OCT 3 1992

NOV 6 1997

Introduction to the Study of Meiofauna

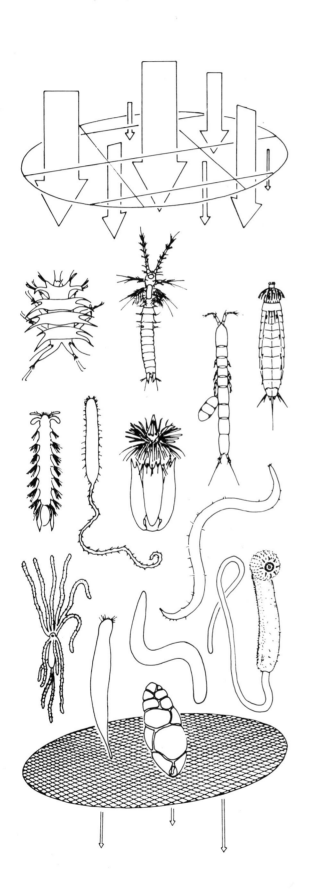

Introduction to the Study of Meiofauna

ROBERT P. HIGGINS

HJALMAR THIEL

Editors

Published by the
Smithsonian Institution Press
Washington, D.C.   London
1988

©1988 by the Smithsonian Institution.
All rights reserved.

Cover design by Robert P. Higgins.

Type set by Ralph Walker and Cathleen Drew.

Library of Congress Cataloging-in-Publication Data

Introduction to the study of meiofauna
Robert P. Higgins, Hjalmar Thiel, editors.
  p. cm.
Includes index.
ISBN 0-87474-488-1
  1. Interstitial fauna. 2. Aquatic invertebrates.
I. Higgins, Robert P. II. Thiel, Hjalmar.
QL 120.I584 1988   88-11336
592.092–dc19   CIP

The paper in this book meets the guidelines for
permanence and durability of the Committee on
Production Guidelines for Book Longevity of the
Council on Library Resources.

Contents

Foreword
List of Contributors

Part 1. Introduction

Chapter 1: Prospectus	11
Chapter 2: History of Meiofaunal Research	14
Chapter 3: Ecology of the Marine Meiofauna	18
Chapter 4: Ecology of the Freshwater Meiofauna	39

Part 2. Methods

Chapter 5: Abiotic Factors	61
Chapter 6: Biotic Factors	79
Chapter 7: Sampling Equipment	115
Chapter 8: Sampling Strategies	126
Chapter 9: Sample Processing	134
Chapter 10: Organism Processing	146
Chapter 11: Culture Techniques	161
Chapter 12: Experimental Techniques	169
Chapter 13: Energetics	181
Chapter 14: Data Processing, Evaluation, and Analysis	197
Chapter 15: The Nanobenthos	232

Part 3. Meiofaunal Taxa

Chapter 16. Taxonomic and Curatorial Considerations	238
Chapter 17. Sarcomastigophora	243
Chapter 18. Ciliophora	258
Chapter 19. Cnidaria	266
Chapter 20. Turbellaria	273
Chapter 21. Gnathostomulida	283
Chapter 22. Nemertina	287
Chapter 23. Nematoda	293
Chapter 24. Gastrotricha	302
Chapter 25. Rotifera	312
Chapter 26. Loricifera	319
Chapter 27. Priapulida	322
Chapter 28. Kinorhyncha	328
Chapter 29. Polychaeta	332
Chapter 30. Aeolosomatidae and Potamodrilidae	345
Chapter 31. Oligochaeta	349
Chapter 32. Sipuncula	355
Chapter 33. Tardigrada	357
Chapter 34. Cladocera	365
Chapter 35. Ostracoda	370
Chapter 36. Mystacocarida	377
Chapter 37. Copepoda	380
Chapter 38. Syncarida	389
Chapter 39. Thermosbaenacea	393
Chapter 40. Isopoda	397
Chapter 41. Tanaidacea	402
Chapter 42. Amphipoda	409
Chapter 43. Cumacea	413
Chapter 44. Halacaroidea	417
Chapter 45. Pycnogonida	423
Chapter 46. Palpigradida	425
Chapter 47. Insecta	428
Chapter 48. Bryozoa	438
Chapter 49. Entoprocta	444
Chapter 50. Brachiopoda	445
Chapter 51. Aplacophora	447
Chapter 52. Gastropoda and Bivalvia	451
Chapter 53. Holothuroidea	457
Chapter 54. Tunicata	461
Index	465

Foreword

My introduction to meiofauna was both public and traumatic. I had given a talk on the benthos of an offshore muddy habitat in the late 1950's, and the subsequent session of questions had gone reasonably well. On leaving the lecture room I was suddenly confronted by a somewhat truculent individual who apparently wished to reopen the discussion on the mesh size of sieves. I felt myself on firm ground here, because although most benthic ecologists had by then abandoned the old 2 mm mesh in favor of the 1 mm size, I regarded even this as too large for work on muds, and preferred an 0.5 mm screen. But my assailant was not appeased, and wanted to know how I handled the sieve filtrate. On being told that, of course, this was wasted over the ship's side, his next question was loud and disturbing, -- "Don't you know that you washed away most of the animals?" I did not know this at the time, but on the next cruise I collected core samples, examined them unpreserved, and thereby discovered a whole new world, fully vindicating the comments of my questioner.

It is of interest that my account, published in the early 1960's, could cite only a handful of earlier papers on the ecology of what everyone now knows as "meiofauna." The changed situation today testifies to the intellectual fascination, scientific relevance, and practical importance of the diverse organisms encompassed by that title. Research in this field, now conducted around the world, is directed to all marine and freshwater habitats from the poles to the tropics, and from the upper intertidal to the abyssal ocean depths. It has stimulated fresh thinking on taxonomy and evolution; raised basic problems in morphology, physiology, behavior, and distribution; brought new ideas to population and community ecology; offered exciting opportunities for experimental and field studies; and it has necessitated the development of innovative sampling and processing techniques appropriate to the unique characteristics of the group. It is also worth noting that far from being an entirely academic pursuit, the study of meiofauna has added to our understanding of the cycling of material in food webs related to commercially exploited species and has provided a valuable tool in the approach to a wide range of pollution problems.

Continued progress in any research field depends on bringing together and maintaining a body of interacting and communicating enthusiasts. It is perhaps in this direction that a significant part of the current strength of meiofaunal research lies. Its practitioners, although scattered around the globe, keep contact through frequent publications, through the Newsletter *Psammonalia,* and, through regular international meetings, always well attended in spite of increasing funding difficulties. They thus exchange data and ideas, and effectively advance and expand their area of science.

This new volume, a further step in that process, needs no detailed preface. Its table of contents indicates the range and diversity of its coverage, while the list of authors confirms the scope of interest. Its editors, contributors, reviewers, and all others concerned with its making are to be congratulated on producing a publication which is worthy of its subject, and which must surely stand as the authoritative text for years to come.

Alasdair D. McIntyre
Aberdeen, Scotland
March, 1988

List of Contributors

Roberto Argano
*Dipartimento Biologia Animale
Università di Roma
Viale Università 32
Rome, Italy*

Patrick M. Arnaud
*Station Marine d'Endoume
Rue Batterie des Lions
13007 Marseille, France*

Ilse Bartsch
*Biologische Anstalt Helgoland
Notkestrasse 31
D-2000 Hamburg 52, FRG*

Susan S. Bell
*Department of Biology
University of South Florida
Tampa, FL 33620, USA*

Dieter Bunke
*Institut für Biologie III
Universität Tübingen
D-7400 Tübingen 1, FRG*

Bryan R. Burnett
*Marine Ecological Consultants
531 Encinitas Blvd., Suite 110
Encinitas, CA 92024, USA*

C. Bradford Calloway
*Museum of Comparative Zoology
Harvard University
Cambridge, MA 02138, USA*

L.R.G. Cannon
*Queensland Museum
P.O. Box 300, South Brisbane
Queensland 4101, Australia*

C. Allan Child
*Department of Invertebrate Zoology
National Museum of Natural History
Smithsonian Institution
Washington, DC 20560, USA*

Bruno Condé
*Université de Nancy I
Musee de Zoologie, Aquarium Tropical
34, Rue Ste Catherine
5400 Nancy, France*

Patricia L. Cook
*Department of Zoology
British Museum (Natural History)
Cromwell Road
London, SW7 5BD, UK*

John O. Corliss
*Department of Zoology
University of Maryland
College Park, MD 20742, USA*

Bruce C. Coull
*Baruch Institute for Marine Biology
and Coastal Research
University of South Carolina
Columbia, SC 29208, USA*

A. Eleftheriou
*Department of Biology
University of Crete
Herakleion 71110
Crete, Greece*

Christer Erséus
*Department of Zoology
University of Göteborg
Box 25059
S-400 31 Göteborg, Sweden*

Richard A. Farris
*Department of Biology
Linfield College
McMinnville, OR 97128, USA*

Anno Faubel
*Institut für Hydrobiologie
und Fischereiwissenschaft
Universität Hamburg
Zeiseweg 9
D-2000 Hamburg 50, FRG*

Robert J. Feller
*Baruch Institute for Marine Biology
and Coastal Research
University of South Carolina
Columbia, SC 29208, USA*

John W. Fleeger
*Department of Zoology and Physiology
Louisiana State University
Baton Rouge, LA 70803, USA*

David G. Frey
*Department of Biology
Indiana University
Bloomington, IN 47405, USA*

Olav Giere
*Zoologisches Institut und Museum,
Universität Hamburg
Martin-Luther-King Pl. 3
D-2000 Hamburg 13, FRG*

Andrew Gooday
*Institute of Oceanographic Sciences
Brook Road, Wormley, Godalming
Surrey, GU8 5UB, UK*

Norbert Greiser
*Institut für Hydrobiologie
und Fischereiwissenschaft
Zeiseweg 9
D-2000 Hamburg 20, FRG*

Eike Hartwig
*Institut für Naturschutz- und
Umweltschutzforschung, (INUF)
des Verein Jordsand
Haus der Natur
2070 Ahrensburg, FRG*

Carlo Heip
*Delta Institute for Hydrobiological Research
Vierstraat 28
4401 EA Yerseke, The Netherlands*

Peter M.J. Herman
*Delta Institute for Hydrobiological Research
Vierstraat 28
4401 EA Yerseke, The Netherlands*

Robert R. Hessler
*Scripps Institution of Oceanography
University of California, San Diego
La Jolla, CA 92093, USA*

Robert P. Higgins
*Department of Invertebrate Zoology
National Museum of Natural History
Smithsonian Institution
Washington, DC 20560, USA*

Alexander D. Huryn
*Department of Entomology
The University of Georgia
Athens, GA 30602, USA*

Dietmar Keyser
*Zoologisches Institut und Museum
Universität Hamburg
Martin-Luther-King Pl. 3
D-2000 Hamburg 13, FRG*

Reinhardt M. Kristensen
*Institute of Cell Biology and Anatomy
University of Copenhagen
Universitetsparken 15
DK-2100 Copenhagen Ø, Denmark*

Susan E. Lenk
*Department of Zoology
University of Maryland
College Park, MD 20742, USA*

Alasdair D. McIntyre
*Marine Laboratory
P.O. Box 101, Victoria Road
Aberdeen AB9 8D8, Scotland, UK*

Claude Monniot
*Laboratoire de Biologie des
Invertebres Marins et Malacologie
Museum National d'Histoire Naturelle
55, Rue de Buffon
75005 Paris, France*

Françoise Monniot
*Laboratoire de Biologie des
Invertebres Marins et Malacologie
Museum National d'Histoire Naturelle
55, Rue de Buffon
75005 Paris, France*

M. Patricia Morse
*Biology Department and
Marine Science Center
Northeastern University
Nahant, MA 01908, USA*

Derek J. Murison
*Marine Laboratory
P.O. Box 101, Victoria Road
Aberdeen AB9 8D8 Scotland, UK*

Claus Nielsen
*Zoologisk Museum
Universitetsparken 15
DK-2100 Copenhagen Ø, Denmark*

Jon L. Norenburg
*Smithsonian Oceanographic Sorting Center
National Museum of Natural History
Smithsonian Institution
Washington, D.C. 20560, USA*

David Pawson
*Department of Invertebrate Zoology
National Museum of Natural History
Smithsonian Institution
Washington, DC 20560, USA*

Robert W. Pennak
*EPO Biology
University of Colorado
Boulder, CO 80309, USA*

Daniel Perlmutter
*Institute of Ecology
University of Georgia
Athens, GA 30602, USA*

Olaf Pfannkuche
*Institut für Hydrobiologie
und Fischereiwissenschaft
Universität Hamburg
Zeiseweg 9
D-2000 Hamburg 50, FRG*

Claude Poizat
*Laboratoire de Biologie Marine
Rue Henri Poincare
13397 Marseille Cedex 13, France*

Günter Purschke
*Spezielle Zoologie
Fachbereich Biologie/Chemie
Postfach 4469
Universität Osnabrück
D-4500 Osnabrück, FRG*

Jeanne Renaud-Mornant
*UA 699 CNRS Biologie des Invertebres Marins
57 Rue Cuvier
75231 Paris Cedex 05, France*

Mary E. Rice
*Smithsonian Marine Station at Link Port
5612 Old Dixie Highway
Fort Pierce, FL 34946, USA*

Franz Riemann
*Alfred-Wegener-Institut für Polar-
und Meeresforschung
D-2850 Bremerhaven 1, FRG*

Edward E. Ruppert
*Department of Biological Sciences
Clemson University
Clemson, SC 29634-1903, USA*

Amelie H. Scheltema
*Woods Hole Oceanographic Institution
Woods Hole, MA 02543, USA*

Jürgen Sieg
*Universität Osnabrück, Abt. Vechta
Fachbereich Naturwissenschaften/Mathematik
Postfach 13 49
D-2848 Vechta, FRG*

Karlien Soetaert
*Marine Biology Section, Zoology Institute
State University of Gent
Ledeganckstraat 35
B-9000 Gent, Belgium*

Wolfgang Sterrer
*Natural History Museum
Flatts FLBX, Bermuda*

Hjalmar Thiel
*Institut für Hydrobiologie
und Fischereiwissenschaft
Universität Hamburg
Zeiseweg 9
D-2000 Hamburg 50, FRG*

David Thistle
*Department of Oceanography
Florida State University
Tallahassee, FL 32306, USA*

John H. Tietjen
*Department of Biology
City College of New York
Convent Avenue at 138th St.
New York, NY 10031, USA*

Paul N. Turner
*Department of Invertebrate Zoology
National Museum of Natural History
Smithsonian Institution
Washington D.C. 20560, USA*

Richard M. Warwick
*Institute for Marine Environmental Research
Prospect Place, The Hoe
Plymouth, PL1 3DH, UK*

Les Watling
*Ira C. Darling Center
University of Maine
Walpole, ME 04573, USA*

John B.J. Wells
*Department of Zoology
Victoria University of Wellington
Private Bag
Wellington, New Zealand*

Wilfried Westheide
*Spezielle Zoologie
Fachbereich Biologie/Chemie
Universität Osnabrück
Postfach 4469
D-4500 Osnabrück, FRG*

1. Prospectus

Robert P. Higgins and Hjalmar Thiel

Elements of the marine and freshwater benthos which we now call meiofauna have been known and studied since the 18th century, but only in this current century have we looked carefully at the myriad of component taxa more analytically and responded with a series of investigations into the nature of this assemblage of microscopic invertebrates. The historical perspectives of the study of meiofauna will be the subject of a separate chapter to follow, yet it is impossible to introduce the rationale of this book without referring to at least some of these landmarks.

Nearly half a century after "meiobenthos" became a word in the scientists's vocabulary (Mare, 1942), its definition is still controversial. Most studies of the benthos of both marine and freshwater communities began with primary attention to organisms large enough to be recognized with relative ease. The biocoenotic concept of Möbius (1877) portrayed a rather static community and conveyed the belief that it was sufficient to consider only larger organisms. Therefore, for many decades to follow, attention was narrowly focused on a convenient size of benthic organisms: those that could be retained on a screen with 1 mm openings, an arbitrary system established by Petersen (1911) for quantitative studies. This assemblage of organisms, whose biomass was clearly significant in aquatic ecosystems, we now call "macrobenthos," and it is accepted that the 1 mm lower limit is not a measure of organism size but sieve mesh size.

The term "meiofauna" is derived from the Greek *meio* meaning "smaller." In this context, it refers to the fauna smaller than what has been defined as the lower size limit for macrofauna, those organisms that are retained on a 1 mm sieve. This does not necessarily infer that the organisms have no single dimension greater than 1 mm, nor does it take the live or preserved state into consideration. Indeed, with reference to this latter instance, the "meiofaunal" species of sipunculids and some other taxa might only pass through a 1 mm mesh sieve in a living, active state. The same procedure of setting a lower size limit to define macrofauna defines meiofauna; 42 µm has been established as the lower limit by most investigators, although sieves with larger mesh sizes have been applied in earlier studies. Thus, benthic biology has been based emperically on the research protocol, on the tools used primarily to separate the fauna from the sediment.

Categories of size, such as macrofauna and meiofauna, generally have little relationship with taxonomic classification, and they have no direct ecological implications. Thus, the terms "meiofaunal taxa" or "macrofaunal taxa of predominantly meiofaunal size" convey the impression that higher taxa fall into definite size categories. This is true for some higher taxa such as the Loricifera, Kinorhyncha, Tardigrada, Gastrotricha, and Gnathostomulida, most, if not all of which, are made up of species whose maximum dimensions at sexual maturity are smaller than 1 mm. However, it is not true for such ecologically important taxa as the Nematoda or the Harpacticoida which contain species whose sexually mature adults may have maximum dimensions well in excess of 1 mm despite the untold numbers remaining much smaller than 1 mm throughout their life history.

A third category of meiofaunal species exists within higher taxa that we normally consider macrofauna; these include representatives of the Brachiopoda, Sipuncula, Ascidacea, Entoprocta, and some even more questionable taxa such as the Holothuroidea and Pycnogonida. These taxa have only a few species approaching meiofaunal dimensions. For the sake of completeness, we have included chapters on these animals because some of their component species have been considered as meiofauna in the literature. Whether or not these taxa should continue to be considered meiofauna is left to the reader's judgment.

The expression "macrofaunal taxa of predominantly meiofaunal size" was used by Hessler (1974) for those species in the deep sea which belong to taxa normally comprised of representatives retained on a 1 mm mesh screen. Classifying the benthos into size categories, based on the methods applied in quantitative assessments, causes taxonomic consideration to lose their relevance.

This is clearly supported by the terms "permanent meiofauna," those species of meiofaunal size throughout their life, and "temporary meiofauna," those species with immature stages falling within the meiofauna size but whose sexually mature stages are of macrofaunal dimensions.

With the notable exception of the insects, considered by freshwater researchers as a major meiofaunal component, we have not included immature stages of taxa which are macrofaunal at sexual maturity. And, as may be concluded from the essay on freshwater meiofauna (Chapter 4), the inclusion of immature insects, especially those which might pass only through a 2 mm sieve, is an inconsistency that requires a careful reevaluation.

Additionally, a strict size definition, prerequisite for quantitative investigations, has not been applied always by taxonomists, especially in the context of "interstitial" species. The term "interstitial fauna," introduced by Nicholls (1935) for animals living in the interstitial water of sand, and its equivalent term "mesopsammon" introduced by Remane (1940), are often confused with the term meiofauna. The "interstitial fauna" or "mesopsammon," and meiofauna are not necessarily interchangable. Organisms living in the interstitial spaces between all types of sediment particles, quartz grains, pieces of broken shell or coral, foraminiferan tests, fecal pellets, or any other conglomerates of particles are properly referred to as interstitial if they move through this habitat with a minimum of disturbance of the constituent particles. The latter part of this definition generally excludes mud-dwelling animals from being classified as interstitial fauna. But not all organisms living in interstitial water are of meiofaunal dimensions, therefore not all interstitial organisms are meiofauna.

Defining meiofauna for quantitative studies requires the establishment of practical limits relevant to size and not to taxonomic classification. Nevertheless, in ecological research it is important to know which size groups are covered when numbers and weights are presented. To our knowledge, a total community, bacteria to megafauna, has never been evaluated ecologically, and it is in this context that well-defined size groups are essential for comparison.

For general ecological studies on abundance and biomass, size groups defined by what is retained on a sieve with a given mesh size during the processing of sediment, for example, are both sufficient and practical. Energetics studies, however, must be based on more exact volume measurements which are too time consuming for the average ecological study.

Thus, meiofauna is defined sufficiently by sieve mesh sizes as summarized by Thiel (1983) and presented in Table 1.1. Size limits have changed historically and from author to author. However, for comparison of faunal size components, it is essential that all researchers use the same mesh sizes for the upper and lower limits, and for subgroups. Sets of sieves with 1000 µm, 500 µm, 250 µm, 125 µm, 62 µm, and 42 µm establish a close to logarithmic system which allows for the most practical comparison of data among investigators. We hope that the definitions presented will be accepted and will promote a more meaningful literature within comparable data sets on the subject of meiofaunal ecology.

The problems inherent with the definition of meiofauna must be recognized, but they should not become an immobilizer of the very research we seek to promote through this book. Together with this introduction, Part I includes a summary of the history of meiofaunal research, showing its development into an independent field of investigation and its bearing on theoretical questions of current importance. Part I concludes with two summarizing chapters of the ecological information derived from meiofauna investigations. In particular, the intent of these latter two chapters is not only to inform, but to guide the prospective student into the myriad of research problems facing this young discipline.

Part II of this book is intended to present the wide range of methods currently used in meiofaunal research. In addition, we have addressed the subject of "those organisms that pass through the 42 µm

Table 1.1.—Meiobenthos and nanobenthos size groups, recommended sample sizes, and processing methods.

Size Group	Mesh Size	(Sub)sample Size	Shipboard Processing	Laboratory Processing
Meiobenthos	42-1000 µm	2-10 cm²	Total preservation	Sieving, staining, sorting with stereo-microscope, microscopic examination
Nanobenthos	2-42 µm	<1->2 cm²	Total preservation	Slide preparation using subvolumes of 0.25 ml, staining, microscopic examination

mesh sieves, namely the "nanofauna." The study of these even more poorly known organisms, suffers from the same constraints as elaborated for the meiofauna.

This book concludes with Part III, an extensive series of chapters on meiofaunal taxa. The classification upon which this section is based generally is in agreement with the contemporary recognition of 40 animal phyla. Meiofaunal organisms are variously represented in 22 of these 40 animal phyla; the extent of coverage of each chapter does not always reflect their relative representation within this meiofauna. These chapters are intended to be merely introductions to the taxa. Each author has included taxonomic characteristics, habitats, collecting, processing, and other methodologies relevant to the taxonomic group. Considerable emphasis is placed on habitus illustrations in order to provide the reader with a broad spectrum of visual recognition. We have purposely excluded taxonomic keys and have tried to minimize the material, especially the details of morphology, that can be found in invertebrate zoology textbooks or other reference literature. These chapters are intended to provide the reader with the information necessary to proceed with the identification of a given taxon through further inquiry into the taxonomic literature, and stimulate the needed studies of the biology and environmental relationships of these taxa.

Nearly every chapter of this book could be a book in itself; perhaps someday we will see meiobenthology developed to this end. If so, we sincerely hope that our efforts have played some role in this achievement.

Acknowledgments

Like all such endeavors, this book is the product of the combined efforts of many persons. We are particularly grateful for the efforts, patience, and cooperation of the many authors who contributed to this book as well as to the following colleagues who reviewed the individual manuscripts: J. Aller, G. Arlt, A. Borror, T. Bowman, R. Brinkhurst, M. Buzas, S. Cairns, C. Clausen, B. Coull, E. Cutler, A. Dinet, J. Eckman, U. Ehlers, A. Eleftheriou, R. Elmgren, K. Fauchald, A. Fricke, P. Gibbs, O. Giere, J. Gilbert, E. Gnaiger, P. Herman, R. Hessler, G. Hicks, D. Hope, W. Hummon, B. Kensley, E. Kirsteuer, L. Kornicker, R. Kristensen, S. Lorenzen, R. Manning, M. MacQuitty, A. McIntyre, A. McLachlan, P. Montagna, W. Noodt, J. Nybakken, M. Palmer, D. Patterson, R. Pennak, M. Petersen, O. Pfannkuche, L. Pollock, K. Reise, J. Renaud-Mornant, F. Riemann, R. Rieger, E. Ruppert, E. Schockaert, P. Schwinghamer, Y. Shirayama, V. Storch, D. Strayer, D. Thistle, G. Uhlig, B. Widbom, J. Winston, and S. Wooden.

In addition, we thank B. Berghahn, C. Drew, D. Fisher, P. Rothman, R. Walker, and M. Wallace, who participated in the technological aspects of the preparation of the computer-generated camera-ready copy and subsequent proofreading.

Finally, we express our appreciation for the financial assistance received from Dr. R. Hoffman, Director, National Museum of Natural History, Smithsonian Institution, and the Sumner Gerard Foundation.

References

Hessler, R.R.
1974. The Structure of Deep Benthic Communities from Central Oceanic Waters. Pages 79-93 in C. Miller, editor, *The Biology of the Oceanic Pacific*. Corvallis: Oregon State University Press.

Mare, M.F.
1942. A Study of a Marine Benthic Community with Special Reference to the Micro-organisms. *Journal of the Marine Biological Association of the United Kingdom*, 25:517-554.

Möbius, K.
1877. *Die Austern und die Austernwirtschaft*. 129 pages. Berlin: Wiegand, Hempel, and Parey.

Nicholls, A.G.
1935. Copepods from the Interstitial Fauna of a Sandy Beach. *Journal of the Marine Biological Association of the United Kingdom*, 20:379-405.

Petersen, C.G.J.
1911. Valuation of the Sea. I. Animal Life of the Sea Bottom. *Reports of the Danish Biological Station*, 20:1-81.

Remane, A.
1940. Einführung in die zoologische Ökologie der Nord- und Ostsee. Volume 1a, pages 1-238 in G. Grimpe and E. Wagler, editors, *Die Tierwelt der Nord- und Ostsee*. Leipzig: Akademische Verlagsgesellschaft.

Thiel, H.
1983. Meiobenthos and Nanobenthos of the Deep Sea. Pages 167-230 in G.T. Rowe, editor, *The Sea*, volume 8. New York: John Wiley and Sons.

2. The History of Meiofaunal Research

Bruce C. Coull and Olav Giere

The study of meiofauna started long before the term meiofauna (or meiobenthos) was coined and the earliest meiofauna studies focused on the discovery and description of new forms. Despite the discovery of many of the higher meiobenthic taxa by the mid-nineteenth century, new meiofaunal groups continue to be described (e.g., Sorberacea (Tunicata) – Monniot et al., 1975; Loricifera – Kristensen, 1983). In fact, of the four animal phyla discovered since the beginning of this century, two have been meiofaunal: the Gnathostomulida (Riedl, 1969) and the Loricifera (Kristensen, 1983). One might think that the systematics of meiofaunal organisms is completely known. However, the task of the meiofauna systematist remains demanding. As more areas are explored, we can be sure that additional descriptions of new families, genera, and species will follow.

Early Meiofaunal Research: Discovery, Taxonomy and Recognition of the Special Characters of Microscopic Invertebrates (1900-1950)

The term "meiobenthos" was introduced and defined in 1942 by Mare in her account of the benthos of muddy substrates off Plymouth, England. But meiobenthic studies preceded her terminology. The earliest work represented regional taxonomic/systematic/morphological descriptions of single groups, e.g., the interstitial opisthobranchs Microhedylidae in the Eastern Mediterranean by Kowalewsky (1901b) or the nematodes populating North American shores in high densities (Cobb, 1914, 1920). However, as early as 1904 Giard recognized in his studies on sandy bottoms of the northern French coast a microfauna so rich "that it would take years to study it." It was Giard who described the first archiannelid (*Protodrilus*) and the gastrotrich *Chaetonotus*.

With the development of more effective sampling techniques for sublittoral habitats (grabs, dredges) between 1911 and 1935, more investigators became aware of the abundance and complexity of fauna in the intermediate size range between classical micro- and macrofauna. Moore (1931) and Rees (1940) in Great Britain as well as Krogh and Spärck (1936) in Denmark quantitatively investigated meiobenthos and enumerated all taxa. But it was Remane, the "father of meiofaunal research," who, on the basis of complete samplings, first recognized the rich populations in intertidal beaches, subtidal sands, and muds and algal habitats as definable ecological assemblages. Remane and his students centered their research on the meiofauna of Kiel Bight (Baltic Sea) and the North Sea shores of Germany. He described *Halammohydra* (1927a), the unique interstitial hydrozoan depicted on the emblem of the International Association of Meiobenthologists, and in 1936, *Monobryozoon*, the curious vagile interstitial bryozoan. In his classic paper (1933) entitled "Verteilung und Organisation der benthonischen Mikrofauna der Kieler Bucht," Remane stressed an essentially new notion that "these biocoenoses differ not only by species abundance and composition, but also in morphological and functional features." Remane wrote extensively on the distribution of the Gastrotricha (1927b), the Rotifera (1929), the Archiannelida (1932), the Kinorhyncha (1936), and other taxa from various shorelines in northern Germany and on the Island of Helgoland; a summary can be found in Remane (1952). His approach has been supplemented and continued by his numerous students and he is creator of the widely acknowledged "German School" of meiobenthology.

Meanwhile, a British scientist, Nicholls, studying the copepods of a Scottish sandy beach (1935), introduced the term "interstitial fauna," but it was Remane who by comparing the fauna from muds, sands, and algae elaborated the adaptive morphology in faunas from specific habitats: In 1952, Remane created the term "Lebensformtypus," where certain morphological features in very different taxa recur as a result of selective pressures for living in a particular habitat. For example, interstitial fauna, regardless of higher taxonomic category, are typically vermiform to allow them to move between the sand grains. Also, during the 1930s – 1950s the Swedish marine biologist Swedmark originated studies on the systematics and ecology of the marine interstitial fauna, particularly the Gastrotricha. After many years of work, primarily in Roscoff (France) and

Kristineberg (Sweden), Swedmark, lovingly called "El Patron" by his students and colleagues, summarized much of his and others' observations in his classic 1964 paper "The Interstitial Fauna of Marine Sand," required reading for all students of meiobenthology.

Independent of marine meiobenthology, a school of freshwater zoologists studied the rich fauna in the sandy bottoms and shores of Eastern European rivers and lakes. This research was initiated in 1927 by Sassuchin in Russia, and during the course of his studies, he introduced the term "psammon" to describe this biocoenosis, characterized by rotifers and protozoans. Sassuchin's papers also include studies on the ecology of limnetic shores, and his work was expanded by Wiszniewski and others to rivers and lakes in Poland. Wiszniewski (1934) created a nomenclature for limnetic shores characterizing the various sandy zones which is still in use in freshwater meiobenthology. Beginning in the 1920's Chappuis from Switzerland studied the subterranean or "phreatic" fauna of river beds and described this as a special biotope, later termed "hyporheic" (Orghidan, 1959).

In America, Pennak gave the first descriptive account of the ecology of freshwater meiobenthos (1939) and in 1940 he published a comprehensive monograph, "Ecology of the Microscopic Metazoa Inhabiting the Sandy Beaches of Some Wisconsin Lakes," which included a comprehensive analysis of the limnetic psammal. In a paper published in 1951, Pennak compared freshwater and marine interstitial fauna. While working on the meiofauna of Massachussetts marine sandy beaches he and Zinn discovered and described (1943) a unique subclass of Crustacea, the Mystacocarida. Also, Delamare Deboutteville from France in his monograph (1960) on subterranean fauna reconciled the often separate pathways in marine and limnetic meiobenthos research underlining the numerous transitions from the littoral to groundwater biotas evident in the Mediterranean area.

Accelerated Interest in Meiofaunal Systematics and Descriptive Ecology (1950s through the mid-1960s)

While Bruce (1928a,b) first detailed the physical micro-environment of sandy beaches, research relating abiotic factors to the occurrence and distribution of meiofauna did not dominate meiobenthology until the 1950s and 1960s. During this period meiofauna studies enjoyed rapid growth due to a variety of factors: more scientists working on meiofauna, a renewed interest in the ecology of the animals, the development of meiobenthic research in North America, better sampling and extracting techniques, new readily available equipment (e.g., electrodes) and better communication amongst meiobenthologists via the "Association of Meiobenthologists" (later changed to the "International Association of Meiobenthologists") and the Association's newsletter *Psammonalia*. The first two-page issue of *Psammonalia* was mimeographed and mailed in November 1966 by R.P. Higgins (a former student of Pennak) and D.J. Zinn to a small list of interested people. After 20 years of its existence, *Psammonalia* now reaches over 300 subscribers in 36 countries. In 1969 the Association sponsored its first International Meiofauna Conference in Tunis, Tunisia. It was from this conference that the predecessor of this present volume, i.e., "A Manual for the Study of Meiofauna," was produced (Hulings and Gray, 1971). Subsequently the Association has held symposia in York, England (1973), Hamburg, Germany (1977), Columbia, South Carolina, USA (1980), Ghent, Belgium (1983), and Tampa, Florida, USA (1986).

During the 1960s, with the advent of meiofauna research in North America and a greater interest in benthic ecology world wide, two general types of meiofaunal studies dominated: (1) descriptive ecology and (2) broadened taxonomic knowledge of various meiofaunal taxa from various habitats. There was much success in correlating meiofaunal taxa with physical factors (such as grain size, water content, tidal exposure). We learned, for example, that (1) certain taxa are restricted to certain sediment types and vertical position in the sediment, (2) the anoxic layers of certain sediments harbor few meiofauna, (3) meiofaunal biomass in estuaries and the deep-sea tend to equal that of the macrofauna, (4) in most shallow areas of the world (<100 m) there are about 10^6 meiofaunal organisms per square meter of sea bottom, and (5) changes in tidal exposure are often the primary factors limiting sandy beach interstitial fauna.

In the field of taxonomy, emphasis on finding, identifying, and classifying new taxa was prevalent during this period. Various investigators all over the world described many new taxa, but, more importantly, they began to consider the role of various meiofaunal organisms in the overall scheme of invertebrate systematics. Ecologically this was a period of routine, but necessary work to determine the abundance and diversity of meiofauna in a habitat. It was during this period that attempts to ascertain what controlled meiofauna distribution and abundance, at least correlatively, were initiated.

Meiofauna in the Laboratory and Species Diversity (late 1960s into the 1970s)

During this era there occurred a surge of

ecological interest in the meiofauna, and in 1969 McIntyre published the first overall review of meiofaunal ecology. The ecological questions of the times centered on what do meiofauna do and how do they do it? A primary emphasis was in ecophysiology and behavior of meiofauna: studies measuring respiration rates, preference and tolerance to various natural environmental parameters (temperature, salinity, anoxia), and attempts to ascertain the functional and life history parameters of meiofauna were conducted primarily in the laboratory. From this physiological approach there developed a great interest in the energetic role of meiofauna in benthic systems: what they ate, how fast (or slowly) they grew. Laboratory experiments, in which specimens were provided with a suite of foods or given one type of food, were used to estimate growth, reproductive potential, and longevity of certain species. Work was initiated to determine what species tolerated and/or required low oxygen or anoxic conditions. Several proposals were advanced on the biochemical mechanisms used by these metazoans to tolerate anaerobic conditions. With the advent of interest in anaerobic metabolism also came various theories on the origins of the Metazoa with several arguing that some anaerobic meiofaunal organism may have been the primitive metazoan, since the early Precambrian atmosphere lacked oxygen.

The analysis of community structure by measuring species diversity was a popular ecological technique during this era, and meiobenthologists also used this community measure. Meiofauna were considered as a good bioassay of community health and rather sensitive indicators of environmental change. Considerable research is still in progress regarding these topics.

Experimental Ecology/Phylogenetic Considerations (1970s into the 1980s)

In ecology, hypothesis testing, whether in the laboratory or in the field, became an important avenue of research and meiofaunal researchers realized that meiofaunal assemblages could be experimentally manipulated. There was significant interest in population dynamics seeking to define controlling factors. The first estimates of *in situ* life histories of meiofauna were produced by short interval field sampling. In addition, a number of investigations were devoted to studying macrofaunal-meiofaunal interactions, including predator/prey interactions (with or without exclusion cages), the role of meiofauna as food for higher trophic levels, the effect of macrobiotic structure on meiofaunal distribution, and the recolonization of meiofauna into disturbed areas. Pollution effects on meiofauna were studied in the laboratory and in the field. In most cases these were true experiments, with a hypothesis and a test (with controls). Such experimental treatments have caused a significant input of statistical rigor to meiofaunal ecology. Much of the present day ecological research is at the stage where hypotheses are proposed, experimentally tested via some manipulation (e.g., addition of predators, addition of a pollutant, change of habitat), and the response quantitatively noted.

First attempts to place meiofauna in a phylogenetic perspective had already begun in the 1960s. In the 1970s–1980s this type of meiofaunal systematic research experienced a broader emphasis, i.e., where to place taxa and how, phylogenetically, to relate one taxon to another? With the popularization and relative accessibility of electron microscopes, many new and unique characters have been discerned and new ideas related to animal phylogeny proposed. We have learned, for example, that most of the Macrodasyoida (Gastrotricha) and the Gnathostomulida have monociliated epithelial cells, whereas in most Turbellaria they are multiciliated. The monociliated/uniflagellate condition is thought to be ancestral to the multiciliated one and thus raises questions concerning the long-held belief that the Turbellaria are the most ancestral bilateral metazoans. This phylogenetic approach to metazoan systematics has been the forte of meiobenthologists and it is becoming rather obvious that understanding meiofaunal evolution is a key to understanding invertebrate evolution.

Summary and Prospect

Meiofauna research has seen a progression from basic taxonomy and descriptive ecology to process-oriented experimental ecology and phylogenetics. The background descriptive phase was, of course, necessary and led to our ability to perform the experimental studies. We are now at a stage that still requires background/descriptive information and we recognize the value of such descriptive studies; they are often requisite to generate data that encourage experimental hypotheses testing. No longer should meiofaunal research restrict itself toward meiobenthic problems alone. Rather, we need to use meiobenthic background data to test basic hypotheses, hypotheses in need of elaboration and amenable to being tested using meiofauna with the experimental method.

Acknowledgments Contribution number 702 from the Bell W. Baruch Institute for Marine Biology and Coastal Research; research partially supported by NSF grant number OCE85-21345 (B.C. Coull and R.J. Feller, principal investigators).

References

Bruce, J.R.
1928a. Physical Factors on the Sandy Beach. Part I. Tidal, Climatic and Edaphic. *Journal of the Marine Biological Association of the United Kingdom*, 15:535-552.
1928b. Physical Factors on the Sandy Beach. Part II. Chemical Changes - Carbon Dioxide Concentration and Sulphides. *Journal of the Marine Biological Association of the United Kingdom*, 15:553-565.

Chappuis, P.A.
1922. Die Fauna der unterirdischen Gewässer der Umgebung von Basel. *Archiv für Hydrobiologie*, 14:1-88.

Cobb, N.A.
1914. North American Free-Living Fresh-Water Nematodes. (Contribution to a Science on Nematology II). *Transactions of the American Microscopical Society*, 33:35-99.
1920. One Hundred New Nemas (Type Species of 100 New Genera). *Contributions to the Science of Nematology*, 9:217-343.

Delamare Deboutteville, C.
1960. *Biologie des eaux souterraines littorales et continentales.* 740 pages. Paris: Hermann.

Giard, A.
1904. Sur une faunule charactéristique des sables à Diatomées d'Ambleteuse. *Compte Rendu des Séances de la Société de Biologie*, Paris, 56:107-165.

Hulings, N.C., and J.S. Gray, editors
1971. A Manual for the Study of Meiofauna. *Smithsonian Contributions to Zoology*, 78:83 pages.

Kowalesky, A.
1901a. Etudes anatomiques sur le genre *Pseudovermis*. *Mémoires de l'Académie des Sciences St. Pétersburg (Sciences Mathematiques, Physique Naturelle)*, 12:1-28.
1901b. Les Hédylides, études anatomiques. *Mémoires de l'Académie des Sciences St. Pétersburg (Sciences Mathematiques, Physique Naturelle)*, 12(6):1-32.

Kristensen, R.M.
1983. Loricifera, A New Phylum with Aschelminthes Characters from the Meiobenthos. *Zeitschrift für Zoologische Systematik und Evolutionsforschung*, 21:161-180.

Krogh, A., and R. Spärck
1936. On a New Bottom-sampler for Investigation of the Microfauna of the Sea Bottom. *Kongelige Danske Videnskabernes Selskab Skrifter, Biologiske Meddelelser*, 13:1-12.

Mare, M.F.
1942. A Study of a Marine Benthic Community with Special Reference to the Micro-organisms. *Journal of the Marine Biological Association of the United Kingdom*, 25:517-554.

McIntyre, A.D.
1969. Ecology of Marine Meiobenthos. *Biological Reviews of the Cambridge Philosophical Society*, 44:245-290.

Monniot, C., F. Monniot, and F. Gaill
1975. Les Sorberacea: une nouvelle classe de Tuniciers. *Archives de Zoologie Expérimentale et Géneral*, 116:77-122.

Moore, H.B.
1931. The Muds of the Clyde Sea Area. III. Chemical and Physical Conditions; Rate and Nature of Sedimentation; and Fauna. *Journal of the Marine Biological Association of the United Kingdom*, 17:325-358.

Nicholls, A.G.
1935. Copepods from the Interstitial Fauna of a Sandy Beach. *Journal of the Marine Biological Society of the United Kingdom*, 20:379-405.

Orghidan, T.
1959. Ein neuer Lebensraum des unterirdischen Wassers: der hyporheische Biotop. *Archiv für Hydrobiologie*, 55:392-414.

Pennak, R.W.
1939. The Microscopic Fauna of the Sandy Beaches. Pages 94-106 in *Problems of Lake Biology*. American Association for the Advancement of Science.
1940. Ecology of the Microscopic Metazoa Inhabiting the Sandy Beaches of Some Wisconsin Lakes. *Ecological Monographs*, 10:537-615.
1951. Comparative Ecology of the Interstitial Fauna of Fresh-water and Marine Beaches. *L'Année Biologique*, 27:449-479.

Pennak, R.W., and D.J. Zinn
1943. Mystacocarida, A New Order of Crustacea from Intertidal Beaches in Massachussetts and Connecticut. *Smithsonian Miscellaneous Collections*, 103:1-11.

Rees, C.B.
1940. A Preliminary Study of the Ecology of a Mud-flat. *Journal of the Marine Biological Association of the United Kingdom*, 24:185-199.

Remane, A.
1927a. *Halammohydra*, ein eigenartiges Hydrozoon der Nord-und Ostsee. *Zeitschrift für Morphologie und Ökologie der Tiere*, 7:643-677.
1927b. Gastrotricha. Pages 1-56 in Grimpe and Wagler, editors, *Die Tierwelt der Nord- und Ostsee*, Part VIId. Leipzig: Akademische Verlagsgesellschaft.
1929. Rotatoria. Pages 1-156 in Grimpe and Wagler, editors, *Die Tierwelt der Nord- und Ostsee*, Part VIIe. Leipzig: Akademische Verlagsgesellschaft.
1932. Archiannelida. Pages 1-36 in Grimpe and Wagler, editors, *Die Tierwelt der Nord- und Ostsee*, part VIa. Leipzig: Akademische Verlagsgesellschaft.
1933. Verteilung und Organisation der benthonischen Mikrofauna der Kieler Bucht. *Wissenschaftliche Meeresuntersuchungen N.F., Kiel*, 21, H. 2:161-221.
1935/ Gastrotricha und Kinorhyncha. Pages 1-55 in H.
1936. Bronn, editor, *Klassen und Ordnungen des Tierreichs*, 4 Bd., 2 Abt., 1 Buch, 2 Teil. Leipzig: Akademische Verlagsgesellschaft.
1936. *Monobryozoon ambulans*, ein eigenartiges Bryozoon des Meeressandes. *Zoologischer Anzeiger*, 113:161-167.
1952. Die Besiedlung des Sandbodens im Meere und die Bedeutung der Lebensformtypen für die Ökologie. *Verhandlungen der Deutschen Zoologischen Gesellschaft, Wilhelmshaven 1951, Zoologischer Anzeiger Supplement*, 16:327-359.

Riedl, R.
1969. Gnathostomulida from America. *Science*, 163:445-452.

Sassuchin, D.N., N.M. Kabanov, and K.S. Neiswestnowa
1927. Über die mikroskopische Pflanzen-und Tierwelt der Sandfläche des Okaufers bei Murom [Russian with German co-title and summary]. *Gidrobiological Zhurnal*, 6:59-83.

Swedmark, B.
1964. The Interstitial Fauna of Marine Sand. *Biological Reviews of the Cambridge Philosophical Society*, 39:1-42.

Wiszniewski, J.
1934. Recherches écologiques sur le psammon et spécialement sur les rotiféres psammoniques. *Archiwum Hydrobiologji i Rybactwa*, 8:149-272.

3. Ecology of the Marine Meiofauna

Bruce C. Coull

The purpose of this chapter is to provide an overview of the state of the art in meiofaunal ecology. This overview is not meant to be exhaustive but rather it attempts to synthesize the major findings and generalities of meiofaunal ecology to date. The reader is referred to Swedmark (1964), McIntyre (1969), Gerlach (1971, 1978), Coull (1973), Fenchel (1978), Coull and Bell (1979), Platt and Warwick (1980), Giere and Pfannkuche (1982), Heip et al. (1982, 1985), Hicks and Coull (1983), Thiel (1983), Coull and Palmer (1984) and Soyer (1985) (and the references cited therein) for more complete reviews of various aspects of meiofaunal ecology.

The meiofauna are by no means a homogeneous ecological group. There is a wide diversity of habitats in which the meiofauna live. Meiofauna occur in both freshwater and marine habitats, from high on the beach to the deepest depths of the water body. Sediments of all kinds from the softest of muds to the coarsest shell gravels, and all those in between, harbor meiofauna. Meiofauna also occupy several "above sediment" habitats including rooted vegetation, moss, macroalgae fronds, sea ice and various animal structures, e.g., coral crevices, worm tubes, echinoderm spines. Still some meiofauna are symbionts living commensally in animal tubes, with bivalves, in association with wood borers or hydrozoan colonies. Thus, it appears that wherever one looks in the aquatic environment meiofauna are likely to be found. Table 3.1 lists recent abundance values for meiofauna from a variety of habitats (primarily sedimentary).

Twenty-two of the 33 metazoan phyla have at least some meiobenthic taxa; the Gastrotricha, Gnathostomulida, Kinorhyncha, Loricifera, and Tardigrada are exclusively meiobenthic. Some animals, usually larvae of the macrofauna, are a part of the meiobenthos only during their juvenile stages (temporary meiofauna), but many taxa have species that are meiobenthic throughout their life cycle (permanent meiofauna). In addition to the five exclusively meiobenthic phyla mentioned above, permanent meiobenthos includes the Mystacocarida and many representatives of Rotifera, Nematoda, Polychaeta, Copepoda, Ostracoda, Turbellaria, Halacaroidea, and some specialized members of the Hydrozoa, Nemertina, Entoprocta, Gastropoda, Aplacophora, Brachiopoda, Holothuroidea, Tunicata, Priapulida, Oligochaeta, and Sipuncula.

Certain taxa are restricted to particular sediment types. Sediments where the median particle diameter is below 125 μm tend to be dominated by burrowing meiofauna. Interstitial groups, e.g., Gastrotricha, Tardigrada, are typically excluded from muddy substrates where the interstitial lacunae are closed. Obviously, the converse is true so that a burrowing taxon, e.g., Kinorhyncha (one exception, *Cateria*), is excluded from the interstitial habitat. In those taxa that have both interstitial and burrowing representatives, e.g., Nematoda, Copepoda, Turbellaria, there are differences in the morphologies of mud and sand dwellers. The sand fauna tends to be slender as it must maneuver through the narrow interstitial openings, whereas the mud fauna is not restricted to a particular morphology but is generally larger. Additionally most interstitial taxa have adhesive glands for attaching to sand grains and they tend to have a low number of eggs (see Swedmark, 1964, for a review of interstitial fauna functional morphology). Epibenthic and phytal forms tend to be larger and often have the ability to swim for short distances. They occupy a variety of above sediment habitats, i.e., just above the surface, on worm tubes, algal fronds, sea grasses, etc., and often occur in great abundances (Hicks 1977).

Abundance and Diversity of Meiofauna

On the average one can expect to find some number times 10^6 meiofaunal organisms per square meter of sediment surface and a standing stock dry weight biomass of 1-2 gm m^{-2} in shallow (<100m) waters. These abundance/biomass values, of course, vary according to season, latitude, water depth, tidal exposure, grain size, habitat, etc. Highest values typically come from intertidal muddy estuarine habitats, lowest values from the deep sea (Table 3.1). In general, sediment grain size is a (the?) primary factor affecting the abundance and species composition of meiofaunal organisms. Different assemblages occur in muddy vs. sandy vs. phytal habitats. Such a generalization, however, must

Table 3.1.—Summary of meiofauna abundance not included in or post-dating similiar tables in McIntyre(1964) and Hicks (1977). Only studies that include number of animals per unit area and either all taxa or total meiofauna are included. Single taxon studies and studies in polluted or perturbed areas are not included. Studies on phytal or epibiontic meiofauna are not included because conversion to a standard unit (10 cm^2 is not possible. (See Coull et al., 1983 for a relatively recent table on phytal studies). (* = mean of 11 years; abundance in number of organisms per 10 cm^2).

Habitat	Location	Nematodes	Copepods	Meiofauna	Author
Freshwater					
Sand (0.5-10 m)	Iraq	126-558	14-71	231-773	Arlt and Saad, 1977
Sand and mud (1-65 m)	Finland	6-186	24-135	76-625	Holopainen and Passivirta, 1977
Sandy/mud (1.2 m)	Finland	435	140	771	Ranta and Sarvala, 1978
Sandy (1 m)	South Carolina, USA	383-3047	—	968-3675	Oden, 1979
Sand and mud (3-15 m)	Canada	—	—	120-1500	Anderson and de Henau, 1980
Muddy sand (11-23 m)	Michigan, USA	146-394	8-25	204-590	Nalepa and Quigley, 1983
Campo	Brazil	42-252	1-172	52-601	Reid, 1984
Marine Intertidal					
Sand	White Sea, USSR	7-239	20-261	319-427	Galtsova, 1971
Sand	Germany	322-1559	34-93	636-1837	Arlt, 1977
Sand	Delaware, USA	57-538	2-8	125-896	Hummon et al., 1976
Sand	New York, USA	1-2262	0-34	4-2293	Martinez, 1975
Sand	South Africa	35-1328	10-502	60-2250	McLachlan, 1977
Sand	South Africa	47-450	2-211	55-584	McLachlan et al., 1977a
Sand	India	158-524	0-224	215-1337	Munro et al., 1978
Sand	Scotland	168-2100	2-1125	203-4262	Munro et al., 1978
Sand	Isle of Man, UK	180-2771	20-353	802-2939	Moore, 1979a
Sand	Alaska	300-2900	86-1248	420-4790	Feder and Paul, 1980
Sand	S. Africa	35-41	54-109	—	Orren et al., 1981
Sand	India	584-4597	102-888	2270-6116	Ansari and Ingole, 1983
Sand	India	161-2046	23-102	245-2195	Kondalarao, 1983
Sand	Martinique	—	—	5-1308	Renaud-Mornant et al., 1983
Sand	Northern Ireland	—	—	260	Boaden and Elhag, 1984
Sand	Poland (Baltic)	—	—	325-2250	Joncyk and Radziejewska, 1984
Mud	Alaska	790-3490	26-351	800-3900	Feder and Paul, 1980
Mud	New Zealand polluted	1-22	0-9	54-210	Coull and Wells, 1981
Mud	New Zealand non-polluted	321-793	26-689	144-723	Coull and Wells, 1981
Mud	India	65-412	3-32	76-413	Rao and Misra, 1983

Table 3.1 (continued)

Habitat	Location	Nematodes	Copepods	Meiofauna	Author
Mud	Plymouth UK	—	—	1400-11,400	Ellison, 1984
Salt marsh muds	Chile	270-1770	0-325	310-2430	Clasing, 1976
Salt marsh muds	South Carolina, USA (high marsh)	87-550	30-192	200-800	Bell, 1979
Salt marsh muds	Louisiana, USA	154-756	12-106	202-814	Fleeger and Chandler, 1983
Salt marsh muds	Louisiana, USA	754-856	11-105	202-914	DeLaune et al., 1984
Salt marsh muds	Louisiana, USA	202-433	25-123	243-627	Smith et al., 1984
Salt marsh muds	South Carolina, USA (low marsh)	325-1475	80-273	525-1700	Palmer and Gust, 1985
Mangrove mud and sand	South Africa	620-4110	10-180	840-5300	Dye, 1983
Mangrove mud and sand	Australia	6-1189	0-37	9-1208	Hodda and Nicholas, 1985
Marine Subtidal (0.5-200 m)					
Sand (0.8-3.0 m)	Germany	8-104	4-23	45-165	Arlt and Holtfreter, 1975
Sand (0.5-3.0 m)	Germany	42-244	8-37	$\bar{x}=197$	Moller et al., 1976
Sand (5-30 m)	Algoa Bay South Africa	47-450	2-211	55-584	McLachlan et al., 1977b
Sand (25 m)	Japan	93-408	7-153	103-526	Ito, 1978
Sand (13-37 m)	Georgia, USA	56-863	6-179	413-1290	Tenore et al., 1978
Sand (20-107 m)	Goa, India	193-654	28-138	361-2295	Ansari et al., 1980
Sand (0-100 m)	Andaman Sea	34-102	—	68-340	Ansari and Parulekar, 1981
Sand (11-96 m)	SE USA Shelf	—	—	$\bar{x}=957$	Coull et al., 1982
Sand (0.5-11 m)	Polynesia	7-64	4-44	39-120	Thomassin et al., 1982
Sand (8-14 m)	Poland (Baltic)	—	—	274-12341	Radziejewski, 1984
Sand (1 m)	South Carolina USA	641*	88*	1240*	Coull, 1985
Mud (6-8 m)	India	769-1107	8-25	1022-1250	Ansari, 1978
Mud (15-50 m)	India	91-500	9-56	142-621	Ansari et al., 1982
Mud (5.5 m)	Beaufort Sea Arctic	5072-5461	204-253	5626-6061	Carey and Montagna, 1982
Mud (8-30 m)	North Adriatic Yugoslavia	94-607	6-90	103-739	Vidakovic, 1984
Mud (1 m)	South Carolina USA	856*	102*	1247*	Coull, 1985
Muddy/sand (6 m)	Germany	105	2	148	Arlt, 1977
Muddy/sand (7 m)	Rhode Island, USA	190-950	10-120	300-1100	Grassle et al., 1981
Muddy/sand (20 m)	Gulf of Mexico USA	—	—	450-1200	Harper et al., 1981

Table 3.1 (continued)

Habitat	Location	Nematodes	Copepods	Meiofauna	Author
Muddy/sand (117-141m)	North Sea	146-280	28-183	156-349	Faubel et al., 1983
Sand and Mud (4-193m)	Kerguelen Archipelago	18-4640	0-424	21-4801	deBovee and Soyer, 1977
Sand and mud, phytal zone (0.5-5 m)	Asko Region Baltic Sea	—	—	800-1400	Kautsky et al., 1981
Sand and mud (18-230 m)	Bay of Bengal India	0-50	0-11	0-127	Rodgrigues et al., 1982
Sand and mud (90-223 m)	Gulf of Mexico	59-346	44-166	181-715	Yingst and Rhoads, 1985
Sea grass sediments (<5 m)	Bothinian Region Baltic Sea	867-3855	90-124	1207-4312	Elmgren et al., 1984
Seagrass sediments (<2 m)	Florida USA	101-191	12-47	185-309	Decho et al., 1985
? (?)	India	191-934	66-243	478-1467	Sarma and Rao, 1980

Marine Deep-sea (>200 m) Depth (m)

2116-2883	Mediterrean	22-106	1-8	28-94	Dinet et al., 1973
1468-3828	North Atlantic	9-187	1-51	29-294	Rachor, 1975
459-1209	Aegean	51-362	2-34	69-415	Dinet, 1976
252-1245	Kergulen Archipelago	166-576	1-26	181-640	DeBovee and Soyer, 1977
400-4000	SE USA	4-942	0-138	74-892	Coull et al., 1977
283-2000	Norwegian Sea	130-905	2-40	139-939	Dinet et al., 1977
1369-4225	Gascogne, France	5-766	0-48	9-978	Dinet and Vivier, 1977
201-4727	NW Indian Ocean	4-465	0-6	11-712	Romano and Dinet, 1978
168-810	Mediterranean	—	—	580-640	Vivier, 1978
2707-3709	Norwegian Sea	9-1127	0-18	9-1158	Dinet, 1979
840	Goa, India	111	0	250	Ansari et al., 1980
501-2000	Andaman Sea	34-302	—	68-438	Ansari and Parulekar, 1981
1750-3000	NW Africa	—	—	440-800	Thiel, 1982
2090-8260	W Pacific (Japan)	31-1195	0-71	37-1315	Shirayama, 1984
4400-4840	Amazon Cone	133-255	4-9	144-278	Sibuet et al., 1984
4626	NW Atlantic	143-197	8-14	153-212	Thistle et al., 1985

Sea Ice

Beaufort Sea		18	2	63	Carey and Montagna, 1982
Frobisher Bay, Canada		6-110	6-110	—	Grainer et al., 1985

be tempered with the consideration of other environmental factors, e.g. temperature, salinity, water movement, O_2 content, seasonality, and others.

Nematodes regularly dominate the meiofauna in sediment biotopes comprising >50% of the total meiofauna. Harpacticoid copepods are usually second in abundance but may dominate in some coarse grained sediments. Occasionally another taxon may rank first or second. As examples at a sandy South Carolina site gastrotrichs were the second most abundant taxon over an 11-year period (Coull, 1985). Galhanó (1970) reported isopods as the dominant interstitial taxon in a Portuguese sandy beach, and Hummon et al. (1976) found ostracods the second most abundant taxon in a Delaware sandy beach.

Table 3.1 summarizes the abundance values for total meiofauna, and the two most common taxa, nematodes and copepods, appearing in the literature subsequent to similar lists in McIntyre (1969) and Hicks (1977). Note that there may be wide ranges given for abundance in a specific location, because ranges include both temporal and spatial variability. Marine shallow water (<100 m) and freshwater values tend to be within the range of 200-3000 meiofauna/10 cm^2, whereas deep sea values typically are an order of magnitude lower. Thiel (1983) provides a very thorough review of meiofauna abundance in the deep sea.

The bewildering array of species data that can result from surveys of meiofaunal assemblages has often prompted investigators to simplify results in a derived species diversity index. Any index involves an inevitable loss of information compared to the data from which it was calculated. In the case of species diversity measures, the information lost includes the identity of the species in the community and the species history via bibliographic background, characteristics of obvious importance in evaluating the consequences of any alteration in community structure. For this reason, diversity indices should never be used alone and must be coupled with population or multivariate analyses, which reflect qualitative community composition. Species richness and evenness are intuitively more meaningful than a diversity index *per se.*

The highest known species diversity for a meiofaunal assemblage is recorded for copepods from algal holdfast communities (Moore, 1973), but shallow water algal frond assemblages and the deep-sea is greatly reduced (Thiel, 1983), there are many different species and dominance by one to several species, a phenomenon common in shallow water systems, does not occur. Diversity of meiofauna (however measured) typically increases into deep-sea environments and may be a function of a variety of factors, e.g., predation/disturbance keeps competitors low in abundance by suppressing dominance or the "homogeneous" deep-sea is really quite heterogeneous due to microbiogenic structures thus allowing more species to coexist. For example, Aller and Aller (1986) found that at 482 m meiofaunal abundance was 2-3x greater at sites where relict burrows were filled with fresh organic matter than at non-burrow sites. That diversity of meiofauna is higher in the deep-sea is known; why that is true is open to interpretation.

In shallow water sedimentary habitats, meiofaunal diversity appears similar world wide with ecologically equivalent species in different parts of the world. These shallow water communities with their characteristic species assemblages and diversities tend to have one to ten dominant species, whether they be in sand or mud. Coull and Fleeger (1977) compared two such subtidal sites (one mud, one sand) within 1 km of each other and found no significant within or between site differences in copepod diversity over a three year period.

This similar diversity within a region and between geographically disjunct regions (compare for example Hartzband and Hummon, 1974; Marcotte and Coull, 1974; Bodin, 1977) suggests comparable diversities can be anticipated in shallow sedimentary biotopes world-wide. While the data base is limited and there are difficulties interpreting diversity data, perhaps there is a standard diversity range which any (most?) shallow water meiofaunal assemblage may be expected to attain.

Zonation

Vertical Distribution.--In almost all meiobenthic studies the majority of the fauna has been found in the upper 2 cm of sediment. Vertical zonation is typically controlled by the depth of the redox potential discontinuity (RPD) level, i.e., the boundary between aerobic and anaerobic sediments. The primary factor or "super parameter" responsible for vertical gradients in the RPD is oxygen, which determines the redox potential as well as the oxidation state of sulfur and various nutrients. When redox potentials drop below +200 mV, metazoan meiofauna densities greatly decrease (McLachlan, 1978). Harpacticoid copepods are typically the most sensitive meiobenthic taxon to decreased oxygen, and generally are confined to oxic sediments (but see Wieser et al., 1974). Some meiofauna appear to be capable of tolerating low or no oxygen conditions and thus penetrate sediments below the RPD (Fenchel and Riedl, 1970; Ott and Schiemer, 1973). There is however debate over how these animals adapt to reduced O_2 levels (Maguire and Boaden, 1975, Powell et al. 1979, 1980; Ott et al., 1983; Mayers et al., 1987) and further, whether, the occupied habitats below the RPD are truly anoxic (Reise and

Ax, 1979, Boaden, 1980).

In muds and sediments heavy in detritus, meiofauna are often restricted to the upper few mm or cm of oxidized sediments (Coull and Bell, 1979). Consequently, most of the research on vertical patterns of meiofauna vertical distribution has been conducted in sandy substrata. In sands the meiofauna can be distributed to the depth of the RPD which on high energy beaches can be 50 cm or more deep. While oxygen content is the ultimate factor controlling vertical distribution of meiofauna in beaches (McLachlan, 1978), desiccation is also important. Meiofauna are known to be sensitive to low pore water content (Jansson, 1968) and as sand dries at low tide, the fauna face desiccation stress regardless of the oxygen content. McLachlan et al., (1977a) found meiofauna migrated downwards on an ebbing tide and upwards on a flooding tide. Vertical migration was less in the winter than in the summer and this appeared to be related to lower winter temperatures and therefore less desiccation at low tide than in the summer. Furthermore, vertical migration was reduced at night, probably in response to cooler night temperatures at low tide and again, less desiccation. Thus the migrations were not entirely dependent on the tides since desiccation varied seasonally and diurnally. Figure 3.1 illustrates the vertical migration of the interstitial polychaete *Hesionides arenaria* as the tide covered an Isle of Sylt (North Sea) sandy beach. Meineke and Westheide (1979) attributed this upward migration to increased interstitial water content of the sand as the tide covered the beach.

Harris (1972a) found no significant change in distribution of copepods at different states of the tide, but this was on a beach where the sand was always saturated, whereas Boaden and Platt (1971) reported that nematodes migrated downward when the tide crossed their sandy beach sampling site and Meineke and Westheide (1979) recorded that some species migrated up at high tide, some migrated down at high tide, some did not migrate vertically in relation to the tide, and yet others migrated horizontally over longer time periods. However, Harris (1972c) demonstrated that the vertical distribution of individual species on a British sandy beach varied seasonally. Since the redox layer is known to migrate in response to temperature (RPD closer to surface when temperatures are high), the vertical distributions reported by Harris (1972c) reflect the upward movement of populations away from the upward migrating RPD layer in the summer (see also Dinet, 1972). Independent of seasonal or diurnal migrations certain species segregate in a vertical plane (Harris, 1972c; Wieser et al., 1974; Mielke, 1976; Hogue, 1978; Warwick and Gee, 1984). The vertical distribution of the dominant nematode species on a sandy mudflat in the Tamar Estuary, southwest England is illustrated in Figure 3.2. Such vertical segregation may be a mechanism to avoid interspecific competition.

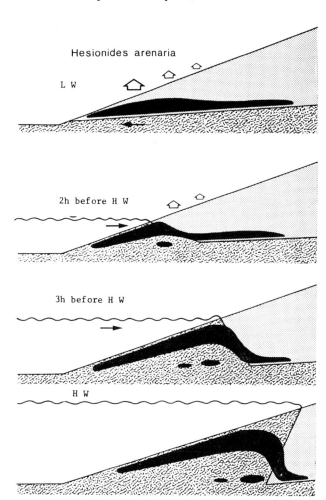

Figure 3.1.--Horizontal migration of the interstitial polychaete *Hesionides arenaria* as the tide floods a sandy beach. The dark striped area represents the *H. arenaria* population (from Meineke and Westheide, 1979).

Horizontal Distribution.--There are many examples of horizontal zonation of meiofauna along salinity gradients in estuaries (e.g., Noodt, 1957, Capstick, 1959, Bilio, 1966, Warwick, 1971, Horn, 1978); across intertidal sandy habitats (e.g., Harris, 1972b, Ott, 1972, Platt, 1977, Moore, 1979b; Rieger and Ott, 1971, Schmidt, 1968, 1969, 1972a,b,; Blome, 1983); across intertidal muddy habitats (e.g., Barnett, 1968, Coull et al., 1979) and with increasing water depth both onto the continental shelf and into the deep sea (e.g., Soyer, 1970, Tietjen, 1976, Coull et al., 1982).

In estuaries there appears to be a distinct relationship between salinity and meiofaunal

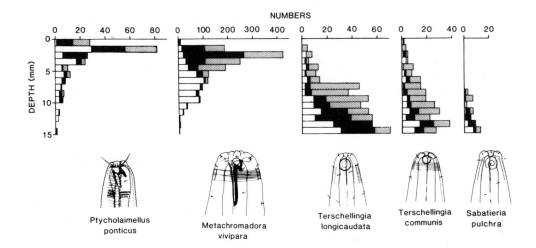

Figure 3.2.—Fine-scale vertical distribution of the dominant nematode species at one site in Tamar Estuary, U.K. Darker shading infers more individuals. Nematode heads are drawn to scale to show buccal structures (from Warwick and Gee, 1984).

assemblages. The relationship is reflected in species composition, abundance and species diversity. Bilio (1967) distinguished four distinct groupings of meiofauna in relation to a salinity gradient, in which he noted a decrease in the number of species proceeding from the "euryhaline Meerestiere" to the "holoeuryhaline Tiere." Reasonably enough lower salinity regions have a preponderance of freshwater forms (Bilio, 1967, Capstick, 1959, Warwick, 1971, Brenning, 1973).

Associated with the switch in fauna, there is usually a decrease in the number of species as brackish water is approached. There is also a general decrease in the number of animals per unit area as one proceeds up an estuary (Capstick, 1959; Brickman, 1972; Van Damme et al., 1980). However, other parameters distort this idealized picture, e.g., Warwick (1971) recorded his lowest nematode species diversity in the middle station of the Tamar Estuary. Numbers of animals usually increase seaward, probably because euryhaline marine animals make up the greatest portion of estuarine faunas and, therefore, most of the changes in faunal distribution are due to the failure of marine animals to penetrate salinities below their adaptive capacities. Truly euryhaline estuarine faunas are rare and euryhaline freshwater forms almost non-existent. Thus, with the preponderance of marine forms responsible for the estuarine assemblage the decreases in population parameters experienced are expected.

Across sandy intertidal habitats, species are often restricted to certain zones. Typically species groups include: (1) a sublittoral fringe guild, (2) eurytopic species, usually with their distribution centered on the lower shore and (3) species confined to the upper shore. Platt (1977) and Blome (1983) report on the zonation of nematodes in such habitats and Moore (1979b) on such distinct harpacticoid assemblages. McLachlan (1980) proposed four strata for the low tide distribution of meiofauna on sandy beaches: (1) a dry sand stratum, near the top of the beach where the upper sand layers are > 50% desiccated. (2) A moist sand stratum, which underlies the dry sand stratum and extends seaward. It reaches to the depth of the permanent water table; desiccation is <50% and O_2 levels are high (>70% saturation). (3) A water table stratum crossing the beach and lying underneath strata 1 or 2. The sand is always moist and oxygen saturation is 40-70%. (4) A low oxygen stratum underneath stratum 3, which may be very deep in high energy beaches, where O_2 tension is <30% saturation.

McLachlan (1980) emphasized that this pattern can be modified by a variety of factors (e.g., change in wave action or sediment grain size, changing beach slope, freshwater seepage, tidal amplitude, temperature as it affects desiccation). Other authors (e.g., Salvat, 1964; Pollock and Hummon, 1971) have reported similar patterns from other areas of the world but have used different controlling factors. Pollock (1970) illustrated the horizontal zonation of tardigrades on a Massachusetts, USA, sandy beach (Figure 3.3). It is obvious that species are zoned on beaches, but clearly there are no unequivocal and universal causative factors.

In muddy substrata, abundance and species also appear to be zoned horizontally. Working with meiobenthic copepods in South Carolina salt marsh muds Coull et al. (1979) reported one species *(Microarthridion littorale)* to cross the entire gradient from high intertidal salt marsh to the subtidal, whereas other species were restricted to certain sub-habitats along the gradient (Figure 3.4). The same genera found by Coull et al., (1979) occupy similar habitats in France (Bodin, 1977; Castel and Lasserre, 1979).

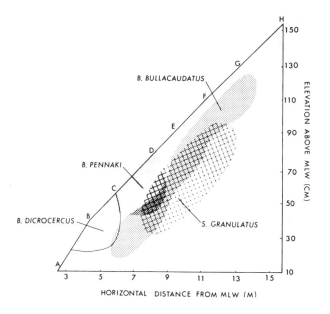

Figure 3.3.--Horizontal distribution of three tardigrade species on a sandy beach at Woods Hole, MA, USA (from Pollock, 1970).

Crossing the continental shelf to the deep sea, the faunal composition is primarily a function of sediment and depth changes. Sandy continental shelves harbor the families and genera traditionally found in sandy substrates. Changes at the taxon level coincide with physical changes in the substratum. Tietjen (1971) found that archiannelid polychaetes, gastrotrichs and halacarids were completely absent and copepod abundance markedly reduced when the sediment changed from sand to silty sand; similar to what one might expect in shallow water. Figure 3.5 illustrates the relative abundance of nematode families from two transects off North Carolina, USA (Tietjen, 1976). Certain families are more abundant at particular depth intervals, and this is also true for the Harpacticoida (Coull et al., 1982; Hicks and Coull, 1983). Additionally, in deep sea areas without current much of the fauna lives on the surface, whereas if currents are present most of the fauna burrows (Thistle et al., 1985). The similarity in fauna with depth, apparently all over the world, led Por (1965) to propose that meiobenthic harpacticoids were pan-bathyal; this also seems to be true for nematodes (Tietjen, 1984). Indeed, it appears that wherever studies have been conducted the same families and/or genera comprise a significant portion of the fauna at similar depths.

In the Mediterranean and the Arctic it appears that many of the "deep-sea" genera are displaced into shallower (<500 m) depths (Soyer, 1970; Montagna and Carey, 1978). In both the Arctic and Mediterranean, sediments typically found in the deep-sea are similarly displaced into shallow water and, of course, shallow Arctic habitats have similar ambient temperatures as the deep-sea elsewhere. There is thus further circumstantial evidence of species composition and distribution being controlled by variables of the physical environment.

Dispersion-Patchiness

Meiofauna are patchy in their distribution (e.g., Vitiello, 1968; Gray and Rieger, 1971; McLachlan, 1978; Thistle, 1978; Findlay, 1981). Causes of such distributions are difficult to define, but Findlay (1981) reviews most of the proposed contributors to micro-scale patchiness. On the large scale (m-km), gradients in physical factors (e.g., salinity, tidal exposure, sediment granulometry, oxygen concentrations) are primarily responsible for variances in abundance, whereas on smaller (cm) scales animal-habitat processes increase in importance (Findlay, 1981). Such animal-habitat interactions may be in response to physical variables (e.g., microtopography of sediments sensu Thistle, 1978, Hogue and Miller, 1981) or biological variables.

Localized food has been suggested as a potential cause of microspatial patchiness (e.g., Gray, 1968; Gray and Johnson, 1970; Gerlach, 1977a; Hicks, 1977; Lee et al., 1977). If food is patchy, predictably, so might be the meiofauna. Macrobenthos (animals and plants) and their associated biogenic structures provide another source of heterogeneity (e.g., Bell et al., 1978; Findlay, 1981). Even in the deep-sea, where the environment has long been thought of as rather homogeneous, biogenic structures, particularly agglutinated foraminiferan tests and polychaete mud balls, are important in providing loci for high meiofauna densities and diversities (Thistle, 1978).

Predation and/or disturbances could cause patchiness by creating "holes" in an otherwise homogeneous background. Thistle (1980) and Reidenauer and Thistle (1981) concluded that since most species were indifferent to "mechanical" disturbances, such events may not be the major factor affecting harpacticoid copepod aggregations. Certainly, we need more information to evaluate predation/disturbance as a factor controlling meiofauna aggregations. Interspecific competition could potentially lead to spatial segregation. Spacing of congeneric population to avoid competition needs further investigation to substantiate this as a mechanism allowing for localized coexistence.

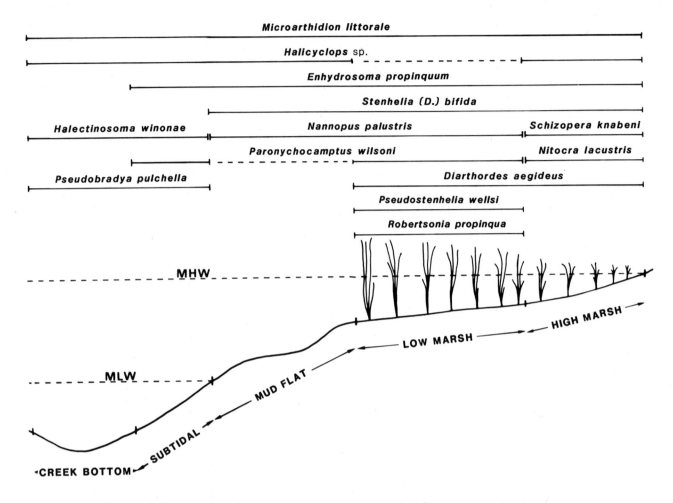

Figure 3.4.—Horizontal distribution of the dominant copepod species in muddy Southeastern United States salt marshes (from Coull et al., 1979).

Reproductive activity and life cycle stage may also affect spatial distribution. Certainly there are ontogenetic differences in feeding preferences and it seems reasonable that reproductively active males and females would need to be near each other.

Differences in "life-style" should also be considered in assessing contagion in populations. Epibenthic species should respond to different niche axes than, for example, interstitial forms. Epibenthic species with their ability to "swim," might therefore be less restricted to a particular microhabitat, and thus less aggregated or occur in aggregations of a larger spatial scale. We presently lack a framework of experimentation and hypotheses which might allow us to test microspatial partitioning of meiofauna. While we know that the species are aggregated, it is not possible to define what induces these aggregations since so many factors can be causative.

Dispersal

Although meiofauna inhabit some of the most dynamic environments imaginable (e.g., exposed shore sediments and algae) these animals have traditionally been considered sedentary. Emphasis has centered on adaptations for remaining in close proximity to the substratum, particularly because pelagic larvae are virtually nonexistent. Development, morphology, and biology all seem designed to ensure that the organism remains in or on the substratum (Swedmark, 1964; Sterrer, 1973). Based on such observations one would expect limited distribution patterns for related taxa. This, however, is not the case as numerous species are cosmopolitan.

Plate tectonics has been invoked as a potential mechanism to describe pan-oceanic and world-wide meiofaunal distributions (Sterrer, 1973; Hicks, 1977; Hagerman and Reiger, 1981). Recognizing plate tectonics as one mechanism, Gerlach (1977b)

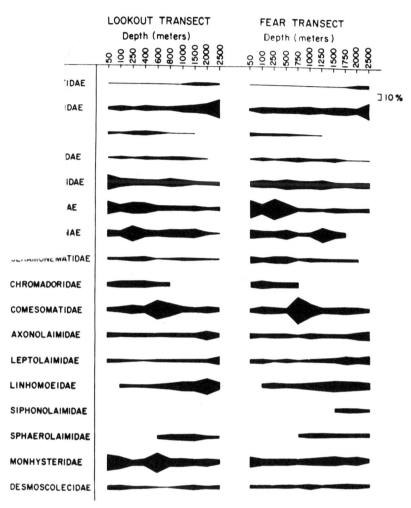

Figure 3.5.--Distribution and relative abundances of the major nematode families between 50 and 2500 meters on two transects off North Carolina, USA (from Tietjen, 1976).

summarized other possible means of meiofauna dispersal: (1) dispersal via airborne animals; (2) rafting on drifting materials; (3) transport in the ballast of sailing vessels; and (4) dispersal by suspension in the water column.

Evidence of widespread dispersal began to accumulate from colonization experiments. Scheibel (1974) revealed that submerged platforms were colonized by harpacticoids within days. It became apparent that colonization could occur much more rapidly when Sherman and Coull (1980) detailed complete recolonization of a mudflat site denuded of meiofauna within one tidal cycle. Such recolonization was suspected to be the result of suspended meiofauna being carried by waves and tides (Hagerman and Rieger, 1981). Palmer (1984) found that under flow conditions in a flume many taxa burrowed (to avoid suspension?) and that those that behaviorally occupied the sediment surface were suspended much more frequently than the burrowers. From the studies in the field Palmer and Gust (1985) reported that transport of meiobenthos was primarily a passive process resulting from mechanical removal due to current scour and that the abundance of meiobenthos in the water column at any one time was a function of the magnitude of the current velocity. As water moves, either as the ebb or flood, there are more meiofauna in the water column (Figure 3.6), thus illustrating the relationship between meiofaunal abundance in the water column and tidal velocity. In sea grass beds meiobenthic copepods actively migrate into the water column (Walters and Bell, 1986). Thus, it appears that while erosion is important in meiofaunal movement, species can also actively leave the sediments. Whether these two modes of excursion into the water column are mutually exclusive requires further study.

Temporal Variability.--Intertidal and shallow water meiobenthos are known to vary seasonally. Typically, maximum abundances occur in the warmer months of the year but there are certainly exceptions. Most studies of the seasonal abundance of meiofauna have been restricted to studies of about a year's duration. While maximum abundance of all meiofauna regularly peak in the warm months, individual taxa or species may reach maximum abundance at other times of the year. Reports of long-term analysis of meiofauna (or individual meiofauna taxa) are limited. McIntyre and Murison (1973) reported on monthly sampling for 12 months and irregular sampling for 9 years on total meiofauna and major taxa; Herman and Heip (1983) on fluctuations of 3 copepod species over 7 years; Coull (1985) on total meiofauna and major taxa over 11 years, and Coull and Dudley (1985) on copepod species over 11 years. In Scotland annual variations in meiofaunal abundance were small (McIntyre and Murison, 1973), but in South Carolina, U.S.A., year to year variations at both a sand and a mud site were greater than the inherent seasonality (Coull, 1985). Herman and Heip (1983) reported that one of their three copepods had a long-term (about 3.5 yr) periodicity, whereas Coull and Dudley (1985) found that no species of 12 investigated had an abundance periodicity of more than a year. Thus long-term periodicity patterns for meiofauna cannot be generalized from the data sets at hand and more long-term studies must be completed to determine if there are indeed any long-term patterns.

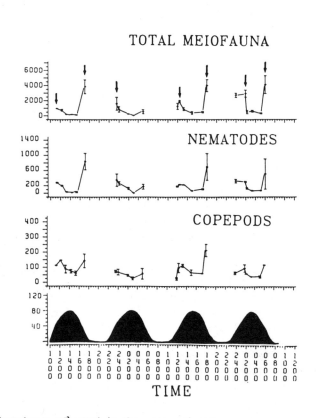

Figure 3.6.--The mean abundance of meiofauna (number x m^{-3} water) in the water column: for two neap and two spring tidal cycles. Stage of the tide is shown as water depth (cm) over the mudflat; 95% confidence intervals are shown (from Palmer and Gust, 1985).

There is no common world-wide pattern of temporal abundance and without reiterating every such study undertaken suffice it to say that before attempting any experimental, manipulative or quantitatively correlative work in a particular area, seasonality of the dominant species should be known. In many parts of the world these data are available in the primary literature, if not, baseline seasonal investigations are recommended on at least the abundance of a dominant species, together with sequential observations on reproductive state, population structure and associated biological/physical variables in that habitat.

The Role of Meiofauna in Benthic Processes

The methodologies to measure meiofaunal processes are outlined in other chapters of this book thus I will not repeat them. In this section I will consider three aspects of meiofaunal ecology as it relates to benthic processes, i.e.: (1) what do meiofauna feed upon *in situ* and what is the role of meiofauna in nutrient remineralization? (2) production and the role of meiofauna as food for higher trophic levels, and (3) the response of meiofauna to benthic disturbances, stress, pollution, etc.

Food of Meiofauna and Remineralization of Organic Matter.--Attempts to discern the food of meiofauna in the field have generally concluded that meiofauna are primarily detrital feeders or indiscriminant feeders on diatoms and bacteria. Since field analyses are mostly based on visual gut content observation, most studies quantifying meiofauna feeding have been conducted in the laboratory. These laboratory studies have found, for example, (1) that turbellarians prey on other meiofauna (Watzin, 1983); (2) up to seven different diatom species are consumed by a single nematode species (Romeyn et al., 1983); (3) nematodes suspected of feeding on bacteria because of their buccal morphology ingested more bacteria than diatoms when given a choice, and those suspected of feeding on diatoms ingested more diatoms than bacteria (Tietjen and Lee, 1977a); and (4) nematodes and copepods are known to elaborate mucus to trap bacteria and the bacterial/mucus mixture is ingested (Riemann and Schrage, 1978; Hicks and Grahame, 1979). Thus, while *in situ* meiofaunal feeding experiments are essentially non-existent because of practical reasons, we have learned much from the laboratory feeding experiments. By radioactively labeling bacteria and diatoms in cores which were replaced in the field, Montagna (1984) found that 3% of the bacteria and 1% of the diatoms were removed per hour. These preliminary field results suggest that meiofauna could maintain bacterial and diatom assemblages in log phase growth and thus correlations of standing stock of food and meiofauna might not be indicative of the rate of feeding.

There is also evidence that meiofauna play an important role in making detritus available to macroconsumers (Tenore et al., 1977). Net incorporation rates of five month-aged eelgrass detritus by the polychaete *Nephtys*, cultured with and without meiobenthos, were nearly doubled in cultures containing meiofaunal organisms. The authors suggested that the observed increase in net incorporation of the aged detritus could be due to a combination of factors, such as "meiofaunal enhancement of microbial activity and subsequent polychaete utilization and/or ingestion of the meiofauna themselves."

Production of Meiofauna.--An important aspect of ecological studies is the energetic role played by the organisms under investigation. Questions such as: how much do they produce? What percentage of the total energy do they use? and what is their ultimate fate? are of particular interest. Gerlach (1971) suggested that for an equivalent biomass, the meiofauna are responsible for about 5 times the total benthic metabolism of the macrofauna. Thus, a macrofauna/meiofauna standing stock biomass ratio of >5:1 is requisite before the total energy requirements of the macrofauna exceed that of the meiofauna. Macrofauna biomass in most systems is usually greater than the 5:1 ratio but in extremely shallow water (mudflats, estuaries) and the deep sea the macro/meiofauna standing stock biomass ratio approximates 1 (Gerlach, 1971; Thiel, 1975; 1983) and thus the meiofauna play a more significant role in benthic energetics in these systems.

Prerequisite for understanding production of meiofauna are estimates of life history parameters such as reproductive potential, fecundity, number of broods, longevity, and rate of development. Extrinsic factors such as temperature, salinity, and nutrition are known to affect life history parameters. Most of our knowledge related to reproductive capabilities of meiofauna comes primarily from species cultured in the laboratory (Tietjen and Lee, 1972, 1977b; Fenchell, 1974; Hicks and Coull, 1983). Attempts to measure such parameters in the field are difficult because of the small size of the organisms and the need for samples to be replicated at close time intervals. Thus, there are relatively few estimates of life history parameters from field collected data. Meiobenthic crustaceans, because they go through a series of distinct larval stages have been the subject of most of the field life history studies (see Tables 6-10 of Hicks and Coull for summary of life history parameters for harpacticoid copepods). Field estimated growth is typically slower than that estimated in the lab and obviously we must have field estimates on a variety of taxa before any such meiofauna generalization can be made. Larval and juvenile stages, of course, must be included and while this is not an easy task for most taxa, life history parameter estimates without them are useless.

Annual turnover rate (life cycle turnover x number of generations), a very difficult measure to quantify, has been estimated to approximate *10* for the meiofauna (McIntyre, 1964; Gerlach, 1971, 1978). Coupling turnover rate with standing crop biomass indicates that in systems dominated by macrofaunal biomass, meiofaunal and foraminiferan production is about equal to deposit feeding macrofauna (Gerlach, 1978), but in systems like the deep sea or shallow water total annual animal production will most certainly be dominated by the meiofauna.

Production is not easy to measure in meiofauna (see Feller and Warwick, Chapter 13). Most production estimates for meiofauna have relied upon indirect methodologies which utilize knowledge of turnover rates of populations. For example, McIntyre (1964) found production of meiobenthos to lie between 2.5–3.8 g C m^{-2} x yr^{-1} assuming a Production to Biomass (P/B) ratio of 10 per year and C to be 40% of dry weight. Ankar and Elmgren (1976) estimated (P/B=10) total meiofaunal production in the shallow Baltic Sea as 2.7 g C m^{-2} x yr^{-1}, Heip (1980) using equations generated from P/B ratios derived production for four benthic copepods as 0.02–0.11 g C m^{-2} x yr^{-1}, Warwick et al., (1979) found a total meiofaunal production of 13.50 g C m^{-2} x yr^{-1} in the Lynher estuary, U.K., and Faubel et al., (1983) 0.16 g C m^{-2} x yr^{-1} at 117–138 m in the North Sea with a P/B ratio of 1.4.

To date there are only four empirically derived production estimates for meiofauna, all crustaceans, which directly measure biomass of various life history stages in the field (Fleeger and Palmer, 1982; Feller, 1982, Herman et al. 1983, 1984). Table 3.2 lists the P/B ratios of these four crustaceans and they are widely variable. Vranken and Heip (1986) estimated biomass turnover per generation at about 3 for laboratory cultured free living marine nematodes and P/B ratios ranging between 4 and 63 for 6 nematode species. The generalized P/B value of *10* obviously needs to be examined as to its continued utility since even within closely related taxa great variability exists. Obviously, more data on *in situ* production/life histories are important if we are ever to truly evaluate the energetic role of the meiobenthos.

Meiofauna as Food for Higher Trophic Levels.—A primary question is what is the ultimate fate of meiofaunal production. There has been much controversy in recent years as to the interaction between meiofauna and higher trophic levels. McIntyre and Murison (1973) and Heip and Smol (1975) suggested that meiobenthic prey species were consumed primarily by meiobenthic predators and thus were not available to higher trophic levels. Lasserre et al. (1975) speculated that there may have been competition for food between species of detrital feeding meiofauna and the European gray mullet, *Chelon labrosus*. McIntyre (1964) and Marshall (1970) concluded that there was competition for food between macrofauna and meiofauna and that meiofauna served primarily as rapid metazoan nutrient regenerators. However, over 50 papers have been published since the early 1970's documenting the presence of meiofaunal prey in the stomach contents of marine fish and invertebrate predators and thus there can no longer be any doubt about the significance of meiofauna as food for higher trophic levels.

While such predation on meiofauna is undeniable, few studies document available prey (i.e., concurrent field abundances along with predator gut contents) and only one (Hicks, 1984) has examined predation effects on population parameters of the prey population. Where both habitat and gut content abundance values are available for prey (e.g., Bodiou and Villiers, 1979; Sibert, 1979; Schmidt-Moser and Westphal, 1981; Alheit and Scheibel, 1982; Carle and

Table 3.2.—Calculated yearly production and production to Biomass (P/B) ratios for four meiobenthic crustaceans determined from field observations, HC = harpacticoid copepod; O = ostracod. (From Herman et al., 1984.)

Species	Production	P/B (yr^2) gcm^2 x yr^1	Reference
Huntemannia jadensis (HC)	0.7–1.7	3.6	Feller, 1982
Microarthridion littorale (HC)	0.6	18.0	Fleeger and Palmer, 1982
Cyprideis torosa (O)	3.68–3.88	2.7	Herman et al., 1983
Tachidius discipes (HC)	0.98–1.01	34.3	Herman et al., 1984

Hastings, 1982; Zander and Hartwig, 1982; Coull and Wells, 1983; de Morais and Bodiou, 1984; Hicks, 1984; Sogard, 1984), benthic copepods are reported to be overwhelmingly selected over other available prey even though copepods are rarely the most abundant taxon in the habitat. Additionally, when species data are available (Bodiou and Villiers, 1979; Sibert, 1979; Alheit and Schiebel, 1982; Coull and Wells, 1983; Hicks, 1984, de Morais and Bodiou, 1984), there appears to be selectivity for certain copepod species – ones that are typically surface floc dwellers and presumably easier to catch.

Meiofauna appears to serve as food for higher trophic levels more in muds than in sands. In most cases where meiofauna are known to be food for higher trophic levels the study has been conducted in a muddy, detrital or phytal substrate. Conversely, those papers which suggest that meiofauna are not food for higher trophic levels are based primarily on work done in sandy environments. In muddy/detrital substrates most of the meiofauna are restricted to the top-most sediment layers and an indiscriminant browser/ingester would inevitably collect the resident meiofauna. However, with all the available interstitial space in sands, meiofauna go deeper into the sediment and are not as susceptible to browsing predation, but epibenthic sand meiofauna are also susceptible to predation. Smith and Coull (1987) have demonstrated experimentally that when given a choice, juvenile fish *(Leiostomus xanthurus)* will preferentially take meiofauna from mud rather than sand. If given no sediment choice the fish fed on meiofauna in mud or sand. Phytal meiofauna are also readily available prey (Hicks and Coull, 1983; Sogard, 1984), and thus while sediment substrate type is important, more important is the relative availability of the prey item to the predator.

Figure 3.7 schematically illustrates a potential meiofaunal food web including macrofauna, meiofauna, swimming predators and meiofaunal food. Meiofauna food is suspected to be diatoms, bacteria, detritus and perhaps protozoans and dissolved

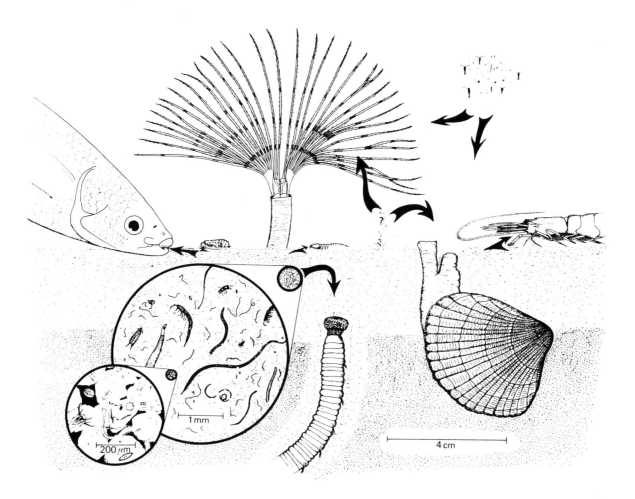

Figure 3.7.--Hypothetical benthic food web, illustrating possible connections between meiofauna and macrofauna (from Platt, 1981).

organic matter. Meiofauna can be eaten by swimming predators (e.g., fish, shrimp, and mysids), deposit feeders or by suspension feeders if the meiofauna is suspended.

Responses of Meiofauna to Perturbations.--Field experimentation as reviewed by Coull and Palmer (1984) along with the methodological outline of Bell (Chapter 12) has provided much insight into responses of meiobenthos to natural or man-made perturbations. Only by determining response of an assemblage before and after an event can one state with statistical surety that the event did or did not change the assemblage. Meiofauna are known to be sensitive indicators of environmental perturbations. Because of their large numbers, relatively stationary life habits, short generation times, benthic larvae and intimate association with sediments known to accumulate various contaminants, meiofauna have also become popular subjects to study in relation to pollution. One must realize however, that meiofauna are notoriously difficult to identify to species and since species should be identified to provide a most complete picture of an effect, the use of meiofauna as perturbation indicators is limited.

Because of the rather difficult taxonomy of meiofauna, many perturbation studies prefer not to include them. Parker (1975) and Raffaelli and Mason (1981) have proposed using the ratio of nematodes to copepods (N/C ratio) as a monitor of pollution or sediment changes that eliminates the need for detailed, time consuming species identification. While this idea is very attractive it has generated considerable controversy (Coull et al., 1981; Raffaelli, 1981; Warwick, 1981; Amjad and Gray, 1983; Vidakovic, 1983; Lambshead, 1984; Platt et al., 1984) and it has not been proven to be an accurate predictor of environmental change. While the technique is simple because no taxonomic experts need be sought, further research is needed to determine the universality of the N/C ratio as a measure of environmental perturbation.

In almost any disturbance there is an immediate decrease in the abundance and diversity of the meiofauna (Coull and Palmer, 1984, Tables 1 and 2), but there is a subsequent recovery. Rates of recovery vary from hours to months, and it is these recovery rates that are of primary interest in colonization studies. The crustacean meiofauna regularly seems to be the most quickly affected by a perturbation and nematodes the least affected. Individual taxa recolonize at different rates and thus, once again, the value in obtaining species information cannot be overempahsized.

Studies on the ecology of meiofauna have progressed from a descriptive mode to one where experimentation is now in vogue. Ecological processes involving meiofauna need to be examined because as we learn more about the role of these ubiquitous organisms it is apparent that they play important roles in aquatic ecosystems. As technology continues to advance, the goal of future meiofaunal studies should be to unravel the mysteries of how meiofauna function. With ingenuity, hard work and some luck, hopefully we can understand just how important meiofauna are.

Acknowledgments

Contribution No. 700 from the Belle W. Baruch Institute for Marine Biology and Coastal Research.

I particularly thank Bettye W. Dudley, Roberta L. Marinelli, Richard A. Eskin, Robert J. Feller, Robert P. Higgins, Hjalmar Thiel, and two anonymous reviewers for suggesting revisions in earlier versions of this manuscript. This research was supported by the Biological Oceanography section of the National Science Foundation, Grant OCE85-21345 (B.C. Coull and R.J. Feller, principal investigators).

References

Alheit, J., and W. Scheibel
1982. Benthic Harpacticoids as a Food Source for Fish. *Marine Biology*, 70:141-147.

Aller, J.Y., and R.C. Aller
1986. Evidence for Localized Enhancement of Biological Activity Associated with Tube and Burrow Structure in Deep-Sea Sediment at the HEBBLE Site, Western North Atlantic. *Deep-sea Research*, 33:755-790.

Amjad, S., and J.S. Gray
1983. Use of the Nematode-copepod Ratio as an Index of Organic Pollution. *Marine Pollution Bulletin*, 14:178-181.

Anderson, R.S., and A.-M. de Henau
1980. An Assessment of the Meiobenthos from Nine Mountain Lakes in Western Canada. *Hydrobiologia*, 70:257-264.

Ankar, S., and R. Elmgren
1976. The Benthic Macro- and Meiofauna of the Askö-Landsort area (Northern Baltic Proper). A Stratified Random Sampling Survey. *Contributions of the Askö Laboratory, University of Stockholm*. Volume 11, 115 pages.

Ansari, Z.A.
1978. Meiobenthos from the Karwar Region (Central West Coast of India). *Mahasagar Bulletin of the National Institute of Oceanography*, 11:163-167.

Ansari, Z.A., and B.S. Ingole
1983. Meiofauna of Some Sandy Beaches of Andaman Islands. *Indian Journal of Marine Science*, 12:245-246.

Ansari, Z.A., and A.H. Parulekar
1981. Meiofauna of the Andaman Sea. *Indian Journal of Marine Science*, 10:285-288.

Ansari, Z.A., A.H. Parulekar, and T.G. Jagtap
1980. Distribution of Sublittoral Meiobenthos off Goa Coast, India. *Hydrobiologia*, 74:209-214.

Ansari, Z.A., C.L. Rodrigues, A. Chatterji, and A.H. Paruleka
1982. Distribution of Meiobenthos and Macrobenthos at the Mouth of Some Rivers of the East Coast of India. *Indian Journal of Marine Science*, 11:341-343.

Arlt, G.
1977. Verbreitlung und Artenspektrum der Meiofauna im Greifswalder Bodden. *Wissenschaftliche Zeitschrift der Universität Rostock, Mathematisch-Naturwissenschaftliche Reihe*, 26:217-222.

Arlt, G., and J. Holtfreter

1975. Verteilung, Zusammensetzung und jahreszeitliche Fluktuationen der Meiofauna im Barther Bodden (Darss-Zingster Boddenkette). *Wissenschaftliche Zeitschrift der Universität Rostock, Mathematisch-Naturwissenschaftliche Reihe*, 6:743-751.

Arlt, G., and M.A.H. Saad
1977. Investigations of the Meiofauna and Sediment of the Shatt Al-Arab near Barash (Iraq). *Freshwater Biology*, 7:487-494.

Barnett, P.R.O.
1968. Distribution and Ecology of Harpacticoid Copepods of an Intertidal Mudflat. *Internationale Revue der gesamten Hydrobiologie*, 53:177-209.

Bell, S.S.
1979. Short- and Long-term Variability in a High Marsh Meiofauna Community. *Estuarine and Coastal Marine Science*, 9:331-350.

Bell, S.S., M.C. Watzin, and B.C. Coull
1978. Biogenic Structure and Its Effect on the Spatial Heterogeneity of Meiofauna in a Salt Marsh. *Journal of Experimental Marine Biology and Ecology*, 35:99-107.

Bilio, M.
1966. Die aquatische Bodenfauna von Salzwiesen der Nord- und Ostsee. II. Ökologische Fauneanalyse: Hydrozoa, Nematodes, Rotatoria, Gastrotricha, Halacaridae, Ostracoda, Copepoda. *International Revue der gesamten Hydrobiologie*, 51:147-195.

1967. Die aquatische Bodenfauna von Salzwiesen der Nord- und Ostsee. III. Die Biotopeinflüsse auf die Faunenverteilung. *International Revue der gesamten Hydrobiologie*, 52(4):487-533.

Blome, D.
1983. Ökologie der Nematoda eines Sandstrandes der Nordseeinsel Sylt. *Mikrofauna des Meeresbodens*, 88:1-76.

Boaden, P.J.S.
1980. Meiofaunal Thiobios and "the *Arenicola* Negation": Case not Proven. *Marine Biology*, 58:25-29.

Boaden, P.J.S., and E.A.G. Elhag
1984. Meiobenthos and the Oxygen Budget of an Intertidal Sandy Beach. *Hydrobiologia*, 118:39-48.

Boaden, P.J.S., and H.M. Platt
1971. Daily Migration Patterns in an Intertidal Meiobenthic Community. *Thalassia Jugoslavica*, 7:1-12.

Bodin, P.
1977. Les peuplements de copépodes harpacticoides (Crustacea) des sédiments meubles de la zone intertidale des côtes Charentaises (Atlantique). *Mémoires du Museum National d'Histoire Naturelle, Series A, Zoologie*, 104:1-120.

Bodiou, J.-Y., and L. Villiers
1979. La prédation de la méiofaune par les formes juvéniles de *Deltentosteus quadrimaculatus* (Teleostei, Gobiidae). *Vie et Milieu*, 28-29(1):143-156.

de Bovee, F., and J. Soyer
1977. Le meiobenthos des Iles Kerguelen Données Quantitatives. I. Le Golfe du Morbihan and II. Le Plateau Continental. In A. Guille and J. Soyer, editors, *Le Benthos du Plateau Continental, Des Iles Kerguelen Résultats Scientifiques*. Centre National Francais Recherches Antarctica (CNFRA), 42:237-258.

Brenning, U.
1973. The Distribution of Littoral Nematodes in the Wismarbucht. *Oikos*, supplement, 15:98-104.

Brickman, L.M.
1972. *Base Food Chain Relationships in Coastal Salt Marsh Ecosystems.* 179 pages. Ph.D. Thesis, Lehigh University.

Capstick, C.K.
1959. The Distribution of Free Living Nematodes in Relation to Salinity in the Upper and Middle Reaches of the River Blyth Estuary. *Journal of Animal Ecology*, 28:189-210.

Carey, A.G., Jr., and P.A. Montagna
1982. Arctic Sea Ice Faunal Assemblage: First Approach to Description and Source of the Under Ice Meiofauna. *Marine Ecology Progress Series*, 8:1-8.

Carle, K.J., and P.A. Hastings
1982. Selection of Meiofaunal Prey by the Darter Goby, *Gobionellus boleosoma* (Gobiidae). *Estuaries*, 5:316-318.

Castle, J., and P. Lasserre
1979. Opportunistic Copepods in Temperate Lagoons of Arcachon Bay: Differential Distribution and Temporal Heterogeneity. *Estuarine and Coastal Marine Science*, 9:357-368.

Clasing, E.
1976. Fluctuaciones anuales de la meiofauna en la marisma de Chinquihue (puerto Montt, Chile). *Studies of the Neotropical Fauna Environment*, 11:179-198.

Coull, B.C.
1973. Estuarine Meiofauna: A Review, Trophic Relationships and Microbial Interactions. Pages 499-511 in L. H. Stevenson and R.R. Colwell, editors, *Estuarine Microbial Ecology*. Columbia: University of South Carolina Press.

1985. Long-term Variability of Estuarine Meiobenthos: An 11 Year Study. *Marine Ecology Progress Series*, 24:205-218.

Coull, B.C., and S.S. Bell
1979. Perspectives of Marine Meiofaunal Ecology. Pages 189-216 in R. J. Livingston, editor, *Ecological Processes in Coastal and Marine Ecosystems*. New York: Plenum Publishing Company.

Coull, B.C., and B.W. Dudley
1985. Dynamics of Meiobenthic Copepod Populations: A Long-Term Study (1973-1983). *Marine Ecology Progress Series*, 24:219-229.

Coull, B.C., and J.W. Fleeger
1977. Long-term Temporal Variation and Community Dynamics of Meiobenthic Copepods. *Ecology*, 58:1136-1143.

Coull, B.C., and M.A. Palmer
1984. Field Experimentation in Meiofaunal Ecology. *Hydrobiologia*, 118:1-19.

Coull, B.C., and J.B.J. Wells
1981. Density of Mud Dwelling Meiobenthos from Three Sites in the Wellington Region. *New Zealand Journal of Marine and Freshwater Research*, 15:411-415.

1983. Refuges from Fish Predation: Experiments with Phytal Meiofauna from the New Zealand Rocky Intertidal. *Ecology*, 64:1599-1609.

Coull, B.C., G.R.F. Hicks, and J.B.J. Wells
1981. Nematode/Copepod Ratios for Monitoring Pollution: A Rebuttal. *Marine Pollution Bulletin*, 12:378-381.

Coull, B.C., S.S. Bell, A.M. Savory, and B.W. Dudley
1979. Zonation of Meiobenthic Copepods in a Southeastern U.S. Salt Marsh. *Estuarine and Coastal Marine Science*, 9:181-188.

Coull, B.C., Z. Zo, J.H. Tietjen, and B.S. Williams
1982. Meiofauna of the Southeastern United States Continental Shelf. *Bulletin of Marine Science*, 32:139-150.

Coull, B.C., R.L. Ellison, J.W. Fleeger, R.P. Higgins, W.D. Hope, W.D. Hummon, R.M. Rieger, W.E. Sterrer, H. Thiel, and J.H. Tietjen
1977. Quantitative Estimates of the Meiofauna from the Deep Sea off North Carolina, U.S.A. *Marine Biology*, 39:233-240.

Decho, A.W., W.D. Hummon, and J.W. Fleeger
1985. Meiofauna-Sediment Interactions Around Subtropical Seagrass Sediments Using Factor Analysis. *Journal of Marine Research*, 43:237-242.

DeLaune, R.D., C.J. Smith, W.H. Patrick, Jr., J.W. Fleeger, and M.D. Tolley
1984. Effect of Oil on Salt Marsh Biota: Methods for Restoration. *Environmental Pollution*, Series A, 36:207-227.

Dinet, A.
1972. Etude écologique des variations quantitative

annelles d'un peuplement de copepodes harpacticoides psammique. *Tethys*, 4(1):95-112.
1976. Étude quantitative du meiobenthos dan le secteur nord de la Mer Egee. *Acta Adriatica*, 18(5):83-88.
1977. Données quatitative sur le meiobenthos bathyal de la Mer de Norvége. Pages 13-14 in *Géochimie Organique des sédiments marins profond, Orgon I, Mer de norvége, Aout 1974*. CEPM-CNEXO, Comite Etudes Geo-Chim. Mar., CNRS.
1979. A Quantitative Survey of Meiobenthos in the Deep Norwegian Sea. *Ambio*, Special Report, 6:75-77.

Dinet, A., and M.-H. Vivier
1977. Le meiobenthos abyssal du Golfe de Gascogne. I. Considérations sur les données quantitative. *Cahiers de Biologie Marine*, 18:85-97.

Dinet, A., L. Laubier, J. Soyer, and P. Vitiello
1973. Resultats biologiques de la campagne Polymede. II. Le meiobenthos abyssal. *Rapports Proces-Verbaux des Reunions, Conseil International pour l'Exploration de la Mer, CIESMN*, 21:701-704.

Dye, A.H.
1983. Vertical and Horizontal Distribution of Meiofauna in Mangrove Sediments in Transkei, Southern Africa. *Estuarine and Coastal Shelf Science*, 16:591-598.

Ellison, R.L.
1984. Foraminifera and Meiofauna on an Intertidal Mudflat, Cornwall, England: Populations; Respiration and Secondary Production; and Energy Budget. *Hydrobiologia*, 109:131-148.

Elmgren, R., R. Rosenberg, A.-B. Andersin, S. Evans, P. Kangas, J. Lassig, E. Leppakoski, and R. Varmo
1984. Benthic Macro- and Meiofauna in the Gulf of Bothnia (Northern Baltic). *Finnish Marine Research*, No. 250:3-18.

Faubel, A., E. Hartwig, and H. Thiel
1983. On the Ecology of Sublittoral Sediments, Fladen Ground, North Sea. I. Meiofauna Standing Stock and Estimation of Production. *"Meteor" Forschungesergebnisse*, 36:35-48.

Feder, H.M., and A.J. Paul
1980. Seasonal Trends in Meiofaunal Abundance on Two Beaches in Port Valdez, Alaska. *Syesis*, 13:27-36.

Feller, R.J.
1982. Emperical Estimates of Carbon Production for a Meiobenthic Harpacticoid Copepod. *Canadian Journal of Fisheries and Aquatic Science*, 39:1435-1443.

Fenchel, T.
1974. Intrinsic Rate of Natural Increase: The Relationship with Body Size. *Oceologia* (Berlin), 14:317-326.
1978. The Ecology of Micro- and Meiobenthos. *Annual Review of Ecology and Systematics*, 9:99-121.

Fenchel, T.M., and R.J. Riedl
1970. The Sulfide System: A New Biotic Community Underneath the Oxidized Layer of Marine Sand Bottoms. *Marine Biology*, 7:255-268.

Findlay, S.E.G.
1981. Small Scale Spatial Distribution of Meiofauna on a Mud- and Sandflat. *Estuarine and Coastal Shelf Science*, 12:471-484.

Fleeger, J.W., and G.T. Chandler
1983. Meiofauna Responses to an Experimental Oil Spill in a Louisiana Salt Marsh. *Marine Ecology Progress Series*, 11:257-264.

Fleeger, J.W., and M.A. Palmer
1982. Secondary Production of the Estuarine Meiobenthic Copepod, *Microarthidion littorale*. *Marine Ecology Progress Series*, 7:157-162.

Galhanó, M.H.
1970. Contribucão para o conhecimento da fauna intersticial em Portugal. *Publicacoes do Institute Zoologia, Dr. A. Nobre, Faculdade de Ciencias do Porto*, 110:1-206.

Galtsova, V.V.
1971. A Quantitative Characteristic of Meiobenthos in the Chupinsky Inlet of the White Sea. *Zoologischeskii Zhurnal*, 50:641-647.

Gerlach, S.A.
1971. On the Importance of Marine Meiofauna for Benthos Communities. *Oceologia* (Berlin), 6:176-190.
1977a. Attraction to Decaying Organisms as a Possible Cause for Patchy Distribution of Nematodes in a Bermuda Beach. *Ophelia*, 16:151-165.
1977b. Means of Meiofauna Dispersal. *Mikrofauna des Meeresbodens*, 61:89-103.
1978. Food Chain Relationships in Subtidal Silty Sand Marine Sediments and the Role of Meiofauna in Stimulating Bacterial Productivity. *Oecologia* (Berlin), 33:55-69.

Giere, O., and O. Pfaunnkuche
1982. Biology and Ecology of Marine Oligochaeta, A Review. *Oceanography and Marine Biology*, Annual Review, 20:173-308.

Grainger, E.H., A.A. Mohammed, and J.E. Lovrity
1985. The Sea Ice Fauna of Frobisher Bay, Arctic Canada. *Arctic*, 38:23-30.

Grassle, J.F., R. Elmgren, and J.P. Grassle
1981. Response of Benthic Communities in MERL Experimental Ecosystems to Low Level Chronic Additions of No. 2 Fuel Oil. *Marine Environmental Research*, 4:279-297.

Gray, J.S.
1968. An Experimental Approach to the Ecology of the Harpacticoid *Leptastacus constrictus* Lang. *Journal of Experimental Marine Biology and Ecology*, 2:278-292.

Gray, J.S., and R.M. Johnson
1970. The Bacteria of a Sandy Beach as an Ecological Factor Affecting the Interstitial Gastrotrich *Turbanella hyalina* Schultze. *Journal of Experimental Marine Biology and Ecology*, 4:119-133.

Gray, J.S., and R.M. Rieger
1971. A Quantitative Study of the Meiofauna of an Exposed Sandy Beach at Robin Hood's Bay, Yorkshire. *Journal of the Marine Biological Association of the United Kingdom*, 51:1-20.

Hagerman, G.M., and R.M. Rieger
1981. Dispersal of Benthic Meiofauna by Wave and Current Action in Bogue Sound, North Carolina. *Publicazioni della Stazione Zoologia di Napoli I, Marine Ecology*, 2:245-270.

Harper, D.E., D.L. Potts, R.R. Salzer, R.J. Case, R.L. Jaschek, and C.M. Walker
1981. Distribution and Abundance of Macrobenthic and Meiobenthic Organisms. Pages 133-177 in B.S. Middleitch, editor, *Environmental Effects of Offshore Oil Production*. New York: Plenum Publishing Corporation.

Harris, R.P.
1972a. The Distribution and Ecology of the Interstitial Meiofauna of a Sandy Beach at Whitesand Bay, East Cornwall. *Journal of the Marine Biological Association of the United Kingdom*, 52:1-18.
1972b. Horizontal and Vertical Distribution of the Interstitial Harpacticoid Copepods of a Sandy Beach. *Journal of the Marine Biological Association of the United Kingdom*, 52:375-388.
1972c. Seasonal Changes in Population Density and Vertical Distribution of Harpacticoid Copepods on an Intertidal Sand Beach. *Journal of the Marine Biological Association of the United Kingdom*, 52:493-506.

Hartzband, D.J., and W.D. Hummon
1974. Sub-community Structure in Subtidal Meiobenthic Harpacticoida. *Oecologia* (Berlin), 14:37-51.

Heip, C.
1980. The Influence of Competition and Predation on Production of Meiobenthic Copepods. Pages 167-177 in K.R. Tenore and B.C. Coull, editors, *Marine Benthic Dynamics*. Columbia: University of South Carolina Press.

Heip, C., and N. Smol.
1975. On the Importance of *Protohydra leuckarti* as a Predator of meiobenthic Populations. *Tenth European Symposium on Marine Biology*, Ostend,

Belgium, 2:285-296.

Heip, C., M. Vincx, and G. Vranken
1985. The Ecology of Marine Nematodes. *Oceanography and Marine Biology Annual Review*, 23:399-489.

Heip, C., M. Vincx, N. Smol, and G. Vranken
1982. The Systematics and Ecology of Free-living Marine Nematodes. *Helminthological Abstracts, Series B, Plant Nematology, Commonwealth Institute of Parasitology*, 51:1-31.

Herman, P.M.J., and C. Heip
1983. Long-term Dynamics of Meiobenthic Popluations. *Oceanologica Acta*, 1983:109-112.

Herman, P.M.J., C. Heip, and G. Vranken
1983. The Production of *Cryprideis torosa* Jones, 1850 (Crustacea, Ostracoda). *Oecologia* (Berlin), 58:326-331.

Herman, P.M.J., C. Heip, and B. Guillemijn
1984. Production of *Tachidius discipes* (Copepoda: Harpacticoida). *Marine Ecology Progress Series*, 17:271-278.

Hicks, G.R.F.
1977. Species Composition and Zoogeography of Marine Phytal Harpacticoid Copepods from Cook Strait, and their Contribution to Total Phytal Meiofauna. *New Zealand Journal of Marine and Freshwater Research*, 11:441-469.

1984. Spatio-temporal Dynamics of a Meiobenthic Copepod and the Impact of Predation-disturbance. *Journal of Experimental Marine Biology and Ecology*, 81:47-72.

Hicks, G.R.F., and B.C. Coull
1983. The Ecology of Marine Meiobenthic Harpacticoid Copepods. *Oceanography and Marine Biology, Annual Review*, 21:67-125.

Hicks, G.R.F., and J. Grahame
1979. Mucus Production and Its Role in the Feeding Behaviour of *Diarthrodes nobilis* (Copepoda: Harpacticoida). *Journal of the Marine Biological Association of the United Kingdom*, 59:321-330.

Hodda, M., and W.L. Nicholas
1985. Meiofauna Associated with Mangroves in the Hunter River Estuary and Fullerton Cove Southeastern Australia. *Austalian Journal of Marine and Freshwater Research*, 36:41-50.

Hogue, E.W.
1978. Spatial and Temporal Dynamics of a Subtidal Estuarine Gastrotrich Assemblage. *Marine Biology*, 49:211-222.

Hogue, E.W., and C.B. Miller
1981. Effects of Sediment Microtopography on Small Scale Spatial Distribution of Meiobenthic Nematodes. *Journal of Experimental Marine Biology and Ecology*, 53:181-191.

Holopainen, I.J., and L. Paasivirta
1977. Abundance and Biomass of the Meiozoobenthos in the Oligotrophic and Mesohumic Lake Pääjarvi, Southern Finland. *Annales Zoologici Fennici*, 14:124-134.

Horn, T.D.
1978. The Distribution of *Echinoderes coulli* (Kinorhyncha) along an Interstitial Tidal Gradient. *Transactions of the American Microscopical Society*, 97:586-589.

Hummon, W.D., J.W. Fleeger, and M.R. Hummon
1976. Meiofauna-macrofauna Interactions: I. Sand Beach Meiofauna Affected by Maturing *Limulus* eggs. *Chesapeake Science*, 17:297-299.

Ito, T.
1978. Meiobenthos of a Shallow-water Sandy Bottom in Ishikari Bay, Hokkaido: A General Account. *Faculty of Science, Hokkaido University, Series VI, Zoology*, 21:287-294.

Jansson, B-O
1968. Quantitative and Experimental Studies of the Interstitial Fauna in Four Swedish Beaches. *Ophelia*, 5:1-72.

Joncyk, E., and T. Radziejewska
1984. Temporal Changes in Sand Meiofauna of a Southern Baltic Beach. *Limnologica* (Berlin), 15:421-423.

Kautsky, H., B. Widbom, and F. Wulff
1981. Vegetation, Macrofauna and Benthic Meiofauna in the Phytal Zone of the Archipelago of Lulea-Bothnian Bay. *Ophelia*, 20:53-77.

Kondalarao, B.
1983. Distribution of Meiofauna in the Gautami-Godavari Estuarine System. *Mahasagar-Bulletin of the National Institute of Oceanography (India)*, 16:453-457.

Lambshead, P.J.D.
1984. The Nematode/Copepod Ratio. Some Anomalous Results from the Firth of Clyde. *Marine Pollution Bulletin*, 15:256-259.

Lasserre, P., J. Renaud-Mornant, and J. Castel
1975. Metabolic Activities of Meiofaunal Communities in a Semi-enclosed Lagoon. Possibilities of Trophic Competition between Meiofauna and Mugilid Fish. *Tenth European Symposium on Marine Biology*, Ostend, 2:393-414

Lee, J.J., J.H. Tietjen, C. Mastropaolo, and H. Rubin
1977. Food Quality and the Heterogeneous Spatial Distribution of Meiofauna. *Helgoländer wissenschaftliche Meeresuntersuchungen*, 30:272-282.

Maguire, C., and P.J.S. Boaden
1975. Energy and Evolution in the Thiobios: An Extrapolation from the Marine Gastrotrich *Thiodasys sterreri*. *Cahiers de Biologie Marine*, 16:635-646.

Marcotte, B.M., and B.C. Coull
1974. Pollution, Diversity and Meiobenthic Communities in the North Adriatic (Bay of Piran, Yugoslavia) *Vie et Milieu*, 24(2B):281-300.

Marshall, N.
1970. Food Transfer Through the Lower Trophic Levels of the Benthic Environment. Pages 52-66 in J. H. Steele, editor, *Marine Food Chains.* Edinburgh: Oliver and Boyd.

Martinez, E.A.
1975. Marine Meiofauna of a New York City Beach, with Particular Reference to the Tardigrada. *Estuarine and Coastal Marine Science*, 3:337-348.

McIntyre, A.D.
1964. Meiobenthos of Sublittoral Muds. *Journal of the Marine Biological Association of the United Kingdom*, 44:665-674.

1969. Ecology of Marine Meiobenthos. *Biological Reviews*, 44:245-290.

McIntyre, A.D., and D.J. Murison
1973. The Meiofauna of a Flatfish Nursery Ground. *Journal of the Marine Biological Association of the United Kingdom*, 53:93-118.

McLachlan, A.
1977. Composition, Distribution, Abundance and Biomass of the Macrofauna and Meiofauna of Four Sandy Beaches. *Zoologica Africana*, 12:279-306.

1978. A Quantitative Analysis of the Meiofauna and the Chemistry of the Redox Potential Discontinuity Zone in a Sheltered Sandy beach. *Estuarine and Coastal Marine Science*, 7:275-290.

1980. Intertidal Zonation of Macrofauna and Stratification of Meiofauna on High Energy Sandy Beaches in the Eastern Cape, South Africa. *Transactions of the Royal Society of South Africa*, 44:213-223.

McLachlan, A., T. Erasmus, and J.P. Furstenberg
1977a. Migrations of Sandy Beach Meiofauna. *Zoologica Africana*, 12:257-277.

McLachlan, A., P.E.D. Winter, and L. Botha
1977b. Vertical and Horizontal Distribution of Sublittoral Meiofauna in Algoa Bay, South Africa. *Marine Biology*, 40:355-364.

Meineke, T., and W. Westheide
1979. Gezeitenabhängige Wanderungen der Interstitialfauna in einem Sandstrand der Insel Sylt (Nordsee). *Mikrofauna des Meeresbodens*, 75:1-36.

Meyers, M.B., H. Fossing, and E.N. Powell
1987. Microdistribution of Interstitial Meiofauna, Oxygen and Sulfide Gradients, and the Tubes of Macroinfauna. *Marine Ecology Progress Series*, 35:223-241.

Mielke, W.
1976. Ökologie der Copepoda eines Sandstrandes der Nordseeinsel Sylt. *Mikrofauna des Meeresbodens*, 59:1-86.

Möller, S., U. Brenning, and G. Arlt
1976. Untersuchungen über die Meiofauna des Barther Boddens unter besonderer Berücksichtigung der Nematoden. *Wissenschaftliche Zeitschrift der Wilhelm-Pieck Universität Rostock, Mathematisch-Naturwissenschaftliche Reihe*, 3:271-291.

Montagna, P.A.
1984. In Situ Measurement of Meiobenthic Grazing Rates on Sediment Bacteria and Edaphic Diatoms. *Marine Ecology Progress Series*, 18:119-130.

Montagna, P.A., and A.G. Carey, Jr.
1978. Distributional Notes on Harpacticoida (Crustacea:Copepoda) collected from the Beaufort Sea (Arctic Ocean). *Astarte*, 11:117-122.

Moore, C.G.
1973. The Kelp Fauna of Northeast Britain. II. Multivariate Classification: Turbidity as an Ecological Factor. *Journal of Experimental Marine Biology and Ecology*, 13:127-163.
1979a. The Distribution and Ecology of Psammolittoral Meiofauna Around the Isle of Man. *Cahiers de Biologie Marine*, 20:383-415.
1979b. The Zonation of Psammolittoral Harpacticoid Copepods Around the Isle of Man. *Journal of the Marine Biological Association of the United Kingdom*, 59:711-724.

de Morais, L.T., and J.Y. Bodiou
1984. Predation on Meiofauna by Juvenile Fish in a Western Mediterranean Flatfish Nursery Ground. *Marine Biology*, 82:209-215.

Munro, A.L.S., J.B.J. Wells, and A.D. McIntyre
1978. Energy Flow in the Flora and Meiofauna of Sandy Beaches. *Proceedings of the Royal Society of Edinburgh*, 76B:297-315.

Nalepa, T.F., and M.A. Quigley
1983. Abundance and Biomass of the Meiobenthos in Nearshore Lake Michigan with Comparisons to the Macrobenthos. *Journal of Great Lakes Research*, 9:530-547.

Noodt, W.
1957. Zur Ökologie der Harpacticoidea (Crustacea, Copepoda) des Eulitorals der Deutschen Meeresküste und der Nagrenzenden Brackewasser. *Zeitschrift für Morphologie und Ökologie der Tiere*, 46:149-242.

Oden, B.J.
1979. The Freshwater Littoral Meiofauna in a Reservoir Receiving Thermal Effluents. *Freshwater Biology*, 9:291-304.

Orren, M.J., G.A. Eagle, A.H. Fricke, W.J. Gedhill, P.J. Greenwood, and H. F-K. O. Hennig
1981. The Chemistry and Meiofauna of Some Unpolluted Sandy Beaches in South Africa. *Water SA*, 7:203-210.

Ott, J.A.
1972. Determination of Fauna Boundaries of Nematodes in an Intertidal Sand Flat. *International Revue der Gesamten Hydrobiologie*, 57:645-663.

Ott, J., and F. Schiemer
1973. Respiration and Anaerobiosis of Free Living Nematodes from Marine and Limnic Sediments. *Netherlands Journal of Sea Research*, 7:233-243.

Ott, J., G. Rieger, R. Rieger, and F. Enderes
1983. New Mouthless Interstitial Worms from the Sulfide System: Symbiosis with Procaryotes. *Pubblicazioni della Stazione Zoologica di Napoli I Marine Ecology*, 3:313-333.

Palmer, M.A.
1984. Invertebrate Drift: Behavioral Experiment with Intertidal Meiobenthos. *Marine Behaviour and Physiology*, 10:235-253.

Palmer, M.A., and G. Gust
1985. Dispersal of Meiofauna in a Turbulent Tidal Creek. *Journal of Marine Research*, 43:179-210.

Parker, R.H.
1975. The Study of Benthic Communities: A Model and Review. *Oceanography*, Series 9. Amsterdam: Elsevier Publishing Company.

Platt, H.M.
1977. Vertical and Horizontal Distribution of Free-living Marine Nematodes from Strangford Lough, Northern Ireland. *Cahiers de Biologie Marine*, 18:261-273.
1981. Meiofaunal Dynamics and the Origin of the Metazoa. Pages 207-216 in P.L. Forey, editor, *The Evolving Biosphere*. Cambridge: Cambridge University Press.

Platt, H.M., and R.M. Warwick
1980. The Significance of Free-living Nematodes to the Littoral Ecosystem. Pages 729-759 in J.H. Price, D.E.G. Irvine, and W.F. Farnham, editors, *The Shore Environment. 2. Ecosystems*. New York: Academic Press.

Platt, H.M., K.M. Shaw, and P.J.D. Lambshead
1984. Nematode Species Abundance Patterns and their Use in the Detection of Environmental Perturbations. *Hydrobiologia*, 118:59-66.

Pollock, L.W.
1970. Distribution and Dynamics of Interstitial Tardigrada at Woods Hole, Massachusetts. *Ophelia*, 7:145-166.

Pollock, L.W., and W.D. Hummon
1971. Cyclic Changes in Interstitial Water Content, Atmospheric Exposure and Temperature in a Marine Beach. *Limnology and Oceanography*, 16:522-535.

Por, F.D.
1965. La faune des Harpacticoides dans les vases profundes de la côte d'Israël. Une faune panbathyale. *Rapports et Proces-verbaux des Reunions de la Counseil International pour l'Exploration de la Mer Mediteranee*, 18:159-162.

Powell, E.N., M.A. Crenshaw, and R.M. Rieger
1979. Adaptations to Sulfide in the Meiofauna of the Sulfide System. I. ^{35}S-sulfide Accumulation and the Presence of Sulfide Detoxification System. *Journal of Experimental Marine Biology and Ecology*, 37:57-76.
1980. Adaptation to Sulfide in Sulfide-system Meiofauna. End Products of Sulfide Detoxification in Three Turbellarians and a Gastrotrich. *Marine Ecology Progress Series*, 2:169-177.

Rachor, E.
1975. Quantitative Untersuchungen über des Meiobenthos der nordatlantischen Tiefsee. *"Meteor" Forschungs-ergebnisse D*, 21:1-10.

Radziejewska, T.
1984. Meiofauna Communities in Organically Enriched Estuarine Environment. *Limnologica* (Berlin), 15:425-427.

Raffaelli, D.
1981. Monitoring with Meiofauna - A Reply to Coull, Hicks and Wells (1981) and Additional Data. *Marine Pollution Bulletin*, 12:381-382.

Raffaelli, D.G., and C.F. Mason
1981. Pollution Monitoring with Meiofauna, Using the Ratio of Nematodes to Copepods. *Marine Pollution Bulletin*, 12:158-163.

Rao, G.C., and A. Misra
1983. Studies on the Meiofauna of Sagar Island. *Proceedings of the Indian Academy of Science, (Animal Science)*, 92:73-85.

Ranta, E., and J. Sarvala
1978. Spatial Patterns of Littoral Meiofauna in an Oligotrophic Lake. *Verhandlungen der Internationalen Vereinigung für Theoretische und angewandte Limnologie*, 20:886-890.

Reid, J.W.
1984. Semiterrestrial Meiofauna Inhabiting a Wet Campo in Central Brazil, with Special Reference to the Copepoda (Crustacea). *Hydrobiologia*, 118:95-111.

Reidenauer, J.A., and D. Thistle

1981. Response of a Soft-Bottom Harpacticoid Community to Stingray (*Dasyatis sabina*) Disturbance. *Marine Biology*, 65:261-267.

Reise, K., and P. Ax
1979. A Meiofaunal "Thiobios" Limited to the Anaerobic Sulfide System of Marine Sand Does Not Exist. *Marine Biology*, 54:225-237.

Renaud-Mornant, J.N., N. Gourbault, and M.N. Helleuet
1983. Prospections meiofaunistiques en Martinique. *Bulletin du Museum National d'Histoire Naturelle*, series 5, 1:221-234.

Rieger, R., and J. Ott
1971. Gezeitenbedingte Wanderungen von Turbellarien und Nematoden eines nordadriatischen Sandstrandes. *Vie et Milieu*, supplement, 22:425-448.

Riemann, F., and M. Schrage
1978. The Mucus-trap Hypothesis of Feeding of Aquatic Nematodes and Implications for Biodegradation and Sediment Texture. *Oecologia* (Berlin), 34:75-88.

Rodrigues, C.L., S.N. Harkantra, and A.H. Parulekar
1982. Sublittoral Meiobenthos of the Northwestern Bay of Bengal. *Indian Journal of Marine Science*, 11:239-242.

Romano, J.-C., and A. Dinet
1978. Relations entre l'abundance du meiobenthos et la biomasse des sediments superciciels estimee par la mesure des adenosines 5'phosphate (ATP, ADP, AMP0. Pages 159-170 in *Geochemie organique des sediments marins profonds*. CNRS, ORGON IV.

Romeyn, K., L.A. Bouwman, and W. Admiraal
1983. Ecology and Cultivation of the Herbivorous Brackish-Water Nematode *Eudiplogaster paramatus*. *Marine Ecology Progress Series*, 12:145-153.

Salvat, B.
1964. Les conditions hydrodynamiques interstitiel des sediments meuble intertidaux et la repartition verticale de la faune endogee. *Comptes Rendus des Séances de l'Academie des Sciences*, (Paris), 259:1576-1579.

Sarma, A.L.N., and D.G. Rao
1980. The Meiofauna of Chilka Lake (Brackish Water Lagoon). *Current Science of India*, 49:870-872.

Scheibel, W.
1974. Submarine Experiments on Benthic Colonization of Sediments in the Western Baltic Sea. II. Meiofauna. *Marine Biology*, 28:165-168.

Schmidt, P.
1968. Die quantitative Verteilung und Populationsdynamik des Mesopsammons am Gezeiten-Sandstrand der Nordseeinsel Sylt I. Faktorengefüge und biologisch Gliederung des Lebensraumes. *Internationale Revue der gesamten Hydrobiologie*, 53:723-779.
1969. Die quantitative Verteilung und populationsdynamik des Mesopsammon am Gezeiten-Sandstrand der Nordseeinsel Sylt. II. Quantitative Verteilung und Populationsdynamik einzelner Arten. *Internationale Revue der gesamten Hydrobiologie*, 54:95-174.
1972a. Zonierung und jahreszeitliche Fluktuationen des Mesopsammons im Sandstrand von Schilksee (Kieler Bucht). *Mikrofauna des Meeresbodens*, 10:353-410.
1972b. Zonierung und jahreszeitliche Flukuationen der interstitiellen Fauna in Sandstränden des Gebiets von Troms. (Norwegen). *Mikrofauna des Meeresbodens*, 12:81-164.

Schmidt-Moser, R., and D. Westphal
1981. Predation of *Pomatoschistus microps* Kroyer and *P. minutus* Pallas (Gobiidae, Pisces) on Macro- and Meiofauna in the Brackish Fjord Schlei. *Kieler Meeresforschungen*, 5:471-478.

Sherman, K.M., and B.C. Coull
1980. The Response of Meiofauna to Sediment Disturbance. *Journal of Experimental Marine Biology and Ecology*, 45:59-71.

Sibert, J.R.
1979. Detritus and Juvenile Salmon Production in the Nanaimo Estuary: II. Meiofauna Available as Food to Juvenile Chum Salmon (*Oncorhynchus keta*). *Journal of the Fisheries Research Board of Canada*, 36:497-503.

Sibuet, M., C. Monniot, D. Desbruyeres, A. Dinet, A. Khripounoff, G. Rowe, and M. Segonzac
1984. Peuplements benthiques et caracteristiques trophiques du milieu dans la plaine abyssale de Demerara. *Oceanologica Acta*, 7:345-358.

Shirayama, Y.
1984. The Abundance of Deep-Sea Meiobenthos in the Western Pacific in Relation to Environmental Factors. *Oceanologica Acta*, 7:113-121.

Smith, C.J., R. DeLaune, W.H. Patrick, Jr., and J.W. Fleeger
1984. Impact of Dispersed and Undispersed Oil Entering a Gulf Coast Marsh. *Environmental Toxicology and Chemistry*, 3:606-616.

Smith, L.D., and B.C. Coull
1987. Juvenile Spot (Pisces and Grass Shrimp) Predation on Meiobenthos in Muddy and Sandy Substrata. *Journal of Experimental Marine Biology and Ecology*, 105:123-136.

Sogard, M.
1984. Utilization of Meiofauna as a Food Source by a Grassbed Fish, the Spotted Dragonet *Callionymus pauciradiatus*. *Marine Ecology Progress Series*, 17:183-191.

Soyer, J.
1970. Bionomie benthique du plateau de la Côte Catalane Française. III. Les peuplements de copepodes harpacticoides (Crustaces). *Vie et Milieu*, 21:337-512.
1985. Mediterranean Sea Meiobenthos. Pages 85-108 in M. Moraitou-Appostolopoulou and V. Kiortsis, editors, *Mediterranean Marine Ecosystems*. New York: Plenum Publishing Corporation.

Sterrer, W.
1973. Plate Tectonics as a Mechanism for Dispersal and Speciation in Interstitial Sand Fauna. *Netherlands Journal of Sea Research*, 7:200-222.

Swedmark, B.
1964. The Interstitial Fauna of Marine Sand. *Biological Review*, 39:1-42.

Tenore, K.R., J.H. Tietjen, and J.J. Lee
1977. Effect of Meiofauna in Incorporation of Aged Eelgrass, *Zostera marina*, Detritus by the Polychaete *Nephthys incisa*. *Journal of the Fisheries Research Board of Canada*, 34:563-567.

Tenore, K.R., C.F. Chamberlain, W.M. Dunstan, R.B. Hanson, B. Sherr, and J.H. Tietjen
1978. Possible Effects of Gulf Stream Intrusions and Coastal Runoff on the Benthos of the Continental Shelf of the Georgia Bight. Pages 577-598 in M.L. Wiley, editor, *Estuarine Interactions*. New York: Academic Press.

Thiel, H.
1975. The Size Structure of the Deep-Sea Benthos. *Internationale Revue der gesamten Hydrobiologie*, 60:575-606.
1982. Zoobenthos of the CINECA Area and Other Upwelling Regions. *Rapports et Proces-Verbaux des Reunion, Conseil International pour l'Exploration de la Mer*, 180:323-334.
1983. Meiobenthos and Nanobenthos of the Deep Sea. Pages 167-230 in G.T. Rowe, editor, *The Sea 8*. New York: John Wiley and Sons, Inc.

Thistle, D.
1978. Harpacticoid Dispersion Patterns: Implications for Deep-Sea Diversity Maintenance. *Journal of Marine Research*, 36:377-397.
1980. The Response of a Harpacticoid Copepod Community to a Small-scale Natural Disturbance. *Journal of Marine Research*, 38:381-395.

Thistle, D., J.Y. Yingst, and K. Fauchald
1985. A Deep Sea Benthic Community Exposed to Strong Near-Bottom Currents on the Scotian Rise (Western Atlantic) *Marine Geology*, 66:91-112.

Thomassin, B.A., C. Jouin, J. Renaud-Mornant, G. Richard, and B. Salvat
1982. Macrofauna and Meiofauna in the Coral Sediments on the Tiahura Reef Complex, Moorea Island (French Polynesia). *Tethys*, 10:392-397.

Tietjen, J.H.
1971. Ecology and Distribution of Deep-sea Meiobenthos off North Carolina. *Deep-Sea Research*, 18:941-957.
1976. Distribution and Species Diversity of Deep-Sea Nematodes off North Carolina. *Deep-Sea Research*, 23:755-768.
1984. Distribution and Species Diversity of Deep-Sea Nematodes in the Venezuela Basin. *Deep-Sea Research*, 31:119-132.

Tietjen, J.H., and J.J. Lee
1972. Life Cycles of Marine Nematodes. Influence of Temperature and Salinity on the Development of *Monohystera denticulata* Timm. *Oecologia* (Berlin), 10:167-176.
1977a. Feeding Behavior of Marine Nematodes. Pages 21-36 in B.C. Coull, editor, *Ecology of Marine Benthos*. Columbia: University of South Carolina Press.
1977b. Life Histories of Marine Nematodes. Influence of Temperature and Salinity on the Reproductive Potential of *Chromadorina germanica* Bütschli. *Mikrofauna des Meeresbodens*, 61:263-270.

Van Damme, D., R. Herman, J. Sharma, M. Holvoet, and P. Martens
1980. Benthic Studies of the Southern Bight of the North Sea and Its Adjacent Continental Estuaries. Progress Report II. Fluctuations of the Meiobenthic Communities in the Westerschelde Estuary. *International Council for the Exploration of the Sea*, 23:131-170.

Vidakovic, J.
1983. The Influence of Raw Domestic Sewage on Density and Distribution of Meiofauna. *Marine Pollution Bulletin*, 14:84-88.
1984. Meiofauna of Silty Sediments in the Coastal Area of the North Adriatic, With Special Reference to Sampling Methods. *Hydrobiologia*, 118:67-72.

Vitiello, P.
1968. Variation de la densité du meiobenthos sur une aire restreinte. *Recueil des Travaux Station Marine d'Endoume*, 43:261-270.

Vivier, M-H.
1978. Conséquences d'un déversement de boue d'alumine sur le méiobenthos profond (Canyon de Cassidaigne, Mediterranee). *Tethys*, 8:249-262.

Vranken, G., and C. Heip
1986. The Productivity of Marine Nematodes. *Ophelia*, 26:429-442.

Walters, K., and S.S. Bell
1986. Diel Patterns of Active Vertical Migration in Seagrass Meiofauna. *Marine Ecology Progress Series*, 34:95-103.

Warwick, R.M.
1971. Nematode Association in the Exe Estuary. *Journal of the Marine Biological Association of the United Kingdom*, 51:439-454.
1981. The Nematode/Copepod Ratio and Its Use in Pollution Ecology. *Marine Pollution Bulletin*, 12:329-333.

Warwick, R.M., and J.M. Gee
1984. Community Structure of Estuarine Meiobenthos. *Marine Ecology Progress Series*, 18:97-111.

Warwick, R.M., I.R. Joint, and P.J. Radford
1979. Secondary Production of the Benthos of an Estuarine Environment. Pages 429-250 in R.L. Jefferies and A.J. Davy, editors, *Ecological Processes in Coastal Environments*. Oxford: Blackwell Scientific Publishing Company.

Watzin, M.C.
1983. The Effects of Meiofauna on Settling Macrofauna: Meiofauna May Structure Macrofaunal Communities. *Oecologia* (Berlin), 59:163-166.

Wieser, W., J. Ott, F. Schiemer, and E. Gnaiger
1974. An Ecophysiological Study of Some Meiofauna Inhabiting a Sandy Beach at Bermuda. *Marine Biology*, 26:235-249.

Yingst, J.Y., and D.C. Rhoads
1985. The Structure of Soft-Bottom Benthic Communities in the Vicinity of the Texas Flower Garden Banks, Gulf of Mexico. *Estuarine and Coastal Shelf Science*, 20:569-592.

Zander, C., and E. Hartwig
1982. On the Biology and Food of Small-Sized Fish from North and Baltic Sea Areas. IV. Investigations on an Eulittoral Mud Flat at Sylt Island. *Helgoländer wissenschaftliche Meeresuntersuchungen*, 35:47-63.

4. Ecology of the Freshwater Meiofauna

Robert W. Pennak

Since the first use of "meiobenthic" in 1942, the concept has been almost exclusively restricted to marine investigations. Indeed, if we search the freshwater literature, we find that there are fewer than 20 titles of journal articles incorporating the root "meio-." This is in striking contrast to the hundreds of "meio-" papers dealing with marine situations. There is, nevertheless, a large number of published freshwater investigations that may be considered meiobenthic studies. It is the purpose of the present chapter to bring together and briefly review the important literature concerned with the ecology of the freshwater meiobenthos.

In general, I am adhering to certain definitions in common use by marine meiobenthic researchers. For example, I shall restrict my discussion of meiobenthic organisms to those freshwater taxa that are less than 2 mm long. I am omitting consideration of protozoans because the freshwater meiobenthic forms are so poorly known. In Hartwig's (1980) bibliography of interstitial ciliates only four papers appear to deal appreciably with the ecology of freshwater forms. Similarly, I am also avoiding consideration of complex algal assemblages (see Davies, 1971).

Nearly all types of aquatic substrates are included, and while some marine studies cover habitats more than a meter below the water-substrate interface, most of the freshwater data are derived from the uppermost few cm, uncommonly extending as deep as 20 cm. As an important exception, however, I am including the hyporheic habitat, which, by my (arbitrary) definition, extends from the substrate surface of lotic situations to a depth of 50 cm. Interstitial water deeper than 50 cm in typical stream substrates I consider as belonging in the phreatic or groundwater regime (see Pennak and Ward, 1986). On exposed shores near flowages, I consider the phreatic water up to 50 cm deep also as a part of the meiobenthon.

Neither am I including situations in caves, springs, seepages, underground flowages, or wells. These are all marginal or very special meiobenthic habitats, even though they may be closely related to adjacent meiobenthic habitats in the long-time evolutionary sense. Indeed, these subterranean habitats may eventually be incorporated into the meiobenthic bailiwick.

The freshwater meiofauna, with few exceptions, consists chiefly of major taxa that are also dominant forms in marine meiobenthic assemblages, including especially Turbellaria, Gastrotricha, Rotifera, Nematoda, Tardigrada, Annelida, Cladocera, Copepoda, Ostracoda, Isopoda, Amphipoda, Hydracarina, and Diptera larvae. In addition, there are several major taxa that are uncommon, rare, or incidental in freshwater meiobenthic habitats. All of these taxa, including dominants and rarities, will be discussed in this chapter as they relate to specific kinds of freshwater situations.

Because freshwater meiobenthic studies are still in the pioneer stages as compared to marine investigations, it is not feasible to write a meaningful state-of-the-art summary. Quantitative data are especially lacking. Indeed, we can characterize the freshwater meiobenthos in a single word: "variability." Population densities vary enormously from place to place and time to time. Local "patchiness" is the rule. Some species are highly restricted geographically; others are widely distributed. Some assemblages consist of many species, e.g., 20 to 30; others consist of two to five species. Some species are euryokous; others are stenokous. Aside from the importance of substrate grain size, meiobenthic species must cope with highly variable ecological conditions such as temperature, water chemistry, interstitial currents, oxygen availability, and food sources. Specific examples of these conditions are cited in the pages which follow.

A Classification of Meiobenthic Habitats

If we use the term "freshwater meiobenthon" in its broad meaning, it includes a wide variety of lotic and lentic situations which range from small to large, covering many m^2 or km^2. Until freshwater meiobenthic studies are more firmly established, I propose to use a simplified physical classification:

Lentic habitats; lakes, reservoirs, and ponds.

Above the water line

1. Sandy margins (eupsammolittoral) up to 3 or 4 m wide and 20 cm deep; includes only damp and wet sand.
2. Gravel and rubble margins; often wave-swept and seldom more than 2 m wide.
3. Wet mud margins.

Below the water line

4. Littoral sand, mud, and gravel substrates; limited wave action; 1 to 15 m wide.
5. Sand, gravel, and rubble substrates; usually 1 to 20 m wide, depending largely on wave action.
6. Mud and sand substrates; may extend from littoral or shores to greatest depths; includes gyttja deposits.

Lotic habitats; small streams to large rivers

Above the water line

7. Sandy margins (eupsammolittoral) up to 4 m wide; usually contain more mud and organic particulates than lake shore margins.
8. Sand, gravel, and rubble margins; intermittently swept by waves and currents.
9. Mud margins; composition variable.

Below the water line

10. Sandy substrates; narrow to broad; usually interspersed with other substrates.
11. Sand, gravel, and rubble substrates; commonly extending across the entire width of the stream.
12. Mud substrates; composition variable; more common in slowly flowing waters.

Although these habitats are listed as though they are "pure" in their composition, it must be borne in mind that there are all sorts of mixtures, e.g., a variable mixture of mud, sand, gravel, and rubble in the near-shore littoral of a lake.

Brief History of Freshwater Meiobenthos Research

No one person can be designated as the original investigator of the freshwater meiobenthos. The amateur and the professional microscopists of the 18th and 19th centuries commonly sought their study materials in pond and stream debris and substrates, and it is from these nebulous sources that many species of substrate micrometazoans were first described. In the broad sense, these pioneers were studying the meiobenthos.

We can probably attribute the real onset of freshwater meiobenthic studies (in the "community" sense) to several Russian investigators of psammolittoral ecology, especially Sassuchin, Kabanov, and Neiswestnova (1927). Sassuchin (1930) wrote of the fauna of air borne sands of the Kirghiz steppes, and in 1931 published a study of conditions in river sands. Unfortunately, these papers are brief and inaccessible. Wilson (1932) first called attention to the harpacticoid copepod populations in the interstitial waters of freshwater and marine beaches. In Poland, Wiszniewski published 12 papers on lake psammolittoral studies between 1932 and 1937; emphasis was on rotifer taxonomy, mostly descriptions of new species, although important ecological measurements were also made (see especially Wiszniewski, 1932, 1934a, 1934b, 1937). Slightly later, Varga (1938) worked on the psammon rotifers of Lake Balaton, Hungary. Pennak, working in Wisconsin, then published a series of papers which covered taxonomy, ecology, and population structure of lake beach interstitial organisms (Pennak, 1939a, 1939b, 1939c, 1940, 1951). Between 1953 and 1962 Ruttner-Kolisko followed with a series of papers on lake and stream psammolittoral, based on locations in Austria, Italy, and Sweden; some of the more important are: Ruttner-Kolisko, 1953, 1954, 1956b, 1962.

In 1935 Karaman marked the beginning of investigations of the interstitial waters of sand, rubble, and gravel shores, and substrates of streams and rivers. But beyond 1950 interests turned sharply toward these hyporheic and phreatic communities, especially in Europe. A discussion of the history of these developments is beyond the scope of this chapter. Danielopol (1982) gives a complete account. The following investigations, however, are especially noteworthy: Angelier, 1953; Chappuis (more than 100 titles); Delamare-Debouteville, 1960; Danielopol, 1976, 1980b; Evans, 1984; Husmann, 1956, 1966; Noodt, 1952, 1955; Picard, 1962; Ponyi, 1960; Rouch, 1968 ; Schwoerbel, 1961a, 1961b, 1967a; Tilzer, 1968. All of these contributions are discussed in subsequent sections of this chapter.

After 1950, papers dealing with lake and stream meiobenthos continued to be published, but most authors contributed only one or two titles. The works of Hummon, Moore, Angelier, Nalepa, Neel, Whitman, Cole, Stanczykowaka and Przytocka-Jusiak, Whiteside and Lindegaard, and Strayer are especially significant, and their contributions are also discussed subsequently.

Physical and Chemical Conditions

Temperature.--As might be expected, the

temperature of a particular meiobenthic habitat generally closely reflects the temperature of the adjacent free water. An important exception is found in the psammolittoral, especially in the summer months when the solar insolation is most intense. At the water's edge the sand temperature is usually close to that of the adjacent open water, but back horizontally from the water's edge the sun often raises the temperature of the damp sand 1-4° C (e.g., at a depth of 3-4 cm). Examples are given in Ruttner-Kolisko, 1953, 1956; Wiszniewski, 1934a; and Pennak, 1940. Similarly, in a vertical sequence, the surface of sandy beaches is usually 1-4° C warmer than at a depth of 8 cm during daylight hours.

In the sandy riffle of a small Texas stream, Whitman and Clark (1984) found that the surface and 30-cm depths were never more than 1.5° C apart. In a rubble mountain stream Pennak and Ward (1986) reported interstitial temperatures at a depth of 50 cm about the same as those of the stream above. Schwoerbel (1961b), however, usually found differences of 1-3° C.

Hydrogen-ion Concentration.--Hydrogen-ion readings of the water and the substrate of meiobenthic habitats are essentially similar at the interface, but below the substrate surface the interstitial water becomes more acid with increasing depth. In psammolittoral waters vertical ranges of more than one pH unit are uncommon, at least to a depth of 15 cm. Pennak (1940) reports typical lake psammon data and Angelier (1953) gives sandy river shoreline substrate values.

Data for gravel, rubble, and wet mud margins of lakes and ponds appear to be lacking, but I presume conditions below the surface to be slightly more acid than those at the surface.

Hydrogen-ion readings for mud substrates in lakes and ponds appear to be incidental, but it is clear that the anaerobic conditions one cm or more below the water-mud interface are progressively more acid, sometimes as much as two pH units. The specific pH readings, however, are variable and depend much on the season of the year, depth, water chemistry, degree of eutrophy, and the chemical nature of the bottom deposits (discussion in Hutchinson, 1957; Hayes, 1964; Mortimer, 1941-1942).

In a more recent paper, Schiff and Anderson (1986) note that in acid lakes the interstitial substrate waters are sometimes less acid than the overlying waters, presumably because of the buffering capacity of the sediment.

Hydrogen-ion data for submerged lotic meiobenthic habitats in sand, gravel, pebble, and rubble substrates are very limited. Whitman and Clark (1984) took interstitial pH readings from Mill Creek, a small stream in Texas, and found that the mean surface pH 6.0 decreased to 5.5 at a depth of 10 cm. Pennak and Ward (1986) found that pH determinations at a depth of 50 cm in a rubble mountain stream were generally within 0.2 of a unit of the stream itself. The small brooks studied by Schwoerbel (1961b) had subsurface readings less alkaline than those of the surface waters, usually by less than 1.0 pH unit. Less pronounced differences were reported by Husmann (1971) and Williams and Hynes (1974). Thus, it may be concluded that lotic meiobenthic habitats are only slightly less alkaline than their overlying flowages, and these differences would appear to be insufficient to be a significant factor in the ecology of the freshwater meiobenthos.

Dissolved Oxygen.--Most studies of the dissolved oxygen content of psammolittoral interstitial water clearly demonstrate reduced concentrations. The most marked horizontal differences in dissolved oxygen were reported for 15 Wisconsin beaches by Pennak (1940). At a depth of 8 cm the average horizontal dissolved oxygen distribution for interstitial water: 25 cm lakeward from the water's edge, at the water's edge, 50 cm shoreward from the water's edge, and 100 cm shoreward were respectively, 8.4, 5.5, 1.0, and 0.4 ppm. Corresponding ranges were 7.4-9.8, 3.4-7.7, 0.1-2.7, and 0.0-0.8 ppm.

These results are corroborated by Wiszniewski (1934a) and by the brief studies of Stangenberg (1934), Neel (1948), Ruttner-Kolisko (1956b), and by the "oxygen availability" study of Enckell (1968).

Pennak (1940) reported black or brown strata in beaches at depths of 6 to 10 cm and usually 150 to 250 cm landward from the water's edge. These were anaerobic fine sand strata where the odor of hydrogen sulfide was easily detected. Such strata contained no active micrometazoans. Neel (1948) found similar "black layers" in Douglas Lake, Michigan, beaches; see his paper for a full discussion of the chemistry of this phenomenon. Gordon (1960) reported similar situations on California marine beaches. These anaerobic habitats, however, are not comparable with the "sulfide system" found in offshore marine deposits with a relatively rich interstitial fauna (Fenchel and Riedl, 1970).

It has long been known that mud substrates in the deeper parts of lakes and ponds are almost always either anaerobic or low in oxygen. Depending on water chemistry, substrate chemistry, time of year, water circulation, and temperature, the level at which anaerobic conditions begin is usually at or a few cm below the water-mud interface (Hutchinson, 1957; Hayes, 1964; Mortimer, 1941-1942). Thus, the meiobenthic assemblage of mud bottoms appears to consist of anaerobic and facultative anaerobic species. Many of these species have highly resistant resting stages (Pennak, 1985).

Only a few dissolved oxygen data for rubble shorelines and rubble littoral areas are available, and we cannot draw meaningful conclusions, but I would assume that subsurface waters in these areas also have lower oxygen concentrations.

Dissolved oxygen conditions in running water substrates are highly variable, depending on local physical and chemical conditions of the particular stream. Ruttner-Kolisko (1961) found less than 1.0 mg/l deeper than 20 cm in the sandy substrates of four streams. But then at greater depths she reported that the oxygen content rapidly increased in the Hollenstein and Langau rivers. Tilzer (1968), working on two alpine streams, reported near-saturation consistently; Pennak and Ward (1986) detected an annual range of 75.6 to 115.6 percent saturation in a Colorado rubble mountain stream at a depth of 50 cm. Schwoerbel (1961b) found that oxygen generally decreased to 50-70 percent saturation at a depth of 30 cm in a series of small brooks. Several workers, on the other hand, report contrary data. Cummins (1975), for example, states that fine sediments are anaerobic at 1 to 10 cm depths, and coarse sediments are anaerobic below 30 cm. Williams and Hynes (1974) seldom found oxygen below 30 cm, and Poole and Stewart (1976) reported less than 1 mg/l at 30 to 40 cm. Whitman and Clark (1982) detected variable oxygen conditions in a Texas sand creek, but generally concentrations decreased from near-saturation at the surface of the sands to 1 to 4.5 mg/l at 30-cm depths.

Free Carbon Dioxide.—Most measurements of the carbon dioxide content of freshwater meiobenthic habitats have been of a limited and casual nature, but it is clear that the higher the CO_2 content, the lower the dissolved oxygen content. Pennak (1940) found a range of 2.0 to 89.0 ppm free CO_2 in the interstitial waters of 15 beaches at a depth of 8 cm, and between the water's edge and 100 cm back from the water's edge. The greater the distance from the lake's edge and the deeper the sample, the greater the CO_2 concentration, although there were many exceptions. Similar but less spectacular results have been reported by Ruttner-Kolisko (1955a, 1956) and Neel (1948).

Free CO_2 determinations for mud rich in organic matter, sand, and gravel habitats in lakes and ponds seem to be rare in the literature. Since such intersititial waters are characterized by low oxygen concentrations, it is logical to assume that the corresponding respiratory processes produce high CO_2 concentrations.

Few data are available for free CO_2 in stream substrates. In three streams Schwoerbel (1961b) found that surface values were 0.0 to 4.4 mg, and at 30 cm the range was 4.8 to 6.5 mg. Other values in the literature are similar.

Freshwater Meiobenthic Communities.

Although an abundance of small metazoans are widely distributed in and on many kinds of freshwater meiobenthic habitats, there are nevertheless a few taxa more or less restricted or typical of one kind of substrate, so that definite communities can be recognized by their combinations of cosmopolitan and endemic taxa. These kinds of communities are briefly characterized in the following sections. Each community type below is identified by the numbers shown earlier in this chapter.

Psammolittoral (1).—The sandy margin (eupsammolittoral) of lakes, reservoirs, and ponds has perhaps the most distinctive and identifiable of all meiobenthic assemblages, especially along waveswept shores. In this habitat the characteristic faunal assemblage is restricted to the strip of damp and wet sand and fine gravel between the water's edge and back as far as 4 m from the water's edge, depending on the slope and exposure of the beach. The psammolittoral community is, for the most part, restricted to damp and wet sand up to 20 cm deep. Below this shallow surface layer there is commonly a variable thick layer of sand, gravel, and clay before the top of the ground water, or phreatic zone, is encountered. The underground phreatic community is therefore separated from the psammon community by a poorly populated ecotone layer that is of little interest from the standpoint of micrometazoans.

Some investigators have worked on the particle size composition of sandy beaches, but it is difficult to say that particular species are restricted to beaches having certain percentages of fine, medium, or coarse sand. In a general way, however, the greatest variety of species and the most dense populations are found in predominant mixtures of medium and coarse sands where there is an abundance of interstitial water and the proper interstitial spaces, as well as some onshore wave action where the interstitial water is frequently replenished. Ruttner-Kolisko (1961) postulated that grain size distribution must consist largely of medium sand grains (250-400 μm) for psammobiotic communities to develop. This produces 35-55 percent interstitial space. In beaches composed of the finest sand the metazoan fauna is almost lacking (Pennak, unpublished).

Morphological adaptations of psammolittoral organisms have been discussed at length by many authors (e.g., Delamare Deboutteville, 1960; Swedmark, 1964; Remane, 1952; and Ax, 1966). As shown in Figures 4.1 and 4.2, however, these organisms are, for the most part, elongated, small, and with reduced appendages. In short, they are especially adapted for navigating the restrictive interstices of wet and damp sand. Movements are produced by writhing, kicking, and ciliary action. Food is obtained by browsing, detritus feeding, and, rarely, by suspension feeding.

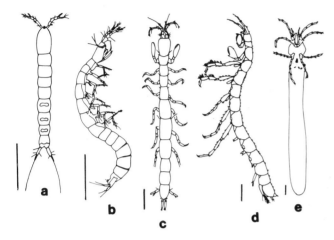

Figure 4.1.--Typical interstitial arthropods, illustrating the prevailing "Lebensform." a, *Parastenocaris* (Harpacticoida, Copepoda); b, *Leptobathynella* (Syncarida); c, *Microcerberus* (Isopoda); d, *Ingolfiella* (Amphipoda); e, *Wandesia* (Hydrachnellae). (From Husmann, 1978.) (Scale = 100 μm.)

Rotifera.--The many species of rotifers found here are for the most part small, elongated, and with the coronal disc oriented in a ventral position (Figures. 4.2, 4.3). Most of our information on genera and species has been supplied by Wiszniewski (1934a, 1934b, 1935, 1936, 1937), Myers (1936), Pennak (1940), Neel (1948), Ruttner-Kolisko (1953, 1954, 1956), Evans (1982), and a few others. Collectively, these papers list more than 220 species reported from beaches, but only about 50 are true "psammobiotic" species that are essentially restricted to beaches and appear only rarely in other (nearby) habitats, and then only fortuitously as stragglers. The other 170 species are found everywhere (a) in or on all kinds of aquatic substrates, or (b) as free-swimming open-water plankton forms whose adults or eggs may be washed ashore where they live only briefly and presumably do not reproduce. Genera having typical psammobiotic species include: *Aspelta, Cephalodella, Dicranophorus, Diurella, Elosa, Encentrum, Euchlanis, Lecane, Lindia, Monostyla, Myersina, Notommata, Pedipartia, Trichotria, Wierzejskiella,* and *Wigrella*.

On typical lake beaches the great majority of rotifers are restricted to a relatively narrow shoreline band, beginning at the water's edge and extending back 100 to 250 cm (seldom 300 cm or more) from the water's edge, depending on the slope of the beach and amount of wave action. Although Pennak (1942) has demonstrated definite horizontal intertidal species preferences for copepods in the marine beach interstitial, there are no such sharp demonstrable preferences in freshwater beaches for rotifers, except, perhaps, for a very few species of the hygropsammon.

Wiszniewski (1934a) recognized three subhabitats in the lake beach habitat: (1) the hydropsammon, including those organisms in continuously submerged sand just below the waterline; (2) the hygropsammon, including those organisms extending from the waterline back for a distance of about 0.5 to 1.0 m; and (3) the eupsammon, including those organisms more than a meter above the water's edge, but usually no more than three meters above. Using this terminology, the horizontal densities of the total rotifer populations may be summarized for a wide variety of beaches (Table 4.1). Population densities are highly variable from one study to another and also at a single station from time to time. Additionally, few rotifers are found in the (submerged) hydropsammon as compared to the hygropsammon and eupsammon.

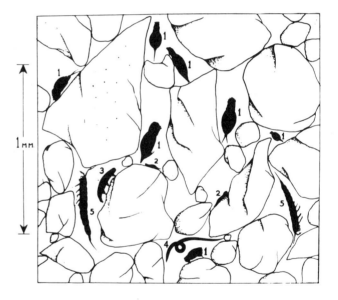

Figure 4.2.--Optical section of a small portion of a sandy beach showing relative sizes of sand grains and some of the common species of micrometazoans. 1, rotifers; 2, gastrotrichs; 3, tardigrades; 4, nematodes; 5, harpacticoid copepods. (From Pennak, 1940.)

The vertical distribution of rotifers is more predictable. Most of the published data clearly show that greatest densities are restricted to the top 2 cm of sand in the hydropsammon and eupsammon, decreasing markedly, so that only a few stragglers (less than 4 per 10 cm^3 of sand) are found as deep as 8 cm. One 10 cm^3 surface sample 100 cm from the water's edge at SW Trout Lake in Wisconsin contained 11,550 rotifers (Pennak, 1940).

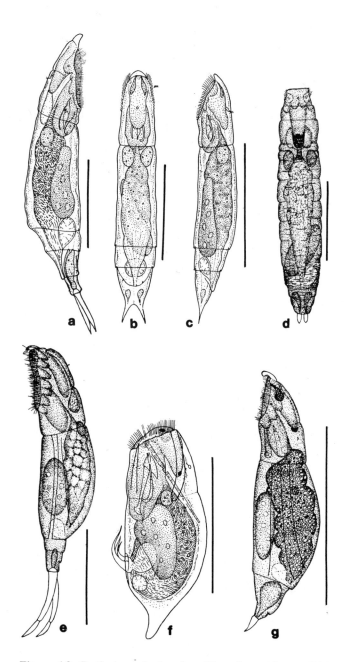

Figure 4.3.--Typical sandy beach rotifers (from Myers, 1936 and Wiszniewski, 1934) (scale = 100 μm.)

Harpacticoid Copepods.--Although by no means as abundant as rotifers, the harpacticoid copepods are nevertheless a taxon characteristic of the psammolittoral. They clearly show a preference for sand that contains less water, and are most abundant between 150 and 300 cm from the water's edge, usually only to a depth of 8 cm. They are therefore a eupsammon group, and although they attain surprising densities, such populations are composed of only one, two, or three species. In Pennak's (1940) monograph on Wisconsin lake beaches only three species were reported. *Parastenocaris brevipes* Kessler was found in all but one beach; *Phyllognathopus paludosus* Mrazek occurred in one beach; and *Parastenocaris starretti* Pennak was found in two beaches. Of 20 beaches, by far the greatest densities occurred at the N. Trout station on Trout Lake. Many 10 cm^3 samples contained more than 300 individuals, the maximum being 878. At the other extreme, 10 of the beaches contained an average of less than 5 harpacticoids per sample. At Lake Monona only 4 individuals were found in 16 samples, and at Weber Lake only 2 individuals in 16 samples. Factors responsible for these wide differences in densities are yet to be elucidated.

Numerical data for other studies are limited, and usually no identifications to genera or species are included. Ruttner-Kolisko (1954) and Enckell (1968) both reported occasional harpacticoids in samples taken in Swedish lake beaches. Evans (1982) reported "Copepods" from Lake Erie beaches, but no numerical data are supplied. Neel (1948) found up to 19 harpacticoids per 10 cm^2 of sand in a few of his eupsammon samples, and listed *Canthocamptus* sp., *Phyllognathopus* sp. The genus *Parastenocaris* is very large and in need of revision. Although a few species, such as *P. brevipes*, are widely distributed, almost all of the others are known from only one, or at most a few, localities on all continents. It is a euryokous genus, and in addition to being found in beaches, it also occurs in springs, caves, wells, moss, mountain brooks, streams, and in phreatic and hyporheic habitats.

A few other genera have been occasionally reported from European psammolittoral studies, especially *Epactophanes, Nitocrella, Canthocamptus, Attheyella,* and *Elaphoidella* (Husmann, 1960).

Nematoda.--Since the early papers of Cobb (1923 et seq), nematodes have been recognized as being common and universally present in all types of wet and damp habitats including sandy beaches. Unfortunately, the nematodes have usually been ignored in psammolittoral studies, as well as in those dealing with other freshwater meiobenthic habitats, even though nematodes are sometimes the most abundant taxon. Few numerical data are given and only seldom are nematodes identified, even to the generic level. Many species are cosmopolitan, but a very large number are undescribed.

Husmann (1962), working on the psammolittoral of "dune lakes," reported 14 species of nematodes in such common genera as *Acrobeles, Plectus, Monhystera, Odontolaimus, Tripyla, Mononchus,* and *Dorylaimus*. Hope (1971) emphasized the need for proper study of marine meiobenthic nematodes, and his comments apply equally as well to freshwater members of this same taxon. Evans (1982), working on three Lake Erie beaches, reported average densities of 10, 0.7, and 3 nematodes per cm^3 in the hydropsammon; 1.0, 0.2, and 0.5 nematodes per cm^3 in the hygropsammon; and 6.0, 0.8, and 0.3 nematodes per cm^3 in the eupsammon. Ruttner-Kolisko (1954) found ranges of 0-12, 1-13, 0-4, 0-3, and 12-50 nematodes per cm^3 in five Swedish lake beaches. Neel (1948) reported from 0-96 nematodes per cm^3 of sand at a depth of 1 to 4 cm in Douglas Lake, Michigan, beaches, although the average was less than 5 per cm^3. Pennak (1940) recorded average densities of 0 (Lakes Michigan and Superior) to 13.6 per cm^3 of sand at SW Trout Lake beach. It is possible that low densities are correlated with small quantities of particulate organic matter in the sand.

Oligochaeta.--Like nematodes, oligochaetes are also generally distributed in damp and wet substrates of all kinds, with many cosmopolitan and otherwise common species. Meiofaunal annelids have likewise received little attention in the psammolittoral. Dense populations are rare, presumably because of the relatively small quantities of organic detritus in waveswept beaches. Pennak (1940) found averages of only 1.0 to 4.2 per 10 cm^3 sand sample, and Ruttner-Kolisko (1956) reported less dense populations in Lake Maggiore beaches, about 1.1 per cm^3. These results are similar to those of Neel (1948) who found an average of about 3 per cm^3 of sand in the eupsammon. Lake Erie beaches (Evans, 1982) contained only negligible numbers. Enckell (1968) recorded greatest concentrations at 50 cm inland from the water line, about 6 per cm^3. Most beach meiofaunal annelids are *Aeolosoma* and species of Naididae, especially *Nais, Pristina,* and *Chaetogaster*.

Tardigrada.--I consider the tardigrades as a fifth taxon that characterizes freshwater beaches, chiefly because they are invariably present if the entire community of additional typical taxa is also present. Where the meiobenthic assemblage is poorly developed, then tardigrades are uncommon to rare. Most psammolittoral papers make only casual mention of tardigrades and seldom contain quantitative data. Ruttner-Kolisko (1953, 1954, 1956) and Enckell (1968) reported only an occasional tardigrade. In the eupsammon of three beaches of Douglas Lake, Neel (1948) seldom found tardigrades, but in three other beaches he sometimes reported large populations, up to 270 per cm^3 of sand, although usually less than 10 per sample. In the Wisconsin study, Pennak (1940) found seven beaches with few or no tardigrades. In the other 13 beaches, populations, though highly variable, sometimes attained astounding densities in the eupsammon, up to 657 per 10 cm^3 in one sample taken at Day Lake. Many other samples contained more than 100 tardigrades. The largest populations were usually recorded (1) from finer grades of sand, (2) from beaches where there was limited wave action, and (3) from beaches of soft-water lakes. On the other hand, I have no explanation for the pronounced month-to-month and year-to-year variations in the tardigrade populations of individual beaches.

Most beach tardigrades are in the genus *Macrobiotus,* a cosmopolitan taxon found everywhere on suitable substrates, including damp mosses and lichens. Schuster et al. (1977) collected *Hypsibius* and *Isohypsibius* in Lake Tahoe beaches, California.

Other taxa.--I consider the five groups discussed in the foregoing paragraphs as being the ecological

Table 4.1.--Horizontal density distribution of rotifers in various psammolittoral studies. Selected data; given variously as ranges and arithmetic means per 10 cm^3 of sand and per 10 cm^2 of sand.

Depth	Number of Beaches	Hydropsammon	Hygropsammon	Eupsammon	Measurement	Author
?	1	28	1130	730-1410	no./10 cm^2	Wiszniewski, 1934
?	1	112	3740		no./10 cm^2	Wiszniewski, 1934
top 3 cm	3	0-100	2-23	0-110	no./10 cm^3	Evans, 1982
top 8 cm	20	0-6	0-130	0-1169	no./10 cm^3	Pennak, 1940
		(2)	(17)	(21)		
?	1	0-15	0-234	0-234	no./10 cm^3	Ruttner-Kolisko, 1953
top 3 cm	6	270	220	220	no./10 cm^3	Neel, 1948
?	1	0	13	32	no./10 cm^3	Ruttner-Kolisko, 1954

dominants in the beach meiofaunal assemblage, namely: Rotifera, harpacticoid Copepoda, Nematoda, Annelida (mostly Oligochaeta), and Tardigrada, -- this in spite of the fact that their numbers vary widely and unaccountably.

Several other taxa play minor roles in this habitat. Meiobenthic turbellarians are occasionally collected, but never in any abundance. Gastrotricha are a common element, especially in the surface sand within 150 cm of the water's edge. Pennak (1940) found an average density of 5 to 10 per cm^3 of sand. *Chaetonotus, Ichthydium,* and *Lepidodermella* are most common. First-instar insect larvae, mostly Diptera, are small enough to be considered meiofauna; they sometimes amount to 2 per 10 cm^3 of sand.

Most washing and preserving procedures destroy the Protozoa, and the only reliable evidences remaining are the resistant tests of Testacea. Often, however, there are abundant ciliates and both green and colorless flagellates, sometimes up to 50,000 per cm^3 of sand. Neel (1948) has a long list of algae found in Douglas Lake, Michigan, beaches.

Nature of the freshwater beach meiofauna.--I close this section with a summary paragraph taken from Pennak (1940): "Perhaps no other (fresh-water habitat) is capable of supporting such a dense and diversified group of microscopic organisms as the sandy beach. If a typical 10 cm^3 sand sample be taken from the surface of the sand at a distance of 150 cm from the water's edge, it will be found to contain 2 to 3 ml of water. Within this small volume will be found 4,000,000 bacteria, 10,000 Protozoa, 400 Rotifera, 40 Copepoda, 20 Tardigrada, and small numbers of other microscopic Metazoa." While this summary may have held true in 1940, more recent studies of the top cm of lake sediments often show similar or greater densities of meiobenthic species (e.g., Strayer, 1985b).

Gravel and Rubble Lake Shores (2).--The interstitial fauna of this type of habitat appears to have been largely neglected. It is usually found on waveswept shores of large lakes so that the interstitial water is frequently renewed. There seem to be no reason to doubt the existence of a meiofaunal assemblage in this habitat, although sampling is difficult.

Muddy Shores (3).--This habitat has also been ignored by meiobenthic researchers. It occurs chiefly as a narrow zone on the shores of smaller lakes and ponds, often where rooted emergent vegetation is well developed and where there is little wave action. The mud is typically saturated with water and consists of more than 50 percent fine organic detritus. Interstitial spaces are therefore greatly restricted, and probably the meiofauna is confined to the top few cm of the wet substrate. I suspect that Annelida, Nematoda, and Protozoa are codominant elements.

Littoral Sand, Mud, and Gravel Substrates (4).--This is a littoral habitat just below the water line of ponds and lakes, and forming a peripheral shoreline band about 1-15 m wide. It is essentially similar to the "hydropsammon," but is wider, and it may or may not support a growth of rooted aquatic plants. Unlike the situation in exposed beaches where waves wash upon the shore and replenish the interstitial oxygenated water supply by gravity, the submerged sand has only diffusion processes and currents. As a result, these sand, mud, and gravel substrates have an adequate oxygen supply only in the top few cm.

Holopainen and Paasivirta (1977) have data for the littoral of Lake Pääjärvi, Finland, where the substrate consists of various mixtures of gravel, mud, and sand with or without hydrophytes. Average densities of the meiobenthos ranged from 400,000 to 630,000 organisms per m^2, depending on the location. Ash-free dry weight ranged from 0.6 to 0.8 grams per m^2. Dominant taxa were Nematoda, Copepoda, Ostracoda, and Chironomidae. Other taxa collectively amounted to only 8 percent of the total numbers and included: Turbellaria, Rotifera, Oligochaeta, Cladocera, Ephemeroptera, Ceratopogonidae, Oribatei, Hydrachnellae, Halacaridae, and Tardigrada. For the lake as a whole, these authors list the common meiobenthic species as follows: 37 Nematoda, 25 Cladocera, 5 Ostracoda, 4 Cyclopoida, 8 Harpacticoida, 15 Chironomidae, and 7 Hydrachnellae.

Extensive biological data are also to be found in Neel (1948), for Douglas Lake, Michigan. His samples were the top 2 to 5 cm of submerged sand bottom at six transects involving stations at various distances lakeward from the water line. Selected data are shown in Table 4.2. Neel listed the Rotifera only as *"Lecane"* and "others," and I suspect that few were true interstitial forms. Note that the nematodes and oligochaetes were more abundant than in the eupsammon, but, more significantly, no copepods were collected. Note further that densities varied unaccountably from one sample to another. Neel's paper is also of considerable interest for its discussion of physical features of submerged sand bottoms. Moffett's (1943) paper also contains a discussion of physical features in sandy shoal waters.

Evans (1982) also has considerable biological data, which are summarized in Table 4.3. He found densities to be much lower than those of Neel, but data are similarly variable.

Littoral Sand, Gravel, and Rubble (5).--Sometimes

lentic habitats have a peripheral zone of sand, gravel, and rubble in various mixtures and extending from the water's edge out for a distance of 1 to 20 m. Such substrates contain little mud and are especially common on waveswept shores and in mountain lakes. This is a true meiobenthic habitat, but it has gone unrecognized. *Between* the coarse particles making up the irregular substrate there is a limited meiobenthic assemblage, but more significantly on the *exposed surfaces* of the larger rubble particles there is a complex community that has, for many years, gone under the name of "periphyton," or, more appropriately, the "lithophyton" or "Aufwuchs," often forming a layer 50 μm thick, but thicker under eutrophic conditions. It is composed of fine inorganic and organic detritus, bacteria, mucus, algae, protozoans, and micrometazoans, as well as larger forms. This is a true meiobenthic community, but, so far as I am aware, there are no major investigations of the meiofauna of the Aufwuchs or interstitial aspects of this situation. Essentially all of the Aufwuchs littoral work has centered around the total organic production, chlorophyll production, and algal populations. However, many macrofaunal studies have been done with the Peterson dredge and similar collecting devices.

Lake Mud and Sand Substrates (6).--Soft muds, mostly gyttja, plus various quantities of sand in the sublittoral, form the predominant substrate in most lakes and reservoirs. Although there are many hundreds of studies of the macrofauna, we know little about the meiofauna. Important contributions are those of Holopainen and Paasivirta (1977), Moore (1939), Nalepa and Robertson (1981), Strayer (1985a,

Table 4.2.—Meiobenthos in sand and gravel substrates of Douglas Lake, Michigan; six transects below the water's edge. Expressed as numbers of organisms per cc of sand. Data selected from Neel (1948).

m From Water's Edge	Rotifers	Gastrotrichs	Nematoda	Tardigrada	Oligochaeta	Ostracoda
-1	5					
-2	4		3			
-3	26		5		5	
-1	32		115	26	12	
-2		26	96	26	26	
-3	13		39			
-1	5	4	6		3	
-2	12		18		12	
-3	3		6	7	10	
-1	27	2	3	23	6	
-2	44	13		19	6	
-3	43		180	13		
-1	18	4	7	11	5	
-5	18	2	6	22	19	
-1	5	8	139			2
-2	16		117		20	10

1985b), Strayer and Likens (1986), Monard (1920), Muckle (1942), Cole (1955), and Whiteside and Lindegaard (1982). Unquestionably, Strayer's (1985b) monograph on the bottom fauna of Mirror Lake is our best single reference. It is a complete review of the gyttja meiobenthos literature, and contains abundant qualitative and quantitative data.

Moore (1939) collected remarkably large numbers of species with his "mud sucker" and small coring apparatus in "muck" and sandy substrates between 4 and 20 m deep. Essentially all of his species are cosmopolitan and consist of: 7 rhizopods, many flagellates, 15 ciliates, 1 hydroid, 6 flatworms, many nematodes, 20 rotifers, 2 gastrotrichs, 8 oligochaetes, 14 cladocerans, 10 hydracarinids, and 1 tardigrade. Quantitatively, his data are extremely variable from station to station and are difficult to interpret. Nematodes, for example, varied from 0 to 2,362 per dm^2 of bottom; *Cyclops agilis* varied from 0 to 104; and *Cypria* (ostracod) from 0 to 364. Some species were rare and were taken only once, including the cladoceran *Ophryoxus gracilis* and the protozoan *Dileptus gigas*.

Strayer (1985b), working on Mirror Lake, New Hampshire, estimated that 322 species of micrometazoans inhabited the lake bottom (mostly gyttja), averaging 1,200,000 per m^2. About half of the animals were in the top cm of sediment. The most abundant groups were chironomids, oligochaetes, chaoborids, nematodes, and copepods, but it should be noted that chironomids and chaoborids are temporary meiofauna. Among the less abundant forms were 23 species of flatworms, 20 or 30 gastrotrichs, "probably" more than 100 species of rotifers, 3 tardigrades, 23 cladocerans, and 4 halacarid mites.

These studies of Moore and Strayer are perhaps our best examples of the complexity of the meiobenthic assemblage on the surface of the gyttja. Essentially all of the taxa represented are common and often cosmopolitan species or species groups (see Frey, 1982) -- a convincing demonstration of adaptability and euryoky. Even in nine relatively barren Canadian mountain lakes, Anderson and DeHenau (1980) found 40 species in the meiobenthic assemblage.

Nalepa and Robertson (1981) sampled the meiobenthos in predominantly sandy substrates at depths of 11 to 23 m in Lake Michigan. Reduced sediments occurred at depths exceeding 5 cm, but only nematodes and tardigrades were collected at such depths. The great majority of Naididae, Cyclopoida, Harpacticoida, Cladocera, Ostracoda, Turbellaria, and Rotifera occurred in the top cm of the sediments. They also found that the meiofauna was more concentrated in the superficial sand where detritus accumulations were greatest.

Kirchner (1975), working on Char Lake, Northwest Territories, Canada, reported that ostracods, nematodes, and oligochaetes were

Table 4.3.—Average densities of meiobenthos for the top 3 cm of sand in the hydropsammon, hygropsammon, and eupsammon. Expressed as numbers of individuals per cm^3. Selected data from Evans (1982).

m From Water's Edge	Rotifera	Nematoda	Turbellaria	Gastrotricha	Oligochaeta
South Bass					
-1	6.8	9.6	2.9	0.7	
0	7.3	1.2	20.9	1.4	0.6
+1	4.9	5.8	51.0	0.9	
Pelee					
-1	2.5	0.7		1.6	1.0
0	8.5	0.2	0.4	3.4	2.1
Kelley's					
-1	1.2	2.7	1.4	3.1	0.4
0	0.6	0.5	0.9	0.5	
+1	3.4	0.3	0.2	1.8	

essentially all restricted to the uppermost three cm of oxidized mud substrate. Similar results were obtained by Särkkä and Paasivirta (1972) for Lake Paijanne, Finland. About 90 percent of the meiofauna was confined to the top 6 cm of sediment and 8.3 percent was confined to the top 2 cm. Cyclopoids, nematodes, and harpacticoids were numerically dominant, with density ranges of 5,500-207,000; 3,000-23,000; and 2,000-13,000 per m^2, respectively. Compared with psammolittoral populations, however, these densities are not particularly high.

Estimates of meiobenthic biomass have been surprising and intriguing. Nalepa and Quigley (1983) measured the biomass in Lake Michigan cores from depths of 11 to 23 m. The standing crop ranged from 0.03 to 0.87 g per m^2, of which 80 percent consisted of nematodes. Overall macrobenthos:meiobenthos biomass ratios ranged from 5:1 to 45:1. Anderson and DeHenau (1980), on the other hand, found that meiobenthic forms accounted for one-third of the total mud biomass, and Strayer and Likens (1986) estimated that about half of the Mirror Lake zoobenthic assimilation is attributable to the meiofauna. Holopainen and Paasivirta (1977) give higher estimates for Lake Pääjärvi, Finland, where the meiofauna:macrofauna biomass ratios were 2:1 and 3:1. All of these ratios must be taken with caution and skepticism, chiefly because of varying methodology from one investigation to another.

One of the regular features of lake gyttja deposits is the reported occurrence of many species of meiofaunal crustaceans that are usually considered free-swimming plankters and littoral species; these are normally taken just above the substrate by whatever sampling device is used, but most of them do actually forage just at the mud surface. Papinska and Prejs (1979), for example, found 17 such species of cyclopoid and calanoid (sic) Crustacea and 12 Cladocera in their study of 24 Polish lakes. The true interstitial forms just below the water-mud interface included the usual assemblage of Harpacticoida, Ostracoda, Annelida, Nematoda, and Chironomidae. Another study by these authors (Prejs and Papinska, 1983) on 20 Polish lakes showed 8,000 to 3,561,000 cyclopoid copepods per m^2 of bottom and near-bottom; biomass for these species ranged from 1.2 to 521.0 g per m^2 (wet weight).

Evans and Stewart (1977) recognize three ecological groups of microcrustaceans inhabiting the bottom deposits and near-bottom waters of southeastern Lake Michigan: (1) euplanktonic species having no affinity for the sediments, (2) epibenthic species usually occurring in the plankton but concentrated in the water just above the sediments, and (3) benthic (meiobenthic) species inhabiting the sediments and only rarely entering the water above the water-sediment interface. All species in the last category were well-known and widely distributed copepods (five species), cladocerans (six species), and ostracods (one species).

The investigation of Tinson and Laybourn-Parry (1986) suggests a mechanism which may account for some of the temporal variations in meiobenthic faunal densities in small lakes. They found that five common species of benthic (meiobenthic) and epibenthic cyclopoid copepods were abundant in gyttja areas of Esthwaite Water (England) during periods of sufficient dissolved oxygen. During summer benthic anaerobiosis, however, these species migrated shoreward into the littoral where the water was well oxygenated.

Several studies on meiobenthic cladoceran populations are notable for their exceptionally high densities. Smirnov (1971), for example, working on Volga River reservoirs, found from 1 to 9 g of cladocerans per m^3 of bottom water (including both meiobenthic forms and near-bottom swimmers). Whiteside et al. (1978) collected up to 2,000,000 cladocerans per m^2 of bottom in Lake Itasca, Minnesota, while Goulden (1971) found highly variable densities of three species in Lake Lacawac, Pennsylvania, with a maximum of 300,000 per m^2 of substrate.

The great majority of qualitative and quantitative meiobenthic samples taken in gyttja by various investigators are notable by the rarity or absence of microturbellarians. A few authors, however, report relatively rich populations, especially Strayer, 1985b, Schwank, 1981, and Rixen, 1961. It is not known whether these investigators just "happened" upon habitats favorable for microturbellarians, or whether they used field and laboratory methods that were especially fruitful. I have seldom encountered meiobenthic microturbellarians in my studies of freshwater meiobenthos.

Sandy Margins of Streams (7).--This eupsammolittoral habitat may extend up to 4 m or more back from the water's edge, and for the most part it contains quantities of organic and inorganic mud. For this reason, interstitial spaces are limited, and the meiofauna is correspondingly much more limited than in lake psammolittoral situations. Also, depending on precipitation, more or less of the width of the habitat is intermittently above the water line. In general, however, physical, chemical, and biological conditions are basically similar to those occuring in the psammolittoral of lakes (e.g. Ruttner-Kolisko, 1961).

The pioneer paper of Sassuchin, Kabanov, and Neiswestnova (1927) dealt with the "microfauna" of Oka River sandy beaches, but it is concerned chiefly with a list of algae and protozoans.

Hummon (1987) worked on shoreline sand substrates (both exposed to air as well as covered with 3 to 20 cm of water) of the headwaters of the Mississippi River, and took samples up to 7 cm deep in the sand. Average meiofaunal densities (numbers per cm^3 of sediment) were: Rotifera 10.8, Tardigrada 4.7, Nematoda 3.4, Oligochaeta 2.0, Gastrotricha 1.9, Diptera 1.3, and Harpacticoida 1.0. These densities are generally comparable with those found for lake beaches. Hummon emphasizes "patchiness" of populations.

At this point it is appropriate to mention the European investigators Kolasa (1982, 1983) and Schwank (1981). They have extensively studied lotic species of microturbellarians and oligochaetes. Collections were made from a variety of substrates, including my habitat categories (7), (8), (9), and (11). Comparable data for North America are lacking but these papers should be consulted by anyone interested in these taxa in lotic situations.

Exposed Sand, Gravel, and Rubble Stream Margins (8).--This substrate consists of variable percentages of sand, gravel, and rubble. It has been studied by many European authors, but most such contributions have dealt with what are called "phreatic" or "ground water." Sampling has often been accomplished with a Bou-Rouch pump or similar apparatus, often at considerable depths. Some workers have simply dug holes into the substrate and sampled the exposed water below the water table.

Since I have arbitrarily set 0.5 m below the top of the ground water table as the lower limit of the meiobenthos for purposes of this chapter, it is often difficult to interpret just what other investigators are discussing, especially for comparisons, and particularly if depths are not specified. Gravel, sand, and rubble margins may be only a meter wide, or they may be extensive exposed bars with a width of 50 m or more. For example, authors often refer simply to "ground water" and one cannot tell how deep or how far from a surface flowage the sample was taken. For practical purpose, however, phreatic waters near the surface of such shores I regard as synonymous with the meiobenthic habitat, at least to a depth of 0.5 m.

One of the few major contributions is that of Husmann (1956) who took a great many phreatic samples from coarse gravel deposits along the shores of four rivers in Germany. None of the species were unusual or endemic. The list includes 33 nematodes, *Troglochaetus*, 18 ostracods, 27 copepods, 2 bathynellids, 10 amphipods (including *Bogidiella*), 1 isopod, and 19 mites. The same author (1957) studied similar habitats in Lower Saxony, and, in general, obtained similar results. Nematodes, copepods, and ostracods were numerical dominants.

There has been a series of phreatic investigations along the shores of the Danube, the most recent and complete being that of Danielopol (1976) in Austria. Sampling was done with the Bou-Rouche pump and by digging holes in riverside gravel deposits. Nematodes, oligochaetes, and cyclopoids were most abundant, representing up to 80 percent of the total fauna. Harpacticoids, isopods, amphipods, limnohalacarids, and other taxa were poorly represented. In general, a depth of 0.5 m in the phreatic layer was sufficiently typical for the sampling station and usually had the most dense populations. It should be emphasized, however, that the numerical vertical distribution of all taxa was extremely irregular and unpredictable. Nematodes often appeared to be most abundant in deposits high in organic matter. Cyclopoid copepods were most abundant where there was little organic matter. Hydrachnellids were absent, in contrast to an abundance reported from upper Danube areas by other investigators. Husmann maintains that hydrachnellids are usually restricted to the uppermost phreatic waters, and limnohalacarids to lower zones.

Dole (1983) worked on shore gravel deposits of the Rhône River, mostly in the top 0.5 m of the phreatic (ground water), and found that the greatest populations were farthest from the water's edge. The compositions of the meiobenthos, however, were highly variable, although the usual groups were present: rotifers, nematodes, oligochaetes, *Troglochaetus,* ostracods, cyclopoid and harpacticoid copepods, amphipods, isopods, hydracarina, and tardigrades. Sometimes cyclopoids were most abundant, sometimes ostracods, and sometimes oligochaetes. From a species standpoint, oligochaetes, with 66 species, were by far the dominant group. There were also 21 species of copepods, 13 amphipods, and 18 nematodes. In fact, this is one of the richest locations to be reported in the literature, even though the great majority of the species are familiar cosmopolitans.

The contribution of Pesce and Maggi (1983) is especially useful for the summarized information it contains on the distribution of the rich phreatic crustacean fauna of southern Europe. Other noteworthy taxonomic and distributional data are contained in such papers as Danielopol (1978, 1980a) on Ostracoda; Lescher-Moutoué (1973) on Cyclopoida; and Rouch (1968) on Harpacticoida.

Danielopol (1980b) presents an important discussion of the role of the limnologist in ground water studies, with special emphasis on the upper phreatic zone. Husmann and other European investigators have become interested in the practical aspects of the role of the phreatic meiobenthos in waste-water gravel beds.

Only one phreatic meiofauna monographic study has been done in North America (Pennak and Ward, 1986). These authors took year-round interstitial water samples from a gravel bar on the South Platte River in Colorado at depths of 15, 30, and 50 cm below the top of the phreatic layer. Their results have produced interesting comparisons with the European literature. Two stations were established: 1-2 m from the water's edge (shore station), and 19 m from the water's edge. All samples were taken with the Bou-Rouch apparatus, and all were sorted and counted under the microscope.

Physical and chemical conditions in the phreatic waters were remarkably similar to those in the nearby river. Crustaceans, insects, "archiannelids" and oligochaetes were the predominant taxa. The "archiannelid" *Troglochaetus beranecki* was especially abundant at the shore station (up to 380 per 5-liter water sample). Only six oligochaete taxa were collected, in contrast to the rich European populations. Nematodes were rare, with only two species being found. *Phyllognathopus viguieri* (harpacticoid copepod) was the most abundant phreatic species of all; some 5-liter water samples contained more then 5,000 individuals. Among the other harpacticoids, *Bryocamptus* was much less abundant, and *Parastenocaris* was uncommon, with only 112 specimens being noted in 58 samples. Among the cyclopoid copepods, only *Acanthocyclops plattensis* and *Microcyclops pumilis* (both new species; Pennak and Ward, 1985a) were found. Both commonly exceeded 20 per 5-liter water sample. Ostracods were rare. *Bathynella riparia* (a new species; Pennak and Ward, 1985b) was the only member of the Bathynellacea; it was especially common in March through September, sometimes exceeding 50 per sample. Amphipods were represented by only two species: *Stygobromus coloradensis* and *S. pennaki* (see Ward, 1977; Ward and Holsinger, 1981). Only 27 samples contained these species, and they seldom exceeded 5 per sample. Aquatic mites were rare, except for a few *Torrenticola* and *Soldanellonyx*. First-instar aquatic insects were also rare.

Thus, there are several striking features which separate the Platte River phreatic meiobenthos from those in Europe, as shown in Table 4.4. Although the Platte River is relatively poor in species, the overall phreatic population density is remarkably greater than those demonstrated for Europe. European phreatic meiobenthic populations are characterized by especially large numbers of Acarina (see, for example, Viets, 1955, 1959; Gledhill, 1982). The same is true for isopods (Coineau, 1971), ostracods (Danielopol, 1978, 1980a), and harpacticoids (Pesce, 1985). Generic differences for the two areas are striking, especially for amphipods, harpacticoids, and cyclopoids.

Exposed Mud Margins of Streams (9).--This type of meiobenthic habitat is characteristic of some reaches of large rivers, especially where the sedimentary load is heavy. I can only speculate that the meiobenthic assemblages are dominated by nematodes. Significant literature seems to be lacking.

Sandy Substrates in Streams (10).--Although many streams have large stretches that consist of relatively clean sand deposits, few investigators have considered this type of meiobenthic habitat, perhaps because of the fact that it is constantly shifting about, driven by the currents and changing discharge, and perhaps also because the sand and fine gravel deposits often are quite shallow, with underlying layers of rock and rubble.

Early brief references to submerged river sands go back to Neiswestnova-Shadina (1935) and Sassuchin et al. (1927) who worked on the Oka River, Behning (1928) on the Volga River, Greze (1953) on the Angara River, and Russev (1974) on the Danube River.

Whitman and Clark (1982) studied the chemistry of the interstitial waters of a submerged sandy riffle in a small stream in Texas. Dissolved oxygen was always near saturation at the surface and dropped off to 1 to 5 mg per liter at a depth of 30 cm, chiefly because of bacterial decomposition coupled with decreased water renewal at that depth. Hydrogen-ion concentrations likewise dropped from pH 6.0 at the surface to near pH 5.5 at 30 cm.

Whitman and Clark (1984) also worked on the ecology of the same riffle. Most of the inhabitants were insect larvae, and the permanent meiofauna were poorly represented; they included only *Hydra*, *Catenula*, three nematodes, some bdelloid rotifers, one tardigrade, four oligochaetes, and four crustaceans, including one *Parastenocaris* (probably the only meiobenthic biont). Although the authors speak of "high" densities of the sand inhabitants, most of them were insect larvae larger than 2 mm and therefore not meiofauna. Populations of the latter category were considerably lower than those of the shoreline psammolittoral.

Evans (1984) studied the rotifer population on the top 4 cm of the sandy bottom of a small brook (1 m) in Ohio. Of the 31 species reported, probably only two (*Lecane paraclosterocera* and *Wierzejskiella* sp.) are definitely known to be interstitial forms. Most or all of the others were cosmopolitan species found on all kinds of freshwater substrates. Densities were very low and highly variable, ranging from 0.0 to 8.6 rotifers per cm^3.

Hummon et al. (1978) studied submerged (unpolluted) sand bars in two small streams in Ohio. Sand samples from surface to 7 cm deep were taken with a small suction syringe, and the sand was washed free of organisms. In general, these substrates had a poor fauna, with only 20 taxa, including nematodes, rotifers, Diptera larvae, microannelids, gastrotrichs, tardigrades, cyclopoids, harpacticoids, and ostracods. Average total numbers under 10 cm^2 of sand surface ranged from 134 to 443.

Schwank (1985), working on the meiofauna of shifting sand bottoms in mountain streams, found a mixture of psammophilic and "hyporheic" species. Although populations were sparse, the number of species found was surprisingly large: nine Turbellaria, 17 Oligochaeta, 16 Nematoda, and 13 Rotifera.

Riemann (1966) did a qualitative study of the interstitial meiobenthos at many stations along the Elbe River both above and below Hamburg. Substrates were chiefly shifting sands, and the freshwater section above Hamburg supported a remarkably impoverished fauna, dominated by nematodes and harpacticoids (five species). Most taxa, e.g., Tardigrada, Cladocera, Ostracoda, and Acarina, were rare.

Whitman's (1984) record of a new harpacticoid, *Parastenocaris texana,* from a small Texas sandy stream is the first North American report of this genus from such a habitat. Specimens were collected at a depth of 20 to 30 cm where the sand was low in dissolved oxygen.

Submerged Lotic Sand, Gravel, and Rubble Substrates (11).--There are really two distinct communities in this category. One consists of the Aufwuchs which is best developed on the exposed surfaces of pieces of rubble. As mentioned earlier, Aufwuchs is a layer consisting of detritus and the associated micro- and macroscopic organisms. Ordinarily, analyses of Aufwuchs in streams show that the largest fraction is inorganic material (clay, silt, fine sand); the next largest fraction consists of organic detritus from many sources; the third largest fraction consists of algae, bacteria, and secreted mucus; and the last is the complicated association of aquatic insects and variable numbers of the usual common meiobenthic taxa (Pennak, 1977, and unpublished). The specific composition of the Aufwuchs layer depends on currents, temperature, geographical location, water chemistry, sedimentary load, etc. Stream Aufwuchs communities are usually more dense and highly developed than lake littoral Aufwuchs communities. Commonly they are 1 to 3 mm thick. Unfortunately, most Aufwuchs studies have been concerned chiefly with (a) total organic content, or (b) with the insect population. Microscopic animal taxa have been mentioned only occasionally and incidentally.

The other community is that of the hyporheic interstitial waters to a depth of 0.5 m below the water-substrate interface. Here there is a typical association of meiobenthic species that has been studied extensively in Europe, but completely neglected in North America until very recently. The preferred collecting apparatus has been the Bou-Rouch pump and similar devices. In general, the hyporheic community is very similar to the shore-phreatic community, but as will be shown below, there are a few important differences.

Perhaps the most significant ecological factor in the hyporheal is dissolved oxygen, and reports in the literature are varied. Poole and Stewart (1976), working on the substrate of the Brazos River in Texas, found that dissolved oxygen ranged from 100 percent saturation to 0.4-0.7 ppm at depths of 30 to 40 cm. Tilzer (1968) found 84 to 63 percent saturation in alpine brooks, and the average for Schwoerbel's (1967a) results on the Danube ranged from 100 percent at the surface to 50 percent saturation at 30 cm depth. Pennak and Ward (1986) found an annual range of 75.6 to 115.6 percent saturation at a depth of 50 cm in the South Platte River. Some of these variations undoubtedly are produced by localized hyporheic currents, in addition to respiratory activities of the hyporheic community. Poole and Stewart (1976) report a negligible flow below 10 cm, but Vaux (1968) emphasizes the localized variations in upwelling and downwelling, depending on the bed profile and localized composition of the substrate, especially with reference to organic particulates. At any rate, it appears that the interstitial organisms are highly adaptable to changing oxygen conditions.

Temperatures in the hyporheic zone are usually one to four degrees below that of the free-flowing water (e.g., Schwoerbel, 1961b, 1967a; Tilzer, 1968), although Pennak and Ward (1986) sometimes found the reverse situation, depending on season and precipitation.

Hydrogen-ion conditions at a depth of 10 to 50 cm are essentially always 0.5 to 1.0 of a pH unit more acid than the overlying water.

Several North American studies have shown that there is a relatively large macrofauna (aquatic insects) at depths of 20 to 40 cm (Williams and Hynes, 1974; Hynes et al., 1976; Poole and Stewart, 1976; and Stanford and Gaufin, 1974).

There are few detailed species lists of meiofauna inhabiting hyporheic deposits (e.g., Schwoerbel, 1961a, 1967b; Tilzer, 1968; Plesa et al., 1964; Ruffo, 1961). Nevertheless, with occasional exceptions, the great majority of species are widely distributed or cosmopolitan. If the results of European studies are compared with those of Pennak and Ward (1986) for the South Platte River, the material will almost duplicate what has been portrayed in Table 4.4. The only important further differences are:

1. Oligochaetes are relatively more abundant in

the South Platte hyporheic than in the phreatic but still were a relatively impoverished group.
2. A few more nematode species were found in the South Platte hyporheic than in the phreatic, but still were a relatively impoverished group.
3. No *Bathynella* were found in the South Platte hyporheic, whereas bathynellids do occur in the European hyporheic.
4. South Platte hyporheic Acarina were more abundant than in the phreatic, but still were an insignificant faunal element.

Microturbellarians are seldom reported from the water-substrate interface and hyporheic habitats of sand-gravel-rubble substrates. Perhaps this situation is a reflection of inadequate collecting (sieving?) methods. Strayer (unpublished personal communication), however, working in southeastern New York State, informs me that patience and great care in sampling such areas, and in laboratory processing, can reveal surprising populations (he has found as many as 31 species).

Pieper (1976) has demonstrated a practical aspect of hyporheic studies. He determined the composition of the meiobenthos of a small German mountain stream that received domestic and brewery waste water. Close to the outfall the fauna was impoverished but at downstream stations the normal community composition was quickly restored.

Mud Substrates in Rivers (12).--Slow-flowing rivers that carry heavy silt loads are likely to have shifting silt and mud bottoms. Although there are many studies of the macrofauna of these situations, the meiofauna has been neglected. Arlt and Saad (1977), however, studied such a river in Iraq and found a meiofauna consisting chiefly of nematodes and oligochaetes; densities ranged from 231 to 773 per 10 cm^2. Kirchengast (1984) found a hyporheic assemblage in an Austrian stream to consist almost entirely of chironomid larvae and nematodes to a depth of 70 cm; water mites were notably absent.

Expression of Population Densities

From the foregoing pages of this chapter, it should be obvious that it is often questionable to compare species abundance and species densities from time to time within the same collecting area because of natural variations in species populations.

It should also be clear that many different methods of expressing population densities are used in the freshwater meiobenthic literature: numbers per

Table 4.4.–A comparison between riverside phreatic meiobenthic faunas of Europe and that of the S. Platte River, Colorado, USA. European data from many sources.

Taxon	S. Platte River	European Rivers
Troglochaetus beranecki	Abundant	Uncommon to rare
Oligochaeta	Four taxa; uncommon	Many spp., common
Nematoda	Few spp., uncommon	Many spp., common
Harpacticoida	*Phyllognathopus, Bryocamptus,* and *Parastenocaris;* extremely abundant	The same plus others; e.g., *Elaphoidella, Moraria, Epactophanes, Nitocra, Maraenobiotus; uncommon*
Cyclopoida	*Acanthocyclops, Mirocyclops;* common	The same plus others; e.g., *Diacyclops, Speocyclops, Graeteriella;* not common
Bathynellacea	*Bathynella riparia;* common	Several other spp. of *Bathynella;* uncommon
Amphipoda	Two spp. *Stygobromus;* common	*Bogidiella, Niphargus, Balcanella, Niphargopsis, Ingolfiella, Metahadzia,* and others; uncommon
Isopoda	Absent	*Microcharon, Angeliera, Proasellus, Stenasellus,* and others; uncommon
Ostracoda	Rare	Many spp.; common
Acarina	Rare	Many spp.; common
Tardigrada	Rare	Rare to common
Rotifera	Rare	Rare to uncommon
Tubellaria	Rare	Occasional

m², per dm², per cm²; to a depth of 3 cm, 30 cm, 50 cm; numbers per cm³, per 10 cm³, per 10 ml, per 5 l, or 10 l of interstitial water; per pump sampling hour; per hole dug into the substrate.

Sieving samples with various mesh sizes and the actual details of using sieves further compound counting errors. Perhaps the most accurate method of assessing meiobenthic densities is to examine large aliquots under the microscope, *without* preliminary sieving. Unfortunately, this process is enormously time-consuming, and few investigators are willing to invest in such a laborious method.

Sometimes results from one sampling technique may be accurately transposed to another technique, but only rarely. It would be most advantageous if freshwater meiobenthologists were to adopt "standard" methods of measuring population densities.

Aside from the problems of trying to compare the results of two or more investigations of the same types of habitat, it is even more fruitless to compare quantitatively the results obtained in the study of two different habitats.

As emphasized in the beginning of this chapter, the chief character of freshwater meiobenthos is *variability*.

Meiobenthic Anomalies and Rarities in Freshwater

Thus far, this discussion has centered around the taxa that are abundant or characteristic of the various kinds of meiobenthic assemblages. In addition, however, there are numerous taxa that (a) have been found only rarely, (b) do *not* occur in the meiobenthos but theoretically and ecologically *should be present,* or (c) have restricted distributions but may be abundant in meiobenthic habitats. Some of these taxa are briefly considered here in the form of concluding remarks.

In the predominantly marine group, the acoel flatworms, two species are notable for their recent invasion into freshwater. One is *Limnoposthia polonica* (Figure 4.4a) from Poland (Kolasa and Faubel, 1974; Faubel and Kolasa, 1978), from sandy substrates. The other is *Oligochoerus limnophilus* (Ax and Dörjes, 1966) which was found on artificial stone embankments of six German rivers. Undoubtedly this species is much more widely distributed in Europe. Both are less than 1 mm long. Strayer (1985b) reported one acoel specimen in the United States (Mirror Lake, New Hampshire).

Ruttner-Kolisko (1955b) found *Marinellina flagellata* in the psammolittoral of the Ybbs River in Austria. Its phylogenetic affinities are uncertain, but it is probably an atypical gastrotrich. The body is 200 µm long, with two posterior bifurcate processes, and a few long cilia. The anterior end bears two adhesive processes, and there is a large pharynx (Figure 4.4b).

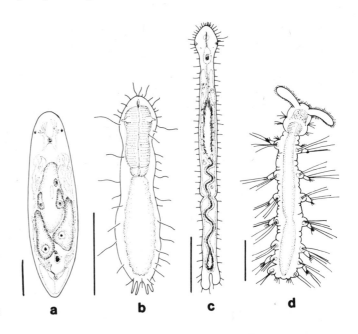

Figure 4.4.--Rare and anomalous vermiform taxa. a, *Limnoposthia polonica*, an acoel flatworm from fresh waters in Poland (from Faubel and Kolaska, 1978); b, *Marinellina flagellata*, from the Ybbs River, Austria (modified from Ruttner-Kolisko, 1955b); c, *Rheomorpha neizwestnovae*, a protooligochaete from European psammolittoral areas (from Ruttner-Kolisko, 1955b); d, *Troglochaetus beranecki*, an archiannelid, widely distributed in Europe and probably also in the USA. (scale = 100 µm).

Rheomorpha neizwestnovae is an unusual protooligochaete known only from the psammolittoral of the Oka River, Poland, the Elbe River, and Lake Maggiore (Ruttner-Kolisko, 1955b). Careful searching in the proper habitats should reveal that it is more generally distributed. This species is minute and has few spicules but neither internal nor external segmentation (Figure 4.4c).

Among the "archiannelida," only *Troglochaetus beranecki* (Figure 4.4d) has been able to successfully invade interstitial freshwater meiobenthic habitats, and it is now known to be widely distributed in Europe. As indicated already in this chapter, however, thousands of specimens have been found in Colorado mountain streams, far from salt water, and it *must* be common elsewhere in North America. The closest morphological relatives of *Troglochaetus* are strictly marine, but *Troglochaetus* is unique in having made a simultaneous cluster of physiological and morphological adjustments to the new environment. Nothing is known of its reproductive habits, or whether it has any special resistant stage in its life history. At any rate, here is *one single* "archiannelid" that has become surprisingly successful in phreatic and hyporheic freshwater. I should also mention that a few very rare "archiannelids" have been reported from subterranean waters in Japan. They are closely related to the marine *Thalassochaetus*.

Although many meiobenthic polychaetes are found along brackish shores, only a very few have penetrated into true freshwater meiobenthic habitats. One example is *Hesionides riegerorum* Westheide, 1979, which occurs in the psammolittoral on the shore of the Chowan River, North Carolina. It is only 1.35 mm long.

To me, it is remarkable that small species of freshwater nemertines only rarely have been reported from meiobenthic habitats. They are highly mobile, microscopic carnivores that feed on a wide variety of small invertebrates, and I should think that they would flourish in the surface layers of freshwater substrates. Although six genera have been reported from widely separated areas on all continents, they are rarely seen and recognized (Gibson and Moore, 1976, 1978). I have collected *Prostoma graecense* only three times, always in the bottom mud of a small pond. Most literature records report freshwater nemertines to be associated with aquatic vegetation.

A most intriguing discovery is the collection of chydorid cladocerans in the phreatic zone on the shores of a small French stream (Dumont, 1983). Two of these small species, *Alona phreatica* and *A. bessei*, are restricted to this meiobenthic habitat and show morphological modifications for locomotion in the water-filled interstices.

Another microcrustacean anomaly is the report of a calanoid copepod of freshwater ancestry, *Notodiaptomus caperatus*, from phreatic waters in Barbuda (Bowman, 1979). Otherwise, all known subsurface copepods are harpacticoids and cyclopoids; indeed, many planktonic and lake-bottom species are sometimes found in phreatic waters, wells, and underground flowages.

Among the Amphipoda, the genus *Bogidiella* (Figure 4.5a) and its relatives are intriguing. They have rarely been found in phreatic waters in Europe (Pesce, 1980), on sandy shores of the Amazon River, in Mexican wells, and in a Texas aquifer. Otherwise they are more characteristic of the marine littoral and interstitial, especially in the Mediterranean area. Apparently they are actively invading freshwater, and their geographic distribution is probably wide and complex. A similar situation exists for the rare and aberrant family Ingolfiellidae (Siewing, 1953, 1963). Most species are marine shoreline interstitial forms but there are a few records of freshwater species from southeastern Europe.

I have previously commented on the occurrence of meiobenthic isopods in the phreatic and hyporheic fresh waters of southern Europe, notably the genera *Microparasellus*, *Microcharon*, *Microcerberus* (Figure 4.1c), and *Proasellus*. A true psammolittoral species, *Proasellus walteri*, is known from Europe (Henry, 1976), and a phreatoicid is know to have invaded Tasmanian sandy beaches (Bayly, 1973). Although North America has an abundance of isopod species living in springs, caves, and underground streams, I have not been able to locate any reports of isopods being collected from meiobenthic assemblages.

Until a few years ago, bathynellids (Figures 4.1b, 4.5b) were thought to be rare, but now we know that there are numerous psammolittoral and phreatic species with restricted ranges on all continents except Antarctica (Pennak and Ward, 1985b), although several European species have wide ranges.

The several species of Stygocaridacea (Syncarida) are restricted to Argentina, Chile, New Zealand, and Australia (Noodt, 1970). They are chiefly meiobenthic forms found in streams and interstitially on shores. More intensive collecting might show a wider geographical distribution and the existence of many additional species, especially in view of their life style.

About nine species of Thermosbaenacea are known from ground waters, shallow wells, and coastal caves in Texas, the Mediterranean area, the Antilles, and Somalia (Zilch, 1972; Chelazzi and Messana, 1982; Maguire, 1965). Ecologically and morphologically, members of this taxon could be considered prime candidates for the invasion of suitable meiobenthic habitats, but presumably the movements away from warm, subsurface, and saline habitats have not yet been surmounted (see Figure 4.5c).

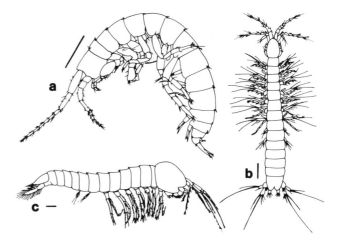

Figure 4.5.--Atypical meiobenthic crustaceans found in fresh and brackish waters. a, female *Bogidiella brasiliensis*, (from Siewing, 1953); b, *Bathynella* (from Gledhill and Gledhill, 1984); c, *Monodella* (Thermosbaenacea) (from Zilch, 1972) (scale = 100 μm).

Although first instar insect larvae are frequently found as elements of the subsurface meiobenthos in freshwater habitats, most of these are "incidental" or "accidental" species normally found on exposed surfaces of substrates. Ferrington (1984), however, found early dipteran larval stages of *Krenosmittia* at a depth of 0.4 m in a Colorado mountain stream and believes that this is the "normal" habitat for this form.

One of the most striking rarities is the occurrence of minute beetles in phreatic waters (Matsumota, 1976). These are not surface species that have accidentally been washed into subsurface waters, nor are they terrestrial cave species. Instead they are actually restricted to the interstitial subsurface meiobenthic habitats. Examples are *Phreatodytes* and *Morimotoa*.

Lastly among insects, some passing attention should be given to the Collembola, or springtails, an order of flightless insects. Even the adults are minute, and all instars are found typically on the surface film in puddles and marshy areas. If they were to lose the effectiveness of the abdominal jumping apparatus and their hydrophobic coat of fine hairs, however, they would seem much more likely members of the meiobenthon. Indeed, when I was collecting in Tasmania a few years ago, I found dense concentrations of Collembola in the interstices of a sandy psammolittoral of a lake at a depth of 5 to 10 cm. In some marine beaches, this apparently is not an uncommon encounter even at depths of 50 cm into the sand (Higgins, personal communication).

In conclusion, then, it appears that only occasional "straggler" species have recently acquired the clusters of (macro?) mutations necessary to: (a) invade the freshwater meiobenthos from marine habitats, (b) change the freshwater life style from a free-swimming existence to substrate or interstitial habitats, or (c) change from a terrestrial (aerial) life style to a meiobenthic life style.

But we have by no means exhausted the opportunities of discovering further meiobenthic anomalies and rarities. I confidently predict that careful searching will uncover many additional "curiosa" which will contribute much toward the problems of ecological evolution and taxonomy.

References

Anderson, R.V., and A.-M. DeHenau
1980. An Assessment of the Meiobenthos from Nine Mountain Lakes in Western Canada. *Hydrobiologia*, 70:257-264.

Angelier, E.
1953. Recherches écologiques et biogéographiques sur la faune des sables submergées. *Archives de Zoologie expérimentale et génerale*, 90:37-161.

Arlt, G., and M.A.H. Saad
1977. Investigations of the Meiofauna and Sediment of the Shatt Al- Arab near Barash (Iraq). *Freshwater Biology*, 7:487-494.

Ax, P.
1966. Die Bedeutung der interstitiellen Sandfauna für allgemeine Probleme der Systematik, Ökologie und Biologie. *Veröffentlichungen des Instituts für Meeresforschung Bremerhaven*, 2:15-66.

Ax, P., and J. Dörjes
1966. Oligochoerus limnophilus. *Internationale Revue der gesamten Hydrobiologie*, 51:15-44.

Bayly, I.A.E.
1973. The Sand Fauna of Lake Pedder: A Unique Example of Colonization by the Phreatoicidea (Crustacea: Isopoda). *Australian Journal of Marine and Freshwater Research*, 24:303-306.

Behning, A.
1928. Das Leben der Wolga, zugleich eine Einführung in die Flussbiologie. *Die Binnengewässer*, 5:1-162.

Bowman, T.
1979. *Notodiaptomus caperatus*, A New Calanoid Copepod from Phreatic Groundwater in Barbuda. *Bijdragen tot de Dierkunde*, 49:219-226.

Chellazi, L., and G. Messana
1982. *Monodella somala* n. sp. (Crustacea, Thermosbaenacea) from the Somali Democratic Republic. *Monitore Zoologico Italiana, N. S. Supplemento*, 16:161-172.

Cobb, N.A.
1913 New Nematode Genera Found Inhabiting Fresh Water and Non-Brackish Soil. *Journal of the Washington Academy of Science*, 3:432-444.

Coineau, N.
1971. Les isopodes interstitiels. Documents sur leur écologie et leur biologie. *Mémoires Museum National d'Histoire Naturelle, Paris, Serie A, Zoologie*, 64:1-170.

Cole, G.A.
1955. An Ecological Study of the Microbenthic Fauna of Two Minnesota Lakes. *American Midland Naturalist*, 53:213-230.

Cummins, K.W.
1975. Macroinvertebrates. Pages 170-198 in B.A. Whitton, editor, *River Ecology*. Berkeley: University of California Press.

Danielopol, D.L.
1976. The Distribution of the Fauna in the Interstitial Habitats of Riverine Sediments of the Danube and the Piesting (Austria). *International Journal of Speleology*, 8:23-51.

1978. Über Herkunft und Morphologie der süsswasserhypogäischen Candoninae (Crustacea, Ostracoda). *Sitzungsberichte Österreichischen Akademie der Wissenschaften, Mathematisch-Naturwissenschaftliche Klasse, Abteilung 1*, 187:1-160.

1980a. On the Carapace Shape of Some European Freshwater Interstitial Candoninae (Ostracoda). *Proceedings of the Biological Society of Washington*, 93:743-756.

1980b. The Role of the Limnologist in Ground Water Studies. *Internationale Revue der gesamten Hydrobiologie*, 65:777-791.

1982 Phreatobiology Reconsidered. *Polskie Archiwum Hydrobiologii*, 29:375-386.

Davies, W.
1971. The Phytopsammon of a Sandy Beach Transect. *American Midland Naturalist*, 86:292-308.

Delamare Deboutteville, C.
1960. *Biologie des Eaux Souterraines Littorales et Continentales*. 740 pages. Paris: Hermann.

Dole, M.J.
1983. Le domaine aquatique souterrain de la plaine alluviale du Rhône a l'est de Lyon. 1. Diversité hydrologique et biocénotique de la dynamique fluviale. *Vie et Milieu*, 33:219-229.

Dumont, H.J.
1983. Discovery of Groundwater-Inhabiting Chydoridae (Crustacea, Cladocera), with the Description of Two New Species. *Hydrobiologia*, 106:97-106.

Enckell, P.H.
1968 Oxygen Availability and Microdistribution of Interstitial Mesofauna in Swedish Fresh-Water Sandy Beaches. *Oikos*, 19:271-291.

Evans, M.S., and J.A. Stewart
1977. Epibenthic and Benthic Microcrustaceans (Copepods, Cladocerans, Ostracods) from a Nearshore Area in Southeastern Lake Michigan. *Limnology and Oceanography*, 22:1059-1066.

Evans, W.A.
1982. Abundances of Micrometazoans in Three Sandy Beaches in the Island Area of Western Lake Erie. *Ohio Journal of Science*, 82:246-261.
1984. Seasonal Abundance of the Psammic Rotifers of a Physically Controlled Stream. *Hydrobiologia*, 108:105-114.

Faubel, A., and J. Kolasa
1978. On the Anatomy and Morphology of a Freshwater Species of Acoela (Turbellaria): *Limnoposthia polonica* (Kolasa et Faubel, 1974). *Bulletin de l'Académie Polonaise des Sciences, Série des Sciences Biologiques, Classe II*, 26:393-397.

Fenchel, T.M., and R.J. Riedl
1970. The Sulfide System: A New Biotic Community Underneath the Oxidized Layer of Marine Sand Bottoms. *Marine Biology*, 7:255-268.

Ferrington, L.C.
1984. Evidence for the Hyporheic Zone as a Microhabitat of *Krenosmittia* spp. Larvae (Diptera: Chironomidae). *Journal of Freshwater Ecology*, 2:353-358.

Frey, D.G.
1982. Questions Concerning Cosmopolitanism in Cladocera. *Archiv für Hydrobiologie*, 93:484-502.

Gibson R., and J. Moore
1976. Freshwater Nemerteans. *Zoological Journal of the Linnean Society of London*, 58:177-218.
1978. Freshwater Nemerteans: New Records of *Prostoma* and a Description of *Prostoma canadiensis* sp. nov. *Zoologischer Anzeiger*, 201:77-85.

Gledhill, T.
1982. Water-mites (Hydrachnellae, Limnohalacaridae, Acari) from the Interstitial Habitat of Riverine Deposits in Scotland. *Polskie Archiwum Hydrobiologii*, 29:439-451.

Gordon, M.S.
1960. Anaerobiosis in Marine Sandy Beaches. *Science*, 132:616-617

Goulden, C.E.
1971. Environmental Control of the Abundance and Distribution of the Chydorid Cladocera. *Limnology and Oceanography*, 16:320-331.

Greze, I.I.
1953. [Hydrobiology of the Lower Part of the Angara River.] *Trudy Vesesoyuznogo Gidrobiologicheskogo Obshchestva*, 5:203-211. [In Russian].

Hartwig, E.
1980. A Bibliography of the Interstitial Ciliates (Protozoa): 1926-1979. *Archiv für Protistenkunde*, 123:422-428.

Hayes, F.R.
1964. The Mud-Water Interface. *Oceanography and Marine Biology, Annual Review*, 2:121-145.

Henry, J.-P.
1976. Remarques sur l'Aselle psammique *Proasellus walteri* (Chappuis, 1948) (Crustacea, Isopoda, Asellota). *International Journal of Speleology*, 8:75-80.

Holopainen, I.J., and L. Paasivirta
1977. Abundance and Biomass of the Meiobenthos in the Oligotrophic and Mesohumic Lake Päajärvi, Southern Finland. *Annales Zoologici Fennici*, 14:124-134.

Hope, W.D.
1971. The Current Status of the Systematics of Marine Nematodes. Pages 33-66 in N.C. Hulings, editor, Proceedings of the First International Conference on Meiofauna. *Smithsonian Contributions to Zoology*, 76:1-205.

Hummon, W.D.
1987. Meiobenthos of the Mississippi Headwaters. Pages 125-140 in R. Bertolani, editor, *Biology of Tardigrades*. Modena: Mucchi.

Husmann, S.
1956. Untersuchungen über die Grundwasserfauna zwischen Harz und Weser. *Archiv für Hydrobiologie*, 52:1-184.
1957. Die Besiedlung des Grundwassers im südlichen Niedersachsen. *Beiträge zur Naturkunde Niedersachsens*, 10:87-96.
1960. Über einige blinde Arthropoden aus dem Mesopsammal bremischer Langsamsandfilter. *Abhandlungen der Naturwissenschaftlichen Vereins zu Bremen*, 35:421-437.
1962. Ökologische und verbreitungsgeschichtliche Studien über limnische Grundwassertiere aus dem künstlichen Mesopsammal der Helgoländer Düneninsel. *Archiv für Hydrobiologie*, 58:405-422.
1966. Versuch einer ökologischen Gliederung des interstitiellen Grundwassers in Lebensbereiche eigener Prägung. *Archiv für Hydrobiologie*, 62:231-268.
1971. Ecological Studies on Freshwater Meiobenthon in Layers of Sand and Gravel. Pages 161-169 in N.C. Hulings, editor, Proceedings of the First International Conference on Meiofauna. *Smithsonian Contributions to Zoology*, 76:1-205.

Hutchinson, G.E.
1957. *A Treatise on Limnology, Volume I, Geography, Physics and Chemistry*. 1015 pages. New York: Wiley and Sons.

Hynes, H.B.N., D.D. Williams, and N.E. Williams
1976. Distribution of the Benthos within the Substratum of a Welsh Mountain Stream. *Oikos*, 27:307-310.

Karaman, S.L.
1935. Die Fauna unterirdischer Gewässer Jugoslawiens. *Verhandlungen, Internationale Vereinigung für theoretische und angewandte Limnologie*, 7:46-53.

Kirchengast, M.
1984. Faunistische Untersuchungen im hyporheischen Insterstitial des Flusses Mur (Steiermark, Österreich). *Internationale Revue der gesamten Hydrobiologie*, 69:729-746.

Kirchner, W.B.
1975. The Effect of Oxidized Material on the Vertical Distribution of Freshwater Benthic Fauna. *Freshwater Biology*, 5:423-429.

Kolasa, J.
1982. On the Origin of Stream Interstitial Microturbellarians. *Polskie Archiwum Hydrobiologii*, 29:405-413.
1983. Formation of the Turbellarian Fauna in a Submontane Stream in Italy. *Acta Zoologica Cracoviensia*, 26:297-354.

Kolasa, J., and A. Faubel
1974. A Preliminary Description of a Freshwater Acoela (Turbellaria): *Oligochoerus polonicus* nov. spec. *Bolletino de Zoologia*, 41:81-85.

Lescher-Moutoué, F.
1973. Sur la biologie et l'écologie des copépodes cyclopides hypogés (Crustacés). *Annales de Spéléologie*, 28:429-502, 581-674.

Maguire, B.
1965. *Monodella texana* n. sp., an Extension of the Range of the Crustacean Order Thermosbaenacea to the Western Hemisphere. *Crustaceana*, 9:149-154.

Matsumota, K.
1976. An Introduction to the Japanese Groundwater Animals with Reference to Their Ecology and Hygienic Significance. *International Journal of Speleology*, 8:141-155.

Moffett, J.W.
1943. A Limnological Investigation of the Dynamics of a Sandy,Wave-Swept Shoal in Douglas Lake, Michigan. *Transactions of the American Microscopical Society*, 62:1-23.

Monard, A.
1920. La faune profonde du Lac de Neuchâtel. *Bulletin de la Neuchâteloise des Sciences Naturelles,* 44:65-236.

Moore, G.
1939. A Limnological Investigation of the Microscopic Benthic Fauna of Douglas Lake, Michigan. *Ecological Monographs,* 9:537-582.

Mortimer, C.H.
1941- The Exchange of Dissolved Substances Between
1942. Mud and Water in Lakes. *Journal of Ecology,* 29:280-329, 30:147-201.

Muckle, R.
1942. Beiträge zur Kenntnis der Uferfauna des Bodensees. *Beiträge zur Naturkundlichen Forschung im Oberrheingebiet,* 7:5-109.

Myers, F.J.
1936. Psammolittoral Rotifers of Lenape and Union Lakes, New Jersey. *American Museum Novitates,* 830:1-22.

Nalepa, T.F., and M.A. Quigley
1983. Abundance and Biomass of the Meiobenthos in Nearshore Lake Michigan with Comparisons to the Macrobenthos. *Journal of Great Lakes Research,* 9:530-547.

Nalepa, T.F., and A. Robertson
1981. Screen Mesh Size Affects Estimates of Macro- and Meio-benthos Abundance and Biomass in the Great Lakes. *Canadian Journal of Fisheries and Aquatic Sciences,* 38:1027-1034.

Neel, J.K.
1948. A Limnological Investigation of the Psammon in Douglas Lake, Michigan, with Especial Reference to Shoal and Shoreline Dynamics. *Transactions of the American Microscopical Society,* 67:1-53.

Neiswestnowa-Shadina, K.
1935. Zur Kenntnis des rheophilen Mikrobenthos. *Archiv für Hydrobiologie,* 28:555-582.

Noodt, W.
1952. Subterrane Copepoden aus Norddeutschland. *Zoologischer Anzeiger,* 148:331-343.
1955. Limnische-subterrane Harpacticoiden (Crust. Cop.) aus Norditalien. *Zoologischer Anzeiger,* 154:78-85.
1970. Zur Eidonomie der Stygocaridacea, einer Gruppe interstitieller Syncarida (Malacostraca). *Crustaceana,* 19:227-244.

Papinska, K., and K. Prejs
1979. Crustaceans of the Near-Bottom Water and Bottom Sediments in 24 Masurian Lakes with Special Consideration to Cyclopoid Copepods. *Ekologia Polska,* 27:603-624.

Pennak, R.W.
1939a. A New Rotifer from the Psammolittoral of Some Wisconsin Lakes. *Transactions of the American Microscopical Society,* 58:222-223.
1939b. A New Copepod from the Sandy Beaches of a Wisconsin Lake. *Transactions of the American Microscopical Society,* 58:224-227.
1939c. The Microscopical Fauna of the Sandy beaches. *Problems of Lake Biology, Publications of the American Association for the Advancement of Science,* 10:94-106.
1940. Ecology of the Microscopic Metazoa Inhabiting the Sandy Beaches of Some Wisconsin Lakes. *Ecological Monographs,* 10:537-615.
1942. Ecology of Some Copepods Inhabiting Intertidal Beaches Near Woods Hole, Massachusetts. *Ecology,* 23:446-456.
1951. Comparative Ecology of the Interstitial Fauna of Fresh-Water and Marine Beaches. *Anneé Biologique,* 27:449-480.
1977. Trophic Variables in Rocky Mountain Trout Streams. *Archiv für Hydrobiologie,* 80:253-275.
1985. The Fresh-Water Invertebrate Fauna: Problems and Solutions for Evolutionary Success. *American Zoologist,* 25:671-687.

Pennak, R.W., and J.V. Ward
1985a. New Cyclopoid Copepods from Interstitial Habitats of a Colorado Mountain Stream. *Transactions of the American Microscopical Society,* 104:216-222.
1985b. Bathynellacea (Crustacea: Syncarida) in the United States, and a New Species from the Phreatic Zone of a Colorado Mountain Stream. *Transactions of the American Microscopical Society,* 104:209-215.
1986. Interstitial Faunal Communities of the Hyporheic and Adjacent Groundwater Biotopes of a Colorado Mountain Stream. *Archiv für Hydrobiologie, Supplement,* 74:356-396.

Pesce, G.L.
1980. *Bogidiella aprutina,* n. sp., a New Subterranean Amphipod from Phreatic Waters of Central Italy. *Crustaceana,* 38:139-144.
1985. The Groundwater Fauna of Italy: A Synthesis. *Stygologia,* 1:15-73.

Pesce, G.L., and D. Maggi
1983. Ricerche faunistiche in acque souterranee freatiche della Grecia meridionale ed insulare e stato attuale delle conoscenze sulla stigofauna di Grecia. *Natura,* 74:15-73.

Picard, J.Y.
1962. Contribution á la connaissance de la faune psammique de Lorraine. *Vie et Milieu,* 13:471-505.

Pieper, H.-G.
1976. Die tierische Besiedlung des hyporheischen Interstitials eines Urgebirgsbaches unter dem Einfluss von allochthoner Nährstoffzufuhr. *International Journal of Speleology,* 8:53-68.

Plesa, C., F. Botea, and G. Racovitja
1964. Recherches sur la faune des biotopes aquatiques souterraines du bassin du Crisul Repede. *Lucrari Instituto Speologia "Emil Racovitza,"* 3:367-396.

Ponyi, J.
1960. Über im interstitialen Wasser der sandigen und steinigen Ufer des Balaton lebende Krebse (Crustacea). *Annales Biologia Tihany,* 27:85-92.

Poole, W.C., and K.W. Stewart
1976. The Vertical Distribution of Macrobenthos within the Substratum of the Brazos River, Texas. *Hydrobiologia,* 50:151-160.

Prejs, K., and K. Papinska
1983. XI. Meiobenthos and Near-Bottom Meiofauna in 20 Lakes. *Ekologia Polska,* 31:477-494.

Remane, A.
1952. Die Besiedlung des Sandbodens im Meere und die Bedeutung der Lebensformtypen für die Ökologie. *Verhandlungen der deutschen Zoologischen Gesellschaft Wilhelmshaven,* 1951:327-359.

Riemann, F.
1966. Die interstitielle Fauna im Elbe-Aestuar. Verbreitung und Systematik. *Archiv für Hydrobiologie, Supplement,* 31:1-279.

Rixen, J.-U.
1961. Kleinturbellarien aus dem Litoral der Binnengewässer Schleswig-Holsteins. *Archiv für Hydrobiologie,* 57:464-538.

Rouch, R.
1968. Contribution à la connaissance des Harpacticides hypogés (Crustacés-Copépodes). *Annales de Spéléologie,* 23:5-167.

Ruffo, S.
1961. Problemi relativi allo studio della fauna interstiziale iporreica. *Bollettino de Zoologia,* 28:271-319.

Russev, B.
1974. [Das Zoobenthos der Donau zwischen dem 845ten und dem 375ten Stromkilometer. III, Dichte und Biomass]. *Academy of Bulgarian Sciences, Bulletin of the Institute of Zoology Museum,* 40:175-194. [In Bulgarian.]

Ruttner-Kolisko, A.
1953. Psammonstudien 1. Das Psammon des Torneträsk in Schwedisch-Lappland. *Sitzungsberichte der Österreichischen Akademie der Wissenschaften, Mathematisch-Naturwissenschaftliche Klasse,* 162:129-161.

1954. Psammonstudien II. Das Psammon des Erken in Mittelschweden. *Sitzungsberichten der Österreichischen Akademie der Wissenschaften. Mathematisch-Naturwissenschaftliche Klasse, Abteilung I*, 163:301-324.

1955a. Einige Beispiele für die unmittelbare Auswirkung des Wetters auf die Lebensbedingungen im feuchten Sand. *Wetter und Leben*, 7:16-21.

1955b. Rheomorpha neiswestnovae und Marinellina flagellata, zwei phylogenetisch interessante Wurmtypen aus dem Süsswasserpsammon. *Österreichsche Zoologische Zeitschrift*, 6:55-69.

1956. Psammonstudien III. Das Psammon des Lago Maggiore in Oberitalien. *Memorie dell'Istituto Italiano di Idrobiologia dott. Marco di Marchi*, 9:365-402.

1961. Biotop und Biozönose des Sandufers einiger Österreichischer Flüsse. *Proceedings of the International Association of Theoretical and Applied Limnology*, 14:362-368.

1962. Porenraum und kapillare Wasserströmung im Limnopsammal, ein Beispiel für die Bedeutung verlangsamter Strömung. *Schweizerische Zeitschrift für Hydrologie*, 24:444-458.

Särkkä, J., and L. Paasivirta
1972. Vertical Distribution and Abundance of the Macro- and Meiofauna in the Profundal Sediments of Lake Paijanne, Finland. *Annales Zoologici Fennici*, 9:1-9.

Sassuchin, D.N.
1930. Materialen zur Frage über die Organismen des Flugsandes in den Kirgisensteppen. *Hydrobiological Journal of the U.S.S.R.*, 9:121-130.

1931. Lebensbedingungen der Mikrofauna in Sandanschwemmungen der Flüsse und im Treibsand der Wüsten. *Archiv für Hydrobiologie*, 22:369-388.

Sassuchin, D.N., N.M. Kabanov, and K.S. Neiswestnova
1927. Über der mikroskopische Pflanzen- und Tierwelt der Sandfläche des Okaufers bei Murom. *Russische Hydrobiologische Zeitschrift*, 6:59-83. [In Russian with German summary.]

Schiff, S.L., and R.F. Anderson
1986. Alkalinity Production in Epilimnetic Sediments: Acidic and Non-Acidic Lakes. *Water, Air, and Soil Pollution*, 31:941-948

Schuster, R.O., E.C. Toftner, and A.A. Grigarick
1977. Tardigrada of Pope Beach, Lake Tahoe, California. *Wasmann Journal of Biology*, 35:115-136.

Schwank, P.
1981. Turbellarien, Oligochaeten und Archianneliden des Breitenbachs und anderer oberhessischer Mittelgebirgsbäche. 1. Lokalgeographische Verbreitung und die Verteilung der Arten in den einzelnen Gewässern in Abhängigkeit vom Substrat. 2. Die Systematik und Autökologie der einzelnen Arten. 3. Die Taxozönosen der Turbellaria und Oligochaeten in Fliessgewässern -- eine synökologische Gliederung. *Archiv für Hydrobiologie, Supplement*, 62:1-85, 86-147, 191-253.

1985. Differentiation of the Coenoses of Helminthes and Annelida in Exposed Microhabitats in Mountain Streams. *Archiv für Hydrobiologie*, 103:535-543.

Schwoerbel, J.
1961a. Subterrane Wassermilben (Acari: Hydrachnellae, Porohalacaridae und Stygothrombiidae), ihre Ökologie und Bedeutung für die Abgrenzung eines aquatischen Lebensraumes zwischen Oberfläche und Grundwasser. *Archiv für Hydrobiologie, Supplement*, 25:242-306.

1961b. Über die Lebensbedingungen und die Besiedlung des hyporheischen Lebensraumes zwischen Oberfläche und Grundwasser. *Archiv für Hydrobiologie, Supplement*, 25:182-214.

1967a. Die stromnahe phreatische Fauna der Donau (hyporheische Fauna). In R. Liepolt, *Limnologie der Donau*, Lieferung 3, Kapitel 5:284-294. Stuttgart, Germany.

1967b. Das hyporheische Interstitial als Grenzbiotop zwischen oberirdischem und subterranem Ökosystem und seine Bedeutung für die Primär-Evolution von Kleinsthöhlenbewohnern. *Archiv für Hydrobiologie, Supplement*, 33:1-62.

Siewing, R.
1953. Bogidiella brasiliensis, ein neuer Amphipode aus dem Küstengrundwasser Brasiliens. *Kieler Meeresforschungen*, 9:243-247.

1963. Zur Morphologie der aberranten Amphipodengruppe Ingolfiellidae und zur Bedeutung extremer Kleinformen für die Phylogenie. *Zoologischer Anzeiger*, 171:76-91.

Smirnov, N.N.
1971. Chydoridae Fauna Mira. *Fauna SSSR. Nov. Ser. 101*, 531 pages.

Stanczykowaka, A., and M. Przytocka-Jusiak
1968. Variations in Abundance and Biomass of Microbenthos in Three Mazurian lakes. *Ekologia Polska*, 16:539-559.

Stanford, J.A., and A.R. Gaufin
1974. Hyporheic Communities of Two Montana Rivers. *Science*, 185:700-702.

Stangenberg, M.
1934. Psammolittoral, ein extrem eutrophes Wassermedium. *Archiwum d'Hydrobiologii i Rybactwa*, 8:273-284.

Strayer, D.L.
1985a. Benthic Invertebrates. Pages 228-234 in G.E. Likens, editor, *An Ecosystem Approach to Aquatic Ecology: Mirror Lake and Its Environment*. New York City: Springer.

1985b. The Benthic Micrometazoans of Mirror Lake, New Hampshire. *Archiv für Hydrobiologie, Supplement*, 72:287-426.

Strayer, D.L., and G.E. Likens
1986. An Energy Budget for the Zoobenthos of Mirror Lake, New Hampshire. *Ecology*, 67:303-313.

Swedmark, B.
1964. The Interstitial Fauna of Marine Sand. *Biological Reviews*, 39:1-42.

Tilzer, M.
1968. Zur Ökologie und Besiedlung des hochalpinen hyporheischen Interstitials im Arlberggebiet (Österreich). *Archiv für Hydrobiologie*, 65:253-308.

Tinson, S., and J. Laybourn-Parry
1986. The Distribution of Benthic Cyclopoid Copepods in Esthwaite Water, Cumbria. *Hydrobiologia*, 131:225-234.

Varga, L.
1938. Vorläufige Untersuchungen über die mikroskopischen Tiere des Balaton-Psammons. *Arbeiten des Ungarischen Biologischen Forschungsinstituts*, 10:101-138.

Vaux, W.G.
1968. Intragravel Flow and Interchange of Water in a Streambed. *Fishery Bulletin*, 66:479-489.

Viets, K.
1955. In subterranen Gewässern Deutschlands lebende Wassermilben (Hydrachnellae, Porohalacaridae und Stygothrombiidae). *Archiv für Hydrobiologie*, 50:33-63.

1959. Die aus dem Einzugsgebiet der Weser bekannten oberirdisch und unterirdisch lebenden Wassermilben. *Veröffentlichungen des Instituts für Meeresforschung Bremerhaven*, 6:303-513.

Ward, J.V.
1977. First Records of Subterranean Amphipods from Colorado with Descriptions of Three New Species of Stygobromus (Crangonyctidae). *Transactions of the American Microscopical Society*, 96:452-466.

Ward, J.V., and J.R. Holsinger
1981. Distribution and Habitat Diversity of Subterranean Amphipods in the Rocky Mountains of Colorado, U.S.A., *International Journal of Speleology*, 11:63-70.

Westheide, W.
1979. *Hesionides riegerorum* n. sp., a New Interstitial Freshwater Polychaete from the United States. *Internationale Revue der gesamten Hydrobiologie und*

Hydrographie, 64:273-280.

Whiteside, M.C., and C. Lindegaard
1982. Summer Distribution of Zoobenthos in Grane Langsø, Denmark. *Freshwater Invertebrate Biology*, 1:2-16.

Whiteside, M.C., J.B. Willliams, and C.P. White
1978. Seasonal Abundance and Pattern of Chydorid Cladocera in Mud and Vegetative Habitats. *Ecology*, 59:1177-1188.

Whitman, R.L.
1984. *Parastenocaris texana*, New Species (Copepoda: (Harpacticoida: Parastenocaridae) from an East Texas Sandy Stream with Notes on its Ecology. *Journal of Crustacean Biology*, 4:695-700.

Whitman, R.L., and W.J. Clark
1982. Availability of Dissolved Oxygen in Interstitial Waters of a Sandy Creek. *Hydrobiologia*, 92:651-658.
1984. Ecological Studies of the Sand-Dwelling Community of an East Texas Stream. *Freshwater Invertebrate Biology*, 3:59-79.

Williams, D.D., and H.B.N. Hynes
1974. The Occurrence of Benthos deep in the Substratum of a Stream. *Freshwater Biology*, 4:233-256.

Wilson, C.B.
1932. The Copepods of the Woods Hole Region, Massachusetts. *Bulletin of the United States National Museum*, 158:1-635

Wiszniewski, J.
1932. Les rotifères des rives sabloneusses du Lac Wigry. *Archives d'Hydrobiologie et d'Ichthyologie*, 6:86-100.
1934a. Recherches écologiques sur le psammon. *Archives d'Hydrobiologie et d'Ichthyologie*, 8:149-271.
1934b. Les rotifères psammiques. *Annales Musei Zoologici Polonici*, 10:339-399.
1935. Notes sur le psammon. II. Rivière Czarna aux environs de Varsovie. *Archives d'Hydrobiologie et d'Ichthyologie*, 9:221-238.
1936. Notes sur le psammon. IV-V. *Archives d'Hydrobiologie et d'Ichthyologie*, 10:235-243.
1937. Différenciation écologique des Rotifères dans le psammon d'eaux douces. *Annales Musei Zoologici Polonici*, 13:1-13.

Zilch, R.
1972. Beitrag zur Verbreitung und Entwicklungsbiologie der Thermosbaenacea. *Internationale Revue der gesamten Hydrobiologie*, 57:75-107.

5. Abiotic Factors

O. Giere, A. Eleftheriou, and D.J. Murison

The abiotic factors to be considered in studies on meiofauna may vary with the objectives of the investigation and the type of the study area. Usually a general characterization of the locality and its external features will be followed by a description of "*in situ*-parameters" characterizing the ambient situation of the meiobenthos. Among these environmental factors, grain size is one of the most important factors since it reflects most other physiographic parameters; fortunately, its assessment is relatively easy and inexpensive.

In contrast, however, the suite of additional substrate- and water-related factors often are much more complex and those familiar with field work will attest to the fact that it is necessary to exercise great discipline in measuring and carefully recording these factors in the field at the time of sampling. Thus one of the first items for consideration is the preparation of a log or a record sheet. Such a log helps to standardize the measurements and establish the sequence in which they should be measured. The format of the record form will depend on the purpose of the investigation and as such will vary, but certain data are essential. These include such items as geographical location, sampling date (and time), and tide and weather conditions. An example of such a data form is shown in Figure 5.1; this example is a form used by the senior author in his field studies on the North Sea tidal shores.

General Description and Environment

In most investigations, it is necessary to give a general physiographical characterization of the study area. The range of required observations will depend on the objectives set at the planning stage, on the physiographic domain of the study area, and on the availability of relevant information from previous investigations. Assessment of ecological relationships between biota and environmental factors usually requires consideration of the following aspects.

Geography.--Latitude and longitude; site name as recorded on map or chart; map (grid) reference, particularly in relation to littoral sampling stations or traverses; location in relation to local (durable) landmarks; physiographic domain (e.g., littoral, continental shelf or slope, oceanic, abyssal, etc.); extent and continuity of study site with adjacent topography, reference to site names/codes in previously published studies in the same region.

Geology.--Substrate; nature of shoreline (erosion-accretion characteristics); sediment source and availability; type and structure of adjacent rock outcrops.

Physiography.--Littoral sites: intertidal and longshore dimensions; beach contours and profile; prominent features such as stream or river courses, rock outcrops, offshore islands, sand bars, tide pools and berms; beach orientation and exposure.

Sublittoral sites: bottom contours and profile; prominent features such as sand ridges, ripple marks, trenches, canyons and seamounts.

Hydrography.--Littoral sites: tidal amplitude and cyclic characteristics (diurnal and longer term); hydrographic effects due to prevailing tidal and wind vectors; effects of local stream and river outfalls.

Sublittoral sites: hydrographic regime as influenced by tidal movements, oceanic currents and terrestrial outfalls.

Light.--Even in eulittoral shores directly exposed to the sun, penetration of light is only in the range of a few millimeters, although in natural coarse quartz sand, this range may extend to 5-10 mm. Thus, light is not a very important factor acting directly on meiofauna; in general, meiofauna are photonegative, but since algae and bacteria may undergo photosynthesis and thereby function in meiofaunal nutrition, light may play an important indirect role.

Penetration of light in sand is, in contrast to conditions in water, governed mainly by refraction (and scattering). This means that in quartz sand the long wave-length infra-red and red light penetrates more deeply than does ultraviolet or blue light. This is of significance mainly for the photosynthetic sediment bacteria which have optimal absorption and highest photosynthetic activity in the longer wave-

Station:

Location:

Date/Time:

Weather conditions:

Temp. (Air):   Tide:

   (Surf. Water):

Depth (cm)	0-1	1-2	2-3	3-4	4-5	5-7.5	7.5-10
Field recordings							
temperature (°C)							
pH							
Eh							
salinity (°/oo S)							
ODR *							
Gwl **							
Sub-Sample taken for							
grain size							
porosity							
permeability							
water saturation and organ. contents							
fauna extraction							

Remarks:

(e.g., previous meteorological conditions)

* oxygen diffusion rate

** ground water level (depth in cm)

Figure 5.1.--Example of a log sheet for field data (from senior author).

length range.

The vertical distribution of diatoms, a major food source for the meiobenthos, is not directly correlated with the penetration of light since many diatoms can regularly be found alive in sediment horizons as deep as 15 cm. Usually, studies on the physiographic milieu of meiobenthos do not refer to light as an environmental factor. Detailed measurements are restricted to special investigations (e.g., Fenchel and Straarup, 1971).

Environmental Factors

The selection of environmental factors to be monitored in a study is a matter of some importance and should be given careful consideration at the planning stage. In general the same elements can be considered in both littoral and sublittoral studies, though due to constraints in sampling technology, the methods of obtaining information may vary considerably. Important abiotic aspects include physiography, hydrography, granulometry, mineralogy, temperature, salinity, oxygen availability, redox potential and substrate porosity, permeability and water saturation. Solar illumination characteristics and turbidity due to particulate matter in the water column overlying subtidal study areas may also be important in relation to primary biotic processes.

The objectives previously defined will dictate the degree of consideration to be given to temporal variations in abiotic characteristics. Where a cyclic (e.g., annual) pattern of fluctuation is anticipated it may be essential to apply a multi-occasion sampling strategy, e.g., on a weekly, monthly, or seasonal basis. Variations associated with tidal or diurnal cycles may be important, particularly in littoral or shallow subtidal situations, and should be investigated accordingly. Where investigations involve comparison of two or more sites, e.g., polluted versus non-polluted areas, temporal aspects may be less important and sampling occasions can be restricted to periods corresponding to maxima and minima for important parameters.

In practice it is seldom possible to avoid a degree of compromise in the design of sampling programs. In some cases the absolute abiotic status of a particular habitat at any point in time may be less important to the biota present than the degree of environmental stability and rate of change. Unfortunately in intertidal investigations, for example, it may be impossible to monitor the study area during and after important, but unpredictable, events such as storms, with consequent loss of crucial information. Greater environmental stability may be anticipated in the deep sea, but important long term fluctuations (e.g., in temperature and dissolved oxygen content of overlying water) should be monitored.

Whenever possible it is best to measure directly in the field, particularly those parameters that may become affected by disturbance of samples or delay in obtaining measurements. The monitoring of environmental parameters should be carried out in close association (the same sample, if possible) with related biotic observations. In most intertidal and many shallow sublittoral situations where SCUBA techniques can be applied this is generally feasible, but in deep sea investigations, using remote sampling techniques, the limitations imposed by suitability or reliability of equipment may render this impossible.

Substrate Related Factors

Sampling Methods.--In soft substrates, sediment sampling devices normally will be identical to those used in meiofaunal collection, and several of these are described in Chapter 7. Clearly, it is of fundamental importance to obtain samples in a minimally disturbed state and by methods which will allow accurate quantification of relevant components.

In the eulittoral, samples are usually obtained as vertical cores using transparent plastic tubes of suitable diameter (1-10 cm) with one end externally beveled to a cutting edge. Tubes may be hand-pushed or hammered (rubber) into the substrate and the core withdrawn following closure (e.g., by a rubber bung) of the top opening. A degree of core compaction ("shortening") may occur depending on the combined effects of internal diameter of the corer, sampler displacement volume, penetration depth, friction, and sediment characteristics; this should be taken into consideration when choosing a particular coring device and in any subsequent subdivision of the sample.

Devices which allow several parameters to be monitored within a single sample have been designed for intertidal studies (e.g., Giere, 1973). Here the Plexiglas box incorporates a series of sampling ports and core sectioning slots, temporarily sealed with adhesive plastic film. Thus, factors such as oxygen availability, redox potential, temperature and salinity may be simultaneously investigated while analysis of particle size and meiofauna population is based on subsamples of the same core fraction.

Sublittoral sediment samples are frequently obtained as subsamples from larger devices such as grabs or box corers. However, the physical disturbance of sediments caused by many of these instruments must cast some doubt on the real value of subsamples so obtained. Similar limitations may be associated with most gravity corers (McIntyre, 1971). Where circumstances permit, sediment cores should be taken by the Craib corer (Craib, 1965 - for single cores) or the SMBA Multiple Corer

(Barnett, Watson, and Conelly, 1984 – up to 12 cores) or by SCUBA diver (Chapter 7).

Grain Size.—Grain size composition is influenced by numerous environmental factors (e.g., exposure, currents, nature and amount of suspended matter) and, in turn, determines many physiographic parameters which are closely related to substrate such as porosity, permeability, oxygen supply and salinity gradients. Since many meiobenthic animals can exploit the interstitial environment of sandy substrates (the mesopsammon), the proportion and distribution of finer sediment particles will influence the degree of accessibility. Thus, sediment structure attains a dominant role in meiobenthic ecology. Because of this and because of the relatively simple analytical techniques involved, grain size characteristics have received more attention than any other abiotic parameter in benthic investigations.

The quantity, distribution, and volume of sediment samples required for particle size analysis will depend on the scope and overall objectives of the investigation. However, because of the large horizontal and vertical variability exhibited by sediments, it may be desirable to use the same, or closely associated, samples for grain size and faunal analysis. Ideally, samples should be taken directly from the field site as undisturbed cores, but it may be necessary to acquire them as subsamples from a grab. As a rough guide, cores of 5 cm diameter (19.6 cm^2) will normally be suitably representative, but smaller samples (e.g., 2 cm diameter) may be adequate for combined meiofaunal/sediment analysis. Vertical division of intervals for sediment cores will depend on the stratification pattern of the sediment, but should conform to those applied to meiobenthos samples. As preservation techniques may alter the grain size distribution (i.e., of organic components), samples should not be subjected to any chemical fixatives and should be analyzed as soon after collection as possible. If preservation is necessary, freezing is the best technique; less reliable, but within reasonable limits, is preservation by 5–10% formalin.

Samples comprising sand, silt, and clay fractions should initially be wet-sieved through a 63 μm-mesh using tap water in order to remove any sea salt and taking care to avoid damage to fecal pellets and other fragile biogenic elements. If not needed, organic matter may be removed from sediment samples by pre-treatment in a 6% solution of hydrogen peroxide (Buchanan, 1984).

The filtrate of the 63 μm-mesh, i.e., the silt-clay fraction, may be further graded by sedimentation analysis after given settling times. Detailed methodology for this type of analysis in marine sediments is given in Buchanan (1984). Although the mechanical procedures involved require only simple laboratory apparatus and techniques, they are, nevertheless, relatively time-consuming and consequently careful consideration should be given to the precise requirements of such analyses. In many meiobenthic ecological investigations it is sufficient to divide the fine fraction into no more than three grades representing coarse silt (63–15.6 μm), fine silt (15.5–3.9 μm), and clay (<3.9 μm). Since the principle of the analysis is that, in a water suspension, large particles will fall faster than small particles, then the size ranges selected (above) may be quantified (weight) by the timed extraction of only three 20 ml pipette samples from the sedimentation column. Because sedimentation rates, and consequently sample extraction times, depend on water temperature, a constant temperature water bath is essential.

Comparatively rapid automatic/semi-automatic analyses of suspended particles based on conductometric (e.g., Coulter Counter) and laser diffraction (e.g., Malvern Autosizer) techniques are increasingly applied, but depend on the availability of sophisticated and correspondingly costly equipment.

Samples comprising sand fractions only, need not be wet-sieved through a 63 μm mesh initially, but before drying (80° C, 24 h), should be cleared of salt (e.g., by repetitive rinsing and decanting with water).

The sand fraction is normally analyzed by passing through a geometric series of test sieves with 0.5 phi (φ) intervals. Phi (φ) = $-\log_2$ of mesh size in mm, see Table 5.1), and consequently, information obtained regarding finer grades will be more detailed than that for the coarser end of the spectrum. For more refined analysis, intermediate apertures (0.25 phi intervals) are available from most manufacturers. If possible, a mechanical sieve shaker should be used and, in order to facilitate comparisons, processing time should be standardized (e.g., 15 min). Excessively prolonged shaking may cause unnecessary damage both to sediment structure and sieves.

On completion of the shaking process the fractionated sediment residues in individual sieves should be weighed (a precision level of 0.01 g will be adequate for a total sample of 100 g or more). For quartz sands a core section of 5 cm diameter and 4 cm vertical depth will provide approximately 100 g dry sediment.

For assessment of characteristic sediment indices, the sediment weight fractions (calculated in percent of the total sample) should be transformed into a cumulative frequency series (Table 5.2) and then plotted as a cumulative frequency curve (Figure 5.2). It is apparent that the use of the phi (φ) notation transforms geometric to arithmetic integer series with consequent advantages in graphical and statistical procedures. From the resulting curve with its roughly sigmoid shape, the median particle diameter, i.e., the φ-value corresponding to the 50% - point of the cumulative scale (Md φ or $\varphi 50$), can be estimated. While this Md φ-value indicates the central tendency or the average in particle size distribution, the upper and lower quartiles ($\varphi 75$ and $\varphi 25 = Q_3 \varphi$ and $Q_1 \varphi$, see Figure 5.1) indicate the spread of the grain size data and, thus, are further characteristic sediment indices. Since it may become difficult to read these and other φ-values (see below) accurately from an S-shaped curve, it is advisable to transform the curve into a straight line by plotting the cumulative percentages (the ordinate) on a probability scale. However, availability of probability scale paper seems to be problematical in many cases. The "Quartile Deviation" (QD) expresses the number of phi units lying between the upper and lower quartile diameters: $QD = \frac{Q_3 \varphi - Q_1 \varphi}{2}$ A sediment with a small spread between the quartiles, i.e., a small QD φ, is regarded as being "well sorted" (see below).

These indices, along with the "quartile skewness" which denotes the asymmetry of the cumulative curve (see below), have been criticized by Folk (1974) on grounds that they ignore both "tails" of the curve, neglecting the fine and coarse sediment fractions. Folk recommends including more φ-points in order to obtain more sensitive (middle 90%) indices.

Graphic means:

$$Md = \frac{(\varphi 16 + \varphi 50 + \varphi 84)}{3}$$

Inclusive Graphic Quartile Deviation:

$$QDI = \frac{\varphi 84 - \varphi 16}{4} + \frac{\varphi 95 - \varphi 5}{6.6}$$

This Quartile Deviation covers over 90% of the total distribution and is, therefore, a better overall measure of sorting. The resulting sorting classes are as follows:

<0.35	very well sorted
0.35-0.50	well sorted
0.50-0.71	moderately well sorted
0.71-1.00	moderately sorted
1.00-2.00	poorly sorted
2.00-4.00	very poorly sorted
>4.0	extremely poorly sorted

Inclusive Graphic Skewness:

$$SkI = \frac{\varphi 16 + \varphi 84 - 2 \varphi 50}{2(\varphi 84 - \varphi 16)}$$

$$+ \frac{\varphi 5 + \varphi 95 - 2 \varphi 50}{2(\varphi 95 - \varphi 5)}$$

A positive skewness indicates preponderance of grain size fractions larger than the median diameter, while in a sediment with a negative skewness the finer fractions prevail.

Table 5.1.—Standard Sieve Series for Determination of Grain Size Distribution.

International Test Sieve Standard R565

ISO 3310-1 Wire Mesh Series

Nominal Aperture		Size Class
Principal Sizes (µm)	Phi Scale Equivalent	(Wentworth Scale)
4.000	-2.00	Granule
2.000	-1.00	Very coarse sand
1.000	0.00	Coarse sand
710	+0.49	
500	+1.00	Medium sand
355	+1.49	
250	+2.00	Fine sand
180	+2.47	
125	+3.00	Very fine sand
90	+3.47	
63	+4.00	Coarse silt
<63	>+4.00	Silt
<2	>9.00	Clay

Table 5.2.—Assessment of a Cumulative Frequency Series from an Exemplary Sediment Sample.

Sieve		Sediment	
µm	phi scale	dry wt. (%)	cum. dry wt. (%)
2000	-1	2.0	2.0
1000	0	7.3	9.3
500	+1	31.9	41.2
250	+2	33.7	74.9
125	+3	16.3	91.2
63	+4	7.9	99.1
<63	>+4	0.9	100.0

Instead of the widely used graphical assessment of the above characteristic sediment indices (Figure 5.2), the same values can be mathematically computed by extrapolation (see Hartwig, 1973). In some cases it is useful to convert phi-values to the grain size diameter in um. The best way of doing this is to construct a conversion chart on one-scale log-paper. A more comprehensive treatise on sediment analytical techniques is given by Buchanan (1984).

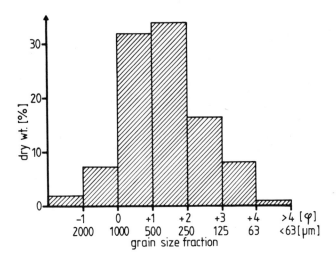

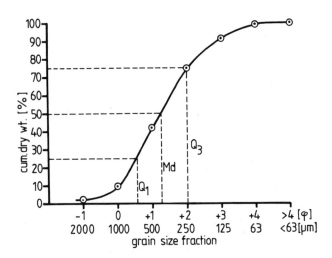

Figure 5.2.--Diagrams of grain size distribution (% values taken from Table 5.2), illustrated as % fractions (above) and as a cumulative frequency curve (below). Position of Median diameter (Md), lower and upper quartiles (Q_1, Q_3) indicated.

Particle Roundness.--By assuming a spherical shape for all particles, grain size analysis will provide a general morphometric description of the interstitial system. However, the natural asymmetry, characteristic of sediment particle structure, is ecologically important both in relation to void configuration and organic encrustation or attachment (Meadows and Anderson, 1966; Webb, 1969; McIntyre and Murison, 1973).

Deviation from sphericity may be used to classify sediment grains on a standard scale. The scale of roundness for quartz particles devised by Powers (1953) is shown in Figure 5.3 and encompasses characteristics of both sphericity and angularity. To determine roundness of particles in a sample, fifty or more grains are assigned to appropriate classes (very angular, angular, etc.) by comparison with the photograph series. The index of average roundness for a sample is obtained by multiplying the number of grains per class by the corresponding geometric mean (0.15 for very angular, 0.2 for angular, etc.) and dividing the sum of products by the total number of grains examined.

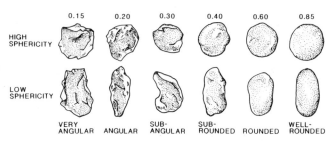

Figure 5.3.--Scale of roundness for quartz grains (from Powers, 1953, in Hulings and Gray, 1971).

Biogenic calcium carbonate fragments may be similarly classified according to the scale of Pilkey, Morton, and Luternauer (1967). Five classes are designated ranging from Class O, which includes whole shells, to Class 4 comprising highly rounded shell fragments. Some biogenic fragments may display greater structural complexity due to internal cavitation (e.g., echinoderm plate fragments).

Porosity.--The porosity of a clastic sediment may be defined as the ratio of void volume to total volume in a given mass of sediment. The porosity value has ecological significance in that it provides a measure of the total interstitial volume which, in the presence of sufficient water, may be available to meiobenthic organisms. The ratio does not, however, convey information regarding dimensional or spatial relationships between discrete particles and voids. For instance, fine sands with correspondingly small mean interstitial aperture generally display higher porosity values than do coarser sands. Also burrows of macrobenthic animals strongly influence sediment porosity.

The accessibility of any given interstitial system to mesopsammic (i.e., locomotion within the interstitial space without apparent displacement of sediment grains as opposed to burrowing) fauna will depend on morphometric relationships between the void network and the particular organisms concerned. In this respect the epoxy resin embedding technique of Williams (1971) is useful in that it enables direct microscopic observation on particle/void spatial and dimensional relationships. The porosity of a given sediment may vary significantly according to the degree of compaction induced by the hydrodynamic forces. It is appropriate, therefore, to relate the range of porosity values associated with a particular sediment system to the prevailing environmental regime. The determination of porosity, P, may be carried out on minimally disturbed cylindrical core samples in the following steps:

1. Calculate sample volume (V = Vs + Vv = $\pi r^2 h$)
2. Weigh washed and dried (80°C) sediment (Ws)
3. Estimate (gravimetrically) the mean density of the sediment (Ds)
4. Calculate volume of solids (Vs = Ws/Ds)
5. Calculate volume of voids (Vv = V - Vs)
6. Calculate porosity of sample (P = Vv/V)
7. Porosity % = P x 100

Ds = density of sediment; h = height of core sample; r = radius of core sample; V = total core volume; Vs = volume of solids; Vv = volume of voids; Ws = weight of sediment

Since the volume of solids (Vs) in a particular sample remains constant, the range of porosity values possible may be determined by calculating core volumes (V) associated with induced states of maximum and minimum compaction. Minimum compaction may be achieved by gradual re-sedimentation of the sample in the water-filled cylinder. Maximum compaction can be obtained by vertical vibration of the cylinder while applying partial vacuum in order to eliminate trapped air bubbles. For detailed discussion of porosity including aspects of ecological significance, the studies of Fraser (1935), Graton and Fraser (1935) and Webb (1958, 1969) should be consulted.

Permeability.--Permeability is an index of the rate that water passes through a given volume of sediment. Although the porosity value defines the total volume of interstitial space available to meiobenthos, it gives no indication of sediment permeability, which may be considerably more important to the organisms concerned. The flow of water through the interstitial network provides a mechanism for transport of both dissolved and suspended substances (particles, solutions, gases). Diffusion and gravitational forces also play an important role in the movement of gases and particles. The permeability of sediments can be measured using the apparatus shown in Figure 5.4.

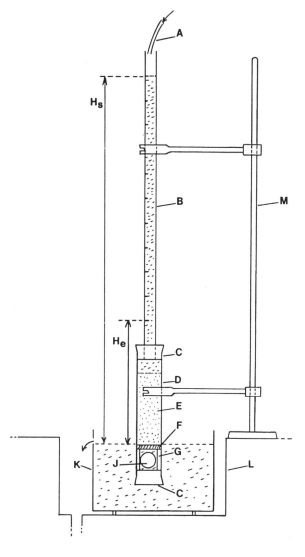

Figure 5.4.--A simple Falling Head Permeameter; a, filtered water supply; b, calibrated acrylic standpipe; c, rubber stopper; d, acrylic coring tube; e, sediment sample; f, porous plastic ("Vyon") filter; g, split-ring plinth; hs, hydraulic head (start); he, hydraulic head (end); j, effluent port (closable); k, constant level bath; l, drained sink; m, clamp stand.

Permeameters, as used in laboratory studies, usually incorporate either a constant level or falling head water reservoir and provide a measure of the core section gravitational flow rate (i.e., coefficient of permeability in cm sec^{-1}) under defined physical conditions (water temperature and viscosity). Laboratory tests carried out on undisturbed vertical core samples collected and measured in the same coring cylinder will provide useful comparative information on the permeability of sediments.

Because of the importance of variables such as compaction and vertical stratification in field situations, measures obtained from homogenized or recomposed samples should be avoided.

Absolute flow rates pertaining to actual field situations are less easy to obtain because of the inherent complexity and variability characterizing many hydrodynamic regimes. Studies on interstitial water movement in intertidal sand using Rhodamine B dye and Gel Cyanogum (Steele et al., 1970) and in permanently submerged *Amphioxus*-sands using fluorescein dye (Webb and Theodor, 1968) provide useful information both on *in situ* techniques and on ecological considerations concerned. Some other aspects of ecological significance are covered in Webb (1958, 1969). A comprehensive study of porosity and permeability in clastic sediments is provided by Fraser (1935).

Erosion-Accretion Dynamics.--Currents and tides, but above all, wave action determine the shape and texture of a beach which at any time is in a dynamic equilibrium with the controlling forces. Any change in this equilibrium is manifested by the transport of sedimentary material in an offshore, onshore or longshore direction accompanied by a change in beach profile. The duration of these movements varies from a few hours during storms, to weeks between neap and spring tides, to months between summer and winter. Not infrequently climate changes or human interference are responsible for much longer cycles lasting several years or in extreme cases even for the elimination of the beach. Shelter promotes stability and beaches which are fully protected do not show such obvious cycles. The frequency and severity of these changes determine the relative stability of the beach. Although most effective in the eulittoral, hydrodynamic forces also influence sediment texture and stability in sublittoral bottoms.

Erosion and accretion processes are accompanied by disruption of the sediment layers, change in the particle size and sorting of the sediment which can cause considerable alterations in the interstitial environment. Most meiobenthic organisms are adapted to such changes provided that these are not too severe and sudden.

Beach processes can be studied in a variety of ways. The simplest, fastest, and most accurate method is by inserting metal stakes into the sediment along transect lines in the intertidal and shallow subtidal. Readings taken at time intervals provide information on the short and long term erosional and depositional processes which allow the construction of sediment budgets for a specific area. The method is effective although as it is open to interference from the public it requires continuous supervision over long periods.

Surveying techniques using staff and theodolite or Abney clinometer measuring the level of the sediment at different positions along transect lines and on successive occasions are also used for constructing beach profiles. On beaches with a very uneven profile including ridges and runnels, readings are taken at close intervals (5 m) but on flat and stable beaches larger intervals can be used. Other simple leveling devices have also been used by many workers (for literature see Eleftheriou and Holme, 1984) with variable degrees of success.

It should be noted that irrespective of the device used for leveling purposes all measurements need to be referred to a fixed point of beach mark located above the beach area. Most of these surveying techniques are accurate enough for ecological purposes; it is, however, essential to remember that most require the cooperation of two persons. Useful information on surveying can be found in Kissam (1956). Komar (1976) discusses in detail the physical processes of beaches. Time series of data on beach processes provide important information on the degree of stability of a particular beach and help to interpret physical and biological data.

Water Related Factors

Sampling Pore Water.--Since many modern physico-chemical analyses require comparatively small water samples, it is often possible to obtain sufficient pore water by centrifuging or pressing a sediment sample (subsample) taken in the immediate vicinity of the meiofauna sample.

In a more precise way, pore water can be obtained directly from the sample to be studied by a suction corer which draws water from a given sediment horizon with negligible disturbance of its structure. In sandy intertidal areas, a normal hypodermic syringe can serve as a simple suction corer. However, in finer, detritus-rich sediments, the needle will tend to get clogged. To overcome this problem, several types of corers with filtering devices have been developed.

Steele et al. (1970) designed probes with filtering "windows" of stainless steel mesh which are pushed by hand to the desired depth. In a similar approach Howes and Wakeham (1985) developed an *in situ* pore water sampler with minimal disturbance of the sediment. Because of its simple and inexpensive construction a modified version is described here: On a thick-walled glass capillary of suitable length a teflon or plastic cap is glued (e.g., the protecting cap of a hypodermic syringe) which has been perforated by a series of small holes (approximately 1 mm diameter). The lumen of this cap is filled with glass wool as a filtering device. At its top end, the glass capillary is tightly sealed with a rubber serum stopper. After pushing the glass corer into the right depth, pore water will passively accumulate in the plastic cap when the corer is left for a while in the sediment. However, water can also immediately be withdrawn by injection of a fitting hypodermic syringe through the rubber stopper into the capillary tube.

For greater convenience the hypodermic syringe can be flexibly connected with the glass rod through a piece of capillary tubing (teflon tubing is permeable for oxygen!) applied to the hypodermic needle on one end and to the needle plus the cut-off adapting stub (Luer stub) of another disposable syringe glued on top of the capillary glass.

Pulling the piston of the syringe, one usually obtains sufficient pore water even from muddy sediments for subsequent sealing in the syringe or transfer to a suitable vial. In stony or much compressed sands it is advisable to push a metal rod into the bottom close to the sampling depth prior to insertion of the fragile glass rod. Using the resulting "entrance channel" it is, however, of prime importance for correct sampling that the last few centimeters of undisturbed sediment are penetrated by the sample rod itself. This glass capillary sampler can be easily used by SCUBA divers for obtaining subtidal pore water in a defined sediment depth.

A more sophisticated, yet similarly effective "pore water lance" was originally developed by Balzer (unpublished). A slightly modified version (Figure 5.5) allows extraction of depth-graduated series of pore water samples from sediment cores: a stainless steel corer (approximately 40 cm long), fortified with a conical steel tip, has an upper lid and a solid handle which is firmly connected with the tip by a central steel axis. From small holes (2 mm diameter), drilled through the wall of the corer at suitable intervals in a spiral arrangement, lead capillary tubes through its inner lumen and lid. Each tube is externally connected to a hypodermic syringe. In order to filter the pore water, each drill hole is covered by a ring-shaped sleeve of porous teflon (5 μm pore diameter) held in position by intermittent steel rings.

After pushing the sampler into the sediment, the syringes are serially "charged" with filtered pore water from controlled depth. Immediately after sampling they can be tightly closed with caps. If used for H$_2$S measurements, the whole system must be pre-flushed with nitrogen. For oxygen measurements, all capillary tubes must be filled prior to sampling with deoxygenated water.

This pore water lance, tested for use in intertidal and subtidal (SCUBA) areas has to be operated by hand. It can easily be adjusted according to individual needs and allows well separated depth gradients to be measured because the suction area of each filtering hole is fairly well defined (tests with rhodamine B). This corer is a simplified version of the automatic, *in situ*, interstitial water sampler described in detail by Barnes (1973) for use in deep water sediments.

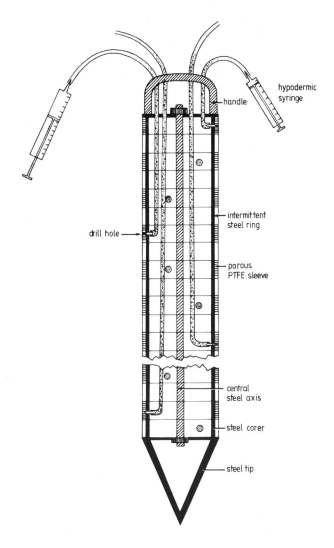

Figure 5.5.--The "pore water lance," a hand-operated suction corer for sampling sediment pore water.

Hydrological Characterization.--The hydrological characterization of a study area will be of primary importance in all benthic investigations. The role of hydrodynamic agencies in the process of transport, sorting, passive sedimentation, active deposition and erosion (see previous section) of clastic sediments is widely recognized. Similarly, the influence of local hydrography on the disposition of soluble inorganic and organic (both soluble and particulate) compounds along with planktonic phases of benthic fauna is of fundamental importance. Further, it has been demonstrated that the faunal characteristics of a beach can be correlated with its degree of exposure to wave action (Eleftheriou and Nicholson, 1975).

Exposure (To Wave Action).--Exposure is a characteristic of a shore line and describes the physical and chemical consequences of more or less intense wave action. It is rather of a conceptual, summative character than a quantifiable parameter controlling sedimentary characteristics such as grain size and sorting. Thus it will determine the parameters of the interstitial environment, e.g., pore space, permeability, oxygen content, pH, temperature and food, essential for an interstitial existence.

The effects of wave action on the macrofaunal organisms of beaches are obvious and direct. A similar effect on meiofauna, however, is not so clearly visible and it could be assumed that it operates through control of the interstitial environment and associated parameters. Thus, it appears that coarse grained beaches exposed to wave action with their rigid physiographical regime have a smaller meiobenthic abundance and biomass than finer sediment in sheltered beaches (Gray and Rieger, 1971).

Wave recorders and thermistors of different types have been used to measure wave action or water movement as indicators of a relative ranking of shorelines according to exposure. However, because of price and maintenance requirements, these methods are beyond the reach of the average worker. Alternative methods used to measure wave action and turbulence (literature for these methods in Eleftheriou and Nicholson, 1975) present inherent difficulties as to calibration and reliability of results and are totally unsuitable for measuring the extremes of the exposure spectrum. As there is a relationship between exposure and particle size of sediments the mean diameter of particles could be used to categorize indirectly beaches with different exposure. An alternative method of using the depth of the reduced layer in the sediment (Seed and Lowry, 1973) as an indicator of exposure is rather too imprecise to be effective.

Exposure scales based on other physical and chemical parameters are too arbitrary, and biologically based scales used for rocky shores are not satisfactory for beaches. The introduction of a compound index that rates a number of parameters such as wave action, surf zone width, particle size, depth of reduced layer and the presence or absence of macrofaunal burrows, combines many of the individual approaches in a unified presentation (McLachlan, 1980). Its universality, however, has not been adequately tested and its effectiveness as a rating system of exposure remains uncertain. The indices and scales suggested here are, as yet, tentative. They should be combined with the experience of the individual worker to assess exposure of a particular shore.

Terrestrial Effluents.--In some respects investigations concerning the role of terrestrial effluents in the marine environment will overlap with general hydrological considerations. However, important differences, both physical and chemical, which characterize waters of marine and terrestrial origin will in many cases necessitate a study of the complex interactions involved. The influence of rivers, streams, and other freshwater effluents such as flood water, industrial waste water, and sewage outfalls, on intertidal and shallow subtidal ecosystems may be substantial. Variability in flow velocity and volume generated by weather conditions (e.g., spate streams) may contribute substantially to erosion/accretion dynamics in high energy sandy beaches, particularly in sites characterized by significant water course deviation.

Water Flow.--Interstitial water movement is controlled by the pore size of the sediments. In intertidal sediments the interstitial water movement during low tide is affected by gravity and has a downward direction; where it is affected by important land drainage a horizontal flow is also evident.

Interstitial water movement can be measured by injection of a solution of a fluorescent dye, Rhodamine B (20 g·l^{-1}) into a core of sand which is placed in the beach. After 12 hours, interstitial water samples are taken from different depths and distances from the original core and concentrations of dye are assessed visually and by means of a fluorometer (Steele et al., 1970). An alternative method is the injection into the sand of Gel Cyanogum 14, a substance soluble in sea water which gels in a predetermined time. After the gelling time the spheroid of gel and sand is excavated and through its acquired shape the horizontal and vertical movement of water can be measured. By relating the movement of the gel to the tidal cycle and to the ratio of interstitial space to sand volume, the water flow through the surface can also be calculated (Steele et al., 1970). In addition to pore space, in shallow subtidal areas water percolation is also affected by pressure due to the wave action (Webb, 1969; Webb and Theodor, 1968).

Water Saturation.--Intertidal beach systems, particularly those of an exposed character, are normally subject to some degree of drying out at least once per day. The upper region of the intertidal zone will be most affected and may show a continuous decrease in pore water content throughout the exposure period. Consequently, water saturation measurements should always be correlated with tidal state observations.

A rapid and comparatively accurate field analysis can be carried out by means of a Speedy Moisture Tester. The instrument is available in two versions handling sediment samples of 20 g and 3 g (wet weight) respectively. The technique measures gas (acetylene) pressure generated when a sample is shaken or pulverized in the presence of an absorbent compound incorporating calcium carbide. The dial of the pressure chamber is calibrated to indicate water content as a percentage of wet weight. Water saturation values for individual samples can be calculated from corresponding porosity and relative density measurements, e.g. a water content of 19.6% represents total saturation in a sediment of 39% porosity and 2.63 density.

A less accurate field method is the displacement technique of Hummon (1969) which measures the total displacement volume (water + sediment) of a sample comprising 5 cm^3 of sediment. Measurement of sediment and associated pore water volume is carried out in a 10 cm^3 calibrated cylinder primed with 5 cm^3 of water. The sediment (plus adsorbed water) is settled in the cylinder until its compacted volume, obtained by frequent agitation, reaches the 5 cm^3 mark. Fully saturated sediment will register an increase of 5 cm^3 whereas completely dry sediment of 40% porosity will raise the water level by 40% less, i.e., $60/100 \times 5 = 3$ cm^3 (0% saturation). Actual porosities of field samples can be tested by oven drying sediment and measuring minimum displacement values (see above).

When field methods prove inappropriate, sediment samples may be stored in air-tight, preferably pre-weighed, tares for laboratory testing of water content, e.g., oven dry (at 80° C for 24 h) weight loss, expressed as water volume, in relation to corresponding porosity value.

Water Table.--The drainage/flood characteristics of a particular intertidal zone will be greatly influenced by the physiography of the beach concern. The water table, or depth below which the sediment pore system is fully saturated, may be important for the vertical distribution of meiobenthos. Position of the ground water table should be ascertained directly from the "sampling hole" right after removal of the corer in order to eliminate the effect of capillary rise (see Figure 5.1). It should always be related to the time of observation. Equally important may be the influence of terrestrial freshwater seepage on both the depth and salinity of the beach water table.

Pore Water Analysis.--The set of physical and chemical factors relevant to the occurrence or distribution of meiofauna is too wide to be covered in this manual in every detail. Although the dominant role of redox processes in the sediment is recognized, details of the underlying physico-chemical principles are omitted here. They are extensively treated in Fenchel (1969). Consideration of parameters conventionally measured in meiofauna studies will be supplemented by descriptions of new techniques for recording oxygen and hydrogen sulfide which are increasingly realized to be decisive for the distribution of meiobenthos. For further reference, the standard manuals for seawater analysis (e.g., Grasshoff et al., 1983) are recommended.

In field work the instrumentation frequently has to meet more rigorous specifications than laboratory experiments. Battery driven, water- and shock-resistant, easily transportable and small recorders are needed. Since one easily forgets to switch off the instrument manually after recording, an automatic cut off triggered by a short period of non-usage should prevent the batteries from early discharging. The instrument's scale should have figures large and bright enough to be read even from a little distance and in full sunlight. While in field measurements a certain inaccuracy is inherent and tolerable, stabilization of the signals without undue fluctuations is very important. Hence, analogue scales have proved in many cases to be superior to the often very erratic digital displays. An alternative facility for either discrete or continuous recording is useful. If equipped with mains-rechargeable power supply and a sufficiently high accuracy (often with a scale divided into several ranges), field instruments are also well suited for lab work, especially if they are to be connected with an automatic chart recorder.

Recent developments in modern electrodes (new materials and chemicals) have greatly facilitated field work. For diver-operated *in situ* measurements in subtidal bottoms, electrodes must be enclosed in solid perspex housings or silicon rubber casts in order to avoid pressure problems. The recording instrument, built in a water-tight pressure housing, allows direct readings under water (see Conti and Wilde, 1972). Some leading manufacturers will build electrodes according to the customers' design and purposes. In every case, working with electrodes still requires understanding of the underlying theoretical and constructional principles and limitations in order to correctly interpret the frequently variable recordings. Electrometry should not be taken as a fool-proof "automatic" method.

Temperature.--As one of the universally important parameters, temperature measured *in situ* at various sediment depths is nowadays usually recorded with thermistors. The classical glass thermometer is less appropriate due to its susceptibility to breakage. The multitude of small and handy electronic recorders precludes the need to recommend specific models. For exact temperature readings on the very surface of the sediment or in the often only mm-thick water layer of residual pools (e.g., on tidal flats), it is advisable to use instruments which alternatively allow correction to surface/touch measurement and/or insertion measurement. Among the wide range of sensor types, robust "needles" of different length and width are best suited for field use. Since in thermistors only the very tip of the needle represents the effective sensor, temperature profiles can be easily read by inserting the probe vertically from the surface or horizontally at selected depths from a hole dug in the sediment. In the latter horizontal method, insertion should be at least 5 cm deep to prevent temperature change by air contact from the hole.

The variety of sensor types requires a thorough guidance by the manufacturer/distributor and careful adjustment to the recording instrument available. Measurement in semi-conductors and thermoelements (e.g., Ni-Cr-Ni) is linear, but thermoelements are quicker to respond, albeit rather inaccurate ($\pm1.5°$ C between -40 and $+40°$ C). On the other hand, the resistance-based, non-linear platinum sensors are slower, but their accuracy is usually much better ($\pm1.0°$ C in the above range). The accuracy indicated in the instruction manuals of the instruments is mostly valid for the recorder only and not for the whole functional system with the sensor connected.

For special studies, instruments can be connected to a series of probes and readings monitored simultaneously. In sublittoral sediments, temperature is usually homogeneous and equal to that of the overlying water layer.

Salinity.--The various methods for measuring salinity share an inherent inaccuracy in coastal, often brackish waters: The salt composition often diverges considerably from that of oceanic water which is constant to a wide degree and, thus, determines "salinity."

Conductivity.--Today, the international standard for salinity is based on conductivity. Consequently, modern oceanographic salinity recorders refer to this electrometrical unit. The electrical conductivity X or σ ($\mu S \cdot cm^{-1}$ or $mS \cdot cm^{-1}$) of a water sample is measured in a field of weak alternating current between two platinum electrodes. Conductivity depends on the total concentration of ions in the water and on the geometrical properties of the electrodes. These are defined by the "cell factor." Additionally, it depends on temperature, so that modern conductivity meters not only allow for adjustment of the cell factors but simultaneously record temperature for subsequent compensation. Conductivity meters and cells exist in numerous versions including battery-powered field recorders.

In more conductive water (brackish or seawater) naked platinum surfaces will suffice for measuring conductivity ranges between 1 and 100 $mS \cdot cm^{-1}$. Additional platination of the Pt-electrodes will enhance the active surface and render a much wider recording range and better accuracy. The convenient "standard cell" with its cell factor = 1.00 can rarely be used for meiobenthic work due to its rather large dimensions. Instead, pin-pointed insertion cells with ring-shaped Pt-electrodes are better suited, particularly since their smaller surfaces reduce the risk of damage. On the other hand, accuracy of the recording will limit the size reduction of the Pt-surface.

For correct results, the Pt-electrodes must be protected from becoming coated by organic substances and drying out. They have to be regularly re-calibrated in solutions of known conductivity. Platinated Pt-electrodes are subject to abrasion of their sensitive "active" surface and must be repeatedly re-platinated according to the manufacturer's prescriptions.

Since the recorded signal depends on the active surface size of the electrodes, reliable conductivity readings can only be obtained if the Pt-plates are completely covered with a water film. Thus, this method may prove inadequate in moist sediment samples from shore regions not saturated with water, and refractometry might be preferred.

Refractometry.--During the last decade, the refractometric determination of salinity has become more and more common. Advantages are its simplicity and independence of substrate nature. For the robust and handy instrument a few drops, drawn with a pipette from a given sediment layer, will suffice. Previously only one American company produced a convenient type of salinity refractometer, albeit at a very high price. More recently, Japanese companies are selling similar products worldwide at competitive prices. All these instruments have a scale already calibrated for direct salinity ($°/_{oo}$ S) readings and automatic temperature adjustment. The first battery-operated instruments with a conveniently large digital display of refraction indices have just entered the market and will, probably, soon be available calibrated for salinity readings. All products, however, have a very wide scale range (0-200$°/_{oo}$ S) limiting the accuracy of salinity measurements to about $1°/_{oo}$S. But, since for most ecological purposes this will be sufficient, many meiofauna specialists prefer this convenient method of salinometry.

Density Determination.--Compared with the

methods mentioned above, salinity measurement by recording the density of water with a set of araeometers is only rarely being done these days because of the fragility of the glass instruments, the amount of water required, and the tedious recording procedure.

Titration.--The classical chemical salinity determination is a very exact, but tedious and, due to its need of silver nitrate, rather expensive laboratory method. Moreover, it requires a volume of water which cannot always be supplied.

Dissolved Oxygen.

Titration.--The classical chemical determination of dissolved oxygen by the "Winkler" iodometric titration method is still in wide use for meiobenthic purposes preferably as a modernized micro-method (e.g., Bryan et al., 1976) which requires only a few ml of water. For field measurements, complete "oxygen kits" are available at a low price, equipped with all reagents and flasks needed. Greater accuracy will be obtained in the laboratory by a potentiometric end-point titration which implies careful stabilization of oxygen as an insoluble precipitate ($Mn(OH)_2$) immediately after sampling.

Electrometry.--This modern and versatile method records the oxygen content polarographically. The electrode acts, in principle, as an oxygen conformer. An actual and comprehensive review on "Polarographic Oxygen Sensors," their various applications and comparison with other methods is given by Gnaiger and Forstner (1983) from which the following general characterization is taken: The operational principle is based on oxygen diffusion to the polarized cathode (mostly platinum or gold), where oxygen becomes reduced (in micromoles O_2 per second) and thereby a corresponding electrical signal (in amperes) is generated. There are several factors which influence the stability of the recordings or interfere with them. The prevalent ones are temperature and salinity for which the readings have to be corrected. The most severely interfering factor, however, is hydrogen sulfide which tends to "poison" the cell quickly (see below).

According to Gnaiger and Forstner (1983): "Variations in the construction of polarographic oxygen sensors are mainly related to the problem how oxygen diffuses to the cathode. In this respect the actual design of every sensor necessarily involves a compromise, since optimizing particular functions, such as sensitivity and response time, detracts from others, such as stability and stirring requirements. The optimal design therefore depends upon the application."

Every oxygen electrode needs careful and repeated calibration which in modern electronic measuring instruments is facilitated by using air as saturation standard. Membranes have to be replaced frequently. Since stabilization may require several minutes, usually a constant (predetermined) reading time has to be applied. Numerous commercial oxygen meters are available with electrodes of varying size, diameter and shape. Regrettably, it is not uncommon to find oxygen electrodes offered at unreasonable prices as "special constructions" and calibrated for special purposes which are in no way better than the classical models at modest costs. Frequently used oxygen electrodes are the "Clarke-electrode type" and the "Rank type." In these, the active surface is relatively large and, because of the thin semipermeable membrane, almost in direct contact with the medium being analyzed. This restricts measurement of oxygen to aqueous media while recordings in sediments are impossible. Moreover, a considerable and constant water flow is required to minimize erroneous recordings. Another problem is the risk of degradation of the Pt-surface in waters of low oxygen with high H_2S-content and high organic load. Recently, a sensor has been manufactured which is totally insensitive to H_2S using a gold cathode which does not tarnish in the presence of S^{2-}.

The thin water layers of shallow pools or the pore water of sediment samples, where steep oxygen gradients often determine the distribution of meiofauna, are now accessible to polarographic sensors since microelectrodes have been developed. Modifying minute sensors used in the physiological studies of tissue respiration, Revsbech and co-workers were mainly responsible for construction of microprobes (Figure 5.6) for field use which allowed them to record oxygen stratifications and oscillations in the mm-range. A detailed description of their electrodes is given by Revsbech and Ward (1983) and Revsbech and Jørgensen (1986).

As shown in Figure 5.5, the sensitive "naked" platinum wire (99.9% pure Pt) is sealed into a glass capillary and shielded by a piece of silicone rubber which, due to its high oxygen permeability, does not falsify the recordings. The glass tip can be made more sturdy using thick-walled micropipettes (Figure 5.6) for easier insertion into the sediment. Even so, these electrodes remain delicate instruments with a limited life-span. Because of their suitability to measure in sediment, or other micro-scale systems, their applicability is much wider than that of most larger polarographic sensors. Furthermore, due to its miniaturization, stirring is not required, which allows reliable analyses of the frequently stagnant pore water. Also stabilization requires only a few seconds. Electro-plating the platinum surface with gold will render the sensor more resistant to chemically interfering factors, e.g., sulfide. This enables the investigator to do *in situ* recordings in muds with their often high concentration of H_2S close to the oxidized surface. As yet, all microelectrodes have to be hand-made in the research laboratories, which requires manual skill and some fine mechanical instruments. Instruction by an experienced colleague can save a lot of time and frustration.

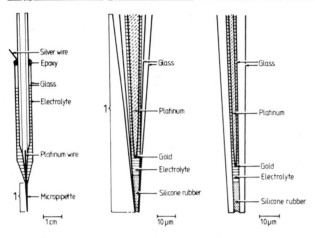

Figure 5.6.--Oxygen microelectrodes with detailed structure of their tip (modified after Revsbech and Ward, 1983).

Oxygen Availability.--This method gives a crude picture of the distribution of oxygen in the sediment. It originates from soil science (Lemon and Erickson, 1952) and is related to the polarographic oxygen measurement (see above). It was Jansson (1967) who first emphasized its use for recordings in marine sediments. At a constant voltage of 0.8 V (batteries) a naked platinum electrode (usually a piece of Pt-wire extending from a glass tube into which it is sealed) reduces as a cathode the oxygen molecules which diffuse onto its surface from the anode, Ag/AgCl reference electrode. The resulting electrical current (recorded by an amperemeter) is proportional to the concentration of oxygen available for diffusion. Values are obtained as g $O_2 \cdot cm^{-2}$ electrode surface $\cdot$ min^{-1} and can be read after several minutes of electrode stabilization. Reference is provided by a normal pH-reference electrode. For reliable readings, the surface of the Pt-wire has to be very clean and completely covered by water which renders measurement in moist sand problematical. The reason why this method is less frequently used today may reflect the increasing application of suitably small oxygen sensors (see above) and the complications introduced by a number of inherent physical factors which make this electrometrical method very unreliable: the naked platinum can be easily "poisoned," mainly by H_2S or by organic coatings. The recording is highly sensitive to temperature changes and particularly to water currents. In areas with considerable ground water or interstitial water flow (e.g., surge zone) it is the pore water velocity rather than its oxygen content which will be measured. Consequently, this electrode can also be used, after pertinent calibration, as a microflowmeter for pore water.

Hydrogen Sulfide.--Not only for thiobiotic animals is hydrogen sulfide one of the most important factors in the benthic ecosystem. Nevertheless, measurement of H_2S has been widely neglected mainly due to difficulties in quantitative recording.

Usually, the characteristic smell of sulfide and often the grayish-black coloration of the sediment caused by iron sulfide indicate the presence of H_2S remarkably well, but odor does not allow for any quantification and coloration may become misleading in sediments poor in ferrous iron (e.g., in calcareous sands) where blackening may be poorly developed. Furthermore, it is in these particular types of sediment that, due to the chemical balance between dissolved H_2S and iron salts, the sulfide concentrations in the pore water can become particularly high. In an alkaline milieu, hydrogen sulfide is predominantly present as HS^- ions, while in the slightly acidic environment of reducing substrates, the undissociated H_2S molecules prevail in solution. These are considered particularly toxic.

Chemical Measurement.--Quantitatively, the H_2S-content of pore water can be measured by potentiometric titration (Fenchel, 1969) or colorimetric reactions (see Cline, 1969; Fonselius, 1983). These methods require careful sampling since sulfide is a very unstable compound. A chemical *in situ* method is given by Reeburgh and Erickson (1982) who constructed a "dipstick sampler" equipped with a polyamide gel cast which was dosed with lead acetate to measure semi-quantitatively depth profiles of sulfide concentrations in the sediment. If appropriately calibrated, the subsequent blackening of the impregnated gel can be evaluated quantitatively.

Electrometrical Measurement.--The sulfide content of water can be measured *in situ* by electrodes now commercially available from leading companies. A silver sulfide electrode is combined with a glass electrode whose leak conductance consists of a silver/silver iodide system. The resulting voltage is a linear function of the logarithm of H_2S-partial pressure.

Recently, Revsbech (1983) manufactured a

sulfide-microelectrode which, due to its thin tip, allows small scale measurements in the sediment. This is of particular relevance since the overlapping zones of oxygen and hydrogen sulfide are often only a few mm thick, but ecologically most important due to its richness in bacteria (Rhoads, 1974; Yingst, 1978; Yingst and Rhoads, 1980). While the problem of sufficient sensitivity in low concentrations of H_2S seems to be solved even in fairly thin electrodes, the rapid abrasion of the sensitive sulfide layer on the silver wire remains a "weak point."

Redox Potential.--The reducing or oxidizing potential of a given water body or associated sediment can be measured with a platinum electrode in combination with a reference electrode (usually a silver/silver-chloride or a calomel system). Combined electrodes with built-in reference are available in varying shapes to accommodate differing sampling procedures. In insertion types the Pt-surface should not be too small in order to avoid long drifting time for the readings; usually 1-3 minutes should be sufficient for stabilization. This is important particularly in transition zones leading to reducing conditions where H_2S could poison the reference electrodes. The signals are recorded in mV and, in order to obtain the conventional Eh value which always refers to a standard hydrogen electrode, they must be corrected by addition of approximately +211 mV (silver reference) or +248 mV (Calomel reference) depending on temperature (the above values are for 20° C; but see remarks which follow).

The resulting mV-range is fairly wide (roughly from -300 to +400) with negative values usually indicating a reducing milieu, and positive values an oxidative environment. Commonly used in benthology as an "indirect," or substitute measure of the oxidative conditions, this method is limited by the fact that all electron transport processes contribute to the resulting redox potential (e.g., $Fe^{2+} \rightarrow Fe^{3+}$; $S^o \rightarrow S^{2-}$), not only those with oxygen involved. This integrative and rather non-specific response causes interpretation problems particularly in the range between +100 mV and -100 mV, where oxygen may or may not be present, while the signal can originate from purely anoxic electron transport. Thus, redox potential can only roughly mirror the general oxygen regime of the sediment, but it is of importance for indicating the Redox Potential Discontinuity (RPD) layer around +0 mV which is of much biological relevance since bacterivorous meiofauna species are often found aggregated here.

An additional problem of measuring Eh in sediment is the low reproducibility in multiple readings which often fluctuate within 10-20 mV due to microniches of decaying matter or animal tubes with entrapped air. If a thin, pointed electrode is inserted swiftly and deep enough into the sediment core, the risk of dragging air from the outside is minimal, so that recordings from the various horizons can be made directly in the core. Multiple probing in each sediment layer will also facilitate statistical treatment and for most biological purposes, the observed scatter will be tolerable considering the wide overall range of the redox scale. A major problem remains the meaningful interpretation of the redox values which requires experience and scientific caution. Consideration of the corresponding pH - milieu (see below) may provide important and helpful additional information.

In all cases, regular calibration of the electrode in a defined calibration buffer (e.g., a ferrocyanide-ferricyanide redox buffer, Zobell, 1946) is essential prior to measuring a sample series. This is particularly important if measurements have been done in sediments containing hydrogen sulfide (color, smell!). A reliable electrode should not deviate from the theoretical standard value by more than +10 mV. Since the commercially available calibration buffers (often based on chinhydrone) give norm readings already corrected for the standard hydrogen electrode, their use for calibration renders further correction of the values unnecessary (see above).

pH.--Measurement of the hydrogen ion concentration (pH) is of relevance mainly in freshwater environments and especially polluted or dystrophic waters where pH can become limiting for many organisms. In the well-buffered marine habitats (around pH 8) only extreme conditions will exert a demonstrable influence on the meiobenthos (e.g., pH 9 in intertidal pools during sunny periods). In strongly reducing marine sediments pH-values can drop to approximately pH 6. However, here the negative impact of anoxia and hydrogen sulfide will exceed the effect of low pH.

The simplest method for quick measurement of pH is with the use of indicator paper or sticks. Contemporary versions of indicator papers or sticks have a non-bleeding indicator which allows an accuracy to within 0.5 pH units.

Mostly, however, pH is measured by an all-glass combination electrode. Selecting the right reference system (usually $Ag/AgCl_2$, also calomel) is very important for reliable pH-readings. In viscous media or samples of low ionic strength, a calomel reference is preferable. About 70% of all faults in pH-measurement are caused by clogging of the junction between reference and measuring unit. This junction is usually a small ceramic plate, a porous PTFE-sleeve (Polytetrafluorethylene, Teflon) or even a platelet of porous wood. Clogging is very often due to internal precipitation of AgS in $Ag/AgCl_2$ internal electrodes after contact with sulfide-containing pore water in reduced sediments.

Blocking of the junction pores by particulate matter (mud) or oil suspensions is also a common source of pH-problems. Thus, a conventional pH-electrode used under field conditions is often subject to relatively long drifting times, to inaccuracies and instabilities due to its electrochemical and constructional properties and it has a restricted life time. Modern versions with new types of glass membranes, chemically altered liquid fillings, more stable reference cells, and better junctions have largely eliminated these problems. These new electrodes are offered by all leading manufacturers in various shapes and sizes (e.g., insertion types with spear tip). In many designs epoxy or PTFE shafts and special heavy duty glass tips have reduced the fragility of the electrode. Models in which gel-fillings render repeated calibration of the electrode unnecessary, are, however, often non-refillable and, thus, relatively expensive. Retractable sleeves and other types of junctions allow for quick replacement after clogging in soils or viscous media. Also special electrodes for soft water, poor in electrolytes, are available. This is important for work on acidifying pollution since in waters of only 10 to 100 $\mu s \cdot cm^{-1}$ conductivity, the current velocity will influence pH-recording. In this case, liquid-filled electrodes would be preferable to solid cells with gel fillings. Measurement should then be done in a sampling jar and not by direct immersion *in situ*.

Most recently, a maintenance-free pH-electrode (a similar one for redox-values) has been constructed with an "aperture junction" that drastically reduces the risk of any clogging, fouling, or contamination, since it simply represents one relatively large hole with a diameter as wide as approximately 1 mm. A viscous polymer hydrogel that contains suspended KCl serves as a filling for the reference electrode and surrounds the pH-electrode. Thus, the outflow of liquid through the junction hole is eliminated and instability of the potential due to loss of KCl minimized. Precision over long periods of time and sturdiness of this new electrode type are claimed to be extraordinary. In other new electrode constructions a corresponding "free-flowing" junction can be flushed clean by pressing down a pumping cap at the head of the glass shaft, thus enabling reliable measurements even in colloids and slurries.

In older pH/Eh recorders, temperature dependent calibration of the pH-electrode was a cumbersome and repetitive procedure. Furthermore, many instruments, although battery-powered for field use, were inadequately water resistant. Modern lightweight, small instruments are water- and shock-resistant in their plastic housings. Fully equipped with microprocessors, they usually combine temperature measurement, recognize in seconds the various calibration buffers, compensate automatically for temperature changes, and store all relevant values in an electronic memory. Thus, in combination with improved electrodes, measurement of pH and Eh in the field has become a simple task and yields more accurate results (two decimal places) than many laboratory recordings before.

Ion-selective Electrometry.--Today, the selective electrometric measurement of particular ions in solution has been developed into a reliable method applicable for field use mainly due to new constructions of electrodes and recorders (basically the same trends as outlined in the discussion of pH). In the response to the accuracy demanded by most ecological studies, many earlier interfering factors and unstable reading characteristics have been eliminated. In freshwater, ecologically important factors such as NO_3^-, SO_4^{2-}, CO_3^{2-}, Cl^-, many heavy metals, Ca^{2+} and Mg^{2+} can be recorded with selective electrodes. In seawater, however, or in moist sediment, the range is more restricted due to major physical and technical problems.

Principally similar to pH-measurement, the ion-selective electrodes usually have either a glass membrane (for sodium), a solid state membrane (e.g., for chloride, sulfide, fluoride), a liquid membrane (e.g., for calcium, nitrate, potassium) or a gas sensing membrane (carbon dioxide, ammonia). Recent developments include exchangeable solid state or glass membrane modules allowing a quick and relatively inexpensive replacement of the sensitive membrane and, thus, enhancing considerably not only the lifetime of the electrode body, but also reliability of readings. Reference electrodes should be one order of magnitude more accurate (± 0.005 pH or better) than routine ones used for pH, in order to produce reliable readings of specific ion concentrations. Their potential should be stable and reproducible under a wide range of sampling conditions. So far micro-electrodes are constructed mainly for ions of medical relevance (e.g., Na^+, K^+); however, first models for recording pollutants are now available. For the recording instruments which have a specific ion-scale instead of the pH- scale, the same modern models with convenient handling are on the market as described above for pH-meters. Using a selector box, several electrodes, sensitive for various ions (pH, NO_3^-, Cl^-, K^+, Na^+), can be monitored simultaneously. Among the more important parameters, so far Mg^{2+}, PO_4^{3-}, and SO_4^{2-} cannot be measured by ion-selective electrodes. In general, development in the field of ion-selective electrometry is still much in progress. For more detailed information see Whitfield (1985).

References

Barnes, R.O.
1973. An In-Situ Interstitial Water Sampler for Use in Unconsolidated Sediments. *Deep-Sea Research*, 20:1125-1128.

Barnett, P.R.O., J. Watson, and D. Conelly
1984. A Multiple Corer for Taking Virtually Undisturbed Samples from Shelf, Bathyal and Abyssal Sediments. *Oceanologia Acta*, 7(4):399-408

Bryan, J.R., J.P. Riley, and Le B. Williams
1976. A Winkler Procedure for Making Precise Measurements of Oxygen Concentration for Productivity and Related Studies. *Journal of Experimental Marine Biology and Ecology*, 21:191-197.

Buchanan, J.B.
1984. Sediment Analysis. Pages 41-65 in N.A. Holme and A.D. McIntyre, editors, *Methods for the Study of Marine Benthos*. IPH Handbook 16. London: Blackwell Scientific Publications.

Bühler, H., and H. Galster
1980. Redoxmessung. Grundlagen und Probleme. 23 pages. Frankfurt: Dr. Ingold KG.

Cline, J.D.
1969. Spectrophotometric Determination of Hydrogen Sulfide in Natural Waters. *Limnology and Oceanography*, 14:454-458.

Conti, U., and P. Wilde
1972. Diver-operated *In Situ* Electrochemical Measurements. *Marine Technology Society Journal, Washington*, 6:17-23

Craib, J.S.
1965. A Sampler for Taking Short Undisturbed Marine Cores. *Journal du Conseil*, 30:34-39.

Eleftheriou, A., and N.A. Holme
1984. Macrofauna Techniques. Pages 140-216 in N.A. Holme and A.D. McIntyre, editors, *Methods for the Study of Marine Benthos*. IPB Handbook 16. London: Blackwell Scientific Publications.

Eleftheriou, A., and M. Nicholson
1975. The Effects of Exposure on Beach Fauna. *Cahiers de Biologie Marine*, 16:695-710.

Fenchel, T.
1969. The Ecology of Marine Microbenthos. IV. Structure and Function of the Benthic Ecosystem, Its Chemical and Physical Factors and the Microfauna Communities with Special Reference to the Ciliate Protozoa. *Ophelia*, 6:1-182.

Fenchel, T., and B.J. Straarup
1971. Vertical Distribution of Photosynthetic Pigments and the Penetration of Light in Marine Sediments. *Oikos*, 22:172-182.

Folk, R.L.
1974. *Petrology of Sedimentary Rocks*. 182 pages. Austin, Texas: Hemphill Publishing Company, 182 pages.

Fonselius, S.
1983. Determination of Hydrogen Sulphide. Pages 73-80 in K. Grasshoff, M. Ehrhard and K. Kremling, editors, *Methods of Seawater Analysis*. Second edition. Basel: Verlag Chemie.

Fraser, H.J.
1935. Experimental Study of the Porosity and Permeability of Clastic Sediments. *Journal of Geology*, 43:910-1010.

Giere, O.
1973. Oxygen in the Marine Hydropsammal and the Vertical Microdistribution of Oligochaetes. *Marine Biology*, 21:180-189.

Gnaiger, E., and H. Forstner, editors
1983. *Polarographic Oxygen Sensors. Aquatic and Physiological Applications*. 370 pages. Berlin: Springer Verlag.

Grasshoff, K., M. Ehrhard, and K. Kremling, editors
1983. *Methods of Seawater Analysis*. Second edition. 419 pages. Basel: Verlag Chemie.

Graton, L.C., and H.J. Fraser
1935. Systematic Packing of Spheres - With Particular Relation to Porosity and Permeability. *Journal of Geology*, 43:785-909.

Gray, J.S., and R.M. Rieger
1971. A Quantitative Study of Meiofauna on an Exposed Sandy Beach at Robin Hood's Bay, Yorkshire. *Journal of the Marine Biological Association of the United Kingdom*, 51:1-19.

Hartwig, E.
1973. Die Ciliaten des Gezeiten-Sandstrandes der Nordseeinsel Sylt. II. Ökologie. *Mikrofauna des Meeresbodens*, 21:1-171.

Howes, B.L., and S.G. Wakeham
1985. Effects of Sampling Technique on Measurements of Porewater Constituents in Salt Marsh Sediments. *Limnology and Oceanography*, 30:221-227.

Hummon, W.D.
1969. *Distributional Ecology of Marine Interstitial Gastrotricha from Woods Hole, Massachusetts, with Taxonomic Comments on Previously Described Species*. Ph.D. Dissertation, University of Massachusetts.

Jansson, B.-O.
1967. The Availability of Oxygen for the Interstitial Fauna of Sandy Beaches. *Journal of Experimental Marine Biology and Ecology*, 1:122-143.

Kissam, P.
1956. *Surveying*. Second edition. 482 pages. New York: McGraw-Hill Book Company.

Komar, P.D.
1976. *Beach Processes and Sedimentation*. 429 pages. Englewood Cliffs, New Jersey: Prentice-Hall, Inc.

Lemon, E.R., and A.E. Erickson
1952. The Measurement of Oxygen Diffusion With a Platinum Microelectrode. *Proceedings - Soil Science Society of America*, 16:160-163.

McIntyre, A.D.
1971. Deficiency of Gravity Corers for Sampling Meiobenthos and Sediments. *Nature*, 231:60.

McIntyre, A.D., and D.J. Murison
1973. The Meiofauna of a Flatfish Nursery Ground. *Journal of the Marine Biological Association of the United Kingdom*, 53:93-118.

McLachlan, A.
1980. The Definition of Sandy Beaches in Relation to Exposure: A Simple Rating System. *South African Journal of Science*, 76:137-138.

Meadows, P.S., and J.G. Anderson
1966. Micro-organisms Attached to Marine and Freshwater Sand Grains. *Nature*, 212:1059-1061.

Pilkey, O.H., R.W. Morton, and J. Luternauer
1967. The Carbonate Fraction of Beach and Dune Sands. *Sedimentology*, 8:311-327.

Powers, M.C.
1953. A New Roundness Scale for Sedimentary Particles. *Journal of Sedimentary Petrology*, 23:117-119.

Reeburgh, W.S., and R.E. Erickson
1982. A Dipstick Sampler for Rapid, Continuous Chemical Profiles in Sediments. *Limnology and Oceanography*, 27:556-559.

Revsbech, N.P.
1983. In-situ Measurement of Oxygen Profiles of Sediments by Use of Oxygen Microelectrodes. Pages 265-273 in E. Gnaiger and H. Forstner, editors, *Polarographic Oxygen Sensors*. Berlin: Springer Verlag.

Revsbech, N.P., and B.B. Jørgensen
1986. Microelectrodes: Their Use in Microbial Ecology. Volume 9, pages 293-352 in K.C. Marshall, editor, *Microbiological Ecology*.

Revsbech, N.P., and D.M. Ward
1983. Oxygen Microelectrode That is Insensitive to Medium Chemical Composition: Use in an Acid Microbial Mat Dominated by *Cyanidium caldarium*. *Applied Environmental Microbiology*, 45:755-759.

Rhoads, D.C.
1974. Organism-sediment Relations on the Muddy Sea Floor. *Oceanography and Marine Biology: An Annual Review*, 12:263-300.

Seed, R., and B.J. Lowry

1973. The Intertidal Macrofauna of Seven Sandy beaches of County Down. *Proceedings of the Royal Irish Academy,* B, 73:217-230.

Steele, J.H., H.L.S. Munro, and G.S. Giese
1970. Environmental Factors Controlling the Epipsammic Flora on Beach and Sub-littoral Sands. *Journal of the Marine Biological Association of the United Kingdom,* 50:907-918.

Webb, J.E.
1958. The Ecology of Lagos Lagoon. V. Some Physical Properties of Lagoon Deposits. *Philosophical Transactions of the Royal Society, London,* Series B, 241:393-419.

1969. Biologically Significant Properties of Submerged Marine Sands. *Proceedings of the Royal Society of London,* Series B, 174:355-402.

Webb, J.E., and J. Theodor
1968. Irrigation of Submerged Marine Sands through Wave Action. *Nature,* 220:682-683.

Whitfield, M.
1985. Ion-selective Electrodes in Estuarine Analysis. Pages 201-277 in P.C. Head, editor, *Practical Estuarine Chemistry: A Handbook.* New York: Cambridge University Press.

Williams, R.
1971. A Technique of Measuring the Interstitial Voids of a Sediment Based on Epoxy Resin Impregnation. Pages 199-205 in N.C. Hulings, editor, Proceedings of the First International Conference on Meiofauna. *Smithsonian Contributions to Zoology,* 76.

Yingst, J.Y.
1978. Patterns of Microfauna and Meiofauna Abundance in Marine Sediments, Measured with Adenosine Triphosphate Assay. *Marine Biology,* 47:41-54.

Yingst, J.Y., and D.C. Rhoads
1980. The Role of Bioturbation in the Enhancement of Bacterial Growth in Marine Sediments. Pages 407-421 in K.R. Tenore and B.C. Coull, editors, *Marine Benthic Dynamics.* Columbia: University of South Carolina Press.

Zobell, C.E.
1946. *Marine Microbiology.* 240 pages. Waltham: Chronica Botanica Press.

6. Biotic Factors

N. Greiser and A. Faubel

In explaining the term "biotic factors" from an ecological point of view, one must realize that there is no precise distinction between "abiotic factors" and the biota. Growth as well as spatial and temporal distribution of organisms are regulated by abiotic factors, e.g., nutrient availability, temperature, salinity, and governed mainly by the redox regime. Organisms compete as herbivores, predators, and parasites and they are involved in the biogeochemical cycles as producers and decomposers.

In this review of the methodology related to meiofaunal biotic factors, we will confine this term to the following parameters: (1) the consumption process of the living cells of organisms, and (2) indicators of organic material (detritus), i.e., "all the particulate organic matter involved in the decomposition of dead organisms" (Odum and La Cruz, 1963), and dissolved organic matter.

It will be the task of the individual investigator to select the most suitable methods for a particular study and to analyze the structural and functional properties of the biotic factors in a given community or ecosystem.

In this chapter we have assembled the parameters related to biotic factors into four categories: (1) methods for measuring total organic matter; (2) methods for measuring dead and living organic matter; (3) methods to determine organic matter of different groups of organisms; and, (4) methods to determine special properties of organic matter. These topics are presented as they relate to the following considerations.

It is standard procedure to determine the standing crop of the meiobenthos by the enumeration of organisms. For many purposes it is desirable to obtain rapid information on the fluctuations of the particulate organic matter (POM) present in different samples; therefore, a method for measuring representative compounds of organic matter is needed. The parameters most commonly used are ash-free dry-weight, organic carbon, protein, carbohydrates, and lipids.

In the analysis of samples where one is to combine different parameters related to organic matter, it is often possible to distinguish between the proportion of dead (detritus) and living biomass. Such parameters may give information about growth and decay within the benthic community.

Some cellular compounds are related to special groups of organisms, e.g., muramic acid to bacteria, and chloroplastic pigments to phytoplankton. As bacteria and phytoplanktonic cells (after sedimentation) are part of the nutrients for the meiobenthic community these parameters may be useful for measuring the seasonal variations of nutrient input and availability.

The last group of parameters chosen is related to special energetic properties of organic matter such as the physiological state of living cells (ATP and energy charge), the special biochemical capabilities (enzyme activity) of organisms in characterizing food and feeding, and, in particular, the estimation of the nature of food and the rate at which digestion occurs.

The problems involved with a particular measurement will be briefly discussed in the introductory remarks dealing with general principles of analytical biases and the expected reliability of the results. Procedures for each measurement are discussed and, in some cases, the methods are compared. No attempt has been made to be comprehensive since this is not within the scope of this chapter. Careful use of the references cited will provide more specific information on specific methods and the interpretation of results.

Methods for Measuring Total Organic Matter

The following parameters can be used either to determine the amount of total organic matter or the proportion of individual groups of organisms present in the samples. For ecological studies and, in particular, for community estimations, it is often advantageous to measure biomass directly rather than indirectly by using abundance values and by conversion factors (Faubel, 1982; Widbom, 1984).

Wet Weight/Dry Weight

Determination of individual wet or dry weight values of meiofauna is an invaluable but very tedious procedure in quantitative ecological investigations, because the faunal items have to be separated from the sediments in the samples. Separation of the fauna has to be done from unpreserved material because weights of formalin-preserved animals exceeded those of unpreserved specimens by as much as 44% for dry weight and 29% for ash-free dry weight (Widbom, 1984). After separation, the organisms may be weighed totally or the abundance figures could be multiplied with average conversion factors for the different taxa. Conversion factors of individual biomass may be estimated by weighing size-depending individual taxa (Faubel, 1982; Widbom, 1984) or by using the formula ($a^2b/1.7$) of Andrassy (1956) in which a = the maximum width and b = the length. The volume multiplied by specific gravity, e.g., 1.13 for all marine nematodes (Wieser, 1960), yields the wet weight. The latter procedure presents the most reliable method for estimation of wet weight biomass but is only practicable for cylindrical and smooth individuals since exact estimates of volume can only be made for such organisms.

Often, one option in the measurement of wet-weight values is to preserve the meiofauna species after the biomass has been determined, thereby making it possible to use the same sample. Another reason for estimating wet weight values as a biomass parameter is the desire to compare meiofaunal with macrofaunal biomass, because wet weight values are commonly used in literature (Wieser, 1960). Therefore, it is necessary to start the weighing procedure by washing the organisms in filtered seawater and subsequent removal of excess liquid by blotting or filtration. Wieser (1960) indicated that determining wet weight of small organisms is particularly difficult because of rapid water evaporation through the body walls of the organisms. The common practice of rolling individuals on filter paper or a cloth is difficult to standardize and can be a source of serious error. Another possible source of error during the determination of dry weight values is the adhesion of mineral or organic particles to the body walls of the organisms.

For the determination of dry weight values it is important to choose temperatures that do not change the organic content of the organisms by losses of volatile substances. Lovegrove (1961) showed that drying at 60° C was comparable to desiccator-drying values and that extended heating did not cause significantly large variations. Note that marine organisms prepared for drying procedures have to be rinsed with distilled water to remove excess salt.

A suitable procedure to determine wet weight and dry weight includes the following steps:

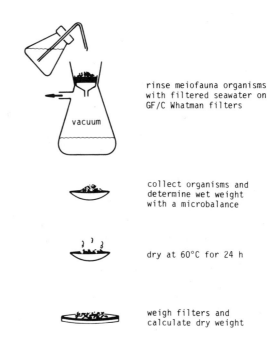

rinse meiofauna organisms with filtered seawater on GF/C Whatman filters

collect organisms and determine wet weight with a microbalance

dry at 60°C for 24 h

weigh filters and calculate dry weight

For comparison and information about the range of wet and dry weight values of different groups of meiofaunal organisms see Wieser (1960), Faubel (1982), and Widbom (1984), where an extended list of corresponding values are given.

Ash-free Dry Weight

This method involves drying at 60° C (see previous discussion) and combustion of the organic matter at high temperatures. The temperature for combustion should be about 450 to 500° C to avoid volatilizing bicarbonates. Hirota and Szyper (1975) pointed out that up to 7 h of combustion no important losses occur at 500° C. Therefore, the loss of weight indicates the amount of total organic matter (TOM) in the samples. Byers et al. (1978), comparing different techniques of determining organic carbon, showed that the recovery rate for glucosamine added to a carbonate free standard sediment amount to 99%.

The following diagram shows the method in detail:

dry sediment sample at 60°C for 24 h and determine dry weight value

ash sample at 475°C for 2 h and calculate ash free dry weight from the loss of weight

Organic Carbon

The determination of organic carbon is widely used for studies of the productivity of pelagic and benthic communities. Generally, it is more difficult carrying out reproducible values from sediment samples than from selected meiofauna organisms. Due to the method used, different sources of bias are responsible for non-reproducible results:

The chemical Chromic Acid Method (see following) is mainly sensitive to chemical interference by different salt ions and ferrous minerals (Wakeel and Riley, 1956) and the chemical reaction depends on the variable nature of the organic matter present (Wakley, 1947).

The use of commercial elemental analyzers (CHN-analyzers) to detect the released organic carbon from combusted samples gives rise to biases due to the properties of the CHN-analyzer used (e.g., Hewlett-Packard, Perkin Elmer, Carlo Erba). The low limits for analysis (e.g., 5 μg C, Hewlett-Packard; Model 185 B; Sharp, 1974) generally require very accurate work to avoid contamination of the apparatus (Sharp, 1974). The most common source of error is carbonates present in the samples. These should be removed by acid treatment or combustion (following).

The Chromic Acid Method.--The method described will determine organic carbon in sediment samples with a limit of detection of about 100 μg C. Therefore, at least 0.15 g of sediment containing an average amount of 1% organic carbon have to be analyzed (Wakeel and Riley, 1956). Gaudette (1974), comparing organic carbon values measured by the chromic acid method and a LECO carbon analyzer, implied a yield of 100% by the chromic acid method in aquatic sediments. Byers (1978) found corresponding yields at about 76%. Holme and McIntyre (1984) pointed out that there is no need for using any conversion factors for comparing chromic acid values with organic carbon values from other methods.

Since chlorides react with chromic acid in sulfuric acid with the formation of volatile chromyl chloride (Wakeel and Riley, 1956), salts have to be removed by washing the sediments with distilled water. The interference of ferrous minerals is only slight unless great amounts of iron are present in a more soluble form (Wakeel and Riley, 1956). Generally, the interference by ferrous ions is reduced if the oxidation procedure is carried out at 100° C and not at 175° C as described by Schollenberger (1927) (Wakeel and Riley, 1956).

The following diagram shows the procedure given by Wakeel and Riley (1956) in detail:

wash sediment samples with distilled water

dry sediment at 60°C for 24 h

homogenize sample in a mortar

weigh out 0.15 - 0.30 g of homogenized sediment into a boiling tube and add 10 ml of chromic acid. Shake the tube gently

heat in boiling water for 15 minutes

cool and pour the content of the tube into 200 ml of distilled water

add one drop of ferrous-phenanthroline indicator and titrate with 0.2 N ferrous ammonium sulphate solution until a pink color just persists

calculate organic carbon: 1 ml of 0.2 N ferrous ammonium sulphate equals 1.15 0.6 mg of carbon

Reagent: *chromic acid.* **Procedure:** dissolve 13 g of *chromium trioxide A.R.* in a minimum amount of water. Add about 900 ml of *concentrated sulfuric acid,* allow to cool, dilute to 1 l with *concentrated sulfuric acid.*

Reagent: *ferrous ammonium sulfate.* **Procedure:** dissolve 39.3 g of the reagent grade solution (0.2 N) salt in 400 ml of water containing 10 ml of concentrated sulfuric acid, dilute to 500 ml.

Reagent: *ferrous-phenanthroline.* **Procedure:** dissolve 0.337 g of o-phenanthroline indicator (0.025 M) monohydrate in 25 ml of 0.695% ferrous sulfate solution.

The Gas-chromatography Method (CHN-analyzer).--The amount of particulate organic carbon (POC) is often overestimated because of variable contents of carbonates present in sediment samples. Some methods for eliminating inorganic carbon are based on washing the samples with chloric acid (Menzel and Vaccaro, 1964; Gordon, 1970). Telek and Marshall (1974) pointed out that the removal of carbonate carbon with acid treatment will be unsatisfactory, because soft treatment will not remove all the carbonates according to different sediment types, and intensive treatment may result in losses of some organic carbon. Roberts et al. (1973) found that acidification removed 9 to 44% of the organic carbon in natural carbonate sediments. Because of these difficulties Hirota and Szyper (1975) introduced a method for separating total particulate carbon in POC and particulate inorganic carbon by combusting the samples at or below 500° C for 4 h between two analyzing steps using a commercial CHN-analyzer.

If the organic carbon content of the samples is low (about 5-200 µg C), it will be important to minimize reagent and gear blanks (Sharp, 1974). Sharp (1974) recommended some modifications in the preparation of the oxygen donor (MnO_2, Cr_2O_3-mixture), substitution of silicone rubber O-rings for Buna-N rubber ones if the CHN-analyzer is equipped with them because of their high C-content. We will present the method of Hirota and Szyper (1975) in detail because of its easy sample preparation and sufficient accuracy. The principle of this method is the independent measurement of TOC and particulate inorganic carbon with calculation of POC by difference. The method requires only two carbon analyzes per sediment at the 20 s flow-mod bypass interval and 1.100° C oxidizer's temperature. For the calculation of correct carbon values some blanks and organic standard samples (e.g., cyclohexanone-2.4-dinitrophenyl-hydrazone) have to be run. Hirota and Szyper (1975) tested their method on reagent grade $CaCO_3$ plus the organic standard.

The following diagram shows the method in detail:

wash sediment samples with distilled water

dry sediment at 60°C for 24 h

homogenize sample in a mortar

1. weigh out parallel samples in silver rods

2. ash one sample at 475°C for 2 h

combust samples at 1.100°C, 20 s-flow mode

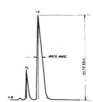

calculate total particulate carbon from sample No. 1

calculate particulate inorganic carbon from sample No. 2

calculate POC by difference considering values of blank and organic standard

Organic Nitrogen

Nitrogen is an essential component of many functionally important compounds in living cells as amino acids (proteins), nucleic acids, and cell wall material (chitin, peptidoglycan). Although nitrogen is the major constituent of the air, only some groups of microorganisms require N-fixing enzyme systems to provide themselves with this essential nutrient.

Therefore, organic nitrogen compounds are broken down and incorporated readily by decomposers if released from dead or living cells. Specialized decomposers convert the organic nitrogen to an inorganic form which enters the organic form merely by the activity of phytoplanktonic cells in aquatic ecosystems. Dortch et al. (1985) described that marine algae can serve as a N-reserve for the planktonic community. Graf et al. (1984) showed the importance of phytoplankton blooms as a nutrient input to benthic communities. Hanson (1982) assumes the organic nitrogen content of detritus to be the primary factor regulating microbial metabolism in benthic communities. Therefore, the determination of organic nitrogen is commonly used in decomposition process studies (Takahashi and Saijo, 1981). In combination with organic carbon measurements, C/N ratios can be calculated to give information about the quality of detritus, e.g., the nutritional value and therefore the state of the decomposition processes (Graf et al., 1984).

The use of commercial elemental analyzers generally combine organic carbon measurements with the detection of organic nitrogen. We will complete the methodological description relating to the above section on CHN-analyzers). As only some laboratories are equipped with CHN-analyzers we especially emphasize the chemical Kjeldahl-method.

The Chemical Kjeldahl-method.--The basic principle of the method is the discovery that boiling of nitrogen-containing organic compounds in concentrated sulfuric acid results in liberation of the nitrogen in the form of ammonium sulfate. Kjeldahl announced this method in 1883 and since then it became widespread in many laboratories, mainly to analyze protein nitrogen content. Polley (1954) found the recovery of nitrogen from various amino acids to be 94% or greater. Sources of errors due to this chemical method are unfavorable digestion conditions and the reproducibility of the ammonia detection technique chosen.

Mann (1963) considered mercury to be the best catalyst for the oxidation procedure whereas Lang (1958) preferred selenium. Kirk (1950) gives an overall view about references that deal with the chemical reaction mechanisms and possible interferences. "The fact that the necessary conditions for reduction of nitrogen must coexist with oxidizing conditions for the remainder of the decomposed organic molecules gives an indication of the narrow oxidation-reduction range in which the decomposition must be carried out. Disregard for this fact has led to numerous unsatisfactory modifications in which strong oxidizing agents were used or other conditions were modified unfavorably" (Kirk, 1950).

The common measurement of ammonia concentration is a colorimetric one after nezzlerization (Lang, 1958). Mann (1963) prefers the formation of phenol-indophenol from ammonia and phenol in the presence of base, sodium hypochlorite, and sodium-ferrinitroso-pentacynide ("nitroprusside") which can be measured photometrically either. This method detects very low concentrations (<15 µg) of ammonia. The disadvantages of this technique are (1) the highly temperature dependent formation of the color by the reactants, and (2) three mixing steps, e.g., three different indicator solutions have to be prepared (the nitroprusside solution was only stable for at most 1 hour). We will describe the method from Lang (1958) in detail, because it is a composite of several conventional techniques and has the following advantages: extreme sensitivity, accuracy, and range, use of inexpensive commonly available equipment, and the possible determination of other elements in the same sample (Lang, 1958). This modification of the Kjeldahl-method may be of value in those laboratories where the use of the classical method has been discouraged because of the expensive elaborate apparatus required as well as the involved manipulation.

Reagent: digestion mixture: potassium sulfate (40 g), selenium oxychloride (2 ml), distilled water (250 ml), sulfuric acid (250 ml).

Reagent: Nessler reagent: use a commercial standard solution (e.g., Merck)

Reagent: ammonium sulfate standard. **Procedure:** This is prepared by dissolving 1.179 g of ammonium sulfate, previously dried for several days in a desiccator, in 250 ml of 0.2N sulfuric acid to give a final concentration of 1000 µg per ml.

Reagent: lysine standard solution. **Procedure:** L-lysinemonohydrochloride recrystallized twice from alcohol-water and dried, and 65.2 mg are dissolved in 100 ml of distilled water to give a final concentration of 100 µg nitrogen per ml. Aliquots of these standard stock solutions are diluted with distilled water to obtain working standards.

The following diagram shows the method in detail supplemented with some modifications according to sediment analysis:

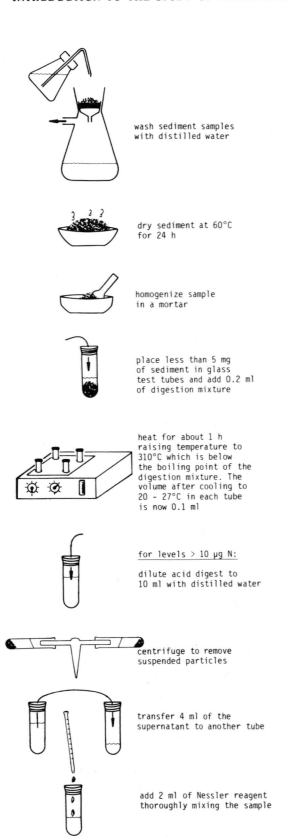

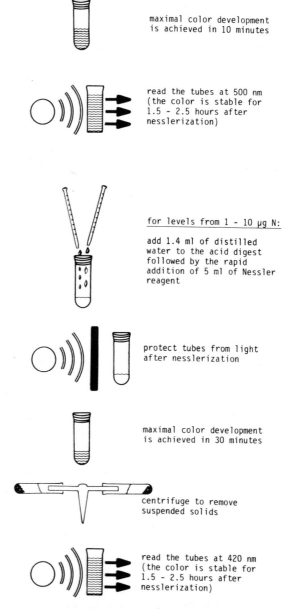

Kjeldahl-method with Gravimetrically Measurement of Nitrogen.--In this technique we will present a special alternative of measuring the organic nitrogen obtained from Kjeldahl digestion: a gravimetrically determination of the released ammonia by precipitation as ammoniumtetraphenylborate (Crane and Smith, 1963). For this method, as for the previous one, no distillation equipment is necessary. Additionally no standard solutions have to be prepared.

Reagent: sodiumtetraphenylborate (TPB) solution.
Procedure: dissolve reagent grade TPB to get a 0.05-0.08 M solution.

Reagent: mercury II sulfate solution (as catalyst).

Procedure: dissolve 50 g mercury II oxide in 250 ml of 6 N sulfuric acid with heating followed by dilution to 500 ml with ammonia free water.

Reagent: sodium sulfate. **Procedure:** use reagent grade sodium sulfate.

Reagent: sodium iodide solution. **Procedure:** dissolve reagent grade sodium iodide to get a 2 M solution.

Reagent: sodium acetate solution. **Procedure:** dissolve reagent grade sodium acetate to get a 3 M solution.

Reagent: ammonium sulfate (use reagent grade).

Reagent: TPB wash solution. **Procedure:** dissolve 500 mg of ammonium TPB and 300 mg of potassium TPB in a few drops of acetone, and then add slowly with stirring about 2 l of distilled water. After equilibrium the mixture should be filtered.

Calculation: The gravimetric factor to calculate organic nitrogen from the precipitate is 0.04154

The complexation procedure has to be performed as follows:

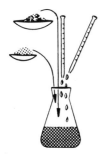

transfer a dry, weighed sample containing 0.1 - 0.5 mequiv of organic nitrogen to a small Kjeldahl flask. Add about 0.25 g of sodium sulphate, 0.3 ml of mercury-II-sulphate solution and 1 - 1.5 ml of concentrated sulphuric acid

heat the mixture gently until frothing ceases. After all black carbon has disappeared, boil the solution for 20 - 30 min

after cooling dissolve the resulting digest in a minimum of freshly boiled (ammonia free) water and transfer it to a 100 ml Erlenmeyer flask

because of interference by the catalyst mercury-II-ion with TPB-ion eliminate the mercury by complexation as tetra-iodo-mercurate-II by adding a 2 M sodium iodide solution dropwise with swirling until the red mercury precipitate dissolves completely, and then a 1 ml excess to complex the mercury quantitatively

dilute the solution with water to 60 - 70 ml and mix well

add 3 drops of 3 M sodium acetate and adjust to pH 2 - 3 with 2 N sodium hydroxide or 2 N sulphuric acid as required

pour a volume of the sodium tetra-phenyl-borate solution, sufficient to provide a 25 - 50% excess of TPB slowly down the side of the vessel with stirring

allow the resulting white precipitate to stand for 20 - 30 minutes at room temperature and then filter quantitatively through a previously weighed glass suction crucible (medium porosity)

test the first portion of the filtrate for complete precipitation by addition of a small volume of dilute ammonium sulphate. If cloudiness results the complete precipitation of ammonium is indicated

wash the precipitate with 3×20 ml of the special wash solution

dry the precipitate for 3 - 6 hours at 50 - 60°C under vacuum

Before calculating the nitrogen content, it is sometimes necessary to deduct any potassium TPB or other precipitable impurities in the ammonium TPB precipitate.

To do so dissolve a known portion of the precipitate in a minimum (10 ml or less) of ethylene glycol, add 2 pellets of sodium hydroxide and warm gently on a hot plate until all ammonia has escaped.

About 10 minutes of heating should suffice to give a negative litmus test. Add a 6-fold excess of water to the cool solution with swirling, let the cloudy precipitate (potassium TPB), if any, stand for 15 - 20 minutes and filter through a weighed glass crucible. Wash and dry the precipitate as described above. Calculate the portion of ammonium TPB by difference.

The Gas Chromatography Method (CHN-analyzer).--The methodological principle has been described in section on organic carbon. The limit for analysis is about 1 µg N using a Hewlett Packard model 158 B CHN-analyzer standardized with acetanilide (Sharp, 1974). The analysis is optimal with samples of 0.5-1.0 mg with about 25-100 µg N. Sharp (1974) reported one possible source of serious error to be improperly constructed sample rods which will give high nitrogen blanks (up to 7 µg N) due to the air trapped in them. Soluble inorganic N like NO_3^- will not contribute much to blanks, because it will be washed away due to the procedure given in detail dealing with organic carbon.

Methods for Measuring Dead and Living Organic Matter

Protein

Proteins are the major constituents of living cells. As organisms have to afford a lot of energy to synthesize proteins, it will be metabolized very soon by microorganisms if released from dead or living cells. Therefore, protein seems to be a good parameter for biomass determination in planktonic and benthic communities (Packard and Dortch, 1976; Christensen and Packard, 1977; Setchell, 1981). Sibuet et al. (1982) showed particulate protein to be an important food resource for sediment feeders, especially Holothurians. These investigations imply that the protein content of oxidized sediments - in anoxic sediments decomposition of proteins will occur much slower (Degens et al., 1964) - represent mainly living organic matter, whereas in anoxic sediments protein comprise dead and living organic matter.

From an analytical point of view the term protein describes a number of very different molecules with different molecular weights and chemical structures. Some proteins are soluble while others are linked to structural components of the cell, e.g., membranes and cell wall material. Therefore, emphasis has to be put on two aspects concerning protein determination: complete extraction, even of structural proteins, and for detection, the sensitivity of the protein binding dyes.

Herein we will describe two different methods of protein-dye-binding, the method of Lowry et al. (1951) and the method of Bradford (1976). The method of Lowry is the most widely used procedure for protein determination. The disadvantages of this method are its lack of specificity due to the amino acid composition of the proteins measured (Chou and Goldstein, 1960), the instability of some reagents (Peterson, 1979), and the interference of many substances like magnesium ions (Kuno and Kihara, 1967), various organic substances (Peterson, 1979), and at least carbohydrates (Lo and Stelson, 1972).

Bradford prepared a method which eliminates many of these disadvantages of the Lowry method. This assay can easily be utilized for processing large numbers of samples and is adaptable for automation.

One possible disadvantage of this method is the considerable variation in extinction with pure proteins (Pierce and Suelter, 1977; Van Kley and Hale, 1977) and protein mixtures (Bio Rad Protein Assay, 1977). Intensive studies with the Bradford technique have been performed by Frauenheim (1984) who determined that for complete extraction of protein from marine sediments, up to seven different extractions from the same sample have to be performed. For practical reasons it is recommended that only two extraction steps per sample are carried out. Standardization with γ-globulin yields up to 75% of the sediment protein content (Frauenheim, 1984).

The Method of Lowry et al. (1951).--The method of Lowry et al. is based on the chemical reaction of proteins with the Folin phenol reagent which contains phosphomolybdic-tungstic mixed acid. These acids are the chromagens whereas tyrosine, tryptophan, and to a lesser extent cystine, cysteine, and hystidine are the chromogenic amino acids of the proteins. As a result of this reaction the mixed acids are reduced by the loss of oxygen atoms from tungstate and molybdate producing a characteristic blue color (λmax 745-750 nm). Copper, which is part of the reagent mixture, too, acts as a catalyst chelating in the peptide structure of the proteins.

Reagents: Folin-Ciocalteu phenol reagent (reagent A). **Procedure:** for use dilute a MERCK standard solution to get a 1 N acid content as follows: add 1 ml of Folin reagent to 30 ml distilled water and titrate with 0.1 N NaOH against phenolphthalein (0.1% solution). On the basis of this titration dilute Folin reagent (about 2-fold) with distilled water to make it 1 N acid.

Reagents: natrium carbonate/sodium-potassium-tartrate solution (reagent B). **Procedure:** dissolve reagent grade Na_2CO_4 and sodium-potassium-tartrate to get a 4%/0.04% (w/v) solution.

Reagent: copper sulfate solution (reagent C). **Procedure:** dissolve reagent grade $CuSO_4 \cdot 5\ H_2O$ to get a 1% (w/v) solution.

Reagent: solution D. **Procedure:** mix equal parts of solution B and solution C. The resulting reagent remains stable for only 1 day. **Standards:** dissolve appropriate amounts of γ-globulin in distilled water to get standard curves at concentrations from 10 to 100 μg per ml. Prepare these standards like the sample protein solution for measurement.

The following figure shows the analyzing steps in detail:

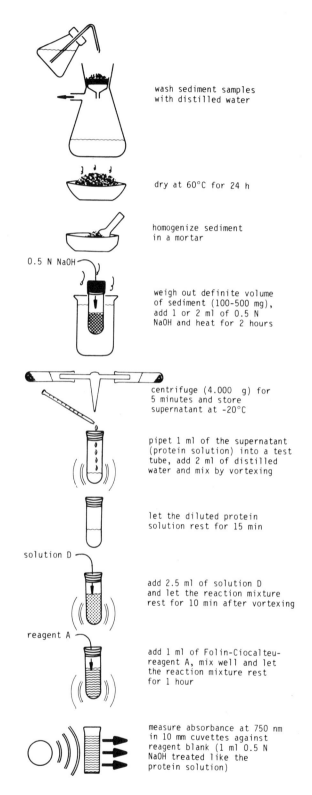

The Method of Bradford.--The method is based on the conversion of the red form of Coomassie Brilliant Blue G-250 to the blue form by binding to proteins. The dye reacts only with proteins of molecular weight greater than 3000. Amino acids and oligopeptides do not affect the reaction (Sedmark and Grossberg, 1977). The blue color develops from the chemical reaction of the SO_3^- ion groups of the dye with the NH_3^+ groups of the proteins (Sedmark and Grossberg, 1977) and is therefore dependent on the pH-value of the reaction mixture (Fazekas de St. Groth et al., 1963).

Reagents: Dye-reagent. Procedure: 40 mg of Coomassie Brilliant Blue G-250 (Sigma) is dissolved in 200 ml of distilled water. To this solution 50 ml of 95% (w/v) ethanol and 100 ml 85% (w/v) phosphoric acid is added. Dilute the resting solution to 1000 ml with distilled water.

Reagent blank: 2 ml of dye-reagent plus 0.3 ml of distilled water (Spector, 1977).

Standards: dissolve appropriate amounts of γ-globulin in distilled water to get final concentrations from 10 to 100 µg per ml. These standards have to be prepared like the sample protein solution for measurement.

The following diagram shows the method in detail:

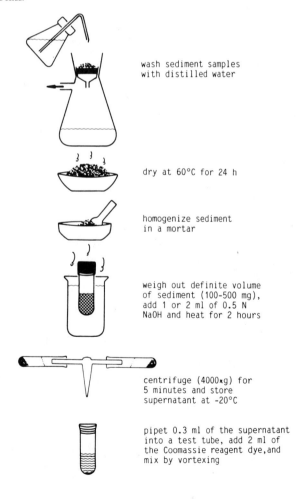

wash sediment samples with distilled water

dry at 60°C for 24 h

homogenize sediment in a mortar

weigh out definite volume of sediment (100-500 mg), add 1 or 2 ml of 0.5 N NaOH and heat for 2 hours

centrifuge (4000×g) for 5 minutes and store supernatant at -20°C

pipet 0.3 ml of the supernatant into a test tube, add 2 ml of the Coomassie reagent dye, and mix by vortexing

measure absorbance at 595 nm after 2 minutes of incubation in 10 mm cuvettes against reagent blank

calculate protein content from γ-globulin standards (10-100 µg) treated like sediment samples

Lipids

The so-called lipids are actually a mixture of very different chemical compounds. Anything that can be extracted from plant, animal, or bacterial cells by organic solvents, such as chloroform, ether, and alcohols, is called a lipid. This can be fatty acids, fats, carotenoids, and steroids. An essential frequent compound of cellular membranes are the phospholipids. By determining the lipid phosphate it is possible to measure the viable membrane biomass of all the microbes (Smith et al., 1982). Hack et al. (1962) examined the lipid pattern of various invertebrates (11 phyla) and microorganisms by chromatography and found specific variations due to the taxa investigated. Therefore, the study of lipids may lead to more differentiated information about the species composition of the living biomass in sediment samples than with other biomass parameters.

As the lipid fraction contains various species of chemical compounds, detection of lipids must be carried out with nonspecific dyes or simple techniques such as weighing. The meaningfulness of lipid values therefore is highly dependent on the extraction and separation method used. In literature various extraction procedures have been described. Ether extraction was used by Giese (1966) for marine invertebrates whereas Freeman et al. (1957) introduced the chloroform-methanol method. This method was investigated carefully by Bligh and Dyer (1959) and applied to invertebrate tissues by Lawrence et al. (1965). Mukerjee (1956) used alcoholic potassium hydroxide for lipid extraction, because the resulting lipid soap can be treated with a cationic dye. The extinction of the resulting blue color can be measured photometrically. An adaptation of this method for marine samples is given by Strickland and Parsons (1972) and will be described later.

For more differentiated lipid determination some simple separation techniques are mentioned. Due to the method of Mukerjee (1956) it is possible to separate lipids into a saponifiable fraction including the triglycerides, phospholipids, fatty acids and plasmalogens, and a non-saponifiable fraction including the sterols, carotenoids, wax-alcohols, and hydrocarbons (Giese, 1966). Freeman et al. (1957) described the separation of polar from non-polar lipids by the use of chromatography. This method was applied by Towle and Giese (1966) for marine organisms. For more sophisticated separation techniques of lipids see Saike et al. (1959), Rapport and Alonzo (1960) and Hack et al. (1962a,b).

The Chloroform-Methanol Extraction Method.--A simple, rapid, and reproducible method for lipid determination is given by Bligh and Dyer (1959). The method combines extraction with chloroform-methanol, separation of the lipid fraction by dilution with chloroform and water, and determination of the lipids by evaporation to dryness and weighing afterwards. A detailed description of the method is given below with some modifications for marine sediment obtained from Jeffries (1969).

homogenize an aliquot of dried sediment containing about 500 mg of organic matter with 30 ml of chloroform/methanol/water (1:2:0.8) in a mortar

add 8 ml of chloroform and homogenize again

add 8 ml of distilled water and homogenize again

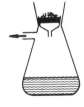

filter the homogenate through a glass fiber filter with slight suction

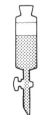

transfer the filtrate into a separator funnel and allow to rest for complete separation of the methanol-water layer and the chloroform layer containing the lipid extract

pour the chloroform layer into a tared flask

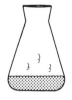

dry the chloroform lipid extract at 40°C in a vacuum oven containing P_2O_5 and purged with nitrogen

weigh out the lipid residue

The Alcoholic Potassium Hydroxide Extraction Method.--This method (Mukerjee, 1956; Strickland and Parsons, 1972) may detect lipids at the ug level quantitatively and therefore does not require great amounts of sediment for sample processing. The suitable volume of sediment sample and the extractant has to be determined by trial and error, because the organic content may vary greatly with the location of the sample and the time of the year.

This method may also be useful in combination with the method described in the discussion of chloroform-methanol extraction (below) The lipid residue extracted with chloroform-methanol can be measured for a sediment sample due to the procedure given below.

Reagent: alcoholic potassium hydroxide solution.
Procedure: dissolve 20 g of analytical reagent grade potassium hydroxide in 50 ml of distilled water and add 50 ml of reagent grade 95% ethanol.

Reagent: bromobenzene (reagent grade monobromobenzene).

Reagent: borate buffer. **Procedure:** dissolve 50 g of analytical reagent grade boric acid and 8.8 g of potassium hydroxide in 2 l of distilled water. Keep the solution well-stoppered to prevent carbonation.

Reagent: Pinacyanol reagent. Procedure: dissolve 16 mg of pinacyanol in 500 ml of borate buffer, dilute 2 l and filter. Keep this solution in the dark in a glass bottle. The solution should have a pH value about 8.5 and remains stable for about two weeks.

Standard solution: dissolve .75 mg of pure stearic acid in 250 ml of 95% ethanol (1 ml of this solution contains 300 µg of stearic acid). Store in a well-stoppered glass bottle. The solution is stable indefinitely. For use add 1 ml of this standard solution into a clean dry 10 ml centrifuge tube and evaporate carefully to dryness. Prepare this standard for measurement due to the treatment of the sediment sample.

The following diagram shows the method in detail:

transfer an aliquot of homogenized dried sediment, containing about 10 - 50 mg of organic matter, into a 10 ml graduated stoppered centrifuge tube

add 1 - 3 ml of alcoholic potassium hydroxide solution

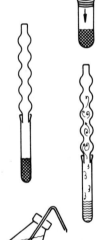
fit a 15 cm air condenser into the centrifuge tube and mix by shaking

heat for 30 minutes at about 100°C so that alcohol is refluxing part way up the condenser

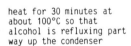
allow to cool to room temperature and remove the condenser. Adjust the volume of liquid inside the tube to 10 ml with distilled water and mix by vortexing

centrifuge the sample (4000×g) and add a 1 ml aliquot of the supernatant to a 60 ml separator funnel containing 40 ml of pinacyanol reagent. Allow to stand for 5 min in the dark

add 5 ml of monobromobenzene and shake the contents in the funnel vigorously for a full minute. Allow the layers to separate for 20 minutes in the dark

withdraw sufficient organic phase from the bottom of the funnel to fill a 10 mm cuvette and measure extinction of the solution immediately at 620 nm against water

calculate lipid content from corresponding standards prepared as indicated above

Total Carbohydrates

Carbohydrates make up a large part of dead and living organic matter. Sugars such as glucose are efficient energy-providing molecules, often stored as starch, whereas cellulose, chitin, and other carbohydrate derivatives are the structural compounds of the cell and body walls of plants, fungi, bacteria, and animals. Ribose and deoxyribose are essential for nucleic acid synthesis, and the acid derivatives of many sugars are used to replenish the amino acid pool of the living cell. Therefore, the determination of carbohydrates may be used as a biomass parameter indicating growth and standing stocks in cultures or natural samples.

For ecological studies concerning the energy flow in a given ecosystem, the release of carbohydrates from algal cells is to be of interest. Burney et al. (1981) reported that carbohydrates accounted for about 30% of the released organic carbon. Data from

Ittekot et al. (1981) show that up to more than 60% of these carbohydrates consisted of glucose.

These investigations lead to the suggestion that besides the particulate carbohydrates, soluble carbohydrates play a distinct role as a nutrient input to benthic communities. The importance of phytoplankton blooms for benthic nutrition has been postulated by several authors (see Graf et al., 1984, for reference). Kormondy (1968) suggested that considerable amounts of decomposed plant material become incorporated into the sediments because of the presence of decay-resistant chemical compounds as cellulose, lignin, and resins. He found cellulose disintegration to be far from complete (only 40 to 50% lost) within nine months in Carolina Bay, near Savannah River (South Carolina, USA). Similar observations were made by Whittaker and Vallentyne (1957) finding high concentrations of sugars, up to 42.4 g per kg dry weight, in lake seston. From the analysis of the carbohydrate content of tissues from marine invertebrates (and vertebrates) in relation to other body constituents, such as proteins or lipids, information on the reproductive cycles and constitution of the organisms become available. For some general notes concerning this field of research see Giese (1966).

Three main methodological approaches have been cited in literature for colorimetric determination of total carbohydrates: The color-producing chemical reaction of sugars and their derivatives with anthrone, phenol-sulfuric acid or 3-methyl-2-benzothiazolinone hydrazone-hydrochloride (MBTH). The more recently established MBTH method has been introduced by Johnson and Sieburth (1977) and Burney and Sieburth (1977) for seawater. The anthrone method was applied as a microdetermination of cellular carbohydrates from yeast by Trevelyan and Harrison (1952), on plant extracts by Yemm and Willis (1954), and has been obtained from Hewitt (1958) for marine samples by Parsons et al. (1984). The phenol-sulfuric acid method obtained from Dubois et al. (1956) has been applied to sediments by Liu et al. (1973).

As the term carbohydrates is a generic one including many different molecular species, each method is more or less specific for individual compounds of the carbohydrate matrix analyzed and non-quantitative for total carbohydrates. Care has to be taken preparing sediment samples for analysis. Drying of test mixtures at 60° C for 20 h induced an average loss of 55% of the monomeric sugars but none of the polymeric carbohydrates which comprise for >98% of the sedimentary carbohydrates (Mopper, 1977).

Carbohydrate Determination by Anthrone Reagent.--The method is based on the reaction of various sugar degradation products (probably the furfural derivatives) formed by digestion with sulfuric acid and anthrone, producing a green color. This method is most sensitive for hexoses. The ketoses react most quickly giving a maximum color development after about 1-5 minutes, the aldoses reach color maximum at about 7 min (Yemm and Willis, 1954). Although the reaction with anthrone is not equal with the different sugars, Yemm and Willis (1954) found satisfactory agreement of carbohydrate values obtained from known sugar mixtures.

The procedure given in detail is based on Parsons et al. (1984), the detection limit is on the ug level. Note that the time of heating is responsible for the color development; extended heating may result in a loss of color intensity. Therefore, the heating period may be varied to achieve optimal color development due to peculiarities of individual samples.

Reagent: Anthrone reagent. **Procedure:** dissolve 0.2 g of anthrone (9,10-dihydro-9- ketoanthracene) in 100 ml of concentrated sulfuric acid (analytical reagent quality). Add 8 ml of ethanol and 30 ml of distilled water. Allow to stand overnight refrigerated and try to protect from light at all times.

Standard solution: dissolve 1 g of glucose in 100 ml of distilled water to prepare a stock solution. This solution remains stable for a few weeks at -5° C. For use dilute 10 ml of the stock solution to 1000 ml to get a final concentration of 100 µg per ml.

Calculation: prepare appropriate standard curves and calculate carbohydrate content as glucose equivalents per dry weight (or volume) of sample analyzed.

The following figures describe the method given by Parsons et al. (1984) modified for sediment samples.

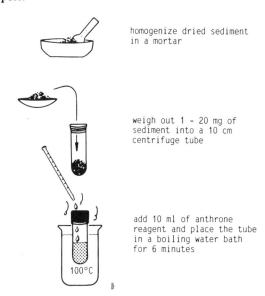

homogenize dried sediment in a mortar

weigh out 1 - 20 mg of sediment into a 10 cm centrifuge tube

add 10 ml of anthrone reagent and place the tube in a boiling water bath for 6 minutes

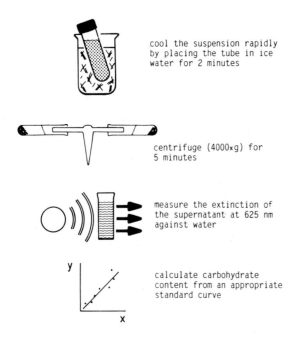

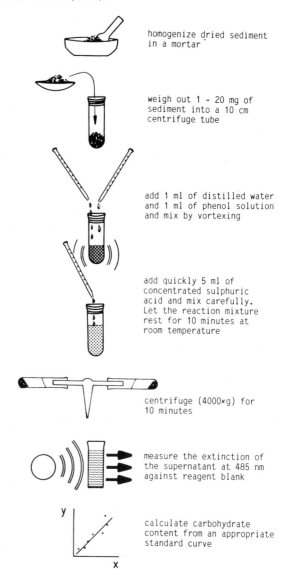

The following diagram shows the procedure given by Liu et al. (1973):

Carbohydrate Determination with Phenol-Sulfuric Acid.--Monosaccharides, oligosaccharides, and their derivatives, including the methylethers with free or potentially free reducing groups, give an orange-yellow color when treated with phenol and concentrated sulfuric acid (Dubois et al., 1956). The advantages of this method are the stability of the produced color, its rapidity, and its specificity for a wide range of carbohydrate compounds. The colorimetric determination of chromatographically separated carbohydrates is also possible with this reagent, because there is no interference with commonly used solvents. The anthrone reagent would, for instance, be rendered useless by traces of phenol containing solvents (Dubois et al., 1956). Liu et al. (1973) adapted this method to sediment samples achieving high reproducibility, and requiring only as little as 1 mg homogenized dry weight of sample.

Reagent: phenol solution. **Procedure:** dilute reagent grade phenol to get a 10% solution, shake well before use.

Reagent: concentrated sulfuric acid.

Standard solutions: dissolve 1 g of soluble starch and 1 g of agar in 100 ml of distilled water on a hot water bath to prepare two different standard stock solutions. These solutions may be stored frozen for weeks. Dilute the required standard to final concentrations at the μg level due to the carbohydrate concentration range of the samples analyzed.

Calculation: calculate the carbohydrate content of the samples from the prepared standard curve as agar (or starch) equivalents per dry weight (or volume) of sediment analyzed.

Carbohydrate Determination with MBTH.--The MBTH method was evaluated, because similar standard curves are obtained from equimolar concentrations of different monosaccharides, and no use of strong acids is required which may cause analytical interferences from non-carbohydrate degrading products (Johnson and Sieburth, 1977). Using the MBTH method, monomeric and polymeric carbohydrates can be analyzed separately from the same sample. This only requires an additional hydrolysis procedure with HCl (Burney and Sieburth, 1977). The methodological principle is the reduction of carbohydrates to sugar alcohols with borohydride and quantitative liberation and detection of formaldehyde from these alcohols by periodate and spectrophotometrically MBTH assay. The method was introduced for seawater in 1977 by the above references and modified by Johnson et al. (1981) to decrease the complexity of the analytical operations.

> **Reagent: borohydride solution. Procedure:** dissolve 100 mg of reagent grade KBH_4 in 5 ml of cold distilled water. Prepare this solution fresh for each batch of samples.
>
> **Reagent: periodic acid solution. Procedure:** dissolve 0.57 g of analytical grade periodic acid in 100 ml of distilled water. This solution should be stored at room temperature protected from light.
>
> **Reagent: sodium arsenite solution. Procedure:** dissolve 3.25 g of sodium arsenite in 100 ml of distilled water, this solution is stable for months.
>
> **Reagent: ferric chloride solution. Procedure:** dissolve 5 g of analytical grade ferric chloride in 100 ml of distilled waters, filter, and store at 5° C.
>
> **Reagent: MBTH solution. Procedure:** dissolve 276 mg of 3-methyl-2-benzothiazolinone hydrazone hydrochloride in 10 ml of 0.1 N HCl with slight heating, filter, and store in an amber bottle. The solution remains stable for about one week.
>
> **Reagent:** reagent grade acetone.
>
> **Standard solution:** dissolve 1 g of soluble starch and 1 g of glucose in 100 ml of distilled water to prepare standard stock solutions for polysaccharides and monosaccharides. Store at 5° C adding 1 ml of saturated mercuric chloride solution as a preservative. Before use dilute stock solutions to the ug level to get appropriate standard curves.

The MBTH method is described in the following diagram with slight modifications for sediment samples.

homogenize dried sediment in a mortar

weigh out 1 - 10 mg of sediment into a glass ampule and add 12 ml of 0.1 N HCl

seal the ampule with a gas burner and hydrolyse at 100°C for 20 hours

centrifuge the sample at 4000×g for 15 minutes

transfer 10 ml of the supernatant into a screw topped test tube and add 2 ml of 0.5 N NaOH for neutralization

to the neutralized hydrolysate add 0.1 ml of freshly prepared borohydride solution

incubate the samples in a water bath at 18°C for 4 h

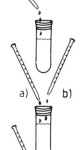
oxidize the excess borohydride with 0.1 ml of 2 N HCl until visible gas evolution ceased

a) add 0.1 ml of periodic acid to the sample and incubate for 10 minutes at room temperature (step a)

b) add 0.1 ml of sodium arsenite solution and incubate for another 10 min (step b)

add 0.2 ml of 2 N HCl and allow the amber color to disappear

add 0.2 ml of the MBTH reagent and place the tightly capped tube in a boiling water bath, remove after 3 minutes and cool to room temperature

add 0.2 ml of the ferric chloride solution and incubate at room temperature in the dark for 30 minutes

add 1 ml of acetone and mix well

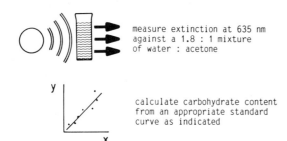

measure extinction at 635 nm against a 1.8 : 1 mixture of water : acetone

calculate carbohydrate content from an appropriate standard curve as indicated

Calculation of carbohydrate content: For analytical controls prepare "reagent blanks" of the samples and standards by carrying out the procedure with one modification: instead of stepwise addition of the periodic acid and the sodium arsenite solution, add 0.2 ml of a pre-mixed (let react for 10 minutes) 1:1 periodic acid/sodium arsenite solution to aliquots of the samples. Continue as for the samples.

These blanks must be run on each sample or standard for correction of turbidity and free aldehydes as well as for the reagents themselves. Subtract the extinction of the blanks from the readings of the samples to get the extinction values for each calculation.

For differentiating between the amount of monomeric and polymeric carbohydrates, run aliquots of the samples with and without the initial hydrolization step. Running the complete procedure will give total carbohydrates (TCH), running the procedure without hydrolysis will give the amount of monomeric carbohydrates (MCH). By subtracting the MCH-value from the TCH- value the amount of the polymeric carbohydrates can be calculated.

DNA/RNA

DNA and RNA represent the genetic material of living organisms and viruses. Whereas DNA is present in definite amounts, the content of RNA varies with the physiological state of living cells. During high metabolic activity a lot of messenger RNA is synthesized via DNA to complete the enzymatic equipment and to facilitate the production of structural proteins for growth. Culture experiments with some phytoplankton species show that RNA can serve as an additional N-reservoir besides protein, amino acids, and inorganic nitrogen to support growth (Dortch et al., 1984).

Early studies of RNA and DNA in microorganisms showed that there is a linear relationship between growth rate and RNA/DNA ratios whereas the DNA/dry weight ratio remains constant (Rosset et al., 1966; Leick, 1968; Brunschede et al., 1977). Dortch et al. (1983) found the highest RNA/DNA ratios in marine samples during phytoplankton blooms, indicating high metabolic activity. Holm-Hansen et al. (1968) tried to use DNA measurements as an indicator of living biomass but found unrealistic high concentrations (4.9 to 11.5%) in seawater in comparison to phytoplankton cultures (0.5 to 2.5%). They proposed high detrital DNA-content to be responsible for such unexpected values. An extended discussion about this problem is given by Dortch et al. (1983).

Båmstedt and Skjoldal (1979) used the RNA/dry weight relationship in zooplankton organisms to get information about their growth rate. Sutcliffe (1965, 1969) demonstrated positive correlation between RNA concentration and growth rate for marine organisms of different taxa. RNA/growth rate relationships obtained from diverse taxa exhibit strong differences. Bamstedt and Skjoldal (1979) suggested that there is no predictive value in determining general relationships in mixed zooplankton samples. Dortch et al. (1985) also concluded that corresponding values would be better interpreted as general indicators for the physiological state of planktonic communities rather than as specific indicators for growth rate. They tried to use protein/DNA and RNA/DNA relationships to estimate the growth rate of plankton populations due to seasonal variations of N-nutrients and possible N-deficiency.

Spectrofluorometric Determination of DNA and RNA.--Le Pecq and Paoletti (1966) described a fluorometric method for determination of nucleic acids. This method was examined and tested for several animal tissues by Prassad et al. (1972), who improved this method making it possible to measure both DNA and RNA on the same sample. Sample extracts are treated with ethidium bromide, a dye which reacts both with DNA and RNA producing a highly fluorescent compound. After measuring total fluorescence from DNA plus RNA, RNAse is added to the sample. A second reading of the fluorescence represents the DNA present in the sample. The RNA content is then calculated by difference. Dortch et al. (1983) adapted this method for marine samples. Sufficient results were obtained grinding the sample for 2 minutes in distilled, deionized water chilled in an ice water bath. Generally, samples for nucleic acid determination should be kept frozen until analyzed or prepared immediately after collection by freeze-drying (Bamstedt and Skjoldal, 1979) to prevent denaturation of the DNA and RNA molecules.

This method described below is based on Prassad (1972) and Dortch et al. (1983) with slight modifications for preparation of sediment samples.

Reagent: ethidium bromide solution. **Procedure:** dissolve 20 mg of 2,7-diamino-9-phenyl-phenanthridine-10-ethyl-bromide in 1 l of distilled water.

Reagent: tris/NaCl buffer solution. **Procedure:** dissolve reagent grade tris-hydroxymethylamino-methane and NaCl in distilled water to get a 0.1 M/0.1 M (w/v) solution. Adjust pH to 7.5 - 8.8.

Reagent: RNAse solution. Procedure: RNAse (Type III-A from bovine pancreas, Sigma Chemical Company) has to be prepared fresh daily with distilled water to a final concentration of 20 mg per milliliter.

DNA/RNA standards: standards should be prepared from Type I DNA and Type III RNA (obtained from Sigma Chemical Company) to get final concentrations at the µg level in 2 ml of the measured solution (1.0 ml ethidium bromide solution + 0.9 ml Tris/NaCl buffer + 0.1 ml DNA or RNA standard). Standard curves have to be prepared daily.

Reagent blank: the fluorometer (Turner fluorometer) is calibrated with reagent blank containing 1.0 ml ethidium bromide solution + 1.0 ml Tris/NaCl buffer solution.

The following diagram shows treatment and measurement of the samples:

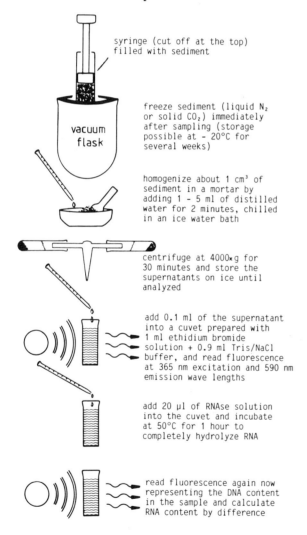

Total Adenylates

During the past 20 years many investigators introduced ATP as a useful parameter for measuring microbial biomass in aquatic ecosystems (Holm-Hansen and Booth, 1966; Christian et al., 1975; Graf et al., 1984). Other authors reported difficulties interpreting their ATP-measurements in relation to biomass (Dahlbäck et al., 1982; Pridmore et al., 1984) because of contributions from meiofauna and protozoa in addition to bacteria (e.g., Yingst, 1978) and because the ATP content per cell varies with stage of growth of the organisms as well as environmental conditions (Dietzler et al., 1974; Hunter and Laws, 1981; Cacciari et al., 1983).

Atkinson (1968, 1971) showed the important role of the adenylate pool (ATP, ADP and AMP) in regulating the metabolic activity in living cells. Shifts in the ATP pool will alter the energy charge (Atkinson and Walton, 1967) of this adenylate system but to a much lesser extent the amount of total adenylates (Chapman et al., 1971; Chapman and Atkinson, 1973). Therefore, the measurement of total adenylates appears to be a better indicator of living biomass than ATP concentration.

Generally, there are three critical points in analyzing the adenylate content of sediment samples:

(1) The metabolism of the living biomass has to be stopped immediately after sampling, because changes in, e.g., oxygen content or temperature will alter the ATP content about 50%.

(2) The adenylates, especially ATP, will be adsorbed to sediment particles, mainly to clay minerals. It is necessary to saturate the free bonds of the sediment particles before extraction (Bulleid, 1978).

(3) During extraction and storage, the adenylate content will be altered due to the properties of the solution used for extraction (e.g., low pH values will force hydrolysis of ATP) and the activity of enzymes (ATPases) present in the samples (Bagnara and Finch, 1972; Tobin et al., 1978; Klinken and Skoldal, 1983). For marine sediments successful attempts (Greiser, 1982) in adenylate determination were made with the methods of Kalbhen and Koch (1967) and Pradet (1967) in combination with the method of Bulleid (1978). The measurement of adenylates is achieved by determining the light output of the luciferin-luciferase bioluminescence reaction with ATP. Firefly-lantern-extracts (FLE-50) were obtained from Sigma Chemical Company. Better purified luciferin-luciferase extracts could be used too but satisfactory reproducibility will be achieved from crude firefly extracts by running two or three times ATP standards parallel to the samples.

To measure total adenylates ADP and AMP must be converted to ATP enzymatically before determining bioluminescence activity. This can be carried out by the enzymatic reaction of pyruvate kinase with ADP and phosphoenolpyruvate (PEP) and the reaction of myokinase (adenylate kinase) with ATP and AMP in combination with the pyruvate kinase reaction (Pradet, 1967).

With the bioluminescence technique very low limits of ATP can be determined using a commercial luminometer (e.g., LKB model 1250), because the luciferin-luciferase system has a very high affinity for ATP. One molecule of ATP gives one quantum of light (Strehler, 1968).

Reagent: phosphate buffer. Procedure: mix appropriate amounts of 0.2 M Na_2HPO_4 buffer and 0.2 M KH_2PO_4 to adjust the resulting solution to the pH value of the sample.

Reagent: glycine buffer. Procedure: dissolve 7.5 g glycine and 5.9 g NaCl in 1 l distilled water. Add about 500 ml 0.1 N NaOH and adjust the resulting solution to pH 9.8.

Reagent: arsenate buffer. Procedure: 0.05 M Na_2HAsO_4 /0.01 M $MgSO_4$-buffer solution adjusted to pH 7.4. Dissolve the Na_2HAsO_4-salt prior to $MgSO_4$ addition!

Reagent: tris-HCL-buffer. Procedure: prepare a 0.2 M solution, pH value should be adjusted to 7.4 with concentrated HCl.

Reagent: K^+4Mg^2-solution. Procedure: dissolve reagent grade KH_2PO_4 and $MgSO_4 \cdot 7 H_2O$ to get a 0.3 M / 0.042 M (w/v) solution.

Reagent: PEP-solution. Procedure: dissolve 0.0235 g of phosphoenolpyruvate (Merck) in 10 ml of distilled water.

Reagent: pyruvate kinase. Procedure: dilute the 10 mg/ml-pyruvate kinase solution (Boehringer-Mannheim) 10-fold (e.g., 0.1 ml + 0.9 ml distilled water).

Reagent: myokinase. Procedure: dilute the 10 mg/ml myokinase solution (Boehringer-Mannheim) 5-fold (e.g., 0.2 ml + 0.8 ml distilled water).

Reagent: sulfuric acid (Procedure: 0.6 N H_2SO_4).

Reagent: firefly enzyme. Procedure: dissolve the content of one vial FLE-50 (Sigma Chemical Company) in 5 ml of distilled water and let the resulting solution stabilize for 1-2 h. Centrifuge the turbid solution for 5 minutes (4000 RPM) to remove undissolved solids before use. Spare enzyme solution can be stored at -20° C for months without significant loss of activity.

ATP-standards: dissolve 55 mg of ATP (Merck) in 100 ml of glycine buffer to get a stock solution of 10 M /100 ml. Dilute this solution to a final concentration of 1-100 M/100 ml (0.1 ml of this solution will contain 1-100 M). It is advisable to measure standards from time to time during ATP-determination of the samples due to the gradually slight decline of the firefly enzyme activity.

The following diagram shows the method mentioned above:

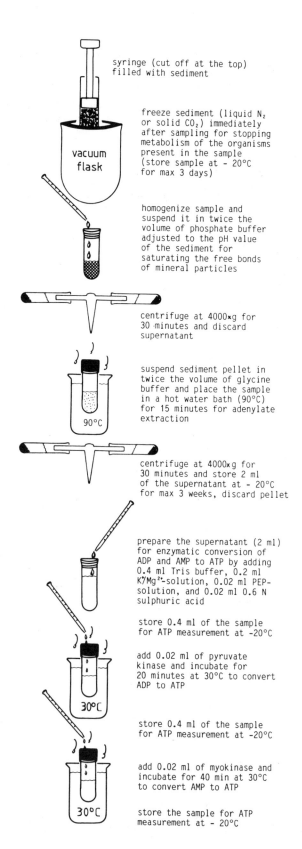

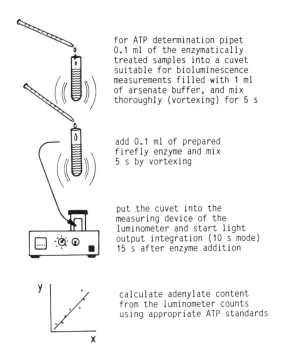

for ATP determination pipet 0.1 ml of the enzymatically treated samples into a cuvet suitable for bioluminescence measurements filled with 1 ml of arsenate buffer, and mix thoroughly (vortexing) for 5 s

add 0.1 ml of prepared firefly enzyme and mix 5 s by vortexing

put the cuvet into the measuring device of the luminometer and start light output integration (10 s mode) 15 s after enzyme addition

calculate adenylate content from the luminometer counts using appropriate ATP standards

Amino Acids

Various investigations indicate that the amino acid content composition of sediments can provide information about biomass of bacteria and endofauna, nutrient availability, and community metabolism.

Hall et al. (1970) showed that changes in the amino acid composition of estuarine suspended solids may reflect the effect of biological breakdown by microorganisms. De La Cruz (1970) reported biochemical changes during decomposition of dead plant material which may lead to an increase in amino acid content exceeding that in living plants. Otsuki et al. (1983) found amino acids to be the major constituent of particulate matter in a representative mesotrophic freshwater pond. As reported by Degens et al. (1970) high levels of amino acids were even found in aerobic sediment layers, probably being by-products of microbial metabolism. Clark et al. (1972) suggested that these upper sediment layers were a major reservoir for amino acids indicated by high values in interstitial water from nearshore samples.

Three main groups of amino acids can be separated from organic matter: (1) the hydrophobic ones (valine, leucine, isoleucine, phenylalanine, and methionine); (2) the hydrophylic quite polar ones (aspartic acid, glutamic acid, asparagine, glutamine, lysine, arginine, and histidine); and (3) the moderate polar but uncharged ones (glycine, alanine, cysteine, serine, threonine, tyrosine, tryptophane, and proline). The various compositions of these amino acid species determine the structural and functional properties of proteins in living organisms. Fluctuations in the amino acid composition of the living cell will have far-reaching effects on the metabolism as a whole, and may be used as an indicator of an organism's constitution reflecting the nutrients incorporated (Cowey and Corner, 1963). As reported by these authors from marine zooplankton studies comparing the cellular and diet amino acid composition throughout the year, zooplankton greatly influence the nitrogen turnover in marine ecosystems by assimilation of non-living particulate nitrogen and excretion of amino acids. Similar results were obtained from Jeffries (1969) and Riley and Segar (1970) suggesting that shifts in the amino acid pool correspond to the ecological pattern of seasonal succession in planktonic communities.

The fluorometric phthalaldehyde method is described here, because it is more commonly used (Grasshoff et al., 1983; Parsons et al., 1984) and because of its low limit of detection at the picomole range. An extended list of other methods with corresponding references with emphasis on possible disadvantages of these methods is given by Dawson and Pritchard (1978).

Early methods for amino acid determination were carried out spectrophotometrically measuring the absorbance due to the color development of the ninhydrin reaction. Troll and Cannon (1953) describe a method used more recently by Andrews and Williams (1971) which produces more reproducible results because of the uniformity in molar color yields with the various amino acids and the low sensitivity to ammonia. Grasshoff et al. (1983) indicated that interference of ammonia may account for only 5% of the corresponding glycine standard values.

Fluorometric Determination of Particulate Amino Acids.--The phthalaldehyde method was evolved by Roth (1971) and permits amino acid analysis from seawater down to the nanomole range. Besides its high sensitivity, the method has the advantage over ninhydrin techniques because there is no need of heating. For analysis of particulate amino acids an initial hydrolization step is required (Riley and Segar, 1970; De La Cruz, 1970). For better reproducibility at low detection limits it is necessary to desalt sample extracts on an ion exchange column (Hall et al., 1970; Parsons et al., 1984). Separation of amino acids prior to analysis due to different groups or single species requires sophisticated chromatographic techniques and is mainly used in combination with an automatic amino acid analyzer. For detailed information about corresponding analytical procedures see Lindroth and Mopper (1979).

The method presented below is based on techniques given in the above references as described by Parsons et al. (1984) for particulate matter with slight modifications for sediment samples.

Reagent: phthalaldehyde solution. **Procedure:** dissolve 200 mg of analytical grade o-phthalaldehyde in 20 ml of 95% ethanol.

Reagent: borate buffer solution. **Procedure:** dissolve reagent grade orthoboric acid in 1 l of distilled water to get a 0.4 M solution. Adjust pH value to 9.5 with 1 N NaOH.

Reagent: reagent solution. **Procedure:** add 1 ml of reagent grade 2-mercaptoethanol to 400 ml of borate buffer and the 20 ml of phthalaldehyde solution. Let stand for 1 hour. The solution remains stable for about 2 weeks in a refrigerator.

Amino acid standard: dissolve 0.751 g of glycine in 1 l of distilled water. Stabilize the solution with 1 ml of toluene. The solution is stable for months and contains 10 µmole glycine per ml. Dilute this stock solution to the concentration range of the samples for preparing appropriate standard curves.

Reagent blank: for amino acid concentrations at the detection limit of this method it is necessary to correct fluorescence values for a reagent blank due to contamination of the reagents and the equipment used. Prepare reagent blank by treating an aliquot of combusted sediment, and for the standard by treating a distilled water sample according to the procedure given below.

The following diagram shows the procedure in detail:

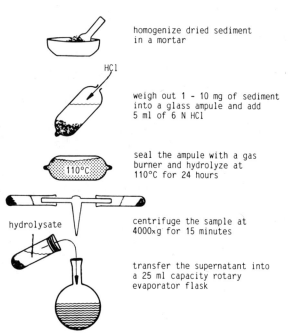
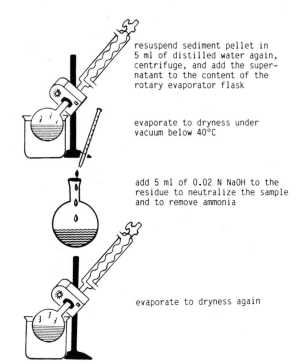
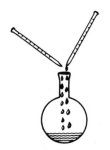

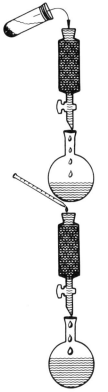

if a desalting step is required:

pass the centrifuged hydrolysate through an ion exchange column (2 x 20 cm in size, cation exchange resin DOWEX 50W-X8 in the H⁺-form after acidification to pH 3) and rinse with about 50 ml of distilled water

elute with 80 ml of 2 N NH₄OH into a rotary evaporator flask and reduce the volume under vacuum below 40°C to about 10 ml
transfer the concentrate to a 25 ml capacity rotary evaporator flask and continue as indicated

Methods to Determine Organic Matter of Different Groups of Organisms

Chloroplastic Pigments

The conversion of inorganic carbon to organic substances facilitated by light energy (primary production) is the most important process (besides release of oxygen) replenishing the energy and food supply of an ecosystem. Studies on primary productivity give information about the energy available passing through the community food web. The relation between the amount of community productivity (P) and respiration (R) may lead to information about the trophic level (P/R ratio) or (seasonal) succession of an ecosystem (Odum, 1971). Many attempts have been made to introduce methods for determining primary production. The measurement of chloroplastic pigments has proven to be a useful parameter to estimate plant biomass and therefore provides an index to its productivity. The determination of, e.g., chlorophyll degrading products (pheophytins, pheophorbids) may lead to information about turnover and decomposition of plant biomass.

Odum (1971) mentioned the possibility of using the ratio between the carotenoids and chlorophylls as an index to the ratio of heterotrophic to autotrophic metabolism in the community as a whole. When photosynthesis exceeds respiration, the chlorophylls predominate, whereas the amount of carotenoids increases as community respiration increases. Züllig (1981) used carotenoid stratigraphy in lake sediments as an indicator of eutrophication processes in lake ontogeny. Similar approaches have been made by several authors (for reference see Sanger and Gorham, 1970). Many authors have used chlorophyll values for estimating phytoplankton biomass (Manuels and Postma, 1973; Leighton, 1981; Riemann, 1983). Difficulties in interpreting corresponding values may arise because of variations in the pigment content of algal cells due to nutrient availability and growth rate (Hunter and Laws, 1981; Lehman, 1981). Steele and Baird (1968) analyzing production on a beach found chlorophyll values to be well correlated with organic carbon due to seasonal variations.

Analyses of the influence of the nutrient input from a phytoplankton bloom to a marine benthic community using chlorophyll measurements in combination with other biomass and bioactivity parameters have been carried out by Thiel (1978), Faubel et al. (1983), Faubel and Meyer-Reil (1983), Graf et al. (1984), and Pfannkuche (1985).

Although the detection of the various types of pigments needs sophisticated chromatographic techniques, the amount of chlorophyll$_a$, $_b$, and $_c$, the degrading products of chlorphyll$_a$, and an approximated amount of the carotenoids, can be determined easily using a commercial spectrophotometer. The fluorometric determination of pigments is more sensitive (5-10 times) but does not allow a simultaneous measurement of the different chlorophylls (Parsons et al., 1984).

Due to the various extraction procedures available caution should be exercised when comparing quantitative values given in the literature. Widely used extractants are ethanol, methanol, and acetone. The latter ones have been generally accepted and the extractant of choice depends on the investigator's special applications. For spectrophotometrical determination Tett et al. (1975) prefers methanol extraction because of its superiority to acetone for the distinction of chlorophyll and pheophytin. Riemann (1978) reported the spectral changes of chlorophyll after acidification to be more moderate in acetone than in methanol. Tett et al. (1975) indicated that chlorophyll seems less stable in methanol than in acetone. Riemann and Ernst, (1982) compared the extraction efficiency of methanol and acetone. They found both methods to be equal in determining the relative amounts of pigments from different samples although methanol gave higher absolute yields.

We will describe the spectrophotometric and fluorometric determination of chloroplastic pigments in detail. Because of the expensive equipment and sophisticated technique only a brief review is given on the high-pressure liquid chromatography (HPLC) method.

The Spectrophotometric Determination.--Due to the spectrophotometric determination (and the fluorometric measurement) optimal sample handling and measuring conditions as well as possible sources of errors have been discussed extensively in literature. Main problems are the reliability of the various extinction coefficients given in the equations used for calculating the pigment concentrations, and the interference of the degrading products with chlorophyll$_a$ measurements. Lorenzen and Jeffrey (1980) compared six spectrophotometric methods by intercalibration, proving that all methods perform satisfactorily for chlorophyll$_a$ in the presence of other chlorophylls but do not differentiate between chlorophyll$_a$ and chlorophyllide$_a$. This may result in overestimating chlorophyll chlorophyll$_a$ concentrations in the presence of high amounts of chlorophyll degrading products. Jacobsen (1982) emphasizes the need for chromatographic separation of chlorophyll$_a$ from the various degradation products to get values comparable to HPLC measurements. Riemann (1982) reported about fairly good agreement of spectrophotometric calculations of pheopigment$_a$ and chlorophyllide$_a$ with corresponding values estimated by paper chromatography.

These readings and other literature cited lead to the conclusion that it is not possible to quantify the chloroplastic pigments exactly by the common methods but it seems mainly possible to determine the correct order of magnitude of pigment concentrations. For comparing corresponding pigment data, detailed knowledge about the chosen methods and the nature of the samples is required. Trees et al. (1985) showed that over- and underestimation of chlorophyll$_a$ concentrations is greatly influenced by the pigment composition of natural samples.

For the extraction procedure of chlorophylls and phaèopigments from sediment samples, the method according to Lorenzen and Jeffrey (1980) with its possible error sources, e.g., interstitial water content of the samples, change of the concentration of the extractant acetone, acidification, influence of light and temperature, was discussed in detail by Wasmund (1984). Therefore, the method described here is based on the sediment sample preparation technique from Wasmund and the spectrophotometric method described by Parsons et al. (1984) using the equations from Jeffrey and Humphrey (1975) for the determination of the chlorophylls. These equations proved to be the most reliable ones due to the intercalibration tests carried out by Lorenzen and Jeffrey (1980). The determination of the pheophytins is performed due to the acidification technique from Riemann (1978) and the equations given by Parsons et al. (1984).

A detailed description of the method, modified by Wasmund (1984) is given as follows:

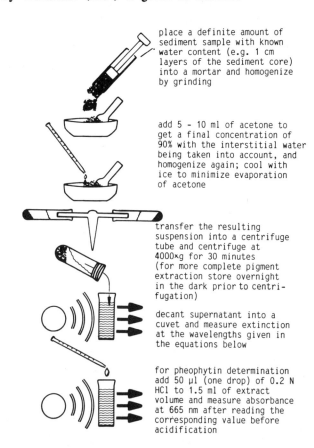

place a definite amount of sediment sample with known water content (e.g. 1 cm layers of the sediment core) into a mortar and homogenize by grinding

add 5 - 10 ml of acetone to get a final concentration of 90% with the interstitial water being taken into account, and homogenize again; cool with ice to minimize evaporation of acetone

transfer the resulting suspension into a centrifuge tube and centrifuge at 4000×g for 30 minutes (for more complete pigment extraction store overnight in the dark prior to centrifugation)

decant supernatant into a cuvet and measure extinction at the wavelengths given in the equations below

for pheophytin determination add 50 µl (one drop) of 0.2 N HCl to 1.5 ml of extract volume and measure absorbance at 665 nm after reading the corresponding value before acidification

Calculation:

1. Measure absorbance at 664, 647, and 630 nm and calculate the chlorophyll$_{a,b,c}$ content from the equations given by Jeffrey and Humphrey (1975):

chlorophyll$_a$ (µg/ml) =

$11.85\ E_{664} - 1.54\ E_{647} - 0.08\ E_{630}$

chlorophyll$_b$ (µg/ml) =

$- 5.43\ E_{664} + 21.03\ E_{647} - 2.66\ E_{630}$

chlorophyll$_c$ (μg/ml) =

$$-1.67\ E_{664} - 7.60\ E_{647} + 24.52\ E_{630}$$

2. Measure absorbance at 665 nm before after acidification for calculating chlorophyll$_a$ in relation to pheopigments:

chlorophyll$_a$ (mg/m^3) =

$$\frac{26.7\ (E_o - E_a)\ x\ V}{V_s\ x\ L}$$

pheopigments (mg/m^3) =

$$\frac{26.7\ (1.7\ [E_a] - E_o)\ x\ V}{V_s\ x\ L}$$

E_o = absorbance before acidification at 665 nm
E = absorbance after acidification at 665 nm
V = volume of water content of the samples plus acetone (100%) added
V_s = volume of sediment sample
L = path length (cm) of the spectrophotometer cell

3. Measure absorbance at 480 and 510 nm to estimate the carotenoid content:

carotenoids (μg/ml) =

$$7.6\ (E_{480} - 1.49\ E_{510})$$

Turbidity blank: correct each extinction for the turbidity blank measured at 750 nm by subtracting the 750 nm value from the 665, 664, 647, and 630 nm values before calculating the pigment content. For calculating carotenoids twice the 750 nm blank has to be subtracted from the 510 nm value and three times from the 480 nm values.

The Fluorometric Determination.--As particulate matter especially from marine phytoplankton samples often has to be concentrated by filtration of large volumes of water, methods have been developed to detect chloroplastic pigments considerably more sensitive than by spectrophotometric analysis. The availability of commercial photomultipliers (e.g., Turner fluorometer) made it possible to detect the fluorescent light emitted from excited pigment molecules. As described for the spectrophotometric method, interference may occur to chlorophyll measurements from other pigments. Lorenzen and Jeffrey (1980) and Gieskes and Kraay (1982, 1983) reported on serious interference from chlorophyll which will be partially calculated as pheopigment. Gowen et al. (1982) separated chlorophyll and pheophytin from chlorophyllide and pheophorbide by thin layer chromatography prior fluorometric determination to allow a more precise measurement. Coveney (1982) introduced a second acidification step to cope with this problem. Trees et al. (1985) found that fluorometric measurements underestimate chlorophyll concentrations from natural samples by an average of 39% compared with the HPLC technique. On the other hand Gieskes and Kraay (1982) reported about fluorometric values to give a good estimate of total chlorophyll phorbids comparing the HPLC and fluorometric method.

As indicated above, fluorometric determination of chloroplastic pigments do not give a precise analysis of the pigment content but its advantage is to provide a simple, sensitive, and rapid method. A general outline of the method described below is given by Yentsch and Menzel (1963) demonstrating some calibration principles. Parsons et al. (1984) recommended calibration of the fluorometer with chlorophyll standards. Due to the fluorometer model used, calibration factors have to be determined for the different sensitivity doors. A second calibration factor is needed for the decrease in fluorescent intensity of pure chlorophyll after acidification. The acidification technique gives an estimate of chlorophyll degrading products. This factor is assumed to be in the range of 1.85/1.95 (Shuman and Lorenzen, 1975). Parsons et al. (1984) reported a factor of 2.2 but indicated its dependence on the fluorometer used. The equations for chlorophyll and pheopigment calculation given below are obtained from Shuman and Lorenzen (1975). For more sophisticated and possibly more precise pigment calculation see Coveney (1982).

Though the fluorometric method discussed is contradictory, the procedure is given as following. For sample preparation and extraction procedure see previous discussion.

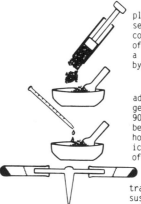

place a definite amount of sediment with known water content (e.g. 1 cm layers of the sediment core) into a mortar and homogenize by grinding

add 5 - 10 ml of acetone to get a final concentration of 90% with the interstitial water being taken into account and homogenize again; cool with ice to minimize evaporation of acetone

transfer the resulting suspension into a centrifuge tube and centrifuge at 4000×g for 30 minutes (for more complete extraction of pigments store overnight prior to centrifugation)

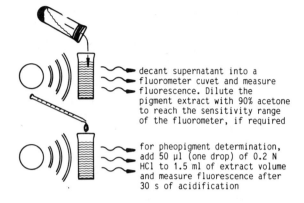

decant supernatant into a fluorometer cuvet and measure fluorescence. Dilute the pigment extract with 90% acetone to reach the sensitivity range of the fluorometer, if required

for pheopigment determination, add 50 μl (one drop) of 0.2 N HCl to 1.5 ml of extract volume and measure fluorescence after 30 s of acidification

calculate chlorophyll$_a$ and pheopigment concentrations from the equations given below

Calculation:

$$\text{chlorophyll}_a \ (\mu g/l) = \frac{\dfrac{k_a}{k_a - 1} \times (k_c) \times (F_o - F_a)}{\text{volume of sediment}}$$

$$\text{pheopigments} \ (\mu g/l) = \frac{\dfrac{k_a}{k_a - 1} \times (k_c) \times [(k_a \times F_a) - F_o]}{\text{volume of sediment}}$$

k_a = acidification factor, computed from F_{Chl}/F_{Chlac}
(F_{Chl} = fluorescence of pure chlorophyll; F_{Chlac} = fluorescence of pure chlorophyll after acidification)
k_c = calibration factor for the sensitivity doors of the fluorometer, if required for the fluorometer used
F_o, F_a = fluorescence before and after acidification

The HPLC Technique.--Lorenzen and Jeffrey (1980) recommended the HPLC technique thin-layer chromatography to be the only method to get precise knowledge about the pigment content and composition of natural samples. Jacobsen (1982) and Trees et al. (1985) suggested that HPLC, compared with standard fluorometric and spectrophotometric methods, should be routinely used for the analysis of chloroplastic pigments in marine samples. For chlorophyll determination in cultures, conventional methods or HPLC gave similar results but for North Sea samples, pigments showing light-absorbing properties like chlorophyll may contribute to 15% of the chlorophyll concentrations causing serious interference (Gieskes and Kraay, 1983).

The use of HPLC eliminates the interferences caused by overlapping absorption and fluorescence bands of the various pigments, because they are separated chromatographically prior to detection. Using mixtures of different solvents, special types of pigment molecules can be isolated and measured. The techniques of separation are sophisticated and rely on the investigator's experience. For detection of all the various pigments and their degradation products the laborious preparation of appropriate standards is needed. This will lead to a restricted analysis of pigment compositions in natural samples, often containing a variety of algal species.

A detailed outline of the standard HPLC method is given by Abaychi and Riley (1979). The more sophisticated reverse-phase HPLC technique is described extensively by Mantoura and Llewellyn (1983). For analysis of sediment pigments see Brown et al. (1981). Bidigare (1985) reported on a shipboard HPLC system which can rapidly and precisely analyze chlorophyll$_{a, b, c}$, and their degradation products from seawater at concentrations as low as 13 pg per injection. The choice between the HPLC technique and conventional methods depends on the questions being addressed. It has to be taken into consideration that HPLC requires expensive equipment and only pays for itself with continuous routine use.

Chitin

Chitin is the major component of the skeleton and body wall of many invertebrate species (e.g., crustaceans, nematodes) and fungi. Although considerable amounts of chitin are produced from planktonic and benthic communities no accumulation of this chemical compound in seawater or sediments has been observed (Goodrich and Morita, 1977a,b). This leads to the suggestion that chitin must be decomposed rapidly if released from organisms. Hock (1941) isolated chitin decomposing bacteria from marine sand, mud, and the intestinal contents of several marine animals. These bacteria proved to be highly specialized, most unable to digest cellulose or to hydrolyze starch. Seki and Taga (1965) found chitin degradation rates of 30 mg per day in the presence of about 10^{10} marine chitin decomposing bacteria. Other authors reported about chitinolytic enzyme activity in the digestive tracts of marine fishes suggesting their important role in the decomposition and recycling of chitin (Fänge et al., 1976, 1979; Goodrich and Morita, 1977a; Danulat, 1986).

The chitinolytic activity is carried out by chitinase, chitobiase, and to a lesser extent by lysozyme being an abundant microbial exoenzyme. Goodrich and Morita (1977b) assumed from their studies on marine fish that the degradation of chitin in the benthos may occur mainly from the chitinoclasts present in the digestive tracts of benthic invertebrates.

As chitin does not seem to accumulate in marine sediments, e.g. Goodrich and Morita (1977) for Yaquina Bay (Newport, Oregon), chitin values must reflect the presence of this compound in living organisms. It is suggested therefore, that chitin is a suitable biomass parameter for chitin bearing invertebrates and fungi.

The Spectrophotometrical Determination of Chitin.

The method for chitin determination has been obtained from Strickland and Parsons (1972) and is based on the colorimetric method of Morgan and Elson (1934).

Reagents: 95% ethanol, 6 N hydrochloric acid, sodium hydroxide solution. **Procedure:** dissolve 250 g of sodium hydroxide in 1 l of distilled water.

Reagent: phenolphthalein indicator. **Procedure:** prepare a 1% (w/v) solution in 95% ethanol.

Reagent: Ehrlich's reagent. **Procedure:** dissolve 0.8 g of p-dimethyl-aminobenzaldehyde in 30 ml of 95% ethanol and 30 ml of concentrated hydrochloric acid. The solution remains stable refrigerated for about one week.

Reagent: acetylacetone reagent. **Procedure:** add 1 ml of pure acetylacetone (2, 4- pentanedione) to 50 ml of a sodium carbonate solution (27 g in 1 l of distilled water). Use within 30 minutes of preparation!

Standard chitin solutions: dissolve 0.3 g of D(+)glucosamine hydrochloride in 100 ml of distilled water. This stock solution contains 3000 μg chitin per 100 ml. Take 5.0 ml of this solution and dilute to exactly 100 ml with 6 N hydrochloric acid. Prepare immediately before use, 1 ml = 150 μg chitin.

Calculation: to obtain mg chitin per dry weight or volume of sediment the measured absorbance has to be corrected by a reagent blank and by a factor F calculated as:

$$F = 300 / E_a - E_b$$

(E_a = extinction of the standard, 2 ml prepared like the sample; E_b = extinction of the reagent blank, 2 ml HCL prepared like the sample. The value of F should be approximately 275 and should be determined each time the period of hydrolysis is varied appreciably.

$$\mu g \text{ chitin/unit of sed.} = \frac{E(\text{sample}) - E_b}{\text{vol. sed. (dwt)}} \times F$$

Note that the result for chitin obtained from the procedure given below will be too great in the presence of high amounts of protein and carbonates. As both react with the reagents affecting the color development, the chitin values should be interpreted only as rough estimates.

The following diagram shows the procedure in detail:

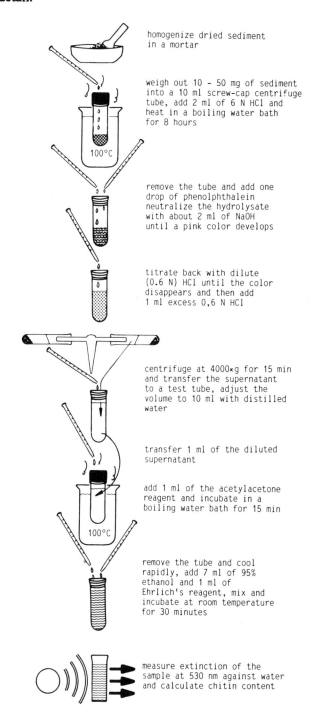

Muramic Acid

Muramic acid is a component of a class of

macromolecules (peptidoglycans) that is of universal occurrence in prokaryotic cell wall material. Muramic acid is a lactyl ether of glucosamine and forms glycan strands alternating with N-acetylglucosamine. The carboxyl group of the lactyl moiety bears a chain of amino acids which may be cross-linked to one another forming a bag-shaped molecular mesh in combination with the glycan strands. This macromolecule of enormous dimensions completely encloses and restrains the protoplast, counterbalancing its turgor pressure (Stanier et al., 1978).

The unique occurrence of muramic acid as an essential compound of the bacterial cell wall leads to its application as a useful biomass parameter (Milar and Casida, 1970; Moriarty, 1975). Classical techniques for microbial biomass determination such as plate counting or direct counting techniques are laborious and give rise to erroneous estimates as compared to each other (Hayes and Anthony, 1959; Jannasch and Jones, 1959). These difficulties increase in trying to quantify the microflora attached to organic detritus and sediment particles (King and White, 1977). Milar and Casida (1970) found muramic acid values at levels 100 to 1,000-fold greater than could be accounted for the bacterial plate count biomass from various soil samples. Estimating bacterial biomass, Milar and Casida (1970) calculated an average muramic acid content of 6.4 +3.3 µg/mg dry weight for these organisms. Moriarty (1977) pointed out, if most of the bacteria are Gram negative or weak Gram positive (with low muramic acid content), a combined value of about 12 µg MA/mg C for these organisms would be sufficiently accurate for estimating biomass. Thus the formula (Moriarty, 1975) should be modified for this environment to C = 1000 MA/ (12 n + 40 p), where n is the proportion of Gram negative or weakly Gram positive bacteria and p is the proportion of strongly Gram positive bacteria.

Although the muramic acid content of the bacterial cell wall may vary by a factor of two (Ellwood and Tempest, 1972) this parameter enables further differentiation of these microbial assemblages in natural samples. King and White (1977) found in samples of estuarine muds and oak litter that the muramic acid content detected by the colorimetric assay, primarily reflects the bacterial component. Corresponding values obtained from enzymatic muramic acid assays, gave values 10- to 20-fold higher.

Hadzija (1974) developed a simple spectrophotometrical method for muramic acid determination which shows high reproducibility and was insensitive to possible interfering substances and varying conditions of the reaction. King and White (1977) adopted this method for marine sediment samples. For more sophisticated techniques involving thin-layer chromatography and gas-liquid chromatography (GLC, HPLC) see Fazio et al. (1979), Findlay et al. (1983).

The Spectrophotometrically Determination of Muramic Acid.-- The principle of the method based on the literature cited above involves the degradation of muramic acid to lactic acid, followed by liberation of acetaldehyde which can be detected photometrically from the reaction with p-hydroxydiphenyl (PHD).

Reagent: 1 N NaOH, concentrated sulfuric acid, copper sulfate solution. **Procedure:** dissolve 4 g of $CuSO_4 \cdot 5\, H_2O$ in 100 ml of distilled water to get a 4% (w/v) solution.

Reagent: PHD solution. **Procedure:** dissolve p-hydroxydiphenyl in 96% ethanol to get a 1.5% (w/v) solution.

Reagent: Standard solution. **Procedure:** prepare 1 mg/ml solution of muramic acid (Sigma Chemical Company) in water. Dilute to get final concentrations of 5-20 µg.

The following diagram shows the procedure in detail:

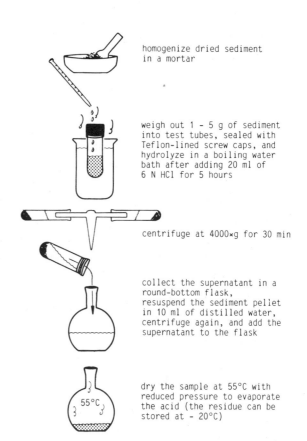

homogenize dried sediment in a mortar

weigh out 1 - 5 g of sediment into test tubes, sealed with Teflon-lined screw caps, and hydrolyze in a boiling water bath after adding 20 ml of 6 N HCl for 5 hours

centrifuge at 4000×g for 30 min

collect the supernatant in a round-bottom flask, resuspend the sediment pellet in 10 ml of distilled water, centrifuge again, and add the supernatant to the flask

dry the sample at 55°C with reduced pressure to evaporate the acid (the residue can be stored at - 20°C)

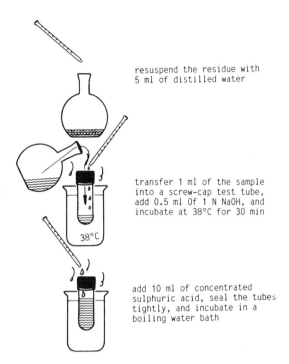
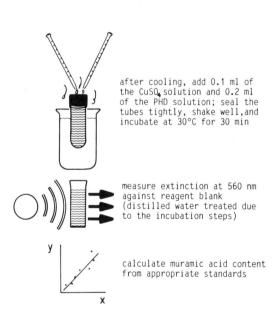

Methods to Determine Special Properties of Organic Matter

Enzymes

The desire to quantify the rates of metabolism, feeding, and growth within the benthic system has given rise to increased applications of biochemical techniques. The detection of exoenzymes involved in the degradation of organic matter and the quantification of their activity is necessary for estimating the breakdown and turnover rates of natural and man-made organic substances. In the decomposition of macromolecules as proteins, ribonucleic acids, cellulose, chitin and lignin exoenzymes are involved, which are secreted from living cells (mainly microorganisms) as well as liberated during the lysis of cells (algae, microorganisms) hydrolyzing the macromolecules to soluble endproducts. The extracellular digestion requires close contact with the substrates and therefore takes place mainly on particles in the water column or aerobic sediment layers. The processes like adsorption, decomposition, and denaturation may greatly influence the efficiency of the enzymes released. Pamatmat and Skjoldal (1974) investigated the microbial activity due to increasing depth of marine sediment layers, finding a corresponding decrease of dehydrogenase activity, comprising a collection of enzymes involved in catabolism. These findings are consistent to the suggestion that high microbial activity in marine sediments is mainly associated with settled detrital material.

In recent years an increased emphasis on activity measurements of specific enzymes has taken place providing a more promising link to processes of interest, e.g., relationships between physiological rates and compounds like ATP, RNA, and DNA of living organisms. The activities of $NADH_2$-dependent dehydrogenase, for instance, are useful indicators of respiratory potential in animals (King and Packard, 1975). The activities must represent a measure of the electron transport system (ETS) of the organism proportional to actual rates of oxygen consumption. Similarly, the digestive enzymes may provide information on the feeding pattern and nutrition of collected animals. The measurement of the activity of digestive enzymes allows acquisition of information on short-term variations or rhythms and on the potential rate at which an animal can digest a given substrate (Boucher and Samain, 1974; Mayzaud and Conover, 1976; Cox, 1981; Faubel, 1984).

Because of the permanent input of organic particles to the sediment surface, information on the degradation potential by diverse decomposers will lead to a better knowledge on the energy flow within benthic ecosystems. To a certain extent, organisms other than bacteria are involved in this degradation of particulate organic matter as a food source (Tenore et al., 1978; Gerlach, 1978; Riemann and Schrage, 1978). Generally, it is known that in lower metazoa the mode of digestion is an intra- and/or extracellular process which takes place in the digestive tract. Recent studies on chitinolytic enzymes in the digestive tracts of marine fishes revealed their important role in the decomposition of (crustacean) chitin (Fänge et al., 1979).

A further field of research is decomposition studies on man-made poisonous products such as pesticides, herbicides, and chemical by-products that accumulate in the biosphere due to environmental pollution. Dagley (1975) gives an extended review on enzymatic mechanisms for biochemical breakdown of these substances and on investigatory approaches on corresponding decomposition studies.

The analysis of enzyme activity from natural samples can be carried out by conventional photometric or fluorometric techniques. The substrates used for enzymatic breakdown bear chromogenic or fluorogenic compounds that are liberated as decomposition proceeds. Chromogenic compounds are p-nitrophenol, phenolphthalein 2,7-di-hydroxynaphthalene, and fluorogenic ones are 3-o-methylfluorescein, β-naphthylamide, 4-methyl-umbelliferyl substrates. Due to the enzyme activity an increase in absorbance or fluorescence can be measured. The special method relies on the type of enzyme investigated (e.g. amylase, pectinase, cellulase, lysozyme, chitinase, peptidase, deoxy/ribonuclease, lipase). For methodological details see Doyle (1968), Petersson and Jansson (1978), Billen et al. (1980), Somville and Billen (1983), and Hoppe (1983). For example the measurement of exoproteolytic activity in natural waters as carried out by Somville and Billen (1983) is described below.

The Measurement of Exoproteolytic Activity.--The method is based on the enzymatic cleavage of aminoacyl-β-naphthylamide as a substrate which gives rise to a fluorescent product (β-naphthylamine) after hydrolysis of the peptide-like bond. The linear increase of fluorescence at 410 nm under excitation at 340 nm is measured over a period of 10-100 minutes and the enzymatic activity is expressed as the amount of β-naphthyl-amine produced per minute of incubation.

The procedure is shown as following:

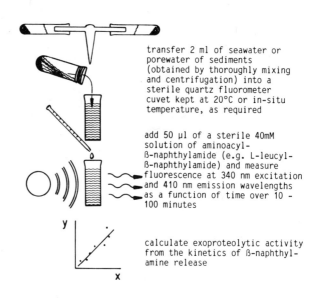

transfer 2 ml of seawater or porewater of sediments (obtained by thoroughly mixing and centrifugation) into a sterile quartz fluorometer cuvet kept at 20°C or in-situ temperature, as required

add 50 µl of a sterile 40mM solution of aminoacyl-ß-naphthylamide (e.g. L-leucyl-ß-naphthylamide) and measure fluorescence at 340 nm excitation and 410 nm emission wavelengths as a function of time over 10 - 100 minutes

calculate exoproteolytic activity from the kinetics of ß-naphthyl-amine release

The Measurement of Digestive Enzyme Activity.--The methods for the measurement of enzymatic activities of digestive protease, amylase, and laminarinase are based on Hassett and Landry (1982) and Head et al. (1984).

Preparation of Organisms.--Individuals have to be collected in adequate numbers from sediment samples and maintained at low temperatures (8-4° C) before processing. The individuals are then preserved by quick-freezing in liquid nitrogen followed by freeze drying. The samples are stored in small vials in desiccant at -40° C. Under these conditions, storage times as long as a year result in no loss of enzyme activity. Samples to be analyzed are brought to room temperature in their vials and then weighed on a micro-balance, if required.

Determination of Soluble Protein.--Before measurement of digestive enzyme activity, the samples are transferred to a micro-tissue grinder along with 10 mM Tris HCl buffer, pH 7.5 (1 ml), and homogenized and kept on ice. The samples are then centrifuged (12.800 g x 5 minutes). The supernatants are decanted off and subsampled for simultaneous measurements of soluble protein and digestive enzyme levels. Soluble protein is measured using the method of Bradford (see section on measuring dead and living organic matter).

Digestive Protease Assay.

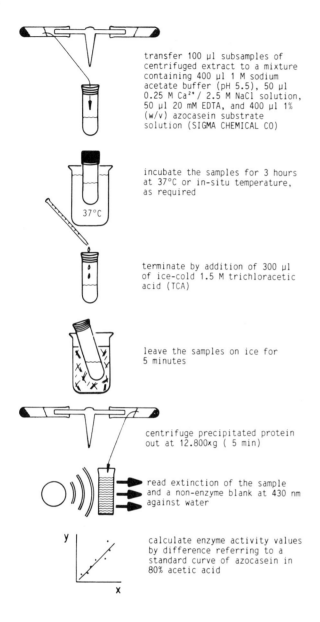

Amylase Assay.

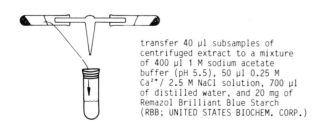

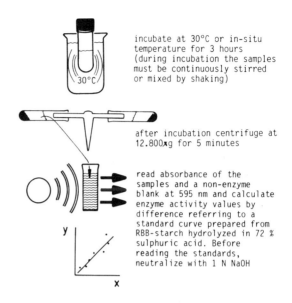

Calculation.—The enzyme activities should be expressed in units of ug substrate h^{-1} or mg soluble protein h^{-1}; units presented should be based on mean values of at least three replicated determinations.

Laminarinase Assay, Step 1:

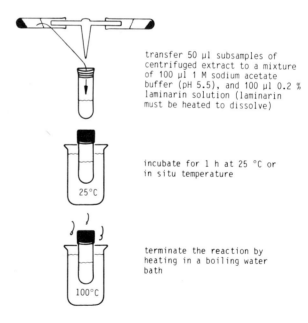

The glucose produced in the first step is then measured using a direct fluorometric assay (Lowry and Passonneau, 1972). This assay which is suitable for glucose, in concentrations from 0.1 to 10 µM, utilizes the fluorescence of NADPH in the reactions:

glucose + ATP $\xrightarrow{\text{hexokinase}}$ glucose-6-P + ADP

glucose-6-P + NADP$^+$ $\xrightarrow{\text{glucose-6-P dehydrogenase}}$ 6-P-gluconolactone + NADPH + H$^+$

Laminarinase Assay, Step 2:

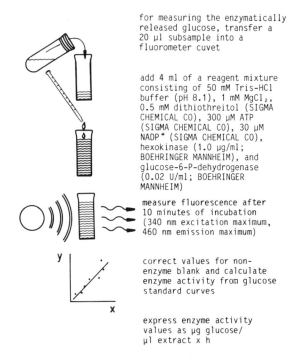

for measuring the enzymatically released glucose, transfer a 20 µl subsample into a fluorometer cuvet

add 4 ml of a reagent mixture consisting of 50 mM Tris-HCl buffer (pH 8.1), 1 mM MgCl$_2$, 0.5 mM dithiothreitol (SIGMA CHEMICAL CO), 300 µM ATP (SIGMA CHEMICAL CO), 30 µM NADP$^+$ (SIGMA CHEMICAL CO), hexokinase (1.0 µg/ml; BOEHRINGER MANNHEIM), and glucose-6-P-dehydrogenase (0.02 U/ml; BOEHRINGER MANNHEIM)

measure fluorescence after 10 minutes of incubation (340 nm excitation maximum, 460 nm emission maximum)

correct values for non-enzyme blank and calculate enzyme activity from glucose standard curves

express enzyme activity values as µg glucose/ µl extract x h

Adenylate Energy Charge

The adenylate energy charge (EC) represents the energy available for metabolism in a living cell (Atkinson, 1971). The EC is calculated from the amount of adenylate nucleotides (ATP, ADP, AMP) by the equation (ATP + 1/2 ADP) / (ATP + ADP + AMP). The formula has been derived from Atkinson and Walton (1967) and Bomsel and Pradet (1968) quantitating the amount of energy yielding P-bonds (two in ATP and one in ADP) in relation to the amount of total adenylates. Mainly by the speed of the adenylate kinase (myokinase) reaction, different cellular adenylate pools are maintained to the same energy level (Bomsel and Pradet, 1968). Values of the EC will be nearly 1 if most of the adenylates are supplied with energy rich bonds and beyond 0.5 if energy consumption exceeds storage from catabolic reactions. EC values therefore indicate the level of conditions (optimal, limiting, severe) for single organisms, populations, or communities as a whole investigated in culture or natural samples. An extended list of EC values in relation to different environmental factors is given by Ivanovici (1980).

Most measurements have been carried out on microorganisms, for reference see Chapman et al. (1971), Ball and Atkinson (1975), and Karl and Holm-Hansen (1978), whereas studies on marine invertebrates are limited to only some species, e.g., crustaceans and mollusks (Skjoldal and Båmstedt, 1977; Ivanovici, 1979; Giesy et al., 1981; Skjoldal et al., 1982). EC values from natural samples gave often ambiguous results. Amblard (1981) found increasing EC values as environmental conditions changed unfavorable for phytoplankton populations. Graf et al. (1982) explained a seasonally unexpected EC decrease by sudden metabolic shifts within the benthic community. Skjoldal and Båmstedt (1976) found the EC of a deep water planktonic community to remain constant at different growth rates which contradicts findings from culture experiments. Souza-Lima et al. (1983) determined low EC values from the nutrient rich surface microlayer of marine samples indicating environmental stress (high light intensities) or accumulation of non-living organic detritus.

As described in the section on methods for measuring dead and living organic matter, environmental stress may lead to severe changes in the ATP pool of living organisms. This requires careful sampling to prevent transitions in temperature, salinity, oxygen supply, and, mostly inevitable for deep sea samples, transitions in pressure. On the other hand varying shares of dead and living organic matter in natural samples will cause misinterpretations of EC values. As outlined by Ivanovici (1980) EC values should therefore be measured in combination with other parameters giving independent estimates of biomass and metabolic activity of organisms. Its useful application on environmental stress or pollution studies has been proved by many investigators.

For further methodological aspects see "Methods for Measuring Dead and Living Matter."

References

Abaychi, J.K., and J.P. Riley
1979. The Determination of Phytoplankton Pigments by High-Performance Liquid Chromatography. *Analytica Chimica Acta*, 107:1-11.

Amblard, C.

1981. Interets du dosage des adénosines 5'-phosphate pour l'étude de la dynamique des populations phytoplanctoniques lacustres (le Parvin-France). *Hydrobiologica*, 85:257-270.

Andrassy, J.
1956. Die Rauminhalts- und Gewichtsbestimmung der Fadenwürmer (Nematoden). *Acta Zoologica Budapest*, 2:1-15.

Andrews, P., and P.J. LeB. Williams
1971. Heterotrophic Utilization of Dissolved Organic Compounds in the Sea. III. Measurement of the Oxidation Rates and Concentrations of Glucose and Amino Acids in Sea Water. *Journal of the Marine Biological Association of the United Kingdom*, 51:111-125.

Atkinson, D.E.
1968. The Energy Charge of the Adenylate Pool as a Regulatory Parameter. Interaction with Feedback Modifiers. *Biochemistry*, 7:4030-4034.
1971. Adenine Nucleotides as Universal Stoichiometric Metabolic Coupling Agents. *Advances in Enzyme Regulation*, 9:207-219.

Atkinson, D.E., and M. Walton
1967. Adenosine Triphosphate Conservation in Metabolic Regulation - Rat Liver Cleavage Enzyme. *Journal of Biological Chemistry*, 242:3239-3242.

Bagnara, A.S., and L.R. Finch
1972. Quantitative Extraction and Estimation of Intracellular Nucleoside Triphosphates of *Escherichia coli*. *Analytical Biochemistry*, 45:24-34.

Ball, W.J., and D.E. Atkinson
1975. Adenylate Energy Charge in *Saccharomyces cerevisiae* during Starvation. *Journal of Bacteriology*, 121:975-982.

Båmstedt, U., and H.R. Skjoldal
1989. RNA Concentration of Zooplankton: Relationship with Size and Growth. *Limnology and Oceanography*, 25:304-316.

Bidigare, R.R.
1985. Rapid Determination of Chlorophylls and Their Degradation Products by High-performance Liquid Chromatography. *Limnology and Oceanography*, 30:432-435.

Billen, G., C. Joiris, J. Wijnant, and G. Gillain
1980. Concentration and Microbial Utilization of Small Organic Molecules in the Scheldt Estuary, the Belgian Coastal Zone of the North Sea and the English Channel. *Estuarine and Coastal Marine Science*, 11:279-294.

Bio Rad Protein Assay
1979. *Bulletin*, 1069 EG.

Bligh, E.G., and W.J. Dyer
1959. A Rapid Method of Total Lipid Extraction and Purification. *Canadian Journal of Biochemistry and Physiology*, 37:911-917.

Bomsel, J.-L., and A. Pradet
1968. Study of Adenosine 5'-mono-, di-, and triphosphate in Plant Tissues. IV. Regulation of the Level of Nucleotides, in vivo, by Adenylate Kinase; Theoretical and Experimental Study. *Biochimica et Biophysica Acta*, 162:230-242.

Boucher, J., and J.F. Samain
1974. L'activité amylasique indice de la nutrition du zooplancton, mise en evidence d'un rhytme quotidien en zone d'upwelling. *Téthys*, 6:179-188.

Bradford, M.M.
1976. A Rapid and Sensitive Method for the Quantitation of Microgram Quantities of Protein Utilizing the Principle of Protein-dye Binding. *Analytical Biochemistry*, 72:248-254.

Brown, L.M., B.T. Hargrave, and M.D. MacKinnon
1981. Analysis of Chlorophyll in Sediments by High-pressure Liquid Chromatography. *Canadian Journal of Fisheries and Aquatic Science*, 38:205-214.

Brunschede, H., T.L. Dove, and H. Bremer
1977. Establishment of Exponential Growth after a Nutritional Shift-up in *Escherichia coli* B/r: Accumulation of Deoxyribonucleic Acid, Ribonucleic Acid, and Protein. *Journal of Bacteriology*, 129:1020-1033.

Bulleid, N.C.
1978. An Improved Method for the Extraction of Adenosine Triphosphate from Marine Sediment and Seawater. *Limnology and Oceanography*, 23:174-178.

Burney, C.M., and J. McN. Sieburth
1977. Dissolved Carbohydrates in Seawater. II. A Spectrophotometric Procedure for Total Carbohydrate Analysis and Polysaccharide Estimation. *Marine Chemistry*, 5:15-28.

Burney, C.M., K.M. Johnson, and J. McN. Sieburth
1981. Diel Flux of Dissolved Carbohydrate in a Salt Marsh and a Simulated Estuarine Ecosystem. *Marine Biology*, 63:175-187.

Byers, S.C., E.L. Mills, and P.L. Stewart
1978. A Comparison of Methods of Determining Organic Carbon in Marine Sediments, with Suggestions for a Standard Method. *Hydrobiologica*, 58:43-47.

Cacciari, I., D. Lippi, S. Ippoliti, and W. Pietrosant
1983. Respiratory Activity, Molar Growth Yields, and ATP Content in an Arthrobacter sp. Ammonium-limited Chemostat Culture. *Canadian Journal of Microbiology*, 29:1136-1140.

Chapman, A.G., L. Fall, and D.E. Atkinson
1971. Adenylate Energy Charge in *Escherichia coli* during Growth and Starvation. *Journal of Bacteriology*, 108:1072-1086.

Chapman, A.G., and D.E. Atkinson
1973. Stabilization of Adenylate Energy Charge by the Adenylate Deamonase Reaction. *Journal of Biological Chemistry*, 248:8309-8312.

Chou, S.C., and A. Goldstein
1960. Chromogenic Groupings in the Lowry Protein Determination. *Biochemical Journal*, 75:109-115.

Christensen, J.P., and T.T. Packard
1977. Sediment Metabolism from the Northwest African Upwelling System. *Deep-Sea Research*, 24:331-343.

Christian, R.R., K. Bancroft, and W.J. Wiebe
1975. Distribution of Microbial Adenosine Triphosphate in Salt Marsh Sediments of Sapelo Island, Georgia. *Soil Science*, 119:89-97.

Clark, M.E., G.A. Jackson, and W.J. North
1972. Dissolved Free Amino Acids in Southern California Coastal waters. *Limnology and Oceanography*, 17:749-758.

Coveney, M.F.
1982. Elimination of Chlorophyll$_b$ Interference in the Fluorometric Determination of Chlorophyll$_a$ and Phaeopigment$_a$. *Archiv für Hydrobiologie*, 16:77-90.

Cowey, C.B., and E.D.S. Corner
1963. On the Nutrition and Metabolism of Zooplankton. II. The Relationship between the Marine Copepod *Calanus helgolandicus* and Particulate Material in Plymouth Sea Water, in Terms of Amino Acid Composition. *Journal of Marine Biology*, 43:495-511.

Cox, J.L.
1981. Laminarinase Induction in Marine Zooplankton and Its Variability in Zooplankton Samples. *Journal of Plankton Research*, 3:345-356.

Crane, F.E., and E.A. Smith
1963. A Non-distillation Kjeldahl Method for Nitrogen Based on the Precipitation as Tetraphenylborate. *Analytica Chemica Acta*, 31:258-267.

Dagley, S.
1975. A Biochemical Approach to Some Problems of Environmental Pollution. *Essays in Biochemistry*, 11:81-138.

Dahlbäck, B., L.A.H. Gunnarsson, M. Hermansson, and S. Kjelleberg
1982. Microbial Investigations of Surface Microlayers, Water Column, Ice and Sediment in the Arctic Ocean. *Marine Ecology*, 9:101-109.

Danulat, E.
1986. Role of Bacteria with Regard to Chitin Degradation in the Digestive Tract of the Cod *Gadus morhua*. *Marine Biology*, 90:335-344.

Dawson, R., and R.G. Pritchard
1978. The Determination of α-amino Acids in Seawater Using a Fluorimeter Analyser. *Marine Chemistry*, 6:27-40.

Degens, E.T.
1970. Molecular Nature of Nitrogenous Compounds in Sea Water and Recent Marine Sediments. Pages 77-106 in D.W. Hood, editor, *Organic Matter in Natural Waters*. Institute of Marine Science, Alaska.

Degens, E.T., J.H. Reuter, and K.N.F. Shaw
1964. Biochemical Compounds in Offshore California Sediments and Sea Waters. *Geochimica et Cosmochimica Acta*, 28:45-66.

De La Cruz, A.A., and W.E. Poe
1970. Amino Acids in Salt Marsh Detritus. *Limnology and Oceanography*, 15:124-127.

Dietzler, D.N., C.J. Lais, and M.P. Leckie
1974. Simultaneous Increases of the Adenylate Energy Charge and the Rate of Glycogen Synthesis in Nitrogen-starved *Escherichia coli* W 4597 (K). *Archives of Biochemistry and Physics*, 160:14-25.

Dortch, Q., J.R. Clayton, S.S. Thoresen, and S.I. Ahmed
1984. Species Differences in Accumulation of Nitrogen Pools in Phytoplankton. *Marine Biology*, 81:237-250.

Dortch, Q., J.R. Clayton, Jr., S.S. Thoresen, J.S. Cleveland, S.L. Bressler, and S.I. Ahmed
1985. Nitrogen Storage and Use of Biochemical Indices to Assess Nitrogen Deficiency and Growth Rate in Natural Plankton Populations. *Journal of Marine Research*, 43:437-464.

Dortch, Q., T.L. Roberts, J.R. Clayton, Jr., and S.I. Ahmend
1983. RNA/DNA Ratios and DNA Concentrations as Indicators of Growth Rate and Biomass in Planktonic Marine Organisms. *Marine Ecology*, 13:61-71.

Doyle, R.W.
1968. Isolation of Enzymes and Other Organic Anions from Freshwater Sediment. *Limnology and Oceanography*, 13:518-522.

Dubois, M., K.A. Gilles, J.K. Hamilton, P.A. Rebers, and F. Smith
1956. Colorimetric Method for Determination of Sugars and Related Substances. *Analytical Chemistry*, 28:350-356.

Ellwood, D.C., and D.W. Tempest
1972. Effects of Environment on Bacterial Wall Content and Composition. *Advances in Microbial Physiology*, 7:83-117.

Fänge, R., G. Lundblad, and J. Lind
1976. Lysozyme and Chitinase in Blood and Lymphomyeloid Tissues of Marine Fish. *Marine Biology*, 36:277-282.

Fänge, R., G. Lundlad, J. Lind, and K. Slettengren
1979. Chitinolytic Enzymes in the Digestive System of Marine Fishes. *Marine Biology*, 53:317-321.

Faubel, A.
1982. Determination of Individual Meiofauna Dry Weight Values in Relation to Definite Size Classes. *Cahiers de Biologie Marine*, 23:339-345.
1984. On the Abundance and Activity Pattern of Zoobenthos Inhabiting a Tropical Reef Area, Cebu, Philippines. *Coral Reefs*, 3:205-213.

Faubel, A., E. Hartwig, and H. Thiel
1983. On the Ecology of the Benthos of Sublitoral Sediments, Fladen-Ground, North Sea. I. Meiofauna Standing Stock and Estimation of Production. "Meteor" *Forschungsergebnisse*, 36:35-48.

Faubel, A., and L.-A. Meyer-Reil
1983. Measurement of Enzymatic Activity of Meiobenthic Organisms: Methodology and Ecological Application. *Cahiers de Biologie Marine*, 24:35-49.

Fazekas de St. Groth, S., R.S. Webster, and A. Datyner
1963. Quantitative Staining for Zone Elektrophoresis. *Biochimica et Biophysica Acta*, 71:377-391.

Fazio St. D., W.R. Mayberry, and D.C. White
1979. Muramic Assay in Sediments. *Applied Environmental Microbiology*, 38:349-350.

Findley, R.H., D.J.W. Moriarty, and D.C. White
1983. Improved Method of Determining Muramic Acid from Environmental Samples. *Geomicrobiology*, 3:135-150.

Frauenheim, K.
1984. *Gesamtprotein im Sediment als Biomasseindikator - Untersuchungen im Auftriebsgebiet vor Nordwestafrika*. 85 pages. Hamburg: Diplomarbeit Institut für Hydrobiologie und Fischereiwissenschaft, Universität Hamburg.

Freeman, N.K., F.T. Lindgren, Y.C. Ng, and A.V. Nichols
1957. Serum Lipide Analysis by Chromatography and Infrared Spectrophotometry. *Journal of Biological Chemistry*, 227:449-464.

Gaudette, H.E., W.R. Flight, L. Tremer, and D.W. Folger
1974. An Inexpensive Filtration Method for the Determination of Organic Carbon in Recent Sediments. *Journal of Sedimentary Petrology*, 44:249-253.

Gerlach, S.A.
1978. Food Chain Relationships in Subtidal Silty Sand, Marine Sediments and the Role of Meiofauna in Stimulating Bacterial Productivity. *Oecologia (Berlin)*, 33:55-69.

Giese, A.C.
1966. Lipids in the Economy of Marine Invertebrates. *Physiological Reviews*, 46:244-298.

Gieskes, W.W., and G.W. Kraay
1982. Comparison of Chromatographic (HPLC and TLC) with Conventional Methods for the Measurement of Chlorophylls in Oceanic Waters. *Archiv für Hydrobiologie*, 16:123.
1983. Unknown Chlorophyll $_a$ Derivates in the North Sea and the Tropical Atlantic Ocean Revealed by HPLC Analysis. *Limnology and Oceanography*, 28:757-766.

Giesy, J.P., S.R. Denzer, C.S. Duke, and G.W. Dickson
1981. Phosphoadenylate Concentrations and Energy Charge in Two Freshwater Crustaceans: Response to Physical and Chemical Stressors. *Verhandlungen der Internationalen Vereinigung für Theroetische und Angewandte Limnologie*, 21:205-220.

Goodrich, T.D., and R.Y. Morita
1977a. Incidence and Estimation of Chitinase Activity Associated with Marine Fish and Other Estuarine Samples. *Marine Biology*, 41:349-353.
1977b. Bacterial Chitinase in the Stomachs of Marine Fishes from Yaquina Bay, Oregon, USA. *Marine Biology*, 41:355-360.

Gordon, D.C., Jr.
1970. A Microscopic Study of Organic Particles in the North Atlantic Ocean. *Deep-Sea Research*, 17:175-185.

Gowen, R.J., P. Tett, and B.J.B. Wood
1982. The Problem of Degradation Products in the Estimation of Chlorophyll by Fluorescence. *Archiv für Hydrobiologie*, 16:101-106.

Graf, G., W. Bengtsson, A. Faubel, L.-A. Meyer-Reil, R. Schulz, H. Theede, and H. Thiel
1984. The Importance of the Spring Phytoplankton Bloom for the Benthic System of Kiel Bight. *Rapports et Proces-Verbaux des Réunions Conseil International pour la Exploration de la Mer*, 183:136-143.

Grasshoff, K., M. Ehrhardt, and K. Kremling
1983. *Methods of Seawater Analysis*. Second Edition. Weinheim, Dearfield Beach, Florida, Basel: Verlag Chemie.

Greiser, N.
1982. *Methodische Untersuchungen zur quantitativen Bestimmung von Adenosintriphosphat und Energy Charge in Prokaryonten und Eukaryonten*. 121 pages. Hamburg: Diplomarbeit Institute für Hydrobiologie und Fischereiwissenschaft, Universität Hamburg.

Hack, M.H., A.E. Gussin, and M.E. Lowe
1962a. Comparative Lipid Biochemistry. I. Phosphatides of Invertebrates (Porifera to Chordata). *Comparative Biochemistry and Physiology*, 5:217-221.

Hack, M.H., R.G. Yeager, and T.D. McCaffery

1962b. Comparative Lipid Biochemistry. II. Lipids of Plant and Animal Flagellates, a Non-motile Alga, an Amoeba and a Ciliate. *Comparative Biochemistry and Physiology,* 6:247-252.

Hadzija, O.
1974. A Simple Method for the Quantitative Determination of Muramic Acid. *Analytical Biochemistry,* 60:512-517.

Hall, K.J., W.C. Weimer, and G. Fred Lee
1970. Amino Acids in an Estuarine Environment. *Limnology and Oceanography,* 15:162-164.

Hanson, R.B.
1982. Organic Nitrogen and Caloric Content of Detritus. II. Microbial Biomass and Activity. *Estuarine and Coastal Shelf Science,* 14:325-336.

Hassett, R.P., and M.R. Landry
1982. Digestive Carbohydrase Activities in Individual Marine Copepods. *Marine Biology Letters,* 3:211-221.

Hayes, F.R., and E.H. Anthony
1959. VI. The Standing Crop of Bacteria in Lake Sediments and its Place in the Classification of Lakes. *Limnology and Oceanography,* 4:299-315.

Head, E.J.H., R. Wang, and R.J. Conover
1984. Comparison of Diurnal Feeding Rhythms in *Temora longicornis* and *Centropages hamatus* with Digestive Enzyme Activity. *Journal of Plankton Research,* 6:543-551.

Hewitt, B.R.
1958. Spectrophotometric Determination of Total Carbohydrate. *Nature,* 182:246-247.

Hirota, J., and J.P. Szyper
1975. Separation of Total Particulate Carbon into Inorganic and Organic Components. *Limnology and Oceanography,* 20:896-899.

Hock, C.W.
1941. Marine Chitin-decomposing Bacteria. *Marine Research,* 2:99-106.

Holme, N.A., and A.D. McIntyre
1984. *Methods for the Study of Marine Benthos.* IBP Handbook 16. 387 pages. Oxford: Blackwell Scientific Publications.

Holm-Hansen, O., and C.R. Booth
1966. The Measurement of Adenosine Triphosphate in the Ocean and Its Ecological Significance. *Limnology and Oceanography,* 11:510-519.

Holm-Hansen, O., W.H. Sutcliffe, Jr., and J. Sharp
1968. Measurement of Deoxyribonucleic Acid in the Ocean and Its Ecological Significance. *Limnology and Oceanography,* 13:507-514.

Hoppe, H.-G.
1983. Significance of Exoenzymatic Activities in the Ecology of Brackish Water: Measurements by Means of Methylumbelliferyl-substrates. *Marine Ecology,* 11:299-308.

Hunter, B.L., and E.A. Laws
1981. ATP and Chlorophyll$_a$ as Estimators of Phytoplankton Carbon Biomass. *Limnology and Oceanography,* 26:944-956.

Ittekot, V., U. Brockmann, W. Michaelis, and E.T. Degens
1981. Dissolved Free and Combined Carbohydrates During a Phytoplankton Bloom in the Northern North Sea. *Marine Ecology,* 4:299-305.

Ivanovici, A.M.
1979. The Adenylate Energy Charge in the Estuarine Mollusc, *Pyrazus ebenius.* Laboratory Studies of Responses to Salinity and Temperature. *Comparative Biochemistry and Physiology,* 66A:43-55.
1980. Adenylate Energy Charge: An Evaluation of Applicability to Assessment of Pollution Effects and Directions for Future Research. *Rapports et Proces-Verbaux des Réunions Conseil International pour La Exploration de la Mer,* 179:23-28.

Jacobsen, T.R.
1982. Comparison of Chlorophyll$_a$ Measurements by Fluorometric, Spectrophotometric and High Pressure Liquid Chromatographic Methods in Aquatic Environments. *Archiv für Hydrobiologie,* 16:35-45.

Jannasch, H.W., and G.E. Jones
1959. Bacterial Populations in Sea Water as Determined by Different Methods of Enumeration. *Limnology and Oceanography,* 4:128-139.

Jeffries, H.P.
1969. Seasonal Composition of Temperate Plankton Communities: Free Amino Acids. *Limnology and Oceanography,* 14:41-52.

Jeffrey, S.W., and G.F. Humphrey
1975. New Spectrophotometric Equations for Determining Chlorophylls a, b, c, c_2 in Higher Plants, Algae and Natural Phytoplankton. *Biochemie und Physiologie der Pflanzen,* 167:191-194.

Johnson, K.M., and J. McN. Sieburth
1977. Dissolved Carbohydrates in Seawater. I. A Precise Spectrophotometric Analysis for Monosaccharides. *Marine Chemistry,* 5:1-13.

Johnson, K.M., C.M. Burney, and J. McN. Sieburth
1981. Doubling the Production and Precision of the MBTH Spectrophotometric Assay for Dissolved Carbohydrates in Seawater. *Marine Chemistry,* 10:467-473.

Kalbhen, D.A., and H.J. Koch
1967. Methodische Untersuchungen zur quantitativen Mikrobestimmung von ATP in biologischem Material mit dem Firefly - Enzymsystem. *Zeitschrift für Klinische Chemie und Klinische Biochemie,* 5:299-304.

Karl, D.M., and O. Holm-Hansen
1978. Methodology and Measurement of Adenylate Energy Charge Ratios in Environmental Samples. *Marine Biology,* 48:185-197.

King, F.D., and T.T. Packard
1975. Respiration and the Activity of the Respiratory Electron Transport System in Marine Zooplankton. *Limnology and Oceanography,* 20:849-854.

King, J.D., and D.C. White
1977. Muramic Acid as a Measure of Microbial Biomass in Estuarine and Marine Samples. *Applied Environmental Microbiology,* 33:777-783.

Kirk, P.L.
1950. Kjeldahl Method for Total Nitrogen. *Analytical Chemistry,* 22:354-358.

Klinken, J., and H.R. Skjoldal
1983. Improvements of Luciferin-luciferase Methodology for Determination of Adenylate Energy Charge Ratio of Marine Samples. *Marine Ecology,* 13:305-309.

Kormondy, E.J.
1968. Weight Loss of Cellulose and Aquatic Macrophytes in a Carolina Bay. *Limnology and Oceanography,* 13:522-526.

Kuno, H., and H.K. Kihara
1967. Simple Microassay of Protein with Membrane Filter. *Nature,* 215:974-975.

Lang, C.A.
1958. Simple Microdetermination of Kjeldahl Nitrogen in Biological Materials. *Analytical Chemistry,* 30:1692-1694.

Lawrence, A.L., J.M. Lawrence, and A.C. Giese
1965. Cyclic Variations in the Digestive Gland and Glandular Oviduct of Chitons (Mollusca). *Science,* 147:508-510.

Lehman, P.W.
1981. Comparison of chlorophyll$_a$ and Carotenoid Pigments as Predictors of Phytoplankton Biomass. *Marine Biology,* 65:237-244.

Leick, V.
1968. Ratios Between DNA, RNA and Protein in Different Microorganisms as a Function of Maximal Growth Rate. *Nature,* 217:1153-1155.

Leighton, G.
1981. Biomass Relationships between Phytoplankton, Zooplankton and Heterotrophic Aerobic Bacteria in a Reservoir. *Verhandlungen der Internationalen Vereinigung für Theoretische und Angewandte Limnologie,* 21:962-966.

LePecq, J.B., and C. Paoletti
1966. A New Fluorometric Method for RNA and DNA Determination. *Analytical Biochemistry*, 17:100-117.

Lindroth, P., and K. Mopper
1979. High Performance Liquid Chromatographic Determination of Subpicomole Amounts of Amino Acids by Precolumn Fluorescence Derivatization with O-phthaldialdehyde. *Analytical Chemistry*, 51:1667-1674.

Liu, D., P.T.S. Wong, and B.J. Dutka
1973. Determination of Carbohydrate in Lake Sediment by a Modified Phenol-sulfuric Acid Method. *Water Research*, 7:741-746.

Lo, C.-H., and H. Stelson
1972. Interference by Polysucrose in Protein Determination by the Lowry Method. *Analytical Biochemistry*, 45:331-336.

Lorenzen, C.J., and S.W. Jeffrey
1980. Determination of Chlorophyll in Seawater. *Unesco Technical Papers in Marine Science*, 35:1-20.

Lovegrove, T.
1961. The Effect of Various Factors on Dry Weight Values. *Rapports et Proces-Verbaux des Réunions*, 153:86-91.

Lowry, O.H., and J.V. Passoneau
1972. *A Flexible System of Enzymatic Analysis*. 291 pages. New York: Academic Press.

Lowry, O.H., N.J. Rosebrough, A.L. Farr, and R.J. Randall
1951. Protein Measurement with the Folin Phenol Reagent. *Journal of Biological Chemistry*, 193:265-275.

Mantoura, R.F.C., and C.A. Llewellyn
1983. The Rapid Determination of Algal Chlorophyll and Carotenoid Pigments and Their Breakdown Products in Natural Waters by Reverse-phase High-performance Liquid Chromatography. *Anales Chimie Analytique*, 151:297-314.

Manuels, M.W., and H. Postma
1973. Measurements of ATP and Organic Carbon in Suspended Matter of the Dutch Wadden Sea. *Netherlands Journal of Sea Research*, 8:292-311.

Mann, L.T.
1963. Spectrophotometric Determination of Nitrogen in Total Micro-Kjeldahl Digests: Application of Phenol-hypochlorite Reaction to Microgram Amounts of Ammonia in Total Digest of Biological Material. *Analytical Chemistry*, 35: 2179-2182.

Mayzaud, P., and R.J. Conover
1976. Influence of Potential Food Supply on the Activity of Digestive Enzymes of Neritic Zooplankton. *Proceedings of the 10th European Symposium on Marine Biology*, 2:415-427.

Menzel, D.W., and R.F. Vaccaro
1964. The Measurement of Dissolved Organic and Particulate Carbon in Seawater. *Limnology and Oceanography*, 9:138-142.

Milar, W.N., and L.E. Casida
1970. Evidence for Muramic Acid in Soil. *Canadian Journal of Microbiology*, 16:299-304.

Mopper, K.
1977. Sugars and Uronic Acids in Sediment and Water from the Black Sea and North Sea with Emphasis on Analytical Techniques. *Marine Chemistry*, 5:585-603.

Morgan, W., and L. Elson
1934. A Colorimetric Method for the Determination of N-acetyl-glucosamine and N-acetyl-chondrosamine. *Biochemical Journal*, 28:988-995.

Moriarty, D.J.W.
1975. A Method for Estimating the Biomass of Bacteria in Aquatic Sediments and Its Application to Trophic Studies. *Oecologia (Berlin)*, 20:219-229.
1977. Improved Method Using Muramic Acid to Estimate Biomass of Bacteria in Sediments. *Oecologia (Berlin)*, 26:317-323.

Mukerjee, P.
1956. Use of Ionic Dyes in the Analysis of Ionic Surfactants and Other Ionic Organic Compounds. *Analytical Chemisty*, 28:870-873.

Odum, E.P.
1971. *Fundamentals of Ecology*. Third Edition. 574 pages. Philadelphia: W.B. Saunders Company.

Odum, E.P., and A.A. de la Cruz
1963. Detritus as a Major Component of Ecosystems. *BioScience*, 13:39-40.

Otsuki, A., T. Miyoshi, T. Unno, and H. Seki
1983. Biochemical Constituents of Particulate Matter in a Mesotrophic Irrigation Pond. *Archiv für Hydrobiologie*, 98:1-14.

Packard, T.T., and Q. Dortch
1975. Particulate Protein-nitrogen in North Atlantic Surface Waters. *Marine Biology*, 33:347-354.

Pamatmat, M.M., and H.R. Skjoldal
1974. Dehydrogenase Activity and Adenosine Triphosphate Concentration of Marine Sediments in Lindaspoliene, Norway. *Sarsia*, 56:1-12.

Parsons, T.R., Y. Maita, and C.M. Lalli
1984. *A Manual of Chemical and Biological Methods for Seawater Analysis*. 173 pages. New York: Pergamon Press.

Peterson, G.L.
1979. Review of the Folin Phenol Protein Quantitation Method of Lowry, Rosebrough, Farr and Randall. *Analytical Biochemistry*, 100:201-220.

Pettersson, K., and M. Jansson
1978. Determination of Phosphatase Activity in Lake Water - A Study of Methods. *Verhandlungen der Internationalen Vereinigung für Theoretische und Angewandte Limnologie*, 20:1226-1230.

Pfannkuche, O.
1985. The Deep-sea Meiofauna of the Porcupine Seabight and Abyssal Plain (NE Atlantic): Population Structure, Distribution, Standing Stocks. *Oceanologica Acta*, 8:343-353.

Pierce, J., and C.H. Suelter
1977. An Evaluation of the Coomassie Brilliant Blue G 250 Dye Binding Method for Quantitative Protein Determination. *Analytical Biochemistry*, 81:478-480.

Polley, J.R.
1954. Colorimetric Determination of Nitrogen in Biological Material. *Analytical Chemistry*, 26:1523-1524.

Pradet, A.
1967. Etudes des adénosine-5'-mono-, di- et triphosphates dans les tissues végétaux. I. Dosage enzymatique. *Physiologie Végetale*, 5:209-221.

Prasad, A.S., E. DuMouchelle, D. Koniuch, and D. Oberleas
1972. A Simple Fluorometric Method for the Determination of RNA and DNA in Tissues. *Journal of Laboratory and Clinical Medicine*, 80:598-602.

Pridmore, R.D., A.B. Cooper, and J.E. Hewitt
1984. ATP as a Biomass Indicator in Eight North Island Lakes, New Zealand. *Freshwater Biology*, 14:75-78.

Rapport, M.M., and N.F. Alonzo
1960. The Structure of Plasmalogens V. Lipids of Marine Invertebrates. *Biological Chemistry*, 235:1953-1956.

Rapport, M.M., N.F. Alonzo, L. Graf, and V.P. Skipski
1958. Immunochemical Studies of Organ and Tumor Lipids. IV. Chromatographic Behavior of the Lipid Hapten of Rat Lymphosarcoma: Dependence of In Vitro Measurements on Lipid Interactions. *Cancer*, 11:1125-1135.

Riemann, B.
1978. Carotenoid Interference in the Spectrophotometric Determination of Chlorophyll Degradation Products from Natural Populations of Phytoplankton. *Limnology and Oceanography*, 23:1059-1066.
1982. Measurement of Chlorophyll$_a$ and its Degradation Products: A Comparison of Methods. *Archiv für Hydrobiologie*, 16:19-24.
1983. Biomass and Production of Phyto- and Bacterioplankton in Eutrophic Lake Tystrup, Denmark. *Freshwater Biology*, 13:389-398.

Riemann, B., and D. Ernst
1982. Extraction of Chlorophyll$_a$ and b from Phytoplankton Using Standard Extraction

Techniques. *Freshwater Biology,* 12:217-223.

Riemann, F., and M. Schrage
1978. The Mucus Trap Hypothesis on Feeding of Aquatic Nematodes and Implication for Biodegradation and Sediment Texture. *Oecologia (Berlin),* 34:75-88.

Riley, J.P., and D.A. Segar
1970. The Seasonal Variation of the Free and Combined Dissolved Amino Acids in the Irish Sea. *Journal of the Marine Biological Association of the United Kingdom,* 50:713-720.

Roberts, A.A., J.G. Palacas, and I.C. Frost
1973. Determination of Organic Carbon in Modern Carbonate Sediments. *Journal of Sedimentary Petrology,* 43:1157-1159.

Rosset, R., J. Julien, and R. Monier
1966. Ribonucleic Acid Composition of Bacteria as a Function of Growth Rate. *Journal of Molecular Biology,* 18:308-320.

Roth, M.
1971. Fluorescence Reaction for Amino Acids. *Analytical Chemistry,* 43:880-882.

Saike, M., S. Fang, and T. Mori
1959. Studies on the Lipid of Plankton. II. Fatty Acid Composition of Lipids from Euphausiacea Collected in the Antarctic and Northern Pacific Ocean. *Bulletin of the Japan Society of Science Fisheries,* 24:837-839.

Sanger, J.E., and E. Gorham
1970. The Diversity of Pigments in Lake Sediments and Its Ecological Significance. *Limnology and Oceanography,* 15:59-69.

Schollenberger, C.J.
1927. A Rapid Method for Determining Soil Organic Matter. *Soil Science,* 24:63-67.

Sedmark, J., and S.E. Grossberg
1977. A Rapid, Sensitive and Versatile Assay for Protein Using Coomassie Brilliant Blue G 250. *Analytical Biochemistry,* 79:544-552.

Seki, H., and N. Taga
1965. Microbiological Studies on the Decomposition of Chitin in Marine Environments. VIII. Distribution of Chitinoclastic Bacteria in the Pelagic and Neritic Waters. *Journal of the Oceanographic Society of Japan,* 21:174-187.

Setchell, F.W.
1981. Particulate Protein Measurement in Oceanographic Samples by Dye Binding. **Marine Chemistry,** 10:301-313.

Sharp, J.H.
1974. Improved Analysis for 'Particulate' Organic Carbon and Nitrogen from Seawater. *Limnology and Oceanography,* 19:984-989.

Shuman, F.R., and C.J. Lorenzen
1975. Quantitative Degradation of Chlorophyll by a Marine Herbivore. *Limnology and Oceanography,* 20:580-586.

Sibuet, M., A. Khripounoff, J. Deming, R. Colwede, and A. Dinet
1982. Modification of the Gut Contents in the Digestive Tract of Abyssal Holothurians. Pages 421-428 in J.M. Lawrence, editor, *International Echinoderms Conference, Tampa Bay.* Rotterdam: A.A. Balkema.

Skjoldal, H.R., and U. Bamsted
1976. Studies on the Deep-water Pelagic Community of Korsfjorden, Western Norway - Adenosine Phosphates and Nucleic Acids in *Meganyctiphanes norvegica* (Euphausiaceae) in Relation to the Life Cycle. *Sarsia,* 61:1-14.

1977. Ecobiochemical Studies on the Deep-water Pelagic Community of Korsfjorden, Western Norway. Adenine nucleotides in Zooplankton. *Marine Biology,* 42:197-211.

Skjoldal, H.R., and S. Barkati
1982. ATP Content and Adenylate Energy Charge of the Mussel *Mytilus edulis* During the Annual Reproductive Cycle in Lindaspollene, Western Norway. *Marine Biology,* 70:1-6.

Smith, G.A., J.S. Nickels, R.J. Bobbie, N.L. Richards, and D.C. White
1982. Effects of Oil and Gas Well-drilling Fluids on the Biomass and Community Structure of Microbiota that Colonize Sands in Running Seawater. *Archives of Environmental Contamination and Toxicology,* 11:17-23.

Somville, M., and G. Billen
1983. A Method for Determining Exoproteolytic Activity in Natural Waters. *Limnology and Oceanography,* 28:190-193.

Souza-Lima de Y., and J.-C. Romano
1983. Ecological Aspects of the Surface Microlayer. I. ATP, ADP, AMP Contents, and Energy Charge Ratios of Microplanktonic Communities. *Journal of Experimental Marine Biology and Ecology,* 70:107-122.

Spector, T.
1978. Refinement of the Coomassie Blue Method of Protein Quantitation. A Simple and Linear Spectrophotometric Assay for ≤0.5 to 50 µg of Protein. *Analytical Biochemistry,* 86:142-146.

Stanier, R.Y., E.A. Adelberg, and J.L. Ingraham
1978. *General Microbiology.* 4th Edition. 871 pages. London: Macmillan Press.

Steele, J.H., and I.E. Baird
1968. Production Ecology of a Sandy Beach. *Limnology and Oceanography,* 13:14-25.

Strehler, B.L.
1968. Bioluminescence Assay: Principles and Practise. *Methods of Biochemical Analysis,* 18:99-181.

Strickland, J.D.H., and T.R. Parsons
1972. A Practical Handbook of Seawater Analysis. 2nd Edition. *Bulletin of the Fisheries Research Board of Canada,* 167:1-310.

Sutcliffe, W.H., Jr.
1965. Growth Estimates from Ribonucleic Acid Content in Small Organisms. *Limnology and Oceanography,* 10:253-258.

1969. Relationship Between Growth Rate and Ribonucleic Acid Concentration in Some Invertebrates. *Journal of the Fisheries Research Board of Canada,* 27:606-609.

Takahashi, M., and Y. Saijo
1981. Nitrogen Metabolism in Lake Kizaki, Japan. II. Distribution and Decomposition of Organic Nitrogen. *Archiv für Hydrobiologie,* 92:359-376.

Tett, P., M.G. Kelly, and G.M. Hornberger
1975. A Method for the Spectrophotometric Measurement of Chlorophyll$_a$ and Pheophytin$_a$ in Benthic Microalgae. *Limnology and Oceanography,* 20:887-895.

Telek, G., and N. Marshall
1974. Using a CHN Analyzer to Reduce Carbonate Interference in Particulate Organic Carbon Analyses. *Marine Biology,* 24:219-221.

Tenore, K.R., W.M. Chamberlain, R.B. Dunstan, B. Hanson, B. Sherr, and J.H. Tietjen
1978. Possible Effect of Gulf Stream Intrusions and Coastal Runoff on the Benthos of the Continental Shelf of the Georgia Bight. Pages 577-599 in M.L. Wiley, editor, *Estuarine Interactions.* New York: Academic Press.

Thiel, H.
1978. Benthos in Upwelling Regions. Pages 124-128 in R. Boje, and M. Tomczak, editors, *Upwelling Ecosystems.* Berlin, Heidelberg: Springer Verlag.

Tobin, R.S., J.F. Ryan, and B.K. Afghan
1978. An Improved Method for the Determination of Adenosinetriphosphate in Environmental Samples. *Water Research,* 12:783-792.

Towle, A., and A.C. Giese
1966. Biochemical Changes During Reproduction and Starvation in the Sipunculid Worm *Phascolosoma agassizi. Comparative Biochemisty and Physiology,* 19:667-680.

Trees, C.C., M.C. Kennicutt, II, and J.M. Brooks
1985. Errors Associated with the Standard Fluorometric Determination of Chlorophylls and Phaeopigments. *Marine Chemistry,* 17:1-12.

Trevelyan, W.E., and J.S. Harrison

1952. Studies on Yeast Metabolism. 1. Fractionation and Microdetermination of Cell Carbohydrates. *Biochemical Journal*, 50:298-303.

Troll, W., and R.K. Cannon
1953. A Modified Photometric Ninhydrine Method for the Analysis of Amino and Imino Acids. *Journal of Biological Chemistry*, 200:803-811.

Van Kley, H., and S. Hale
1977. Assay for Protein by Dye Binding. *Analytical Biochemistry*, 81:485-487.

Wakeel el, S.K., and J.P. Riley
1956. The Determination of Organic Carbon in Marine Muds. *Journal du Conseil International pour l'Exploration de la Mer*, 22:180-183.

Walkley, A.
1947. A Critical Examination of a Rapid Method for Determining Organic Carbon in Soils - Effect of Variations in Digestion Conditions and of Inorganic Soil Constituents. *Soil Science*, 63:252-256.

Wasmund, N.
1984. Probleme der spektrophotometrischen Chlorophyllbestimmung. *Acta Hydrochimica et Hydrobiologica*, 12:255-272.

Whittaker, J.R., and J.R. Vallentyne
1957. On the Occurence of Free Sugars in Lake Sediment Extracts. *Limnology and Oceanography*, 2:98-110.

Widbom, B.
1984. Determination of Average Individual Dry Weights and Ash-free Dry Weights in Different Sieve Fractions of Marine Meiofauna. *Marine Biology*, 84:101-108.

Wieser, W.
1960. Benthic Studies in Buzzards Bay. II. The Meiofauna. *Limnology and Oceanography*, 5:121-137.

Yemm, E.W., and A.J. Willis
1954. The Estimation of Carbohydrates in Plant Extracts by Anthrone. *Biochemical Journal*, 57:508-514.

Yentsch, C.S., and D.W. Menzel
1963. A Method for the Determination of Phytoplankton Chlorophyll and Phaeophytin by Fluorescence. *Deep-Sea Research*, 10:221-231.

Yingst, J.Y.
1978. Patterns of Micro- and Meiofaunal Abundance in Marine Sediments, Measured with the Adenosine Triphosphate Assay. *Marine Biology*, 47:41-54.

Züllig, H.
1981. On the Use of Carotenoid Stratigraphy in Lake Sediments for Detecting Past Developments of Phytoplankton. *Limnology and Oceanography*, 26:970-976.

7. Sampling Equipment

John W. Fleeger, David Thistle, and Hjalmar Thiel

Practical meiofauna work usually begins with sampling. Almost all investigators have considered the problems associated with sampling, and many have arrived at individualized solutions to their problems. However, certain concepts apply to all sampling methods. In this chapter we shall discuss these principles and describe the more common sampling methods and equipment for use in different habitats. We hope that a discussion of the principles behind good sampling technique will direct the choice of equipment in future investigations, and that a comparison of methods will facilitate the exchange of data.

Ideally, samples should always accurately reflect the populations from which the sample was taken. Such samples would be unbiased and quantitative, and should serve as the basis of studies of meiofaunal abundance, biomass, production and community structure. Unbiased samples are not always required, however. Qualitative collections are those in which a device or method captures fauna in an unpredictable or variable fashion. They are useful to formulate a partial and biased faunistic list or to collect specimens of specific taxa for culturing, physiological, or systematic studies. Qualitative and quantitative sampling methods vary with habitat and sediment type, and whether sampling is done with visual contact, i.e., with close inspection of the sampling site, or by remote means in which equipment is deployed without visual contact.

Qualitative Sampling

Qualitative collection methods differ between air-exposed and water-covered sediments, and may be visual or remote. In the supralittoral, any method of digging or scraping sediments may be sufficient to collect meiofauna. The washings of coarser substrates, including stranded materials and algae may also be used.

A method frequently used in beaches is the concentration of meiofauna from the water that seeps from the sides of a pit. This water may be scooped up and filtered through a net or sieved with mesh of the appropriate size (see meiofauna definition, Chapter 1), or the animals may be stirred into the water and filtered with a small hand net. Similarly, the Bou-Rouch pump has been used to concentrate meiofauna from coarse river sands. It has a perforated lower hollow column that allows water and associated fauna to seep into it. The collected water is then pumped up and sieved (Bou, 1974). Meiofauna may be isolated and concentrated from large-volume sand samples by decantation after anesthetization with 6% $MgCl_2$ (i.e., 73.2 g/l) or a weak solution of formaldehyde. In muds, meiofauna can be concentrated by sieving hand-scooped superficial sediments through a coarse screen to remove large detritus particles and a smaller screen, of appropriate size, to retain fauna. Living meiofauna can be extracted from this retained material by sucrose flotation or by the use of phototactic responses. For example, some harpacticoid species will swim from such material into the overlying water of a bucket toward a beam of light (cool light from a fiber-optics illuminator is best because it does not cause a temperature change in the water), and can be captured by aspiration. Subtidal qualitative collection may be achieved with any of several types of grabs or corers (see below), from which surface scrapings or sediment subsamples may be extracted to obtain meiofauna. Much of the equipment described by Eleftheriou and Holme (1984) for macrofauna sampling is appropriate for meiofauna qualitative sampling.

Several researchers (e.g., Higgins, 1964; Ockelmann, 1964; Bieri and Tokioka, 1968) have used small dredges (Figure 7.1) for remote collection of meiofauna. Such a dredge should be constructed from light-weight materials and the bottom or skids should be relatively large so that the gear remains on the sediment surface. The cutting edge of the dredge mouth suspends sediment, and the bag captures and sieves the material. The upper sediments are skimmed off, but do not always flow freely into the dredge bag because sediment accumulates in front of the dredge. With the resulting increase in wire tension, the dredge is frequently lifted from the sediment surface, and it jumps from bite to bite. Dredge samples cannot be related to sediment area, and are therefore not quantitative. Nevertheless, dredges may be useful for the collection of larger and temporary meiofauna, which often occur in densities too low to be sampled in large enough numbers with corers.

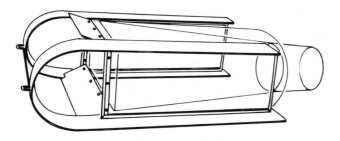

Figure 7.1.—Meiobenthic sled. Note: cod-end untied (open) in illustration (see Higgins, 1964).

Quantitative Sampling

Meiobenthologists take quantitative samples primarily for two reasons. They wish to describe the temporal or spatial distribution of meiofaunal taxa, or they wish to test hypotheses. In both cases, sampling is critical to the quality of the answer obtained. For most ecological work, unbiased, quantitative samples are required. Unfortunately, the efficiency and sources of bias of most sampling techniques are poorly studied. Nevertheless, efforts are continually made to understand sampler bias and to design gear that takes the best samples possible.

We summarize below quantitative sampling techniques according to major habitat types.

Sediments.—In sediments, coring is the best quantitative sampling or subsampling technique. Corers are devices with a known surface area. Most are cylindrical, and can be made from tubing or piping of any available rigid material (e.g., Cubit, 1970). The diameter chosen depends on the volume and depth of sample required. Corers should have a smooth internal surface to facilitate sediment penetration and core removal. Corer inner diameters of 2-4 cm have been used in many habitats, and provide a meiofaunal sample that can be sorted in its entirety. Smaller corers (1 cm diameter or less) are desirable in sediments where densities are unusually high, or to determine small-scale distributions. Tubes of clear plastic allow the core to be viewed through the walls. The lower end of the tube should be beveled to facilitate sediment penetration. Hand-held corers may also be made by cutting the needle end from a plastic, disposable syringe (e.g., Chandler and Fleeger, 1983) for use in sampling low-tide intertidal sediments or for subsampling (discussed later). For best results the plunger of this piston-style corer initially should be near the level of the tube opening (Figure 7.2). The corer tube should then be pushed slowly into the sediment to the desired depth while holding the plunger in place. The plunger provides the suction necessary to retain the core while the corer is carefully removed from the sediment. These modified syringes can be re-used, but if so, syringe plunger heads made of rubber, rather than plastic, are best because they retain their flexibility and suction.

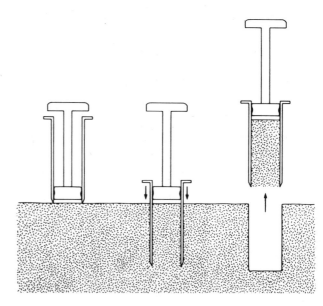

Figure 7.2.—Proper technique for using a piston-style corer for visual sampling and subsampling. The plunger of the corer should be located near the sediment surface (a) and the wall of the corer pushed downward while holding the plunger in place (b). Suction from the plunger retains the sediment while the corer is removed (c).

When used with care, corers are excellent samplers because they collect a known area or volume of sediment with all depths equally represented, and all animals present before sampling are captured. Three general problems arise with core sampling: a bow-wave-induced reduction in abundance, effects on population parameters due to the underlying distribution of the fauna, and sample distortion due to core compaction. Care should be taken to minimize these problems, and we discuss below some ways to accomplish this.

The bow wave arises because water can flow more easily around a corer than through it due to the friction water exerts against the walls of the corer. As the corer approaches the sediment, surface material and attendant meiofauna may be caught in this flow around the corer and washed out of the region to be sampled, thereby biasing the sample. However, if the surface of the sediment lacks a resuspendable layer, the bow wave may be irrelevant. The investigator can minimize it by taking cores slowly and by using flow-through corer designs.

Meiofauna are well known for their spatial patchiness, and the size of the corer can interact with the underlying spatial distribution, or dispersion, of the fauna. Estimates of interspecific association and population parameters such as density and patch size may be affected by this interaction. Many small cores (samples or subsamples) provide better density estimates (and tighter confidence limits) than a small number of large cores (Elliot, 1977), and this approach is recommended for general use in meiofaunal studies. However, there is a limit on how small a corer should be, and Green (1979) cautions that the corer should be sufficiently large so that the ratio of the area of the organism to the area of the sample unit be 0.05 or less. Too little is now known of the small-scale distribution of meiofauna to suggest an optimal corer size for all situations (but see Findlay, 1982, for an approach to the problem). Preliminary or theoretical studies should be done to select an appropriate core size before a study of dispersion or species associations is begun. Corer sizes that coincide with any scale of patch size should be avoided (Green, 1979).

Compaction may occur as the corer is forced into the sediment (Williams and Pashley, 1979). Friction between the sediment and the walls of the corer can compress the contained sediment, so the volume of the core does not represent the volume of sediment sampled. This effect can bias the reconstruction of vertical profiles from cores divided into layers (see below). A related problem is the drawing down of individuals from upper layers into lower layers along the wall as the corer is inserted into the sediment. Some sediments resist compaction, in particular, sands and muds of low-water content. Compaction and draw-down of material can be minimized by increasing the diameter of the corer, which also minimizes the fraction of the volume of the core affected by the walls. For subsampling, a second solution is to use a corer with a piston (the syringe corer discussed above or the corer of Cubit, 1970).

Intertidal.—Intertidal sampling has typically been done at low tide to allow easy access to the study site. This advantage is tempered by the need to avoid disturbing the location to be sampled. Intertidal habitats vary from fine muds to gravels. In most sediment types, tubes with the characteristics described above may serve as a sampler for the upper (< 10 cm) portion of the sediment. A tight-fitting stopper secured in the upper end of the coring tube or the piston of the syringe will provide suction to hold the sediment in place while the corer is removed. Another stopper may secure the lower end of the tube for transport.

These techniques will not work well if the fauna extends deep into the sediment. In the supralittoral of beaches, which can be inhabited to depths of greater than 2 m (Fenchel, 1978), special demountable corers have been developed, e.g., Renaud-Debyser (1957). Alternatively, a pit can be dug to the required depth and core samples taken from the vertical face (Pollock, 1970).

Subtidal.—If at all possible, subtidal samples should be taken by SCUBA divers. Divers usually obtain superior samples because they are able to position the samplers with care and insert the corer slowly (see McIntyre, 1971). Also, the presence of the investigator will often yield important insights about the ecology of the site or practical aspects of sampling. Similarly, cores from submersibles and remotely controlled vehicles (Thiel and Hessler, 1974; Thistle, 1978) may be taken carefully, and with visual observation. As an aid in diver coring, Jensen (1983) devised a three-tube-corer with a circular handling plate to allow careful positioning with the diver some distance from the actual sampling site. McIntyre (1971) and Holopainen and Sarvala (1975) used subsampled diver cores to suggest that larger corers (> 8 cm^2) are more efficient than smaller corers in estimating meiofaunal abundance. The higher estimates in larger, but subsampled, diver-collected cores could partly be due to a concentration of meiofauna in the center of the core, rather than a truly higher sampling efficiency (Rutledge and Fleeger, 1988).

A variety of remote samplers have been developed for use in shallow water. Pole samplers (Figure 7.3) with closing mechanisms that minimally block the flow of water through the corer are likely to take unbiased samples because entrance velocities will be low and the bow wave should be small. The device of Frithsen et al. (1983) consists of a tube mounted on a long, light pole, and can be used in depths up to 4 m. The top lid is held open during descent and sediment penetration. A plate triggers the lid release mechanism on full penetration, and the tube is well sealed to secure the core in the tube. In the Hamburg pole corer (Figure 7.3), the closing mechanism is triggered by a line running up the pole. This system has the advantage that closure is independent from penetration depth. In contrast, free-fall corers (e.g., traditional geological gravity corers) enter the sediment at a relatively high velocity and have a severe bow wave (McIntyre, 1971). Therefore, they are unlikely to sample the surface of the sediment or a flocculent layer well, and their use should be avoided.

Several new, relatively inexpensive corers are light enough to be used on smaller research vessels that cannot accommodate a box corer (see below), and yet take samples yielding densities similar to those of diver-collected cores. Ankar and Elmgren (1976) described a simple one-tube corer which free falls only the last 2 m. The devices of Plocki and Radziejewska (1980) and Raisanen et al. (1981) were designed for coarse subtidal sands, and may serve as alternatives to grab samplers in such conditions. The Kajak corer (Hakala, 1971) works well in muddy sediments and takes multiple cores (Holopainen and Sarvala, 1975; Jensen, 1983; Chandler et al., 1988). Good design features include slow sediment penetration, large, flow-through tubes and a trip mechanism that does not interfere with water flow through the tube core or disturb the sediment before penetration.

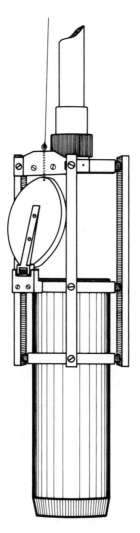

Figure 7.3.--The Hamburg pole corer. A pin holds the spring loaded lid open during descent. The pin is pulled away to release the lid after sediment penetration.

Environments with a well-developed flocculent layer or an easily resuspended sediment surface are the most difficult to sample. Deliberate corer samples are the least disturbed in these circumstances (McIntyre and Warwick, 1984), and are therefore the most highly recommended for remote sampling. These wire-lowered corers consist of a supporting frame and a movable sampling unit. The frame arrives on the bottom first. When it stops, the sampling unit is positioned some 30 cm above the sediment surface. It is released and slowly descends to penetrate the sediment with one or several collecting tubes. Their penetration is slow, retarded by the action of a piston. Flocculent material is retained, and the water above the sample is truly a near-bottom water sample. Craib (1965) describes a single-corer device for use from a small boat in shallow water; Barnett et al. (1984) have devised a version that takes 12 cores simultaneously (Figure 7.4), but requires the gear-handling facilities of a large vessel.

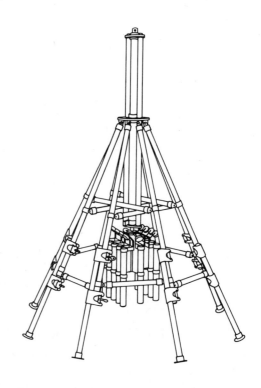

Figure 7.4.--The Scottish Marine Biological Laboratory corer, a 12-place deliberate corer after Barnett et al. (1984). The framework is 3.5 m high and 2.4 m across at its widest point. Some details, e.g., bottom core catchers, are not shown.

Cores, especially from sandy sediments, will sometimes be lost during retrieval if the tube is not closed at its lower end; upper lid suction may not be strong enough to hold the core in place. Various styles of core "catchers" are available to aid in core retention. Many are internal to the corer (e.g., the orange-peel type), and typically disturb the surface and furrow the sides of the core. External core catchers are therefore recommended. Mills (1961) describes a catcher consisting of two half lids screwed to phosphor bronze spring strips (Figure 7.5a). The lids are held open by the tube. At corer penetration, the lids are triggered, move down the tube on retrieval, and the springs pull the lids into the closing position. Fenchel (1967) used a ball on a rubber band design (Figure 7.5b). Core penetration releases the ball, it moves to the tube opening, covers it and stays in position by the force of the partially released rubber band. The multiple-corer design of Barnett et al. (1984) has two lids; the lower is mounted on a long lever that falls onto the sediment after triggering and swings in closing position when the tubes are raised above the sediment surface.

Box (spade) corers frequently have been used successfully in the sampling of large areas of sediment (Thiel, 1971; Dinet, 1973; Coull et al., 1977; Fleeger et al., 1983). First described by Reineck (1958) as the "Kastengreifer," the box corer has been enlarged as the USNEL corer, Hessler and Jumars, (1974) and modified by Thiel (1983). Subsequent unpublished modifications have been made (Hessler, personal communication) (Figure 7.6).

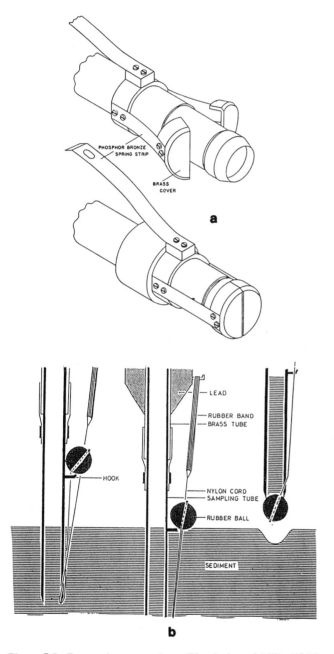

Figure 7.5.--External core catchers. The design of Mills (1961) with a two-sided closing lid (a), and the design of Fenchel (1967) using a ball and rubber band (b). (Courtesy Ophelia Publications.)

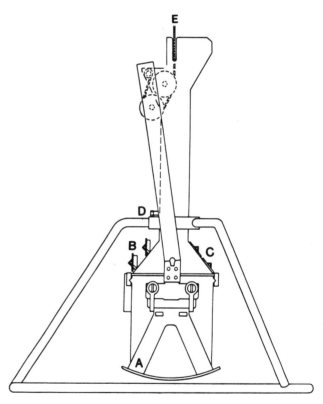

Figure 7.6.--A box corer. The corer is shown in the closed position. A, is the detachable spade, which allows the core to be removed without inserting a bottom plate. B, shows the positions of the flaps, open for descent. C, shows the flaps closed for ascent. D, is the safety pin. E, is the cable to the ship. The details of the firing mechanism and the pre-trip device have been omitted. This design is after unpublished modifications of the Hessler and Jumars (1974) USNEL corer. (Courtesy Ocean Instruments, Inc.)

Box corers weigh from 150 to 750 kg empty, and collect a block of sediment from 0.02 to 0.25 m^2 in surface area, and from 20 to 50 cm in depth. In effect, the corer is an open box which penetrates the sediment under its own weight after reaching the sea floor (after the frame of the box corer comes to rest on the bottom, the open box is released and is slowly pushed into the sediment). As the corer is pulled from the sediments, a lever is turned that causes a spade to cut through the sediment below the box. Upper lids close off the top of the corer, and the sediment and overlying water are firmly held in place while the box corer is retrieved.

Because the box enters the sediment at a moderate speed, and because in recent designs, water flow through the corer is relatively unobstructed (the upper opening totals 50% of the corer surface area), the bow wave is modest. Thistle (1983) and Thistle and Sherman (1985) detected no evidence of bow-wave-induced bias in samples of harpacticoid copepods and nematodes respectively at a site with little or no flocculent layer. However, the box corer probably does not routinely take samples with a quality equal to the deliberate corer. Thiel et al. (personal communication) simultaneously employed a box corer and the corer of Barnett et al. (1984). The deliberate corer regularly collected a layer of phytodetritus on the sediment surface which was missing in box corer samples. This suggests that very flocculent materials are inadequately sampled by the box corer.

Sometimes a relatively large volume of water may be retained in the box corer, e.g., when sampling sandy sediments that limit penetration. This water overlying the sediment can percolate through the sample and disturb the faunal distribution and may even erode the sediment surface. Currents may cause the box to penetrate at an angle, or may not allow the corer to be pulled from the sediment in an upright manner; either effect will cause the surface of the sample to be disturbed. Box-corer samples biased by any of these means should be rejected.

Although grabs have been used for quantitative meiofaunal sampling, many workers have severe reservations about the quality of such samples (e.g., Elmgren, 1973; Heip et al., 1977). Grabs are usually constructed from two compact claws which can create strong bow waves and cause the loss of surface material (Ankar, 1977). Additionally, claw penetration and closure disturb and compress the sediment. This is especially true for grabs closed by internal chains or other structural parts (Petersen, Okean-50 grabs). Even with open space inside the grab (Van Veen grab, Figure 7.7), disturbance is strong. Grabs often do not close properly (due to construction defects or to shell or debris becoming wedged between the claws), and overlying water with fauna will drain from the grab during recovery. Therefore, grab samples are usually too disturbed for quantitative meiofaunal subsampling, and they should be avoided if possible.

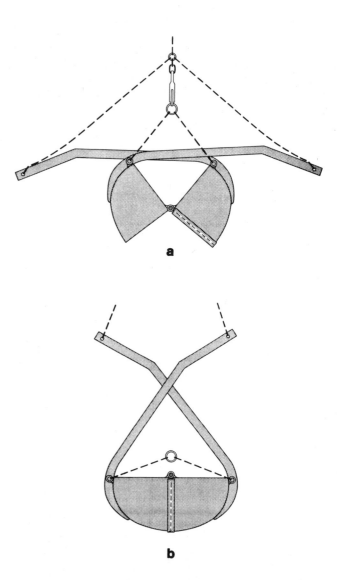

Figure 7.7.--The Van Veen grab sampler in open (a) and closed (b) position.

To sample larger, temporary meiofauna, Muus (1964) designed the "mousetrap" (Figure 7.8). On contact with the bottom, a stationary plate and a frame with a plankton net are positioned opposite to each other. The frame penetrates the sediment a few centimeters, and moves to the sealing plate, scraping the surface sediment (200 cm^2) into the net.

Vertical Profiles.--The vertical distribution of meiofauna is frequently of interest. The appropriate method to determine vertical profiles varies with the sediment type. When the deposit is such that compaction is not a concern (e.g., coarse sand), the core can be extruded from below and sliced off in appropriate layers (e.g., Thiel, 1971; Thistle, 1978). If the surface layer is poorly consolidated and cannot be extruded without loss, as with soft muds, then the core can be allowed to slip down the core tube by loosening the top stopper. Careful manipulation will allow the core to be cut into appropriate depth intervals, avoiding additional compaction that may occur in extrusion. Markings at appropriate intervals on the corer or the extruder help identify the thickness to be sliced. Cores should be processed immediately because meiofauna are known to migrate in standing sediments. When the vertical profile is very deep, specialized corers (Renaud-Debyser, 1957) may be necessary. An approach to this problem for a beach has been discussed above. Very small-scale profiles can be measured by the technique of Joint et al. (1982) in which a micrometer is attached to a core tube, allowing 1-mm layers to be extruded. The "meiostecher" (Figure 7.9a) of Thiel (1966) allows for good profiling at cm-intervals from subsamples by using a plate support (Figure 7.9b). Although cores have been fast frozen in liquid nitrogen in the field to preserve the vertical profile (Bell and Sherman, 1980; Chandler and Fleeger, 1983), convective flows within the core during freezing disrupt the profile greatly (Rutledge and Fleeger, 1988), destroying the utility of this technique. Slow freezing causes less disturbance but also permits meiofauna the chance to migrate before freezing takes place.

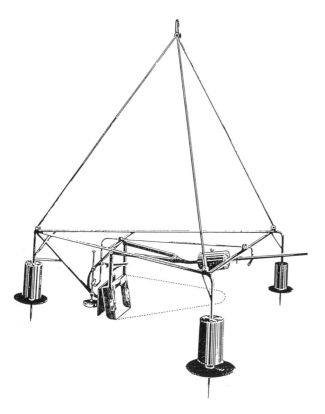

Figure 7.8.--The Muus (1964) "mousetrap," a sampler designed to collect a large area of superficial sediments. (Courtesy Ophelia Publications.)

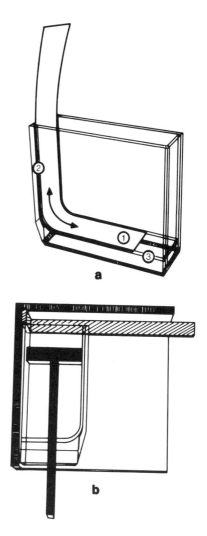

Figure 7.9.--The meiostecher after Thiel (1966). The device itself takes subsamples (a), and includes (1) a plastic tongue to cut the subsample, (2) a slot for the tongue, and (3) grooves to guide the tongue horizontally. A Plexiglas support allows sectioning of the subsample (b). A plunger pushes the core up and a sliding bar pulls away single sediment layers.

For larger samples in soft sediments, Hagge (personal communication) used a corer in which half the tube from the upper 1-cm had been cut off. The remaining half supported the core when pushed up, and a rounded plate fit to the inside of the tube was used to separate the layers of the core (Figure 7.10).

Plants.--Macroalgae (Hicks, 1977) and seagrasses

(Novak, 1982; Bell et al., 1984) have meiofauna living on them. Quantitative sampling usually involves removing the entire above-ground portion of the plant or some piece of it (e.g., an algal blade) and then extracting the fauna. Samples taken while the plant is exposed by the receding tide need no special treatment. They are usually fixed in the field in individual containers so animals are not lost (Gunnill, 1982). When samples are taken underwater, it is necessary to contain the fauna present on the structure. For example, the plant can be placed in a bag underwater (Hicks, 1977). Care must be taken so that disturbances resulting from placing the structure in a sample container and then detaching it do not result in a loss of animals, and simultaneously that contaminants do not enter the sample container from the water column. In data analysis, abundances should be normalized, for example, to plant weight or surface area.

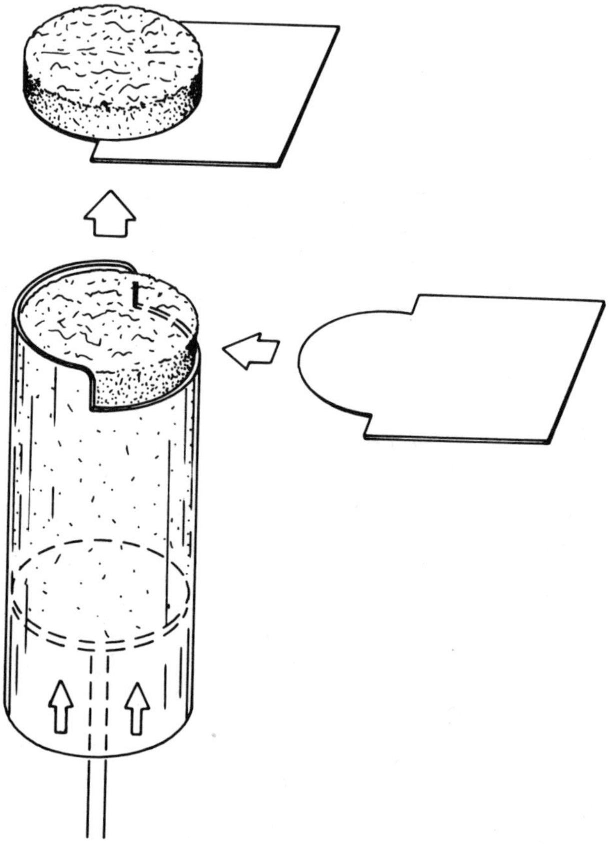

has been done on the meiofauna of aufwuchs, hard substrates, or hyperbenthos (near-bottom water). Suction devices are described (e.g., Tanner et al., 1977) that might be useful to collect aufwuchs or meiofauna from hard substrates, although most samples are taken by scraping a known area. Meiofauna that reside in near-bottom waters or those which leave the sediment have been sampled by pumps (Sibert, 1981; Palmer and Gust, 1985), and by emergence traps (Alldredge and King, 1980; Fleeger et al., 1983; Walters and Bell, 1986) (Figure 7.11). No clear consensus is available on the

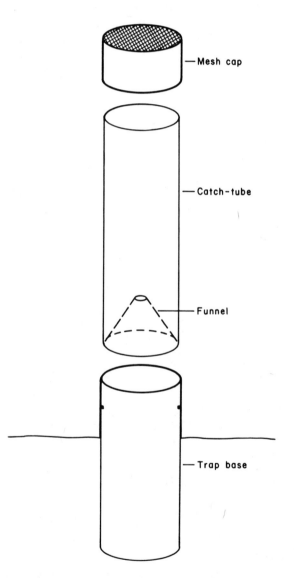

Figure 7.10.--The Hagge corer for vertical sectioning. One half of the circumference of the upper part of the corer is removed and a plate fit to the dimensions of the corer is used to section and remove layers of sediment.

Figure 7.11.--The emergence trap of Walters and Bell (1986). This Plexiglas trap captures meiofauna that emerge from the sediment and enter the catch tube through the funnel. A mesh cover allows movement of water but retains the meiofauna.

Other Habitats.--Relatively little quantitative work

efficiency of such sampling devices (but see Youngbluth, 1982, for comparisons of emergence traps).

Subsampling

By subsampling, we refer to procedures in which a sample is removed from a larger sample, for example, a small core taken from a box core. The necessity for subsampling arises when more than one type of analysis is to be done on each sample or when the sample is too large to be processed in its entirety (e.g., box cores or cores from deliberate corers). Deep-sea investigations are frequently limited in sample number, and many subsamples (for grain size and chemical and faunal analyses) from one box core must be taken. When surface or near-surface values are of interest, subsampling should be done before the sample is removed from the sediment, because during recovery, water motion in the sampler may resuspend the superficial sediment layers. The flocculent-layer and superficial sediment meiofauna can be redistributed in the core, and the natural variability present when the core was taken will be replaced by a variability created by the mixing during recovery (Rutledge and Fleeger, 1988). "In situ" subsampling techniques that avoid this problem have been developed. For example, a small corer may be inserted on the site to be sampled before a larger diameter corer is used to collect it and the surrounding sediment. In subtidal work, subcorers of an appropriate size should be mounted within larger samplers, so both sample and subsample are taken simultaneously (Bacescu, 1957; Jumars, 1975; Burnett, 1979). If coring and recovery result in a truly undisturbed core (the deliberate corers approach this goal), careful subsampling should result in an unbiased sample.

Data Standardization

Faunal densities and biomass values from sediments are frequently reported in terms of a standard surface area to facilitate comparisons between studies using different sampling techniques. Although this procedure will ease comparisons among reports, statistical tests must be run on raw, non-standardized data to assure correct variance estimates (Elliot, 1977). By convention, the number of individuals (or biomass) per 10 cm^2 (i.e., 10 square centimeters not 10 centimeters squared) are reported. Figure 7.12 clarifies this issue by illustrating the exact area of 10 square centimeters. A 10 centimeter squared area is equivalent to a square 10 cm on a side, or 100 cm^2.

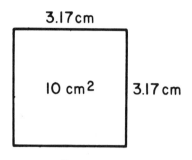

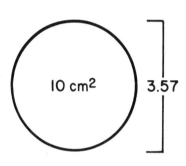

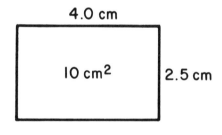

Figure 7.12.--Areas of 10 cm^2 (i.e., 10 square centimeters) drawn to original size in a square, circle and rectangle. In contrast, 10 centimeters squared (10 cm x 10 cm) is 100 cm^2 in surface area.

References

Alldredge, A.L., and J.M. King
1980. Effects of Moonlight on the Vertical Migration Patterns of Demersal Zooplankton. *Journal of Experimental Marine Biology and Ecology*, 44:133-156.

Ankar, S.
1977. Digging Profile and Penetration of the Van Veen Grab in Different Sediment Types. *Contributions of the Askö Laboratory, University of Stockholm*, 16:1-22.

Ankar, S., and R. Elmgren
1976. The Benthic Macro- and Meiofauna of the Askö-Landsort Area (Northern Baltic Proper). A Stratified Random Sampling Survey. *Contributions of the Askö Laboratory, University of Stockholm*, 11:1-115.

Băcescu, M.
1957. Apucatorul-sonda pentru studiul cantitativ al organismelor de fundun aparat mixt pentru colectarea simultana a macro-si microbentosulia. *Buletinul Institutului de Cercetari Piscicole,* 16:69-82.

Barnett, P.R.O., J. Watson, and D. Connelly
1984. A Multiple Corer for Taking Virtually Undisturbed Samples from Shelf, Bathyal and Abyssal Sediments. *Oceanologica Acta,* 7:399-408.

Bell, S.S., and K.M. Sherman
1980. A Field Investigation of Meiofaunal Dispersal: Tidal Resuspension and Implications. *Marine Ecology Progress Series,* 3:245-249.

Bell, S.S., K. Walters, and J.C. Kern
1984. Meiofauna from Seagrass Habitats: A Review and Prospectus for Future Research. *Estuaries,* 7:331-339.

Bieri, R., and T. Tokioka
1968. Dragonet II, An Opening-closing Quantitative Trawl for the Study of Micro-vertical Distribution of Zooplankton and Meio-epibenthos. *Publications of the Seto Marine Biology Laboratory,* 15:373-390.

Bou, C.
1974. Les methodes de recolte dans les eaux souterraines interstitielle. *Annales of Speleology,* 29:611-619.

Burnett, B.R.
1979. Quantitative Sampling of Microbiota of the Deep-sea Benthos. II. Evaluation of Technique and Introduction to the Biota of the San Diego Trough. *Transactions of the American Microscopical Society,* 98:233-242.

Chandler, G.T., and J.W. Fleeger
1983. Meiofaunal Colonization of Azoic Estuarine Sediment in Louisiana: Mechanisms of Dispersal. *Journal of Experimental Marine Biology and Ecology,* 69:175-188.

Chandler, G.T., T.C. Shirley, and J.W. Fleeger
in press The Tom-tom Corer: A New Design of the Kajak Corer for Use in Meiofauna Sampling. *Hydrobiologia,* (in press).

Coull, B.C., R.L. Ellison, J.W. Fleeger, R.P. Higgins, W.D. Hope, W.B. Hummon, R.M. Rieger, W.E. Sterrer, H. Thiel, and J.H. Tietjen
1977. Quantitative Estimates of the Meiofauna from the Deep Sea off North Carolina, USA. *Marine Biology,* 39:233-240.

Craib, J.S.
1965. A Sampler for Taking Short Undisturbed Marine Cores. *Journal du Conseil Permanent International pour l'Exploration de la Mer,* 30:34-39.

Cubit, J.
1970. A Simple Piston Corer for Sampling Sand Beaches. *Limnology and Oceanography,* 15:155-156.

Dinet, A.
1973. Distribution quantitative du meiobenthos profond dans la region de la dorsale de Walvis (Sud-Ouest Africain). *Marine Biology,* 20:20-26.

Eleftheriou, A., and N.A. Holme
1984. Macrofauna Techniques. Pages 140-216 in N.A. Holme and A.D. McIntyre, editors, *Methods for the Study of Marine Benthos.* London: Blackwell Scientific Publications.

Elliott, J.M.
1977. *Some Methods for the Statistical Analysis of Samples of Benthic Invertebrates.* Freshwater Biological Association Scientific Publication No. 25. 156 pages.

Elmgren, R.
1973. Methods of Sampling Sublittoral Soft Bottom Meiofauna. *Oikos,* Supplement, 15:112-120.

Fenchel, T.
1967. The Ecology of Marine Microbenthos. I. The Quantitative Importance of Ciliates Compared with Metazoans in Various Types of Sediments. *Ophelia,* 4:121-137.
1978. The Ecology of Micro- and Meiobenthos. *Annual Review of Systematics and Ecology,* 9:99-121.

Findlay, S.E.G.
1982. Influence of Sampling Scale on Apparent Distribution of Meiofauna on a Sand Flat. *Estuaries,* 5:322-324.

Fleeger, J.W., G.T. Chandler, G.R. Fitzhugh, and F.E. Phillips
1984. Effects of Tidal Currents on Meiofauna Densities in Vegetated Salt Marsh Sediments. *Marine Ecology Progress Series,* 19:49-53.

Fleeger, J.W., W.B. Sikora, and J.P. Sikora
1983. Spatial and Long-term Temporal Variation of Meiobenthic-Hyperbenthic Copepods in Lake Pontchartrain, Louisiana. *Estuaries and Coastal Shelf Science,* 16:441-453.

Frithsen, J.B., D.T. Rudnick, and R. Elmgren
1983. A New Flow-through Corer for the Quantitative Sampling of Surface Sediments. *Hydrobiologia,* 99:75-79.

Green, R.H.
1979. *Sampling Design and Statistical Methods for Environmental Biologists.* 257 pages. New York: John Wiley and Sons.

Gunnill, F.C.
1982. Macroalgae as Habitat Patch Islands for *Scutellidium lamellipes* (Copepoda: Harpacticoida) and *Ampithoetea* (Amphipoda: Gammaridae). *Marine Biology,* 69:103-116.

Hakala, I.
1971. A New Model of the Kajak Bottom Sampler, and Other Improvements in the Zoobenthos Sampling Technique. *Annales Zoologici Fennici,* 8:422-426.

Heip, C., K.A. Willems, and A. Goossens
1977. Vertical Distribution of Meiofauna and the Efficiency of the Van Veen Grab on Sandy Bottoms in Lake Grevelingen (The Netherlands). *Hydrobiological Bulletin* (Amsterdam), 11:35-45.

Hessler, R.R., and P.A. Jumars
1974. Abyssal Community Analysis from Replicate Box Cores in the Central North Pacific. *Deep-Sea Research,* 21:185-209.

Hicks, G.R.F.
1977. Species Composition and Zoogeography of Marine Phytal Harpacticoid Copepods from Cook Strait, and Their Contribution to Total Phytal Meiofauna. *New Zealand Journal of Marine and Freshwater Research,* 11:441-469.

Higgins, R.P.
1964. Three New Kinorhynchs from the North Carolina Coast. *Bulletin of Marine Science of the Gulf and Caribbean,* 14:479-493.

Holopainen, I.J., and J. Sarvala
1975. Efficiencies of Two Corers in Sampling Soft-bottom Invertebrates. *Annales Zoologici Fennici,* 12:280-284.

Jensen, P.
1983. Meiofaunal Abundance and Vertical Zonation in a Sublittoral Soft Bottom, with a Test of the Haps Corer. *Marine Biology,* 74:319-326.

Joint, I.R., J.M. Gee, and R.M. Warwick
1982. Determination of Fine-scale Vertical Distribution of Microbes and Meiofauna in an Intertidal Sediment. *Marine Biology,* 72:157-164.

Jumars, P.A.
1975. Methods for Measurement of Community Structure in Deep-sea Macrobenthos. *Marine Biology,* 30:245-252.

McIntyre, A.D.
1971. Deficiency of Gravity Corers for Sampling Meiobenthos and Sediments. *Nature,* 231:260.

McIntyre, A.D., and R.M. Warwick
1984. Meiofauna Techniques. Pages 217-244 in N.A. Holme and A.D. McIntyre, editors, *Methods for the Study of Marine Benthos.* London: Blackwell Scientific Publications.

Mills, A.A.
1961. An External Core Retainer. *Deep-Sea Research,* 7:294-295.

Muus, B.
1964. A New Quantitative Sampler for the Meiobenthos. *Ophelia,* 1:209-216.

Novak, R.
1982. Spatial and Seasonal Distribution of the Meiofauna in the Seagrass *Posidonia oceanica*. *Netherlands Journal for Sea Research*, 16:380-388.

Ockelmann, K.W.
1964. An Improved Detritus-sledge for Collecting Meiobenthos. *Ophelia*, 1:217-22.

Palmer, M.A., and G. Gust
1985. Dispersal of Meiofauna in a Turbulent Tidal Creek. *Journal of Marine Research*, 43:179-210.

Plocki, W., and T. Radziejewska
1980. A New Meiofauna Corer and Its Efficiency. *Ophelia*, Supplement, 1:231-233.

Pollock, L.W.
1970. Distribution and Dynamics of Interstitial Tardigrada at Woods Hole, Massachusetts, U.S.A. *Ophelia*, 7:145-166.

Raisanen, P., O. Timola, and T. Valtonen
1981. A New Corer for Sampling Sand and Marine Bottom Meiofauna. *Annales Zoologici Fennici*, 18:133-137.

Reineck, H.E.
1958. Kastengreifer und Lotröhre "Schnepfe," Geräte zur Entnahme ungestörter, orientierterr Meeresgrundproben. *Senckenbergiana lethaea*, 39:42-48; 54-56.

Renaud-Debyser, J.
1957. Description d'un carottier adapte aux prelevements des sables de plage. *Revue de l'Institut Francais du Petrole*, 12(4):501-502.

Rutledge, P.A., and J.W. Fleeger
1988. Laboratory Studies on Core Sampling With Application to Subtidal Meiobenthos Collection. *Limnology and Oceanography* 33:276-282.

Sibert, J.R.
1981. Intertidal Hyperbenthic Populations in the Nanaimo Estuary. *Marine Biology*, 46:259-265.

Tanner, C., M.W. Hawkes, and P.A. Lebednik
1977. A Hand-operated Suction Sampler for the Collection of Subtidal Organisms. *Journal of the Fisheries Research Board of Canada*, 34:1031-1034.

Thiel, H.
1966. Quantitative Untersuchungen über die Meiofauna des Tiefseebodens. *Veröffentlichungen des Instituts für Meeresforschung, Bremerhaven*, Supplement 2:131-147.

1970. Bericht über die Benthosuntersuchungen während der "Atlantischen Kuppenfahrten 1967" von F. S. "Meteor". *"Meteor" Forschungsergebnisse, Reihe D.*, 7:23-42.

1971. Häufigkeit und Verteilung der Meiofauna im Bereich des Island-Färör-Rückens. *Berichte der Deutschen wissenschaftlichen Kommission für Meeresforschung*, 22:99-128.

1980. Benthic Investigations of the Deep Red Sea. Cruise Reports: R.V. "Sonne" - MESEDA I (1977), R.V. "Valdivia" - MESEDA II (1979). *Courier Forschungsinstitut Senckenberg*, 40:1-35

1983. Meiobenthos and Nanobenthos of the Deep Sea. Pages 167-230 in G. Rowe, editor, *Deep-Sea Biology*. New York: John Wiley and Sons.

Thiel, H., and R. Hessler
1974. Ferngesteuertes Unterwasserfahrzeug erforscht Tiefseeboden. *UMSCHAU in Wissenshaft und Technik*, 74:451-453.

Thistle, D.
1978. Harpacticoid Dispersion Patterns: Implications for Deep-sea Diversity Maintenance. *Journal of Marine Research*, 36:377-397.

1983. The Response of a Harpacticoid Copepod Community to a Small-scale Natural Disturbance. *Journal of Marine Research*, 38:381-395.

Thistle, D., and K.M. Sherman
1985. The Nematode Fauna of a Deep-sea Site Exposed to Strong Near-bottom Currents. *Deep-Sea Research*, 32:1077-1088.

Walters, K., and S.S. Bell
1986. Diel Patterns of Active Vertical Migration in Seagrass Meiofauna. *Marine Ecology Progress Series*, 34:95-103.

Williams, J.D.H., and A.E. Pashley
1979. Lightweight Corer Designed for Sampling Very Soft Sediments. *Journal of the Fisheries Research Board of Canada*, 36:241-246.

Youngbluth, M.J.
1982. Sampling Demersal Zooplankton: A Comparison of Field Collections Using Three Different Emergence Traps. *Journal of Experimental Marine Biology and Ecology*, 61:111-124.

8. Sampling Strategies

David Thistle and John W. Fleeger

Meiofaunal ecology has only recently made the transition from a qualitative to a quantitative science. As a result, many workers in the field have little training in statistics and are at a disadvantage when faced with designing sampling programs (Coull and Palmer, 1984). Below we introduce what we feel to be fundamental issues in sampling design. We also describe some statistical procedures in order to point out problems that arise in analysis. We do not attempt to duplicate the coverage of a formal text. Rather, we concentrate on principles critical to the collection and interpretation of sample data. In our presentation, we have attempted to minimize use of the vocabulary of statistics and to confine ourselves to relatively simple situations. In some of our comments, we have deliberately chosen to be conservative. Our presentation is not intended to substitute for the many textbooks on this subject, and the reader is encouraged to consult such resources for a more complete understanding.

Concepts at the Start

Statistical analyses are based on the assumption that the samples have been collected in such a way that they represent, without bias, a defined population. (We use "population" in the statistical sense, to mean a group of objects or events (Tate and Clelland, 1957), not in the biological sense.) Only if the samples meet this criterion do they allow logical inferences to be made about the population from which they are drawn. Therefore, the population must be carefully specified before the sampling design is planned. In some circumstances, the population will be easy to define. For example, all the nematodes between mean low water and mean high water (i.e., in the eulittoral zone) on a beach is a well-defined population in the statistical sense (Figure 8.1). Properly taken samples would allow inferences about the nematodes of the eulittoral region of that beach. However, they would not allow statistical inferences about the nematodes of all beaches, of a similar adjacent beach, or even of the non-eulittoral portion of the beach being studied. Stated more abstractly, specification of the population defines the universe that must be sampled and the universe within which inferences are strictly valid. Although extrapolation of results to other, similar beaches may be ecologically useful, it is still extrapolation.

In practice, specifying a population is not always simple. In some circumstances, the question being asked or logistic considerations may make it

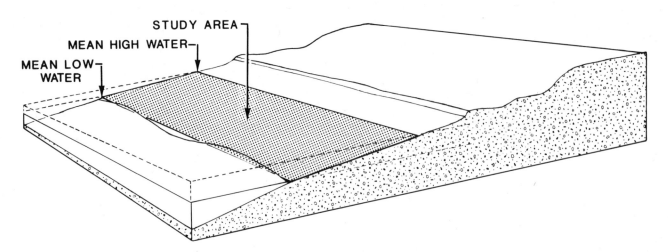

Figure 8.1.--A three-dimensional representation of the beach described in the text, showing the rectangular sampling area in stipple.

impractical to study a naturally bounded population. For example, a beach may extend along shore for kilometers without a natural break. It is likely to be impractical to study the entire beach. A region of any convenient shape can be selected for study and arbitrarily bounded. In these circumstances, although the population is bounded arbitrarily, it still exists. It is the population subject to sampling and to which inferences strictly apply.

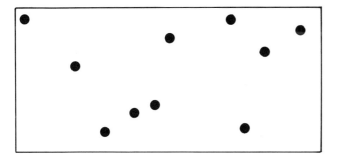

Figure 8.2.--A diagram of the sampling area as seen from above. The locations of randomly placed samples are shown. The placement was achieved by imposition of a Cartesian coordinate system on the beach with the along-shore axis (x axis) running from 0 to 100 and the axis perpendicular to shore (y axis) running from 0 to 50. Pairs of numbers were drawn from a random-number table and used as coordinates in the placement of samples. The procedure is as follows (see also Elliott, 1977). From the circumstances, one can decide the number of digits that are needed to make up a coordinate. The random-number table is entered at the beginning or where one previously left off (Snedecor and Cochran, 1967), and successive entries (using as many digits as necessary) are examined. If a value falls out of the range of values that could serve as coordinates (e.g., is larger than 50 for the y-axis), then the entry is not used. Successive entries are examined until the required number of coordinates is obtained.

The investigator takes samples to obtain unbiased information about a population. The way to accomplish this goal is to take samples randomly from the population (Tate and Clelland, 1957; Elliott, 1977). In the simplest case, the samples are taken such that every unit of the population has an equal opportunity of being selected (Snedecor and Cochran, 1967). To assure this condition, a mechanical randomization procedure, such as the use of a random-number table (see Elliott, 1977), must be involved in the selection of locations to be sampled. For example, if we wished to estimate the average (mean) density of nematodes living on the eulittoral beach described above, we would have to sample at randomly selected locations. This sample placement could be achieved by imposition of a coordinate system on the eulittoral region (the population of interest) (Figure 8.2); x and y coordinates could be selected by means of a random-number table to specify the location of each sample (see Elliott, 1977; Thistle, 1980; Ravenel and Thistle, 1981).

Survey Sampling

A variety of circumstances can arise in which an investigator would wish simply to estimate a parameter of a population. In our example, the density of nematodes might be critical in estimating the biological productivity of the beach. No test of a hypothesis is involved; merely a description is required, but a population must still be identified and randomly sampled.

Simple Random Sampling.--The scheme of sample placement described above for estimating nematode abundance on a beach by random selection of sample locations is an example of simple random sampling. It has the virtue of being easy to use, but it works best when the range of values of the parameter of interest (in our example nematode density) is small. In the example, if the number of nematodes per unit area were not too variable, then, even if the number of samples were small, a simple random sampling scheme would give an estimate of the mean density nearly equal to the true value for the population. However, if nematode density were markedly variable, then simple random sampling might do a poor job. Usually meiofaunal workers analyze small numbers of samples (5-20). If these are placed in a simple random manner, they may not sample the highs and lows of the distribution of nematode densities on the beach in their true proportions. The use of simple random sampling is legitimate in that every location on the beach has an equal probability of being sampled, but, in the face of much variability (as is usually the case with variables of interest to meiofauna ecologists (Fenchel, 1978)), it will estimate a mean value close to the true value only if many samples are taken.

Stratified Random Sampling.--An alternative sampling scheme that works better when density (or another variable of interest) is markedly variable is stratified random sampling. In this scheme (Figure 8.3), the population is divided into regions (strata) before sampling. Because the scheme imposes a regularity on the sample coverage, the irregularities in the distribution of the parameter being measured (i.e., nematode density) are more likely to be sampled in their true proportions by a small number of samples. To present this approach, we chose an example in which the strata are of equal area. They need not be, as discussed below. In our beach example, preliminary data may suggest that nematode concentrations are much greater in the offshore portion than in the onshore region. Suitable strata would be bands of equal onshore-offshore dimension that paralleled the shore; each would receive the same number of samples. When the strata are of equal area and samples are apportioned in this manner, the calculation of the mean and variance are unaffected by the stratification.

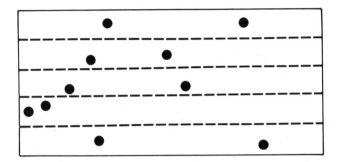

Figure 8.3.—A diagram of the sampling area as in Figure 8.2. Strata are indicated by horizontal lines. Sample placement was random with the constraint that the same number (two in this case) of samples be placed in each stratum. The procedure follows that described in Figure 8.2. Additionally, the pairs of coordinates must be examined for cases in which a stratum would receive more than its share of samples. Such coordinates are discarded, as are coordinates that would place a sample exactly on the boundary between two strata.

Designs with strata of equal area constitute a special case of the general scheme of proportional stratified-random sampling, in which samples are assigned to each stratum in proportion to its area. When the strata are unequal, the mean can be calculated as the average of the resulting values, but the variance (and other statistics) require new formulae, which are given by Elliott (1977) and more completely by Snedecor and Cochran (1967).

Finally, in survey sampling and elsewhere, there is a tendency to fix attention on the mean because it is the value that is used to represent the population and is used in statistical tests for differences. For both ecological and statistical reasons, however, the variability around the mean is also important. If the variability is small, then most samples would yield values similar to the mean, so the mean characterizes the population well. If the variability is large, few of the samples will yield values similar to the mean, so the mean does not characterize the population in the same way; in this case, it is much more a statistical abstraction. Populations with large variability are likely to be heterogeneous in biologically important ways. Statistically, high variability decreases the investigator's ability to discern differences among populations. For both reasons, it is necessary that estimates of variability accompany reports of mean values. Elliott (1977) describes the calculation and use of several measures of variability, including the variance, the standard deviation, the standard error, and confidence intervals.

Sampling in Support of Hypothesis Testing

In testing a hypothesis, the investigator wishes to determine whether some treatment brings about a change in the population. Statistical techniques that allow such questions to be answered in a rigorous fashion require erection of a null hypothesis appropriate to the situation. The null hypothesis takes the form "the treatment has no effect." If such a null hypothesis is true, the values measured in samples drawn at random from the portion of the population that was exposed to the treatment and those drawn from the portion that was not exposed will be the same except for natural (that is, chance) variation. Because samples only *estimate* the true population values, the two sets of samples will differ even if the treatment has no effect. Statistical procedures allow one to examine the size of the observed difference and to state how probable it is that such a difference would arise if the null hypothesis were true. Generally, if a difference as large as or larger than the observed difference would arise by chance only 1 time in 20 or fewer (that is, at the 5% significance level), it is conventional to reject the null hypothesis of no difference and to accept the alternative hypothesis that the treatment has an effect. (See Green, 1979, for more discussion of the formation of appropriate null and alternative hypotheses.) Tests of hypotheses occur under two fundamentally different circumstances. In one case, the treatment is a condition that has been identified by an investigator; in the other, the treatment is applied by an experimenter.

Tests for Differences among Conditions.—From theory, observations, or preliminary data, an ecologist may suspect that two (or more) regions differ on a variable of interest. We introduce techniques appropriate to testing such a supposition using an example (see also Cochran, 1983).

Assume that preliminary samples suggest that nematode density varies with tide level on the eulittoral portion of a beach. To test this working hypothesis, the ecologist will compare densities in four along-shore bands (Figure 8.4). If the ecologist is confident, based on preliminary data, that the variation in nematode density within bands is small compared to the differences among the bands, then simple random sampling is appropriate. Equal numbers of samples are assigned to each band. Within each band, the samples are located at random, so that every location within a band has an equal probability of being sampled (Figure 8.4). It is not appropriate to take the samples from a single location or a small number of adjacent locations within each band if inferences are to be made about the differences between the bands over the entire eulittoral portion of the beach.

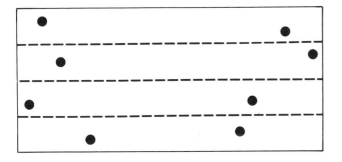

Figure 8.4.—A diagram of the beach as in Figure 8.2 but with the four regions that are the treatments in the example shown as horizontal bands. Samples are placed at random, equal numbers in each band.

The data from this simple random sampling scheme are appropriate for analysis by one-way analysis of variance. Analysis of variance is a statistical procedure that allows one to test for differences among treatments, for example, the differences in nematode density among the bands described above. Either a parametric or a nonparametric test can be used. (When only two treatments are to be compared, Student's T test can be used.) Parametric statistical procedures entail a number of assumptions (see Sokal and Rohlf, 1969), and these merit checking; statistics texts (e.g., Sokal and Rohlf, 1969, 1981) describe procedures. If the data do not satisfy the assumptions, the analysis may still be able to produce reliable tests (see Green, 1979; Underwood, 1981), or it may be possible to transform the data to meet them (see Sokal and Rohlf, 1969; Green, 1979; Underwood, 1981).

These data could also be analyzed by means of nonparametric statistics. Nonparametric procedures differ from parametric in that they require no particular assumptions about the form of the population distributions (Tate and Clelland, 1957), making this approach particularly suitable for the type of data usually available to meiofaunal ecologists (i.e., small numbers of samples from populations with unknown distributions). The nonparametric equivalent of the one-way analysis of variance is the Kruskal-Wallis test (Sokal and Rohlf, 1969). The nonparametric equivalent of Student's T test is the rank test (Tate and Clelland, 1957), which is also called the Wilcoxon T test and the Mann-Whitney U test (Sokal and Rohlf, 1969).

Meiofaunal ecologists usually wish to know which conditions (or treatments) differ from which others (in the example, which along-shore bands differ from which others in nematode density). The results from the analysis of variance (except in the two-treatment case) will not provide this answer directly. If the meiofaunal ecologist has planned to test for a specific difference among treatments before the experiment is carried out, the procedures for *a priori* testing are appropriate (see Sokal and Rohlf, 1969, for methods). However, it is often the case that neither theory nor preliminary data are available to suggest specific comparisons of interest. In these circumstances, an *a posteriori* multiple-comparison procedure should be used to decide which treatments differ. For parametric analyses, a variety of procedures exists, many of which are given by Sokal and Rohlf (1969). For nonparametric analyses, the multiple comparisons procedure based on the Kruskal-Wallis test is available (Hollander and Wolfe, 1973).

If both along-shore and onshore-offshore differences were of interest, a different sampling scheme would be needed, one in which there were two treatments. In the beach example, to assess along-shore differences, the investigator should divide the beach into equal bands perpendicular to the shore. The result is a grid of sampling units (Figure 8.5). Samples are located at random within each cell. This design is appropriate for two-way analysis of variance, a statistical procedure that allows the effects of two treatments to be tested simultaneously. Although a two-way analysis of variance can be done without replication, a good design will have two or more samples per cell in order to permit the calculation of the interaction between the two treatments. A discussion of interaction and other complexities that arise in the two-way analysis of variance is beyond the scope of this chapter. Pitfalls exist, however, and a thorough familiarity with the procedure should be achieved before this design is employed (see, for example, Sokal and Rohlf, 1969; Underwood, 1981).

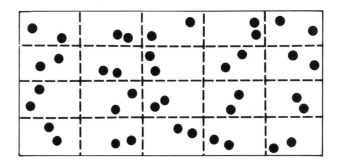

Figure 8.5.—A diagram of the beach as in Figure 8.4 but with the five along-shore bands that constitute the second condition in the example indicated as vertical bands. Equal numbers of samples are placed at random within each sampling unit.

There are many phenomena that can only be

studied by means of testing for differences among conditions, particularly those involving the impact of pollution on the environment. However, because the conditions are chosen rather than applied by the investigator, results can be influenced by variables not part of the study that could contribute to the detected difference (Hurlbert, 1984). For example, if fuel oil from a spill washed half-way up the beach, in a subsequent impact study, any differences detected between the oiled lower part and the non-oiled upper part might be caused by the pollution, but they could be caused by any other differences that existed between the two regions. In our example, the mean density of nematodes on the beach decreases onshore. This difference between regions will exist together with the difference caused by the oil. So, although the oil might cause a major decrease in the density of nematodes, the effect might not be detected if the upper portion of the beach was used as an unaffected, comparison site because the reduction on the lower part of the beach might only be to densities approximating the ordinarily low density on the upper portion of the beach. Green (1979) and Hurlbert (1984) discuss this problem. Stewart-Oaten et al. (1986) provide an approach for cases when it is possible to sample before and after the condition of interest (e.g., the oil spill) occurs.

Tests for Differences among Experimentally Applied Treatments.--A variety of criteria exist that guide and regulate the design of manipulative experiments. We present those we feel to be most important to meiofaunal ecologists by using an example. Imagine the following hypothetical experiment. On the beach described above, an experimenter wishes to determine the impact on nematode recolonization of disturbances of different sizes. Assume the investigator knows from preliminary work that raking a plot will create a nematode-free patch. The experiment planned has four treatments: unraked plots of 0.25 m^2 and 1.0 m^2 and raked plots of 0.25 m^2 and 1.0 m^2. Five plots of each type will be laid out, making a total of 20 plots. The population about which inferences are to be made is to be the entire eulittoral portion of the beach, so the plots must be located at random in this region to make each location equally likely to be included in the experiment.

The plots must also be independent of each other; that is, observations made on one plot must not be influenced by proximity to another plot. For example, plots must be sufficiently far apart that raking a plot does not increase or decrease the availability of nematodes to colonize nearby plots. Further, the plots should not affect one another during the experiment. For example, if two 1.0-m^2 raked plots were close together, the nematode population in the surrounding sediment available for colonization could be decreased below that normally available while the plots are being recolonized. As a result, these plots could recover at a lower rate than they would have if they had been isolated (i.e., independent). (Underwood, 1981, gives a more general account of this issue.) An actual determination of the minimum spacing required for independence would require a preliminary experiment. This effort is seldom made. Most workers rely on a "feel" for a minimum spacing and decide during the planning that, in assigning plot locations at random, if the random numbers drawn would place a plot less than a specified minimum distance from a plot already located, then a new set of random coordinates will be drawn. Although this intuitive procedure is better than no consideration at all of the independence of the plots, it does not guarantee independence.

The requirement for independence suggests that plots be located as far as possible from one another. However, if environmental conditions that can affect the outcome of the experiment (e.g., number of nematodes available for recolonization) are distributed with large differences in the study site, then the random placement of plots may not efficiently detect differences among treatments because of the added variation contributed by this background heterogeneity. Specifically, if variations in the density of nematodes are large, then the rate of recolonization of the plots could be affected by this variability as well as by that caused by differences in plot size. To minimize the effect, the experimenter can lay the plots out in a randomized complete-block arrangement.

The motivation for the randomized complete-block design is to reduce the impact of background variability. In this procedure, "blocks" of experimental material are chosen that are expected to be relatively homogeneous. In most field situations, the expectation is that places close together will be more similar to each other than locations that are far apart, so a block is typically an area that is large enough to accommodate one plot of each treatment type. Each block must be located at random on the eulittoral portion of the beach, so that the beach remains the population. Block orientation should be randomized. However, if a strong gradient is known to exist, then blocks can be oriented in a consistent manner relative to that gradient so that trends in the population will affect each block similarly (Figure 8.6). Within each block, the treatments are assigned to plots at random.

In the example, the two sizes of unraked plots and two sizes of raked plots are the treatments. One plot of each treatment will be present in each block. If 5 replicates of each plot are to be set up, 5 blocks will be created and located at random on the beach. Blocks should be separated sufficiently to be independent. Given the known onshore-offshore gradient on beaches, the blocks are probably best oriented parallel to shore with the 4 treatment plots side by side separated by a distance sufficient to assure their independence. Within each block, the relative positions of the treatment plots should be randomized. If at the end of the experiment a single sample is taken at random from each plot, the data can be analyzed as a randomized, complete-blocks design (Sokal and Rohlf, 1969) for the parametric case. The Friedman rank-sums procedure (Hollander and Wolfe, 1973) is the equivalent nonparametric procedure. If two or more samples are taken from each plot, the data can be analyzed by two-way analysis of variance (Sokal and Rohlf, 1969), an approach advocated by Underwood (1981).

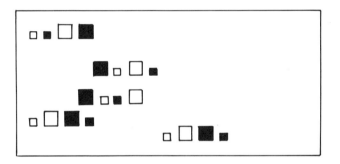

Figure 8.6.--A diagram of the beach as in Figure 8.2 showing a possible layout of an experiment designed as a randomized complete-block design. Squares indicate the locations of plots. Filled squares indicate the raked plots; unraked plots are open squares. Blocks are groups of four adjacent plots, one of each treatment. The placement of treatments within a block is randomized. Blocks are placed at random on the beach, but separated from each other by a minimum distance to assure independence.

The treatment plots in the randomized complete-block design are "interspersed" in Hurlbert's (1984) terms. He argues that interspersion in space or time is a requirement of good experimental design because it protects the experimenter from confusing treatment effects with natural variability in experimental material or changes that occur in experimental conditions during the experiment that only affect a subset of the experimental units. As an example of how such effects could arise, assume in the above experiment that the plots for each treatment type were grouped in different areas of the beach. If a flock of shore birds arrived on the beach, fed in the area that happened to contain the small raked plots, and then were frightened away, they could have affected the nematode densities in the small raked plots and potentially changed the results of the experiment (by artificially slowing the recolonization). If the plots had been interspersed, an effect on only one treatment could not have occurred. Hurlbert (1984) points out that simple random designs can result in an arrangement of treatments among plots that is largely segregated. Blocked designs circumvent this problem.

Contrast the above design with a similar one in which four plots are placed at random locations, each receiving one of the treatments (chosen at random), and where five samples are taken at random locations from each plot (Figure 8.7), to yield a total of 20. The population has not changed, but if the null hypothesis to be tested is the original one of no difference among the treatments, such a design is "pseudoreplicated" (Hurlbert, 1984) because differences among the locations cannot be separated from differences among treatments. Only additional applications of the treatments can allow the differences that necessarily exist among locations to be separated from differences caused by the treatments. More generally, it is the number of repetitions of the manipulation, not simply the number of samples, that is of importance in statistical tests.

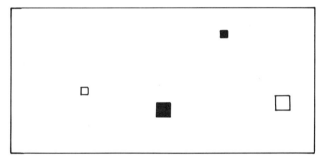

Figure 8.7.--A diagram of the beach as in Figure 8.2 with symbols as in Figure 8.6. The plots were located at random on the beach. In this pseudoreplicated design, five samples are taken from each plot.

Conclusions based on manipulative experiments are less likely to be incorrect than those based on unmanipulated experiments because the random assignment of treatments to locations results in an equal (on the average) contribution of any extraneous variables to each treatment. However, Hurlbert (1984) describes situations where even well-designed manipulations would yield incorrect answers.

Power

Consider a simplified version of the beach example of Figure 8.3 in which, rather than five onshore-offshore bands, only two bands are considered. A simple random sampling scheme followed by a one-way analysis of variance (or a t test) would allow the testing of the null hypothesis that the bands did not differ in nematode density. Test results indicating a significant difference at the usual (5%) significance level mean that the probability that as extreme or more extreme a result would arise by chance alone is 1 in 20. By convention, this level of confidence that chance is not causing the observed differences is considered adequate to reject the null hypothesis of "no difference between the bands."

If the result were not significant, the null hypothesis would not be rejected. This result is much less useful because it is not legitimate to claim that in the absence of a statistically *significant* difference there is *no* difference between the bands (Cochran, 1983). The reason that accepting a null hypothesis is of limited utility is that the sensitivity of the statistical test affects the ability of the test to detect differences. The sensitivity (power) of the test is a function of the size of the difference to be detected, the significance level chosen, the number of samples taken, and the variability present. If the difference to be detected or the number of samples taken is small or the variability is large, then the power of the test may not be sufficient to detect a difference that really exists (the null hypothesis will not be rejected even though it is false).

Questions of power should be considered when a sampling scheme is designed. In particular, good procedure would include an attempt to determine the ability of the design to detect the likely differences. It is best to obtain preliminary samples to get rough estimates of the difference between treatments and the variability of the population. With this information, one can use power tables or statistical procedures (see Elliott, 1977, or Green, 1979) to estimate the number of samples required to detect expected differences.

It may be important to be able to assert that the treatments differ negligibly. As discussed above, the lack of a significant difference between the treatments does not provide statistical support for such an assertion. However, such support can be obtained by means of a procedure described by Rotenberry and Wiens (1985). Briefly, one can estimate the smallest difference between treatments that would have had a 95% chance of being detected given the design; any undetected difference must be smaller. If this smallest detectable difference is sufficiently small in biological terms, then the fact that no difference was detected implies that any difference that might exist is also small in biological terms. Therefore, one can assert, with statistical support, that there is no biologically important difference between the treatments. However, if the smallest detectable difference is large in biological terms, then the fact that no difference was detected is uninformative because an undetected difference could be large, and no assertion can be made about the difference or lack of one between treatments.

Multiple Testing

Circumstances arise frequently in which a number of hypotheses are tested in the same data set. For example, if 20 species are counted from the samples and their abundances correlated with five environmental measures, 100 correlation coefficients can be calculated. Some of the correlations may be significant at the 5% level. In fact, from the definition of the significance level, it can be anticipated that, even if there were no relationship between any species and any environmental variable, on the average, 5 of the 100 coefficients would be significant. That is, even if the original data were thrown away and the data set reconstructed by drawing values from a random-numbers table, on the average 5% of the correlations would be significant. As a consequence, when a group of tests is performed, the significance level of the tests taken as a group differs from the nominal 5% level, and precautions must be taken in order to avoid increasing the probability of rejecting a true null hypothesis (see Underwood, 1981). In analysis of variance, the *a posteriori* multiple-comparison procedures correct for the multiple testing done, for example, when all possible treatments are compared. When procedures other than analysis of variance are used and multiple tests are made, the Bonferroni procedure (Brown and Hollander, 1977) should be followed. That is, the nominal significance level should be divided by the number of tests to yield a corrected (but conservative) significance level for the group of tests (see Thistle, 1980, and Taghon, 1982). For example, if one wanted a 5% significance level for a group of 10 tests, each individual test would be examined for significance at the 0.05/10 = 0.005 level.

Comments at the Finish

Designing a proper sampling scheme is an exacting, time-consuming process. However, the time spent in planning is very cost effective if it saves the investigator from an after-the-fact salvage job, from having uninterpretable results, or from being mentioned in reviews of improper design or analysis (e.g., Hurlbert, 1984). Green (1979) gives recommendations. We would emphasize two. First, if at all possible, consult a statistician about the proposed design. Second, the information obtainable from preliminary samples can be crucial to arriving at a proper design. For example, one can establish that the sampling gear works as anticipated (see Green, 1979:31, for a sobering example). One can obtain estimates of means and variances with which to gauge the number of samples required to obtain a desired degree of precision (see Elliott, 1977, and Green, 1979). One can obtain information on the minimum spacing required for treatments to be independent. The results of such preliminary investigations should permit adjustment of the design to increase the resolving power for the effort invested.

Acknowledgments

The manuscript has been improved by the comments of T. Chandler, P. Culley, F. Dobbs, P. Jumars, D. Meeter, K. Sherman, A. Thistle, and R. Varon.

References

Brown, B.W., Jr., and M. Hollander
1977. *Statistics: A Biomedical Introduction*. 456 pages. New York: John Wiley.

Cochran, W.G.
1983. *Planning and Analysis of Observational Studies*. 145 pages. New York: John Wiley.

Coull, B.C., and M.A. Palmer
1984. Field Experimentation in Meiofaunal Ecology. *Hydrobiologia*, 114:1-19.

Elliott, J.M.
1977. *Some Methods for the Statistical Analysis of Samples of Benthic Invertebrates*. 156 pages. Cumbria: Freshwater Biological Association Scientific Publication No. 25.

Fenchel, T.M.
1978. The Ecology of Micro- and Meiobenthos. *Annual Review of Ecology and Systematics*, 9:99-121.

Green, R.H.
1979. *Sampling Design and Statistical Methods for Environmental Biologists*. 257 pages. New York: John Wiley.

Hollander, M., and D.A. Wolfe
1973. *Nonparametric Statistical Methods*. 503 pages. New York: John Wiley.

Hurlbert, S.H.
1984. Pseudoreplication and the Design of Ecological Field Experiments. *Ecological Monographs*, 54:187-211.

Ravenel, W.S., and D. Thistle
1981. The Effect of Sediment Characteristics on the Distribution of Two Subtidal Harpacticoid Copepod Species. *Journal of Experimental Marine Biology and Ecology*, 50:289-301.

Rotenberry, J.T., and J.A. Wiens
1985. Statistical Power Analysis and Community-wide Patterns. *American Naturalist*, 125:164-168.

Snedecor, G.W., and W.G. Cochran
1967. *Statistical Methods*. 593 pages. Ames, Iowa: The Iowa State University Press.

Sokal, R.R., and F.J. Rohlf
1969. *Biometry*. 776 pages. San Francisco: W.H. Freeman.
1981. *Biometry*. Second edition, 859 pages. San Francisco: W.H. Freeman.

Stewart-Oaten, A., W.W. Murdoch, and K.R. Parker
1986. Environmental Impact Assessment: "Pseudoreplication" in Time? *Ecology*, 67:929-940.

Taghon, G.L.
1982. Optimal Foraging by Deposit-feeding Invertebrates: Roles of Particle Size and Organic Coating. *Oecologia*, 52:295-304.

Tate, M.W., and R.C. Clelland
1957. *Nonparametric and Shortcut Statistics*. 171 pages. Danville, Illinois: Interstate.

Thistle, D.
1980. The Response of a Harpacticoid Copepod Community to a Small-scale Natural Disturbance. *Journal of Marine Research*, 38:381-395.

Underwood, A.J.
1981. Techniques of Analysis of Variance in Experimental Marine Biology and Ecology. *Oceanography and Marine Biology Annual Review*, 19:513-605.

9. Sample Processing

Olaf Pfannkuche and Hjalmar Thiel

This chapter outlines methods for the treatment of sediment samples from the splitting into subsamples through extraction, concentration, fixation, preservation, staining, and enumeration. These procedures are usually the most laborious and time consuming in meiofaunal research. The techniques discussed in this chapter range from those used to effectively remove the meiofauna from large quantities of sediments in order to obtain living meiofauna for experimental work, to the removal of organisms from smaller samples for strictly quantitative investigations. We have attempted to include the more common techniques both simple and inexpensive as well as those that require more sophisticated solutions. Regardless of the kind of study, the efficient extraction of the animals from the sediment is extremely important.

Maintaining Samples

When it is not possible to process the sediments immediately after their collection, it is necessary to maintain the samples in such a condition so that living meiofauna can be extracted later. This can be accomplished by storing the sediments at low temperatures, below ambient but without freezing. When organisms are required for experimental work, it is preferable to keep samples *at* ambient temperature in insulated containers. These should be filled only to about 50% capacity and covered by a thin layer of water to facilitate better aeration. Fine sediments, stored in this manner, can be maintained for 24 to 48 hours; sandy sediments can be stored even longer.

Meiofauna in supra- or eulittoral sands can be kept in good condition for several weeks and still be used for laboratory experiments employing a method proposed by Giere and Pfannkuche (1978). This technique requires that sediments be spread out in a layer 20-30 cm thick, in large plastic troughs placed obliquely to imitate the angle of the sampled beach in order to create physiographic gradients (e.g., moisture). This enables specimens to select zones of preferred environmental conditions. Samples should be maintained at the ambient temperature measured in the ground water.

Sediments collected for quantitative studies can be held only for short periods (hours) before they must be processed. The short generation times of certain meiofaunal taxa, especially protists, and the occurrence of predacious organisms, e.g., certain turbellaria or nematodes, necessitate preservation of samples or live extraction as soon as possible after collection. If the vertical distribution of meiobenthos is to be studied, it is essential to divide samples into the appropriate sections immediately after collection, since the maintenance of such samples - usually taken with a corer or a device such as a "Meiostecher" (Thiel, 1966) - can produce changes in abiotic factors which elicit a corresponding response in the vertical distribution of the fauna, especially of motile forms. If it is not possible to split such samples immediately after their collection, we recommend that such samples be frozen. Cores should be transferred into a $-20°$ C deep freezer and can be stored in this manner for long periods. Deep freezing of cored samples or subsamples from silty sediments also facilitate sectioning, however, freezing can result in distortion of sediment layers (see Chapter 7), and the methods of processing must be adjusted to the questions asked. From our experience with silty deep-sea sediments, we found slicing of a deep-frozen core sample superior to sectioning in natural conditions; however, as with all techniques there may be some elements of the meiofauna that are not as well processed as others; in this case, protists and "soft" meiofaunal taxa may be lost or damaged by this freezing process (Schwinghamer, personal communication).

Fixation and Preservation of Samples

Fixation of meiofauna is normally done by bulk fixation of the sediment samples. However, the fixation technique must be appropriate both for the taxonomic group of interest and for the purpose of the study.

Formalin.--Bulk fixation usually involves the use of formalin. It is important to remember that *formaldehyde* is a gas produced by the oxidation of methyl alcohol whereas *formalin* is a saturated solution of this gas in water (containing 37-40% formaldehyde by weight). Thus, a 10% solution of formalin is the equivalent of a 4% solution of formaldehyde. For bulk fixation 10% formalin still represents the most popular fixative as most meiofaunal taxa such as nematodes, copepods, kinorhynchs, and mites ("hard meiofauna") remain identifiable whereas the so called "soft meiofauna" (e.g., some gastrotrichs, turbellarians, nemertines, etc.) are hardly identifiable and tend to disintegrate after treatment with formalin. For these taxa we recommend live extraction and special fixation techniques (see Chapter 10). Other authors e.g., Schwinghamer (1983), suggest the use of 25% formalin as a fixative. If live extraction and subsequent preservation are not possible immediately after sampling, the material can be frozen until it can be further processed.

Formalin to be used as a fixative should be buffered at a minimum pH of 8.2 because calcium carbonate will dissolve at a lower pH. Commercial formalin (generally purchased as 37-40% aqueous formaldehyde) must be buffered with either 200 g borax or 200 g hexamethylene tetramine (hexamine-urotropine) per liter. If a lower pH is required, sodium glycerophosphate (up to 400 g per liter) may be added. Marble chips also represent a good buffer when added to the stock aldehyde in excess (Schwinghamer, personal communication). If the buffer is in excess the formalin is to be decanted from the residue.

For marine organisms seawater should be used as dilutant for either fixatives or preservatives. Freshwater may be detrimental as a dilutant especially for volumetric determinations of the specimens. Thus seawater-formalin has an increased osmotic pressure, and soft meiofauna may be damaged. Seawater must be filtered in order to prevent contamination of the sample. Generally the use of a 42 μm-mesh sieve (this size representing the generally accepted lower size limit of meiofauna) will be sufficient. For working with size classes <42 μm, seawater should be filtered through corresponding membrane filters.

For fixation of meiofauna, a 10% (or 25%) formalin solution usually is prepared from a buffered 40% stock solution of formaldehyde. However, for large (bulk) samples, including sediments, one must take into consideration the total volume of sediment (and/or specimens), not just the volume of the supernatant fluid. To make sure that the correct volume of fixative is applied, use the volume of the container to calculate the volume of stock solution needed to achieve the desired strength of formalin, add this amount, then fill the jar with seawater and invert several times to mix the fixative, water, and sediment. An extensive review of formaldehyde fixation and preservation of marine zooplankton and the chemistry of fixation and preservation with aldehydes was presented by Steedman (1976) and is highly recommended for special questions.

Formalin fixation of samples should last at least 7 to 10 days but samples can be stored much longer in this fixative. For preservation of meiofaunal specimens (without sediment) a 5% solution of buffered formalin may be sufficient. A note of caution: for health safety, the exposure to formalin should be restricted to an absolute minimum. Handling of formalin should always be carried out in an open area or in a fume hood. Although there are some alternative ways of preserving samples, formalin is still the most reliable and easily manageable fixative.

Microwave Fixation.--This is a well established technique for histological examination of marine invertebrates (Berg and Adams, 1984, and references therein) and may be a promising treatment prior to special fixation for soft meiofauna and protozoa. Specimens can be fixed in a bath of ambient water either after extraction or directly in corer-type subsamplers without any treatment. First tests performed in our laboratory gave promising results for fixation of protozoans from cultures and of sediments in medical syringes converted to corers (Chapter 7).

Glutaraldehyde.--Bulk fixation can also be accomplished with the use of glutaraldehyde. This is particularly effective for the fixation of protists and "soft" metazoan meiofauna. Samples should be narcotized before fixation with an equal volume of magnesium chloride as suggested below. After 1-2 hr, the sample is made to 2% glutaraldehyde using only fresh aldehyde which has been kept refrigerated. The chemical must be clear, not yellow, and must have no sweet odor of esters indicative of deterioration. Samples thus fixed must be kept cold. They can later be transferred after extraction to a suitable preservative with little distortion as long as isotonic conditions are maintained. Buffer is not necessary unless the samples are stored for months. In this case marble chips can be added as buffer to the stock aldehyde.

Storage of Specimens

Most (but not all: see Part III) meiofauna can be stored either in 70-80% ethanol or in "Steedman's mixture," a mixture of propylene phenoxetol and propylene glycol in a 5% concentration (Steedman, 1976). To prepare a stock solution, 100 ml propylene phenoxetol is mixed with 1000 ml propylene glycol. For use as a preservative, 5.5 ml stock solution are added under constant stirring to 94.5 ml filtered sea or distilled water. The pH of the solution is slightly acidic, therefore either borax or sodium glycerophosphate should be added. During the last few years, in our laboratory we have used Steedman's mixture as a preservative and we do all sorting in it with good results. However, we have no experience concerning the long term effects of Steedman's mixture on preserved specimens. According to Steedman (1976) the mixture keeps animals in a good state for at least 10 years at room temperature. For specimens stored for museum collections we still recommend preservation in 70-80% ethanol with glycerin added to prevent accidental desiccation.

Sample Sieving

Sediment samples may have to be sieved to extract a certain size fraction or remove the fine sediment before organism concentration and processing. Sieving must be performed as carefully as possible since rough handling will damage the animals, and should always follow a consistent procedure for better comparability. Large quantities of sediment, especially silts or clays, should be split into small portions and washed without forcing through the sieve. One method is to place sediment into a bucket of water (10% sediment to the volume of the bucket) and to break down the resistant sediment or cohesive lumps of clay into small particles by constant stirring. The disaggregated sediment is then passed through sieves. When using the deck water supply on a ship the water must always be filtered through a plankton net of maximum 42 μm mesh size to prevent contamination with planktonic specimens.

With smaller samples a pre-treatment to improve sieving efficiency is possible. Thiel et al. (1975) used ultrasonic treatment to break up adhesive sediments without any apparent damage to the meiofauna. However, the effect of sonification should be checked carefully as it could damage or destroy protists and other taxa such as kinorhynchs and other more fragile segmented meiofauna. One should test the effect of this procedure by subjecting half of a single sample to ultrasonic treatment and leave the other half untreated. McIntyre and Warwick (1984) suggested the use of the softening agent "Calgon" by adding 100 g Calgon to 500 ml sample volume; Higgins (personal communication) has found that commercial wetting agents, such as those used to avoid water-spots on drying film negatives, are especially useful in sieving fine sediments.

It is always essential to discriminate between the sieving of preserved or live material, because of the ability of certain motile species to pass through the meshes actively and because of the higher flexibility of living and dead but unpreserved specimens. Fractionation of unpreserved material results in different size class distributions from those obtained using preserved material. In case of live sieving Dybern et al. (1976) proposed to check the loss of small specimens either with a sieve finer than the lowest mesh size limiting the investigated size spectrum or by treating the filtrate with formalin and then re-sieving through the finest mesh employed.

When sieving is performed, sediment should be passed through a set of sieves of different mesh sizes in order to facilitate further processing by working up separate size fractions and to compare counts with other surveys where different mesh sizes were used. We propose the following set of sieves preferably with metal gauze (stainless steel, brass, or bronze) or with Nytex nylon gauze which can be glued into plastic cylinders:

- 1000 μm – representing the upper size limit of meiofauna.
- 500 μm – sometimes used as upper size limit of meiofauna; is also recommended by several authors as lower size limit for macrofauna especially in limnology; represents a fraction mainly consisting of juvenile macrofaunal organisms (temporary meiofauna).
- 250 μm
- 125 μm
- 63 μm – used and still in use by some investigators as the lower limit for the retention of meiofauna.
- 42 μm – the generally accepted lower limit for the retention of meiofauna.

If it is one's specific objective to achieve a better comparison with older data, such as in the case of the lower limit of macrofauna of the deep-sea studies in the Pacific Ocean by Hessler and Jumars (1974), a 300 μm mesh sieve can be used. It might be also advisable to include other mesh sizes, preferably in a logarithmic scale, especially when determining biomasses in relation to size categories.

Sieves should be regularly checked for damage to the mesh and for clogging. The latter can be prevented by frequent cleaning using a softening agent such as Calgon and an ultrasonic water-bath.

Organism Concentration and Extraction of Specimens

The concentration of organisms generally refers to their extraction from the bulk sediment. Extraction techniques vary widely according to sediment type, to whether quantitative or qualitative results are required, whether the study is specific to certain taxa only, and whether the organisms are living or preserved.

Some common extraction techniques are listed in Table 9.1. They fall into two broad categories: the processing of preserved or unpreserved sediments. These are divided again into techniques involving either coarse or fine/silty sediments.

Table 9.1.–Methods for Extracting Meiofauna from Sediments.

Live Material

Coarse sediments	Fine/silty sediments
Deterioration techniques (change of temperature, depletion of oxygen, only motile organisms, not quantitative)	Deterioration techniques (only for certain taxa, not quantitative)
	Oxygen depletion technique (only quantitative for certain taxa)
Decantation (with anesthetic)	Bubbling technique (only certain taxa, not quantitative)
Elutriation techniques	
Seawater-ice technique (only quantitative for certain taxa)	Sieving and examination of residue (preferably through a set of different mesh sizes, only recommended for extremely silty sediments)
	Centrifugation in silica sol-sorbitol mixture

Preserved Material

Coarse Sediments	Fine/silty sediments
Decantation	Sieving and examination of the residue
Elutriation techniques	Flotation or centrifugation in colloidal silica (Ludox) or other media creating density gradients (e.g., sugar)

Qualitative Extraction.--Such methods are preferably used to extract living organisms for experimental use or for taxonomic purposes. One of the simplest methods of extracting living meiofauna from either coarse or fine sediments is to allow a sediment sample, covered with ambient water, to rest at a slightly higher temperature than measured *in situ* (depletion technique). Many organisms will come to the surface due to the depletion of oxygen and the rise of temperature, especially in muds, or because of a phototactic reaction.

An equally simple method, most suitable for fine sediments, is that of bubbling air into the sediment (bubbling technique); this technique is useful in the concentration of kinorhynchs and small crustaceans which possess a hydrophobic cuticle. Such animals, when entrapped by the surface tension of seawater, can be lifted from the surface by repeatedly placing a piece of ordinary paper (typing or xerographic copy paper) on the surface momentarily, lifting it from the surface film, washing the attached ("blotted") organisms into a fine-mesh net using a wash-bottle with seawater until sufficient quantities are collected, then transferring the concentrated specimens into a dish for live examination or into a container to which sufficient formalin is added to achieve proper fixation (Higgins, 1964). Like the oxygen depletion technique, this technique has its limitations and may remove only certain organisms effectively.

In most instances, live extraction is facilitated by the use of an anesthetic, since many interstitial meiofauna are attached to sand grains. The most common method for marine organisms is the treatment of sediments with an isotonic solution of magnesium chloride: 7.5 g $MgCl_2$ $6H_2O$ dissolved in 100 ml distilled water. When using magnesium chloride as an anesthetic, it should be added to the sample in equal volume and the sediment should be stirred gently into suspension in a container suitable for subsequent decantation or elutriation. The anesthetic should be allowed 10 minutes to react before further processing of the sample.

Good results can also be obtained by adding sufficient ethanol to either freshwater or marine sediment samples in order to effect a 5-10% solution, allowing 3-4 minutes to anesthetize the meiofauna before further processing. Both freshwater and marine meiofauna also may be anesthetized by the introduction of weak formalin in a similar manner.

Another simple method, akin to anesthetization, which has proven effective in the case of samples of medium to coarse marine sediments, is that described by Kristensen and Higgins (1984) which involves the momentary addition of freshwater into the sample. This technique requires careful procedure, but has been extremely effective in removing such taxa as the Tardigrada and Loricifera from sediment. A few handfuls of the sand sample are placed in a bucket with only enough freshwater to allow its immediate and brief (10-15 seconds!) suspension before decanting the suspended material into a fine-mesh net. The sand is reintroduced to the freshwater a second time, stirred, and decanted; the material in the fine-mesh net is then returned to filtered seawater. If this process is completed rapidly, there appears to be very little prolonged effect on the meiofauna, even on some of the notoriously delicate ones such as the Gastrotricha or some Turbellaria.

Quantitative Extraction.--Several methods have been described for quantitative isolation of meiofauna from the sediments. However, there is no single technique to remove all the specimens of all taxa from a given sediment sample, short of a most time-consuming and extremely careful examination of the sediment sample, small portions at a time, under the highest (normally 50x) magnification of a stereomicroscope. All methods have shortcomings depending on sediment type and animal group. Combinations of different methods may overcome the limitations of single ones, and it is strongly recommended to test the errors introduced with specific procedures. In most cases the application of a standardized method corrected for inherent error can yield quantitative results even through losses occurring during extraction. Quantitative extraction of meiofauna can be grouped into the following methods:

1. Those such as decantation, elutriation, and flotation in various media which rely on differential sinking rates of sediment particles, detritus and meiofauna. They are suitable for both live and fixed samples.
2. Those relying on the reaction of organisms towards environmental gradients (e.g., seawater-ice technique, or oxygen depletion technique), suitable only for live extraction and only for certain taxa.

Extractions from Coarse Sediments.--Medium and coarse sands have a higher specific gravity than the inhabiting animals. The simplest technique without great expenditures is the decantation method, which can be employed for both preserved and live materials. Filtered water should always be used to avoid contamination. For live extractions it is preferable to use *in situ* water together with an anesthetic. Procedure is as follows:

1. Place about 150-200 ml of sand into a glass cylinder of 30-40 cm height and pour in 800-1000 ml of water.
2. Put a stopper on top of the cylinder and invert it several times.
3. Leave the cylinder to let the sand particles settle out (time depends on the nature of the sediment, but should not exceed 60 seconds).
4. Decant the supernatant water through a sieve of 42 µm or preferably through a set of sieves (see above).
5. For quantitative extraction repeat the procedure several times (about 3-5 times).
6. If using a set of sieves, gently wash some more water through the sieves in order to facilitate a good sorting into size fractions.

Elutriation of sediments is also based on the hydrodynamic equivalent of the diameter of the particles, but represents an advanced and more effective method of animal concentration. The method was introduced by Boisseau (1957) and has been modified by many meiobenthologists (cf. Uhlig et al., 1973, and references therein; Ankar and Elmgren, 1976; Fricke, 1979). A simple experimental set up using standard "Quickfit" laboratory glassware has been presented by Hockin (1981).

As a standard procedure (Figure 9.1), sediment is placed in a separatory funnel. A constant stream of water enters for about 15 minutes from below, which washes animals out of the whirled up sediment through the top outflow and onto a sieve. For preserved material an open water supply can be used whereas for unpreserved samples a pump-driven water circuit of preferably ambient water should be employed in combination with use of a narcotic agent (e.g., magnesium chloride). The elutriator's outflow should be drained onto a fine sieve (preferably 42 µm) so that the water meets the sieve surface at an oblique angle. Many authors also claim the advantage of a second sieve. Elutriation separates most of the lighter organisms such as nematodes, harpacticoids, and turbellarians from coarse sediments. However, heavier, shell-bearing organisms such as foraminiferans, ostracods, and molluscs are not quantitatively separated from the sediment. A check of the residue under the stereomicroscope is always necessary. Elutriation is only applicable to medium and coarse sands. Gravel and stones should be picked out prior to an elutriation in order to avoid damage to organisms and glassware.

Tiemann and Betz (1979) evaluated the physical background and suggested methodological improvements of elutriation. They constructed a separatory funnel larger and narrower (about 40 cm height) than those used in earlier described systems (cf. Uhlig et al., 1973). The shape of the funnel is critical for reducing turbulence, enhancing separation efficiency, and reducing sediment accumulation on the collecting sieve. Tiemann and Betz (1979) also claimed a statistically significant higher meiofaunal sorting efficiency for their newly designed separator with a sorting efficiency of 95 ($\pm$5)% for a sediment sample containing different meiofaunal taxa against 84 ($\pm$7)% when opposed to conventional separators.

Uhlig (1964) developed a separation technique based on the behavioral reactions of organisms in sediments exposed to melted water from seawater-ice. Besides a possible temperature effect, organisms are mainly influenced by two physiographic parameters: (1) the formation of salinity and temperature gradients from low to high salinities and temperatures caused by the melting process, and (2) the slow perfusion of water down through the sediment column.

For the standard procedure, about 50 cm^3 of sediment are placed in a plastic tube of 15 cm length and 5 cm cross-sectional diameter. The bottom of the tube is covered with a tightly fitting nylon gauze of about 100-150 μm mesh size. The sediment is covered with a layer of cotton wool before crushed seawater ice is added into the tube, thus avoiding direct contact between sediment and ice. A Petri dish filled with filtered seawater is adjusted below the tube so that the nylon gauze is in contact with the water through surface tension. The Petri dish is placed in a second larger one to catch the overflowing water (Figure 9.2).

(1973) for the extraction of ciliates.

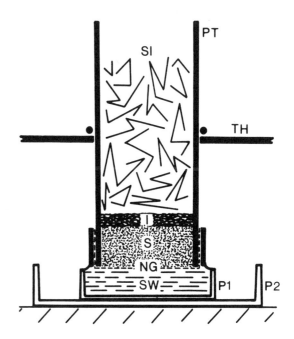

Figure 9.2.–Seawater-ice extractor: I, insulation material; NG, nylon gauze; P1+2, Petri dishes; PT, plastic tube; S, sediment; SI, seawater-ice; SW, sea water; TH, tube holder (redrawn from Uhlig et al., 1973).

The seawater-ice method is most effective with sediments bearing a capillary structure, especially sands. However, Uhlig et al. (1973) also recommend its use for pre-partitioned sediment fractions sieved from silty sediments. According to Riemann (personal communication) even deep-sea sediments can be successfully treated with this method.

Most motile taxa are driven out of the sediment with the seawater-ice method and concentrate in the collecting petri dish. Other taxa, especially nematodes, are not extracted quantitatively. The technique is only well suited for the extraction of motile soft meiofauna such as turbellarians and gastrotrichs and also for ciliates and flagellates. According to Sopott (1973), 97-99% of proseriate turbellarians were driven out of a sandy sediment with a single seawater-ice extraction. The effectiveness can be enhanced if necessary by repeated extractions or by combining with subsequent decantation, elutriation, or hand sorting.

Poizat (1975) described a modification of the seawater-ice technique to treat large volumes of sediment (up to 5000 cm^3) to separate interstitial gastropods from coarse sediments.

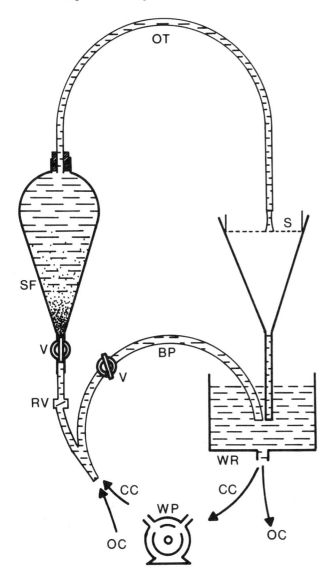

Figure 9.1.–Elutriation apparatus combined for closed and open system: BP, by-pass; CC, closed circuit; OC, open circuit; OT, overflow tube; RV, reflex valve; S, sieve; SF, separation funnel; V, valve; WP, water pump; WR, water reservoir (redrawn from Uhlig et al., 1973).

Modifications of Uhlig's (1964) apparatus were introduced by Schmidt (1968). Plastic tubes of different size and diameter were used by Hartwig

Extractions from Fine to Silty Sediments: Live Specimens.--The extraction of live organisms from fine and muddy sediments is generally more of a problem than the extraction from coarse sands, due to the small size and low settling velocity of the sediment particles and the high proportion of detritus which hamper further processing.

Living material can also be extracted from sediments using sieving. The selection of mesh sizes depends mainly on whether only a certain taxon or a complete quantitative survey is planned. In the latter case a sieve of 42 µm or even of a lower mesh size must be used (see above).

A method involving a flotation medium to extract living organisms from muds was described by Schwinghamer (1981). It employs the use of a "Percoll"-sorbitol mixture which, in contrast to other flotation media such as "Ludox" (see below), is non-toxic to meiofauna. The "Percoll"-sorbitol mixture after Price et al. (1978) is prepared by dissolving 91.1 g sorbitol, 2.64 g Tris-HCl and 4.03 g Tris base in 500 ml "Percoll." To this mixture 1.43 g $MgCl_2$ $6H_2O$ dissolved in 100 ml filtered seawater is added. Finally the volume is made up to 1 liter with "Percoll." The mixture should be filtered and stored in a refrigerator to prevent growth of bacteria and diatoms. The density of the mixture (1.15 g ml^{-1}) should be checked regularly. The extraction procedure runs as follows:

1. Place 5 ml sediment sample and 20 ml "Percoll"-sorbitol mixture in a 50 ml centrifuge tube (preferably with round bottom). A treatment with 7% $MgCl_2$-solution as a narcotic agent is advisable.
2. Mix sediment and flotation mixture by inversion or on a vortex mixer.
3. Centrifuge at 2000-3000 rpm for 15 to 30 min. Spinning speed and time should be adjusted to the sediment type so that sufficient separation between a compact sediment plug and a clear supernatant is achieved but allowing an easy re-suspension of the plug for a further extraction.
4. Decant the supernatant or remove it by vacuum on a screen of at least 42 µm mesh size (preferably 10 µm).
5. Re-suspend the residue of the first extraction in 20 ml fresh "Percoll"-sorbitol mixture and follow steps 2 to 4 to get an adequate representation of all taxa present.

Generally two extractions have been found sufficient for all organisms removable with this method. The method does not damage soft meiofauna or protozoa. "Percoll" is expensive and may be used for specific questions only.

An alternate method of centrifugation, useful for both live and preserved material, involves the use of a saturated sucrose solution (Higgins, 1977). The solution is prepared by dissolving 680 gm sucrose in sufficient tapwater to produce 1 liter. Two portions of sucrose solution mixed with one portion of sieved sediment (to suspend the particulate matter) are placed in 100 ml centrifuge tubes. Subsequent centrifugation is nearly identical to the method described below. After pouring the supernatant through a fine mesh sieve, the sieve is flushed immediately by a seawater rinse. Most organisms recover within a few minutes when subsequently transferred to a Petri dish with seawater.

A method based on oxygen depletion and the corresponding behavioral reactions of organisms was recently presented by Armonies and Hellwig (1986). Mud samples of 2-10 cm^2 surface are placed in plastic tubes and are covered with a 2 cm layer of clean dry sand of a grain size between 500 to 1000 um. Filtered ambient water is rinsed into the tube until the sand layer is just moistened. The tubes are closed and stored in the dark. After a period of days, oxygen depletes in the cores. The organisms in the mud follow the oxygen gradient to higher concentrations and accumulate in the sand layer. This layer can be removed periodically and renewed. Animals concentrated in the sand can be further extracted by decantation or elutriation. According to Armonies and Hellwig (1986) turbellarians, polychaetes, oligochaetes, copepods, and ostracods were extracted quantitatively, whereas nematodes and mites, which are fairly resistant to anaerobic conditions, were not completely driven out of the original sediment.

Extractions from Fine to Silty Sediments: Preserved Specimens.--Sieving and examination of the residue is an adequate method to separate many organisms from preserved mud and silty sand samples, especially deep-sea samples. Most of the mineral sediment is washed through a 42 µm screen so that only a small sieve residue has to be examined.

If samples are of greater volume or consist of a large amount of residue even after sieving, animals can be extracted by density gradient separation. Various media have been used, including zinc chloride (Lydell, 1936), sodium chloride (Lyman 1943), carbon tetrachloride (Dillon, 1964), and a variety of sugar solutions (Anderson, 1959; Hallas, 1975; Heip et al., 1974; Higgins, 1977, described above; Kajak et al., 1968; Schwoerbel, 1980). However, especially soft meiofauna may be damaged by the high osmotic pressure created with sugar solutions.

The flotation media mentioned above have been replaced recently in most laboratories by a colloidal silica polymer called "Ludox," which was introduced into meiobenthic research by De Jonge and Bouwman (1977). They suspended sediment samples of 2 cm^3 in a glass beaker with 300 ml 25% (V/V) Ludox-TM (specific gravity 1.39) made up with distilled water.

The use of distilled water is important as Ludox gels immediately in seawater. The surface of the Ludox suspension was then gently flooded with distilled water to prevent desiccation and formation of insoluble crystals of the Ludox. After 16 h of settlement, the Ludox containing the meiofauna was sucked off and washed onto a 35 μm screen. Organisms were washed with distilled water to remove adherent Ludox gel before further processing. The Ludox method is applicable to samples of coarse as well as fine sediments.

The use of Ludox-AM (specific gravity 1.21) in combination with centrifugation of a Ludox-AM-sediment suspension as a more rapid method of extraction was described by Nichols (1979). In contrast to Ludox-TM, Ludox-AM shows a slower gel formation rate in the presence of divalent cations and is therefore more suitable for use in marine sediments.

The following method has been developed at our laboratory and at the Zoological Institute of the University of Hamburg (Giere, personal communication) for both freshwater and marine muds:

1. A sediment sample of 15 cm^3 is placed in a 100 ml centrifuge tube. 60 ml pre-filtered Ludox-AM are added and mixed with the sediment into a homogeneous mash.
2. Centrifuge tubes are left for 5 min to allow sedimentation of heavier particles.
3. Tubes are transferred into the centrifuge and spinning speed is increased stepwise within 5 min to 1500 rpm. This speed is maintained for another 5 min.
4. After an immediate stop using the magnetic break, the supernatant containing the organisms is decanted through a 42 μm screen and washed thoroughly with distilled water.
5. The sediment plug is homogenized with fresh Ludox-AM and procedures 2-4 are repeated two more times.

According to our experience, the separation of 98-100% of many taxa is achieved with 3 centrifugations, but this should be checked for each sediment type.

A thorough washing of organisms after extraction is essential to prevent gel formation, especially on animals with appendages and spines. Glassware and sieves used in the procedure should also be cleaned thoroughly, preferably by washing with dilute NaOH. Ludox irritates the skin, especially mucous membranes. Rubber gloves should be worn when handling Ludox. Ludox should be stored tightly closed in the dark to prevent desiccation and growth of photosynthetic organisms such as diatoms. One disadvantage of Ludox is that, at present, it is sold only in quantities of 210 liters.

Sample Splitting

Rather large sediment samples are sometimes required for quantitative work because of the small scale patchy distribution of meiofauna. Sub-sampling of the original sample or of the residue left after extraction by splitting can reduce the time required for sorting. Subsamples can then be selected for further processing. The type of selection (e.g., random, systematic) should be appropriate to the previous treatment and intent of subsequent analysis. Some sediments, especially of silty character, can be divided before organism extraction or, alternatively, sieve residues, preferably fractioned by a set of sieves, can be split into subunits.

Sample splitting can be carried out with simple techniques. An easy but usually non-quantitative method is to place a sample into a beaker and to fill it up with tap water to a known volume. The sample is then vigorously stirred into suspension and a subsample is rapidly drawn out with wide mouthed pipette or preferably with a subsampler of known volume such as the "Stempel Pipette" or "Hensen Pipette."

Some sample splitters designed for plankton research have also been successfully used in meiobenthic research, e.g., the "Folsom" plankton splitter (McEwen et al., 1954) by Juario (1974). However, plankton splitters are only useful for lighter fractions of sediment samples. A splitter better fitted for the heavier fractions is the "microsplitter" used by Olsson (1975).

The "Askö" splitter (Elmgren, 1973) was designed specifically for meiobenthic studies. The apparatus (Figure 9.3) consists of a tightly closed Plexiglas cylinder with the lower part divided radially into eight equal chambers. A sediment sample is mixed in the converted cylinder with 1 liter of water by vigorous agitation. It is then left in the position with the chambers at the bottom for one hour to allow the particles to settle out. The water overlying the splitting chambers is slowly drained off until it reaches the level of the dividing walls. The chambers can be emptied separately by removing rubber stoppers in the bottoms of the chambers.

A more advanced meiofauna splitter which we call the "Jensen" splitter (Jensen, 1982) is more efficient than the "Askö" splitter and also requires less time. The "Jensen" splitter is constructed of PVC and consists of two cylindrical chambers (Figure 9.4): an upper mixing chamber with a central opening (7 mm cross-sectional diameter) which leads into a lower splitting chamber. The top of the mixing chamber is closed by an O-ring bearing lid, and the hole leading into the splitting chamber is closed by a rubber stopper. The splitting chamber is fitted to the mixing chamber with a watertight joint and has a

2 mm hole near the top for pressure equilibration.

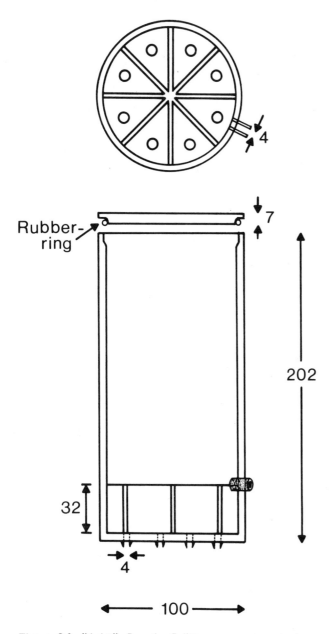

Figure 9.3.--"Askö" Sample Splitter, measurements in mm (redrawn from Elmgren, 1973).

The splitting chamber has a central rod with a conical top and is divided by 8 radial compartments (58 ml volume) with 64 mm-high walls. Each compartment has an individual drainage hole in its bottom sealed with a rubber stopper. The splitter is used as follows:
1. The sample is washed into the mixing chamber and the volume is made up to 275 ml.
2. The mixing chamber is closed with the lid and vigorously shaken.
3. The mixing chamber is fitted to the splitting chamber and its rubber stopper is removed.
4. The suspension running over the cone of the splitting chamber is split into the eight compartments and the mixing chamber is rinsed with water to remove any remaining organisms.
5. The mixing chamber is removed.
6. The subsamples to be examined are collected by removing the rubber stoppers and rinsing the samples into containers.

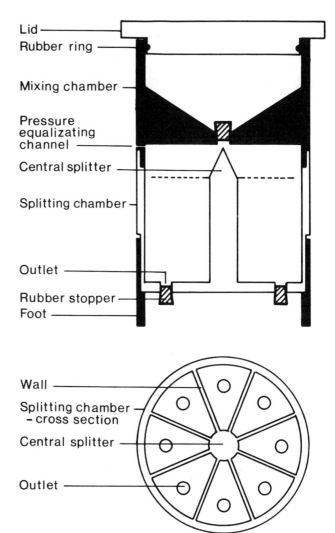

Figure 9.4.--"Jensen" Sample Splitter (redrawn from Jensen 1982).

The "Jensen" splitter has been successfully used in our laboratory with some modification (Giere, personal communication) by giving the bottom of the dividing chambers a concave shape to facilitate a better drainage.

The methods described do not allow for

recognition of small scale distribution patterns in the meiofauna. This can be achieved only by collecting and processing smaller core samples individually.

Sample Staining

Sorting of meiofauna from sediments or from extraction residues is hampered, especially in preserved material, because smaller organisms are not easily detectable when not in motion. To facilitate animal detection entire samples can be stained before further processing. Rose Bengal, a protein stain, commonly has been used by many meiobenthologists for this purpose. It can be employed either at the time of preservation or on samples already processed to a certain extent. For staining in combination with fixation in formalin or alcohol, 10 ml of 1% Rose Bengal solution (1 g Rose Bengal in 1 liter of 10% formalin) are added to a sample volume of 1 liter, which will create a sufficient stain after some days. Samples already processed to a certain extent, such as extraction residues, can be placed on a sieve into a bath of 1% Rose Bengal solution. Staining for 10-15 minutes is sufficient for some animals but as much as 48 hours is required to insure the complete staining of animals whose exoskeletons are notoriously difficult to penetrate (e.g., Halacarida, Kinorhyncha). The addition of 5 g of phenol to the stock solution (Thiel, 1966) is advisable for pH adjustment, as Rose Bengal stains best at a pH of 4-5.

A method of separating meiofauna from sediments employing fluorescent dyes in combination with longwave ultraviolet light was described by Hamilton (1969). He found that, of nine different fluorescent dyes, 0.1% Rhodamine B in tapwater was the most effective stain for living organisms, whereas for formalin preserved organisms 0.1% dye in 70-98% ethanol was more suitable. Freshwater invertebrates stained a bright pink and fluoresced a brilliant orange when scanned under longwave ultraviolet light.

Vital stains such as Neutral Red have been successfully used with living meiofauna such as rotifers (Berghahn, 1979).

Sample Examination and Counting of Specimens

Meiofauna should be examined and counted under a stereomicroscope at a minimum of 25x magnification. The smallest meiofauna, among them the tardigrades, loriciferans, and juvenile stages of these and many others, are often only seen using 50x magnification. Enumeration of preserved specimens stained with Rose Bengal remains the most accurate method for total fauna counts. Specially designed counting trays or Petri dishes marked with parallel lines or squared paper grids on the bottom facilitate accurate scanning. Most identifications to the species level and sometimes to even higher taxonomic levels require that the specimens be removed and mounted on slides for examination under a high power microscope (Chapter 10, and Part III).

Schwinghamer (personal communication) uses an inverted plankton microscope for meiofauna examination and counting. Extracted subsamples are placed directly in settling chambers for counting, sizing (using image analysis), and some identification. According to his experience the counting process is much quicker than under a stereomicroscope and small organisms are not missed. Due to the higher range of magnification there may be less need to pick and mount specimens for individual examination, depending on the purpose of the study.

The extraction of the tiny specimens from the samples may be done with hooked needles (especially for elongated specimens) or with fine loops of thin stainless steel, called Irwin loops (Chapter 10). These methods assure a safe transfer of specimens into preservation vials and small observation dishes, or onto slides. The use of thin pipettes is also possible, but specimens, especially when still alive, easily stick to the tube and may get lost. Coating of pipettes by dipping them into hot silicone reduces this risk. Transfer by mouth-pipette for living specimens allows for a better adjustment to the motion of organisms and more careful handling of soft meiofauna. The ultimate criteria for the effectiveness of the extraction and transfer technique is the *absence* of the specimen from the transfer tool and the *presence* of the specimen in the new medium; usually confirmation of the first tenet of this is sufficient.

Faunal counts should be expressed in numbers of specimens per unit area or per unit volume at a specified depth horizon. The generally accepted unit of area in meiofaunal studies is 10 cm^2 (equivalent of an area measuring 3.16 cm x 3.16 cm) (Chapter 7) as opposed to studies of macrofauna where faunal counts are expressed as numbers of organisms per m^2. However, the objectives of the sample collection and analysis should be taken into account before deciding on units. Where more than one sample per location has been taken, the published data should include the following information: (1) numbers of replicate cores and distances between them; (2) numbers of subsamples per core; (3) numbers of specimens in total and per taxon in each sample; and (4) calculated mean values and standard deviations.

References

Anderson, R.O.
1959. A Modified Flotation Technique for Sorting Bottom Fauna Samples. *Limnology and Oceanography*,

Ankar, S., and R. Elmgren
1976. The Benthic Macro- and Meiofauna of the Askö-Landsort Area (Northern Baltic Proper). *Contributions of the Askö Laboratory*, 11:1-115.

Armonies, W., and M. Hellwig
1986. Quantitative Extraction of Living Meiofauna from Marine and Brackish Muddy Sediments. *Marine Ecology Progress Series*, 29:37-43.

Berg, C.J., and N.L. Adams
1984. Microwave Fixation of Marine Invertebrates. *Journal of Experimental Marine Biology and Ecology*, 74:195-199.

Berghahn, R.
1979. Schädigung limnischer Rotatorien bei der Passage durch das Kühlsystem eines Kohlekraftwerkes. *Archiv für Hydrobiologie*, Supplement, Elbe-Ästuar, 4:225-235.

Boisseau, J.-P.
1957. Technique pour l'étude quantitative de la faune interstitielle des sables. *Comptes Rendus du Congrés des Sociétés Savantes de Paris et des Départements*, 1957:117-119.

Dillon, W.P.
1964. Flotation Technique for Separating Fecal Pellets and Small Marine Organisms from Sand. *Limnology and Oceanography*, 9:601-602.

Dybern, B.I., H. Ackefors, and R. Elmgren
1976. Recommendations on Methods for Marine Biological Studies in the Baltic Sea. *Baltic Marine Biologists*, 1:1-98.

Elmgren, R.
1973. Methods of Sampling Sublittoral Soft Bottom Meiofauna. *Oikos*, Supplement, 15:112-120.

Fricke, A.H.
1979. Meiofauna Extraction Efficiency by a Modified Oostenbrink Apparatus. *Helgoländer Wissenschaftliche Meeresuntersuchungen*, 32:436-443.

Giere, O., and O. Pfannkuche
1978. An Ecophysiological Approach to the Microdistribution of Meiobenthic Oligochaeta. II. *Phallodrilus monospermathecus* (Tubificidae) from Boreal Brackish-water Shores in Comparison to Populations from Subtropical Beaches. *Kieler Meeresforschungen*, Sonderheft, 4:289-301.

Hallas, T.E.
1975. A Mechanical Method for the Extraction of Tardigrada. Pages 153-157 in R.P. Higgins, editor, Proceedings of the First International Symposium on Tardigrades. Memorie dell'Istituto Italiano di Idrobiologia Dott. Marco de Marchi, 32 (Supplement).

Hamilton, A.L.
1969. A Method of Separating Invertebrates from Sediments Using Longwave Ultraviolet Light and Fluorescent Dyes. *Journal of the Fisheries Research Board of Canada*, 26:1667-1672.

Hartwig, E.
1973. Die Ciliaten des Gezeiten-Sandstrandes der Nordseeinsel Sylt II. Ökologie. *Mikrofauna des Meeresbodens*, 21:1-171.

Heip, C., N. Smol, and W. Hautekiet
1974. A Rapid Method for Extracting Meiobenthic Nematodes and Copepods from Mud and Detritus. *Marine Biology*, 28:79-81.

Hessler, R.R., and P.A. Jumars
1974. Abyssal Community Analysis from Replicate Box Cores in the Central North Pacific. *Deep-Sea Research*, 21:185-209.

Higgins, R.P.
1964. A Method for Meiobenthic Invertebrate Collection. *American Zoologist*, 4:291.
1977. Two New Species of *Echinoderes* (Kinorhyncha) from South Carolina. *Transactions of the American Microscopical Society*, 96:340-354.

Hockin, D.C.
1981. A Simple Elutriator for Extracting Meiofauna from Sediment Matrices. *Marine Ecology Progress Series*, 4:241-242.

Jensen, P.
1982. A New Meiofauna Sample Splitter. *Annales Zoologici Fennici*, 19:233-236.

Jonge, V.N. de, and L.A. Bouwman
1977. A Simple Density Separation Technique for Quantitative Isolation of Meiobenthos Using the Colloidal Silcia Ludox-TM. *Marine Biology*, 42:143-148.

Juario, J.V.
1974. *Artenzusammensetzung der Nematodenfauna und jahreszeitliche Fluktuation einer sublitoralen Meiofauna-Gemeinschaft in der Deutschen Bucht.* 76 pages. Ph.D. Dissertation, Universität Hamburg.

Kajak, Z., K. Dusoge, and A. Prejs
1968. Application of the Flotation Technique to Assessment of Absolute Numbers of Benthos. *Ekologia Polska*, Series A, 16:607-620.

Kristensen, R.M., and R.P. Higgins
1984. A New Family of Arthrotardigrada (Tardigrada: Heterotardigrada) from the Atlantic Coast of Florida, U.S.A. *Transactions of the American Microscopical Society*, 103:295-311.

Lydell, W.R.S.
1936. A New Apparatus for Separating Insects and Other Arthropods from Soil. *Annals of Applied Biology*, 23:862-879.

Lyman, F.E.
1943. A Pre-impoundment Bottom Water Study of Watts Bar Reservoir Area (Tennessee). *Transactions of the American Fisheries Society*, 72:52-62.

McEwen, G.F., M.W. Johnson, and T.R. Folsom
1954. A Statistical Analysis of the Performance of the Folsom Plankton Splitter, Based upon Test Observations. *Archiv für Meteorologie, Geophysik, Bioklimatologie*, Serie A, 7:502-527.

McIntyre, A.D., and R.M. Warwick
1984. Meiofauna Techniques. Pages 217-244 in A.D. McIntyre, and N.A. Holme, editors, *Methods for the Study of Marine Benthos*. IBP Handbook 16. Oxford: Blackwell Scientific Publications.

Nichols, J.A.
1979. A Simple Flotation Technique for Separating Meiobenthic Nematodes from Fine-grained Sediments. *Transactions of American Microscopical Society*, 98:127-130.

Olsson, I.
1975. On Methods Concerning Marine Benthic Meiofauna. *Zoon*, 3:49-60.

Poizat, C.
1975. Technique de concentration des gastéropods opistobranches mésopsammiques en vue d'études quantitatives. *Cahiers de Biologie Marine*, 16:475-481.

Price, C.A., E.M. Reardon, and R.R.L. Guillard
1978. Collection of Dinoflagellates and Other Marine Microalgae by Centrifugation in Density Gradients of a Modified Silicia Sol. *Limnology and Oceanography*, 23:548-553.

Schmidt, P.
1968. Die quantitative Verteilung und Populationsdynamik des Mesopsammon am Gezeiten-Sandstrand der Nordseeinsel Sylt I. Faktorengefüge und biologische Gliederung des Lebensraums. *Internationale Revue der gesamten Hydrobiologie*, 53:723-779.

Schwinghamer, P.
1981. Extraction of Living Meiofauna from Marine Sediments by Centrifugation in a Silicia Sol-Sorbitol Mixture. *Canadian Journal of Fisheries and Aquatic Sciences*, 38:476-478.
1983. Generating Ecological Hypotheses from Biomass Spectra Using Causal Analysis: A Benthic Example. *Marine Ecology Progress Series*, 13:151-166.

Schwoerbel, J.
1986. *Methoden der Hydrobiologie, Süsswasserbiologie.* Third edition, 301 pages. Stuttgart: Gustav Fischer Verlag.

Sopott, B.

1973. Jahreszeitliche Verteilung und Lebenszyklen der Proseriata (Turbellaria) eines Sandstrandes der Nordseeinsel Sylt. *Mikrofauna des Meeresbodens*, 15:253-258.

Steedman, H.F.
1976. *Zooplankton Fixation and Preservation*. 350 pages. Paris: UNESCO Press.

Thiel, H.
1966. Quantitative Untersuchungen über die Meiofauna des Tiefseebodens. *Veröffentlichungen des Instituts Meeresforschung Bremerhaven*, Sonderband, 2:131-147.

Thiel, H., D. Thistle, and G.D. Wilson
1975. Ultrasonic Treatment of Sediment Samples for More Efficient Sorting of Meiofauna. *Limnology and Oceanography*, 20:472-473.

Tiemann, H., and K.-H. Betz
1979. Elutration: Theoretical Considerations and Methodological Improvements. *Marine Ecology Progress Series*, 1:227-281.

Uhlig, G.
1964. Eine einfache Methode zur Extraktion der vagilen mesopsammalen Mikrofauna. *Helgoländer wissenschaftliche Meeresuntersuchungen*, 11:178-185.

Uhlig, G., H. Thiel, and J.S. Gray
1973. The Quantitative Separation of Meiofauna. *Helgoländer wissenschaftliche Meeresuntersuchungen*, 25:173-195.

10. Organism Processing

Wilfried Westheide and Günter Purschke

Examination of Living Material

Microscopic Methods

The examination of living meiofauna without the use of a microscope is, of course, impossible. The quality of research results is to a large extent directly dependent on the quality of microscope and light sources, and also on their proper application.

Stereomicroscopes and Light Sources.--The stereomicroscope is the most widely employed instrument for routine observation. For sorting, stereomicroscopes are recommended with at least a maximum total magnification of 40x. Working with minute specimens like chaetonotoid gastrotrichs or rotifers make higher magnifications necessary; up to 240x is desirable, e.g., for dissecting of copepods or ostracods. Besides the total magnification obtainable, the working distance should influence the choice of the type of stereomicroscope: the longer the distance the better the working conditions using all kinds of culture dishes and for all procedures of dissection and preparation.

The type of illumination that one uses is a question of habit. However, one should be aware that both incident light and transmitted light have advantages and disadvantages. Inclined incident light emphasizes details, structures, and colors of the surface of animals but makes recognition of animals between sediment particles sometimes more difficult than transmitted light that is directed through the animals and thus sets off their dark appearing figures against the light background. Incident light coming from ordinary lamps or low voltage lamps (if more light is required with halogen bulbs) strongly heats up the culture medium and may damage or kill the animals. A glass heat filter may be used. Cold light sources with one or two flexible light guides are highly recommended; they also offer highest brightness and best focusing possibility for observation with high magnification. Danger of heating is less with transmitted light, but this illumination needs specific stands. Equipment is available for several types of stereomicroscopes that allows a combination of incident and transmitted light. Stereomicroscopes with swinging arm stands are recommended, e.g., for the observation of delicate cultures which might be disturbed by handling the dishes.

Compound Microscopes and Special Techniques of Illumination.--For the standard analysis of details a compound microscope with a set of objectives including a 100x oil objective has to be available. Most living microscopic specimens are lacking in contrast compared with stained histological objects. This can best be remedied by using optical contrast methods, such as dark-field, phase contrast or differential interference contrast in addition to the ordinary bright-field illumination.

With the dark-field illumination objects which scatter light appear bright against a dark background. Linear structures, especially like cracks, edges, flagella, hairs, and bacteria, are made visible even when their thickness is below the resolving power of the objective. However, the objects might not be represented with absolute accuracy. Specific mirror condensers are required for dark-ground investigations.

Different structures may have practically the same transmission within the visible region of the spectrum, so brightness of color differences cannot be perceived, though the phase of light waves that passed through them is changed. Those phase differences are converted into clearly visible brightness differences by the phase contrast method. Ideal phase objects are as thin as possible (5 µm and thinner) and have no major differences of refraction – mainly surface structures of living squeezed animals or unstained sections. Their images appear darker against a less dark background if their refractive index is higher than their surroundings, brighter if the objects have a lower refractive index. Extensive structures especially are invariably surrounded by halos around dark image details and dark fringes around bright details, that often restrict the application of this method and do not allow exact measurements.

Only very fine details, like cilia, tips of chaetae, and small bacteria, are largely free from halation. Contrast can be considerably influenced by the immersion medium: physiological media should be used for unstained material, media of a refractive index smaller than that of Canada Balsam, e.g., Zeiss W15, a water miscible mounting medium, for permanent mounts. The phase contrast illumination can only be carried out with a special condenser and special objectives; however, these can also be used for bright-field illumination.

Usually interference contrast illumination is the method chosen. A phase displacement between two waves passing closely through adjacent areas of the object is made visible with polarizing aids as a difference in brightness. It produces a conspicuous three-dimensional image of all unstained transparent objects, including thin details and organs inside thick objects. The images are free from halos and appear being illuminated obliquely, which allows especially good photographs. Equipment for interference contrast is expensive, consisting of polarizer, special condenser with Wollaston prisms, and analyzer. Fluorite objectives are necessary.

Inverted Microscopes.--For the examination of living objects in chambers, microtest plates and various kinds of culture dishes (see Figure 10.1a-h), an inverted microscope for transmitted light is highly recommended. It should have a receptacle for a reflex camera housing. Dark-field, phase contrast, and differential interference contrast equipment is available for this type of microscope, the advantages of which are retention of image sharpness even at higher magnifications and a large working distance that allows the use of high dishes on the stage. Together with an inverted microscope the technique of microinjection into living animals with micromanipulator and specific microinjection equipment should be considered as a method of great promise for meiofauna research. Uhlig and Heimberg (1981) devised a versatile compression microchamber (Figure 10.2), especially for inverted microscopes (see also page 148').

Observation of animals, eggs, cocoons, and other items in culture dishes is also possible with a normal compound microscope, equipped with submersible objectives which, however, are rare and expensive. It is easier to use objectives covered with a "diving cap," that may be self-made from a PVC tube, closed with a coverslip (Heunert, 1973) (Figure 10.1d).

Photographic and Graphic Documentation

Photomacrography

Though sometimes highly desirable for documentation of both sediments and animals, macrography is a "Cinderella" of photographic methods in many meiofauna laboratories. The equipment is expensive; for technical reasons results

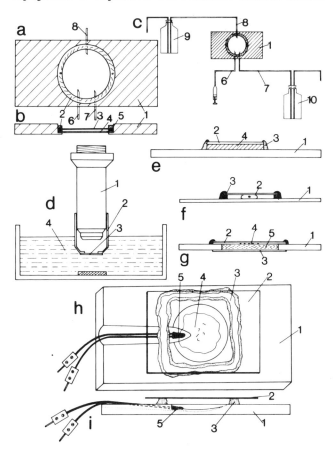

Figure 10.1.--a-c, Chamber for continuous flow system: a,b, details of chamber. 1, body of chamber; 2, thread ring to close the chamber; 3,4, upper and lower coverslip; 5, silicone ring; 6, cannula to suck out air bubbles; 7, drain cannula; 8, inflow cannula; 9,10, inflow and drain bottles. d, Submersible objective: 1, objective; 2, PVC-cap; 3, coverslip; 4, petri dish. e, Agar chamber: 1, slide; 2, coverslip; 3, vaseline sealing; 4, agar. f, Sealed chamber with coverslip support (3): 1, slide; 2 coverslip. g, Paraffin chamber: 1, slide with hole; 2,3, upper and lower coverslip; 4, object; 5, paraffin oil. h,i, Chamber with NTC feeler (5): 1, slide with concave depression; 2, coverslip; 3, vaseline sealing and support; 4, medium with animals. (a-g, after Heunert, 1973.)

are often unsatisfactory and the resolving power may be inadequate. Certain macroscopes allow photos with a field diameter between 0.7 mm and over 25 mm, thus covering a range where bright-field microscopes can only work with difficulty or not at all. These instruments have a built-in phototube that accepts different kinds of photographic systems, cine and video cameras. Instead of expensive modular photomicrographic camera systems, normal 35 mm reflex cameras with an automatic exposure control system are recommended, together with a photoflash group controlling flash emission directly through the camera and allowing fully automatic photography of moving objects. A series of different stands is available for various kinds of illumination. The specific illumination technique to be used is dependent on the specific object and may require experimentation or experience. Modern macroscopes are easy to obtain especially for video macrographic recording of precise measurements of activities like feeding, egg laying, transmission of sperm, molting, and locomotion. Recording of life activities with higher magnification in culture dishes should be made with an inverted microscope (see page 147).

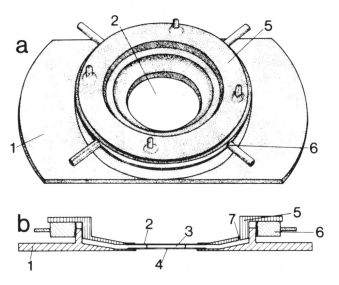

Figure 10.2.--**Roto-compression chambers**; a, View from above; b, Cross section: 1, chamber body; 2, upper coverslip; 3, medium with organisms; 4, lower coverslip; 5, rotor; 6, compression ring; 7, air hole. (a, after Heunert, 1973; and b, after Uhlig and Heimberg, 1981.)

Photomicrography

All modern high quality microscopes can be equipped with a phototube suitable for photomicrography. Built-in ordinary lamps or 6V 5W low-voltage illuminators can be used for black-and-white photomicrography. For color photos, the microscope must be equipped with a 6V 15W illuminator, also desirable for dark-ground, phase contrast, and interference contrast. If resolution of finest structures and correction of chromatic aberration is essential, expensive plano objectives should be chosen, in which the curvature of field is also corrected, so that the image will be sharp throughout the entire photographic format. Fluorite systems are less expensive, and especially in black and white reproduction all normal achromats can be successfully employed.

If whole-page reproductions and a very high quality is needed one should choose a large format camera. However, for economical reasons it is more practical to use a 35 mm format since it is also favorable for color photos when highly-resolving films are used.

Most microscope manufactures offer attachment camera systems for automatic photomicrography, which, however, are expensive and cannot be used in the field. A normal 35 mm reflex camera housing with an automatic exposure control may also comply with the requirements - eventually used together with a camera motor. However, such use requires a firm microscope stand to minimize vibration effects of the shutter. Flash photomicrography is highly desirable especially for taking photos of living material. Unfortunately, suitable commercially available equipment is scarce and not always satisfactory. Microflashes can be employed together with the automatic microscope camera equipment but generally do not permit the use of the automatic exposure system. Therefore, flash photomicrography needs an indirect light-measuring procedure utilizing a photometer. With certain qualifications normal reflex camera housings provided for TTL flash systems can be used together with microflashes.

Several conditions are essential for good micrographs, e.g., flawless slides, correct coverglasses (0.17 mm thickness) in dry systems, a clean optical system, centered illumination, avoidance of air bubbles and dirt particles in the object chamber, accommodation of the eye with a focusing eyepiece, and a careful attention to correct Köhler illumination - adjusted for each objective employed. Further details may be found in instruction guides of the various microscope manufacturers.

Usually black-and-white films are satisfactory for the living meiofauna specimens which rarely show striking colored structures. Black and white material film is also recommended for all kinds of stained whole-mounts and histological sections. In the majority of cases panchromatic films are chosen, which render it possible to change the gray-tone reproduction of certain colors by means of color filters in the illuminating beam. A green filter is particularly used for photomicrography, that permits only those rays to reach the objective for which it is best corrected. The green filter reduces blue, red and orange, which become lighter in the film and, therefore, darker in the paper print. Different colored filters have specific applications depending on what effects are desired.

Slow speed films with fine grains (25-32 ASA) are preferred because they produce high contour sharpness and high contrast, which is especially desirable in bright-field illumination. For slowly moving objects, however, films with higher sensitivity may be necessary, except when a microflash is used.

For taking color photographs - transparencies will be mostly needed for demonstrations - artificial-light films and daylight films are available. Daylight films can be easily used for flash microphotography. Exposure to low-voltage microscope illumination needs artificial-light films, because of its lower color temperature. For details see the manufacturer's instructions and special literature on photomicrography.

Image and Morphometric Analysis

Microscopical investigations should not be done without taking linear measurements of the objects. This is usually accomplished by the use of an eyepiece micrometer whose scale division appears together with the image of the object in the eyepiece. The micrometer value, different with each objective/eyepiece combination, has to be determined by calibration against a stage micrometer (glass slide with an engraved scale, usually 2 mm divided into 200 intervals). The size of an object can then be measured in scale units which can be mathematically converted. A focusing eye-piece is necessary for the focusing of the reticule.

On a binocular tube one may use the measuring eyepiece with the better eye. Details to be measured should be moved into the center of the field of view, the distortion error being here at a minimum. In order to have scales in photomicrographs one should take a photo of the scale on the stage micrometer with each objective. When tracing with a camera lucida one should add a scale to the figure on the sheet made from a stage micrometer at the same magnification as used for the tracing. Rough measurements can also be carried out with mechanical stages graduated in mm in both axes. For the most accurate measurements under the microscopes screw micrometer eyepieces are available.

Extensive quantitative analyses, e.g., of sediment particles, microscopic specimens, and various structures in light- and electronmicrographs may require semi-automatic image analyzers. They offer a direct evaluation of all kinds of images for length and area measurements, particle volume and stereological analysis. Programs are available, e.g., to determine frequencies and cumulative frequencies of the measured diameters; calculation of the statistical parameters may be called and documented, as well as main values, standard deviations and other statistical data of the measured results. Trends in image analysis tend towards the use of fully automatic recording and computation of geometric and densitometric parameters to considerably reduce evaluation time of images.

For further information on morphometric methods see Weibel (1979).

Mounting Specimens for Live Observations

The Mounting Process

Successful microscopic studies of living small organisms depend essentially on suitable methods to slow them down. For the appropriate method one has to differentiate between short-term observations and long-term studies.

For short-term observations specimens are usually placed on a normal flat slide in a small drop of water, preferably from the same habitat, and then covered with a coverslip. Transferring of specimens can be carried out with either a disposable glass Pasteur pipette equipped with a small rubber bulb, a silicone tube with a small mouth piece (Figure 10.3b), or an Irwin loop (Figure 10.3c,d). In the case of the silicone tube and small mouth piece, the manipulation of a specimen is regulated by controlling air pressure within the tube by the mouth. The diameter of the narrow opening end of the pipette may be reduced by drawing it out by hand in the flame of a Bunsen-burner to an opening depending on the size of the animals investigated. Adhesion of many animals in pipettes is a serious problem, which especially affects the slow working beginner. To avoid this the inside of pipettes may be coated with silicone. The advantage of the Irwin loop (nickel-chromium wire) in sorting is it transfers very little fluid with the specimen and when used properly, insures that a tiny specimen, e.g., a tardigrade, kinorhynch, or loriciferan, actually is transferred from one medium to another. The specimen should be observed as it is lifted on the loop. Use of a needle for placing specimens on a slide should be restricted to "hard-bodied" animals like copepods or nematodes.

The coverslip is set with one side on the slide, and then allowing it to carefully glide down along a needle onto the water with the specimen. Thus, the animal is not damaged and no air bubbles are trapped under the coverslip (Figure 10.3a). Water may be gradually removed with a small piece of filter paper until the animal is unable to move. Studies of the internal organization may require an increase in squeezing by removing most of the liquid or applying pressure to the coverslip with a needle. Preferably, one or more drops of anesthetic may be added at one side of the coverslip, then drawn under the slip with filter paper from the opposite side until the animal is immobilized (Figure 10.3a).

A less recommended alternative procedure immobilizes the animals by adding the narcotizing agents to the watch glass prior to the squeeze preparation. A similar effect can be brought about by removing most of the liquid from the edge of the coverslip, either by capillary pipette or an absorptive paper placed in contact with the edge of the coverslip. Especially for larger individuals and for soft-bodied species the edges of the coverslip should be supported by small flecks of bee's wax or plastiline (scraped off directly with the edges of the coverslip, or alternatively, by a dissecting needle). Preparations of this kind may not be parallel with the objective's surface and thus will result in some aberration noticeable in photographic documentation, especially with phase-contrast or interference contrast illumination. For this reason minute hard structures such as glass micro-beads, "glimmer plates," cut pieces of nylon threads or monofilament thread, or even pieces of broken coverslips may offer a better supporting mechanism of coverslip support.

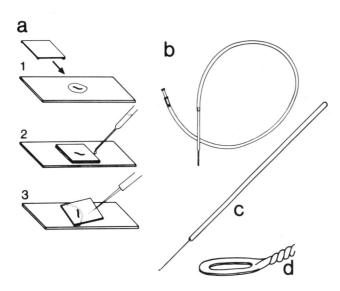

Figure 10.3.--a, Specimen mounting: 1, object and coverslip with support; 2, changing or adding fluids; 3, setting coverslip onto object with the aid of a preparation needle. b, pipette with silicone tube and mouth piece. c-d, Irwin loop: c, entire loop; d, tip of loop (diameter approximately 1 mm). (c,d, courtesy of R.P. Higgins.)

Storage of slide-coverslip preparations of the kind described is possible for a short period of time in a Petri dish with top that is transformed into a moisture chamber by some wet pieces of filter paper. Microscopic long-term studies for life activity observations, cine- or video-recordings have to be carried out in special chambers which do not allow evaporation of the fluid. The most simple chamber can be constructed by using a 1:1 mixture of vaseline and paraffin to support the coverslip. A chamber of this kind should not be completely filled with water, an air bubble will help to avoid the rapid development of anoxic conditions (Figure 10.1f).

Glass slides with concave depressions allow larger chambers. They are generally unsuitable for observations with higher magnifications but can be used for various experiments, e.g., when equipped with a built-in NTC feeler for temperature measurements (Figure 10.1h) (Westheide and von Basse, 1978; Purschke, 1981). Roto-compressor chambers (Uhlig and Heimberg, 1981) are expensive, but useful high precision instruments which make it possible - under exact control - to slow organisms, to arrest them at a specific place under the coverslip, to turn them from one side to another and to gently squeeze them without damaging them (Figure 10.2). They also allow exact determinations of body thickness by focusing the upper and lower coverslip area of the instrument and observing the distance on the dial indicator of the focusing wheel of the microscope.

For objects such as developing eggs that require a higher amount of oxygen one may build a so-called paraffin chamber (Heunert, 1973) (Figure 10.1g). A hole is cut into a glass slide with a diameter of 15-17 mm. This hole should be covered by a coverslip on one side (glue, e.g., "Eukitt"). The formed and sealed chamber is to be filled with paraffinum liquidum. The object is then placed with a small amount of water on a coverslip that is carefully positioned onto the paraffin oil with the drop on the underside and sealed with vaseline. Paraffin oil "binds" oxygen.

For more fragile organisms one may use chambers through which a streamline flow of a medium is possible. Figure 10.1a-c shows a system of this kind with an observation chamber made out of stainless steel (for further details see Ax et al., 1966). Rotocompressor models can also be transformed into flow systems. In any case only animals that have been extracted very carefully out of the sediment should be used for long-term observations. The sea water ice method - not to be used in arctic and some tropic localities - is especially recommended for getting specimens of high vivacity (Chapter 9). Various micro-agar plates used commonly by microbiologists are also useful for the culture and live observation of certain taxa, e.g., copepods (George, 1975). Figure 10.1e shows a simple glass slide with a piece of agar covered with a coverslip and sealed with vaseline.

None of these chambers create optimal conditions for most of the meiobenthic species; thus, various animals will not respond with a natural behavior or locomotion. Unfortunately, the utilization of some of the sediment from the animal's habitat in any chamber will partly or completely limit adequate observation and subsequent documentation. In order to simulate natural conditions artificial sand systems have been proposed. Coineau and Coineau (1979) constructed a transparent model based on small resin casts made from molds which in turn were prepared from blocks commonly used in the printing industry (Figure 10.4). Giere and Welberts (1985) constructed a similar but more natural environment by photographic transfer of the normal sand grains to a plastic cliche using modern block-making techniques. Microscopic observation and photography are among its possible applications.

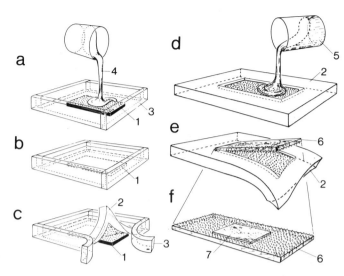

Figure 10.4--Production of artificial sand system; a-c, Casting of a mold of silicone rubber; d-e, production of the epoxy resin cast; f, example for use of the artificial sand system: 1, aluminium or zinc cliché; 2, silicone mold; 3, casting frame; 4, silicone; 5, epoxy; 6, epoxy cast-resin; 7, coverslip. (after Coineau and Coineau, 1979.)

Narcotization

Some form of narcotization usually is desirable (see microscopic methods, photographic and graphic documentation) for the observation of living organisms under a compound microscope, especially, if documentation by drawing or photography is planned. For histological and ultrastructural investigations narcotization should precede fixation when working with slender elongated specimens that would otherwise curl up during the first seconds of the fixation process. For most marine animals a $MgCl_2$ solution isotonic to seawater (about 7.5% $MgCl_2 \times 6H_2O$ in distilled water) is an excellent anesthetic. Preferably, the specimens should not be transferred directly into the narcotizing solution; instead, this solution should be added to in such a manner as to slowly (drop by drop) replace the original fluid to the extent that the animal is relaxed. The amount of $MgCl_2$ solution necessary for relaxation may vary greatly depending on the species involved. Usually, the specimens must remain in the narcotizing solution for 10 to 15 min before being transferred into a fixative (see histological and ultrastructure processing).

Heparin or a freshwater shock (see Chapter 9) is recommended for animals with adhesive glands. Freshwater animals may be narcotized by a 2% solution of urethane (D. Bunke, personal communi-cation). Further narcotics recommended include: MS 222, chloral hydrate, propylene phenoxetol (1% in seawater), lidocaine, nembutal (1% in seawater), and menthol (crystals added to seawater).

Vital Staining

Intravital staining may be used to sort both living and dead meiofauna organisms, e.g., nematodes (Shepherd, 1962; Ogiga and Estey, 1974) and copepods (Dressel et al., 1972). Vital staining can also be applied for the recognition or elucidation of organs. Some of these structures may be species specific and thus help to distinguish species in life observations. Especially the distribution pattern, number and shape of epidermal glands can be conspicuously set off by vital stains, e.g., the paired dorsal glands in the gastrotrich genus *Tetranchyroderma* (Nixon, 1976), and the species specific transverse rows of epidermal glands in intertidal enchytraeids (Kossmagk-Stephan, 1984). However, the staining capacity of some of these organs may depend on the physiological condition of the individual animal and interpretation of results should be carried out carefully.

Various water soluble dyes are available for vital staining, e.g., the basic dyes Neutral Red, Methylene Blue, Methylene Azure, Toluidine Blue, Nile Blue Sulfate, Bismarck Brown, Methyl Violet and the acid dyes Pyrrol Blue and Trypan Blue. Methylene Blue and especially Neutral Red are mainly used because they obviously are the least harmful substances. According to Nixon (1976) the death rate of gastrotrichs was only higher for animals when stained with Neutral Red (concentration of the stain $5.7 \times 10^{-5}\%$) if they were left in stained seawater for long periods of time. Application is usually by placing the animals into seawater containing the stain, or by pouring stained seawater into the sediment samples followed by washing out the stain after a certain time. For special purposes it may be advantageous to feed animals with a stained food to prevent precipitation of dyes in seawater (Reisinger, 1936).

Recording Data

A "check-list" of characters should be set up for every taxon worked with either in a living or dead state. This means a complete list of characters to be observed, arranged according to their optimal visibility during the different stages of examination. This allows for a more efficient recording of a maximum amount of data from each specimen observed.

Generally, observations of living animals should start with the characterization of their normal behavior, which might give useful information as to their identity, e.g., from their adhesive behavior, and feeding and sexual behavior. Other important data may include: color; size; sexual state, including number and size of eggs, presence of sperm; notes on epidermal glands and ciliary structures, contents of the gut; and hard structures (e.g., jaws, chaetae, stylets). Most of these data can only be obtained from living animals and only partially compensated for by other techniques, thus it may be very important to consider this as part of one's research protocol. In many meiofaunal taxa, determination of species, as well as their description from fixed material alone, is insufficient.

Preparation of Permanent Mounts

Fixation

The method of fixation of specimens for permanent mounts often varies according to the taxon involved (see Part III). In most cases 10% formalin (see Chapter 9) or Bouin's fluid give satisfactory results. A buffered solution of formalin is absolutely necessary for organisms with calcareous structures (e.g., foraminiferans, molluscs, most crustaceans). Suitable buffers, borax being one of the best, have been discussed in Chapter 9 as have other useful fixatives such as 2% glutaraldehyde.

Mounting and Staining

"Soft-bodied" animals are usually stained prior to mounting; "hard" fauna may remain unstained or stained. Recommended water-soluble mounting media include: 10% glycerin in 95% ethanol; Hoyer's or Faure's media (see page 159); and "Zeiss W15" (highly recommended because of its high refractive index of nD = 1.515). Mounting media for xylene- or toluene-processed specimens include: "Depex," "Malinol," "Technicon Mounting Medium," and "Euparal." Most of the mountants serve as clearing agents. If additional clearing is desirable, warm lactic acid (50° C) may be used. Pepsin, used to dissolve body tissues usually does not work in formalin-fixed specimens. Soft parts may be preferentially dissolved in aqueous 0.5-1.0% sodium or potassium hydroxide, or even by the judicious use of household bleach (sodium hypochlorite).

One of the mounting procedures suggested is that recommended by Erseus (personal communication) who has used it specifically for oligochaetes but is equally useful for other "soft-bodied" taxa. Pre-mount staining with a solution of boraxcarmine or paracarmine (see page 158) for 15-30 min is carried out in snap cap bottles (10-30 ml). To avoid loss, specimens are kept in modified Beem capsules (see page 154 and Figure 10.7). Differential staining is effected by use of acid alcohol (4 drops of concentrated HCl in 100 ml 70% ethanol). Differentiation should be observed with a stereomicroscope so as to control the amount of staining; a subsequent careful rinsing (several changes) with 70% ethanol is necessary to rinse out the HCl. The specimens are then mounted immediately in a water-soluble medium (see above) or dehydrated in a series of 80%, 90%, 100% ethanol (1h each or less). Then they are brought into a 1:1 solution of ethanol/xylene (or toluene) for 1 h, and then into 100% xylene (or toluene) where they should be subjected to at least two 30-min changes before the specimens are removed from the capsules and transferred to glass slides and mounted in one of the xylene/toluene-soluble mounting media noted previously. A coverslip should be carefully lowered to the mounting medium in a manner so as to avoid the development of air bubbles (Figure 10.3a). Fixation in Zenker's fluid (page 158), dehydration to 80% ethanol and subsequent staining in an aqueous solution of Mayer's muchaematein (page 158) for 25-30 seconds produces a marked staining of ciliary structures or ciliated epithelia (Jägersten, 1952).

Staining procedure may alternatively be carried out in square watch glasses, concave or standard slides, if the specimens are tiny, allowing a precise control of the procedure by observing the process under a stereomicroscope. The fixative, e.g., 5-10% formalin on the slides may be slowly infused with a solution of 10% glycerin in 96% ethanol and covered by a coverslip (Kristensen, 1983; Kristensen and Higgins, 1984). This solution evaporates to glycerin. Riemann (Chapter 23) provides a detailed description of whole mount and transfer procedures for nematodes which is probably applicable to many other similar taxa. The viscous whole-mount preparations may be sealed by "Araldite," "Murrayite," "Zut," and many other suitable commercially available sealants, including epoxy paint or even fingernail polish.

The Cobb metal slide frame preparation (Figure 10.5) holds a double coverglass mounted preparation and allows the microscopic examination of an organism from either of the two surfaces. Figure 10.5a-c demonstrates the assembly of a Cobb slide frame: A 25 mm square #1 coverglass (2) is placed over the 20 mm diameter opening of the metal slide frame (4), two plastic spacers, also 25 mm, are placed on either side, and the upward bent edges of the frame are "crimped" tightly over the plastic spacers, holding the spacers tightly pressed against the lateral edges of the square coverglass. A drop of mounting medium (7) is placed on the upper surface of the square coverglass (2), the specimen (5) is added and oriented with a proper tool. The 12 mm #0 circle coverglass is placed on top of the mounting medium and allowed to settle into place. The organism may be manipulated further during this phase before the coverglass is sealed (6).

A recent variation of the Cobb slide frame is the

H-S slide (R.P. Higgins and Y. Shirayama; Figure 10.5d,e). It consists of two standard microslide-sized pieces of plastic fused into a single unit the same thickness as a standard microslide. The lower plastic element has a 16 mm hole which is overlain by an upper element with an 18 mm hole (9). Cement is added to the 1 mm wide ring-like shelf created by the lower piece of plastic, and an 18 mm #0 circle coverglass is set into the shelf (Figure 10.5d,e). No crimping or other preparation is needed. The mounting process is the same as noted for the Cobb slide preparation. The H-S slide mount allows for a more precise centering of the specimen at a level equidistant between the upper and lower surface of the finished preparation than does the Cobb slide preparation and is easier to prepare. The lateral upper right and left surfaces will accommodate a slide label or can be written on directly using ink or pencil lead.

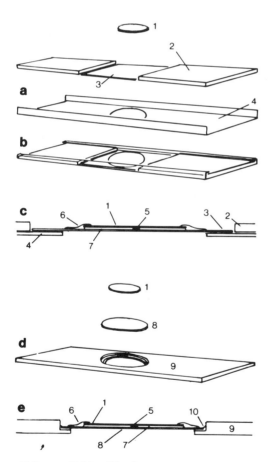

Figure 10.5--a-c, Cobb slide frame: a, component parts; b, assembled preparation; c, optical view of assembled preparation; d,e, H-S slide frame: d, component parts; e, optical view of assembled preparation; 1, 12 mm #0 circle coverglass; 2, 25 mm square plastic spacers; 3, 25 mm #1 square coverglass; 4, slide frame; 5, specimen; 6, sealing medium; 7, mounting medium; 8, 18 mm #0 coverglass; 9, H-S slide frame; 10, cement. (Courtesy of R.P. Higgins.)

Individual taxa require dissections (see Part III), preferably carried out in glycerin-containing viscous media. Finely pointed needles are considered to be generally needed dissecting instruments. For this purpose insect needles or thin tungsten wires (about 0.2 mm diameter) sharpened by dipping into molten sodium nitrite are fixed on various holders, e.g., wooden kabob skewers, glass capillaries, etc.

Chromosome Preparations

Chromosome staining may be used in species description and identification, e.g., in oligochaetes. The following procedure has been successfully used for small polychaetes. It is carried out in square watch glasses or on slides. One or several specimens are transferred into a colchicine solution (0.1% colchicine in Ringer's solution (page 158), diluted 1:1 with seawater (or 0.5% colchicine in seawater) for 4 h so as to arrest the dividing cells at metaphase. After replacing the colchicine solution by a hypotonic solution of 0.075 M KCl (3-4 min), the specimens are fixed in absolute ethanol/glacial acetic acid (3:1, 3 changes) for 3 min to 12 h. The fixative is replaced by 50% acetic acid (3-15 min); if the fixation has been carried out in watch glasses, specimens are transferred onto slides. Fixed specimens are gently picked to pieces with a preparation needle. One should take care that the drop of acetic acid does not roll away. The material dries out in concentric circles and then has to be stained with one drop of 1% orcein in 100% ethanol/100% lactic acid (1:1). Alternatively 1% aceto-orcein (1% orcein in 45% acetic acid) may be used. The slides are covered by a coverslip and have to be studied using a phase contrast microscope. Slides may be stained in staining dishes (same solution, 30-60 min) dehydrated in ethanol (100%, changes, 30 min) and embedded in Euparal.

Histological Processing

The following method is suggested for the preparation of meiofaunal organisms for histological examination. For other methods, the reader is referred to comprehensive manuals on the subject such as those of Romeis (1968) or Grimstone and Skaer (1972).

Fixatives

The most satisfactory fixatives for histological preparations include: buffered 10% formalin, Bouin's fluid (see page 158), Duboscq-Brasil, an alcoholic Bouin's fluid, which is a rapidly penetrating fixative especially useful for arthropods (see page 158). It is necessary not to reduce the ratio fixative volume to sample volume below 9:1.

A possible, but yet seldom practiced, alternative to chemical fixatives is a procedure using microwave irradiation such as generated by household microwaves ovens. This fixation, compared with formalin-fixation, provides good results with marine invertebrates, especially in taxa with cuticular structure which do not allow rapid penetration of chemicals (Berg and Adams, 1984).

Procedure

Narcotization prior to fixation is highly recommended. Relaxed specimens should be transferred with as little fluid as possible into a snap cap bottle (e.g., 10-ml size) containing the fixative. Without narcotization Bouin's fluid may be used at 30–60° C to avoid curling and contraction. However, this may give poorer results. The fixation may last 2 h or longer; the specimens can be stored for a considerable time in Bouin's. The entire process may be carried out with an agitator, whose gentle rotary motions accelerate penetration and ensure uniform exposure to the fluids. The specimens may also be fixed in a modified Beem capsule as described below (see Figure 10.7) or in glass tubes (Padawer, 1951) to avoid loss of material during the processing.

Dehydration is done in an ethanol series: 30%, 40%, 55%, 70%, 80% (5 min each), 90% (3 h, or overnight), 95% (15 min), 100% (3 times, 30 min each). A few drops of 0.5% aqueous eosin may be added to the 95% ethanol in order to make tiny specimens visible in paraplast blocks. From ethanol, specimens are transferred into methylbenzoate (3 steps, overnight), which is subsequently replaced by butanol (change every 20 min, 4 steps to 100% butanol, then another 3 changes, each 20 min, or over night).

Infiltration starts in an oven (60° C) with butanol/paraffin (3:1, 2 times, 20-30 min; 3:2, 2 times, 20-30 min each) followed by 100% paraffin. Paraffin is replaced by the commercial Paraplast (2 changes, 60 min each). Embedding of specimens is carried out in tissue embedding molds (durable aluminum alloy). A method of orientating specimens during embedding has been described by Antonius (1965): specimens are stained with borax carmine and embedded in celloidin in special vessels with a V-shaped depression. This leads to a longitudinal orientation. The celloidin rods produced this way are examined under a compound microscope to determine the exact position of the specimens. This makes graphic reconstruction easier. The posterior and anterior end of the specimen are marked by scratches which are visible in the ribbon of section. Then the celloidin rod is embedded in paraffin. Further processing is similar to that described below; the stain recommended is azocarmine – Pasini after Kohashi (see Romeis, 1968).

Free floating ova or small developing stages may be fixed on a millipore filter where they adhere mechanically and by Van der Waal's forces. This allows fixation, dehydration, and embedding without changing containers (Rinaldi et al., 1966).

After cooling the Paraplast, the blocks are positioned on a chuck (microtome block). To get entire series of sections a rotary microtome is recommended. Sections are mounted on clean glass slides (ethanol; numbered with a diamond marker) which are covered by a very thin film of Mayer's albumen (a small drop is rubbed evenly over the slide with a clean finger). The slide is flooded with a few drops of distilled water. The ribbon of sections is placed on the slide which is placed on a slide-warming table at 40° C in order to spread the sections; they are arranged in their final position with a needle and then the water is drained off. The slides are dried over night in an oven at 30–40° C. Staining is carried out in glass staining dishes with lids (each solution requires a separate dish) and the slides are positioned in a staining tray with slots. Heidenhain's stain is suggested: Dissolving of paraffin in xylene (or toluene) (3 times: 5 min, 1 min, few seconds), xylene/100% ethanol (1:1 for 1 min) and then sections are brought to water as follows: 100% ethanol (2 min), 90% ethanol (2 min), 70% ethanol (2 min), 40% ethanol (2 min), distilled water (2 times: 3 min, 1 min). Staining starts with carmine-alum (10–20 min, see page 158), distilled water (few seconds), phosphotungstic acid (5 min), distilled water (few seconds), Aniline blue and Orange G (5–10 min, see page 158). Sections are then rinsed very briefly in distilled water, followed by 80%, 90% and 100% ethanol, ethanol/xylene (1:1, 1 min), xylene (2 times, 1 min each) and xylene where the sections can be stored. Mounting media such as those mentioned earlier may be used.

Semi-thin section technique uses plastic embedded specimens (as used for electron microscopy, see below). Sections (0.5–1 µm) are prepared with an ultramicrotome or specific rotary-microtome. The sections are cut with glass (or diamond) knives and picked up individually with a fine glass needle, transferred into a small drop of water on a slide. To prevent spreading of the water, the following slide-coating procedure is recommended: slides are cleaned in ethanol and dipped into a solution of 0.5% gelatin and 0.05% $KCr(SO_4)_2$ in distilled water. Adhesion of sections is accelerated by use of a slide warming table (60° C). Sections may be stained with a drop of 1% methylene blue in 1% borax by heating until the stain begins to steam. Another common dye is toluidine blue borax. After washing with distilled water and drying, the sections may be mounted in immersion oil. Serial sections, i.e., ribbons of sections, may be obtained by painting top and bottom of blocks with a toluene soluble contact cement as described by Smith and Tyler (1984).

Preparation for Ultrastructure Study

Fixatives.--Several fixatives and methods for their use are common in ultrastructure research, and most of them may be applied for meiofauna organisms. Because of their small size usually an entire meiofaunal specimen can be fixed. Narcotization prior to fixation is recommended. However, keeping the time for narcotization short may be an advantage since some narcotics may create artifacts. Different narcotics (e.g., $MgCl_2$, MS 222, and urethane) may result in different appearances of certain structures; thus, testing of the effects of the narcotization procedure is a prerequisite. Most investigators use $MgCl_2$. To ensure that the initial entry of the fixative causes as little disturbance as possible, it has to be buffered to a physiological pH (approximately 7.2-7.4) and maintained at a suitable osmolality (for marine invertebrates about 900 milliosmol). Aldehydes are normally employed in conjunction with osmium tetroxide for ultrastructural fixation.

The following fixatives and buffers are recommended for marine meiofauna: (1) SPAFG, a modification of Bouin's fluid, introduced by Ermak and Eakin (1976) (see page 158); (2) 2-3% glutaraldehyde in phosphate buffer (see page 158), 0.45 M of sucrose), or cacodylate buffer (see page 158), 0.45 M of sucrose). For rinsing after fixation the respective buffer is used (0.45 M sucrose). Kristensen (personal communication) recommends the use of trialdehyde as a fixative (Kalt and Tandler, 1971).

Procedure

The anesthetized specimens are removed with a Pasteur pipette or Irwin loop (with as little fluid as possible) into a small vial (e.g., snap-cap bottles of 5 ml) that contains the fixative (room temperature). Animals have to be completely free from grit and sand grains. Fixation should last 1-2 h at 4° C followed by rinsing with buffer for 2.5 h (or longer) at 4° C (5 changes minimum). Fixative and buffer are removed with a Pasteur pipette observed under a stereomicroscope leaving the specimens always covered by a small amount of fluid. Another method of changing fluids is given below. After rinsing in buffer, specimens are post-fixed in 1% aqueous OsO_4 solution (dilute equal volumes of 2% OsO_4 with 0.2 M buffer containing 0.9 M sucrose) for 1 h at 4° C. After rinsing in buffer (5 min) specimens are dehydrated in an ethanol series (30% - 5 min, 50% - 5 min, 70% - 5 min, 95% - 2 x 10 min, 100% - 2 x 10 min). In 100% ethanol specimens are brought to room temperature.

For transmission electron microscopy (TEM) an infiltration medium (propylene oxide/100% ethanol 1:1-10 min, propylene oxide 2 x 5 min) is used, from which specimens are carefully removed with a Pasteur pipette into a square watch glass (40 x 40 mm) containing a 1:3 mixture of the embedding medium (Epon/Araldite, see page 159) and propylene oxide. A cover glass is placed on the watch glass so as to leave a small opening in order to allow the infiltration medium to evaporate overnight. The specimens are then transferred with a preparation needle or a flame polished glass capillary tube into a drop of embedding medium on a microscope slide and immediately put for 5 min into an oven (60° C). After three changes specimens are embedded into cube-shaped silicone molds (self made or available as EM accessories). After staying for 6 h under a hood at room temperature specimens can be oriented with a preparation needle. For another method of orientation see Ryuntyu (1980). Finally, the embedments should be placed into an oven for approximately 2 days. For further processing (e.g., ultramicrotomy and staining) see specific literature (e.g., Reid 1975).

Rieger and Ruppert (1978) described a method using resin embeddings of quantitative meiofauna samples. The advantage of this method is to have one single method for observation, identification, preservation, and archiving as well as for histology and transmission electron microscopy. Moreover, this method allows all meiofauna groups to be treated simultaneously, and large numbers of quantitative samples can be handled at the same time.

This procedure involves the following steps as shown in Figure 10.6: 1, after quantitative extraction of the meiofauna (see Chapter 9) the specimens are placed into a petri dish; 2, the specimens are anesthetized with $MgCl_2$(7.5% $MgCl_2 \cdot 6H_2O$); 3, the $MgCl_2$ solution is removed by a Pasteur pipette covered with a 20 or 35 μm nylon net to avoid loss of organisms; 4, the dish with the specimens is then flooded with phosphate buffered 2.5% glutaraldehyde (2-4 h at 4° C); 5, the specimens are removed from the dish using a Pasteur pipette; 6, the specimens are then placed into a modified number 00 Beem capsule standing in a square watch glass (40 x 40 mm); 7, after all the specimens are in the capsule, the fluid is removed from the bottom of the watch glass and added through the capsule. Note: care should be taken not to fill the capsule further than to the upper edge and not to empty it completely to avoid organisms being lost during processing as well as exposure of specimens to the air which would result in serious artifacts. Also, a tissue processor may be used for this process.

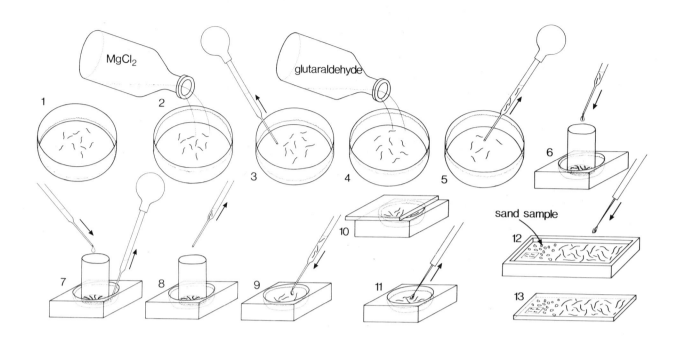

Figure 10.6--Steps in making slide-shaped transparent blocks for light and electron microscopic study of many specimens: 1, specimens in Petri dish; 2, addition of magnesium chloride; 3, removal of magnesium chloride; 4, addition of glutaraldehyde; 5, removal of specimens; 6, specimens placed in Beem capsule strainer (see Figure 10.7); 7, changing fluids; 8, removal of specimens; 9, transfer of specimens to watch glass containing propylene oxide and Epon/Araldite; 10, evaporation of propylene oxide; 11, transfer of specimens to slide-shaped embedding mold; 12, specimens in slide-shaped embedding mold; 13, completed slide-shaped transparent block, ready for microscope use. (See Rieger and Ruppert, 1978: figure 1.)

After specimens have been in propylene oxide, 8, they are removed from the capsule; 9, the specimens are transfered with a Pasteur pipette into a square watch glass containing Epon/Araldite and propylene oxide (1:3); 10, the propylene oxide is allowed to evaporate slowly (overnight); 11, the specimens are removed with a preparation needle; 12, the specimens are then placed into slide-shaped embedding molds (made from silicone rubber using glass slides as negatives, or available as polyethylene resin molds) that contain Epon/Araldite; a small sand sample may be processed in a similar manner – a label or some identifier should be added; 13, after polymerization (60° C, 2 d) the embedments can be removed from the mold and studied in the same manner as microslides using a stereomicroscope or a compound microscope (phase contrast is highly recommended). If specimens are needed for transmission electron microscopy, or semi-thin sections for light microscopy, they may be cut out and re-embedded in a standard embedding mold as already described (if a block-shaped piece can be cut out, re-embedding is not necessary).

The construction of the microstrainer from a Beem capsule is as follows (Figure 10.7): the pyramid-shaped bottom and the cap are cut away, a nylon mesh (30–40 μm) heat-sealed onto one end, or the top of the cap may be cut off leaving a rim only which can then be used to clamp the net by slipping the rim over a piece of nylon mesh (see Flood, 1973). Similar vessels may be prepared with gold mesh (Christie and Edwarson, 1968), or from capillary-tubes (Galey and Dunn, 1971).

The entire fixation procedure may also be carried out in microstainer-syringes as described by Marks and Briarty (1970) or Baker (1972). The specimens, including fluid, are drawn into a syringe. Then a strainer cap is fitted to the syringe to prevent the specimens being ejected during subsequent fluid changes. These caps are made from disposable hypodermic needles and nylon gauze: the needle is cut off from the shank and the nylon net is heat-sealed with a miniature soldering iron or a heated needle. For solution changes, the original solution is ejected and the replacement fluid is drawn in and flushed through 2-3 times. An air space may be maintained between the syringe plunger and the solution which allows almost complete fluid change and minimizes the risk of crushing the sample. This syringe is used until the specimens are passed to the epoxy-resin-propylene oxide mixture, then the cap is removed and the specimens are ejected. The advantages of this method lie in the ease and rapidity of changing fluids, in its cleanness, and in measuring volumes of solution by the calibrated syringe.

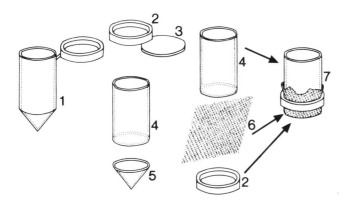

Figure 10.7--Steps in making a microstrainer from a Beem capsule for use in fixation, staining, critical point drying: 1, Beem capsule; 2, rim of lid; 3, top of lid; 4, remaining tube; 5, pyramid-shaped bottom; 6, nylon net; 7, completely assembled microstainer (note: the nylon net may also be sealed by heating one end of the Beem capsule - touching the rim to a hot plate).

Very small specimens may be embedded in coagulated albumin (e.g., Shands, 1968; Sawicki and Lipetz, 1971) or agar. The latter may either be used for free floating cell suspensions (Winters and Slade, 1971) or isolated cells (De Haller et al., 1961); e.g., ova, ciliates. They are brought into 45° C-agar (2-4%) (fixed or unfixed) which is poured onto a glass microscope slide. Single cells may be arranged to the desired position under the microscope. After cooling, blocks of about 1 mm side length containing the cells are cut out from the agar and treated as already described.

For scanning electron microscopy (SEM), which is especially useful for the study of surface structures such as chaetae, cilia, gland openings, the same fixation procedure may be used as in transmission electron microscopy until the specimens are passed to 100% ethanol. Dehydration can also be carried out by acetone which may give even better results with particularly fragile meiofauna, e.g., ciliates, tardigrades, loriciferans. Certain taxa may require special procedures, e.g., nematodes (see Green et al., 1975). To minimize loss of very small animals such as larvae or developing stages, they are attached to a coverslip as follows: The coverslip is cleaned ultrasonically and then heated in the flame of a Bunsen burner until the flame becomes colored. The cooled coverslip is placed in an aqueous solution of 1% poly-L-lysine (use within 10 min after drying). The animals are released in a small drop of buffer from a Pasteur pipette onto this glass where they should remain for 2 min without any disturbance. The same amount of 5% glutaraldehyde solution is added (20 min without moving the coverslip). After washing and dehydration, the coverslip with its attached specimens is placed in a critical point drying apparatus using CO_2.

For critical point drying, larger specimens should be transferred into a porous capsule with a cap or a modified Number 00 Beem capsule covered with nylon net on both ends (Figure 10.7). Specimens or coverslips, respectively, are then mounted on aluminum stubs which have been covered by a double sided sticky tape, conductive carbon cement, thermoplastic adhesives, or other glues. Higgins (personal communication) uses an alternate method of fixing specimens to a coverslip, or directly to an aluminum stub, or a rivet having the same shank diameter as a TEM stub. A thin coat of "Elmer's glue" is allowed to "almost" dry on the surface of the stub or coverslip, leaving a slightly sticky "membrane" to which a specimen can be attached. This type of preparation requires immediate transfer into a drying chamber to keep the specimen from absorbing moisture from the glue. Some laboratories believe that the use of this glue causes less technical problems in the maintenance of the SEM.

After drying, specimens are extremely lightweight and subjected to the effects of static electricity. To avoid damage and loss, specimens are best transferred by a small camel's hair brush or a single eye brow hair. However, a better orientation is achieved with fine forceps. Best results in the scanning electron microscope are obtained if the stub angle to the electron beam is 45°; therefore a particular orientation of the specimens on the stub is necessary: A simple holder (holding stubs at 45°) for mounting specimens has been described by Pulsifer (1975). Care should be taken to mount specimens with sufficient space between them. For survey pictures specimens may be mounted singly on small nails or a fine wire bent to an s-shape and stuck on the stub with specimens on top of the nail. This mounting allows specimens to be photographed without photographing the aluminum stub (see Eisenbeis and Wichard, 1985). The use of a rivet (head rounded) similarly offers the advantage of improved beam access (wider angles and closer working distance) to the specimen. The mounted specimens are sputter coated with gold or gold-platinum before examination in the scanning electron microscope. The dried specimens can be stored for a longer period in a dessiccator above silica gel or P_4O_{10}.

Meiofaunal animals, because of their small size, are excellent subjects for documenting position and three dimensional reconstruction of single cells, complex organs, or even entire animals at the ultrastructural level. They are sufficiently small to enable serial sectioning for TEM studies with an acceptable amount of effort, so that, for instance, nerve cells and their processes can be followed in their entirety in order to describe them correctly. For this purpose ribbons of ultrathin sections have to be collected on single slot grids, coated with support films, and complete series of electron micrographs have to be taken of the structures concerned. Serial sectioning requires ultramicrotomy with diamond knives. This method allowed the reconstruction of the entire nervous system – comprising 68 cells – of the dwarf male of the polychaete *Dinophilus gyrociliatus* out from a total of about 700 sections (Windoffer and Westheide 1988a, 1988b). Ware et al. (1975) used an electron microcope serial section cinematographic procedure to reconstruct the nerve ring in the nematode *Caenorhabditis elegans*. Computer aided reconstructions may be a future benefit to morphology.

Autoradiography

Autoradiography is used for studies of synthesis, transport, localization and turnover of any substance which is retained in tissue sections and for which a radioactively labeled precursor (β-particles) is available, e.g., amino acid uptake by meiofaunal organisms (Tempel and Westheide, 1980). It can be used at the light microscope and electron microscope level. The labeled substances are made visible by covering sections with photographic emulsion, which acts as a radiation indicator. After exposure the emulsion is developed, fixed and studied in the microscope. The silver grains, which mark the site of the radioactive decay, can be related to underlying structures of the section (see Grimstone and Skaer, 1972; Kleinig and Sitte, 1986).

Histochemistry

The most promising future methodological approaches to morphology and physiology of meiofaunal organisms will probably be based on fluorescence microscopical techniques, e.g., the glyoxylic acid fluorescence histochemical method as a sensitive method for visualization of biogenic catecholamines, indolamines, and structurally related compounds (Axelsson et al., 1973), or immunofluorescence methods for whole mounts to attain extremely clear pictures of the nervous system (Reuter et al., 1986). Instructions for histochemical methods of this kind as well as for electron-microscopy, however, are beyond the scope of this manual and should be taken from the special publications or from laboratory guides, e.g., Plattner and Zingsheim (1986).

Appendix

Borax carmine: 3 g carmine, 4 g borax, 100 ml distilled water, boiled for 30 min, after cooling an equal volume of 70% ethanol is added, solution is filtered before use.

Paracarmine: 1 g carmine acid, 0.5 g aluminum chloride, 4 g calcium chloride are dissolved in 100 ml 70% ethanol under slight heating (caution!), cooled solution is filtered.

Zenker's fluid: 5g $HgCl_2$ are dissolved in 100 ml of a solution made up of 2.5 g $K_2Cr_2O_7$, 1 g Na_2SO_4, and 100 ml distilled water, 5 ml glacial acetic acid are added immediately before use.

Mayer's muchaematein: 0.2 g haematein are mixed with a few drops of glycerin, 0.1 g $AlCl_3$, 40 ml glycerin, and 60 ml distilled water are added.

Ringer's solution (insect): 210 ml NaCl (29.2 g in 1000 ml), 20 ml KCl (37.3 g in 1000 ml), 30 ml $NaH_2PO_4 \cdot 2 H_2O$ (78 g in 1000 ml), 70 ml $Na_2HPO_4 \cdot 12 H_2O$ (89.5 g in 1000 ml), 20 ml $CaCl_2 \cdot 6 H_2O$ (21.9 g in 1000 ml), 20 ml $MgCl_2 \cdot 6 H_2O$ (20.3 g in 1000 ml), distilled water to make 1000 ml, pH should be 6.8.

Bouin's fluid: 15 parts saturated aqueous solution of picric acid, 5 parts 37% aqueous solution of formaldehyde, 1 part glacial acetic acid.

Duboscq-Brasil: 1 g picric acid dissolved in 150 ml 80% ethanol, 15 ml glacial acetic acid, 60 ml 37% aqueous solution of formaldehyde.

Carmine-alum: 1 g carmine acid in 1000 ml distilled water, 20 g potassium alum, 1 g salicylic acid, boiled for 10-20 min, filtered before use.

Aniline blue and Orange G: 0.5 g Aniline blue, 2 g Orange G, 100 ml water, 8 ml glacial acetic acid.

SPAFG: 20 g para-formaldehyde, 150 ml double-filtered saturated picric acid (aqueous solution), heated to 60°C, drops of 2.5% aqueous NaOH are added until the solution is clear. The cooled solution is filtered and made up to 1000 ml with phosphate buffer (3.31 g $NaH_2PO_4 \cdot H_2O$, 33.77 g $Na_2HPO_4 \cdot 7 H_2O$ dissolved in double distilled water to a final volume of 1000 ml). This solution (PAF) is very stable and may be stored for several months. 15.5 g sucrose are dissolved in this solution to final volume of 50 ml. Equal volumes of this solution and 6% glutaraldehyde are combined to give SPAFG (0.45 M sucrose, pH 7.3).

Phosphate buffer: (a) see SPAFG, dilute 1:1 with double distilled water for rinsing.
(b) for Sørensen buffer, prepared from stock solutions (double distilled water): *solution A:* 0.2 M $NaH_2PO_4 \cdot H_2O$; *solution B:* 0.2 M $Na_2HPO_4 \cdot 7 H_2O$, 23 ml *solution A* and 77 ml *solution B* diluted to 200 ml (double distilled water) produces 0.1 M buffer (pH 7.3).

Cacodylate buffer: 0.2 M buffer: 4.28 g sodium cacodylate $[Na(CH_3)_2AsO_2 \cdot 3 H_2O]$ in 100 ml double distilled water, 0.2 M HCl is added until a pH of 7.4 is achieved (Approximately 5.5 ml).

Sodium tetraborate (borax): 9.534 g sodium tetraborate $(Na_2B_4O_7 \cdot 10 H_2O)$ in 1000 ml double distilled CO_2-free water; 18.8 ml 0.1 N HCl and 50 ml of the borax solution are diluted to 100 ml (pH 8.2).

Embedding mixture: (a) 10 ml (10.0 g) DDSA (Dodecenylsuccinic anhydride EM), 6.2 ml (7.6 g) Poly Bed 812, 8.1 ml (9.0 g) Araldite resin grade 506, 0.75 ml (0.8 g) Dibutylphthalate, 27 drops (0.85 g) DMP-30 [2,4,6-tri(dimethylaminomethyl)phenol] are carefully mixed with a laboratory stirrer (at least 15 min) before use.
(b) Mixture A: 62 ml Epon 812 (Poly Bed 812) and 100 ml DDSA. Mixture B: 100 ml Epon 812 and 85 ml Nadic methyl anhydride (NMA/MNA) A and B may be stored in a refrigerator. A and B are mixed 1:1 and 1.5% DMP 30 are added prior to use.

Hoyer's mounting medium (for light microscopy): 30 g extremely fine pulverized gum arabic are dissolved in 50 ml distilled water, 20 ml glycerin are added; in this solution 125 g chloral hydrate are dissolved. *[Note: the original formula contains 200 g chloral hydrate. Higgins' (1983) modification reduces this amount to 125 g or even 100 g to prevent over-clearing of meiofaunal specimens, and further recommends the addition of 2 g iodine (crystals) and 1 g potassium iodide to the above formula].* Filter through cotton wool. Media made up from the same substances are Faure's, Berlese's, and Farrant's mounting media (see Adam and Czihak (1964) for additional recipes).

References

Adam, H., and G. Czihak
1964. *Arbeitsmethoden der makroskopischen und mikroskopischen Anatomie.* 583 pages. Stuttgart: Gustav Fischer.

Antonius, A.
1965. Methodischer Beitrag zur mikroskopischen Anatomie und graphischen Rekonstruktion sehr kleiner zoologischer Objekte. *Mikroskopie,* 20:145-153.

Ax, W., U. Kaboth, and H. Fischer
1966. Immunbiologische Studien am Omentum. I. Mitt.: Mikrokinematographische Beobachtungen an kultivierten Mäuse-Omenten; Nachweis gebildeter Antikörper. *Zeitschrift für Naturforschung,* 21b:782-788.

Axelsson, S., A. Björklund, B. Falck, O. Lindwall, and L.- A. Svensson
1973. Glyoxylic Acid Condensation: A New Fluorescence Method for the Histochemical Demonstration of Biogenic Monoamines. *Acta Physiologica Scandinavica,* 87:57-62.

Baker, V.
1972. A Microstrainer-syringe Combination for Handling Specimens in the 1 mm^3 Range in Fluid Changes During Staining Processes. *Stain Technology,* 47:105-106.

Bé, A.W.H., and O.R. Anderson
1976. Preservation of Plankton Foraminifera and other Calcareous Plankton. Pages 250-258 in H.F. Steedman, editor, *Zooplankton Fixation and Preservation.* Paris: UNESCO Press.

Berg, C.J., Jr., and N.L. Adams
1984. Microwave Fixation of Marine Invertebrates. *Journal of Experimental Marine Biology and Ecology,* 74:195-199.

Christie, S.R., and J.R. Edwarson
1968. A Disassemblable Microstainer for Biological Specimens. *Stain Technology,* 43:122-123.

Coineau, Y., and N. Coineau
1979. Une nouvelle technique d'observation des animaux interstitiels: les modèles de réseaux interstitiels microscopiques transparents. *Mikroskopie,* 35:319-329.

De Haller, G., C. Ehret, and R. Naef
1961. Technique d'inclusion et ultramicrotomie destinée á l'étude de développement des organelles dans une cellule isolée. *Experientia,* 17:524-526.

Dressel, D.M., D.R. Heinle, and M.C. Grote
1972. Vital Staining to Sort Dead and Live Copepods. *Chesapeake Science,* 13:156-159.

Eisenbeis, G.B., and W. Wichard
1985. *Atlas zur Biologie der Bodenarthropoden.* 434 pages. Stuttgart, New York: Fischer.

Ermak, T.H., and R.M. Eakin
1976. Fine Structure of the Cerebral and Pygidial Ocelli in *Chone ecaudata* (Polychaeta: Sabellidae). *Journal of Ultrastructural Research,* 54:243-260.

Flood, P.R.
1973. A Simple Technique for the Prevention of Loss or Damage to Planktonic Specimens During Preparation for Transmission and Scanning Electron Microscopy. *Sarsia,* 54:67-74.

Galey, F.R., and R.F. Dunn
1971. Capillary-tube Units for Individual Block Preparation in Light and Electron Microscopy. *Journal of Microscopy* (London), 94:191-193.

George, J.D.
1975. The Culture of Benthic Polychaetes and Harpacticoid Copepods on Agar. *10th. European Symposium on Marine Biology in Belgium,* 1:143-159.

Giere, O., and H. Welberts
1985. An Artificial "Sand System" and its Application for Studies on Interstitial Fauna. *Journal of Experimental Marine Biology and Ecology,* 88:83-89.

Green, C.D., A.R. Stone, R.H. Turner, and S.A. Clark
1975. Preparation of Nematodes for Scanning Electron Microscopy. *Journal of Microscopy,* 103:89-99.

Grimstone, A.V., and R.J. Skaer
1972. *A Guide Book to Microscopical Methods.* 134 pages. Cambridge: Cambridge University Press.

Heunert, H.H.
1973. Präparationsmethoden für Vitalbeobachtungen an Microorganismen. *Zeiss Information,* 81:40-49.

Jägersten, G.
1952. Studies on the Morphology, Larval Development, and Biology of *Protodrilus. Zoologiska Bidrag fran Uppsala,* 29:427-511.

Kalt, M.R., and B. Tandler
1971. A Study of Fixation of Early Amphibian Embryos for Electron Microscopy, *Journal of Ultrastructure Research,* 36:633-645.

Kleinig, H., and P. Sitte
1986. *Zellbiologie.* 528 pages. Stuttgart, New York: Fischer.

Kossmagk-Stephan, K.J.
1984. A Method of Identifying Immature Specimens of Marine Enchytraeidae (Oligochaeta) by Vital Staining of Epidermal Glands. *Hydrobiologia,* 115:55-58.

Kristensen, R.M.
1983. Loricifera, a New Phylum with Aschelminthes Characters from the Meiobenthos. *Zeitschrift für Zoologische Systematik und Evolutionsforschung,* 21:163-180.

Kristensen, R.M., and R.P. Higgins
1984. A New Family of Arthrotardigrada (Tardigrada: Heterotardigrada) from the Atlantic Coast of Florida, U.S.A. *Transactions of the American Microscopical Society,* 103:295-311.

Marks, I., and L.G. Briatty
1970. A Syringe-filter for Processing Specimens through Fluids Used in Plastic Embedding. *Stain Technology,* 45:36-38.

Nixon, D.E.
1976. Dynamics of Spatial Pattern for the Gastrotrich *Tetranchyroderma bunti* in the Surface Sand of High Energy Beaches. *Internationale Revue der gesamten Hydrobiologie,* 61:211-248.

Ogiga, I.R., and R.H. Estey
1974. The Use of Meldola Blue and Nile Blue A, for Distinguishing Dead from Living Nematodes. *Nematodologica,* 20:271-276.

Padawer, I.
1951. Handling Small Numbers of Marine Eggs through Dehydration and Embedding. *Stain Technology,* 26:103-104.

Plattner, H., and H.P. Zingsheim
1986. *Elektronenmikroskopische Methodik in der Zell- und Molekularbiologie.* 335 pages. Stuttgart, New York: Fischer.

Pulsifer, J.
1975. Some Techniques for Mounting Copepods for Examination in a Scanning Electron Microscope. *Crustaceana,* 28:101-105.

Purschke, G.
1981. Tolerance to Freezing and Supercooling of Interstitial Turbellaria and Polychaeta from a Sandy Tidal Beach of the Island of Sylt (North Sea). *Marine Biology,* 63:257-267.

Reid, N.
1975. Ultramicrotomy. In A.M. Glauert, editor, *Practical Methods in Electron Microscopy,* Part II. 353 pages. Amsterdam, New York: Elsevier.

Reisinger, E.
1936. Zur Exkretionsphysiologie von *Ophryotrocha puerilis* Claparède & Metschnikoff. *Thalassia,* 2(4):1-24.

Reuter, M., M. Wikgren, and M. Lehtonen
1986. Immunocytochemical Demonstration of 5-HT-like and FMRF-amide like Substance in Whole Mounts of *Microstomum lineare* (Turbellaria). *Cell and Tissue Research,* 246:7-12.

Rieger, R.M., and E. Ruppert
1978. Resin Embedments of Quantitative Meiofauna Samples for Ecological and Structural Studies - Description and Application. *Marine Biology,* 46:223-235.

Rinaldi, R.A., V.M. Pickel, and R.E. Stephans
1966. A Filter Method for Processing Small Samples of Marine Ova through Fixation, Dehydration and Embedding without Changing Containers. *Stain Technology,* 41:197-200.

Romeis, B.
1968. *Mikroskopische Technik.* 757 pages. München, Wien: Oldenbourg.

Ryuntyu, Y.M.
1980. A Technique of Orientated Embedding of Phytonematodes for Electron Microscopy. *Zoologiceskii Zhurnal,* 59:934-935 [in Russian].

Sawicki, W., and J. Lipetz
1971. Albumen Embedding and Individual Mounting of One or Many Mammalian Ova on Slides for Fluid Processing. *Stain Technology,* 46:261-263.

Shands, I.W.
1968. Embedding Free-floating Cells and Microscopic Particles; Serum Albumin Coagulum-epoxy Resin. *Stain Technology,* 43:15-17.

Shepherd, A.M.
1962. New Blue R, a Stain That Differentiates Between Living and Dead Nematodes. *Nematodologica,* 8:201-208.

Smith, I.P.S., and S. Tyler
1984. Serial-sectioning of Resin-embedded Material for Light-microscopy: Recommended Techniques for Micro-metazoans. *Mikroskopie,* 41:259-270.

Tempel, D., and W. Westheide
1980. Uptake and Incorporation of Dissolved Amino Acids by Interstitial Turbellaria and Polychaeta and their Dependence on Temperature and Salinity. *Marine Ecology Progress Series,* 3:41-50.

Uhlig, G., and S.H.H. Heimberg
1981. A New Versatile Compression Chamber for Examination of Living Microorganisms. *Helgoländer Meeresuntersuchungen,* 34:251-256.

Ware, R.W., D. Clark, K. Crossland, and R.L. Russell
1975. The Nerve Ring of the Nematode *Caenorhabditis elegans:* Sensory Input and Motor Output. *Journal of Comparative Neurology,* 162:71-110.

Weibel, E.R.
1979. *Stereological Methods. Practical Methods for Biological Morphometry.* Volume 1, 415 pages. London, New York: Academic Press.

Westheide, W., and M. von Basse
1978. Chilling and Freezing Resistance of Two Interstitial Polychaetes from a Sandy Tidal Beach. *Oecologia,* 33:45-54.

Windoffer, R., and W. Westheide
1988a. The Nervous System of the Male *Dinophilus gyrociliatus* (Annelida: Polychaeta). I. Number, Types and Distribution Pattern of Sensory Cells. *Acta Zoologica* (Stockholm), 69:55-64.

Windoffer, R., and W. Westheide
1988b. The Nervous System of the Male *Dinophilus gyrociliatus* (Annelida: Polychaeta). II. Electron Microscopical Reconstruction of Nervous System and Effector Cells. *Journal of Comparative Neurology.*

Winters, C., and M. Slade
1971. Embedding Free-floating Cells in Stained Agar for Rapid Scanning and Subsequent Ultramicrotomy. *Stain Technology,* 46:161-162.

11. Culture Techniques

John H. Tietjen

In the past two decades quantitative information on the growth rates, fecundity, nutrition, physiological ecology, and energetics of several meiofauna species has added significantly to our understanding of the functional role of meiobenthos in aquatic ecosystems. Much of the knowledge gained in the foregoing areas has come from experimental studies of animals maintained in laboratory culture and, in fact, could only have come from such studies. While techniques for culturing many species (especially sediment-dwelling ones) remain to be discovered, the successful establishment of laboratory cultures of Protozoa, Turbellaria, Nematoda, Gastrotricha, Harpacticoida, Ostracoda, Oligochaeta, Polychaeta, and other taxa has marked a significant achievement in the study of meiofauna. Many of the successful techniques have been summarized in a treatise on the cultivation of marine animals by Kinne (1977) and several co-authors (Provasoli, 1977; Levandowsky, 1977; Bernhard, 1977). An excellent bibliography included by these authors covers references to 1975.

The purpose of this paper is to review meiofauna culture techniques and to illustrate the types of information that can be gained from the successful culturing of meiofauna.

Goals of Culturing

Types of Cultures.--The aim of a particular study, and the reasons for isolating animals into laboratory culture, must be carefully defined before commencing any culturing effort. Laboratory cultures may be *agnotobiotic* (crude mixtures of the desired species in the presence of other, unknown species) or *gnotobiotic* (where the desired species is maintained in a defined medium either with or without other known species). If it is known that the presence of unknown associated species will not significantly affect the parameters being observed, agnotobiotic cultures are preferred. They are relatively easy to establish and, because they may be "closer to nature" than gnotobiotic cultures, may be better than the latter for short-term experiments. Agnotobiotic cultures are far less labor-intensive to maintain and can be re-established quickly when they fail. However, one never knows when they may fail because agnotobiotic cultures are generally undependable on the time scale from weeks to months (but see Vranken et al., 1984).

Classification of Cultures.--Gnotobiotic cultures can be classed as *polyxenic* (with many known associated species), *tri-*, *di-*, or *monoxenic* (with three, two, or one known species) or *axenic* (without any associated species). Gnotobiotic cultures, even polyxenic ones, are generally much more difficult to establish than agnotobiotic cultures. The usual goal of establishing a gnotobiotic culture is to maintain the desired species in a medium as simple as possible. Once such a culture has been established the types of experiments that can be conducted with the species expands enormously, and the results obtained will, in most cases, be more reproducible than those obtained from similar experiments employing agnotobiotic cultures. For studies of the nutritional requirements or energetics of a species monoxenic or axenic cultures are a necessity.

Field Collection and Preliminary Sample Processing

A good laboratory culture begins with careful field collections. Many meiofauna are easily damaged or killed by rough handling or excessive processing. Phytal samples are the easiest to obtain. Samples of macrophytes may be placed in plastic buckets, gently rinsed by hand and decanted through sieves to remove the large debris; contact with the sieves (preferably nylon or stainless steel) should be as brief as possible. The choice of mesh size of the sieves will depend on the nature of the sample; sieves with mesh sizes of 5.00, 1.00, and 0.50 mm, respectively, have proven excellent for removal of macrophytic material, larger animals, and large sedimentary particles. If sediment samples are being obtained, the top 1-2 cm of sediment should be used because in most sedimentary environments the majority of meiofauna occur in the uppermost 2 cm. Well-drained sandy beaches may be different; from such habitats the top 10 cm should probably be used. Rinsed phytal samples or sediments should be placed in a wide-mouth, screw cap polyethylene jars containing sea water. Plastic bags may also be used as containers. No more than 20% of the jar or bag should be filled with sediment or phytal material. The remainder of the containers are filled with sea water and they are transferred immediately to an ice chest. If necessary, cooling packs are used to keep the samples a few degrees below ambient temperature; an elevated temperature will either kill the animals directly or result in anoxia and rapid bacterial growth which will also cause death.

As soon as the collections can be brought to the laboratory they are transferred to incubation vessels. These may include glass or plastic Petri dishes of different sizes, glass stacking dishes (100 x 40 mm; Carolina Number 74-1004, Carolina Biological Supply Co., Burlington, NC 27215, USA), or 150 cm^2 plastic tissue culture flasks (Corning Number 25120, Corning Glass Works, Corning, NY 14830, USA). These offer the advantage of reducing evaporation loss because they have screw caps. If stacking culture dishes are used care should be taken to use rubber dish seals that bind the stacks together. Petri dishes should be covered at all times to avoid evaporation and contamination by airborne microbes. Erlenmeyer or other wide-bottom flasks can also be used for cultures that require a relatively large volume of water.

There is no standard inoculum size that should be added to the incubation vessels; the volume of material depends upon the richness of the sample. However, the culture vessels should contain samples that are sufficiently dilute to permit the observation of individual animals. Larger competitive or predatory organisms may be removed immediately from the dishes. At this time individual animals (for example, gravid females) can be removed and isolated into separate cultures. Several meiofauna species migrate toward the surfaces of the dishes where they can easily be collected by pipette and isolated.

Extraction of animals from sediments can be especially difficult. Two techniques have been developed to aid this problem. The first is the Uhlig (1964) Seawater-Ice-Technique (see Chapter 9), which is very useful for the extraction of ciliates, flagellates, turbellarians, gastrotrichs, and archiannelids, less so for nematodes and most microcrustaceans. Silica gels (Ludox TM and Percoll) have also been used successfully (Schwinghamer, 1981; Alongi, 1986). Both papers should be consulted for details, but briefly the technique involves inserting sediment samples into 30 ml centrifuge tubes containing 5 ml of silica gel. The samples are vortexed for 1-2 minutes, allowed to stand for 1 hr, and centrifuged for 20 minutes at 490 x G. The supernatent is then poured into a petri dish for microscopic examination and isolation of individual animals.

The samples to be incubated are overlain either with filtered (0.20 µm) autoclaved sea water or one of several growth media (see section on Growth Media). Cultures are incubated at the temperature of collection in light, dark or light/dark cycle conditions. Crude agnotobiotic cultures are generally incubated either in the light or on a programmed light/dark cycle (Lee et al., 1970). However, if algal growth is to be discouraged, the cultures should be covered with aluminum foil. If possible, cultures should be incubated in a controlled environmental chamber to reduce chances of aerial contamination and wide temperature fluctuations.

These crude agnotobiotic cultures are examined daily for evidences of "blooms" of certain animals. Should such "blooms" occur, the animals can be isolated into separate cultures, preferably into small (25 cm^2) tissue culture flasks (Corning Number 25100). To decrease the chances of contamination, individual specimens of the desired species should be rinsed in sterile sea water, sometimes with antibiotic mixtures (see "Use of Antibiotics"). The washed animals are then inoculated into fresh media with potential food organisms obtained from various sources (see "Selection of Food Organisms").

The choice of transfer methods depends on the resistance of the animals to mechanical damage. Foraminifera can be brushed free of adhering material by using glass needles or a camel's hair brush. Other protozoa are transferred using micropipettes. Nematodes can be transferred with micropipettes, but some species either stick to the pipette walls or are broken by sudden pressure changes associated with suction. For these species a curved glass needle is better. Harpacticoida, other microcrustacea and gastrotrichs can be transferred by pipette with no apparent harm. Irwin loops are also excellent for transfer of meiofauna. Trial and error is the only method of discovering the best manner of transfer for a certain species, but I suggest micropipettes as a start.

Repeated washing of many individuals in the above manner will result in an inoculum of relatively clean specimens that can be used to start a single species culture. The size of the inoculum varies with the species being cultured. In general, an inoculum of at least 50-75 individuals is recommended if a 25 cm^2 tissue culture flask is used. However, if gravid females can be obtained, fewer individuals are obviously needed. Recently, Vranken et al. (1985) described the use of small (35 mm inside diameter) vented Petri dishes for culture of nematodes. These dishes appear to be useful for the culture of nematodes and should be tried for other animals. Choice of the proper growth medium and food organisms are the remaining decisions to be made.

Growth Media and Food Organisms

The choice of a growth medium is a function of several factors: the type of culture desired, the trophic characteristics of the desired species (herbivore, bacteriovore, predator, omnivore), the sensitivity of the species to antibiotics and to various abiotic variables, the ability of the animal to co-exist with other species in the same culture vessel and finally, ever present intangibles that can't be anticipated. For these reasons, as many replicate dishes as possible should be maintained to reduce chances of culture loss.

Growth Media.--Meiofauna have been maintained in liquid and solid media. Nematodes have been maintained on solid (Chitwood and Murphy, 1964; Tietjen and Lee, 1972, 1975, 1977a; Marchant and Nicholas, 1974; Romeyn et al., 1983; Vranken et al., 1984) and liquid (Tietjen and Lee, 1977b, 1984; Alongi and Tietjen, 1980) media. In addition, fungal mats have been used to trap and culture nematodes (Hopper and Meyers, 1966; Meyers and Hopper, 1966). Harpacticoid copepods are generally maintained in liquid media (Hicks and Coull, 1983), as are gastrotrichs and most protozoa (Lee et al., 1970; Lee, 1974; Röttger, 1972; Hummon, 1974; Persoone and Uyttersprot, 1975; Parker, 1979). Techniques for the culturing of annelids, which may involve several complex procedures, are described by Giere and Hauschildt (1979), Reish (1980), Redman (1985), and papers cited by these authors.

The best liquid medium for most cultures is Nuclepore filtered sea water from the collection site. To encourage the growth of bacteria several materials can be added to the water. Plant materials that have been used include various species of *Ulva, Enteromorpha, Gracilaria, Zostera, Thalassia* and *Spartina* (Findlay, 1982). Rieper (1978) maintained cultures of the harpacticoid, *Tisbe holothuriae,* on mixtures of herring, mussel meat, crab, sea urchin gonads, and sea stars. Non-marine materials that have been used include lemon rind (Droop and Doyle, 1966), baby cereal (Alongi and Tietjen, 1980; Findlay, 1982), cottage cheese and commercial fish food (Rieper, 1978), powdered rat food, wheat and soya flour, garden vegetables, and chicken droppings (see Hicks and Coull, 1983). Whatever its source, the "detritus" is generally leached in sea water (either by boiling for several hours or naturally for 3-4 days), dried, ground to a fine powder (fine enough to pass through a 250 µm sieve) and inoculated into culture flasks containing Nuclepore-filtered (0.20 µm), autoclaved sea water. The amount of detritus to be added will vary with its nutritional quality (see Findlay, 1982), but a good starting point is approximately 1-3 mg dry weight per cm^2 of surface area (Alongi and Tietjen, 1980).

Erdschreiber (soil extract) is a good culture medium for meiofauna that feed on microalgae. To prepare the soil extract, autoclave equal weights of a good quality forest soil and tap water in a stainless steel vessel and, after cooling, filter it through cheesecloth and filter paper (Whatman Number 1). The filtrate is frozen, thawed, and refiltered through Whatman Number 1 paper. This process may be repeated several times. The extract is finally filtered through a Nuclepore filter (0.20 µm), diluted to a pale straw color and frozen until used. As a growth medium 5 ml of Erdschreiber is added to 95 ml sea water, along with 5 mg% of $Na_2PO_4 \cdot 5 H_2O$ and 10 mg% of KNO_3 (Lee et al., 1970). Erdschreiber medium may also be used to maintain algae in culture as potential food organisms.

In addition, various artificial sea water media have been developed to support the growth of meiofauna and their food (Provasoli et al., 1957; Lee et al., 1970, 1975; Tietjen and Lee, 1975; Vranken et al., 1984). Good monoxenic cultures can be obtained in most cases using either Erdschreiber or one of the detritus-based media cited above. Artificial media are generally reserved for organisms with more sensitive nutritional requirements.

Use of Antibiotics.--If possible, antibiotics should not be used, mainly because the bactericidal concentrations of the antibiotics may be quite close to the lethal concentrations for the animals (Tietjen and Lee, 1973). By maintaining the medium at a sufficiently dilute concentration the growth of bacteria and fungi can usually be controlled within acceptable limits. However, if bacteria-free cultures are desired, antibiotics will have to be used. The necessary concentrations will have to be determined empirically; for cultures of the nematode *Chromadorina germanica,* Alongi and Tietjen (1980) used a mixture of penicillin (10,000 Units ml^{-1}), Fungizone (25 µg ml^{-1}) and Streptomycin (10,000 µg ml^{-1}). Other antibiotic mixtures are described by Lee et al. (1970, 1975) and Vranken et al. (1984).

Transfer Rates.--The rates at which culture media should be changed depend upon the type of culture and the reproductive rates of the animals. For agnotobiotic cultures renewal of the medium should occur at 2-3 day intervals, especially when cultures are first established. Gnotobiotic cultures generally require less transfer (weekly to monthly). Because diet and temperature have been shown to be important factors in determining the growth and reproductive rates of meiofauna in culture (Tietjen and Lee, 1972, 1977b; Gerlach and Schrage, 1971; Feller, 1980; Alongi and Tietjen, 1980; Hicks and

Coull, 1983; Vranken et al., 1984; Redman, 1985), no general rule of thumb can be applied. Experience alone will determine the rate of transfer of cultures being maintained on different combinations of diet and temperature.

Selection of Food Organisms.--Trophic studies using laboratory-maintained animals have revealed that meiofauna may be selective feeders (Lee et al., 1966, 1976; Muller and Lee, 1969; Tietjen and Lee, 1977a; Rieper, 1978; Alongi and Tietjen, 1980; Tietjen, 1980; Hicks and Coull, 1983; Heip et al., 1985; and papers cited by these authors). Among the factors that have been shown to contribute to selectivity include food type (bacteria vs. algae, diatoms vs. chlorophytes), food size (and perhaps shape), method of food presentation (attached to a surface vs. suspended in water), and life stage of the animal (Jensen, 1982, 1984). It is not the purpose of this paper to discuss the trophic ecology of meiofauna, but merely to indicate that (1) certain species of algae and bacteria have been found to be excellent sources of nutrition for a variety of animals while others have not; (2) certain meiofauna can be maintained in bacteria-free culture while others cannot; (3) some meiofauna change their feeding preferences as they progress through their life cycles while others do not; and (4) all of the preceding are not necessarily consistent within a given animal family or genus.

Selection of microbial species that may serve as food for protozoans, nematodes, gastrotrichs, copepods, ostracods, polychaetes and other algal or bacterial feeders may take several routes. Generally, it is not possible for the single meiofauna researcher to isolate potential food organisms from the animals' habitats and maintain a stock collection of algal and bacterial cultures. The numbers of different microbial species to be maintained on several different media is sufficiently large to warrant a full-time person who does little else. Given this problem, how can a person interested in maintaining a gnotobiotic culture of meiofauna achieve this?

Most bactivorous meiofauna can be established by isolating the strains of bacteria that develop on the detritus added initially to the culture medium (a relatively easy process following standard microbiological procedures), growing the bacteria in culture, and experimenting with combinations of bacteria until the correct strain or strains are found that will sustain the meiofauna species. The technical advice of a microbiologist should obviously be sought.

Feeding rates of several meiobenthic species on bacteria are known (Lee et al., 1966; Tietjen et al., 1970; Tietjen and Lee, 1977a; Marchant and Nicholas, 1974; Rieper, 1978, 1982; Schiemer et al., 1980; Admiraal et al., 1983; Montagna, 1984) and can be used to establish guidelines for the concentrations of bacteria needed for successful cultures. An initial concentration of $10^7 - 10^8$ bacteria per ml appears to be sufficient for most bactivorous meiofauna, but certain species might require higher or lower inocula of food organisms (Schiemer, 1983; Vranken et al., 1984).

Selection of species of algae to serve as food organisms for herbivorous meiofauna is a more difficult task. Ideally, the food organisms should come from the environment that the animal was isolated from. However, there may be hundreds of species of diatoms, chlorophytes, and other groups of algae to choose from, assuming that all species of microphytes could be isolated and maintained. Procedures for the isolation, identification, and maintenance of microalgae are given by Lee et al. (1975); an examination of this paper will reveal the enormity of the task involved in isolating and maintaining axenic cultures of algae.

If agnotobiotic cultures of herbivorous meiofauna are desired, the numerous species of algae that develop in the Erdschreiber medium will probably suffice. It is a good idea to maintain agnotobiotic cultures of algae to serve as necessary stock cultures. It will also be possible to separate diatoms from chlorophytes (if time permits) by removing individuals of each group and isolating them into separate cultures. Repeated transfer of the algae under aseptic conditions will help to keep the bacterial densities sufficiently low. Maintenance of meiofauna under these conditions is relatively easy and suitable for most short-term experiments, but it must be remembered that most agnotobiotic cultures may crash at any time.

Sufficient knowledge exists that certain types of meiofauna prefer one major group of algae to another, that certain algal species are good foods for a variety of animals, and that some algal species are better foods than other members of the same genus. As early as 1959 Provasoli et al. identified such algae as *Isochrysis galbana*, *Rhodomonas lens*, and *Monochrysis lutheri* as excellent foods for *Tigriopus* and *Artemia*; these algae are also good foods for other animals, including oysters (Dean, 1957). Algal feeding nematodes grow better on diatom species of the genera *Nitzschia*, *Cylindrotheca*, *Achnanthes*, *Fragillaria*, and *Phaeodactylum* than they do on most chlorophytes, although within the latter group *Nanochloris* species appear capable of sustaining nematode growth for longer periods of time than do *Dunaliella* species (Tietjen and Lee, 1977a; Alongi and Tietjen, 1980). On the other hand, certain species of *Dunaliella* are excellent food sources for foraminifera (Muller and Lee, 1969) and harpacticoid copepods (Provasoli et al., 1959; Hicks and Coull, 1983). Even within a given algal genus (*Dunaliella*, for example), certain species will sustain reproduction of animals while others will not (Muller and Lee, 1969; Tietjen and Lee, 1977a).

The ability of a single algal species, or a combination of more than one species, to sustain meiofauna in the absence of bacteria appears to vary. Muller and Lee (1969) discussed the apparent indispensability of bacteria for successful culturing of foraminifera, and several authors (Provasoli et al., 1959; Brown and Sibert, 1977; Rieper, 1978; Hicks and Coull, 1983) have indicated how several harpacticoid copepod species are aided by the presence of bacteria in cultures. Similarly, it is well known that growth and reproduction of most animals are better on mixed rather than on single species algal diets, especially when the algae are bacteria-free (Muller and Lee, 1969; Smol and Heip, 1974; Heinle et al., 1977; Hicks and Coull, 1983; and others). However, Sellner (1976) found that the copepod, *Thompsonula hyaenae*, grew better on pure cultures of the diatom *Navicula pelliculosa* than on a diet of mixed diatoms.

The factors that determine the suitability of particular microbes as food organisms for particular meiofauna are as yet largely unknown. Food size and shape have been shown to be of some importance to nematodes (Tietjen and Lee, 1977a; Alongi and Tietjen, 1980; Jensen, 1987) and harpacticoids (Marcotte, 1977; Hicks and Coull, 1983). The method of food presentation is also important. For example, a nematode equipped with teeth for rasping organic material off a substratum may not ingest bacteria in suspension, but will eat the same bacteria if they can be scraped off a hard surface (Alongi and Tietjen, 1980). Similarly, Rieper (1978) found that the ability of harpacticoid copepods to ingest bacteria improved significantly when the bacteria were presented as a paste rather than in suspension.

After ingestion, factors such as the nutritional quality of the food (calorific value, chemical composition) and the ability of the animal to digest and assimilate the food become important. For example, Tietjen et al. (1970) observed that, while large numbers of the chlorophytes *Nanochloris* sp. and *Dunaliella parva* were ingested by the nematode, *Rhabditis marina*, the algae passed through the gut undigested. Deutsch (1978) demonstrated significant differences in gut ultrastructure between the algal-feeding nematode, *Chromadorina germanica*, and the bacteriovore, *Diplolaimella* sp. A discussion of the biochemical factors influencing the suitability of various substrates (algae, bacteria, detritus) as food for marine detritovores is given by Phillips (1984).

Density of food is another factor to be considered. Romeyn et al. (1983) have shown that populations of the brackish water nematode, *Eudiplogaster paramatus*, grow better at diatom densities of 10^6 cells ml^{-1} than lower densities. Tietjen and Lee (1977a) and Alongi and Tietjen (1980) employed similar algal densities in their studies, and Lee et al. (1966) showed that algal concentrations of $10^5 - 10^6$ cells ml^{-1} were necessary for foraminifera to feed consistently. The effects of different levels of food supply on growth, reproduction and respiration of two non-marine species of nematodes have been elegantly shown by Schiemer (1982a,b; 1983).

If possible, axenic cultures of algae should be used for herbivores. Axenic cultures may possibly be obtained from an institution that maintains culture collections. Unfortunately, the availability of these cultures is rapidly decreasing in the face of reduced financial support for culture maintenance. However, if one can obtain bacteria-free cultures of some benthic diatoms mentioned above, their maintenance is relatively easy. Because many of these algae have proven to be excellent food sources for a broad spectrum of animals, I recommend trying them as a first step. Any expert in algal ecology or physiology should be aware of the places where algal cultures are currently being maintained.

Use of Radioactive Isotopes

Tracer feeding techniques have contributed much to our knowledge of meiofauna feeding habits (Lee et al., 1966; Chia and Warwick, 1969; Tietjen et al., 1970; Tietjen and Lee, 1973, 1975, 1977a; Marchant and Nicholas, 1974; Brown and Sibert, 1977; Rieper, 1978, 1982; Lopez et al., 1979; Schiemer et al., 1980; Tempel and Westheide, 1980; Guidi, 1984; Montagna, 1984; Carman and Thistle, 1985). Because of the complexity of the subject matter and the variety of techniques employed, the papers cited above should be consulted for details regarding technique. From radioisotopic studies we have (1) obtained quantitative feeding rates of meiofauna belonging to several phyla; (2) seen that most species tested are selective in their feeding habits; (3) observed that several animals may change their feeding preferences throughout their life cycles; (4) been able to identify certain foods that appear to be excellent nutritional sources for a variety of taxa; (5) been able to study uptake of dissolved organic substances by meiofauna; and (6) studied the energetics of several species. Radioisotopic tracers can be used effectively in studies of natural populations of organisms (Montagna, 1984; Carman and Thistle, 1985) as well as in the laboratory, and no doubt will extend our knowledge of the feeding habits of meiofauna forward in the next few years.

Maintenance of Gnotobiotic Cultures

Given the effort involved in their establishment, gnotobiotic cultures should be carefully maintained to avoid contamination by airborne bacteria and fungi as well as by other aquatic organisms (especially algae, bacteria, and protozoa). Flagellates and ciliates are notorious for their ability to contaminate a culture and quickly ruin it. Some simple precautions will prevent contamination and enable a gnotobiotic culture to be maintained for years.

First, use only autoclaved sea water, Erdschreiber, or whatever growth medium is being employed. Sterility of the medium becomes more important as the richness of the medium increases. Second, use disposable pipettes and autoclave all glass and plastic culture vessels to be used. Due to the escalating cost of glass and plastic ware more laboratories are re-using them. Re-use of such materials is possible if proper sterility conditions are maintained. Use of standard microbiological sterility tests of all media and culture vessels is strongly recommended.

All transfers of cultures should be made under aseptic conditions. If a culture transfer room is not available, transfers should at least be made under one of the many good (and inexpensive) transfer hoods currently available. Several replicate cultures of each species should be maintained in at least two locations to prevent culture loss in the event of an air-conditioning or other power failure. If possible, cultures of animals on more than one species of food (either di- or trixenic cultures), or several different monoxenic cultures should also be maintained to provide insurance should one culture fail.

Finally, stock gnotobiotic cultures should be maintained at minimal population densities in dilute media. This will prevent overcrowded conditions from occurring in the cultures, which can cause death by anoxia and by poisonous metabolites excreted by the animals themselves. Dilute cultures will also reduce the frequency of transfer necessary for culture maintenance.

Gnotobiotic cultures are necessary if information on the growth and reproduction rates, nutrition, and energetics of the animals on different foods is desired. They have also proven useful in the examination of interactions among closely related nematode species under defined diet conditions (Alongi and Tietjen, 1980). The use of nematodes as a bioassay for assessing the health of sediments and the effects of specific toxicants can also be aided by maintaining the animals in gnotobiotic culture (Tietjen and Lee, 1984; Vranken et al., 1984). However, Romeyn et al. (1983) illustrate the excellent quality of information that can be obtained on growth, feeding behavior, and physiological ecology of meiofauna under agnotobiotic conditions.

Concluding Remarks

As indicated at the beginning, the goals of a particular study will determine the type of laboratory culture needed. There is much information on the growth, development, fecundity, survivorship, and other life history parameters that can be obtained from maintaining animals under unknown conditions. However, the results of experiments conducted under such conditions will probably not be reproducible under sets of other unknown conditions. It is the nature of experimental work to know, and to isolate, all factors that can significantly affect the outcome of a particular experiment. The milieu of the meiobenthos is a complex one in terms of the number of meiofauna species, their distributions in time and space, and the influence of other organisms and sediment geochemistry directly on the meiofauna and on other species with which the meiofauna interact. To date, almost all of the meiofauna thus far successfully cultured have been eurytopic, surface-dwelling species; few true interstitial species have been established in continuous culture. A major remaining goal of meiofauna culturing is the development of techniques for the maintenance of interstitial species. Once this goal is achieved experimental work on meiofauna can expand much further than at present.

References

Admiraal, W., L.A. Bouman, L. Hoekstra, and K. Romeyn
1983. Quantitative Interactions between Microphytobenthos and Herbivorous Meiofauna of an Estuarine Mudflat. *Internationale Revue der Gesamten Hydrobiologie*, 68:175-191.

Alongi, D.M.
1986. Quantitative Estimates of Benthic Protozoa in Tropical Marine Systems Using Silica Gel: A Comparison of Methods. *Estuarine and Coastal Shelf Science*, 23:443-450.

Alongi, D.M., and J.H. Tietjen
1980. Population Growth and Trophic Interactions Among Free-living Marine Nematodes. Pages 151-166 in K.R. Tenore and B.C. Coull, editors, *Marine Benthic Dynamics*. Columbia: University of South Carolina Press.

Bernhard, M.
1977. Chemical Contamination of Culture Media: Assessment, Avoidance and Control. Pages 1459-1499 in O. Kinne, editor, *Marine Ecology*, Volume III, Part 3, Cultivation. New York: Wiley-Interscience.

Brown, T.J., and J.R. Sibert
1977. Some Food for Harpacticoid Copepods. *Journal of the Fisheries Research Board of Canada*, 34:1028-1031.

Carman, K.R., and D. Thistle
1985. Microbial Food Partitioning by Three Species of Benthic Copepods. *Marine Biology*, 88:143-148.

Chia, F.S., and R.M. Warwick
1969. Assimilation of Labelled Glucose From Sea Water by Marine Nematodes. *Nature*, 224:720-721.

Chitwood, B.G., and D.G. Murphy
1964. Observations on Two Marine Monhysterids - Their

Classification, Cultivation and Behavior. *Transactions of the American Microscopical Society*, 83:311-329.

Dean, D.
1957. The Experimental Feeding of Oysters. Ph.D. Thesis, Rutgers University, New Brunswick, NJ, USA.

Deutsch, A.
1978. Gut Structure and Digestive Physiology of the Free-living Marine Nematodes, *Chromadorina germanica* (Bütschli, 1874) and *Diplolaimella* sp. *Biological Bulletin*, 155:317-335.

Droop, M.R., and J. Doyle
1966. Ubiquinone As a Protozoan Growth Factor. *Nature*, 212:1474-1475.

Feller, R.J.
1980. Development of the Sand-dwelling Meiobenthic Harpacticoid Copepod *Huntemannia jadensis* Poppe in the Laboratory. *Journal of Experimental Marine Biology and Ecology*, 46:1-15.

Findlay, S.E.G.
1982. Effect of Detrital Nutritional Quality on Population Dynamics of a Marine Nematode (*Diplolaimella chitwoodi*). *Marine Biology*, 68:223-227.

Gerlach, S.A., and M. Schrage
1971. Life Cycles in Marine Meiobenthos. Experiments at Various Temperatures with *Monhystera disjuncta* and *Theristus pertenuis* (Nematoda). *Marine Biology*, 9:274-280.

Giere, O., and D. Hauschildt
1979. Experimental Studies on the Life Cycle and Production of the Littoral Oligochaete *Lumbricillus lineatus* and Its Response to Oil Pollution. Pages 113-122 in E. Naylor and R.G. Hartnoll, editors, *Cyclic Phenomena in Marine Plants and Animals*. Oxford: Pergamon Press.

Guidi, L.D.
1984. The Effect of Food Composition on Ingestion, Development, and Survival of a Harpacticoid Copepod, *Tisbe cucumariae* Humes. *Journal of Experimental Marine Biology and Ecology*, 84:101-110.

Heinle, D.R., R.P. Harris, J.F. Ustach, and D.A. Flemer
1977. Detritus As a Food For Estuarine Copepods. *Marine Biology*, 40:341-353.

Heip, C., M. Vincx, and G. Vranken
1985. The Ecology of Marine Nematodes. *Oceanography and Marine Biology, Annual Review*, 23:399-489.

Hicks, G.R.F., and B.C. Coull
1983. The Ecology of Marine Meiobenthic Harpacticoid Copepods. *Oceanography and Marine Biology, Annual Review*, 21:67-175.

Hopper, B.E., and S.P. Meyers
1966. Aspects of the Life Cycle of Marine Nematodes. *Helgoländer wissenschaftliche Meeresuntersuchungen*, 13:444-449.

Hummon, W.D.
1974. Effects of DDT on Longevity and Reproductive Rate in *Lepidodermella squammata* (Gastrotricha, Chetonotida). *American Midland Naturalist*, 92:327-339.

Jensen, P.
1982. Diatom-feeding Behaviour of the Free-living Marine Nematode *Chromadorita tenuis*. *Nematologica*, 28:71-76.
1984. Food Ingestion and Growth of the Diatom-feeding Nematode *Chromadorita tenuis*. *Marine Biology*, 81:307-310.
1987. Feeding Ecology of Free-living Aquatic Nematodes. *Marine Ecology Progress Series*, 35:187-196.

Kinne, O.
1977. Cultivation of Animals-research Cultivation. Pages 579-1293 in O. Kinne, editor, *Marine Ecology*. New York: Wiley-Interscience.

Lee, J.J.
1974. Toward Understanding the Niche of Foraminifera. Pages 207-260 in R.H. Hedley, editor, *Foraminifera*, Volume 1. New York: Academic Press.

Lee, J.J., M. McEnery, E.M. Kennedy, and H. Rubin
1975. A Nutritional Analysis of a Sublittoral Diatom Assemblage Epiphytic on *Enteromorpha* from a Long Island Salt Marsh. *Journal of Phycology*, 11:14-49.

Lee, J.J., M. McEnery, S. Pierce, H.D. Freudenthal, and W.A. Muller
1966. Tracer Experiments in Feeding Littoral Foraminifera. *Journal of Protozoology*, 13:659-670.

Lee, J.J., J.H. Tietjen, and J.R. Garrison
1976. Seasonal Switching in the Nutritional Requirements of *Nitocra typica*, a Harpacticoid Copepod from Salt Marsh Epiphytic Communities. *Transactions of the American Microscopical Society*, 95:628-637.

Lee, J.J., J.H. Tietjen, R.J. Stone, W.A. Muller, J. Rullman, and M. McEnery
1970. The Cultivation and Physiological Ecology of Members of Salt Marsh Epiphytic Communities. *Helgoländer wissenschaftliche Meeresuntersuchungen*, 20:136-156.

Levandowsky, M.
1977. Multispecies Cultures and Microcosms. Pages 1399-1458 in O. Kinne, editor, *Marine Edology*, Volume III, Part 3, Cultivation. New York: Wiley-Interscience.

Lopez, G., F. Riemann, and M. Schrage
1979. Feeding Biology of the Brackish Water Oncholaimid Nematode *Adoncholaimus thalassophygas*. *Marine Biology*, 54:311-318.

Marchant, R., and W.L. Nicholas
1974. An Energy Budget for the Free-living Nematode *Pelodera* (Rhabditidae). *Oecologia*, 16:237-252.

Marcotte, B.M.
1977. An Introduction to the Architecture and Kinematics of Harpacticoid (Copepoda) Feeding: *Tisbe furcata* (Baird, 1837). *Mikrofauna Meeresboden*, 61:183-196.

Meyers, S.P., and B.E. Hopper
1966. Attraction of the Marine Nematode, *Metoncholaimus* sp., to Fungal Substrates. *Bulletin of Marine Science*, 16:143-150.

Montagna, P.A.
1984. *In situ* Measurement of Meiobenthic Grazing Rates on Sediment Bacteria and Edaphic Diatoms. *Marine Ecology Progress Series*, 18:119-130.

Muller, W.A., and J.J. Lee
1969. Apparent Indispensibility of Bacteria in Foraminiferan Nutrition. *Journal of Protozoology*, 16:471-478.

Parker, J.G.
1979. Toxic Effects of Heavy Metals upon Cultures of *Uronema Marinum* (Ciliophora: Uronemaidae). *Marine Biology*, 54:17-24.

Persoone, G., and G. Uyttersprot
1975. The Influence of Inorganic and Organic Pollutants on the Rate of Reproduction of a Marine Hypotrichous Ciliate: *Euplotes vannus* Muller. *Revue Internationale d'Océanographie Médicale*, 37:125-151.

Phillips, N.W.
1984. Role of Different Microbes and Substrates as Potential Suppliers of Specific, Essential Nutrients to Marine Detritivores. *Bulletin of Marine Science*, 35:283-298.

Provasoli, L.
1977. Axenic Cultivation. Pages 1295-1398 in O. Kinne, editor, *Marine Ecology*, Volume III, Part 3, Cultivation. New York: Wiley-Interscience.

Provasoli, L., J.J.A. McLaughlin, and M.R. Droop
1957. The Development of Artificial Media for Marine Algae. *Archiv für Mikrobiologie*, 25:392-428.

Provasoli, L., K. Shiraishi, and J.R. Lance
1959. Nutritional Idiosyncrasies of *Artemia* and *Tigriopus* in Monoxenic Culture. *Annals of the New York Academy of Science*, 77:250-261.

Redman, C.
1985. Effect of Temperature and Salinity on the Life History of *Capitella capitata* (Type I). Ph.D. Thesis, City University of New York, New York, NY, USA.

Reish, D.J.
1980. Use of Polychaetous Annelids as Test Organisms for Marine Bioassay Experiments. Pages 140-154 in A.L. Buikema, Jr., and J. Cairns, Jr., editors, *Aquatic*

Invertebrate Bioassays. Philadelphia: ASTM Publications.

Rieper, M.
1978. Bacteria as Food for Marine Harpacticoid Copepods. *Marine Biology,* 45:337-345.
1982. Feeding Preferences of Marine Harpacticoid Copepods for Various Species of Bacteria. *Marine Biological Progress Series,* 7:303-307.

Romeyn, K., L.A. Boumann, and W. Admiraal
1983. Ecology and Cultivation of the Herbivorous Brackish-water Nematode *Eudiplogaster paramatus. Marine Ecology Progress Series,* 12:145-153.

Röttger, R.
1972. Die Kultur von *Heterostegina depressa* (Foraminifera, Nummulitidae). *Marine Biology,* 15:150-159.

Schiemer, F.
1982a. Food Dependence and Energetics of Freeliving Nematodes. I. Respiration, Growth and Reproduction of *Caenorhabditis briggsae* (Nematoda) at Different Levels of Food Supply. *Oecologia,* 54:108-121.
1982b. Food Dependence and Energetics of Freeliving Nematodes. II. Life History Parameters of *Caenorhabditis briggsae* (Nematoda) at Different Levels of Food Supply. *Oecologia,* 54:122-128.
1983. Comparative Aspects of Food Dependence and Energetics of Freeliving Nematodes. *Oikos,* 41:32-42.

Schiemer, F., A. Duncan, and R.Z. Klekowski
1980. A Bioenergetic Study of a Benthic Nematode, *Plectus palustris* de Man 1880, throughout Its Life Cycle. II. Growth, Fecundity and Energy Budgets at Different Densities of Bacterial Food and General Ecological Considerations. *Oecologia,* 44:205-212.

Schwinghamer, P.
1981. Extraction of Living Meiofauna from Marine Sediments by Centrifugation in a Silica Sol-sorbitol Mixture. *Canadian Journal of Fisheries and Aquatic Science,* 38:476-478.

Sellner, B.
1976. Survival and Metabolism of the Harpacticoid Copepod *Thompsonula hyaenae* (Thompson) Fed Different Diatoms. *Hydrobiologia,* 50:233-238.

Smol, N., and C. Heip
1974. The Culturing of Some Harpacticoid Copepods from Brackish Water. *Biologisch Jarrboek,* 42:159-169.

Tempel, D., and W. Westheide
1980. Uptake and Incorporation of Dissolved Amino Acids by Interstitial Turbellaria and Polychaeta and Their Dependence on Temperature and Salinity. *Marine Ecology Progress Series,* 3:41-50.

Tietjen, J.H.
1980. Microbial-Meiofaunal Interrelationships: A Review. *Microbiology,* 1980:335-338.

Tietjen, J.H., and J.J. Lee
1972. Life Cycles of Marine Nematodes. Influence of Temperature and Salinity on the Development of *Monhystera denticulata* Timm. *Oecologia,* 10:167-176.
1973. Life History and Feeding Habits of the Marine Nematode *Chromadora macrolaimoides* Steiner. *Oecologia,* 12:303-314.
1975. Axenic Culture and Uptake of Dissolved Organic Substances by the Marine Nematode, *Rhabditis marina* Bastian. *Cahiers de Biologie Marine,* 16:685-694.
1977a. Feeding Behavior of Marine Nematodes. Pages 21-35 in B.C. Coull, editor, *Ecology of Marine Benthos.* Columbia: University of South Carolina Press.
1977b. Life Histories of Marine Nematodes. Influence of Temperature and Salinity on the Reproductive Potential of *Chromadorina germanica* Butschli. *Mikrofauna Merresboden,* 61:263-270.
1984. The Use of Free-living Nematodes As a Bioassay for Estuarine Sediments. *Marine Environmental Research,* 11:233-251.

Tietjen, J.H., J.J. Lee, J. Rullman, A. Greengart, and J. Trompeter
1970. Gnotobiotic Culture and Physiological Ecology of the Marine Nematode *Rhabditis marina* Bastian. *Limnology and Oceanography,* 15:535-543.

Uhlig, G.
1964. Eine einfache Methode zur Extraktion der vagilen, mesopsammalen Mikrofauna. *Helgoländer wissenschaftliche Meeresuntersuchungen.* 11:178-185.

Vranken, G., D. Van Brussel, R. Vanderhaegen, and C. Heip
1984. Research on the Development of a Standardized Ecotoxicological Test on Marine Nematodes. I. Culturing Conditions and Criteria for Two Monhysterids, *Monhystera disjuncta* and *Monhystera microphthalma.* Pages 159-184 in G. Persoone, E. Jaspers, and C. Claus, editors, *Ecotoxicological Testing for the Marine Environment.* State University of Gent and Institute for Marine Scientific Research, Bredene, Belgium.

Vranken, G., R. Vanderhaegen, and C. Heip
1985. Toxicity of Cadmium to Free-living Marine and Brackish Water Nematodes (*Monhystera microphthalma, Monhystera disjuncta, Pellioditis marina*). *Diseases of Aquatic Organisms,* 3:1-10.

12. Experimental Techniques

Susan S. Bell

Research on organisms of meiofaunal size has been conducted in a vast array of habitats and for a large number of taxa. Much progress in research on the biology of meiofauna has been attributed to the development of suitable sampling devices, increased systematic sophistication, and a substantial rise in research effort compared to earlier years (see Coull and Bell, 1979). Field studies on meiofauna have been primarily descriptive in nature (see Chapter 14) while laboratory efforts have struggled with developing adequate techniques for working with small, sediment associated organisms. Some advances in the tools for meiofauna research have emerged, but, except in most recent years (e.g., Coull and Palmer, 1984), limited emphasis has been placed on utilizing experimental methodology. Such a method employs significance testing where an investigator makes some experimental modification of a system (=experimental treatment) and compares it to a characteristic set of outcomes if the modification was entirely without effect (=controls). Numerous synthetic papers which summarize experimental techniques or state of the art exist (e.g., Wieser, 1975; Lasserre, 1976; Vernberg and Coull, 1980; Heip et al., 1982; Coull and Palmer, 1984) and generally agree that experimentation poses a formidable challenge to meiobenthologists. Collection of basic biological information coupled with experimental methodology, however, are requisite for understanding dynamics of communities and populations as well as behavior and physiology of meiofauna. Below, representative experimental techniques extracted from the meiofaunal literature and methodologies are presented.

Using Meiofauna as Experimental Organisms

Meiofaunal organisms possess some characteristics which are attractive for experimental work. Animals are readily obtainable in large numbers and generation times of meiofauna are generally short, compared to other benthic taxa. Such features enhance the use of meiofauna in community and population level research. Both the advantages and disadvantages as well as the logistical limitations of working with meiofauna from a particular habitat will dictate whether experimental efforts are feasible. For some better studied, easily collected or cultured taxa, experimental methodology would seem suitable whereas other systems pose a challenge given current techniques.

Because meiofaunal organisms are so small and visually unobservable in natural settings, direct observation, one of the most powerful tools for generating testable ideas for experimentation, is lacking for meiobenthologists. As a result, basic knowledge of animal movement, feeding, or reproduction has been mainly inferential. The lack of information about basic biological traits prevents the testing of some integral questions about field assemblages of meiofauna, impedes designing manipulative experiments and evaluating independence of treatment sites, and raises questions about laboratory studies on organisms sustained on food items not necessarily representative of natural conditions. It is clear that current advancements in critical areas of feeding and behavior will accompany the emergence of readily designed and interpretable experimental studies. A second problem which further hinders the application and interpretation of experimental methodology is the reporting trends of major taxa only and the failure of many researchers to identify species in data analyses. This is especially disconcerting if two different species show opposite and equal responses to experimental treatments, so that at the major taxon level no response is recorded (see, e.g. Bell, 1980). The inability to resolve species specific responses to experimental treatments remains a formidable hurdle in this domain of study.

Two fundamental questions need be addressed in designing experiments: (1) should a laboratory or field setting be used? and (2) what kind of experimental methodology (investigator imposed treatment) should be employed? Both questions require evaluation in context of research goals, logistical limitations, and statistical considerations (see Chapter 14). Arguments for the use of either laboratory or field settings for experimental studies can be offered. Laboratory conditions allow for more controlled conditions than field experimentation. Field studies may contain a host of unmeasured or unknown factors, not controlled in experiments, which can influence organismal responses, possibly nullifying important treatment effects. Moreover field sites must be free from outside interference and experiments easily retrievable, which may not be true in a variety of habitats.

Selection of field sites for experimental trials may be plagued by marked variability in abundance, species composition and/or age structure of populations within visually similar and adjacent areas (Eskin and Coull, 1984; Kern and Bell, 1984) thus presenting problems for establishing replicability of treatments or "blocks" for experimental purposes. Laboratory studies on the other hand may contain artifacts due to removal of organisms from natural settings, a condition often avoided in field studies. A possible solution to the dilemma may be to use a combination of both lab and field studies to serve as checks on experimental artifacts. Certainly the use of micro- or mesocosms provide a promising avenue for experimental manipulation and their utility continues to be examined (e.g., Elmgren and Frithsen, 1982).

Experimental Considerations

Due to the small size and biological characteristics of meiofaunal organisms it is critical to employ (1) appropriate experimental methods for working on a small scale and (2) a feasible experimental design (see Chapter 8) given limitations of working with meiofaunal organisms. Experimental design and sampling strategies can be determined once logistical questions about time scale (how long should experiments run? how many times should the experiment be replicated?) and level of analyses (do species need to be identified?) are considered.

Often because of the large amount of manpower involved in processing meiofaunal samples it may not be feasible to construct a sophisticated experimental design and thus the usefulness of experimental methodology must be questioned. Meiofaunal organisms generate problems in experimental design because (1) often it is impossible to collect a sufficient number of samples necessary to detect differences between treatments, (2) knowledge or control of extraneous factors is lacking and, (3) measurement error and patchy distribution of organisms, singly and/or combined, hinder assessment of treatment effects. Given these drawbacks, it may be necessary to continually re-evaluate and pretest a number of experimental methods before one suitable for the research question is found. The value of background information is apparent. Even if problems of experimental methodology can be addressed, progress may be slow. If a null hypothesis is erected and not rejected by the experimental results it does not necessarily follow that the null hypothesis is proven. Given failure to reject a null hypothesis, alternative tests should be conducted or a better experiment may be mandated. Thus experimentation can be a very time consuming and often frustrating approach. The answers well designed experiments can provide, however, often supersede experimental deterrents.

Statistical and sampling procedures are not the only decisions in experimental investigations. In most experimental studies on meiofauna to date, results are reported from short term investigations, and analyzed to either major taxa level (for the majority of organisms) and/or perhaps species level for one taxon (see Coull and Palmer, 1984), often harpacticoid copepods. Clearly improvements in experimental interpretation will coincide with better taxonomy. The problem in ignoring species level responses to treatments has been cited before. Moreover there is a distinct need to replicate experiments (not treatments) and/or conduct long term investigations to test generality of results.

Although numerous considerations must be addressed when utilizing meiofaunal organisms in experimental work, many workers have gleaned valuable information from experimental studies. Most experimental field studies have been directed at assessing dynamics of meiofaunal populations or assemblages. Comprehensive surveys of existing experimental work indicate that harpacticoid copepods received the bulk of attention while nematodes, gastrotrichs, archiannelids, and turbellarians have been studied less frequently. Eight topics predominate experimental work: recolonization, biological interactions, disturbance, associations with biogenic structure, pollution, feeding (including demography), preference (behavior), and tolerance. Below, representative techniques used in experimental work to address questions about meiofaunal assemblages in the field and lab are outlined and methodology of one experimental study in each category is discussed in detail.

Representative Experimental Methodology

Colonization and Recolonization.--Information on rate and mode of meiofaunal recolonization has been acquired from studies on habitat selection and meiofaunal movement into defaunated sediment patches or islands. Colonization experiments involving the insertion of virgin substrate near areas inhabited by meiofauna have been conducted for numerous taxa, in both bottom sediment and water column locations, at a variety of water depths and for a diversity of substrates. Many studies utilize substrata prepared by the researcher, while some studies have been coupled with investigations on defaunation caused by biological or physical agents present in the field (e.g., Table 12.1). The popularity of these experimental techniques is probably attributable to the relative ease of preparing defaunated sediments or artificial substrates and placing them into the field. Freezing, autoclaving, or extraction of fauna by mechanical stirring and/or the addition of $MgCl_2$ for narcotization have been used to prepare azoic substrates. Such substrate modification may alter organic content or microalgal abundance which may be important. Furthermore special recolonization chambers have been devised for use in the field (Chandler and Fleeger, 1983) and laboratory (Hockin, 1982) and discrete patches can be easily retrieved for analyses. Plant structure is easily defaunated as well and can be manipulated simply for investigation on phytal forms. The importance of time scale for sampling is of extreme importance in the above studies since meiofauna may appear in treatments within hours of initiation of the experiment. Long term experiments may not be possible for recolonization studies due to problems with maintaining experimental treatments over extended time intervals.

Colonization of sediments of varying particle size by harpacticoid copepods was studied by Hockin (1982) under field conditions. Rather than utilizing natural sediments of non-uniform composition and irregularity, soda-glass ballotine beads with diameters of 0.485, 0.367, 0.267, and 0.147 mm were employed. Particles of each size were placed inside a small latticed plastic tray (volume = 7 dm^3). Both lids and bases of traps were lined with .054 mm polyester netting to keep artificial sediments intact and prevent invasion of natural sediments. Traps were placed at two different tidal heights and sampled with a 2.5 cm corer every 5 days for 14 weeks thus designed to look at long-term colonization events.

Samples were analyzed for all copepod fauna. In addition, a complementary laboratory experiment was conducted employing techniques similar to preference experiments (see below) whereby animals in laboratory microcosms were allowed a choice of four different substrata. These experiments were run for 3 months and it was necessary to add bacterial nutrient supplies to and circulate seawater through the laboratory microcosm. Hockin's (1982) use of both laboratory and field experiments were helpful in interpreting his results.

Biological Interactions and Disturbance.-- Numerous studies have explored the phenomena of biological interactions between meiofaunal assemblages and other, generally larger, taxa using experimental techniques (Table 12.2). Many efforts have been aimed at elucidating the effect of predation on meiofaunal communities either by introducing or excluding predators of meiofauna (Figure 12.1a). Cages made with screening with a variety of mesh sizes have been used to both exclude and include large predators such as fish, shrimp, and crabs and subsequently monitor meiofaunal abundance and composition in cage treatments compared to treatments with no alteration of predator activity (uncaged areas) (=controls).

Most studies have been done in intertidal areas although some subtidal work exists. Because of possible cage artifacts (Virnstein, 1978), cage controls (e.g., sideless, topless cages) which assess possible effects of cage structure other than alteration of predator activity must be included in experimental design and it is desirable to have corroborative evidence from gut contents to verify predator activity. In order to readily interpret experimental findings a good knowledge of predator feeding activities is requisite as is information on other members of the benthic community which may

Table 12.1 Representative studies using experimental techniques to study (re) colonization.

Taxa Studied	Experimental Techniques	Authors
Copepods	transplanted defaunated algae onto rock substrate in field	Gunnill (1982)
Copepods	placed artificial monometric sediment into microcosms in field and lab	Hockin (1982)
Nematodes	placed baited sand in small cages in field	Gerlach (1977)
Major taxa; copepods	azoic sediment in containers placed into field	Scheibel and Rumohr (1979)
Major taxa; copepods	transplanted "azoic" fecal mounds and monitored (re) colonization in field	Thistle (1980)
Major taxa; copepods	defaunated sediment placed into traps with different openings to assess pathways of invasion in field	Chandler and Fleeger (1983)
Turbellarians	constructed aggregates of Tellinid bivalves 110 cm below the sediment surface and monitored colonization	Reise (1983b)

Table 12.2 Representative studies using experimental techniques to study predation and disturbance.

Taxa Studied	Experimental Techniques	Authors
Major taxa; copepods	predator exclusion cages (field)	Bell (1980)
Major taxa	predator exclusion cages (field)	Reise (1979)
Major taxa; copepods	predator exclusion cages (field)	Fleeger et al. (1982)
Major taxa; copepods	predator introduction (field)	Bell and Coull (1978)
Major taxa	predator introduction (field)	Dethier (1980)
Major taxa; copepods	predator introduction (field)	Castel and Lasserre (1982)
Copepods	predator introduction (field)	Warwick et al. (1982)
Copepods	predator introduction (lab and field)	Coull and Wells (1983)
Turbellarians	predator feeding trials (laboratory)	Watzin (1985)
Major taxa; copepods	disturber exclusion cages (field)	Bell and Woodin (1984)
Major taxa; nematodes	mimic of disturbance (field)	Sherman and Coull (1980)

respond to investigator imposed treatments. Often it is impossible logistically to monitor all biotic members of sediment systems and if predator effects are mediated by responses of non-meiofaunal organisms (e.g., microbes, macroinfauna), experimental results may be misinterpreted. Manipulation of predators in microcosms has been performed successfully (Bell and Coull, 1978; Castel and Lasserre, 1982) as have experiments on predators feeding on meiofauna in laboratory settings using small aquaria (Coull and Wells, 1983). In laboratory situations where predator activity can be controlled, behavior may be altered relative to that displayed in the natural system so field verification of results should be attempted. Many of the critiques aimed at experimental work with predators on large fauna (e.g., Dayton and Oliver, 1980) equally apply to meiofaunal organisms.

In a series of laboratory experiments, Coull and Wells (1983) investigated fish predation on phytal harpacticoid copepods and how the structural complexity of algal substrata influenced predator efficiency. Copepods were extracted from *Corallina* algae with isotonic $MgCl_2$ (73.2 g/l), concentrated on a 63 μm sieve, and placed back into a seawater solution to make a concentrated suspension of copepods. The copepod suspension was then divided into equal subsamples with a Folsom plankton splitter (but an autopipet may work equally well) and subsamples introduced into plastic aquaria. Various kinds of algal or artificial structure were introduced into aquaria as were fish predators which were previously starved. Replicated aquaria of each structure treatment were included in the study as was an aquaria with prey only which served to assess natural mortality.

At the end of 24 hours all animals and structure were retrieved from the aquaria and enumerated. Comparison of predator removal of harpacticoid prey under various structure treatments was determined by examining the number of prey remaining in all aquaria at the end of the experiments. In addition a field manipulation where predators were introduced into tide pools and feeding assessed served as a comparison of laboratory vs. field results.

Interest in understanding intra-meiofaunal interactions has generated attempts to experimentally manipulate meiofaunal populations in the field and laboratory. Manipulations of organismal density or species composition in the laboratory, especially in cultures (e.g., Fava and Crotti, 1979), have been performed to assess possible competitive effects. Some field attempts have been performed as well to increase densities of meiofauna and measure responses by other members of the community (Bell, 1983, Service and Bell, 1987) although the logistics of manipulating assemblages in the field are severely problematic. Manipulation of meiofaunal size predators such as turbellarians has been accomplished by adding predators, previously extracted from field samples, to experimental chambers stocked with field-collected sediments and various prey densities and monitoring meiofaunal changes when turbellarian predators were added (Watzin, 1983). In some cases temporary meiofauna, i.e., juvenile macrofauna, have been included as categories in experimental studies, but high spatial variability and temporal unpredictability of larval settlement of macrofauna often hinder controlled experimental work.

Although meiofauna may not be eaten by all

higher trophic levels they may be extremely susceptible to disturbance, or physically imposed mortality, by agents such as hemichordates, stingrays, and horseshoe crabs, which turn over sediment. Effects of disturbance have traditionally been evaluated by comparing meiofaunal assemblages in treatments with and without disturbance agents, or by investigating the effect of disturbance by raking over sediment (Figure 12.1b). Mimics of disturbance may not be a perfect imitation of disturbance created by sediment movers and thus conclusions may need tempering when extrapolating results. Disturbance studies combined with investigations on recolonization (Table 12.1) serve as a powerful tool for assessing dynamics of community and population responses in the field. In addition to measuring the ability of meiofauna to recover from physically induced mortality, successful understanding of disturbance effects on meiofaunal assemblages requires knowledge of the area effected by disturbance activities and the frequency of disturbances.

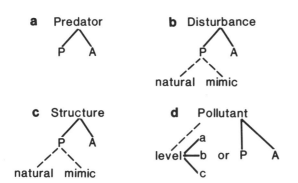

Figure 12.1.–Schematic diagram of types of experimental comparisons utilized in studies on, a, predation; b, disturbance; c, response to structure; and d, pollution. Presence (+) vs. absence (-), natural vs. a mimic, or different levels or concentrations (a,b,c) of treatments have been used in meiofaunal investigations.

Structure Effects.--As tools for small-scale sampling of meiofauna have emerged and studies aimed at examining spatial relationships of meiofauna have increased, it has become apparent that meiofaunal organisms show differential distributions near plant or animal produced structures. Researchers have recorded the associations of meiofaunal taxa with biogenic structure such as plant stems, roots, burrows, and tubes (e.g., Bell et al., 1978; Reise and Ax, 1979; Eckman, 1983). The explanations for meiofaunal exploitation of structures are limited. Suggestions of utilization of structures as sites of (1) aggregated food items or increased oxygen content, (2) predator/disturber refuges, or (3) refuges from physical stress have been purported but have received little experimental assessment (Table 12.3). A variety of experimental approaches have been employed to discern the nature of meiofauna-structure relationships including removing or adding natural structure, or adding mimics of structure in field situation (Figure 12.1c) and subsequently charting meiofauna changes to structure manipulation. The overlap of structure manipulations with colonization studies is common (e.g., Reise, 1983a,b).

Experimental studies on structures and meiofauna seem to be a fruitful avenue for pursuit due to the relatively large size of tubes, burrows, or plants which make them easy to manipulate. As with disturbance studies the use of structural mimics must be carefully scrutinized to determine if they are adequate simulations of natural conditions. Straws, special plastics, or polystyrene ribbons have been used previously to imitate plant and animal structures. Often experimental timetables must be tailored to the hardiness of manipulated structure or mimics and a good knowledge of natural turnover or persistence of structures in the field should accompany experimental studies. Both short and long time scales may be required to unravel the precise nature of associations between biogenic structure and organisms. Laboratory observations with microscopy on organismal utilization of structure can provide valuable information when designing field treatments and interpreting results.

Bell and Coen (1982) performed experiments to assess the influence of the macroalgae *Ulva* sp. upon the abundance of meiofauna living/associated with tube-caps of the polychaete, *Diopatra*, which decorates its cap with the macroalgae. Their experiment was conducted in an easily accessible, shallow mudflat over a 3-week period. Within experimental plots two kinds of treatments were established: tube-caps which had algae removed daily and tube-caps which were similarly handled by investigators but on which algae was left intact. At the end of the experiment tube-caps of both treatments were clipped at the sediment surface and placed into vials with formalin. In addition, algae alone from tube-caps were retrieved from the field. Meiofauna were enumerated from all preserved tube-caps and algal samples. Abundances were compared to determine the effect of algal presence on meiofaunal taxa. The structure here was readily manipulated in the field and the animals could be easily collected from the structure – a distinct advantage of working with epibenthic or phytal meiofauna. For studies utilizing organisms around structures, additional sediment sampling would be necessary.

Pollution.--Studies aimed at assessing effects of pollutants such as PCP, oil, diflubenzuron, and barite have been conducted for meiofauna under laboratory and field situations. Pollution studies have employed

Table 12.3 Representative studies using experimental techniques to study structure - meiofaunal associations. All studies were conducted in the field.

Taxa Studied	Experimental Techniques	Authors
Major taxa; copepods	manipulated algal cover on polychaete tube-caps	Bell and Coen (1982)
Polychaetes; turbellarians	manipulated algal cover in field sediments	Reise (1983a)
Major taxa; copepods	manipulated seagrass structure (mimics)	Hall and Bell (in ms)
Major taxa	manipulated seagrass structure (mimics)	Thistle et al. (1984)
Major taxa; copepods	manipulated mimics of sea pansies	Creed and Coull (1984)

experimental methodology to assess either direct mortality of meiofauna or sublethal stress as impact on reproduction and growth (Table 12.4). As Coull and Palmer (1984) point out, many "experimental" studies in the field have employed situations where pollutants are introduced into an area (either by accident or design) and the affected area compared to (1) the same area before, (2) similar areas in different geographic locations, or (3) adjacent sites not affected by pollutants (Figure 12.1d).

Because of the difficulty in assuming similarity of meiofaunal assemblages in field sites before treatments (e.g., Eskin and Coull, 1984) or in assuming pollution as the only different feature between sites or over time, comparison of treatment sites must be carefully considered (see also Stewart-Oaten et al., 1986). Laboratory studies using more controlled conditions eliminate the problem of isolating pollution effects and have been employed to study toxicity on those taxa which readily survive in culture or establish themselves in microcosms/mesocosms (e.g., Elmgren and Frithsen, 1982). Laboratory conditions allow for monitoring of physiological parameters from experimental treatments, and lethal dosages of pollutants can be directly tested (e.g., Ustach, 1979). Often, laboratory investigations are confined to a single species but the work by Cantelmo et al. (1979), Elmgren and Frithsen (1982), and Frithsen et al. (1985) using "natural" communities with experimenter controlled addition of pollutants into mesocosms improves upon this limitation.

Lehtinen et al. (1984) have extended strict studies on toxic effects to include the effect of temperature on the toxicity of TiO_2 for a harpacticoid copepod, *Nitocra spinipes*. In a laboratory experiment, Lehtinen

Table 12.4 Representative studies of experiments on pollution.

Taxa Studied	Experimental Techniques	Authors
Harpacticoid copepod	cultured harpacticoid copepod in dishes with suspensions of oil	Dalla-Venezia and Fossata (1977)
Whole meiofaunal assemblages	added different concentrations of PCB to aquaria with flow-thru system	Cantelmo and Rao (1978)
Whole meiofaunal assemblages	added barite to aquaria with flow-thru system	Cantelmo et al. (1979)
Harpacticoid copepod (*Nitocra*)	placed oil-water mixture into cultures	Ustach (1979)
Sand assemblages	added PO_4 and NO_3 domestic sewage	Wormald and Stirling (1979)
Harpacticoid copepod (*Tisbe*)	added detergent to cultures of *Tisbe*, combined with effects of crowding	Fava and Crotti (1979)
Whole assemblages	used mesocosms and added #2 fuel oil to pelagic and benthic systems	Elmgren et al. (1980) Frithsen et al. (1985)
Whole assemblages	added oil to sediment	Fleeger and Chandler (1983)
Harpacticoid copepod (*Nitocra*)	added waste water with TiO_2 to cultures	Lehtinen et al. (1984)
Harpacticoid copepod (*Tigriopus*)	added diflubenzuron, dissolved in acetone to water	Antia et al. (1985)

et al. (1984) collected waste water from outlet tubes from the pollution point source and returned the waste water to the laboratory for subsequent evaluation. Acute toxicity tests were performed at three different temperatures (4, 7, 21° C) with a series of waste water concentrations. Ten copepods were placed into test tubes with 10 ml of diluted artificial seawater into which specific waste water concentrations were added. After 96 hr of exposure the number of dead individuals was recorded. This study is representative of others where toxic effects are examined under laboratory conditions. Coincident field samplings of the copepod and measurement of field temperatures and TiO_2 levels can provide additional important information on the possibility of pollutant effects.

Feeding.--Information that exists on feeding of meiofauna has been primarily gathered from examination of guts from both field and/or laboratory animals raised on selected food. Experimental studies on meiofaunal feeding are scarce but those that do exist generally utilize one of two approaches: (1) feeding a consumer different kinds of food and measuring population growth, reproduction, or physiological efficiencies, or (2) examining the performance of a consumer species on different levels of the same food. Experiments conducted in the laboratory employ organisms extracted from field sediments or grown in culture. As such, much information on feeding coincides with culturing techniques (see Chapter 11). Removal of organisms from field conditions may cause experimental artifacts, and alternative methods for measuring *in situ* feeding activities are presently being evaluated (see Montagna, 1984; Carmen and Thistle, 1985). Recent efforts have been made to employ "natural" food items in experimental studies (e.g., Ustach, 1982) in the laboratory to better simulate field conditions.

Because meiofauna are suggested to feed via a number of different modes (uptake of DOC; ingestion of algae, bacteria, or detritus; predation) a variety of techniques to investigate feeding behavior have been developed (Table 12.5). Copepods, especially those maintained well in cultures, have received a large amount of attention in feeding studies (see Hicks and Coull, 1983, for review). Food uptake or feeding rates on various food sources have been measured by labeling substrates with ^{14}C and comparing incorporation of label with appropriate controls (Montagna, 1983). Often measurements of feeding have been combined with measurements of respiration or assessed under different salinity or temperature regimes (e.g., Gaudy et al., 1982). Improved information on copepod feeding mechanisms and selection may be provided through the use of microvideo recordings such as has been used for calanoid copepods (Friedman and Strickler, 1975; Marcotte, 1977).

Nematodes which obtain nutrition through numerous methods have been employed in experimental work. Culturing or laboratory studies using food labeled with ^{32}P and ^{14}C have provided insight into feeding dynamics and physiological characteristics. Often experiments have been conducted in small glass containers, although Findlay and Tenore (1982b) employed a specially designed flow through system to monitor the amount of organic ^{14}C mineralized to $^{14}CO_2$ by nematodes from different sizes and types of detritus. Much work is needed in the field of meiofaunal feeding, especially if experimental work is to become more sophisticated and complete. The work of Findlay and Tenore (1982) using differential labeling of microbes and plant substrates in polychaete feeding studies offers

Table 12.5 Representative studies using experimental techniques to study meiofaunal feeding.

Taxa Studied	Experimental Techniques	Authors
Nematodes	^{14}C label of dissolved organic carbon and microbes	Lopez et al. (1979)
Nematodes	^{14}C tracer to follow C mineralization	Findlay and Tenore (1982b)
Nematodes	fed bacteria under different salinity/temperature regimes	Tietjen and Lee (1972)
Nematodes	raised nematodes on different densities of food and at different temperatures	Schiemer et al. (1980)
Nematodes	population growth monitored with two different species of algae and bacteria	Alongi and Tietjen (1980)
Copepods	population monitored under different foods, temperature and salinity conditions	Gaudy et al. (1982)
Copepods	raised 1 species of copepod on different foods collected from the field	Ustach (1982)
Copepods	offered variety of strains of bacteria to 4 copepod species	Rieper (1982)
Copepods	offered bacteria and ^{14}C labeled alga in different mixtures	Vanden Berghe and Bergmans (1981)

some possibilities to meiofaunal systems.

Trotter and Webster (1984) employed "cafeteria" experiments to investigate the feeding of nematodes collected from macroalgae (*Macrocystis*) on bacteria and diatoms. Nematodes utilized in feeding experiments were removed from algae and kept in cultures in tissue flasks with autoclaved filtered seawater, pieces of macroalgae, and detritus. Isolated bacterium and diatom species from detritus and scrapings from algal blades served as food items in the experiments. Petri dishes were filled with agar and wells created within the agar. A selected bacterium or diatom species was introduced into each of the wells and nematodes were put in a center well. The contents of the petri dish were then covered with a 0.5 cm deep layer of sterile seawater. Plates were left for 72 hours after which the number of nematodes present in each food well was enumerated. Such cafeteria experiments allow for the precise positioning of food items and thus feeding preference can be directly evaluated. Whether this technique would be useful for detecting feeding preference of sediment dwelling forms is not clear although coupling of these techniques with field observations and gut analyses would be instructive.

Studies on population biology or demography of meiofaunal organisms, while often nonexperimental in nature, can be combined with laboratory culturing and feeding studies to provide excellent insight into life history traits of organisms (e.g., Jensen, 1983). Bergmans' (1984a) study on *Tisbe* (Copepoda: Harpacticoida) is a most appropriate example. However, the demographic literature has been fraught with bias and misinterpretation of demographic parameters (Bergmans, 1984b) and the reader is directed to Mertz (1971) and Bergmans (1984a and references within) for important discussions of analytical and interpretive techniques.

Preference.--Of primary concern to meiofauna is selection of sediments by different fauna, and a classical set of laboratory investigations of meiofaunal preference for different sediment types or characteristics (size, angularity) has been conducted by Gray (1966, 1968) (Table 12.6). Other sediment preference experiments have been done more recently in the field by offering defaunated sediments for colonization (see Table 12.1) (e.g., Conrad, 1976; Hockin, 1982) and in the laboratory by introducing organisms into chambers with different sediment types and recording their preferences (e.g., Ravanel and Thistle, 1981). Such studies often complement or overlap studies on colonization or distributional patterns.

Giere (1979) outlines experimental methodology for investigating meiofaunal response (or preference) to *gradients* of salinity, temperature, or moisture rather than examining responses to discrete treatments or levels (see Hicks, 1980, below). For example, to test animal preference for moisture content of sediments, a Perspex trough, measuring 100 cm x 10 cm, with walls 3 cm high at one end and 9 cm high at the other was constructed. The trough was filled with 2 cm of sterile sand and positioned at an oblique angle with the shorter side elevated. A predetermined amount of seawater was added to the trough and a moisture gradient established with absence of moisture at the short-walled, elevated end to full water saturation at the longer-walled end. Meiofauna of interest could then be introduced into the apparatus and the trough covered with a tight lid. Troughs with uniform moisture content of sediments and similar introduced fauna would serve as an appropriate control for the study. This relatively simple design could be modified to investigate salinity and temperature gradients as well (Giere, 1979).

Limited other studies on behaviorally directed movement of meiofauna are reported in the literature. Chemotaxis of meiofaunal copepods was investigated by Hicks (1977) who utilized a Y maze to evaluate preference of copepods to chemicals released by various algae. Phototaxis of harpacticoid copepods has been assessed in the laboratory by introducing them into containers with sediment and recording their movement (preference) to completely darkened or lighted areas (Gray, 1968, Palmer, 1984). Additional information on meiofaunal movement or

Table 12.6 Representative studies using techniques to study meiofaunal preference.

Taxa Studied	Experimental Techniques	Author
Copepods	used multiple preference experiments to establish response to light, gravity, grain size, gregariousness, and sediment bacterial coating	Gray (1968)
Copepods	used Y maze to measure algal preference	Hicks (1977)
Copepods	used salinity and temperature troughs to establish gradients, looked at organism preference for physical regime	Giere (1979)
Copepods	used preference chambers to establish selection of distinct sediment types	Ravanel and Thistle (1981)

Table 12.7. Representative studies using experimental techniques to study meiofaunal tolerance to abiotic factors.

Taxa Studies	Experimental Technique	Author
Mites	Laboratory study on tolerance to 8 different salinities at one temperature; animals placed into bowls	Ganning (1970)
Gastrotrichs	Laboratory study on respiration rates using microrespirometer at one temperature; upper or lower salinity lethal limits assessed by putting 20 individuals in petri dishes	Hummon (1974)
Copepods	Laboratory study of temperature and salinity tolerance in combination with tolerance to anaerobiosis, bubbled 100% N into flasks and monitored activity of animals under different salinities and temperatures	Vernberg and Coull (1975)
Nematodes	Laboratory study of upper lethal temperature by subjecting 20-30 animals in vials to elevated temperatures for 1-10 hours	Wieser and Schiemer (1977)
Copepods	Laboratory study to measure survival of copepods in small vials under different combinations of temperature and salinity	Hicks (1980)

phototaxis in the field has been collected using specially designed emergence traps (e.g., Fleeger et al., 1984; Walters and Bell, 1986) during daytime and nighttime. Advances in understanding meiofaunal behavior will be aided by combining field and laboratory studies, and improving observational techniques of organisms in natural settings, microcosms, or flumes.

Tolerance.--Tolerance experiments on a variety of meiofauna taxa have been performed to detect the impact of specific salinity, temperature, or oxygen levels on survival (Table 12.7). Experiments have been limited to laboratory conditions using small chambers and animals from the field or cultures. A number of thorough reviews of methodologies for physiological studies on tolerance are available (Wieser, 1975; Lasserre, 1976) which discuss techniques which can be used in experimental studies. Vernberg and Coull (1980) provide a helpful overview of results of meiofaunal physiology.

A study by Hicks (1980) illustrates a multiple factor approach to studying tolerance of meiofauna. Temperature and salinity tolerances of the harpacticoid copepod *Zaus spinatus spinatus* were simultaneously tested using single animals in 4 ml vials with water of a desired salinity (ranging from 8-42°/oo) maintained at 5 different temperatures (0, 5, 10, 15, and 20° C). Twenty five replicates for each temperature-salinity combination were employed and animals were checked daily to observe movement or lack of same (mortality). These results were then graphically examined to evaluate survivorship over time of the copepods under different temperature and salinity levels. This study took advantage of copepods which were easily collected from phytal habitats and performed well in the laboratory.

Physiological studies are logical complements to feeding and behavior studies, especially investigations geared towards energetics (see Chapter 13). Physiological studies have been limited to laboratory conditions, however, and extrapolation of results must proceed cautiously. A combination of field studies and laboratory experiments can provide supportive evidence for physical factors limiting the distribution of organisms and thus are of importance for understanding control of meiofaunal assemblages and geographic distribution. Krebs (1984) outlines such procedures coordinating information on distributions of organisms in the field with physiological tolerance studies, and such approaches have been used to answer questions on what factors limit the vertical and horizontal distribution of fauna within a habitat (Wieser, 1975).

As in feeding studies, cultured organisms often serve as investigated subjects in physiological studies although organisms extracted from field collected sediments have also been used. Problems may arise if organisms from either cultures or readily extracted from sediments are used in physiological tests without access to sediment particles (see Vernberg et al., 1977) and a check on this potential problem probably should accompany experimental efforts. Moreover it has been noted that differences in respiration rates of some taxa (oligochaetes) existed depending on whether the organisms' respiration was measured alone or in combination with other species (Lasserre, 1976). Although single species studies have been invaluable, determination of such parameters as energy budgets of organisms in artificial laboratory conditions should consider acclimation and individual seasonal variation (Lasserre, 1976).

Conclusion

As one reviews the techniques and procedures employed in field and laboratory experimentation with meiofauna, it is evident that although progress has been made, a greater attention need be directed toward developing suitable methodologies for studying these small organisms. Experimentation with meiofauna is relatively young, yet has produced new insight into the biology of meiofauna. Thus as more sophistication is achieved and experimentation becomes more commonplace among research efforts, basic information will be forthcoming. Such efforts will be a necessary pathway to gain knowledge about dynamics of populations as well as behavior and physiology of meiofauna.

References

Alongi, D.M., and J.H. Tietjen
1980. Population Growth and Trophic Interactions Among Free-Living Marine Nematodes. Pages 151-167 in K.R. Tenore and B.C. Coull, editors, *Marine Benthic Dynamics*. Columbia: University of South Carolina Press.

Antia, N.J., P.J. Harrision, D.S. Sullivan, and T. Bisalputra
1985. Influence of the Insecticide Diflubenzuron (Dimilin) on the Growth of Marine Diatoms and a Harpacticoid Copepod in Culture. *Canadian Journal of Fisheries and Aquatic Science*, 42:1272-1277.

Bell, S.S.
1980. Meiofauna-Macrofauna Interactions in a High Salt Marsh Habitat. *Ecological Monographs*, 50:487-505.
1983. An Experimental Study of the Relationship between Below-ground Structure and Meiofaunal Taxa. *Marine Biology*, 76:33-39.
1985. Habitat Complexity of Polychaete Tube-caps: Influence of Architecture on Dynamics of a Meioepibenthic Assemblage. *Journal of Marine Research*, 43:647-671.

Bell, S.S., and L.P. Coen
1982. Investigations on Epibenthic Meiofauna. II. Influence of Microhabitat and Macroalgae on Abundance of Small Invertebrates on *Diopatra cuprea* Tube-caps in Virginia. *Journal of Experimental Marine Biology and Ecology*, 61:175-188.

Bell, S.S., and B.C. Coull
1978. Field Evidence that Shrimp Predation Regulates Meiofauna. *Oecologia*, 35:141-148.

Bell, S.S., M.C. Watzin, and B.C. Coull
1978. Biogenic Structure and Its Effect on the Spatial Heterogeneity of Meiofauna in a Salt Marsh. *Journal of Experimental Marine Biology and Ecology*, 35:99-107.

Bell, S.S., and S.A. Woodin
1984. Community Unity: Experimental Evidence for Meiofauna and Macrofauna. *Journal of Marine Research*, 42:605-632.

Bergmans, M.
1984a. Life History Adaptation to Demographic Regime in Laboratory-cultured *Tisbe furcata* (Copepoda, Harpacticoida). *Evolution*, 38:292-299.
1984b. Critique of Some Practices of Life History Studies with Special Reference to Harpacticoid Copepods. *Australian Journal of Marine and Freshwater Research*, 35:375-383.

Cantelmo, F.R., and K.R. Rao
1978. Effect of Pentachlorophenol (PCP) on Meiobenthic Communities Established in an Experiment System. *Marine Biology*, 46:17-22.

Cantelmo, F.R., M.E. Tagatz, and K. Rango Rao
1979. Effect of Barite on Meiofauna in Flow Through Experimental System. *Marine Environmental Research*, 2:301-309.

Carman, K.R., and D. Thistle
1985. Microbial Food Partitioning by Three Species of Benthic Copepods. *Marine Biology*, 88:143-148.

Castel, J., and P. Lasserre
1982. Regulation biologique du meiobenthos d'un ecosysteme lagunaire par un alevinage experimental en soles (*Solea vulgaris*). Pages 243-251 in SCOR/IABO/UNESCO, *Actes Symposium* International sur les lagunes côtières, Bordeaux. Oceanologica Acta.

Chandler, G.T., and J.W. Fleeger
1983. Meiofaunal Colonization of Azoic Estuarine Sediment in Louisiana: Mechanisms of Dispersal. *Journal of Experimental Marine Biology and Ecology*, 69:175-188.

Conrad, J.E.
1976. Sand Grain Angularity as a Factor Affecting Colonization by Marine Meiofauna. *Vie et Milieu*, 26:181-198.

Coull, B.C., and S.S. Bell
1979. Perspectives of Marine Meiofaunal Ecology. Pages 189-216 in R.J. Livingston, editor, *Ecological Processes in Coastal Marine Systems*. New York: Plenum Press.

Coull, B.C., and M.A. Palmer
1984. Field Experimentation in Meiofaunal Ecology. *Hydrobiologia*, 118:1-19.

Coull, B.C., and J.B.J. Wells
1983. Refuges from Fish Predation: Experiments with Phytal Meiofauna from The New Zealand Rocky Intertidal. *Ecology*, 64:1599-1609.

Creed, E.L., and B.C. Coull
1984. Sand Dollar, *Mellita quinquiesperforata* (Leske) and Sea Pansy, *Renilla reniformis* (Cuvier) Effects on Meiofaunal Abundance. *Journal of Experimental Marine Biology and Ecology*, 84:225-234.

Dalla-Venezia, L., and V.U. Fossato
1977. Characteristics of Suspensions of Kuwait Oil and Corexit 7664 and Their Short- and Long-term Effects on *Tisbe bulbisetosa* (Copepoda: Harpacticoida). *Marine Biology*, 42:233-237.

Dayton, P.K., and J.S. Oliver
1980. An Evaluation of Experimental Analysis of Population and Community Patterns in Benthic Marine Environments. Pages 93-120 in K.R. Tenore and B.C. Coull, editors, *Marine Benthic Dynamics*. Columbia: University of South Carolina Press.

Dethier, M.N.
1980. Tidepools as Refuges: Predation and the Limits of the Harpacticoid Copepod *Tigriopus californicus* (Baker). *Journal of Experimental Marine Biology and Ecology*, 42:99-111.

Eckman, J.E.
1983. Hydrodynamic Processes Affecting Benthic Recruitment. *Limnology and Oceanography*, 28:241-257.

Elmgren, R., and J.B. Frithsen
1982. The Use of Experimental Ecosystems for Evaluating the Environmental Impact of Pollutants: A Comparison of an Oil Spill in the Baltic Sea and Two Long-term, Low Level Oil Addition Experiments in Mesocosms. Pages 153-165 in G.D. Grice and M.R. Reeve, editors, *Marine Mesocosms*. New York: Springer-Verlag.

Elmgren, R., G.A. Vargo, J.F. Grassle, J.P. Grassle, D.R. Heinle, G. Langlois, and S.L. Vargo
1980. Trophic Interactions in Experimental Marine Ecosystems Perturbed by Oil. Pages 774-800 in J.P. Gievy, Jr., editor, *Microcosms in Ecological Research*. DOE Symposium Series 52, Conf-781101. National Technical Information Services, Springfield, Virginia.

Eskin, R.A., and B.C. Coull
1984. *A priori* Determination of Valid Control Sites: An Example Using Marine Meiobenthic Nematodes. *Marine Environmental Research,* 12:161-172.

Fava, G., and E. Crotti
1979. Effetto di un detersivo commerciale e dio uno dei soui componenti, LAS, sulla produzione di nauplii in *Tisbe holothuria* (Copepoda, Harpacticoida) in condizioni di alto e basso affollamento. *Accademia Nazionale dei Lincei,* 66:223-231.

Findlay, S., and K. Tenore
1982a. Effect of a Free-living Marine Nematode (*Diplolaimella chitwoodi*) on Detrital Carbon Mineralization. *Marine Ecology Progress Series,* 8:161-166.
1982b. Nitrogen Source for a Detritivore: Detritus Substrate Versus Associated Microbes. *Science,* 218:371-373.

Fleeger, J.W., and G.T. Chandler
1983. Meiofauna Responses to an Experimental Oil Spill in a Louisiana Salt Marsh. *Marine Ecology Progress Series,* 11:257-264.

Fleeger, J.W., S.A. Whipple, and L.L. Cook
1982. Field Manipulations of Tidal Flushing, Light Exposure and Natant Macrofauna in a Louisiana Salt Marsh: Effects on the Meiofauna. *Journal of Experimental Marine Biology and Ecology,* 56:87-100.

Fleeger, J.W., G.T. Chandler, G.R. Fitzbugh, and F.E. Phillips
1984. Effects of Tidal Currents on Meiofauna Densities in Vegetated Salt Marsh Sediments. *Marine Ecology Progress Series,* 19:49-53.

Friedman, M.M., and J.R. Strickler
1975. Chemoreceptors and Feeding in Calanoid Copepods (Arthropoda: Crustacea). *Proceedings of the National Academy of Sciences, USA,* 72:4185-4188.

Frithsen, J.B., R. Elmgren, and D.T. Rudnick
1985. Responses of Benthic Meiofauna to Long-term, Low-level Additions of No. 2 Fuel Oil. *Marine Ecology Progress Series,* 23:1-14.

Ganning, B.
1970. Population Dynamics and Salinity Tolerance of *Hyadesia fusca* (Lohman) (Acarina, Sarcoptiformes) from Brackish Water Rockpool, with Notes on the Microenvironment inside *Enteromorpha* tubes. *Oecologia,* 5:127-137.

Gaudy, R., J.P. Guerin, and M. Moraiton-Apostolopoutou
1982. Effect of Temperature and Salinity on the Population Dynamics of *Tisbe holothuriae* Humes (Copepoda: Harpacticoida) Fed on Two Different Diets. *Journal of Experimental Marine Biology and Ecology,* 57:257-271.

Gerlach, S.A.
1977. Attraction to Decaying Organisms as a Possible Cause For Patchy Distribution of Nematodes in a Bermuda Beach. *Ophelia,* 16:151-165.

Giere, O.
1979. Some Apparatus for Preference Experiments with Meiofauna. *Journal of Experimental Marine Biology and Ecology,* 41:125-131.

Gray, J.S.
1966. The Behaviour of *Protodrilus symbioticus* Giard in Temperature Gradients. *Journal of Animal Ecology,* 34:455-461.
1968. An Experimental Approach to the Ecology of the Harpacticoid *Leptastacus constrictus* Lang. *Journal of Experimental Marine Biology and Ecology,* 2:278-292.

Gunnill, F.C.
1982. Macroalgae as Habitat Patch Islands for *Scutellidium lamellipes* (Copepoda Harpacticoida) and *Ampithoe tea* (Amphipoda Gammaridae). *Marine Biology,* 69:103-116.

Heip, C., M. Vincx, N. Smol, and G. Vranken
1982. The Systematics and Ecology of Free-living Marine Nematodes. *Helminthological Abstracts* Series B, 51:1-31.

Hicks, G.R.F.
1977. Observations on Substrate Preference of Marine Phytal Harpacticoids (Copepoda). *Hydrobiologia,* 56:7-9.
1980. Seasonal and Geographic Adaptation to Temperature and Salinity in the Harpacticoid Copepod *Zaus spinatus spinatus* Boodsir. *Journal of Experimental Marine Biology and Ecology,* 42:253-266.

Hicks, G.R.F., and B.C. Coull
1983. The Ecology of Marine Meiobenthic Harpacticoid Copepods. *Annual Review of Oceanography and Marine Biology,* 21:67-175.

Hockin, D.C.
1981. An Apparatus to Study the Colonization of Sediments by the Meiofauna. *Estuarine and Coastal Shelf Science,* 12:119-120.
1982. The Effect of Sediment Particle Diameter Upon the Meiobenthic Copepod Community of an Intertidal Beach: A Field and Laboratory Experiment. *Journal of Animal Ecology,* 51:555-572.

Hummon, W.D.
1974. Respiratory and Osmoregulatory Physiology of a Meiobenthic Marine Gastrotrich *Turbanella ocellata* Hummon 1974. *Cahiers de Biologie Marine,* 16:255-268.

Jansson, B.O.
1967. The Importance of Tolerance and Preference Experiments for Interpretation of Mesopsammon Field Experiments. *Helgoländer wissenschaftliche Meeresuntersuchungen,* 15:41-58.

Jensen, P.
1983. Life History of the Free-living Marine Nematode *Chromadorita tenuis* (Nematoda: Chromadorida). *Nematologia,* 29:335-345.

Kern, J.C., and S.S. Bell
1984. Spatial Heterogeneity in Size Structure of Meiofaunal-sized Invertebrates on Small-spatial Scales (Meters) and Its Implications. *Journal of Experimental Marine Biology and Ecology,* 78:221-235.

Krebs, C.J.
1984. *Ecology: The Experimental Analysis of Distribution and Abundance.* 3rd Edition. 678 pages. New York: Harper and Row.

Lasserre, P.
1976. Metabolic Activities of Benthic Microfauna and Meiofauna. Recent Advances and Review of Suitable Methods of Analysis. Pages 95-142 in I.N. McCave, editor, *The Benthic Boundary Layer.* New York: Plenum Press.

Lehtinen, K.J., B.-E Bengtsson, and B. Bergstrom
1984. The Toxicity of Effluents from a TiO_2 Plant to the Harpacticoid Copepod, *Nitocra spinipes* Boeck. *Marine Environmental Research,* 12:273-283.

Lopez, G., F. Riemann, and M. Schrage
1979. Feeding Biology of the Brackish-water Oncholaimid Nematode *Adoncholaimus thalassophagus. Marine Biology,* 54:311-318.

Marcotte, B.M.
1977. The Ecology of Meiobenthic Harpacticoids (Crustacea: Copepoda) in West Lawrencetown, Nova Scotia. Ph.D. Dissertation. 212 pages. Dalhousie University.

Mertz, D.B.
1971. Life History Phenomena In Increasing and Decreasing Populations. Pages 361-399 in G.P. Patil, E.C. Pielou, and W.E. Waters, editors, *Statistical Ecology, Volume 2. Sampling and Modeling, Biological Populations and Population Dynamics.* University Park: Pennsylvania State University Press.

Montagna, P.A.
1983. Live Controls for Radioisotope Tracer Food Chain Experiments Using Meiofauna. *Marine Ecology Progress Series,* 12:43-46.
1984. *In situ* Measurement of Meiobenthic Grazing Rates on Sediment Bacteria and Edaphic Diatoms. *Marine Ecology Progress Series,* 18:119-130.

Palmer, M.A.
1984. Invertebrate Drift Behavioral Experiments with Intertidal Meiobenthos. *Marine Behavior and Physiology,* 10:235-253.

Ravenel, W.S., and D. Thistle
1981. The Effect of Sediment Characteristics on the Distribution of Two Subtidal Harpacticoid Copepod Species. *Journal of Experimental Marine Biology and Ecology,* 50:289-301.

Reise, K.
1979. Moderate Predation on Meiofauna by the Macrobenthos of the Wadden Sea. *Helgoländer wissenschaftliche Meeresuntersuchungen,* 32:453-465.
1983a. Sewage, Green Algal Mats Anchored by Lugworms and the Effects on Turbellaria and Small Polychaeta. *Helgoländer wissenschaftliche Meeresuntersuchungen,* 36:151-162.
1983b. Biotic Enrichment of Intertidal Sediments by Experimental Aggregates of the Deposit-feeding Bivalve *Macoma balthica. Marine Ecology Progress Series,* 12:229-236.

Reise, K., and P. Ax
1979. A Meiofaunal "Thiobios" Limited to the Anaerobic Sulfide System of Marine Sand Does Not Exist. *Marine Biology,* 54:225-237.

Rieper, M.
1982. Feeding Preferences of Marine Harpacticoid Copepods for Various Species of Bacteria. *Marine Ecology Progress Series,* 7:303-307.

Scheibel, W., and H. Rumohr
1979. Meiofaunaentwicklung auf Künstlichem Weichboden in der Kieler Bucht. *Helgoländer wissenschaftliche Meeresuntersuchungen,* 32:305-312.

Schiemer, F.
1982. Food Dependence and Energetics of Freeliving Nematodes. II. Life History Parameters of *Caenorhabditis briggsae* (Nematoda) at Different Levels of Food Supply. *Oecologia,* 54:122-128.

Schiemer, F., A. Duncan, and R.Z. Klelowski
1980. A Bioenergetic Study of a Benthic Nematode, *Plectus palustris* de Man 1880, throughout its Life Cycle. *Oecologia,* 44:205-212.

Service, S.
1986. Manipulative Studies of the Effects of Harpacticoid Copepod Density on Dispersal. M.S. Thesis, 36 pages, University of South Florida, Tampa.

Service, S.K., and S.S. Bell
1987. Density-influenced Active Dispersal of Harpacticoid Copepods. *Journal of Experimental Marine Biology and Ecology,* 114:49-62.

Sherman, K.M., and B.C. Coull
1980. The Response of Meiofauna to Sediment Disturbance. *Journal of Experimental Marine Biology and Ecology,* 45:59-71.

Stewart-Oaten, A., W.W. Murdock, and K.R. Parker
1986. Environmental Impact Assessment: "Pseudoreplication" in Time? *Ecology,* 67:929-940.

Tenore, K., J.H. Tietjen, and J.J. Lee
1977. Effect of Meiofauna on Incorporation of Aged Eelgrass, *Zostera marina,* Detritus by the Polychaete *Nepthys incisa. Journal of the Fisheries Research Board of Canada,* 34:563-567.

Thistle, D.
1980. The Response of a Harpacticoid Copepod Community to a Small-scale Natural Disturbance. *Journal of Marine Research,* 38:381-395.

Thistle, D., J.A. Reidenauer, R.H. Findlay, and R. Waldo
1984. An Experimental Investigation of Enhanced Harpacticoid (Copepod) Abundances Around Isolated Seagrass Shoots. *Oecologia,* 63:245-299.

Tietjen, J.H., and J.J. Lee
1972. Life Cycles of Marine Nematodes. *Oecologia,* 10:167-176.

Trotter, D.B., and J.M. Webster
1984. Feeding Preferences and Seasonality of Free-living Marine Nematodes Inhabiting the Kelp *Macrocystis integrifolia. Marine Ecology Progress Series,* 114:151-157

Ustach, J.F.
1979. Effects of Sublethal Oil Concentrations on the Copepod, *Nitocra affinis. Estuaries,* 2:273-276.
1982. Algae, Bacteria and Detritus as Food for the Harpacticoid Copepod, *Heteropsyllus pseudonunni* Coull and Palmer. *Journal of Experimental Marine Biology and Ecology,* 64:203-214.

Vandenberghe, W., and M. Bergmans
1981. Differential Food Preferences in Three Co-occurring Species of *Tisbe* (Copepoda: Harpacticoida). *Marine Ecology Progress Series,* 4:215-219.

Vernberg, W.B., and B.C. Coull
1975. Multiple Factor Effects of Environmental Parameters of the Physiology, Ecology, and Distribution of Some Marine Meiofauna. *Cahiers de Biologie Marine,* 41:721-732.
1980. Meiofauna. Pages 147-177 in F.J. Vernberg, editor, *Functional Adaptations of Marine Organisms.* New York: Academic Press.

Vernberg, W.B., B.C. Coull, and D.D. Jorgensen
1977. Reliability of Laboratory Metabolic Measurements of Meiofauna. *Journal of the Fisheries Research Board of Canada,* 34:164-167.

Virnstein, R.W.
1978. Predator Caging Experiments in Soft-sediments: Caution Advised. Pages 261-274 in M.L. Wiley, editor, *Estuarine Interactions.* New York: Academic Press.

Walters, K., and S.S. Bell
1986. Diel Patterns of Active Vertical Migration in Seagrass Meiofauna. *Marine Ecology Progress Series,* 34:95-103.

Warwick, R.M., J.T. Davey, J.M. Gee, and C.L. George
1982. Faunistic Control of *Enteromorpha* Blooms: A Field Experiment. *Journal of Experimental Marine Biology and Ecology,* 56:23-31.

Watzin, M.C.
1983. The Effects of Meiofauna on Settling Macrofauna: Meiofauna May Structure Macrofaunal Communities. *Oecologia,* 59:163-166.
1985. Interactions Among Temporary and Permanent Meiofauna: Observations on the Feeding and Behavior of Selected Taxa. *Biological Bulletin,* 169:397-416.

Wieser, W.
1975. The Meiofauna as a Tool in the Study of Habitat Heterogeneity: Ecophysiological Aspects: A Review. *Cahiers de Biologie Marine,* 16:647-670.

Wieser, W., and F. Schiemer
1977. The Ecophysiology of Some Marine Nematodes from Bermuda: Seasonal Aspects. *Journal of Experimental Marine Biology and Ecology,* 26:97-106.

Wormald, A.P., and H.P. Stirling
1979. A Preliminary Investigation of Nutrient Enrichment in Experimental Sand Columns and Its Effect on Tropical Intertidal Bacteria and Meiofauna. *Estuarine and Coastal Marine Science,* 8:441-453.

13. Energetics

Robert J. Feller and Richard M. Warwick

The study of energetics is fundamental to our understanding of ecology and evolution at all levels of biological organization from individual animals to ecosystems. The fluxes and fates of ingested materials and flows of energy through aquatic organisms are measured to provide raw data for the eventual development of ecosystem or landscape models (e.g., Moller et al., 1985; de Wilde et al., 1984). Before such models can become useful as management tools or for generating new, testable hypotheses about ecological interactions or for comparing productivities among parallel units of ecosystems, we must often rely on the reductionist approach to acquire the necessary data.

Energetics studies, whether conducted on the energy-budget level for individual species or at some higher level of community organization, are notoriously difficult, time-consuming, and subject to a myriad of vagaries. These difficulties are usually exacerbated in studies of meiofaunal energetics – primarily because of sampling problems and the small size of individuals. What follows is intended as an introduction to the literature on energetics of meiofauna. This introduction often entails methods and examples that do not pertain strictly to meiofauna. Meiobenthologists conceptually still rely heavily upon techniques derived by workers on macrobenthos, zooplankton, and fishes. We intend this to be more of a "where to find out how to" measure energetics and production parameters than a comprehensive "how to do it" manual. Key citations are utilized rather than a complete review of the literature, and we also try to point out difficulties where appropriate.

Determination of Standing Stock

Biomass Determination.--When large numbers of meiofaunal organisms are available, or when individual animals are towards the top end of the meiofaunal size range, dry weights can be determined directly on balances of appropriate sensitivity. More usually we need to determine the biomass of small numbers of small species, and estimates of weight must be extrapolated from volume calculations.

1. Gravimetric Methods.--The most useful type of balance for meiofauna work is an electrobalance sensitive to ±0.1 µg, such as the Cahn automatic electrobalance. These are convenient and easy to use, but relatively expensive. A cheaper possibility would be to build and calibrate a simple quartz fiber balance of equal sensitivity. Instructions for this are given in Lowry (1941). Prior to weighing on an electrobalance, the animals are oven-dried to constant weight at temperatures of 60-100° C and stored in a dessicator. Most investigators dry their specimens at 60° C for 24 hr. For the quartz fiber balance animals are transferred to a small drop of water on a balance hook and oven dried on this hook (to which they hopefully adhere).

Additional biomass estimates are given by Widbom (1984), who calculated average individual dry weights and ash-free dry weights for different sieve fractions of major meiofaunal taxa. Weights of formalin-preserved animals exceeded those of unpreserved specimens by as much as 44% for dry weight and 29% for ash-free dry weight (Widbom, 1984).

2. Volumetric Methods.--For taxa such as nematodes which can be assumed to have a regular cross-section (in this case circular), the volume can be determined from scale drawings made under the camera lucida or from photomicrographs. The animal shape can be approximated to geometrical figures (Andrassy, 1956), or by using Simpson's first rule, taking equally spaced coordinates along the body length (Warwick and Price, 1979). Depending on the degree of accuracy required, a rough estimate of volume (V) can be obtained by measuring the body length (l) and maximum width (W) under the camera lucida and applying a formula for an "average" nematode such as $V = lW^2/16 \times 10^5$, where V is in nl and W and l in µm (Andrassy, 1956) or $V = 530\, lW^2$, where V is in nl and W and l in mm (Warwick and Price, 1979). Measurements from simple drawings or outlines can be taken with dividers for straight lines or a good quality opisthometer (map measurer) for curves and applying an appropriate calibration from a stage micrometer slide. Alternatively, an eyepiece micrometer can be used directly for straight measurements. Volume can be approximately translated to dry weight by assuming a specific

gravity of 1.13 (Wieser, 1960), with estimates of dry weight for nematodes varying between 20 and 25% of wet weight (Myers, 1967; Wieser, 1960). Specific gravity and wet:dry weight ratios have not been determined for most meiofaunal groups, but the values for nematodes should give reasonable approximations. Specific gravity can be determined by flotation in different mixtures of kerosene and bromobenzene (Wieser, 1960) or in a gradient column of bromobenzene and xylene (Low and Richards, 1952).

To determine the body volume of irregularly shaped "hard" meiofauna, such as copepods, it may be necessary to make models based on camera lucida drawings or photographs from different aspects to determine the relationship between body dimensions and volume. Models can be made of some type of modeling clay and their volume determined by immersion in a measuring cylinder of water, or using a eureka can. Figure 13.1 shows the factors used to convert lW^2 to body volume derived from such models. In Table 13.1 we present some very approximate conversion factors for other meiofaunal groups.

Table 13.1.-- Approximate conversion factors (C) in the equation $V = L \times W^2 \times C$, where V is body volume in nl, L is length and W maximum width (both in mm). To convert V to dry weight in micrograms, multiply by specific gravity (e.g., 1.13) and dry/wet weight ratio (e.g., 0.25).

Taxon	Conversion Factor (C)
Nematodes	530
Ostracods	450
Mites	399
Kinorhynchs	295
Turbellarians	550
Gastrotrichs	550
Tardigrades	614
Hydroids	385
Polychaetes	530
Oligochaetes	530
Tanaids	400
Isopods	230

Copepods (depends on shape: see Figure 13.1)

The body volume of "soft" meiofaunal species can be determined by measuring the area of specimens under the camera lucida when gently squashed to uniform thickness in a microcompression chamber such as that described by Uhlig and Heimberg (1981). The distances between the glasses of the chamber (i.e., the thickness of the specimen) can be determined by measuring the focal distance from the top to bottom of the specimen using the calibration on the fine focus knob of a good compound microscope.

As an alternative to volume estimation based on measurements, Holter (1945) has described an ingenious method of direct estimation. The animal is drawn up into a capillary tube of known bore in a column of dye solution of known optical density. The length of the dye column is measured and the dye emptied into a known volume of water. The dye concentration is determined colorimetrically in a microcuvette using a spectrophotometer and compared with a standard curve made from solutions of dye columns alone. Automatic calculation of volumes is now possible using algorithms for two-dimensional projected images of three-dimensional objects using digitizing tablets coupled to desk-top computers. Numerous companies market these products, with costs for the digitizer software and hardware ranging from about US $1,800 and upwards.

Carbon-Nitrogen Analysis.--Models for the flow of energy and materials through ecosystems, and also budgets for individual species, may use many different currencies, but principally dry weight, carbon, nitrogen or energy (now measured in joules, not calories). Again, depending on the accuracy required, rough conversions to these last three units can be made from biomass data, or they can be determined empirically. As a rough approximation, carbon is about 40% of dry weight and nitrogen about 10%.

Exact procedures for Carbon-Hydrogen-Nitrogen analysis will vary with the make and model of CHN analyzer, and it is neither possible nor appropriate to generalize these instructions here. Most CHN analyzers will give reliable results with single animals of the very largest meiofaunal species with dry weights of more than about 10 µg. However, an average sized nematode or harpacticoid copepod weighs about 0.5 µg, in which case 20 or more specimens would be required. In any case, one is generally limited by the sensitivity of the electrobalance when initially weighing material for combustion. See Chapter 6 for more details on biomass determinations.

Calorimetry.--As a rough approximation, 1 mg ash-free dry weight is equivalent to 25 joules. For more accuracy, direct determinations of calorific content of dried tissue can be made with a micro-bomb calorimeter. The most widely used is that designed by Phillipson (1964). Crisp (1984) details

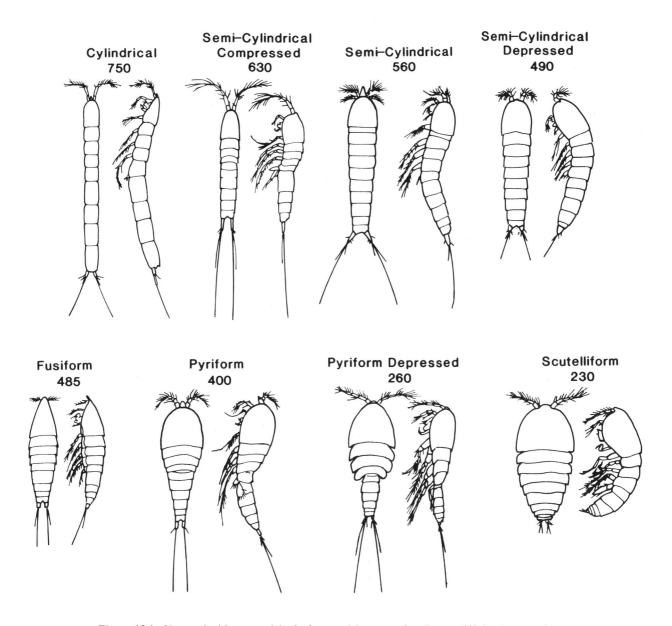

Figure 13.1.--Harpacticoid copepod body forms with conversion factors (C) in the equation: body volume (nl) = length (mm) x width^2 (mm) x C (from Warwick and Gee, 1984).

some more recent refinements of the technique. This apparatus takes samples of 5-10 mg, which would require on the order of 10,000 average sized adult nematodes or harpacticoid copepods! Except for the very largest species, or for species that can be grown in large numbers in laboratory culture and extracted cleanly from the culture medium, this technique is inappropriate for meiofauna. It is possible to calculate calorific values from CHN analysis (Gnaiger and Bitterlich, 1984), and since less material is required for the latter, this technique might be more appropriate for meiofauna. However, the method is not without difficulties, and "Trouble Shooting Instructions" (a cyclostyled newsletter) should be obtained from Erich Gnaiger before embarking upon it.

Special Problems.--Most of the techniques used to determine meiobenthic standing stocks are quite standard. The only real problem is the small size of the animals. When weighing animals that have been dried, static electricity may cause them to disappear from the weighing pan. Before animals are loaded into it, the weighing pan should be degaussed by touching the pan with a grounding wire.

Trophic Aspects

Mechanisms of Food Intake (How Meiofauna Feed).--Incorporation of food by meiofauna entails a wide variety of feeding mechanisms, including the general categories of suspension-feeding (filtering food particles from the water above the sediments), deposit-feeding (ingesting food particles or sediment particles lying in the sediments), predation (the act of killing and ingesting live prey), scavenging (eating dead or decomposing prey), and absorption (the uptake of dissolved nutrients from the water across the body wall). However, the mouthparts of many meiofaunal species are highly specialized to deal with one particular type of food particle, e.g., prizing open or puncturing diatoms. Based on the types of food eaten by meiofauna, taxa may be defined variously as carnivorous, herbivorous, detritivorous, or omnivorous. The terms bacterivore and fungivore have also been used. Some authors choose to indicate the general nature of the food type eaten by describing organisms as phytophagous (eating algae or plants), zoophagous (eating animals), or euryphagous (eating most anything!). Others have invoked mechanistic descriptions of the feeding process by using such terminology as browsing, sand-filing, and plane sweeping (e.g., Marcotte, 1977). Regardless of the classification scheme employed, meiofauna do all these things (Figure 13.2) and some species even cultivate their own algal food in a so-called gardening technique in which slime trails promote the growth of algae and/or bacteria so that animals may turn around and regraze the trail left earlier (Riemann and Schrage, 1978; Warwick, 1980).

Following classic descriptions of nematode feeding modes by Wieser (1953) and by Staarup (1970) for turbellarians, the first reasonably comprehensive review of meiofaunal trophic relationships was published by Coull (1973). Aside from a few detailed descriptions of feeding observed by patient microscopists viewing live organisms in the laboratory, we have not progressed very far since then in our basic understanding of how meiofauna feed. Their small size, rapid movements, our inability to see animals eating while they are within the sediment fabric, and the small size of the food particles ingested - all these factors have inhibited progress. Recent accounts of feeding in meiofauna are nicely summarized by Hicks and Coull (1983) for harpacticoid copepods, by Heip et al. (1985) for nematodes, and by Vernberg and Coull (1981) and Watzin (1985) for other taxa. Watzin's study is indicative of the great amount of effort required (500 hr of observation with a microscope) to attain a modest idea of the potential food web for just a few species. The nutritional status and ecological importance of mouthless nematodes and gutless oligochaetes are relatively recent subjects of study (e.g., Giere and Langheld, 1987). The role of endosymbiotic bacteria and uptake of dissolved organic matter by meiofauna is also not well known, but if uptake is as widespread among meiofaunal taxa as it is for other soft-bodied benthic and planktonic marine organisms (e.g., Manahan and Crisp, 1982), our abilities to describe trophic relationships among meiofauna are obviously still limited. Given the high surface-to-volume ratio of soft-bodied meiofauna, the probability of dissolved organic matter uptake as a significant mechanism of nutrition in the group is high.

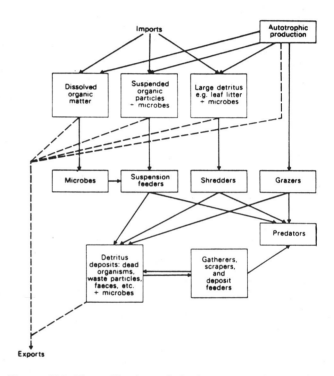

Figure 13.2--The utilization of food resources in aquatic ecosystems (from Barnes and Mann, 1980).

Methods for Stomach or Gut Contents Analysis.--Despite obvious differences between fishes and meiofauna, it is constructive to peruse the literature on dietary analysis methods for these larger organisms simply because similar questions are asked about diet and feeding among these and other animal groups. Particularly helpful are Kiritani and Dempster (1973), Berg (1979), and Hyslop (1980).

Examination and analysis of material contained within the digestive tract of an organism is called either stomach contents analysis or gut contents analysis rather than the incorrect terms "stomach analysis" or "gut analysis" which refer to analysis of stomach or gut tissue, not their contents. Basic gut contents analysis is typically performed to identify food items present, to quantify the size of a meal (as weight or volume), or to make quantitative measures of feeding rates. Data requisite for these purposes may be obtained in a number of ways, with each method having characteristic drawbacks.

1. Visual Examination.--Gut material, whether fresh or preserved, may be extruded for examination by gently squeezing the organism, by making an incision in the organism's body wall through which material will leak out, or by squashing whole mounts of the organism with a cover slip onto a microscope slide. Once material is available, microscopical examination with direct illumination on a dark or light background at magnifications of not less than 50x will be necessary to identify anything present. Compound microscopy at magnifications of 100-500x is almost always necessary. Unless an organism's body wall is cleared, it is usually not possible to identify material inside the gut with any degree of accuracy. About the best one could expect would be to qualitatively ascertain whether there was food present or not. Since gut material extruded onto a slide will rapidly dry out, it is best to use coverslips or prefabricate a small humidified chamber in which the slide can be placed for observation. Addition of water to the sample will of course prevent evaporation, but sample dilution may not always be desirable. One can, however, still obtain reliable measures of dry weight after such dilution.

Problems: Material in the gut may be unidentifiable, existing only as amorphous pieces, partially or completely digested. Gut volume is difficult to measure because gut fullness is only subjectively known and may not be maximal in the specimens under observation. Incomplete removal of gut contents or contamination of the sample with body tissues may occur when trying to estimate gut contents biomass or volume. Preservatives (formalin or alcohol) may alter the original biomass or volume of gut contents. Enzymes in the gut contents of fresh or frozen organisms will continue digestive breakdown during storage or observation.

2. Tracer Techniques.--Tracer techniques in trophic ecology are commonly construed to mean that some sort of inert or radioactive element is used to label food that is to be ingested by an organism. Some time after ingestion of the labeled food has taken place, the predator's body is either solubilized and the quantity of radioactivity taken up measured with a scintillation counter, or the inert particles are identified from the gut contents. Radioisotopes commonly used to label foodstuffs are 14-C, 32-P, and 3-H glucose. Details of radiotracer methodology are in Tietjen et al. (1970) and Montagna (1984). Although incorporation of radioactive labels into natural food items is preferable to obtain "natural" estimates of feeding rates, it is also possible to use inert foodstuffs as a means of detecting predation in meiofauna, e.g., glass beads, polystyrene particles, latex beads, fluorescent particles, or the like. These ingested particulate tracers must be counted or observed in some way, an obvious drawback compared to the relative rapidity with which scintillation counting of radioisotopes may be performed.

The use of stable isotopes as tracers of the past feeding history of organisms is somewhat new to meiofauna ecology, primarily because samples weighing tens of milligrams are required for the analyses of the ratios between 13-C and 12-C or other pairs of stable isotopes such as 15-N/14-N, or 34-S/32-S. Rounick and Winterbourn (1986) provide an interesting overview of this biogeochemical technique.

Problems: As with any tracer technique, there may be incomplete or selective labeling of the food causing bias in estimation of ingestion rates. Incomplete or partial labeling of food generally causes underestimation of true ingestion rates, whereas selective ingestion of labeled versus non-labeled food can cause overestimation. A special permit or license for handling and disposing of radioactive material may be required. The requirement for equipment (scintillation counter, fluorescence microscope for counting bacteria, mass spectrometer for stable isotope analysis, special chemicals) is expensive. Since the cycling rates of radioisotopes or stable isotopes within an organism may not be known, the potential exists for additional bias in estimating true uptake rates.

3. Gut Chlorophyll Analysis.--Chlorophyll and its degradation products (pheopigments) fluoresce naturally. The concentration and, if stomach contents volume is known, the quantity of fluorescing material within an herbivore's stomach may be measured with a fluorometer. Caution must be taken to ensure that freshly-collected samples are analyzed, else one might measure chlorophyll degradation products rather than the material that was originally ingested. Standards of known chlorophyll content are used to calibrate the technique. Key references for this method are: Mackas and Bohrer (1976), Parsons et al. (1984), Downs and Lorenzen (1985), Baars and Helling (1985), Schneckenburger et al. (1985), and Conover et al. (1986). Adoption of flow cytometry as a standard tool for the quantitative detection of similarly sized algae from different taxonomic groups makes this technology attractive for studies of selective feeding by meiofaunal organisms that ingest discrete particles (Cucci et al., 1985). A combination of fluorescence and immunological techniques

(immunofluorescence) also holds promise for feeding studies involving ingestion of microbes.

Problems: Degradation of chlorophyll may occur during sample storage. Some interference is possible from non-chlorophyll containing substances in the gut contents. This method will require pooling or compositing of samples due to the small size of meiofaunal herbivore stomachs.

4. Serology.--This method takes advantage of the highly specific nature of antigen-antibody interactions and utilizes antibodies to recognize antigenic determinants on soluble proteins of prey organisms inside the gut of a predator (Figure 13.3). Antiserum is produced by a mammal - usually a rabbit, chicken, goat, or rat - in response to injection of a purified preparation or whole-organism extract of soluble proteins of the prey organism one desires to detect. The method is most suitable for situations in which prey are rendered visually unidentifiable by their predator, e.g., by chewing and grinding, or if the meal is somehow taken in liquid form. Once a specific antiserum is available, the gut contents of an organism are assayed with the antiserum using diffusion in gel or immuno-electrophoretic methods which allow precipitation of the antigen-antibody complex. If antisera are made to several specific types of prey, the gut contents of a predator may be assayed for several prey simultaneously. Details of the methods for preparing antibodies and immunization techniques are found in numerous textbooks, but the articles by Boreham and Ohiagu (1978) and Warr (1982) are especially useful. Details for the ecological application and interpretation of serological diet data may be found in Feller (1984), and an extension of the basic methodology used as a screening device to check large numbers of predators for the presence of particular prey is described by Grisley and Boyle (1985).

Problems: Antibodies which recognize proteins from one species may also recognize proteins from closely-related species (known as a cross-reaction), thus requiring purification of the antiserum. Quantification of gut contents is possible only under certain conditions, hence serological methods are best used qualitatively for presence/absence determinations (see Feller and Ferguson, in press). Small prey (like meiofauna!) cannot be examined individually - lumping of samples is necessary. Collection and purification of target prey for preparation of antibodies may be difficult and laborious, as several hundred milligrams of wet weight are required as an immunogen to produce an antibody with sufficiently high titer. Cannibalism cannot be detected serologically.

5. Other Methods.--Besides the basic approaches described above, it may also be possible to adopt methodologies used to study calanoid copepod feeding which measure the concentrations of specific digestive enzymes (e.g., laminarinase, carbohydrase) induced as a result of feeding activity. Bakker et al. (1985) provide an excellent review of these and other methods.

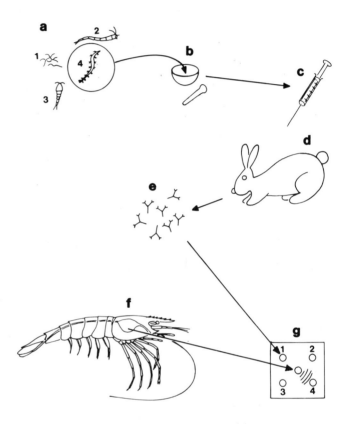

Figure 13.3--Serological analysis of a shrimp's diet. (a) a target prey organism (4) likely to be in the diet is selected and starved to remove foreign protein contamination; (b) several clean target organisms, all of the same species (if a species-specific antiserum is desired), are solubilized in buffered saline with mortar and pestle and centrifuged to remove particulates; the soluble antigenic proteins are injected (c) into a mammalian host (d) for production of antiserum (e); after checking for titer and specificity, the antiserum to each target prey species (1-4 in g) is placed into wells surrounding a central well containing the unknown mixture of proteins from the shrimp's gut contents (f); the template for the wells is supported on a bed of agarose through which antigens from the central well containing gut contents and antiserum from the surrounding wells diffuse and form precipitin lines (antigen-antibody complexes) for those target prey present in the unknown mixture from the shrimp's stomach contents. A positive reaction of four precipitin lines indicates presence of only prey species 4 in this example (g).

The wide availability of HPLC (high-performance liquid chromatography, also called high-pressure liquid chromatography) makes this technology a prime candidate for use in the study of substances present in food that might be ingested by meiofauna. The study of dissolved organic matter cannot be done adequately without using HPLC, and based on the fact that so many meiofaunal taxa utilize detritus as food, HPLC will probably also be useful in studying the nutritional quality of detrital degradation products (Alvarez et al., 1981; Roman and Tenore, 1984).

Measuring Consumption Rates (C).--The measurement of consumption or ingestion rates requires that the food material ingested be quantifiable in some manner, either as numerical counts of particles, some gravimetric measure (weight, volume), or as some other biochemical attribute (e.g., chlorophyll concentration, ratios of constituents of the food, counts of radioactive decay products, etc.). Most work has been done in the laboratory using culture dishes or other microcosm designs and, again, many of the techniques used for measuring feeding in meiofauna derive from methods originally developed for other animal groups in the benthos or zooplankton.

1. Laboratory Measurements.--The use of radioactive tracers is the most common technique employed to measure ingestion rates in meiofauna (e.g., Tietjen and Lee, 1977; Carman and Thistle, 1985). The basic approach is to let the food, usually a mixture of photoautotrophs and bacteria, incorporate the radioactive label until it reaches some uniform concentration and then feed this labeled material to the consumer. The animal is allowed to feed for a period of time and is then killed and homogenized so that the amount of label taken up can be measured by a scintillation counter. With appropriate controls (measures of label uptake in microcosms containing non-feeding animals), one can easily calculate ingestion rates of the labeled material and then convert the measurement to other units such as carbon or calories.

The isotopic composition of elements within an organism may impart information about its past feeding history, provided something is known about the isotopic composition of its potential sources of food. The stable isotope technique of food web analysis is still in its infancy with respect to meiofauna, but basic principles are detailed in Fry and Sherr (1984) and Simenstad and Wissmar (1985). Since this method requires only that animals be collected for analysis in a mass spectrometer, it is attractive for feeding studies in a qualitative sense, but its utility for estimating feeding rate dynamics of organisms has yet to be demonstrated. Analytical costs are high with stable isotope analyses.

Standard grazing experiments developed for zooplankton have also been used to measure ingestion rates for meiofauna (see Hicks and Coull, 1983). These simply measure the number of particles removed from a known volume of water per unit time by individual organisms. In principle the same type of grazing experiment could be conducted in sediment were it possible to count discrete food particles on sediments. Yamamoto and Lopez (1985) relate some of the difficulties and potential sources of error involved with counting bacteria in marine sediments. The suitability of standard equations for calculating grazing rates of meiofaunal organisms must be viewed with caution, as the manner in which food concentrations are calculated can seriously bias the estimates of feeding rates (Marin et al., 1986).

The uptake of dissolved organic compounds is another potential source of nutrition for meiofauna (Chia and Warwick, 1969; Jensen, 1987), but some authors (Siebers, 1982) believe that this is an unlikely source of food given the relative abundance of bacteria available within the sediment. New methods, some using HPLC, developed for measuring the uptake of DOM, are outlined in Stephens (1982) and Stephens and Manahan (1984).

Problems: Label experiments suffer several major difficulties, primary among them the possibility that incompletely labeled food was ingested (one must check for uniform labeling), that feeding took place during a period so long as to allow metabolism and excretion, hence recycling, of the label to occur (results in an underestimation of uptake), and that the experiments must be performed in the laboratory (can results be extrapolated from the laboratory to the field?).

2. Field Measurements.--One can quibble endlessly whether it is legitimate to consider an experiment which brings an intact piece of the natural environment into the laboratory for study is actually a field experiment. Such arguments are not constructive. Regardless, the only "field study" of meiofaunal ingestion rates was done by Montagna (1984) who incubated sediment cores with radioactive label and measured subsequent uptake by various meiofaunal taxa in the cores. His treatment of control experiments for radiotracer studies is worthy of note (Montagna, 1983).

Special Problems.--Regardless of whether feeding experiments themselves are done in the laboratory or in the field, we should be able to relate these rates in an ecologically meaningful way to potential supplies of food in nature. This capability has been heavily exploited by zooplankton ecologists, as feeding rate models have been incorporated into larger ecosystem models having predictive capabilities. Models of feeding rates similar to those developed by Cammen (1980) for macrofaunal deposit feeders and detritivores should be developed for meiofaunal taxa.

Energy Budgets

Definitions.--Assuming steady state conditions, the components of an animal's energy budget can be described by the balanced energy equation of Winberg (1960):

$$C = P + R + F + U$$

where C = food intake or consumption, P = production in terms of somatic growth (P_g) and reproductive output (P_r), R = respiration, F = egested feces, and U = excretion, all values expressed in the same energy equivalents. Absorption (Ab) is the portion of consumed energy not egested as feces (Ab = C - F or P + R + U), whereas assimilation (A) is the portion of consumed energy used for production and respiration (A = C - F - U or P + R). Since consumption (C) has already been dealt with (see Trophic Aspects: Measuring Consumption Rates), this section deals with the ways in which we can measure how the ingested ratio is apportioned among the components on the right hand side of the energy budget equation. Such information is essential if we are to understand the energetic role played by the meiobenthos in marine food webs.

Production (P).

1. *Growth and Survivorship.*--The best estimates of growth rate for an organism are obtained under natural field conditions for individuals of known age at the start of the period of interest. A variety of mark-recapture methods has been used on large organisms to make such measurements (Poole, 1974). Meiofaunal organisms, however, can be neither marked nor aged with any degree of precision long enough to use these methods. Hence we are constrained to use methods which by their nature may introduce bias into the measurements. Specifically, animals are typically held and fed in the laboratory under semi-natural conditions. If it is possible to isolate individuals, then one can estimate growth rates in terms of meristic characters such as length or width as they change over time in the culture dish. Alternatively, large numbers of animals may be cultured together (usually with an "unlimited" food supply) and individuals sampled at random from the population. The advantage of this method over random collections of individuals from the field is that one has complete control over immigration (migration into the population), emigration (migration of individuals out of the population), and what sizes of individuals are initially placed into the culture dishes. Recruitment to laboratory populations may also be controlled by the investigator. Inability to control recruitment, immigration, or emigration in the field creates the potential for bias even when estimating growth of field populations using modal changes in length- or size-frequency histograms over time. For animals with distinct developmental stages (e.g., the separate naupliar and copepodite stages of harpacticoid copepods), changes in the stage-frequency histograms can be used to measure growth of populations in the field. This method is also subject to the same biases mentioned above regarding recruitment, immigration, and emigration.

Growth of an individual organism may be measured as its rate of change of biomass (or length or some other meristic character) over some period of time, whereas growth of a population can be measured as the rate of change of average biomass or average body size over discrete time intervals. Various growth function models have been proposed, especially for fish populations (Gulland, 1969), but for meiofauna the simplest models assume either linear growth over short time intervals or exponential growth. Since most methods for measuring body sizes of individuals are likely to be destructive (e.g., dry weight, biomass as carbon or nitrogen, energy content), growth is typically measured on random samples of individuals from a larger population.

Survivorship is a measure of the proportion of individuals remaining alive in a population after some interval of time, where the proportion is based upon the number of individuals alive at the beginning of the period. The opposite of survivorship is mortality, wherein the proportion dying during an interval is measured. Both survivorship and mortality rates are subject to measurement error depending upon whether (again!) recruitment, immigration, and emigration are accounted for in the population. Note that individuals within a population do not have survivorship or mortality rates. Such rates are characteristic of populations, not individuals, and can only be calculated for populations of individuals, be they from the field or the laboratory.

Estimates of survivorship or mortality are important in the study of energetics of populations because it is possible, under certain assumptions, to calculate production from these data. This has not yet been done for meiofaunal populations, but the methodology and assumptions are detailed in Allan et al. (1976). Pyke and Thompson (1986) discuss statistical methods for comparing survivorship curves between different populations or between similar populations raised under different conditions.

2. *Methods of Calculation.*--The rate of production of biomass (and energy) by meiofaunal organisms can be calculated by a variety of methods

(Heip et al., 1982). The different methods all require quantitative estimates of organism abundance and weight and, depending upon whether reproduction is simultaneous (the most desirable situation) or continuous for the population in the area of study, estimates of growth rate may also be required. Calculations of production sum either successive growth increments of all members of the population over a period of time or the amount of biomass lost from the population due to all combined sources of mortality. A large number of publications concerning the calculation of production exist, and the literature may seem very confused since so many variations of the basic calculation methodologies are used. This confusion, however, is a natural consequence of the great variety of life histories exhibited by marine organisms, for example, whether females carry eggs, how many generations per year are produced, or whether all or only a few developmental stages of an organism can be identified and counted. Unfortunately, published production measurements for meiofauna are far fewer in number relative to those made for macrofauna and zooplankton (Downing and Rigler, 1984; Crisp, 1984), hence those interested in making production measurements on meiofauna will likely be modifying existing methods.

Basic references for calculating production include Edmondson and Winberg (1971), Winberg (1971), Edmondson (1974), Waters (1977), Crisp (1984), and Omori and Ikeda (1984). None of these deals specifically with meiofauna. Summaries of what little we know about meiofaunal production are contained in Hicks and Coull (1983), Gray (1981), Heip et al., (1985), and Herman and Heip (1985).

Cohorts Identifiable.--When reproduction occurs simultaneously among all members of a population over a short period of time, this group of same-aged newborn individuals is called a cohort. If reproduction occurs within a very narrow period of time, we refer to this as "knife-edged recruitment", indicating that the reproductive event happened all at once. This is common in some fish populations but has not been observed for any meiofaunal taxa. The productivity of the cohort, i.e., cohort production, can be estimated in a number of ways under the following assumptions: samples of the population are collected at intervals during which growth and mortality take place, but no additional births occur, i.e., there is no recruitment to the population. Gains and losses to the population via immigration and emigration must be accounted for or else assumed to balance each other or to be negligible and thus ignored. Cohort production is not the same as annual production unless the cohort takes exactly one year to reproduce again, i.e., its generation time is one year. Assuming quantitative data are available on population abundance and average weights of individuals in the population (usually as total population biomass/total number of individuals in the population), Waters and Crawford (1973) give four basic methods briefly described as follows.

(1) *Increment-summation,* in which the product of average abundance on two successive dates and the change in average weight of an individual is production for the interval; (2) *removal summation,* in which the change in abundance between two sampling periods is multiplied by the average individual weights on the two dates to give production lost during the interval -- the sum of production lost in the interval and the change in remaining biomass is total production for the interval; (3) *instantaneous growth,* in which the product of average individual biomass on successive dates and the instantaneous growth rate of average individual biomass is production between the dates; and (4) *mean weight-survivorship or the Allen curve,* in which the area beneath the curve on a graph of population abundance (ordinate) and average weight of an individual in the population (abscissa) as they change over time is cohort production.

The first three methods assume that growth of population or individual biomass is exponential, whereas the Allen curve method makes no assumptions about the shape of the growth curve. Comparisons among the various mathematical calculations for these four methods are detailed in Gillespie and Behnke (1979), and Waters and Crawford (1973) compare results of the four methods applied to a single data set.

Cohorts Unidentifiable.--When distinct cohorts cannot be identified due either to co-existence of overlapping generations or to continuous reproduction in the population, production methods require data on incremental increases in weight of individuals from the time they are born until they die. Production is then simply the sum of weight increments or addition of biomass to the population by each developmental stage in an interval of time. This summation method assumes that population size remains constant and that the population age distribution remains uniform during the interval in question (Edmondson, 1974). Since data required include the individual biomass and duration of each developmental stage of the organism, the organism must be grown in laboratory culture with all attendant artifacts which may adversely influence growth. The need for production estimates of continuously reproducing populations should be very great before this approach is taken, as laboratory culture data are extremely difficult to generate. Herman et al. (1984) describe "short-cut methods" which apply to populations with continuous reproduction. These are indirect methods assuming constant proportionality between respiration and production or between adult size and the annual production/biomass ratio. For instance, by measuring respiration alone, one could then estimate production under these assumptions. Vranken and Heip (1986), however, caution against using a single constant value of the P/B ratio.

Special Problems.--Regardless of the method used for calculating production for a population, one will be faced with the problem of what to do with the estimate obtained. Production estimates by themselves serve no real purpose. It is only when enough estimates are available for comparisons between species, between habitats for the same species, or when the influence of food or other factors on levels of production is known that such estimates become interesting. When sufficient data become available, it will then be possible to integrate them into models of energy flow for an area (e.g., Bodin et al., 1984) or taxon of interest or to generate theory or hypotheses that may be tested with the accumulation of more data (Gray, 1981). Sufficient data on meiofauna are not yet available for this purpose. For that matter, one should not attempt to estimate production without having some valid hypothesis in mind ahead of time - it simply requires too much work! Although comparative studies of meiofaunal energetics exist (e.g., Banse, 1982; Schiemer, 1983), directly comparable productivity data are relatively sparse in both the meiobenthic and macrobenthic literature. Some methods for calculating confidence intervals about estimates of production (necessary for quantitative, statistically valid comparisons) are detailed in Krueger and Martin (1980) and Feller (1982). A particularly interesting and relevant account of the consequences of changes in sample size (number of replicate cores per sampling time) for calculations of production of aquatic insects is given in Resh (1979).

In cases for which distinct cohorts may be identified for the population of interest, there are additional shortcuts one can take to estimate secondary production. LeBlond and Parsons (1977) illustrate that one may not need to obtain individual weights for each age or stage of the organism and allow that estimates of population biomass alone will suffice. When individual weights are required for estimates of biomass, an obvious shortcut is to generate a length-weight regression for the animal so that subsequent weights may be estimated from measurements of length alone. Bird and Prairie (1985) present guidelines and pitfalls for the use of such length-weight regressions.

For some meiofaunal taxa, animals that reach adult size may stop growing or stop producing eggs. Using growth increment methods for calculating production, it might appear that such individuals are no longer productive. This is of course not true, as they are still assimilating food and releasing energy in the form of excretory products and respiration. Such energy expenditures should be accounted for in the method you select to measure production. Much the same can be said for those fauna that produce exuviae (or mucus) - either include the exoskeleton in the measurement of individual biomass or include the biomass of exoskeletons in the fraction of population biomass that is lost to the environment over time. There is also a tendency to neglect the production by naupliar stages of copepods because of their small size. Conflicting evidence as to the magnitude of naupliar production exists (Herman et al., 1984), and since there are so few studies, it is our hope that future investigators will attempt to measure this component of production.

Last in this brief section on problems, we must mention the ever present difficulty that all scientists have with the extrapolation of data derived in the laboratory to natural field populations. We have no solution to this except to suggest that when making such an extrapolation, treat the data accordingly and make sure that existence or use of the extrapolation is clearly stated. Another potentially serious error can occur easily when attempting to estimate productivity of natural populations - failure to sample the same population in the field through time. This difficulty is obviously great in planktonic populations which are subject to advective motions of the surrounding medium; since benthic populations may exhibit a variety of spatial scales of patchiness within the same general location, it is advisable to check for similarities in age structure or some other population parameter amongst replicate samples (Feller, 1980; Kern and Bell, 1984). Clearly, if the same population is not sampled through time, then productivity of that population cannot be calculated.

Production/Biomass Ratio (P/B).--Since it is difficult and laborious to measure the growth rates, mortality, and age composition of natural populations, it is conceptually pleasing to utilize a simpler method for estimating the production of organisms (Waters, 1977). The ratio between annual production and mean annual biomass (P/B ratio) appears to be relatively predictable, depending upon whether it is scaled to lifespan (Robertson, 1979) or to adult body size (Banse and Mosher, 1980; Heip et al., 1982). Hence it is much easier to simply measure population biomass over time, derive an estimate of mean average biomass, and extrapolate, from the P/B ratio of similar fauna, to the production of the population of interest.

The P/B ratio, however, is influenced by several factors, especially important being whether cohort production or annual production is under consideration and the number of generations an organism has per year (Benke, 1979; Waters, 1979). Obviously, animals with more than one generation per year will replace or turnover their biomass more frequently (have a higher P/B ratio) than those having only one generation, for instance, every two years. It is very important to note the time interval for which P/B ratios are calculated before using them to estimate production for other populations. Even with these and other suitable precautions, calculations of production based on extrapolations from P/B ratios should not be considered as anything more than best guesses necessitated by or resulting from an absence of life-history data for meiofauna (Warwick, 1980). Compared to production and biomass data for macrobenthos and zooplankton, it is clear why P/B ratios are not sufficiently known for meiofauna (Figure 13.4). Since few authors estimate confidence intervals about their estimates of the P/B ratio, ratios are very difficult to compare among taxa except in a qualitative way. Brey (1986) presents a method for calculating P/B ratios for macrobenthos using length-frequency data, but his method has not yet been utilized for meiofaunal species.

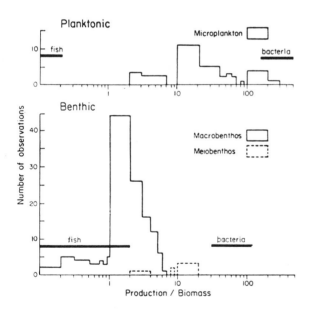

Figure 13.4--Summary of P/B ratio data available for planktonic and benthic species, including a few freshwater species (from I. Valiela, 1984).

Reproductive Output (P_r).--The biomass of reproductive products (eggs, sperm, egg cases) also contributes to the productivity of organisms. It is generally not possible to observe egg-bearing individuals in the field, hence laboratory observations of gravid individuals is necessary in most cases. Animals which reproduce only once (semelparity) and carry their eggs before they are released or hatch represent the easiest situation for estimating fecundity, because one can establish when breeding periods occur in the field by taking samples of the population at intervals less than the development time of the eggs. The proportion of females in the population which are carrying eggs will typically change whenever reproductive periods commence or terminate, but actual fecundity (number of eggs per female per unit time) may change during reproductive periods, as fewer eggs may be produced in successive clutches or due to changes in food supply, salinity, or other factors. Females which reproduce over an extended period of time (iteroparity) represent the most difficult situation. In such cases about the only thing one can do is bring the animal into the laboratory for observation of its egg-laying characteristics. It may be tempting to maintain gravid individuals in laboratory culture and count the number of offspring which successfully hatch as a measure of fecundity. This will likely be an underestimate of the animal's actual fecundity, especially if there is any egg mortality or if any hatchlings die before they can be observed and counted. Our guesses today about the number of generations per year for meiofaunal taxa are a direct result of our inability to measure fecundity successfully.

Estimates of instantaneous birth rates (b) for females may be obtained using the so-called egg-ratio technique. This requires information on the number of eggs per female and the development time of the eggs and other assumptions (see Paloheimo, 1974). These data are essential for calculations of instantaneous rates of population increase (r), as r = b - d, provided one also has data for calculating instantaneous death rates (d). The population dynamics of meiofauna have not been explored to the same extent as for planktonic crustacea, nor have life tables (mathematical models of population growth rates) been generated for meiofauna. This research avenue is explained well in Wilson and Bossert (1971).

Respiration and Metabolism (R).--For aerobic meiofauna, the rate of oxygen consumption is the usual measure of metabolic activity. However, it should be born in mind that certain species, particularly those characteristic of the deeper layers of fine sediments, may employ anaerobic pathways either totally (obligate anaerobes) or partially (facultative anaerobes).

1. Oxygen Consumption.--The usual manometric methods of determining oxygen consumption with

Warburg or Gilson respirometers are not sufficiently sensitive for meiofauna unless large numbers of animals are available, for example, from laboratory cultures. Scaled down versions of this type of apparatus such as the differential volumetric respirometer described by Dixon (1979), have, however, been used successfully in our (RMW's) laboratory for cultured nematodes (*Diplolaimelloides bruciei*) and meiofaunal polychaetes (*Ophryotrocha* spp.). Reference should be made to Dixon's paper for a description of the technique, which because of its relative simplicity would be our first choice of method if sufficient material is available.

More sensitive techniques must be employed for single animals or small numbers of animals. Lasserre (1974) gives a useful review of microrespirometers available for use with meiofauna. Cartesian divers of the standard or stoppered type have been most widely used. Klekowski (1971) has provided an excellent step-by-step practical account of the manufacture, operation and calculations involved in the use of stoppered divers which provides a useful lead-in to this rather difficult and delicate technique. We would recommend that beginners spend some time in the laboratory of an experienced user of the system to learn the finer points of technique. It is possible to simplify Klekowski's system somewhat: for example, we have replaced the rather elaborate coarse and fine pressure control valve with a simple blow-and-suck mouthpiece, and we have done away with the glass spacers in the flotation chambers, which are difficult to make and seem to serve no obvious function.

The non-Cartesian gradient diver is in many ways less exacting to use and is more sensitive than the Cartesian diver, but each diver can be used only once and the technique has rarely been employed for meiofauna. Likewise, the use of oxygen electrodes has been limited, mainly due to the problem of constructing chambers sufficiently small to do the experiments. Atkinson and Smith (1973), however, have devised an oxygen electrode microrespirometer which has been used successfully for relatively large nematodes (Atkinson, 1973). Probably the most sensitive technique of all is the reference diver of Scholander et al. (1952). This method is, however, not particularly suitable for active meiofaunal animals which will interfere with the reference bubble, a bubble which is attached to a hydrophobic weight and floats in the same chamber as the animal. Nevertheless, Lasker et al. (1970) successfully used the technique with the harpacticoid copepod *Asellopsis*.

2. Calorimetry.--A direct measurement of metabolic activity can be made by determining the rate of heat dissipation of biological processes. Calorimetry is a non-specific method in that it does not depend on the nature of the metabolic pathways employed (aerobic or anaerobic), and so may be more appropriate for many meiofaunal species which might be facultative or obligate anaerobes. Another advantage of the method is that the organisms can be maintained under more natural open-flow conditions, rather than being confined to small ampoules as in Cartesian divers. Gnaiger (1983) and Knudson et al. (1983) describe such systems. But there are disadvantages too. The only suitable microcalorimeters presently available (manufactured by LKB) currently (1986) cost about US $50,000. They can also only be used for relatively large, or large numbers of small, meiofaunal individuals. A useful review of the application of calorific methods to ecological studies is given by Widdows (1987), which provides useful references to methodological papers.

Feces Production (F) and Excretion (U).--Here we begin to enter virgin territory: so far as we are aware, no direct measurements of these two parameters of the energy budget have been determined for any meiofaunal species. They are usually estimated by difference (to neatly balance the budget!). The best we can do is recommend that attempts be made to scale down some of the techniques which have been developed for larger organisms, as outlined by Crisp (1984). However, the chemical composition of the nitrogenous excreta (ammonia, urea?) is for the most part unknown, so it is difficult to know exactly what we are trying to measure. Mention should be made here of two other potentially important components of the energy budget which are usually considered under the heading of "excreta," namely shed cuticle and the production of mucus. Both might be particularly important in certain meiobenthic taxa. Clearly, this is a wide open field which is ripe for future research.

Ecological Efficiencies.--The term "ecological efficiency" is generally employed as a measure of the efficiency of energy transfer from one trophic level to the next, and is calculated as the fraction of the energy consumed by one trophic level that is exploited by predators at the next higher trophic level. Slobodkin (1962) concluded that in most cases this value is around 5-20%, and 10% is a widely used figure, but we really have no idea how this applies to trophic steps involving the meiofauna. Its value depends on various physiological efficiencies such as the production efficiency (P/P+R) and assimilation efficiency (C-F/C). Some authors have suggested that the former is unusually high for meiofauna, a large part of the assimilated ratio being devoted to growth rather than respiration, but others have disputed this. The question will not be settled until more empirical data for the energy budgets of meiofaunal species are obtained. Only then will we be able to quantify, for example, the role played by the meiofauna in the transfer of primary food sources (diatoms and bacteria) up the food chain to juvenile fish, or the extent to which meiofauna are an energy sink in the system, respiring much of what they eat.

Acknowledgments

Contribution No. 701 from the Belle W. Baruch Institute for Marine Biology and Coastal Research. Partially supported by NSF grants OCE-8110148 to RJF and OCE-8521345 to RJF and B.C. Coull.

References

Allan, J.D., T.G. Kinsey, and M.C. James
1976. Abundances and Production of Copepods in the Rhode River Subestuary of Chesapeake Bay. *Chesapeake Science*, 17:86-92.

Alvarez, V.L., C.A. Roitsch, and O. Henriksen
1981. High-pressure Liquid Chromatography of Proteins and Peptides. Pages 83-103 in I. Lefkovits and B. Pernis, editors, *Immunological Methods*, Volume 2. New York: Academic Press.

Andrassy, I.
1956. Die Rauminhalts-und Gewichtsbestimmung der Fadenwürmer (Nematoden). *Acta Zoologica Academiae Scientarum Hungaricae*, 2:1-15.

Atkinson, H.J.
1973. The Respiratory Physiology of the Marine Nematodes *Enoplus brevis* (Bastian) and *Enoplus communis* (Bastian). I. The Influence of Oxygen Tension and Body Size. *Journal of Experimental Biology*, 59:255-266.

Atkinson, H.J., and L. Smith
1973. An Oxygen Electrode Microrespirometer. *Journal of Experimental Biology*, 59:247-253.

Baars, M.A., and G.R. Helling
1985. Methodical Problems in the Measurement of Phytoplankton Ingestion Rate by Gut Fluorescence. *Hydrobiological Bulletin*, 19:81-88.

Bakker, C., R.D. Gulati, and K. Kersting, editors
1985. The Measurement of Ingestion of Phytoplankton by Zooplankton: Techniques, Problems and Recommendations. *Hydrobiological Bulletin*, 19:1-105.

Banse, K.
1982. Mass-scaled Rates of Respiration and Intrinsic Growth in Very Small Invertebrates. *Marine Ecology Progress Series*, 9:281-297.

Banse, K., and S. Mosher
1980. Adult Body Mass and Annual Production/Biomass Relationships of Field Populations. *Ecological Monographs*, 50:355-379.

Barnes, R.S.K. and K.H. Mann, editors
1980. *Fundamentals of Aquatic Ecosystems*. 299 Pages. Oxford: Blackwell Scientific Publications.

Benke, A.C.
1979. A Modification of the Hynes Method for Estimating Secondary Production with Particular Significance for Multivoltine Populations. *Limnology and Oceanography*, 24:168-171.

Berg, J.
1979. Discussion of Methods of Investigating the Food of Fishes, with Reference to a Preliminary Study of the Prey of *Gobiusculus flavescens* (Gobiidae). *Marine Biology*, 50:263-273.

Bird, D.F., and Y.T. Prairie
1985. Practical Guidelines for the Use of Zooplankton Length-Weight Regression Equations. *Journal of Plankton Research*, 7:955-960.

Bodin, P., D. Boucher, J. Guillou, and M. Guillou
1984. The Trophic System of the Benthic Communities in the Bay of Douarnenez (Brittany). Pages 361-370 in *Proceedings of the Nineteenth European Marine Biology Symposium, Plymouth, Devon, U. K.* London: Cambridge University Press.

Boreham, P.F.L., and C.E. Ohiagu
1978. The Use of Serology in Evaluating Invertebrate Prey-Predator Relationships: A Review. *Bulletin of Entomological Research*, 68:171-194.

Brey, T.
1986. Estimation of Annual P/B Ratio and Production of Marine Benthic Invertebrates from Length-Frequency Data. *Ophelia*, Supplement, 4:45-54.

Cammen, L.M.
1980. Ingestion Rate: An Empirical Model for Aquatic Deposit Feeders and Detritivores. *Oecologia* (Berlin), 44:303-310.

Carman, K.R., and D. Thistle
1985. Microbial Food Partitioning by Three Species of Benthic Copepods. *Marine Biology*, 88:143-148.

Chia, F.-S., and R.M. Warwick
1969. Assimilation of Labelled Glucose from Seawater by Marine Nematodes. *Nature*, 224:720-721.

Conover, R.J., T. Durvasula, S. Roy, and R. Wang
1986. Probable Loss of Chlorophyll-derived Pigments During Passage through the Gut of Zooplankton, and Some of the Consequences. *Limnology and Oceanography*, 31:878-887.

Coull, B.C.
1973. Estuarine Meiofauna: A Review: Trophic Relationships and Microbial Interactions. Pages 499-512 in L.H. Stevenson and R.R. Colwell, editors, *Estuarine Microbial Ecology*. The Belle W. Baruch Library in Marine Science, Number 1. Columbia, S.C.: University of South Carolina Press.

Crisp, D. J.
1984. Energy Flow Measurements. Pages 284-372 in N.A. Holme and A.D. McIntyre, editors, *Methods for the Study of Marine Benthos*. Oxford: Blackwell Scientific Publications.

Cucci, T.L., S.E. Shumway, R.C. Newell, R. Selvin, R.R.L. Guillard, and C.M. Yentsch
1985. Flow Cytometry: A New Method for Characterization of Differential Ingestion, Digestion and Egestion by Suspension Feeders. *Marine Ecology Progress Series*, 24:201-204.

de Wilde, P.A.W.J., E.M. Berghuis, and A. Kor
1984. Structure and Energy Demand of the Benthic Community of the Oyster Ground, Central North Sea. *Netherlands Journal of Sea Research*, 18:143-159.

Dixon, D.R.
1979. A Differential Volumetric Micro-respirometer for Use With Small Aquatic Organisms. *Journal of Experimental Biology*, 82:379-384.

Downing, J.A., and F.H. Rigler, editors
1984. *A Manual on Methods for the Assessment of Secondary Productivity in Fresh Waters*. 2nd Edition. 501 pages. Oxford: Blackwell Scientific Publications.

Downs, J.N., and C.J. Lorenzen
1985. Carbon:Pheopigment Ratios of Zooplankton Fecal Pellets as an Index of Herbivorous Feeding. *Limnology and Oceanography*, 30:1024-1036.

Edmondson, W.T.
1974. Secondary Production. *Mitteilungen, Internationale Vereinigung für Theoretische und Angewandte Limnologie*, 20:229-272.

Edmondson, W.T., and G.G. Winberg
1971. *A Manual on Methods for the Assessment of Secondary Productivity in Fresh Waters*. 358 pages. IBP Handbook No. 17. Oxford: Blackwell Scientific Publications.

Feller, R.J.
1980. Quantitative Cohort Analysis of a Sand-dwelling Harpacticoid Copepod. *Estuarine and Coastal Marine*

Science, 11:459-476.

1982. Empirical Estimates of Carbon Production for a Meiobenthic Harpacticoid Copepod. *Canadian Journal of Fisheries and Aquatic Sciences*, 39:1435-1443.

1984. Serological Tracers of Meiofaunal Food Webs. *Hydrobiologia*, 118:119-125.

Feller, R.J., and R.B. Ferguson.
1987. Quantifying Stomach Contents Using Immunoassays: A Critique. In C.M. Yentsch, F.C. Mague, and P.K. Horan, editors, *Immunochemical Approaches to Estuarine, Coastal and Oceanographic Questions*. Berlin: Springer-Verlag, Coastal Lecture Note Series.

Fry, B., and E.B. Sherr
1984. $\delta 13C$ Measurements as Indicators of Carbon Flow in Marine and Freshwater Ecosystems. *Contributions in Marine Science*, 27:13-47.

Giere, O., and C. Langheld
1987. Structural Organisation, Transfer and Biological Fate of Endosymbiotic Bacteria in Gutless Oligochaetes. *Marine Biology*, 93:641-650.

Gillespie, D.M., and A.C. Behnke
1979. Methods of Calculating Cohort Production from Field Data - Some Relationships. *Limnology and Oceanography*, 24:171-176.

Gnaiger, E.
1983. The Twin-flow Microrespirometer and Simultaneous Calorimetry. Pages 134-136 in E. Gnaiger and H. Forstner, editors, *Polarographic Oxygen Sensors. Aquatic and Physiological Applications*. Berlin: Springer Verlag.

Gnaiger, E., and C. Bitterlich
1984. Proximate Biochemical Composition and Caloric Value Calculated from CHN Determination: A Stoichiometric Concept. *Oecologia* (Berlin), 62:289-298.

Gray, J.S.
1981. *The Ecology of Marine Sediments: An Introduction to the Structure and Function of Benthic Communities*. Cambridge Studies in Modern Biology: 2. 185 pages. Cambridge: Cambridge University Press.

Grisley, M.S., and P.B. Boyle
1985. A New Application of Serological Techniques to Gut Contents Analysis. *Journal of Experimental Marine Biology and Ecology*, 90:1-9.

Gulland, J.A.
1969. *Manual of Methods for Fish Stock Assessment*. FAO Manuals in Fishery Science, Vol. 4, Part 1, 154 pages.

Heip, C., P.M.J. Herman, and A. Coomans
1982. The Productivity of Marine Meiobenthos. *Academiae Analecta (Klasse der Wetenschappen)*, 44:1-20.

Heip, C., M. Vincx, and G. Vranken
1985. The Ecology of Marine Nematodes. *Oceanography and Marine Biology Annual Review*, 23:399-489.

Herman, P.M.J., and C. Heip
1985. Secondary Production of the Harpacticoid Copepod *Paronychocamptus nanus* in A Brackish-water Habitat. *Limnology and Oceanography*, 30:1060-1066.

Herman, P.M.J., G. Vranken, and C. Heip
1984. Problems in Meiofauna Energy-flow Studies. *Hydrobiologia*, 118:21-28.

Hicks, G.R.F., and B.C. Coull
1983. The Ecology of Marine Meiobenthic Harpacticoid Copepods. *Oceanography and Marine Biology Annual Review*, 21:67-175.

Holter, H.
1945. A Colorimetric Method for Measuring the Volume of Large Amoebae. *Comptes-rendus des Travaux du Laboratoire de Carlsberg, Serie Chimie*, 25:156-157.

Hyslop, E.J.
1980. Stomach Contents Analysis - A Review of Methods and Their Application. *Journal of Fish Biology*, 17:411-429.

Jensen, P.
1987. Feeding Ecology of Free-living Aquatic Nematodes. *Marine Ecology Progress Series*, 35:187-196.

Kern, J.C., and S.S. Bell
1984. Short-term Temporal Variation in Population Structure of Two Harpacticoid Copepods, *Zausodes arenicolus* and *Paradactylopodia brevicornis*. *Marine Biology*, 84:53-63.

Kiritani, K., and J.P. Dempster
1973. Different Approaches to the Quantitative Evaluation of Natural Enemies. *Journal of Applied Ecology*, 10:323-330.

Klekowski, R.Z.
1971. Cartesian Diver Respirometry for Aquatic Animals. *Polskie Archiwum Hydrobiologii*, 18:93-114.

Knudsen, J., P. Famme, and E.S. Hansen
1983. A Microcalorimeter System for Continuous Determination of the Effect of Oxygen on Aerobic-anaerobic Metabolism and Metabolite Exchange in Small Aquatic Animals and Cell Preparations. *Comparative Biochemistry and Physiology*, 74A:63-66.

Krueger, C.C., and F.B. Martin
1980. Computation of Confidence Intervals for The Size-frequency (Hynes) Method of Estimating Secondary Production. *Limnology and Oceanography*, 25:773-777.

Lasker, R., J.B.J. Wells, and A.D. McIntyre
1970. Growth, Production, Respiration and Carbon Utilization of the Sand-dwelling Harpacticoid Copepod *Asellopsis intermedia*. *Journal of the Marine Biological Association of the United Kingdom*, 50:147-160.

Lasserre, P.
1974. Metabolic Activities of Benthic Microfauna and Meiofauna. Recent Advances and Review of Suitable Methods of Analysis. Pages 95-142 in I.N. McCave, editor, *The Benthic Boundary Layer*. New York: Plenum Press.

LeBlond, P.H., and T.R. Parsons
1977. A Simplified Expression for Calculating Cohort Production. *Limnology and Oceanography*, 22:156-157.

Low, B.W., and F.M. Richards
1952. The Use of Gradient Tube for The Determination of Crystal Densities. *Journal of the American Chemical Society*, 74:1660-1666.

Lowry, O.H.
1941. A Quartz Fiber Balance. *Journal of Biological Chemistry*, 140:183-189.

Mackas, D., and R. Bohrer
1976. Fluorescence Analysis of Zooplankton Gut Contents and an Investigation of Diel Feeding Patterns. *Journal of Experimental Marine Biology and Ecology*, 25:77-85.

Manahan, D.T., and D.J. Crisp
1982. The Role of Dissolved Organic Material in The Nutrition of Pelagic Larvae: Amino Acid Uptake by Bivalve Veligers. *American Zoologist*, 22:635-646.

Marcotte, B.M.
1977. An Introduction to the Architecture and Kinematics of Harpacticoid (Copepoda) Feeding: *Tisbe furcata* (Baird, 1837). *Mikrofauna des Meeresbodens*, 61:183-196.

Marin, V., M.E. Huntley, and B.W. Frost
1986. Measuring Feeding Rates of Pelagic Herbivores: Analysis of Experimental Design and Methods. *Marine Biology*, 3:49-58.

Moller, P., L. Pihl, and R. Rosenberg
1985. Benthic Faunal Energy Flow and Biological Interaction in Some Shallow Marine Soft Bottom Habitats. *Marine Ecology Progress Series*, 27:109-121.

Montagna, P.A.
1983. Live Controls for Radioisotope Tracer Food Chain Experiments Using Meiofauna. *Marine Ecology Progress Series*, 12:43-46.

1984. In situ Measurement of Meiobenthic Grazing Rates on Sediment Bacteria and Edaphic Diatoms. *Marine Ecology Progress Series*, 18:119-130.

Myers, R.F.
1967. Osmoregulation in *Pangrellus redivivus* and

Aphelenchus avanae. Nematologica, 12:579-586.

Omori, M., and T. Ikeda
1984. *Methods in Marine Zooplankton Ecology.* 332 pages. New York: John Wiley and Sons.

Paloheimo, J.E.
1974. Calculation of Instantaneous Birth Rate. *Limnology and Oceanography,* 19:692-694.

Parsons, T.R., Y. Maita, and C.M. Lalli
1984. *A Manual of Chemical and Biological Methods for Seawater Analysis.* 173 pages. Oxford: Pergamon Press.

Phillipson, J.
1964. A Miniature Bomb Calorimeter for Small Biological Samples. *Oikos,* 15:130-139.

Platt, H.M.
1981. Meiofaunal Dynamics and The Origin of The Metazoa. Pages 207-216 in P.L. Forey, editor, *The Evolving Biosphere.* London: British Museum (Natural History), Cambridge University Press.

Pomeroy, L.R.
1980. Detritus and Its Role As a Food Source. Pages 84-102 in R.S.K. Barnes and K.H. Mann, editors, *Fundamentals of Aquatic Ecosystems.* Oxford: Blackwell Scientific Publications.

Poole, R.W.
1974. *An Introduction to Quantitative Ecology.* 532 pages. New York: McGraw-Hill Book Co. 532 pages.

Pyke, D.A., and J.N. Thompson
1986. Statistical Analysis of Survival and Removal Rate Experiments. *Ecology,* 67:240-245.

Resh, V.H.
1979. Sampling Variability and Life History Features: Basic Considerations in The Design of Aquatic Insect Studies. *Journal of the Fisheries Research Board of Canada,* 36:290-311.

Riemann, F., and M. Schrage
1978. The Mucus-trap Hypothesis on Feeding of Aquatic Nematodes and Implications for Biodegradation and Sediment Texture. *Oecologia* (Berlin), 34:75-88.

Robertson, A.I.
1979. The Relationship between Annual Production:Biomass Ratios and Lifespans for Marine Macrobenthos. *Oecologia* (Berlin), 38:193-202.

Roman, M.R., and K.R. Tenore
1984. Detritus Dynamics in Aquatic Ecosystems: An Overview. *Bulletin of Marine Science,* 35:257-260.

Rounick, J.S., and M.J. Winterbourn
1986. Stable Carbon Isotopes and Carbon Flow in Ecosystems. *BioScience,* 36:171-177.

Schiemer, F.
1983. Comparative Aspects of Food Dependence and Energetics of Freeliving Nematodes. *Oikos,* 41:32-42.

Schneckenburger, H., B.W. Reuter, and S.M. Schoberth
1985. Fluorescence Techniques in Biotechnology. *Trends in Biotechnology,* 3:257-261.

Scholander, P.F., C.L. Claff, and S.L. Sveinsson
1952. Respiratory Studies of Single Cells. I. Methods. *Biological Bulletin,* 102:157-177.

Siebers, D.
1982. Bacterial-invertebrate Interactions in Uptake of Dissolved Organic Matter. *American Zoologist,* 22:723-733.

Simenstad, C.A., and R.C. Wissmar
1985. C^{13} Evidence of The Origins and Fates of Organic Carbon in Estuarine and Nearshore Food Webs. *Marine Ecology Progress Series,* 23:141-152.

Slobodkin, L.B.
1962. Energy in Animal Ecology. *Advances in Ecological Research,* 1:69-101.

Staarup, B.J.
1970. On the Ecology of Turbellarians in A Sheltered Brackish Shallow Water Bay. *Ophelia,* 7:185-216.

Stephens, G.C.
1982. Recent Progress in The Study of "Die Ernährung der Wassertiere und der Stoffhaushalt der Gewässer." *American Zoologist,* 22:611-619.

Stephens, G.C.. amd D.T. Manahan
1984. Technical Advances in The Study of Nutrition of Marine Molluscs. *Aquaculture,* 39:155-164.

Tietjen, J.H., and J.J. Lee
1977. Feeding Behavior of Marine Nematodes. Pages 21-35 in B.C. Coull, editor, *Ecology of Marine Benthos.* The Belle W. Baruch Library in Marine Science. Columbia: University of South Carolina Press.

Tietjen, J.H., J.J. Lee, J. Rullman, A. Greengart, and J. Trompeter
1970. Gnotobiotic Culture and Physiological Ecology of the Marine Nematode, *Rhabditis marina* Bastian. *Limnology and Oceanography,* 15:535-543.

Uhlig, G., and S.H.H. Heimberg
1981. A New Versatile Compression Chamber for Examination of Living Microorganisms. *Helgoländer Meeresuntersuchungen,* 34:251-256.

Valiela, I.
1984. *Marine Ecological Processes.* 546 pages. New York: Springer-Verlag.

Vernberg, W.B., and B.C. Coull
1981. Meiofauna. Pages 147-177 in F.J. and W.B. Vernberg, editors, *Functional Adaptations of Marine Organisms.* New York: Academic Press, Inc.

Vranken, G., and C. Heip
1986. The Productivity of Marine Nematodes. *Ophelia,* 26:429-442.

Wang, A.-C.
1982. Methods of Immune Diffusion, Immunoelectrophoresis, Precipitation, and Agglutination. Pages 139-161 in J.J. Marchalonis and G.W. Warr, editors, *Antibody as A Tool: The Applications of Immunochemistry.* New York: John Wiley and Sons.

Warr,. G.W.
1982. Preparation of Antigens and Principles of Immunization. Pages 21-58 in J.J. Marchalonis and G.W. Warr, editors, *Antibody as A Tool: The Applications of Immunochemistry.* New York: John Wiley and Sons.

Warwick, R.M.
1980. Population Dynamics and Secondary Production of Benthos. Pages 1-24 in K.R. Tenore and B.C. Coull, editors, *Marine Benthic Dynamics.* The Belle W. Baruch Library in Marine Science. Columbia: University of South Carolina Press.

Warwick, R.M., and J.M. Gee
1984. Community Structure of Estuarine Meiobenthos. *Marine Ecology Progress Series,* 18:97-111.

Warwick, R.M., and R. Price
1979. Ecological and Metabolic Studies on Free-living Nematodes from an Estuarine Mud-flat. *Estuarine and Coastal Marine Science,* 9:257-271.

Waters, T.F.
1977. Secondary Production in Inland Waters. *Advances in Ecological Research,* 10:91-164.
1979. Influence of Benthos Life History upon The Estimation of Secondary Production. *Journal of the Fisheries Research Board of Canada,* 36:1425-1430.

Waters, T.F., and G.E. Crawford
1973. Annual Production of a Stream Mayfly Population: A Comparison of Methods. *Limnology and Oceanography,* 18:286-296.

Watzin, M.C.
1985. Interactions Among Temporary and Permanent Meiofauna: Observations on The Feeding and Behavior of Selected Taxa. *Biological Bulletin,* 169:397-416.

Widbom, B.
1984. Determination of Average Individual Dry Weights and Ash-free Dry Weights in Different Sieve Fractions of Marine Meiofauna. *Marine Biology,* 84:101-108.

Widdows, J.
1987. Application of Calorimetric Methods in Ecological Studies. Pages 182-215 in A.M. James, editor, *Thermal and Energetic Studies of Cellular Biological Systems.* Bristol: Hilger.

Wieser, W.

1953. Die Beziehung zwischen Mundhöhlengestalt, Ernährungsweise und Vorkommen bei freilebenden Marinen Nematoden. Eine ökologisch-morphologische Studie. *Arkiv för Zoologie,* Series II. 4:439-484.
1960. Benthic Studies in Buzzards Bay. II. The Meiofauna. *Limnology and Oceanography,* 5:121-137.

Wilson, E.O., and W.H. Bossert
1971. *A Primer of Population Biology.* 192 pages. Stamford, Connecticut: Sinauer Associates, Inc.

Winberg, G.G.
1960. The Metabolic Intensity and Food Requirements of Fish. *Translation Series of the Fisheries Research Board of Canada,* 194:1-202.
1971. *Methods for The Estimation of Production of Aquatic Animals.* 161 pages. (A. Duncan, translator.) London: Academic Press.

Yamamoto, N., and G. Lopez
1985. Bacterial Abundance in Relation to Surface Area and Organic Content of Marine Sediments. *Journal of Experimental Marine Biology and Ecology,* 90:209-220.

14. Data Processing, Evaluation, and Analysis

Carlo Heip, Peter M.J. Herman, and Karlien Soetaert

This chapter aims at introducing a number of methods for the post-hoc interpretation of field data. In many ecological studies a vast amount of information is gathered, often in the form of numbers of different species. The processing of these data is possible on different levels of sophistication and requires the use of tools varying from pencil and paper to a mini computer. However, some "number crunching" is nearly always necessary and with the increasing availability and decreasing prices of the personal computer, even in developing countries, there is little point in avoiding the more advanced statistical techniques. There is also a more fundamental reason: ecological data are often of a kind where the application of simple statistics is not only cumbersome, but also misleading and even incorrect. The basic assumptions on which uni- and bivariate statistics are based are often violated in the ecological world: concepts such as homogeneity of variance, independence of observations etc. do not belong to the "real" world of the ecologist.

All this is not to say that ecology may be reduced to statistics or that numbers convey anything that was not already there. However, structure and relationships in the data, and their significance, are in most cases best studied using the appropriate statistics.

The increasing availability of computers and the software required to perform most of the more powerful eco-statistical methods have at least one important drawback: this technology entrains many possible pitfalls for the user who is not fully aware of the assumptions underlying the methods used (e.g., are the data of a form suitable for that particular analysis?). Thus, the user may misinterpret or fail to interpret the results that computers offer so generously (Nie et al., 1975). This chapter was written with the software package user especially in mind.

The chapter is structured as follows: following this introduction is part 1 on the subject of data storage and retrieval. It is intended to introduce the reader to the software that exists and how it can be accessed. The chapter then continues with the methods used to study spatial pattern (part 2), structure (classification and ordination) in part 3, species-abundance distributions in part 4, diversity and evenness indices in part 5, and temporal pattern in part 6. Each of these topics is discussed independently and each topic can be consulted without having to read the others. For this reason, each part will be preceded by its own introduction.

Part 1. Data Storage and Retrieval

The existence of many software packages, useful in ecology, will not be unknown to most ecologists. Many computer centers provide their users with routines which facilitate the input, transformation and appropriate output of the data, and the linking of available software packages into one flexible system. As the bulk of mathematical treatments of ecological data is already available, it is far more efficient to use the existing software from data centers or to implement software packages on a small computer than to try to program the necessary analyses oneself, especially when one envisages that different techniques are to be applied to the same data (Legendre and Legendre, 1979).

Software Packages

SPSS (*Superior Performing Software Systems*).--Includes: data transformations, arithmetic expressions, basis statistics, distribution statistics, regression, correlation, contingency and association analysis, analysis of variance and covariance (principal component analysis), factor analysis, and discriminant analysis. See Nie et al., 1975.

BMD and BMDP.--Includes: encoding of data and subsequent calculations, transposition of data matrices, regression, correlation, contingency and association analysis, dispersion matrix/missing values, statistical tests, cumulative frequencies, analysis of variance and covariance, data transformation, principal component analysis, discriminant analysis, harmonic analysis and periodic regression, and spectral analysis. See Dixon, 1973, and Dixon and Brown, 1977.

CEP (*Cornell Ecology Programs*).--Includes: DECORANA (Hill, 1979a): reciprocal averaging/

detrended correspondence analysis; TWINSPAN (Hill, 1979b): hierarchical classifications by two-way indicator species analysis; ORDIFLEX (Gauch, 1977): weighted averages, polar ordination, principal component analysis, reciprocal averaging; COMPCLUS (Gauch, 1979): non-hierarchical classifications by composite clustering; CONDENSE (Singer and Gauch, 1979): reading data matrices and copying this into a single, standardized format (efficient for computer processing); DATAEDIT (Singer, 1980): data matrix-edit options.

CLUSTAN.--Includes: classification methods, and principal component analysis as an option. See Wishart, 1978.

NT-SYS (Numerical Taxonomy System) and CLASP.--These are classification program packages, mainly featuring agglomerative polythetic techniques. See Rohlf et al., 1976.

SAS.--This is an integrated software system providing data retrieval and management, programming, and reporting capabilities. Statistical routines include descriptive statistics, multivariate and time series analyses. All analytical procedures utilize a consistent general linear model approach. See SAS Institute, 1985.

Data Bases

A *data base* is a computerized record keeping system designed to record and maintain information (Date, 1981). A *field* is the smallest named unit of data stored in a data base. A *record* is a collection of associated fields. In most systems, the record is the unit of access of the database. A *file* is a collection of records.

In multidisciplinary research, or when several ecologists work together, a varied and large amount of information concerning the ecosystem studied becomes available. It can be advantageous to group all these data in one file, store it in a data base, and let every researcher access parts of the file which are useful to him. In this way the validation of common data has to be performed only once, and, as everybody has access to all data contained in the file, the extension of one's own data with data gathered by other workers is made easy (Frontier, 1983).

In many data centers, there exists a data base management system (DBMS); that is, a layer of software between the data base itself (i.e., the data as actually stored) and the user of the system. The DBMS thus shields the data base-user from hardware-level detail and supports user-operations. These user-operations can be performed on a high level (simplified statements) (Date, 1981).

SIR (*Scientific Information Retrieval*).--This is one of several DBMS in common usage in North America. It furnishes a direct interface with SPSS and BMDP and is well documented by a users guide (Robinson et al., 1979) and a pocket guide (Anderson et al., 1978).

Dbase III + and LOTUS 123.--These are powerful packages that interface with each other and with several word processors. With the increasing availability of personal computers, it is to be expected that these and similar data bases and spreadsheets will play a major role in the storage and exchange of data for all but the largest data sets.

Standardization

In order to exchange data between different research institutes or data centers, certain standards should be adhered to in the representation of the data. At the moment, several systems are being developed that involve the standardized coding of different parameters.

The Intergovernmental Oceanographic Commission General Format 3 (IOC GF-3).--This is a system for formatting oceanographic data onto magnetic tape. The ICES (International Council for the Exploration of the Sea) Oceanographic Data Reporting System (hydrochemical and hydrographic data) has been based on this system.

The Biological Data Reporting Format (Helcom System).--This format is being developed by ICES, and is designed to interface with the ICES Oceanographic Data Reporting System. A data file consists of four types of records, structured hierarchically (Figure 14.1): (1) The File Header Record (called Biomaster), is the file header for all the biological data obtained at one station. It contains 80 columns, the first 27 of which are identical to the first 27 columns of the cards used in the ICES Oceanographic Data Reporting System. In this way the relevant hydrographic, hydrochemical, and meteorological data may be obtained easily. The Biomaster further identifies the parameters that have been measured at that station (e.g., phytoplankton, meiobenthos) and how many Series Header Records have been prepared for each parameter. (2) The Series Header Record (called Type Master), defines the method that has been used to obtain the data for the parameter (e.g., cores, box-corers etc.). (3) The Data Cycle Record contains the data obtained for the parameter with the method encoded by the Type Master at the station defined by the Biomaster.

For instance, it may contain information on the number or biomass of each species present at the station. (4) Finally, there is a Plain Language Record that may be used at any level in the format to insert comments which are relevant to the interpretation of the data.

The system is not complete at the time of writing this chapter but has been described rather extensively since other systems are often quite similar and it has been adopted by the ICES Benthos Ecology Working Group for cooperative programs.

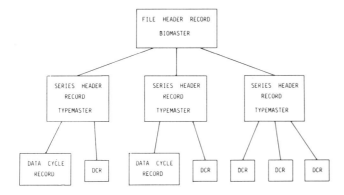

Figure 14.1.--Scheme of the structure of "Helcom system" data base files. See text for an explanation of the terms and a description of the contents of the different records.

NODC (National Oceanographic Data Center, USA.).--This agency has a format for use with benthic organisms. A file contains seven types of 80-character records: (1) The Header record defines the cruise (date, investigator, etc.). (2) There is a Station Header Record for dredge tows that defines the station and describes the tow (type of dredge). (3) There also is a Station Header Record for point sampling that defines the station and describes the pointsampler (e.g., grabs, corers). (4) The Environmental Record describes surface environmental conditions for each station. (5) The Bottom Characteristics Record describes environmental conditions of the bottom and core information. (6) The Taxonomic Data Record gives the number and weight of each taxon collected within a sample. (7) In addition, there is a Text Record.

Taxonomic Codes.--For data exchange, a single system of taxonomic coding would be preferable. However, there are several such systems: (1) For field reporting, the user requires a system that is simple and easy to use, and which minimizes the risk of errors. This implies a mnemonic code based preferably on the scientific name. When using standard abbreviation principles (e.g., Rubin Code: first four letters of genus name, space, first three letters of species name, see Osterdahl and Zetterberg, 1981), the risk of duplicate codes exists. The full latin name, on the other hand, is long and subject to typing and spelling errors, and the software to handle full scientific names is more complicated.

For data exchange in computer compatible form, structured numeric codes are more useful (e.g., the NODC code). These codes can be allocated, taking into account the taxonomic position of the species.

As one should aim at establishing one single coding system, a mechanism to relate the various taxonomic codes to a single common system is needed, such that translation to the common system could be carried out automatically by the computer.

For more information, the reader should contact the Intergovernmental Oceanographic Commission, Place de Fontenoy, Paris, France, or the International Council for the Exploration of the Sea, Palaegade 2-4, DK-1261 Copenhagen K, Denmark.

Part 2. Spatial Pattern

Patterns in the distribution of meiofaunal populations may be studied at length scales that range from that of an individual to the global scale of biogeographic studies. We will restrict our discussion to the small-scale pattern of a deme, the distribution in space of the potentially interbreeding individuals of the population. Examples may be a nematode species on a particular beach or a copepod species in an estuarine creek. Within such large areas, where environmental gradients may or may not exist (but usually will), individual animals occupy a much smaller space during their life. In this smaller space, individual processes and interactions occur: feeding, reproduction, mortality, etc. Small-scale pattern refers to these length scales, which for meiofauna may be between 10^{-2} and 10^2 m, probably mostly below 10 m. Analysis of this pattern may generate hypotheses on the environmental gradients or factors which influence the species. Also, the information obtained from spatial pattern analysis may be used in the design of an adequate sampling scheme.

In meiofauna ecology the sample is mostly taken by a core of some size varying from drinking straws to 10 cm^2 surface area cores. These cores may be contiguous or discontiguous. One should always be aware of the importance of core size on the analysis, and cores should not be too small, i.e., they must be much larger than an individual (see below).

Random Patterns

The Poisson Distribution.--At a particular scale the distribution of animals may be random or not; the detection and analysis of randomness or non-randomness is a starting point for further investigation of the factors involved.

A random pattern implies that the probability of finding an individual at a point in the area sampled is the same for all points. When this probability is very low, e.g., when p <0.01, it may be calculated using the Poisson distribution. Alternatively, a Poisson distribution is appropriate when the probability of finding an individual in a sample of the size of an individual is very small. In these cases the Poisson distribution is the standard to which randomness is checked. It should be noted that the mean number of animals per sample is irrelevant in deciding whether or not a Poisson distribution is appropriate: a Poisson distribution applies if the maximum possible number that could occur in a sample is much higher.

The variance of the Poisson distribution is equal to the mean, and the distribution is thus determined by only one parameter. It may be calculated from the arithmetic mean $\bar{x}$ of the observed frequency distribution of the number of individuals per sample. The probability that a sample contains 0, 1, 2, ..., p, ... individuals is given by:

$$e^{-\bar{x}}, \; \bar{x}e^{-\bar{x}}, \; \frac{\bar{x}}{2}e^{-\bar{x}}, \ldots, \; \frac{\bar{x}^p}{p!}e^{-\bar{x}}, \ldots$$

in which each term can be calculated easily from the previous one. The agreement between the observed frequency distribution and the Poisson distribution can then be calculated as X^2 with n-2 degrees of freedom (Pielou, 1969).

The $s^2/\bar{x}$ Ratio.--Since in the Poisson distribution the variance equals the mean, the criterion most frequently used to detect departure from randomness is $s^2/\bar{x} = 1$. Indeed, almost all criteria are in fact based on calculation of the first and second moments of the observed frequency distribution (see Greig-Smith, 1983, for an extensive review). When the ratio is significantly larger than one, the pattern is assumed to be aggregated. When the ratio is significantly smaller than one the pattern is considered to be regular (this has never been observed for meiofauna and will not be considered further). The significance of departure from one can be tested using the fact that $(n-1)s^2/\bar{x}$ is distributed as X^2 with n-1 degrees of freedom.

The Binomial Distribution.--If the number of individuals actually occurring in a sample approaches the maximum possible, the use of the Poisson distribution becomes inappropriate and the frequencies of the density per sample will approximate to a binomial distribution obtained by the expansion of $(p+q)^n$. In this formula p is the probability of finding an individual in the sample, q = 1-p and n is the maximum number possible per sample. Since n is hard to calculate in meiofauna data, the binomial distribution is difficult to apply. A hypothetical meiofauna animal of 1 mm length and 50 µm width has a surface of 0.05 mm^2. With a density of 1000 per 10 cm^2, 5% of the area is covered by the species and that is the chance of finding one if a corer of 0.05 mm^2 surface area is used. At such high densities, the use of the Poisson distribution may thus become doubtful.

Aggregated Patterns

Contagious Frequency Distributions.--When the variance of the observed frequency distribution is significantly larger than the mean, aggregation is inferred. There may be several mechanisms that generate this, and many possible theoretical distributions describe such situations. They are usually called contagious distributions. Examples are the Neyman and Thomas distributions, which are combinations of two Poisson distributions (Neyman, 1939, described clusters of insect eggs and number of eggs per clusters as being both Poisson-distributed; Thomas, 1949, described randomly distributed colonies with individual populations following a Poisson distribution).

Most frequently used is the Negative Binomial Distribution: it may arise in several ways, e.g., when groups are randomly distributed and the number per group follows a logarithmic distribution, or, when groups are randomly distributed with means that vary according to X^2. It may also arise from true contagion, when the presence of an individual increases the chance that another will be there.

The negative binomial distribution is completely defined by two parameters, the arithmetic mean $\bar{x}$ and a positive exponent k. The probability that a sample contains x individuals is given by:

$$p_x = \frac{(k+x-1)!}{(k-1)!x!} \cdot \frac{(\bar{x}/k)^x}{(1-(\bar{x}/k))^{k+x}}$$

The variance of the Negative Binomial Distribution is equal to $s = \bar{x} + \bar{x}^2/k$. From this it follows that:

$$k = \frac{\bar{x}}{(s^2/\bar{x})-1}$$

To obtain an unbiased estimation of k one has to use the maximum likelihood method described by Bliss and Fisher (1953). A computer program for this is given by Davies (1971).

The value of k is a measure of the degree of aggregation (contagion): the smaller it is, the larger the degree of aggregation. Comparisons can only be made for equal sample sizes.

Taylor's Power Law.--Taylor (1961) observed that often a relationship between the variance and the mean exists, such that:

$$s^2 = a \bar{x}^b$$

The exponent b is an index of aggregation. If b=0, the pattern is regular (s^2=a); if b=1, the pattern is random (if a=b=1, one obtains the Poisson distribution); if b>1, the pattern is aggregated.

A plot of log s^2 against log $\bar{x}$ yields a straight line:

$$\log s^2 = \log a + b \log \bar{x}$$

Plotting log ($s^2/\bar{x}$) versus log $\bar{x}$ also yields a straight line:

$$\log (s^2/\bar{x}) = \log a + (b-1) \log \bar{x}$$

The advantage of using Taylor's power law lies in the fact that the b-value is independent of sample size. If b=2, $s^2 = a\bar{x}^2$ and $s^2/\bar{x} = a\bar{x}$. A linear relationship between $s^2/\bar{x}$ and $\bar{x}$ has been found by Heip (1975) for harpacticoid copepods. In this case there exists a general relationship between s^2, x, and k, such that $\bar{x}=\sqrt{ks}$, relating density, variability and the degree of aggregation (Heip, 1975).

Description of Patches

If the distribution is contagious, then, whatever the exact mechanism producing it, individuals are found in patches with higher or lower abundance. Information on the size and shape of these patches is clearly desirable. Some attempts to measure patch size in benthic data have been published (Heip and Engels, 1977; Findlay, 1982; Hogue, 1982), but they are not entirely satisfactory. Methods used in plant ecology are described by Greig-Smith (1983). They are based on plotting variance against increasing sample size; a peak in such a plot corresponds to a patch and indicates its size. This method depends on contiguous samples.

Crowding.--The degree of crowding experienced by an individual animal may be estimated by the Index of Mean Crowding (Lloyd, 1967) as:

$$x^* = \bar{x} + (s^2/\bar{x} - 1)$$

The ratio of mean crowding to mean density was called patchiness (Lloyd, 1967).

Location of Patches.--To utilize the information contained in the geographical location of the patches, several methods have been proposed. Krishna Iyer (1949) proposed to classify samples into two categories, one with values higher than the mean, one with values lower than the mean. The expected number of connections between samples with high density values if they are located at random in space can be calculated (see Pielou, 1969 for an extensive description) and compared with the number of connections actually observed. This method has been used by Heip (1976) and Heip and Engels (1977) to describe spatial patterns of the ostracod *Cyprideis torosa* and of six species of harpacticoid copepods.

Another possibility is to use the ratio of the autocovariance to the sample variance weighted as a function of the distance between the samples as a measure of spatial autocorrelation (Jumars et al., 1977). Two statistics proposed by Cliff and Ord (1973) may be used for this purpose:

$$I = (\frac{n}{w}) \sum_i^n \sum_j^n w_{ij} z_i z_j / \sum_i^n z_i^2$$

$$C = (\frac{n-1}{w}) \sum_i^n \sum_j^n w_{ij} (x_i - \bar{x})^2 / \sum_i^n z_i^2$$

where x_i is the variate value in sample i, n is the number of samples, $z = x_i - \bar{x}$ and w_{ij} are the weights as a function of spatial distance between the samples:

$$w = \sum_i \sum_j w_{ij}$$

The selection of weights is discussed by Jumars et al. (1977). Often used is:

$$w_{ij} = (\text{distance})^{-2}$$

The use of I, c and $s^2/\bar{x}$ is illustrated by Jumars and Eckman, (1983). If the pattern is random, $I \sim 1/(n-1)$ and $C \sim 1$. Significant departures from these values can be calculated according to the formulae found in Cliff and Ord (1973) or Jumars et al. (1977).

Spectral Analysis

The method of spectral analysis, normally applied to temporal series, may also be applied to spatial data. Hogue and Miller (1981) used the autocorrelation of the data to detect periodicity in nematode density data.

Part 3. Classification and Ordination

Classification (clustering) and ordination are two sets of techniques capable of synthesis and ordering of the data collected to describe communities. Classification involves arranging objects (in benthic ecology usually the samples or stations) into groups setting them apart from the members of other groups, and a typical product of classification is a graph called a dendrogram. Ordination attempts to place objects in a space defined by one or more axes in such a way that knowledge of their position relative to the axes conveys the maximum information about the objects. This information may relate to species composition of samples (stations) or to occurrence of species in samples (stations).

Ordination and classification both start from a data matrix, which in benthic ecology takes the form of a n x p matrix with p stations (samples) as columns and n species as rows. More generally, the columns of the data matrix represent the objects (independent variables), the rows represent the descriptors used to describe these objects (dependent variables). The entries in the matrix are a measure of the abundance y_{ij} of species i in station j.

Some measure is then used to compare all the rows, two by two, or all the columns, two by two, of the data matrix. In this way two association matrices may be derived from the data matrix. These association matrices are square matrices with particular properties which make them apt for further analysis. In the Q-mode, the p x p association matrix Q is derived from the comparison between objects (stations), in the R-mode, the n x n association matrix R is derived from the comparison between descriptors. The choice of a suitable association measure depends on whether the analysis is in the Q- or the R-mode. Numerous measures have been proposed; the following account is broadly based on the lucid review of Legendre and Legendre (1979) which should be consulted for more detail.

Association Measures - Q-Mode

Distance Measures.--The association between the p stations (objects) can be conceptualized as resulting from ordering the different stations in a n-dimensional space in which the n axes are formed by the n species (descriptors). The most obvious way to measure association is to measure the distance between the stations (objects). In a two-dimensional space (only two species are used to describe the stations) the distance between the two stations is given by the familiar formula for the hypothenusa of a rectangular triangle derived by Pythagoras (Figure 14.2):

$$D = \sqrt{(y_{22}-y_{21})^2 + (y_{12}-y_{11})^2}$$

This is readily generalized to n dimensions as:

$$D_1 = \sqrt{\sum_i (y_{i1}-y_{i2})^2}$$

However, there are several problems in applying the Euclidean distance: it has no upper limit but increases with the number of species, and it depends on the scale of the descriptors. Consequently, numerous other distance measures have been proposed. Their calculation is shown in Table 14.1. First, one may reduce the effect of the number of species by dividing by n, to find an average distance:

$$D_2 = \sqrt{\frac{1}{n} \sum_i (y_{i1}-y_{i2})^2}$$

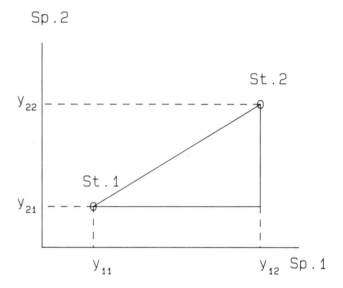

Figure 14.2.--Calculation of the distance between two stations (St. 1 and St. 2) in two-dimensional space (i.e., each station is described by two descriptors, one along each axis). The distance between the two points can easily be calculated by the Pythagorean formula.

Orloci (1967) proposes the chord distance D_3, varying between 0 and $\sqrt{n}$, which is an Euclidean distance after standardization of the descriptors (standardization: mean divided by the standard deviation; the objects will then be situated on a hypersphere with radius one). This may be applied to abundance data:

$$D_3 = \sqrt{2(1 - \frac{\sum_i y_{i1} y_{i2}}{\sqrt{\sum_i y_{i1}^2 \sum_i y_{i2}^2}})}$$

CHAPTER 14: DATA PROCESSING, EVALUATION, AND ANALYSIS 203

Table 14.1. -- A practical example of the calculation of the different distance, similarity, and dependence measures given in the text. The data for this example are given at the start of the table. The steps necessary for the calculation of the distance and similarity between stations 1 and 2 are exemplified next, together with the results for all the measures defined. The third section exemplifies the calculation of the dependence measures for metric and ordered descriptors. Finally the calculation of two dependence measures for binary descriptors are shown.

1. Data Matrix

	St.1	St.2
Spec. 1	13	10
Spec. 2	1	4
Spec. 3	0	2
Spec. 4	1	0
Spec. 5	2	1
Spec. 6	0	2

2. Distance and Similarity Between Stations 1 and 2

	Sp.1	Sp.2	Sp.3	Sp.4	Sp.5	Sp.6	Σ	Σy^2
Stat. 1								
y	13	1	0	1	2	0	17	175
y/Σy	0.76	0.06	0.00	0.06	0.12	0.00	1.00	
pres./abs.	1	1	0	1	1	0	4	
Stat. 2								
y	10	4	2	0	1	2	19	125
y/Σy	0.53	0.21	0.11	0.00	0.05	0.11	1.00	
p/a	1	1	1	0	1	1	5	

Calculations

		Sp.1	Sp.2	Sp.3	Sp.4	Sp.5	Sp.6	
1.	y1-y2	3	-3	-2	1	1	-2	4
2.	[y1-y2]	3	3	2	1	1	2	12
3.	(y1-y2)^2	9	9	4	1	1	4	28
4.	y1+y2	23	5	2	1	3	2	36
5.	(y1+y2)^2	529	25	4	1	9	4	572
6.	y1·y2	130	4	0	0	2	0	136
7.	%y1-%y2	0.24	0.15	0.11	0.00	0.07	0.11	0.72
8.	min(y1,y2)	10	1	0	0	1	0	12
9.	min(%y1,%y2)	0.53	0.06	0.00	0.00	0.05	0.00	0.64
10.	[y1-y2]/(y1+y2)	0.13	0.60	1.00	1.00	0.33	1.00	4.06
11.	a (p/p)	1	1			1		3
12.	b (p/a)				1			1
13.	c (a/p)			1			1	2

Table 14.1 (continued)

Distance Coefficients

$D1 = \sqrt{28} = 5.29$

$D2 = \sqrt{28/6} = 2.16$

$D3 = \sqrt{2(1 - \frac{136}{\sqrt{175 \cdot 125}})} = 0.401$

$D4 = \arccos(1 - \frac{0.401^2}{2}) = 23.15$

$D6 = 12$

$D7 = 12/6 = 2$

$D8 = (1/2)(0.72) = 0.36$

$D8' = 2(1 - 0.64) = 0.72$

$D9 = 4.06$

$D10 = 12/36 = 0.33$

Similarity Coefficients

$S1 = \frac{3}{3+1+2} = 0.50$

$S2 = \frac{6}{6+1+2} = 0.67$

$S3 = \frac{3}{3+2+4} = 0.33$

$S4 = \frac{2 \cdot 12}{17+19} = 0.67$

$S5 = \frac{1}{2}(\frac{12}{17} + \frac{12}{19}) = 0.67$

---

3. Dependence Between Stations

	Sp.1	Sp.2	Sp.3	Sp.4	Sp.5	Sp.6	Σ	Σy²	ȳ
Station 1									
y	13	1	0	1	2	0	17	175	2.8
y-ȳ	10.2	-1.8	-2.8	-1.8	-1.8	-2.8			
(y-ȳ)²	103.4	3.4	8.0	3.4	0.7	8.0	126.9		
R	1	3.5	5.5	3.5	2	5.5			
p/a	1	1	0	1	1	0			
Station 2									
y	10	4	2	0	1	2	19	125	3.2
y-ȳ	6.8	0.8	-1.2	-3.2	-2.2	-1.2			
(y-ȳ)²	46.7	0.7	1.4	10.0	4.7	1.4	64.9		
R	1	2	3.5	6	5	3.5			
p/a	1	1	1	0	1	1			
Calculations									
(y1-ȳ1)(y2-ȳ2)	69.5	-1.5	3.3	5.8	1.8	3.3	82.2		
d_j	0	1.5	2	2.5	3	2			
d_j^2	0	2.25	4	6.25	9	4	25.5		
a (p/p)	1	1			1		3		
b (p/a)				1			1		
c (a/p)			1			1	2		

Metric descriptors

Covariance $s_{ki} = \frac{82.2}{5} = 16.4$

Variance y1: $s^2_y = \frac{126.9}{5} = 25.4$

Variance y2: $s^2_y = \frac{64.9}{5} = 13.0$

Correlation $r_{12} = \frac{16.4}{25.4 \times 13.0} = 0.903$

CHAPTER 14: DATA PROCESSING, EVALUATION, AND ANALYSIS

Table 14.1 (continued)

Ordened descriptors

Spearman's rank correlation coefficient:

$$R_s = 1 - \frac{6 \times 25.5}{6^3 - 6} = 0.271$$

Kendall's rank correlation coefficient:

$$\tau = \frac{S}{\sqrt{(p^2-p-T_1)(p^2-p-T_2)}}$$

a) calculation of S:

y1 (ordened):	1	2	3.5	3.5	5.5	5.5
y2	1	5	2	6	3.5	3.5

N higher ranks   5   1   3   0   0   0   N = 9 to the right in y2 (i.e., first value has rank 1, higher ranks are 5, 2, 6, 3.5 and 3.5, total 5. Second value has rank 5, higher value is 6, total 1. Third value has rank 2, higher values are 6, 3.5 and 3.5, etc.)

$S = 4N - p(p-1) = 4 \times 9 - 6 \times 5 = 6$

b) calculation of correction term for ties:

$T = t(t-1)$, summation over all groups of ties, with t the number of values (ranks) in each tie.

$T_1 = 2 \times 1 + 2 \times 1 = 4$ (two groups of ties with 2 members each in y1)

$T_2 = 2 \times 1 = 2$ (one group of ties with 2 members in y2)

c) final calculation:

$$\tau = \frac{6}{\sqrt{(36-6-4)(36-6-2)}} = 0.222$$

---

4. Binary descriptors (presence/absence)

Point correlation coefficient:

$$r = \frac{ad - bc}{\sqrt{(a+b)(a+c)(b+c)(b+d)}}$$

$$r = \frac{3 \times 0 - 1 \times 2}{\sqrt{(3+1)(3+2)(1+2)(1+0)}} = -0.258$$

Chi square:

$$X^2 = \frac{p((ad-bc) - (p/2))^2}{(a+b)(a+c)(b+c)(b+d)}$$

$$= \frac{6(-2-3)^2}{(4)(5)(3)(1)} = 2.5$$

It may be transformed into the geodesic metric as shown below. This measures the length of an arch on the surface of a hypersphere with radius one. An example for two species is shown in Figure 14.2.

$$D_4 = \arccos\left(1 - \frac{D_3^2}{2}\right)$$

Another general distance measure has been proposed by Minkowski:

$$D_5 = \left(\sum_i |y_{i1} - y_{i2}|^r\right)^{1/r}$$

It reduces to the Euclidean distance for r = 2 and several other variants may be deduced from it. For r = 1 one obtains the Manhattan-metric:

$$D_6 = \sum_i |y_{i1} - y_{i2}|$$

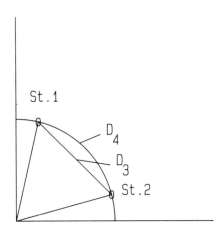

Figure 14.3.--Graphic representation of the difference between the distance measures D_3 and D_4 for two stations in two-dimensional space. In two dimensions, the stations are situated on a circle with radius 1 after standardization. D_3 is then the length of the chord (straight line) between the two points. D_4 is the length of the arch between lines.

The Manhattan-metric has the same disadvantages as the Euclidean distance. One therefore derives:

$$D_7 = \frac{1}{n}\sum_i |y_{i1} - y_{i2}|$$

The index of association of Whittaker is derived as:

$$D_8 = \frac{1}{2}\sum_i \left| \frac{y_{i1}}{\sum_i y_{i1}} - \frac{y_{i2}}{\sum_i y_{i2}} \right|$$

or as:

$$D_8' = 2\left[1 - \sum_i \min\left(\frac{y_i}{\sum_i y_i}\right)\right]$$

It is also applied to species abundances.

A variant of the Manhattan-metric is the Canberra-metric (Lance and Williams, 1966), quite popular in benthic ecology:

$$D_9 = \sum_i \frac{|y_{i1} - y_{i2}|}{(y_{i1} + y_{i2})}$$

Also widely used is the index proposed by Bray and Curtis:

$$D_{10} = \frac{\sum_i |y_{i1} - y_{i2}|}{\sum_i (y_{i1} + y_{i2})}$$

Distance and Similarity.--In the Q-mode association measures are of two forms: distance and similarity. Similarity S is maximal for two identical objects, distance D is maximal for two completely dissimilar objects. Between similarity and distance simple relationships exist such as $D = 1-S$, $D = \sqrt{1-S}$ or $D = \sqrt{1-S^2}$. Williams and Dale (1965) urged as a minimum requirement that the distance measure used should be metric (M), i.e., should have the following properties: (a) $D \geq 0$; (b) $D(a,b) = D(b,a)$; (c) $D(a,c) \leq D(a,b) + D(b,c)$ (inequality of the triangle). When this condition is not satisfied the distance measure is called semimetric (SM), when both conditions (a) and (c) are not satisfied it is called non-metric (NM). D1 to D9 are metric distance coefficients, D10 is semimetric.

Similarity Coefficients.--Similarity coefficients are used to measure the association between stations (objects). They are never metric and may not be used to position objects in a metric space as in ordination. However, the complements of several similarity coefficients are metric and may be used in ordination.

Similarity may be measured from simple presence and absence of species, based on the following table:

		Station 1	
		present	absent
Station 2	present	a	b
	absent	c	d

in which a, b, c and d are the numbers of species present in both stations, in station 1 or 2 only, or in none of the stations. In ecological applications, the double absence of a species is normally not taken into account. The following coefficients are among the more frequently used:

$$S_1 = \frac{a}{a+b+c} \quad \text{(Jaccard) (M)}$$

$$S_2 = \frac{a}{2a+b+c} \quad \text{(Sorensen) (SM)}$$

$$S_3 = \frac{a}{a+2b+2c} \quad \text{(Sokal and Sneath) (M)}$$

Similarity coefficients based on abundance but with exclusion of the double zeros may be derived from the binary coefficients shown above by calculating the amount of concordance, which may be based on presence or absence but also on abundance classes.

$$S = \text{concordance}/(n\text{-double zeros})$$

When abundance data are available, other coefficients use this information better. The following of frequent use are based on raw abundance data arranged as in the following example:

	Station 1	Station 2	Minimum
Sp. 1	13	10	10
Sp. 2	1	4	1
Sp. 3	0	2	0
Sp. 4	1	0	0
	A = 15	B = 16	W = 11

$$S_4 = \frac{2W}{A+B} \quad (=22/31 = 0.71) \quad \text{(Kulczynski) (SM)}$$

$$S_5 = \frac{1}{2}\left(\frac{W}{A} + \frac{W}{B}\right) \quad (=0.71) \quad \text{(Steinhaus) (SM)}$$

Coefficients of Dependence - R-Mode

In the R-mode (clustering and ordination of species) the primary goal is to determine the relationship between the species, although the association matrix may be used for an ordination of objects as well in certain cases (Principal Component Analysis). The coefficients that measure dependence between species (descriptors) are different according to the nature of the variables (metric, ordened non-metric, non-ordened).

Metric descriptors may be classified or ordinated with parametric dependence coefficients: the covariance s_{ki} and Pearson's correlation coefficient r_{ki} between species k and i.

$$s_{ki} = \frac{1}{p-1} \sum_j (y_{kj} - \bar{y}_k)(y_{ij} - \bar{y}_i)$$

$$r_{ki} = \frac{s_{ki}}{\sqrt{s_k^2 s_i^2}}$$

Note that the correlation coefficient is the covariance of the centered standardized variables.

Ordened non-metric descriptors (e.g., ranks, abundance classes) may be classified using the non-parametric correlation coefficients of Spearman R_s and Kendall τ:

$$R_s = 1 - \frac{6 \Sigma d_j^2}{p^3 - p}$$

in which d_j is the difference in rank of the two species in sample (station) j.

$$\tau = \frac{2S}{p^2 - p}$$

To calculate S, the two samples (stations) are ordened according to increasing rank of the species in one of the samples. One then adds one point for all $p(p-1)/2$ pairs of species of the other sample between which there is also an increase and substracts one point when there is a decrease. When there are ties (equal ranks), the formula has to be corrected (see Table 14.1 for an example of calculation).

Spearman's R_s is founded on the difference in rank occupied by the same object in the two series that have to be compared. Kendall's τ is based on the number of ranks of higher and lower order in the two series.

For non-ordened descriptors, the analysis is based on the use of contingency tables and the basic coefficient is X^2. Alternatively, the analysis may be based on the amount of information common to both descriptors B and the amount of information A and C exclusive to one of them, as calculated by the Shannon information function (see Legendre and Legendre, 1979).

Analysis of Species Abundances.--Since species abundances are normally metric, they may be classified using the parametric dependence measures described above. If the data are normalized to ln $(1+y)$, covariance and correlation coefficients are appropriate. However, a large number of double zeros will affect the analysis. This may be remedied by excluding rare species or double zeros. Another problem is that covariance and correlation measure the correlation between abundance fluctuations, the degree to which species fluctuate together. Since one can define associations on the basis of co-occurrence as well, binary coefficients such as Jaccard and Sorensen, or the point correlation-coefficient, which is based on presence (1) and absence (0) (Daget, 1976), may be used instead.

Classification or Clustering

Classification or clustering aims at grouping objects that are sufficiently similar according to some criterion and the usual product is a graph called a dendrogram.

In benthic ecology, classification is nearly always based on hierarchical methods, i.e., the ensemble of samples (stations) is subdivided or the samples (stations) are grouped successively so that members of groups of inferior rank are also members of the groups with higher rank. Non-hierarchical systems exist as well (see Lance and Williams, 1967, for a review).

Two further choices of strategy are required. Firstly, the ensemble may be progressively subdivided into groups of diminishing size (divisive strategy) or a hierarchy may be constructed by fusing individual stations progressively into groups of increasing size (agglomerative strategy). Secondly, there is a choice to be made between monothetic and polythetic strategies. In the first strategy, groups are formed based on the presence or absence of a single descriptor (species). In the polythetic strategy the dichotomies are based jointly on a number of attributes so that groups are defined in terms of the overall similarity of their members.

Divisive Methods.--Except for agglomerative-monothetic methods, the three other possible combinations are potentially useful. Divisive methods are in principle based on all the available data. The best known divisive-monothetic procedure is that of Association Analysis (Williams and Lambert, 1959). The analysis is based on binary descriptors and one looks at the species that is most associated with the others by calculating X^2_{ik} between all possible pairs of descriptors (species) i and k from 2x2 contingency tables using the familiar formula:

$$\chi^2_{ik} = \frac{p(ad-bc)^2}{(a+b)(a+c)(b+c)(b+d)}$$

with a,b,c and d the numbers in the four cells of the contingency table.

After this, the sum of X^2_{ik} for all descriptions of each descriptor k is calculated:

$$\sum_i \chi^2_{ik} \quad i \neq k$$

The largest sum corresponds to the species that is most strongly correlated with the others. The first division of the ensemble follows the descriptions of that descriptor. One forms a first group of stations coded 0 and a second group coded 1. The descriptor is then eliminated from the analysis and calculations are redone for each of the two groups separately, and so on.

Agglomerative-Polythetic Methods.--Benthic data matrices commonly have a large proportion of empty cells so that a large proportion of the data from a particular station are negative, i.e., absence of species. Treatment of benthic data is usually based on agglomerative methods. Agglomerative methods involve the successive fusion of those separate groups which are more similar according to some criterion at each stage. Two decisions have to be made in the selection of the procedure: the similarity measure, already discussed, and the strategy of fusion.

Starting with the matrix of between-group distances, the first operation is always to fuse the two most similar (nearest) stations or, if several pairs have the same smallest distance, to fuse the members of each pair. Two stations fused initially form a group and it is necessary to define the distance of other stations from this group. Lance and Williams (1964) considered five possible definitions:

(1) Nearest-neighbor: the distance between two groups is defined as the shortest distance between each possible pair of stations, one from each group.

(2) Furthest-neighbor: the distance between two groups is defined as the greatest distance between each possible pair of stations, one from each group.

(3) Centroid: a group is replaced on formation by the coordinates of its centroid, i.e., by the same number of stations each having the average composition of the group. The distance between two groups is the distance between the centroids.

(4) Median: a new group is formed as in centroid sorting but is placed midway the positions of the two groups forming it.

(5) Group-average: the distance between two groups is defined as the average distance between all possible pairs of stations, one from each group. A modification, weighted-average sorting exists, in which the group average distances of a third group to each of the two fusing groups are weighted equally.

Lance and Williams (1966) considered three properties to be important in fusing strategies:
(1) Compatible versus incompatible: measures calculated later in the analysis are of exactly the same kind as the initial measures versus not.
(2) Combinatorial strategies: consider two groups i and j forming a new group k, with n_i and n_j stations per group and intergroup distance d_{ij}. Consider a third group h. If the distance d_{hk} of groups h and k can be calculated from d_{hi}, d_{ij}, n_i, and n_j, the strategy is said to be combinatorial. Lance and Williams postulated:

$$d_{hk} = \alpha_i d_{hi} + \alpha_j d_{hj} + \beta d_{ij} + \gamma (d_{hi} - d_{hj})$$

In the following table the values for the coefficients are given:

	α_i	α_j	β	γ
Near. neighbor	1/2	1/2	0	-1/2
Furth. neighbor	1/2	1/2	0	1/2
Centroid	n_i/n_k	n_j/n_k	$-\alpha_i/\alpha_j$	0
Median	1/2	1/2	-1/4	0
Group Average	n_i/n_k	n_j/n_k	0	0
Weight. Average	1/2	1/2	0	0
Flexible Sorting	$1/2(1-\beta)$	$1/2(1-\beta)$	β	0

(3) Space-conserving or space-distorting: the initial distances between the separate stations may be regarded as defining a space with known properties. By forming groups this space may be conserved or not. When not, it may be contracted when there is a tendency for stations to join an already existing group rather than act as a nucleus for a new group, resulting in a chained hierarchy of little ecological interpretation, as in nearest-neighbor. On the other hand, with furthest-neighbor groups appearing to move away from some or all of the remaining stations and space is dilated.

The potential value of a varying degree of space distortion led Lance and Williams to propose a flexible sorting strategy with the constraints:

$$\alpha_i + \alpha_j + \beta = 1 \; ; \; \alpha_i = \alpha_j \; ; \; \beta < 1 \; ; \; \gamma = 0$$

This strategy is combinatorial and compatible for Euclidian distance (and Sorensen's coefficient) but not for the correlation coefficient. As β approaches unity, the system becomes increasingly space-contracting, as β falls to zero and becomes negative it becomes increasingly space-dilating. In most ecological practice, flexible sorting with a coefficient $\beta = -0.25$ appears to be satisfactory.

Which Strategy to Use?.--For the reader not wishing to go into all the details of this overview: in benthic ecology a now often used and satisfactory procedure is group average sorting of stations based on the Bray-Curtis similarity index calculated on double root transformed densities of the species.

Ordination

When dealing with a large number of species (descriptors) or stations (objects) the simple application of uni- or bivariate statistical methods not only is cumbersome but may be misleading as well. Multivariate methods have become indispensable tools in ecology since the general availability of large computers made their use practical. They have two basic roles: (1) to discover structure in the data, and (2) to summarize the data objectively. In contrast with classical statistics, which are concerned with hypotheses testing, ordination tries to elicit some internal structure from which hypotheses can be generated (Williams and Gillard, 1971).

Ordination is simply an operation by which objects are placed along axes that correspond to relationships of order, or on graphs formed by two or more of these axes. The relationships may be metric or not. Ordination tries to reduce the number of dimensions in which the dispersion of stations or species is represented so that the great tendencies of variability in the sample for the ensemble of all descriptors are distinguished. The dispersion of stations (objects) is first represented in a mulidimensional graph with as many axes as species (descriptors). One then looks at the projections in planes of these multi-dimensional graphs which are of most interest. These planes are defined by new axes that permit representation of the variability in the data in an optimal way in a space with reduced dimensions. Therefore, the end product of an ordination is a graph, usually two-dimensional, in which similar species or samples (or both) are near each other and dissimilar ones are far apart.

Many software packages exist in which the most important ordination techniques are included, among which BMD, SPSS, CLUSTAN and NTSYS are widely available. Other programs have been published by Blackith and Reyment (1971), Davies (1971), Orloci (1978), and Hill (1974, 1979a).

Principal Component Analysis (PCA).--The most powerful of ordination techniques, Principal Component Analysis (PCA), is most readily visualized when the stations are described by only two species (Figure 14.4). The stations may then be represented in a two-dimensional graph with the two species as axes. The relationships between the stations can be represented as well by any two other axes in the same plane. PCA in its original form works on centered data, meaning that first the origin of the axes is moved to the centroid of the data (the point representing the mean densities of the species over all stations). The analysis then proceeds by projecting the stations onto a line through the centroid and so oriented that the sum of the squared distances of

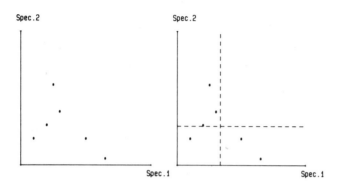

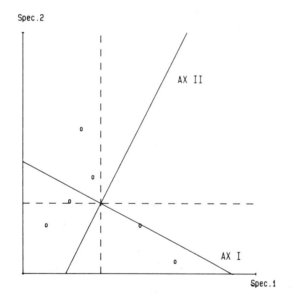

Figure 14.4.--The basic operations of Principal Component Analysis, exemplified for a two-dimensional space. Stations are plotted in this space (top left), axes are centered (top right), and rotated (bottom). The figure is fully explained in the text.

the stations to the line is minimized, or, what amounts to the same, the sum of the squared projections of the stations on the line is maximized. This line, obtained as a rotation of the original axes around the centroid, is the first principal component or axis. In the two species case, the second principal component is perpendicular to the first in the original plane. If there are more than two species, the second axis could be placed in any direction perpendicular to the first. It is so orientated that a maximum amount of the variability not explained by the first axis is explained by the second.

PCA in itself does not produce a reduction in dimensionality, since the number of components is equal to the original number of species. However, by changing from the original axes to a new orthogonal set of axes, it concentrates the variability in the data in the successively derived axes. If the first few axes extracted are accepted as adequate to display the information in the original data, then a reduction in dimensionality has been achieved.

The new axes (z) derived from PCA may be represented as algebraic functions of the old original axes (y):

$$z_i = a_{i1}y_1 + a_{i2}y_2 + \ldots + a_{in}y_n$$

for the i-th new axis in a system containing n species. The a's are constants calculated in such a way that the station's distances from the z-axes are minimized: they are called the components loadings. These loadings represent the characteristic of a particular species along a particular axis and species with similar distributions will thus have similar loadings.

In general, one has to resort to matrix algebra and a computer in order to solve for the new axes and the position of the stations. The starting point is an association matrix A of similarities between stations based on correlation coefficients or covariances/variances. Such a matrix has N eigenvalues which are the values on the diagonal of the matrix when all other elements have been made zero. Associated with each eigenvalue λ is an eigenvector V. The eigenvectors give the loadings of the n species on the p axes and thus define the principal axes. When normalized to unity they permit one to find the principal components. The value of λ for a component is proportional to the total variability accounted for by that component. When covariances are used, $\Sigma \lambda$ is equal to the total variance, when correlation coefficients are used, $\Sigma \lambda$ is equal to the number of species.

Transformation.--PCA has been developed in relation to analyses of psychological tests where centering and standardization of the data are appropriate and sometimes necessary. The use of covariances and correlations implies such transformation. However, it is not always clear in an ecological context whether these explicit or implied transformations are valid. From an extensive discussion by Noy-Meir (1973) and Grieg-Smith (1983) we retain that among several

possibilities the following may be noted:
(1) Raw data: y_{ij} : product-moment correlation.
(2) Centered data $y_{ij} - \bar{y}$: covariance.
(3) Standardized by total y_{ij}/ny_i.
(4) Standardized and centered: $y_{ij} - \bar{y}_i/s_i$: correlation.

When untransformed data are used, stations will be weighted by species richness for presence/absence data or total density for quantitative data. Centering will alter the weighings.

Other Ordination Methods

PCA.--PCA can only validly be used on cross-product similarity coefficients and relationships between stations may be more validly expressed by distance measurements. Principal coordinate analysis (Gower, 1966) permits the analysis of every association matrix Q based on a metric distance coefficient.

The calculations are as follows:
(1) The starting point is a matrix of metric distances D_{ij}.
(2) This matrix is transformed into a new matrix A by defining $a_{ij} = -\frac{1}{2}D^2_{ij}$.
(3) The new matrix A is centered to form a matrix α by calculating $\alpha_{ij} = a_{ij} - \bar{a}_i - \bar{a}_j + \bar{a}$.
(4) The eigenvalues and eigenvectors of α are calculated; the eigenvectors are standardized to the square root of their eigenvalue.
(5) The standardized eigenvectors are placed in columns: the rows in such a table are the Principal Coordinates of the stations (objects).

If an environmental gradient exists there often is a non-linearity between similarity measures and the between-station distance along the underlying gradient. This phenomenon is at the basis of most of the unsatisfactory features of PCA. Ordination based on ranking of similarity seems attractive and has been developed by Kruskal (1964) in a technique called nonmetric multidimensional scaling.

The calculations are as follows:
(1) The starting point may be every matrix of similarity or distance or even non-metric ordered scores. From this a matrix of distances D_{ij} (pxp) is calculated.
(2) A number of axes, t, thought to be sufficient is chosen a priori (often t=2).
(3) The configuration of the p objects in the reduced t-dimensional space is determined: starting from a random order, a matrix Δ with distances δ_{ij} is calculated in reduced space using an appropriate distance measure.
(4) The dispersion of distances D_{ij} in relation to distances δ_{ij} is graphed and a regression between them is calculated.
(5) The regression coefficients aid in the calculation of an approximation $\hat{D}_{ij}$ of the value necessary to conserve the distance relationship monotonously. The values form a new matrix $\hat{D}$.
(6) One then calculates the "stress" S between the matrices $\hat{D}$ and Δ and repeats steps (4) to (6) until a minimum value of this stress is found.

$$S = \sqrt{\sum_i \sum_j (\delta_{ij} - \hat{D}_{ij})^2 / \sum_i \sum_j (\delta_{ij} - \bar{\delta})^2}$$

Reciprocal Averaging and Detrended Correspondance Analysis.--Reciprocal Averaging (Hill, 1974) is a form of PCA in which the species ordination scores are averages of the station ordination scores and vice versa. It amounts to a double standardization and has considerable advantages in that being non-centered it is efficient with heterogeneous data but has two faults. It shows the arch or horseshoe effect: a linear gradient of composition is expressed as an arch in two dimensions of the ordination. The second fault is that equivalent differences in composition are not represented by the same differences in first axis position (see Gauch, 1984, for an extensive discussion).

The modified version of reciprocal averaging, detrended correspondance analysis (DCA) (Hill, 1979a) is now one of the most widely used ordination methods in benthic ecology. The arch effect is removed by adjusting the values on the second axis in successive segments by centering them to zero mean. The variable scaling on the axes is corrected by adjusting the variance of species cores within stations to a constant value. The software for this ordination technique is available in the package DECORANA.

The calculation is as follows:
(1) Each abundance value y_{ij} of the matrix is standardized according to rows and columns:

$$\frac{y_{ij}}{\sqrt{r_i c_j}}$$

in which

$$r_i = \sum_i y_{ij} \text{ and } c_j = \sum_j y_{ij}$$

(2) The matrix of covariances S is then calculated.
(3) The eigenvalues and eigenvectors of S are extracted.

The direct iteration method (Hill, 1974) is an efficient algorithm for obtaining one to several axes. One starts by assigning arbitrary species ordination scores. The weighted averages are used to obtain sample scores from these species scores. The second iteration produces new species scores by weighted averages of the sample scores, and so on. Iterations are continued until the scores stabilize. The scores converge to a unique solution, not affected by the initial, arbitrary scores.

Ordination - Space Partitioning

A set of very powerful methods has been developed in which the data are first subjected to an ordination and then to a divisive classification. They are explained fully in Pielou (1984) on which this account is based.

The Minimum Spanning Tree.--When the p stations are plotted in a n-dimensional species space, the points of the swarm may be linked by a minimum spanning tree, a set of line segments linking all the n points in the swarm in such a way that every pair of points is linked by one and only one path. None of the paths form closed loops. The length of the tree is the sum of the lengths of its constituent line segments. The minimum spanning tree of the swarm is the tree with the minimum length. When this tree is formed, it is divided by cutting in succession first the longest link in the tree, then the second longest link etc.

This is done starting from a pxp matrix of distances. The first segment in the tree is the shortest distance in the matrix. Next, the shortest distance linking a third point to either one of the first two points is found. Next, the shortest distance linking a fourth point to either one of the first three points is found, etc. (this procedure amounts to nearest neighbor clustering).

Lefkovitch's Partitioning Method.--This method consists in first ordering the data in n-space by means of a centered PCA. Then the first division is made by breaking the first axis at the centroid, the second by breaking the second axis at the centroid, etc.

TWINSPAN.--This method consists of carrying out a one-dimensional Reciprocal Averaging ordination and breaking the axis at the centroid so as to divide the data points into two classes. Each of the two classes is then split in the same way, after a RA ordination, etc. (Hill, 1979b).

Part 4. Species-Abundance Distributions

If one records the abundances of different species in a community, invariably one finds that some species are rare, whereas others are more abundant. This feature of ecological communities is independent of the taxonomic group(s) or the area(s) investigated.

An important goal of ecology is to be able to describe consistent patterns in different communities, and explain them in terms of biotic and abiotic interactions.

Although "community" is defined as the total set of organisms in an ecological unit (biotope), it must be specified as to the actual situation. No entities exist within the biosphere with absolutely closed boundaries, i.e., without interactions with other parts. Some kind of arbitrary boundaries should always be drawn. Pielou (1975) recommends to specify explicitly the following features: (1) The spatial boundaries of the area or volume containing the community and the sampling methods, (2) the time limits between which observations were made, (3) the taxocene (i.e., the set of species belonging to the same taxon) treated as constituting the community.

The results of a sampling program of the community take the form of species lists, indicating for each species a measure of its abundance (usually number of individuals per unit surface, although other measures, such as biomass, are possible). Many methods are used to plot these data. The method chosen often depends on the kind of model one wishes to fit to the data. Different plots of the same (hypothetical) data set are shown in Figure 14.5.

It can readily be seen that a bewildering variety of plots is used. They yield quite different visual pictures, although they all represent the same data set. Figure 14.5a-d are variants of the Ranked Species Abundance (*RSA*) curves. The S species are ranked from 1 (most abundant) to S (least abundant). Density (often transformed to percentage of the total number of individuals N) is plotted against species rank. Both axes may be on logarithmic scales. It is especially interesting to use a log-scale for the Y-axis, since then the same units on the Y-axis may be used to plot percentages and absolute numbers (there is only a vertical translation of the plot).

In so-called "k-dominance" curves (Lambshead et al., 1983) (Figure 14.5e-f) the cumulative percentage (i.e., the percentage of total abundance made up by the kth-most dominant plus all more dominant species) is plotted against rank k, or log rank k. To facilitate comparison between communities with different numbers of species S, a "Lorenzen curve" may be plotted. Here the species rank k is transformed to $(k/S) * 100$. Thus the X-axis always ranges between 0 and 100 (Figure 14.5g).

The "collector's curve" (Figure 14.5h) addresses a different problem. When one increases the sampling effort, and thus the number of animals N caught, new species will appear in the collection. A collector's curve expresses the number of species as a function of the number of specimens caught. Collector's curves tend to flatten out as more

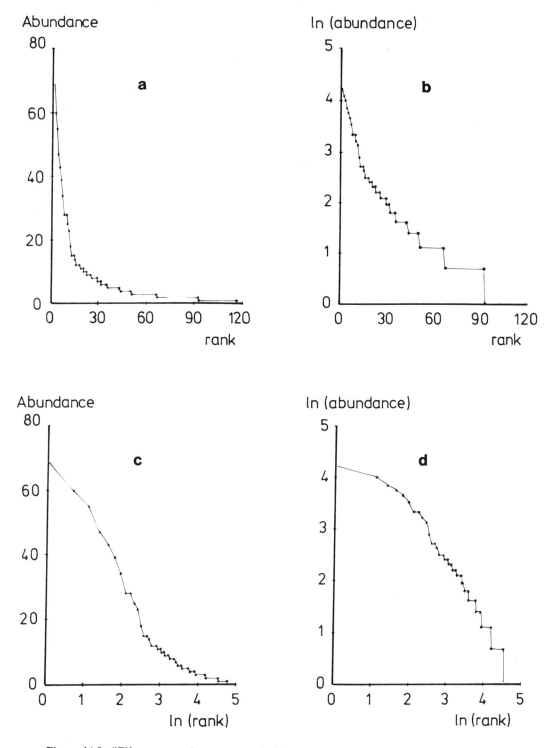

Figure 14.5.—"Fifty ways to lose your reader" in representing species-abundance data. The different figures represent the same species-abundance data. The curves are explained in detail in the text. a-d, Ranked Species Abundance curves with none, one, or both axes on a log scale.

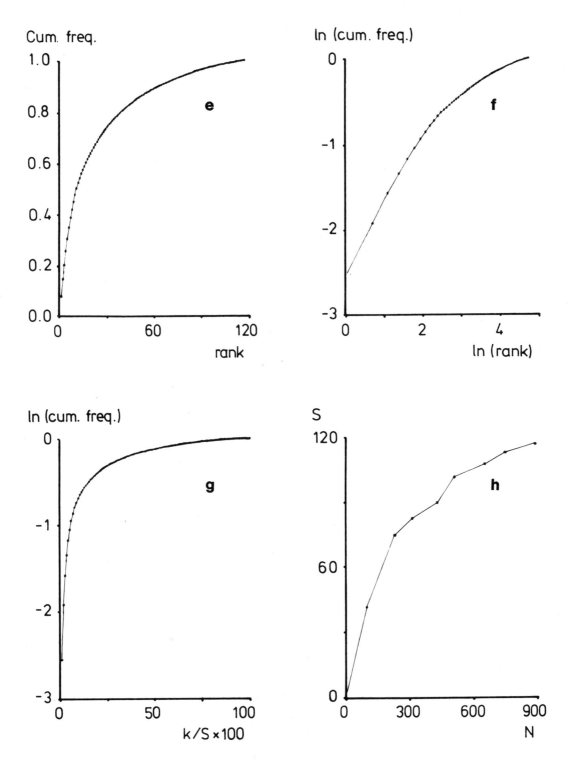

Figure 14.5 (continued).-- e-f, k-dominance curves; g, "Lorenzen curve"; h, collector's curve.

CHAPTER 14: DATA PROCESSING, EVALUATION, AND ANALYSIS 215

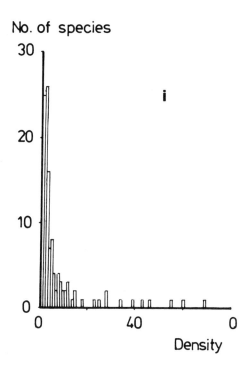
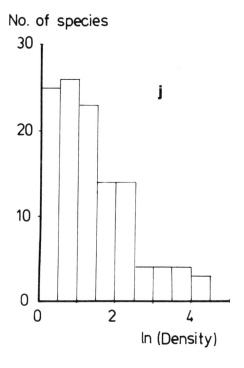
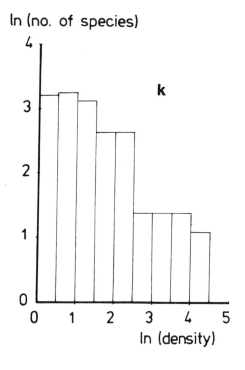
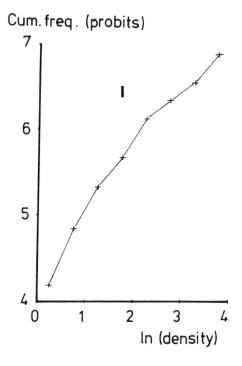

Figure 14.5 (continued).—i-k, species abundance distributions; l, cumulative species abundance distribution on probit scale.

specimens are caught. However, due to the vague boundaries of ecological communities they often do not reach an asymptotic value: as sampling effort (and area) is increased, so is the number of slightly differing patches.

The plots in Figure 14.5i-k are species abundance distributions. They can only be drawn if the collection is large, and contains many species (a practical limit is approximately $S>30$). Basically, a species-abundance distribution (Figure 14.5i) plots the number of species that are represented by $r = 0, 1, 2, \ldots$ individuals against the abundance r. Thus, in Figure 14.5i there were 25 species with 1 individual, 26 species with 2 individuals, etc. More often than not, the species are grouped in logarithmic density classes. Thus one records the number of species with density e.g., between 1 and $\exp(0.5)$, between $\exp(0.5)$ and $\exp(1)$, etc. (Figure 14.5j). A practice, dating back to Preston (1948), is to use logarithms to the base 2. Thus one has the abundance boundaries 1, 2, 4, 8, 16, etc. Although these so-called "octaves" are still used, they have two disadvantages: the class boundaries are integers, which necessitates decisions as to which class a species with an abundance equal to a class mark belongs; the theoretical formulation of models is "cluttered" (May, 1975) by factors $\ln(2)$, which would vanish if natural logs were used. The ordinate of species-abundance distributions may be linear or logarithmic. Often one plots the cumulative number of species in a density group and all less abundant density groups on a probit scale (Figure 14.5l).

Species-Abundance Models

Two kinds of models have been devised to describe the relative abundances of species. "Resource apportioning models" make assumptions about the division of some limiting resource among species. From these assumptions a ranked abundance list or a species-abundance distribution is derived. The resource apportioning models have mainly historical interest. In fact, observed species abundance patterns cannot be used to validate or discard a particular model, as has been extensively argued by Pielou (1975, 1981). One should consult these important publications before trying to validate or refute fortuitously a certain model!

"Statistical models" make assumptions about the probability distributions of the numbers in the several species within the community, and derive species-abundance distributions from these.

The Niche Preemption Model (Geometric Series Ranked Abundance List).--This resource apportioning model was originally proposed by Motomura (1932). It assumes that a species preempts a fraction k of a limiting resource, a second species the same fraction k of the remainder, and so on. If the abundances of the species are proportional to their share of the resource, the ranked-abundance list is given by a geometric series:

$$k, k(1-k), \ldots, k(1-k)^{(s-2)}, k(1-k)^{(s-1)}$$

where S is the number of species in the community. May (1975) derives the species abundance distribution from this ranked abundance list (see also Pielou, 1975).

The geometric series yields a straight line on a plot of log abundance against rank. The communities described by it are very uneven, with high dominance of the most abundant species. It is not very often found in nature. Whittaker (1972) found it in plant communities in harsh environments or early successional stages.

The Negative Exponential Distribution (Broken Stick Model).--A negative exponential species abundance distribution is given by the probability density function:

$$\Psi(y) = S\, e^{-Sy}$$

Stated as such, it is a statistical model, an assumption about the probability distribution of the numbers in each species. However, it can be shown (Webb, 1974) that this probability density function can be arrived at via the "broken-stick model" (MacArthur, 1957).

A limiting resource is compared with a stick, broken in S parts at $S-1$ randomly located points. The length of the parts is taken as representative for the size of the S species subdividing the limiting resource. If the S species are ranked according to size, the expected size of species i, y_i, is given by:

$$E(y_i) = \frac{1}{S} \sum_{x=1}^{S} \frac{1}{x}$$

The negative exponential distribution is not often found in nature. It describes a too even distribution of individuals over species to be a good representation of natural communities. According to Frontier (1985) it is mainly appropriate to describe the right-hand side of the rank frequency curve, i.e., the distribution of the rare species. As these are the most poorly sampled, their frequencies depend more on the random elements of the sampling, than on an intrinsic distribution of the frequencies.

Pielou (1975, 1981) showed that a fit of the negative exponential distribution to a field sample does not prove that the mechanism modeled by the broken-stick model governs the species-abundance pattern in the community. Moreover, the broken-stick model is not the only mechanism leading to this distribution. The same prediction of relative abundance can be derived by at least three other models besides the niche partitioning one originally used (Cohen, 1968; Webb, 1974).

The observation of this distribution does indicate (May, 1975) that some major factor is being somewhat evenly apportioned among the community's constituent species (in contrast to the lognormal distribution, which suggests the interplay of many independent factors).

The Log-series Distribution.--The log-series was originally proposed by Fisher et al. (1943) to describe species abundance distributions in large moth collections. The expected number of species with r individuals, E_r, is given as:

$$E_r = \alpha \frac{X^r}{r}$$

(r = 1,2,3,....). α ($\alpha > 0$) is a parameter independent of the sample size (provided a representative sample is taken), for which X ($0 < X < 1$) is the representative parameter. The parameters α and X can be estimated by maximum likelihood (Kempton and Taylor, 1974), but are conveniently estimated as the solutions of:

$$S = -\alpha \ln (1-X)$$

and

$$N = \frac{\alpha X}{1-X}$$

The parameter α, being independent of sample size, has the attractive property that it may be used as a diversity statistic (see further). An estimator of the variance of $\hat{\alpha}$ is given as:

$$\hat{var} (\hat{\alpha}) \cong \frac{\hat{\alpha}}{-\ln X(1-X)} \quad \text{(Anscombe, 1950).}$$

Kempton and Taylor (1974) give a detailed derivation of the log-series distribution. It was fitted to data from a large variety of communities (e.g., Williams, 1964; Kempton and Taylor, 1974). However, it seems to be generally less flexible than the log-normal distribution. In particular, it cannot account for a mode in the species-abundance distribution, a feature often found in a collection. According to the log-series model, there are always more species represented by 1 individual than there are with 2. The truncated log-normal distribution can be fitted to samples with or without a mode in the distribution.

Caswell (1976) derived the log-series distribution as the result of a neutral model, i.e., a model in which the species abundances are governed entirely by stochastic immigration, emigration, birth and death processes, and not by competition, predation or other specific biotic interactions. He proposes to use this distribution as a "yardstick," with which to measure the occurrence and importance of interspecific interactions in an actual community. Other models have been proposed to generate the log-series distribution (Boswell and Patil, 1971) but they all contain the essentially neutral element as to the biological interactions. However, the proof that any form of biological interaction will yield deviation from the log-series is not given. Neither is it proven that "neutral" communities cannot deviate from the log-series. Therefore, we think that the fit of this distribution cannot be considered as a waterproof test for species interactions.

The Log-normal Distribution.--Preston (1948) first suggested the use of a log-normal distribution for the description of species-abundance distributions. It was shown by May (1975) that a log-normal distribution may be expected, when a large number of independent environmental factors act multiplicatively on the abundances of the species (see also Pielou, 1975).

When the species-abundance distribution is log-normal, the probability density function of y, the abundance of the species, is given by:

$$\Psi(y) = \frac{1}{y \sqrt{2\pi V_z}} \exp \left(\frac{-(\ln y - \mu_z)^2}{2 V_z} \right)$$

The mean and variance of y are:

$$\mu_y = \exp (\mu_z + \frac{V_z}{2})$$

and

$$V_y = (\exp(V_z) - 1) \exp(2\mu_z + V_z)$$

where μ_y and V_z are the mean and variance of $z = \ln(y)$.

If the species abundances are lognormally distributed, and if the community is so exhaustively sampled that all the species in the community (denoted S^*) are represented in the sample, the graph of the cumulative number of species on a probit scale (Figure 14.5l) against log abundance will be a straight line. This is not normally the case.

In a limited sampling a certain number of species S^*-S will be unrepresented in the sample (S being the number of species in the sample). The log-normal distribution is said to be truncated. In the terminology of Preston (1948) certain species are hidden behind a "veil line" (see insets in Figure 14.6), it follows that it is not a good practice to estimate the parameters of the lognormal distribution from a cumulative plot on probit scale. In fact if one does not estimate the number of unsampled species, it is impossible to estimate the proportion of the total number of species in a particular log density class. Species abundances that are lognormally distributed will not yield straight lines if one takes into account only the species sampled (see Figure 14.6). Note also that the normal regression analysis is not applicable to highly correlated values such as cumulative frequencies. (If the frequencies are replaced by evenly distributed random numbers, their cumulative values on probit scale still yield very "significant" correlations with log abundance). In order to fit a lognormal distribution it is absolutely necessary to estimate the number of unrepresented species, S^*-S.

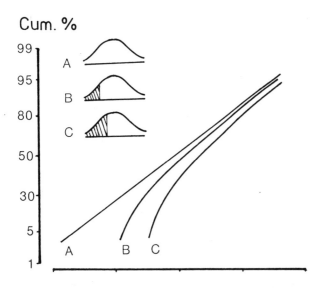

Figure 14.6.--Effect on the probability plot of cumulative % of species vs. abundance of truncating the log-normal distribution (A), by 15% (B), and 30% (C). The hatched areas in the insets represent the unsampled portion of the species (i.e., the species hidden behind the "veil line"). Scale of x-axis arbitrary. After Shaw et al. (1983).

In fitting the log-normal two procedures are used (apart from the wrong one already discussed). The conceptually most sound method is to regard the observed abundances of species j as a Poisson variate with mean λ_j, where the λ_j's are lognormally distributed. The probability, P_r, that a species contains r individuals is then given by the Poisson log-normal distribution (see Bulmer, 1974). P_r can be solved approximately for $r > 10$, but must be integrated numerically for smaller values of r. Bulmer (1974) discusses the fitting to the data by maximum likelihood. Pielou (1975) argues that the fitting of the Poisson lognormal, though computationally troublesome, is not materially better than the alternative procedure, consisting in the direct fitting of the continuous lognormal. The complete procedure in recipe-form is given in Pielou (1975).

The Mandelbrot Model.--This relatively flexible model was derived in information science to model rank-frequency curves of messages in complicated systems (e.g., words in a natural language). It describes the frequency of a species with rank r as:

$$f_r = f_o(r+\beta)^{-\gamma}$$

in which $-\gamma$ is the slope of the asymptote towards which the curve approximates; β is related to the deviation at the left-hand side of the curve (Figure 14.7). This model is extensively discussed in Frontier (1985), where useful references may be found. The model is particularly useful to describe the rank-frequencies of the dominant species in a community. However, for the rare species large deviations may be found.

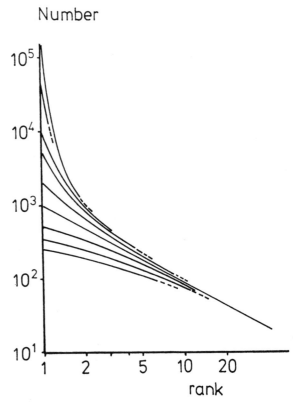

Figure 14.7.--Curves for the Mandelbrot model on a log-log scale, with $N_0 = 1000$, $\gamma = 1$, and the following values for β (from top down): -1, -0.98, -0.8, -0.5, 0 (straight line), 1, 2, 3. After Frontier (1985).

On Fitting Species-abundance Distributions.--
Ever since Fisher et al. (1943) used the log-series, and Preston (1948) proposed the log-normal to describe species-abundance patterns, ecologists have been debating about which model is the most appropriate. Especially the log-normal and the log-series have (had) their fan-clubs, recently also among benthologists (e.g., Shaw et al., 1983; Gray, 1983, and other papers). In our opinion, these debates are spurious. As Pielou (1975) remarked, the fact that, e.g., the log-normal fits well in many instances, tells us more about the versatility of the log-normal than about the ecology of these communities. Although most of the distributions have a kind of biological rationale (to make them more appealing to a biological audience?) the fact that they fit does not prove that the "biological" model behind them is valid in the community.

The fitting of a model to field data is meaningful if the parameter estimates are to be used in further analysis. This is analogous to the use of the normal distribution in ANOVA: in order to perform an ANOVA, the data should be normally distributed. Of course this must be checked, but only as a preliminary condition. No one draws conclusions from the fit or non-fit of the normal distribution to experimental data, but from the test performed afterwards. Similarly, if a particular model fits reasonably well to a set of field data, the parameter estimates can be used, e.g., in respect to the diversity of the communities.

Indices of Diversity and Evenness

It is common practice among ecologists to complete the description of a community by one or two numbers expressing the "diversity" or "evenness" of the community. For this purpose a bewildering *diversity* of diversity indices have been used or proposed.

Two different aspects are generally accepted to contribute to the intuitive concept of diversity of a community. These are (following the terminology of Peet, 1974) *species richness*, a measure somehow related to the total number of species in the community (note that the actual number of species in the community is usually unmeasurable), and *equitability*, which expresses how evenly the individuals are distributed among the different species. Some indices, called *heterogeneity* indices by Peet (1974), incorporate both aspects.

It has been clearly demonstrated (May, 1975) that no single diversity index can summarize the species-abundance distribution in a community. However, this is seldom a goal in itself in ecology. Usually, one tries to show how some characteristic feature(s) of the ecosystem may change in relation to evironmental variables. Depending on the situation, several indices may give a good indication of these relations. Anyway, it is useful to keep in mind that a complete specification of the species-abundance relationship contains more information than a single index.

Indices Derived from Species-abundance Distributions

Historically, the first diversity measure was derived by Fisher, Corbet, and Williams (1943) as a result of the derivation of the log-series distribution. As mentioned earlier, the parameter α of the log-series distribution is independent of sample size. From the first equation under the topic of log-series distribution, it is easily seen that α is the only parameter describing the relationship between number of individuals and number of species in the sample. Thus, this parameter describes the way in which the individuals are divided among the species, which is a measure of diversity. Note that, in adopting the log-series model for the species-abundance distribution, the equitability is already specified, so that α only measures the relative species richness of the community. α, as determined by the fitting of the log-series model to the sample, is only valid as a diversity index when the log-series fits well to the data. The same reasoning can be extended to the log-normal distribution. Preston (1948) expressed the diversity as the (calculated) total number of species in the community, S^*.

The use of the log-series α was taken up again, and extended by Kempton and Taylor (1974). Taylor et al. (1976) showed that, when the log-series fits the data reasonably well, α has a number of attractive properties. The most important of these was that (compared to the information statistic H' and Simpson's index; see below) it provided a better discrimination between sites, it remained more constant within each site (all sites were sampled in several consecutive years), it was less sensitive to density fluctuations in the commonest species, and it was normally distributed. On the other hand, when the data deviate from the log-series, α is more dependent on sample size than the other indices.

Kempton and Taylor (1976), and Kempton and Wedderburn (1978) extended this approach, by noting that for the log-series distribution the parameter α is the asymptotic expectation of Q, a "mid-range statistic," defined as:

$$Q = (S^*/2) / \log(P_{S^*/4} / P_{3S^*/4})$$

where S^* is the total number of species in the community and the proportional abundances P_i ($i = 1, 2, ... S^*$) are arranged in descending order of size.

For discrete data this index is estimated from the sample statistic:

$$a_Q = \frac{\frac{1}{2}n_{R_1} + \sum_{r=R_1+1}^{R_2-1} n_r + \frac{1}{2}n_{R_2}}{\log(R_2/R_1)}$$

where the sample quartiles, R1 and R2 are chosen such that:

$$\sum_{r=1}^{R_1-1} n_r < \frac{S^*}{4} \leq \sum_{1}^{R_1} n_r$$

$$\sum_{r=1}^{R_2-1} n_r \leq \frac{3S^*}{4} < \sum_{1}^{R_2} n_r$$

and n_r is the number of species with abundance r. It can be seen from Figure 14.8 that Q depends mostly on the abundance of the moderately abundant species. Q may either be estimated directly from the data or alternatively from the parameter estimates

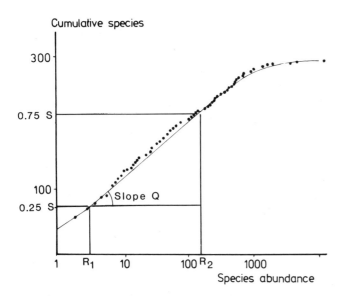

Figure 14.8.—Graphic determination of the "mid-range statistic" Q from a plot of cumulative number of species vs. log (species abundance). Q can be seen to approximately equal the slope of this curve for the moderately abundant species. After Kempton and Wedderburn (1978).

of a fitted species-abundance distribution. For the log-series, $\hat{Q} = \alpha$; for the log-normal, $\hat{Q} = 0.371\, S^*/\hat{\sigma}$ provides a reasonable approximation if more than 50 % of the species are represented in the sample. If Q is estimated in a series of samples by fitting the same model to each of the data sets, it is an index of species richness (or, if the fit is too bad or the sample too small, of nothing anymore). However, if Q is estimated directly from the data, it incorporates both species richness and equitability.

Rarefaction

An obvious index of species richness is the number of species in the sample. However, it is clear that this measure is highly correlated with sample size, an undesirable property. Sanders (1968) proposed a method to reduce samples of different sizes to a standard size, so as to make them comparable in terms of the number of species. The formula used by Sanders (1968) was corrected by Hurlbert (1971), who showed that the expected number of species in a sample of size n is given by:

$$E(S) = \sum_{i=1}^{S} \left[1 - \binom{N-N_i}{n} / \binom{N}{n} \right]$$

where N_i is the number of individuals in the i-th species in the full sample, which had sample size N and contained S species. Alternatively, random samples can be drawn by computer from the original sample (Simberloff, 1972).

Hill's (1973) Diversity Numbers

Hill (1973) provided a generalized notation that includes, as a special case, two often used heterogeneity indices. Hill defined a set of "diversity numbers" of different order. The diversity number of order a is defined as:

$$N_a = \left(\sum_i p_i^a \right)^{1/(1-1)}$$

where p_i is the proportional abundance of species i in the sample.

N_0 can be seen to equal S, the number of species in the sample.

N_1 is undefined by the first of these equations. However, defining

$$N_1 = \lim_{a \to 1} (N_a)$$

it can be shown that

$$N_1 = \exp\left(-\sum_i p_i (\ln p_i)\right)$$

$$= \exp(H')$$

where H' is the well known Shannon-Wiener

diversity index:

$$H' = -\sum_i p_i (\log p_i)$$

Note that in the usual definition of the Shannon-Wiener diversity index, logarithms to the base 2 are used. Diversity then has the peculiar units "bits ind^{-1}". The diversity number N_1 is expressed in much more natural units. It gives an equivalent number of species, i.e., the number of species S' that yields N_1 if all species contain the same number of individuals, and thus if all $p_1 = 1/S'$. This can be seen in the last equation which in this case reverts to:

$$N_1 = \exp(-\ln(1/S')) = S'$$

An additional advantage of N_1 over H' is that it is approximately normally distributed. It has been argued (see e.g., Pielou, 1975) that for small, fully censused communities the Brillouin index should be used. This index is given by:

$$H = \frac{1}{N} \log \frac{N!}{\prod_i N_i!}$$

We do not recommend this index for meiofaunal assemblages. The information theoretical argument for its use should be regarded as allegoric: it has no real bearing to ecological theory. Furthermore, finite collections that are non-destructively sampled do not occur in meiobenthic research. The most compelling argument, however, is given by Peet (1974), who shows with an example that the Brillouin index has counterintuitive properties: depending on sample size, it can yield higher values for less evenly distributed communities.

The next diversity number, N_2, is the reciprocal of Simpson's dominance index, which is given by:

$$\lambda = \sum_i p_i^2$$

for large, sampled, communities. If one samples at random and without replacement, 2 individuals from the community, Simpson's index expresses the probability that they belong to the same species. Obviously, the less diverse the community is, the higher is this probability.

In small, fully censused communities, the correct expression for Simpson's index is:

$$\lambda = \sum_i \frac{N_i(N_i-1)}{N(N-1)}$$

where N_i = number of individuals in species i, and N is the total number of individuals in the community. Pielou (1969) shows that for large communities which are sampled, the first equation is a biased estimator of λ. An unbiased estimator is given by the last equation, where λ and N are then sample values, not parametric values as in the case of fully sampled communities.

In order to convert Simpson's dominance index to a diversity statistic it is better to take reciprocals, as is done in Hill's N_i, than to take $1-\lambda$. In that way the diversity number is again expressed as an equivalent number of species.

The diversity number of order + infinity ($N_{+\infty}$) is equal to the reciprocal of the proportional abundance of the commonest species. It is the "dominance index." May (1975) showed that it characterizes the species-abundance distribution "as good as any, and better than most" single diversity indices. Recently it has received some attention in the context of pollution monitoring (Shaw et al., 1983.

Hill (1973) showed that the diversity numbers of different orders probe different aspects of the community. The number of order $+\infty$ infinity only takes into account the commonest species. At the other extreme, $N_{-\infty}$ is the reciprocal of the proportional abundance of the rarest species, ignoring the more common ones. The numbers N_0, N_1, and N_2, are in between in this spectrum. N_2 gives more weight to the abundance of common species (and is, thus, less influenced by the addition or deletion of some rare species) than N_1. This, in turn, gives less weight to the rare species than N_0, which, in fact, weighs all species equally, independent of their abundance. It is good practice to give diversity numbers of different order when characterizing a community. Moreover, these numbers are useful in calculating equitability (see below).

The Subdivision of Diversity

Hierarchical Subdivision. -- In the calculation of diversity indices, all species are considered as different, but equivalent: one is not concerned with the relative differences between species. However, in nature some species are much more closely related to some other species than to the rest of the community. This relation may be considered according to different criteria, e.g., taxonomic relationships, general morphological types, trophic types, etc. Therefore, it may be desirable to subdivide the total diversity in a community in a hierarchical way. Pielou (1969) shows how the Shannon-Wiener diversity H' can be subdivided in a hierarchical way. The species are grouped in genera, and the total diversity H'_t equals:

$$H'_T = H'_g + H'_{wg}$$

where

$$H'_g = -\sum_i q_i \log q_i$$

is the between-genera diversity, and

$$H'_{wg} = \sum_i q_i \left(-\sum_j r_{ij} \log r_{ij}\right)$$

is the average within-genus diversity.

The same procedure may be repeated to partition the between-genera diversity into between-families and average within-family diversity.

This approach was generalized by Routledge (1979) who showed that the only diversity indices that can be consistently subdivided are the diversity numbers of Hill (1973) (of which H' can be considered a member, taking into account the exponential transformation). The decomposition formulae are:

$$\left(\sum_i \sum_j t_{ij}^a\right)^{1/(1-a)} = \left(\sum_i q_i^a\right)^{1/(1-a)} *$$

$$* \left\{\left(\sum_i q_i^a \sum_j r_{ij}^a\right) / \sum_i q_i^a\right\}^{1/(1-a)}$$

for $a \neq 1$

and

$$\prod_i \prod_j t_{ij}^{-t_{ij}} = \prod_i q_i^{-q_i} \prod_i \left(\prod_j r_{ij}^{-r_{ij}}\right)^{q_i}$$

for $a = 1$

In these equations, q = the proportional abundance of the group (e.g., genus) i, r_{ij} = the proportional abundance of species j in group i, and t_{ij} = the proportional abundance of species j (belonging to group i) relative to the whole community:

$$t_{ij} = q_i r_{ij} \quad \text{and} \quad q_i = \sum_k t_{ik}$$

It can be seen that the community diversity is calculated as the *product* of the group diversity and the average diversity within groups, weighted by the proportional abundance of the groups. Note that this is consistent with Pielou's formulae noted in the previous equations since $N_1 = \exp(H')$.

The hierarchical subdivision of diversity may be useful to study the differences in diversity between two assemblages. Is a higher diversity in one assemblage mainly attributable to the addition of some higher taxa (suggestive of the addition of new types of niches), or rather to a diversifying of the same higher taxa that are present in the low-diversity assemblage?

It may also be useful to study groups other than taxonomic ones. Natural ecological groupings, such as the feeding types of nematodes (Wieser, 1953) or the body types of harpacticoids (Hicks and Coull, 1983) may be particularly interesting. Heip et al. (1984, 1985) used as a "trophic diversity index" to describe the diversity in feeding types of nematodes, the index:

$$\Theta = \sum_{i=1}^{4} \vartheta_i^2$$

where q_i is the proportion of feeding type i in the assemblage. This index can be seen to be the reciprocal of N_2 (for estimation purposes, it would be better to use the equation for Simpson's index although the bias is small when 200 nematodes are sampled). This approach can be naturally extended, using last equation to (1) diversity indices of other orders, and (2) achieve a more complete description of the assemblage, by a subdivision of the total diversity.

Spatio-Temporal Diversity Components.--All ecological communities are variable at a range of spatio-temporal scales. Thus if one examines a set of samples (necessarily) taken at different points in space, and possibly also in time, and calculates an overall diversity index, it is unclear what is actually measured. Whereas diversity may be small in small patches at a particular instant, additional diversity may be added by the inclusion in the samples of diversity components due to spatial or temporal patterns.

Following Whittaker (1972) one often discerns for the spatial component: (1) *Alpha*-diversity – the diversity within a uniform habitat (patch); (2) *Beta*-diversity – the rate and extent of change in species composition form one habitat to another (e.g., along a gradient); and (3) *Gamma*-diversity – the diversity in a geographical area (e.g., the intertidal range of a salt marsh).

The subdivision of total diversity H' in ecological components is discussed by Allen (1975). He treats a sampling scheme where S species are sampled in q sites, each consisting of r microhabitats. The problem is different from a hierarchical subdivision, since the same species may occur in different microhabitats and sites (it can, of course, only belong to one genus, one family, etc., in hierarchical subdivision). Allen (1975) presents two solutions. One can treat the populations of the same species in different microhabitats as the fundamental entities. Total diversity is then calculated on the basis of the proportional abundance (in relation to the total abundance in the study) of these populations. This total diversity can then be subdivided hierarchically.

Alternatively, one can subdivide the species diversity in the total study in average within microhabitat diversity, average between microhabitat (within site) diversity, and average between site diversity components. The latter computations are generalized for Hill's (1973) diversity numbers by Routledge (1979).

Equitability

Several equations have been proposed to calculate equitability (evenness) from heterogeneity measures. The most frequently used measures, which converge for large samples (Peet, 1974) are:

$$E = (D-D_{min})/(D_{max}-D_{min})$$

and

$$E = D/D_{max}$$

where D is a heterogeneity index, and D_{min} and D_{max} are the lowest and highest values of this index for the given species number and the sample size. To this class belongs Pielou's $J = H'/H'_{max} = H'/\log S$.

As discussed by Peet (1974) these measures depend on a correct estimation of S^*, the number of species in the community. It is quasi impossible to estimate this parameter. Substituting S, the number of species in the sample, makes the equitability index highly dependent on sample size. It also becomes very sensitive to the chance inclusion-exclusion of rare species in the sample.

Hill (1973) proposed to use ratios of the form:

$$E_{a:b} = N_a/N_b$$

as equitability indices (where N_a and N_b are diversity numbers of order a and b respectively). Note that $H' - H'_{max} = \ln(N_1 N_0)$ belongs to this class, but that $J' = H'/H'_{max}$ does not. These ratios are shown to possess superior characteristics, compared with J'. Hill (1973) also showed that in an idealized community, where the hypothesized number of species is infinite and the sampling is perfectly random, $E_{1:0}$ is always dependent on sample size. $E_{2:1}$ stabilizes, with increasing sample size, to a true community value. However, in practice all measures depend on sample size.

Heip (1974) proposed to change the index $E_{1:0} = e^{H'}/S$ to $(e^{H'}-1)/(S-1)$. In this way the index tends to 0 as the equitability decreases in species-poor communities. Due to a generally observed correlation between equitability and number of species in a sample, $E_{1:0}$ tends to $1/S$ as both $e^{H'} \longrightarrow 1$ and $S \longrightarrow 1$.

In general, one cannot attach too much importance to equitability indices. Species-abundance distributions show more information about the equitability than any single index.

The Choice of an Index

Lambshead et al. (1983) have noted that, whenever two k-dominance curves do not intersect, all diversity indices will yield a higher diversity for the sample represented by the lower curve. Equivocal results arise as soon as the k-dominance curves intersect. Different measures of diversity are more sensitive to either the commonest or the rarest species (see Hill's diversity numbers). An elegant approach to the analysis of this sensitivity is provided by the response curves of Peet (1974).

In order to summarize the diversity characteristics of a sampled community, it is advisable to provide the diversity numbers N_0, N_1, N_2, possibly also N_{+00}, the dominance index. If permitted by the sampling scheme, one can use these indices in a study of hierarchical and/or spatio-temporal components of diversity. Equitability indices should be regarded with caution. Hill's (1973) ratios $E_{a:b}$ seem preferable if an index is desired. However, one should rather use species-abundance plots to study equitability.

In any case, it should be remembered that the indices depend on sample size, sample strategy (e.g., location of the samples in space and time), spatio-temporal structure of the community, and sampling error.

Although formulae for the estimation of the variance of H' have been proposed, these do not include all these sources of error. Frontier (1983) estimated the background noise in a time series of H' in plankton samples as roughly one unit (bit/ind.). However, in a species-poor community of meiobenthic copepods, Heip and Engels (1974) found that the variance of H' was conservatively estimated by the formula of Hutcheson (1970). Typical coefficients of variation (s/x) were about 0.3. No index was normally distributed.

The diversity statistics derived from species-abundance patterns are less sensitive to the inclusion/ exclusion of rarities, and to the density variations of the most abundant species. They may be more useful in the description of inherent diversity characteristics of communities, and in the discrimination between sites (see Kempton and

Wedderburn, 1978). On the other hand, large samples are required for their estimation. Thus, a relatively large number of environmental patches may be pooled, without the possibility of separation.

Finally, we should stress the possibilities and limitations of diversity indices. A diversity index must be regarded as a summary of a structural aspect of the assemblage. As has been stressed throughout this chapter, different indices summarize slightly different aspects. In comparing different assemblages, it is useful to compare several indices: this will indicate specific structural differences.

The variability of an index within one assemblage cannot be estimated very well. As has been indicated, the estimate of the variance of an index (Hutcheson, 1970) should be used with caution. Spatio-temporal variability of the assemblage should preferably be included in the study, in order to evaluate the variability of the index. Indices may then be transformed (if necessary) so that they are approximately normally distributed, and compared using standard statistical techniques.

A diversity index summarizes the structure, not the functioning of a community. Thus, it is very well possible that two assemblages have a similar diversity, whereas the mechanisms leading to their structures are completely different (e.g., Coull and Fleeger, 1977). Often these functional aspects cannot readily be studied by observing resultant structures, and may require an experimental approach.

Part 6. Time Series Analysis

The analysis of ecological time series is a topic of both practical and theoretical interest. In pollution monitoring studies, pollution effects can only be detected as a divergence between the observed state of some ecological variable, and the range predicted for undisturbed conditions. In the absence of knowledge on the dynamics of a system, prediction is possible when it is based on a long enough time series (Poole, 1978).

From a theoretical point of view, a wide variety of temporal patterns in populations and ecosystems is predicted by many models. Population dynamics models show that populations can exhibit types of behavior in time which range from extreme stability to apparently chaotic behavior (May and Oster, 1976), including periodic and pseudoperiodic cycles. In the field of systems theory, the dynamics of non-linear open systems far from equilibrium are obviously appealing for theoretical ecology. The essence of these systems' dynamics is an unstable but coherent behavior consisting of periodicities and cycles. Finally, some methods of time series analysis may be applicable to studies of spatial structure, in particular where some environmental characteristic varies in an oscillatory way along a gradient (e.g., ripple marks: Hogue and Miller, 1981).

Time series can either be studied in the time domain or in the frequency domain. In the time-domain a probabilistic model is fitted to the data. This model can be used for prediction. The core of the frequency domain studies is spectral analysis. It is analogous to an analysis of variance: the total variance of the series is attributed to oscillations of different periodicities. Whereas the time domain studies are usually superior for prediction purposes, the study in the frequency domain is rather a probe into the structure of the time series. It may lead to an identification of processes acting on the variable. The methods for time series analysis have mainly been developed in the physical and economic sciences. They have a number of features which may represent serious limitations for their application in ecology, and which should be kept in mind when analyses of this kind are planned.

The main body of theory has been developed for the analysis of single series, although extension to a limited number of simultaneous series is possible. For the analysis of the evolution of community structure in time, where a relatively large number of species densities may be recorded at any moment, the ecologist must recur to methods based on ordination or classification.

Many methods in time series analysis are only applicable to relatively long series (more than 100 data points is often considered as constituting a "short series"), which may involve a lot of work if e.g., species densities must be recorded on all dates. Some of the methods discussed below, however, may be used for shorter series. The data should be equispaced for almost all analyses. This is a very important point to consider before starting the sampling. Although in some cases it may be possible to interpolate missing values, one should try to keep the sampling interval as constant as possible from the start onwards. Again this restriction may be weakened for some simple, robust analyses (see e.g., Herman and Heip, 1986). Also in another way the sampling strategy determines the possible outcomes of the analyses. When looking for systematic oscillations in the data (as e.g., in spectral analysis) it is impossible to detect oscillations with a period longer than the time span sampled. Usually a more restrictive factor is given, and anyhow oscillations with a period longer than 1/4 of the time span sampled should be treated with caution. On the other hand, the shortest period that can be resolved is 2 Δt (where Δt is the sampling interval). If important oscillations are missed because the sampling interval is too long, these may still show up in the results of the analyses and yield erroneous conclusions, a phenomenon known as aliasing.

Diagnostic Tools

An essential first step in time series analysis is to plot the data as a function of time. Several features may be apparent: the presence of trend, seasonality, outliers, turning points (where a rising trend changes into a declining one or vice versa), etc. It is often apparent from inspection of the data if a transformation is desirable.

A time series is called stationary if its mean, variance, and higher moments are time-independent. Most analyses require the series to be stationary; this should be checked early in the analysis, usually at the plotting stage. In practice one is usually concerned with non-stationarity of the mean and variance. The mean is non-stationary when, e.g., an increasing or decreasing trend, or an obvious periodic ("seasonal") fluctuation is present.

Biological data, especially population densities, often need transformation. Transformation may stabilize the variance when, in the original series, it varies with the mean. Moreover, it can make seasonal variation additive instead of multiplicative. Several transformations are used in ecology. Legendre and Legendre (1986) give an instructive overview of their properties.

Trend and seasonal effects may be described and/or removed by several techniques (see below). A relatively simple but effective test for the presence of trend makes use of Kendall's correlation coefficient (Kendall, 1976). Consider a series $x_1,...,x_n$, and count the number of times $x > x_i$ for $j > i$. Call this number P. Its expected value in a random series is $n(n-1)/4$, i.e., half the number of comparisons made. P is related to Kendall's τ by the formula:

$$\tau = \frac{4P}{n(n-1)} - 1$$

The significance of τ can be tested by comparison to tabulated values. A positive significant τ points to a rising trend, a negative value to a falling trend.

An important diagnostic tool in time series analysis is the correlogram. This is a plot of the autocorrelation function with lag k against k. The autocorrelation function with lag k expresses the correlation between observations k units apart in time. It is estimated (for reasonably large N, and k not greater than about $N/4$) by:

$$r_k = \frac{\sum_{t=1}^{N-k} (x_t - \bar{x})(x_{t+k} - \bar{x})}{\sum_{t=1}^{N} (x_t - \bar{x})}$$

Defining the (sample) autocovariance function C_k as:

$$c_k = \frac{1}{N} \sum_{t=1}^{N-k} (x_t - \bar{x})(x_{t+k} - \bar{x})$$

it can be seen that $r_k = c_k/c_o$. This is also the computational formula.

An approximate 5% confidence interval for r_k is $\pm 2\sqrt{N}$. In a purely random series 5% of the values are expected to lie outside this range.

The correlogram reveals much of the structure of a time series: trend, periodic oscillations, alternation of points (zig-zag type of time series). Typical patterns in the correlogram point to suitable models to fit to the series. In analogy to the autocovariance function, a cross-covariance function of two simultaneous series is defined as:

$$c_{xy}(k) = \begin{cases} \sum_{t=1}^{N-k} (x_t - \bar{x})(y_{t+k} - \bar{y})/N \\ \qquad [k=0,1,..,(N-1)] \\ \sum_{t=1-k}^{N} (x_t - \bar{x})(y_{t+k} - \bar{y})/N \\ \qquad [k=-1,..,-(N-1)] \end{cases}$$

The cross-covariance plot reveals at which time-lag(s) the two series show a maximum correspondence.

Description and Removal of Trend

The presence of trend is most often the reason for the non-stationarity of the mean in time series. Trend must be removed from the series for most further analyses, but it can be of interest in itself. For example, Heip and Herman (1985) conclude from the near absence of trend in copepod respiration (as opposed to trend in the densities of the separate species, and in "structural" characteristics of the copepod assemblage) that the assemblage is continually changing in composition and structure, but is functionally stable.

Trend may be described (and removed) by different methods. A function (e.g., linear, quadratic, exponential, Gompertz, hyperbolic,...) may be fitted by least squares to the data. This is analogous to a regression calculation and yields entirely valid estimates of the parameters. However, when applying a regression program for these calculations, one should not use the tests of significance, nor the estimates of the standard error of the parameters which are usually provided in the program output. Due to the serial correlation usually present in time series, the basic assumption of independence of the Y's in regression is violated.

An advantage of curve-fitting is that the trend may be summarized in a few parameters. Disadvantages are that updating is computationally cumbersome (for every data point added to the series, the whole fitting must be redone), and that the estimate of trend in, e.g., the first year of the time series is continually changing with each data point added to the series, even 10 years later.

The fitting of moving averages avoids these problems. In essence, one fits a polynomial of order p (to the choice of the investigator) to the first 2m+1 data points in order to estimate the trend term in the (m+1)th point. Then a polynomial of the same order is fitted to the 2nd through (2m+2)th points to estimate the trend in the point m+2 etc.. In practice this procedure reverts to the calculating of linear combinations of the data points with tabulated coefficients. These differ with the order of the polynomial and the length 2m+1 of the data segment. The coefficients, and guidelines for a proper choice of the order p of the polynomial, can be found in Kendall (1976) and Kendall and Stuart (1968).

Inevitably, the fitting and distraction of a moving average distorts the cyclical and random elements of the remainder of the series. Cyclical terms with periods longer than the length of the segment taken into account for the computation of the moving average, will tend to be incorporated into the trend. Terms with shorter periods will remain in the residuals. Moreover, the distraction of a moving average from a purely random series will induce non-zero autocorrelations in the remainders, and thus give the impression that a systematic pattern is present in the data. This phenomenon is known as the Slutzky-Yule effect. It is well studied, and the period of the induced oscillation can be well approximated (Kendall, 1976; Legendre and Legendre, 1986).

Finally, an effective method to remove non-stationarity due to trend from the data is to difference the series. For a linear trend, define the difference operator ∇, such that the differenced series:

$$y_t = \nabla x_{t+1} = x_{t+1} - x_t$$

Occasionally second-order differencing is used, where the differenced series is differenced again. This method is most often used in the fitting of ARIMA models to a series (see below).

Seasonal Effects

Seasonal effects are often prominently present in ecological data. Their causes are usually well known, and are often not the incentive for the long-term study of the system. They may therefore be hindering the interpretation of other, more interesting features.

Seasonal effects may be removed in several ways. A simple, robust method is described by Kendall (1976). A 1-year moving average is calculated, and subtracted from the data. The residuals for all Januaries, all Februaries, etc. (in the case of monthly data) are averaged over all the observations for that month. These are the raw estimates of the seasonal effects. The final estimates are calculated by scaling the seasonal effects such that they sum to zero. The seasonal effects are then subtracted from the data.

Alternatively, a seasonal differencing may be used. This is done with the seasonal difference operator ∇^d, where $\nabla^d x_t = x_t - x_{t-d}$.

Time Domain Studies

Time domain studies use a class of models called ARIMA models (Autoregressive Integrated Moving Average). This class of models was developed for an integrated modeling strategy by Box and Jenkins (1976). Basically, ARIMA models contain several "building blocks":

--An autoregressive process of order p is defined as:

$$x_t = \phi_1 x_{t-1} + \phi_2 x_{t-2} + \ldots + \phi_p x_{t-p} + a_t$$

where the error terms a_t are independently, identically distributed normal random variables with zero mean and variance σ^2_a, and the series is corrected for a (stationary) mean. This equation has a structure similar to a multiple regression, but the "regression" is not on independent variables, but on the series itself. This explains the term "autoregression."

--A moving average process of order q is:

$$x_t = a_t + \theta_1 a_{t-1} + \ldots + \theta_q a_{t-q}$$

A MA is a model of processes where random events affect not only the present state of the variable, but also have repercussions on its future state.

AR and MA processes are related to each other: AR processes of finite order can be written as infinite-order MA processes and vice versa. For parameter parsimony it is often useful to combine both processes in a "ARMA (p,q)" process:

$$x_t = \phi_1 x_{t-1} + .. + \phi_p x_{t-p} + a_t + \theta_1 a_{t-1} + .. + \theta_q a_{t-q}$$

ARMA processes are stationary, and applicable only to stationary series.

An ARMA model, fitted to a differenced series, is called an ARIMA model of the original series. It is called "integrated" because it has to be summed or integrated to provide a model of the non-stationary (undifferenced) series.

The fitting of an ARIMA model is performed in steps. The data are transformed and differenced to obtain a normally distributed, stationary series. The autocorrelation function is plotted and compared with theoretical autocorrelation functions of ARIMA processes. A suitable model is selected and fitted. The residuals are checked for systematic deviations from the model. If necessary the model is reformulated, fitted, etc., until a model has been selected that provides the best fit with the least possible number of parameters. In practice a suitable computer program package is necessary, and the model fitting requires a good deal of experience. The reader is referred to Chatfield (1976) for an introductory text and useful references. Detailed descriptions of the method are provided by Box and Jenkins (1976).

Examples of the use of ARIMA models in an ecological context are Poole (1978) and Keller (1987).

Spectral Analysis

The power spectral density function, or spectrum of a discrete stationary time series is a function $f(\omega)$ of frequency ω, with the following physical interpretation: $f(\omega)d\omega$ is the contribution to the variance of the series by components with frequencies in the range $(\omega, \omega + d\omega)$. When $f(\omega)$ is plotted against ω, the surface under the curve equals the total variance in the series. A peak in the spectrum indicates a frequency with a particularly important contribution to the explanation of the variance.

Thus the spectrum of a time series with a clear seasonal oscillation (e.g., temperature in a temperate climate) will have a high peak on a frequency of 1 yr^{-1}.

The spectrum is a parametric function (in the same way as μ is usually defined as the parametric mean of a set of data). Its estimation from the data is the aim of spectral analysis.

The interpretation of the spectrum becomes clear if we look at the derivation of the periodogram, which is closely related to it.

According to a fundamental result of Fourier analysis a finite time-series $\{x_t\}(t = 1,2,...,N)$ can be decomposed in a sum of sine and cosine functions of the form:

$$x_t = a_o + a_{N/2}\cos \pi t +$$
$$+ \sum_{p=1}^{N/2-1} \{a_p \cos(2\pi pt/N) + b_p \sin(2\pi pt/N)\}$$

where $a_o = \bar{x}$

$$a_{N/2} = \Sigma (-1)^t x_t / N$$

$$a_p = 2\{\Sigma x_t \cos(2\pi pt/N)\}/N$$

$$b_p = 2\{\Sigma x_t \sin(2\pi pt/N)\}/N$$

$$[p=1,..,N/2-1]$$

Note that the first equation has N parameters to describe N observations. It fits the data exactly. The component with frequency:

$$\omega_p = 2\pi p/N$$

is called the p-th harmonic. the amplitude of the p-th harmonic is given as:

$$R_p = \sqrt{(a_p^2 + b_p^2)}$$

It can be shown that:

$$\Sigma (x_t - \bar{x})^2/N = \sum_{p=1}^{N/2-1} R_p^2/2 + a_{N/2}^2$$

This equation expresses how the total variance of the series is divided over the harmonic components. If we assume that the series has a continuous spectrum, we can regard $R^2_p/2$ as the contribution to the explanation of the variance of all components with frequencies in the range $\omega_p \pm \pi/N$.

The periodogram is a plot of $I(\omega)$ against ω, where

$$I(\omega) = N R_p^2/4\pi$$

for $\omega_p - \pi/N < \omega \leq \omega_p + \pi/N$

and

$$I(\omega) = N a_{N/2}^2/\pi$$

for $\pi(N-1) < \omega \leq \pi$

It can be seen that the surface under the plot of $I(\omega)$ in the range $\omega_p \pm \pi/N$ equals

$$\frac{N R_p^2}{4\pi} \frac{2\pi}{N} = \frac{R_p^2}{2}$$

The periodogram is an unbiased estimator of the spectrum, but it is not consistent: its variance does not decrease as N becomes infinitely large, and neighboring values of $I(\omega)$ are asymptotically uncorrelated. Several methods are devised to make the estimate of the spectrum more consistent. In essence, these methods revert to some form of smoothing of the periodogram.

Smoothing may be performed directly on the periodogram. This possibility has become popular since the development of the Fast Fourier Transform algorithm. Calculation of the periodogram with the "classical" Fourier transform takes about N^2 operations. This is drastically reduced with FFT, to $2 \log_2(N)$. Therefore, calculation of the periodogram has become a reasonable possibility, and for long series ($N > 10^4$) it is the only feasible method.

In fact, all methods for the calculation of the spectrum are mathematically equivalent to a smoothing of the periodogram. In this smoothing operation a compromise should always be reached: smoothing over a broad range reduces the variance of the spectrum estimates, but increases the "bandwidth": a peak in the spectrum is spread over a rather broad range of frequencies. Inversely, reducing the bandwidth increases the variance, and in general: bandwidth x variance = constant.

Spectral analysis can be extended to the study of bivariate processes, where two simultaneous time series are studied.

The cross-spectrum is the finite Fourier transform of the cross-covariance function. However, in contrast to the case of a single time series, this cross-spectrum is complex. In order to study the cross-spectrum, several functions are defined from it. These express the real and imaginary parts, and various combinations thereof. The subject will not be treated in this text. For an introduction and useful references, see Chatfield (1976).

MESA

Maximum Entropy Spectral Analysis is a recently developed method for spectral analysis, which has been especially designed for the analysis of short series. An excellent account of the method and an annotated program listing are provided by Kirk et al. (1979). Burg (1967) has proposed the method on the basis of the following reasoning. In using windows to estimate the spectrum, one makes implicit assumptions on the unavailable data, i.e., the data of the time series before the observations have started, and after they have ended. Given a set of autocorrelation values, with the condition that the spectrum be non-negative definite, there usually exist infinitely many power spectra which are consistent with the given data. MESA selects among these the most random spectrum, i.e., the spectrum with the maximum entropy.

Equivalently, one can say that the autocorrelation values are extrapolated beyond the length M ($<=N$) in the most random way. The method thus corresponds to making the least stringent assumptions possible on the unavailable data.

It can be shown (see Kirk et al., 1979, for a summary) that this method is equivalent to the estimation of the spectrum from the least squares fitting of autoregressive (AR) model of order M to the data. The spectrum is then directly estimated from the coefficients of the AR model, as are the extrapolated autocorrelation values.

MESA is especially suited for short time series. It provides a better resolution in the low frequency range, does not produce sidebands in the spectrum, and can predict periods in the same order as the length of the time series (Kirk et al., 1979). Major drawbacks of the method are the computational effort required, and especially the problems in choosing an appropriate order M of the AR filter. This problem is as yet unsolved. MESA is therefore not very well suited for long series. However, for short series, as are almost all ecological time series, it is a powerful method. The method has been applied in meiobenthic research by Herman and Heip (1984) and Heip and Herman (1985) for the spectral analysis of 7-year series in a copepod community.

Which Method to Choose?

It is difficult to guide the choice of an analytical method, as it depends on a large number of factors. For the analysis of short, irregularly spaced series a simple analysis of trend and seasonal components may suffice (see Herman and Heip, 1986). When longer, more or less equispaced data series are available, the choice becomes more complicated. For the aim of prediction, statistical (ARIMA) models usually are superior to spectral analysis. Ecological series often show pseudoperiodicity: swings in the data are more or less regular, but contain phase shifts and changes in amplitude. Extending "harmonic components" into the future for the purpose of prediction, cannot take such features into account.

On the other hand, spectral analysis does reveal on which time scales the most important variability is to be found in the series. This in itself is an important structural feature of the series, which can yield scientific insight into the structure and functioning of the system.

The specific method chosen for spectral analysis will depend on the length of the time series (MESA may be superior for shorter series, FFT is the only feasible method for very long series). Often it will suffice to plot a periodogram without pursuing the analysis further, e.g., if it is the purpose to show the influence of tides, seasonal influences, etc.

The computer programs available are also an important factor to take into account. It is impossible to perform any of these analyses without a computer. It is also advisable to use standard software packages.

Acknowledgments

The authors acknowledge the relevant criticism of two anonymous reviewers. The three authors were supported through grants from the Belgian Fund for Scientific Research (NFWO) an through grant FKFO 32.9007.82 from the Belgian Fund for Collective Fundamental Research.

References

Allen, J.D.
1975. Components of Diversity. *Oecologia*, 18:359-367.
Anderson, G.D., E. Cohen, W. Gazdzic, and B. Robinson
1978. *User's Pocket Guide to SIR with Sections on SPSS and BMDP*. 108 pages. Evanston, Illinois: S.I.R., Incorporated.
Anscombe, F.J.
1950. Sampling Theory of the Negative Binomial and Logarithmic Series Distributions. *Biometrika*, 37:358-382.
Blackith, R.E., and R.A. Reyment
1971. *Multivariate Morphometrics*. 412 pages. London: Academic Press.
Boswell, M.T., and G.P. Patil
1971. Chance Mechanisms Generating the Logarithmic Series Distribution Used in the Analysis of Number of Species and Individuals. Pages 100-130 in G.P. Patil, E.C. Pielou, and W.E. Waters, editors. *Statistical Ecology*. Volume I. University Park: Pennsylvania State University Press.
Box, G.E.P., and G.M. Jenkins
1976. *Time-series Analysis, Forecasting and Control*. 553 pages. San Francisco: Holden-Day.
Bulmer, M.G.
1974. On Fitting the Poisson Lognormal Distribution to Species-abundance Data. *Biometrics*, 30:101-110.
Burg, J.P.
1967. *Maximum Entropy Spectral Analysis*. Paper presented at the 37th Annual International Meeting, Society of Exploratory Geophysicists, Oklahoma.
Caswell, H.
1976. Community Structure: A Neutral Model Analysis. *Ecological Monographs*, 46:327-354.
Chatfield, C.
1976. *The Analysis of Time Series: Theory and Practice*. 263 pages. London: Chapman and Hall.
Cliff, A.D., and J.K. Ord
1973. *Spatial Autocorrelation*. 178 pages. London: Pion.
Cohen, J.E.
1968. Alternate Derivations of a Species-abundance Relation. *American Naturalist*, 102:165-172.
Coull, B.C., and J.W. Fleeger
1977. Long-term Temporal Variation and Community Dynamics of Meiobenthic Copepods. *Ecology*, 58:1136-1143.

Daget, J.
1976. *Les Modeles Mathematiques en Ecologie*. 172 pages. Paris: Masson.
Date, C.J.
1981. *An Introduction to Database Systems*. Third edition. 574 pages. Reading, Massachusetts: Addison-Wesley Publishing Company.
Davies, R.G.
1971. *Computer Programming in Quantitative Biology*. 492 pages. London: Academic Press.
Dixon, W.J., editor
1973. *BMD Biomedical Computer Programs*. 773 pages. Los Angeles: University of Calififornia Press.
Dixon, W.J., and M.B. Brown, editors
1977. *BMDP-77 Biomedical Computer Programs. P-series*. 880 pages. Berkeley: University California Press.
Findlay, S.E.G.
1982. Influence of Sampling Scale on the Apparent Distribution of Meiofauna on a Sand-flat. *Estuaries*, 5:322-324.
Fisher, R.A., A.S. Corbet, and C.B. Williams
1943. The Relation Between the Number of Species and the Number of Individuals in a Random Sample of an Animal Population. *Journal of Animal Ecology*, 12:42-58.
Frontier, S., editor
1983. *Strategies d'Echantillonnage en Ecologie*. 494 pages. Paris: Masson.
1985. Diversity and Structure of Aquatic Ecosystems. *Oceanography and Marine Biology Annual Review*, 23:253-312.
Gauch, H.G., Jr.
1977. *ORDIFLEX - A Flexible Computer Program for Four Ordination Techniques: Weighted Averages, Polar Ordination, Principal Components Analysis, and Reciprocal Averaging*. 185 pages. New York: Ecology and Systematics, Cornell University.
1979. *COMPCLUS - A FORTRAN Program for Rapid Initial Clustering of Large Data Sets*. 59 pages. New York: Ecology and Systematics, Cornell University.
1984. *Multivariate Analysis in Community Ecology*. 298 pages. Cambridge: Cambridge University Press.
Gower, J.C.
1966. Some Distance Properties of Latent Roots and Vector Methods Used in Multivariate Analysis. *Biometrika*, 53:325-338.
Gray, J.S.
1983. Use and Misuse of the Log-normal Plotting Method for Detection of Effects of Pollution - A Reply to Shaw et al. (1983). *Marine Ecology Progress Series*, 11:203-204.
Greig-Smith, P.
1983. *Quantitative Plant Ecology*. 359 pages. Oxford: Blackwell Scientific Publications.
Heip, C.
1974. A New Index Measuring Evenness. *Journal of the Marine Biological Association of the United Kingdom*, 54:555-557.
1975. On the Significance of Aggregation in Benthic Marine Invertebrates. Pages 527-538 in H. Barnes, editor, *Proceedings of the 9th European Marine Biology Symposium*. Aberdeen: Aberdeen University Press.
1976. The Spatial Pattern of *Cyprideis torosa* (Jones, 1850) (Crustacea, Ostracoda). *Journal of the Marine Biological Association of the United Kingdom*, 56:179-189.
Heip, C., and P. Engels
1974. Comparing Species Diversity and Evenness Indices. *Journal of the Marine Biological Association of the United Kingdom*, 54:559-563.
1977. Spatial Segregation in Copepod Species from a Brackish Water Habitat. *Journal of Experimental Marine Biology and Ecology*, 26:77-96.
Heip, C., and P.M.J. Herman
1985. The Stability of A Benthic Copepod Community. Pages 255-263 in P.E. Gibbs, editor, *Proceedings of the 19th European Marine Biology Symposium*. Cambridge: Cambridge University Press.

Heip, C., R. Herman, and M. Vincx
1984. Variability and Productivity of Meiobenthos in the Southern Bight of the North Sea. *Rapports et Proces-verbaux des Reunions. Conseil Permanent International pour l'Exploration de la Mer*, 183:51-56.

Heip, C., M. Vincx, and G. Vranken
1985. The Ecology of Marine Nematodes. *Oceanography and Marine Biology Annual Review*, 23:399-490.

Herman, P.M.J., and C. Heip
1984. Long-term Dynamics of Meiobenthic Populations. *Oceanologica Acta. Proceedings of the 17th European Marine Biology Symposium:* pages 109-112.
1986. The Predictability of Biological Populations and Communities: An Example from the Meiobenthos. *Hydrobiologia*, 142:281-290.

Hicks, G.R.F., and B.C. Coull
1983. The Ecology of Marine Meiobenthic Harpacticoid Copepods. *Oceanography and Marine Biology Annual Review*, 21:67-175.

Hill, M.O.
1973. Diversity and Evenness: A Unifying Notation and Its Consequences. *Ecology*, 54:427-432.
1974. Correspondence Analysis: A Neglected Multivariate Method. *Applied Statistics*, 23:340-354.
1979a. *DECORANA - A FORTRAN Program for Detrended Correspondence Analysis and Reciprocal Averaging*. 52 pages. New York: Ecology and Systematics, Cornell University.
1979b. *TWINSPAN - A FORTRAN Program for Arranging Multivariate Data in an Ordered Two-Way Table by Classification of the Individuals and Attributes*. 90 pages. New York: Ecology and Systematics, Cornell University.

Hogue, E.W.
1982. Sediment Disturbance and the Spatial Distributions of Shallow Water Meiobenthic Nematodes on the Open Oregon Coast. *Journal of Marine Research*, 40:551-573.

Hogue, E.W., and C.B. Miller
1981. Effects of Sediment Microtopography on Small-scale Spatial Distribution of Meiobenthic Nematodes. *Journal of Experimental Marine Biology and Ecology*, 53:181-191.

Hurlbert, S.H.
1971. The Nonconcept of Species Diversity: A Critique and Alternative Parameters. *Ecology*, 52:577-586.

Hutcheson, K.
1970. A Test for Comparing Diversities Based on the Shannon Formula. *Journal of Theoretical Biology*, 29:151-154.

Jumars, P.A., D. Thistle, and M.L. Jones
1977. Detecting Two-Dimensional Spatial Structure in Biological Data. *Oecologia (Berlin)*, 28:109-123.

Jumars, P., and J.E. Eckman
1983. Spatial Structure within Deep-sea Benthic Communities. Pages 399-451 in G.T. Rowe, editor, *Deep-Sea Biology*. New York: John Wiley and Sons.

Keller, A.
1987. Modeling and Forecasting Primary Production Rates Using Box-Jenkins Transfer Function Models. *Canadian Journal of Fisheries and Aquatic Sciences*, 44:1045-1052.

Kempton, R.A., and L.R. Taylor
1974. Log-series and Log-normal Parameters as Diversity Discriminants for the Lepidoptera. *Journal of Animal Ecology*, 43:381-399.
1976. Models and Statistics for Species Diversity. *Nature* 262:818-820.

Kempton, R.A., and R.W.M. Wedderburn
1978. A Comparison of Three Measures of Species Diversity. *Biometrics*, 34:25-37.

Kendall, M.G.
1976. *Time-series*. 2nd edition. 197 pages. London: Charles Griffin and Company.

Kendall, M.G., and A. Stuart
1968. *Advanced Theory of Statistics*. Volume 3, 2nd edition. 552 pages. London: Charles Griffin and Company.

Kirk, B.L., B.W. Rust, and W. Van Winkle
1979. *Time Series Analysis by The Maximum Entropy Method*. ORNL-5332. 218 pages. Oak Ridge, Tennessee: Oak Ridge National Laboratory.

Krishna Iyer, P.V.
1949. The First and Second Moments of Some Probability Distributions Arising from Points on a Lattice and Their Application. *Biometrika*, 36:135-141.

Kruskal, J.B.
1964. Nonmetric Multidimensional Scaling: A Numerical Method. *Psychometrika*, 29:115-129.

Lambshead, P.J.D., H.M. Platt, and K.M. Shaw
1983. The Detection of Differences Among Assemblages of Marine Benthic Species Based on an Assessment of Dominance and Diversity. *Journal of Natural History*, 17:859-874.

Lance, G.N., and W.T. Williams
1966. A Generalized Sorting Strategy for Computer Classifications. *Nature*, 212:218.
1967. A General Theory of Classificatory Sorting Strategies. II. Clustering Systems. *Computer Journal*, 10:271-277.

Legendre, L., and P. Legendre
1986. *Ecologie Numerique*. Tome 1, 192 pages; Tome 2, 247 pages. Quebec: Masson, Les presses de l'Universite du Quebec, 2eme edition.

Lloyd, M.
1967. Mean Crowding. *Journal of Animal Ecology*, 36:1-30.

MacArthur, R.H.
1957. On the Relative Abundance of Bird Species. *Proceedings of the National Academy of Science of the United States of America, Washington*, 43:293-295.

May, R.M.
1975. Patterns of Species Abundance and Diversity. Pages 81-120 in M.L. Cody and J.M. Diamond, editors, *Ecology and Evolution of Communities*. Cambridge, Massachusetts: Belknap Press.

May, R.M., and G.F. Oster
1976. Bifurcations and Dynamical Complexity in Simple Ecological Models. *American Naturalist*, 110:573-599.

Motomura, I.
1932. Statistical Study of Population Ecology. [In Japanese.] *Doobutugaku Zassi*, 44:379-383.

Neyman, J.
1939. On a New Class of Contagious Distributions, Applicable in Entomology and Bacteriology. *Annals of Mathematical Statistics*, 10:35-57.

Nie, N.H., C.H. Hull, J.G. Jenkins, K. Steinbrenner, and D.H. Bent
1975. *-SPSS- Statistical Package for The Social Sciences*. Second edition. 675 pages. New York: McGraw-Hill Book Company.

Noy-Meir, I.
1973. Data Transformations in Ecological Ordination. I. Some Advantages of Non-centering. *Journal of Ecology*, 61:329-341.

Orloci, L.
1967. An Agglomerative Method for Classification of Plant Communities. *Journal of Ecology*, 55:193-205.
1978. *Multivariate Analysis in Vegetation Research*. 451 pages. The Hague: Junk.

Osterdahl, L., and G. Zetterberg
1981. *Rubin-Species Codes and Species Numbers*. Stockholm: National Swedish Environment Protection Board Report 1427.

Peet, R.K.
1974. The Measurement of Species Diversity. *Annual Review of Ecology and Systematics*, 5:285-307.

Pielou, E.C.
1969. *An Introduction to Mathematical Ecology*. 286 pages. New York: John Wiley and Sons.
1975. *Ecological Diversity*. 165 pages. New York: John Wiley and Sons.
1981. The Broken-stick Model: A Common Misunderstanding. *American Naturalist*, 117:609-610.
1984. *The Interpretation of Ecological Data*. 263 pages. New York: John Wiley and Sons.

Poole, R.W.
1978. The Statistical Prediction of Population Fluctuations. *Annual Review of Ecology and Systematics,* 9:427-448.

Preston, F.W.
1948. The Commonness and Rarity of Species. *Ecology,* 29:254-283.

Robinson, B., G.D. Anderson, E. Cohen, and W. Gazdzic
1979. *S.I.R., Scientific Information Retrieval.* 474 pages. Evanston, Illinois: SIR, Incorporated.

Rohlf, F.J., J. Kispaugh, and D. Kirk
1972. *NT-SYS Numerical Taxonomy System of Multivariate Statistical Programs.* Stony Brook, N.Y.: State University of New York.

Ross, G.J.S., F.B. Lauckner, and D. Hawkings
1976. *CLASP Classification Program.* Harpenden, England: Rothampsted Experimental Station.

Routledge, R.D.
1979. Diversity Indices: Which Ones Are Admissible? *Journal of Theoretical Biology,* 76:503-515.

Sanders, H.L.
1968. Marine Benthic Diversity: A Comparative Study. *American Naturalist,* 102:243-282.

SAS Institute, Incorporated
1985. *SAS User's Guide: Statistics.* 584 pages. Cary, North Carolina: SAS Institute, Incorporated.

Shaw, K.M., P.J.D. Lambshead, and H.M. Platt
1983. Detection of Pollution - Induced Disturbance in Marine Benthic Assemblages with Special Reference to Nematodes. *Marine Ecology Progress Series,* 11:195-202.

Simberloff, D.
1972. Properties of the Rarefaction Diversity Measurement. *American Naturalist,* 106:414-418.

Singer, S.B.
1980. *DATAEDIT - A FORTRAN Program for Editing Data Matrices.* 42 pages. New York: Ecology and Systematics, Cornell University.

Singer, S.B., and H.G. Gauch, Jr.
1979. *CONDENSE - Convert Data Matrix from Any Ordiflex Format into A Condensed Format by Samples.* 7 pages. New York: Ecology and Systematics, Cornell University.

Taylor, L.R.
1961. Aggregation, Variance and the Mean. *Nature,* 189:732-735.

Taylor, L.R., R.A. Kempton, and I.P. Woiwod
1976. Diversity Statistics and the Log-series Model. *Journal of Animal Ecology,* 45:255-272.

Thomas, M.
1949. A Generalization of Poisson's Binomial Limit for Use in Ecology. *Biometrika,* 36:18-25.

Webb, D.J.
1974. The Statistics of Relative Abundance and Diversity. *Journal of Theoretical Biology,* 43:277-291.

Whittaker, R.H.
1972. Evolution and Measurement of Species Diversity. *Taxon,* 21:213-251.

Wiser, W.
1953. Die Beziehung zwischen Mundhoehlengestalt, Ernaehrungsweise und Vorkommen bei freilebenden marinen Nematoden. *Arkiv för Zoologie,* 4:439-484.

Williams, C.B.
1964. *Patterns in the Balance of Nature.* 324 pages. New York: Academic Press.

Williams, W.T., and M.B. Dale
1965. Fundamental Problems in Numerical Taxonomy. *Advances in Botanical Research,* 2:35-68.

Williams, W.T., and P. Gillard
1971. Pattern Analysis of a Grazing Experiment. *Australian Journal of Agricultural Research,* 22:245-260.

Williams, W.T., and J.M. Lambert
1959. Multivariate Methods in Plant Ecology. I. Association-analysis in Plant Communities. *Journal of Ecology,* 47:83-101.

Wishart, D.
1978. *CLUSTAN Users Manual.* 3rd edition. Edinburgh: Edinburgh University.

15. The Nanobenthos

Bryan R. Burnett and Hjalmar Thiel

Organisms that make up the nanobenthos are between 2 and 42 µm in size (Thiel, 1983). This segment of the benthos has received little attention, mainly because of technical difficulties in their study. Most studies directly link bacteria with higher trophic levels and fail to consider whether or not nanobenthos represent an intervening link.

Prior to the introduction of "nano-" (Thiel, 1983), terms applied to these organisms had the prefix "micro-." However, as Thiel (1983) pointed out "micro-" is overly broad in its application and generates confusion as to what organisms are being considered. Therefore, references to "microbenthos" should be avoided when dealing with organisms in the size range of the nanobenthos.

Evaluation of nanobenthos in biotopes with large sediment grains has been by direct microscopical observation of organisms in a dish, with sediment or without sediment, using the seawater-ice technique. The extraordinary work of Fenchel (1967, 1968, and 1969) uses these techniques. The seawater-ice technique (Uhlig, 1964; Chapter 9) depends upon both the motility of organisms and their negative response to the melting of seawater ice. Nonmotile organisms or those of limited motility cannot be evaluated, and thus observation of the nanobenthos is biased toward motile, responsive organisms.

The seawater-ice technique has even less value with fine-grained sediments, which typically lack the interstitium that would allow the organisms to move. Although Uhlig (1973) succeeded in extracting ciliates from such sediments collected at a depth of more than 3000 m, obtaining actual estimates of abundance of the nanobenthos from a fresh or preserved fine-grained sediment sample in a Petri dish slurry is impossible because of clouding by suspended sediment. Thus, abundance estimates of nanobenthic organisms in fine-grained sediments, such as those found in most sea bottoms, has had limited success until recently. Burnett (1973, 1977, 1979, 1981) has developed a technique which is more effective in the quantitative evaluation of the nanobenthos of fine-grained sediments than any previous studies. Although this technique has only been applied to deep-sea sediments of the San Diego Trough (1200 m), San Clemente Basin (1130 m), and the central North Pacific (5800 m), there is no reason why the technique could not be used with shallow fine-grained sediments. Alongi (1986, 1987) applied a density gradient centrifugation method for extraction of nanobenthos from unpreserved samples.

Organisms of the Nanobenthos

The nanobenthos represents an extraordinary diverse group of organisms. One has to become familiar with both prokaryote and eukaryote taxa to become proficient with the nanobenthos. Typically, a mounted preparation contains bacterial clumps, large prokaryotes, flagellates, amoebae, testate protozoans, ciliates, and occasional small metazoans. Many identifications are tentative, and in rare cases even a decision as to kingdom can be questionable.

Prokaryotes usually represent a significant portion of the nanobenthos. They can be single-celled and relatively large (Figure 15.1a). They also occur in colonial unicells (Figure 15.1b,c). Flexibacteria (Figure 15.1d) are often found. In the central North Pacific benthos, much of the nanobenthos consists of clumps of cells that appear to be members of the genus *Hyphomicrobium* (Figure 15.1e,f), although it has been suggested that they are thraustochytrid fungi (Riemann, personal communication). Staining characteristics suggest affinity to the prokaryotes. It must be pointed out that most bacteria do not stain with the technique described in this chapter.

Fungi (Figure 15.1g) and yeast-like cells (Figure 15.1h–j) are frequently observed, especially in the sediments from the San Diego Trough (Burnett, 1981). These cells usually stain heavily and are frequently observed with buds.

Flagellates (Figure 15.1k–o) are often encountered in the sediments of the San Diego Trough. Numerous specimens of a multiflagellate (Figure 15.1o) were observed from one sample of the San Diego Trough (Burnett, 1981). In the flagellates, nuclear staining is usually dark and the staining of the cytoplasm variable. The cytoplasm is typically difficult to discern at the magnification used for scanning slides.

Sarcodines are also regularly observed. They usually are small and have a single lobopod (Figure

15.1p). Infrequently, leptomyxid (Figure 15.1q) or chlamydomyxid (Figure 15.1r) sarcodines may be found (e.g., San Diego Trough, 1200 m). Testate sarcodines (Figure 15.2a-c) are numerous in sediments from both the San Diego Trough and the central North Pacific. (See Sarcomastigophora).

Ciliates (Figure 15.2d-g) are rarer in deep-sea fine-grained sediments than are other nanofauna. (The Ciliophora are discussed in Chapter 18.)

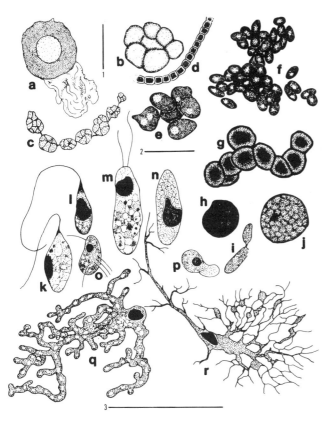

Figure 15.1.--Deep-sea nanobenthos. Most of these are redrawn from the original published micrograph or drawing. a, Microspheroid with associated ghost (from Snider et al., 1984); b,c, Colonial unicells (from Snider et al., 1984); d, Part of a flexibacterial colony (from Burnett, 1979); e,f, Large prokaryote colonies, probably hyphomicrobial bacteria from the central North Pacific, 5800 m; g, Fungus (from Snider et al., 1984); h-j, Yeast-like cells (from Burnett, 1979); k-n, Mastigophora (from Burnett, 1979); o, Multiflagellate from the San Diego Trough, 1200 m), p, Amoeboid sarcodine (from Burnett, 1979); q, Leptomyxid sarcodine from the San Diego Trough, 1200 m; r, Chlamydomyxid sarcodine from the San Diego Trough, 1200 m. (Scales = 10 μm; scale line 1: a-c, g; scale line 2: e,f,h-p; scale line 3: q,r.)

Technique for the Quantification of Nanobenthos in Fine-grained Sediments

Slide Preparation Technique.--Snider et al. (1984) succeeded in generating data on the nano-, meio-, and macrobenthos as well as in estimating the porosity and granulometry of the same samples. A flow scheme for the handling of samples is shown in Figure 15.3. (See Snider et al., 1984, for additional details.)

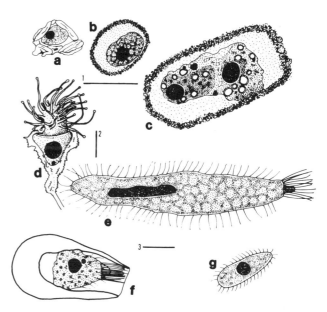

Figure 15.2.--Deep-sea nanobenthos (continued). a-c Testate sarcodines (from Burnett, 1979); d, Suctorian that was found attached to the operculum of a gastropod (from Burnett, 1973); e, Trachelophyllid (?) ciliate, San Diego Trough, 1200 m; f, Tintinnid ciliate, Sand Diego Trough, 1200 m.; g, Ciliate from the abyssal central North Pacific (from Burnett, 1977). (Scales = 10 μm; scale line 1: a-c; scale line 2: d; scale line 3: e-g.)

Sampling.--The first quantitative nanobenthic samples were taken using 0.95 cm (inner diameter), soft plastic tubes (Burnett, 1977, 1981). Side friction severely limited the depth of sampling. These tubes were later replaced by 1.9 cm (inner diameter) acrylic tubes (Snider et al., 1984). The hard hydrophobic surface of the acrylic tubes reduces friction and permits much deeper sediment sampling with little sample displacement.

Each tube sample was divided into a top-water sample (overlying water plus a variable amount of surface sediment) and 5 mm plugs of sediment (Snider et al., 1984). The sediment plugs were extruded from the leading edge of the sampler with a plunger (Figure 15.3), and then preserved.

Fixation.--Rapid fixation, as soon as possible after collection is important in processing of nanobenthos samples. Protozoa are probably more sensitive than metazoa to temperature fluctuations beyond their normal experience (see Burnett, 1979). Therefore, if fixation does not immediately follow sampling, then the sample should be kept at its ambient temperature.

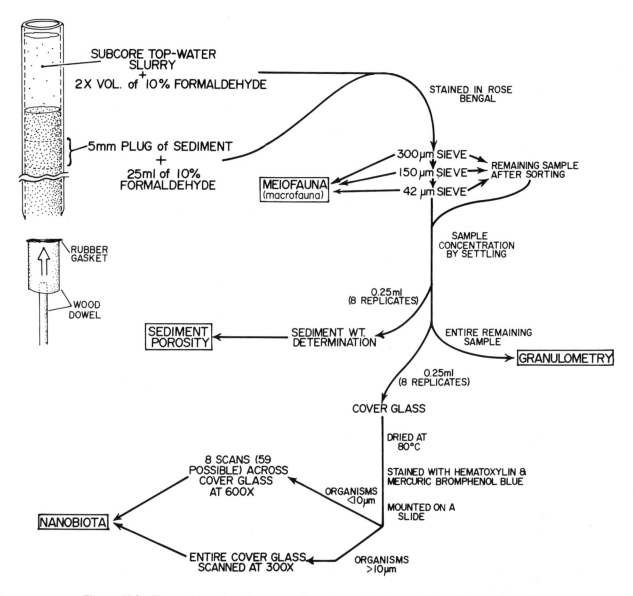

Figure 15.3.—Flow scheme for the processing of samples from 1.9 cm i.d. acrylic tubes (Snider et al., 1984). Major processing steps are indicated at the side of the flow lines. See Snider et al. (1984) for the technique of determining sediment porosity and granulometry.

The penetration of the fixative is another problem. The sediment plug, especially from lower levels of the core, needs to be broken up for adequate fixative penetration. This should be done by gently shaking the sample in the fixative solution until it appears that the entire clump has broken up. The fixative should be kept at the sample temperature for 20 to 30 minutes after introduction.

An 8% formalin solution is the most effective fixative for nanobenthos samples. It both penetrates tissue and fixes rapidly. But formalin does have drawbacks. It will eventually break down, the acidic product dissolving tissue. Thus, formalin fixative needs to be buffered, usually with sodium borate. The pH of the sample should be checked frequently, and buffer as necessary. With an adequate monitoring program, samples should be valuable for at least five years.

The buffered formalin should be filtered prior to use. Organic buffers may interfere with processing nanobenthos samples, and are not recommended. Storing samples in alcohol should also be avoided because alcohol can extract protein from the organisms.

Steedman (1976, see Chapter 9) proposed a mixture of propylene glycol and propylene phenoxetol in distilled water, which is in use for meiofauna and should be checked for nanobenthos preservation.

For bulk fixation for TEM processing of nanobenthos, picric acid formaldehyde (Stefanini et al., 1967) is recommended.

Sample Dilution.--A sample is usually too concentrated with sediment per volume of seawater for further processing. To dilute the sample, first mark the volume of each sample on the containers. With a Pasteur pipette, remove a measured, well-mixed subsample and add it to a known amount of filtered, buffered formalin. The dilution of the subsample should depend on both the quantity of sediment in the sample and its particle-size composition.

Cover-glass Preparation.--Spread a thin layer of Mayer's albumin on a clean 22 x 22 mm cover glass using the narrower side of a glass Pasteur pipette. Avoid touching the cover glass or the albumin, as doing so will leave epithelial cells in the preparation. The albumin coat should be smooth; carefully wipe off any excess at the edge of the cover-glass with a lintless tissue. Place the albuminized cover glass face up on a *cold* slide warmer. Measure and mark 0.25 ml on a glass Pasteur pipette. When removing material from the container, be sure the sediment mixture is well homogenized by rapidly moving the mixture in and out of the pipette. Quickly remove the pipette with 0.25 ml from the sample jar, to minimize loss of heavy particles, and apply this volume to one albuminized cover-glass surface. Try for as even a distribution of the sediment on the cover-glass surface as possible. The preparation of six replicate cover-glass subsamples is recommended. Turn on the slide warmer and warm to 80º C. Dry the cover-glass preparations until 30 minutes after all indication of moisture is gone. The dried preparations can be stored for at least two weeks, probably much longer.

Staining.--The key to the technique for finding nanobenthos among the sediment is differential staining. Protein specific stains, such as hematoxylin and mercuric bromophenol blue, stain organisms and not sediment.

The cover-glass preparations are stained and rinsed in Columbia jars. Each Columbia jar holds four cover glasses. Use great care while removing or inserting the cover-glass preparations in the jar slots. Quick movement through the solution meniscus will disrupt the preparations.

Place the dry cover glasses directly into Ehrlich's hematoxylin (Humason, 1967) and stain for 15 minutes. After every batch of four cover glasses, shake the Columbia jar. After four batches (16 cover glasses), replace the hematoxylin. Good Ehrlich's hematoxylin is deep brick red and somewhat translucent. If the hematoxylin changes to a dark purple, it will over-stain.

Carefully remove each stained cover glass from the Columbia jar and slowly slip it into 70% ethanol for 3 minutes. Change the ethanol after every 2 batches (8 cover glasses). Do a second rinse of 70% ethanol.

The cover-glass preparation now appears dark blue because both organisms and surrounding sediment stain. In order to be able to see the organisms, the sediment has to be rid of the hematoxylin. A destaining step accomplishes this. Destain one cover glass per Columbia jar in acid-ethanol (2.5 ml HCl in 1 liter of 70% ethanol) for 2.5 to 4.5 minutes, depending upon the sample. Replace the acid-alcohol after each cover glass. Experiment with test samples to determine the proper destaining time for your sample. Following destaining, carefully slip the preparation into 95% ethanol for 2 minutes.

Place the cover glasses in mercuric bromphenol blue (10 g mercuric chloride and 100 mg bromphenol blue into 100 ml 95% ethanol) for 15 minutes, followed by 95% ethanol (saturated with sodium borate) for 2 minutes. Replace the sodium-borate ethanol solution after each use. Complete dehydration with two changes of 100% ethanol, one change of 100% ethanol/toluene, and two changes of toluene. Mount on a glass slide with permount. Allow to dry at 40–45º C for four to five days before analysis.

Control Cover-glass Preparations.--It is easy to overstain as well as understain the preparations. In overstained ones, sediment aggregations often give the appearance of organisms. In understained preparations, however, few or no organisms can be discerned. To obtain proper destaining time, it is necessary to have a cover-glass preparation with known organisms present. This means that for every three preparations, one is a control. The controls are made by adding protozoa from a culture to a subsample of the original sediment. It is absolutely necessary always to have such a control with each batch of cover-glass preparations.

Counting the Organisms.--If organisms smaller than 10 µm are being evaluated, the cover glass is scanned eight times in a stratified random manner at 600x (Burnett, 1977). If nanobiota larger than 10 µm are being evaluated, then the entire cover glass is scanned at 300x (Snider et al., 1984).

Specific Problems

A wide variety of organisms (Figures 15.1 and

15.2) have been observed using this technique. Preservation, staining, and mounting processes result in substantial shrinkage of most organisms (Figure 15.4). In an estimation of the nanobenthos biomass of a sample, organisms without an observable cell wall were found to shrink an average of 70.3% (Burnett, 1981). In eukaryotes the average was 60% (Snider et al., 1984). Prokaryote shrinkage can range from none to as much as 93% and probably depends upon the robustness of the cell wall.

Drastic changes occur in the structure of the organism (e.g., disruption of the cirri in the *Euplotes* sp., Figure 15.4 and Burnett, 1979). Many organisms show little or no change in shape with this technique. However, in the case of most flagellates and many ciliates, distortion can be such as to make an identification from a single individual impossible. With enough specimens this problem can be overcome, and in many cases a taxonomic determination can be made at least to the family level.

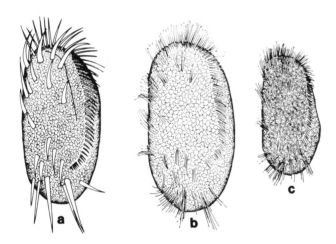

Figure 15.4.--Comparison drawings of a marine *Euplotes* sp. in different stages of processing. a, Live individual; b, Fixation in formalin; c, Mounted and stained. Nuclear staining was unusually light and cirral disruption occurred in this species. Drawn from photomicrographs from Burnett (1979).

Another problem is the sediment grain size. Large particles (>50 μm) tend to fall off the cover glasses during processing through the solutions. Such loss often disrupts the entire preparation on the cover glass.

Venrick (1971) showed that variance estimates increase exponentially with subsampling. With as many as three subsamplings (dilution of the original tube sample by part transfer to another sample container; subsampling the sample on the cover glass; and, scanning the cover glass for organisms smaller than 10 μm), this can be a significant problem with which the investigator should be aware. Any processing should minimize subsampling as much as possible.

The amount of time that it takes to scan and record the organisms from one cover-glass can be more than five hours. Thus, collection, preparation, and evaluation of nanobenthos requires a significant commitment of time.

Density Gradient Centrifugation Technique

Centrifugation has been used for the separation of meiofauna from sediments (Chapter 9) and it was successfully tested by Alongi (1986) on mud samples from mangrove and (1987) from bathyal sites. An advantage to this method is that it is applicable to unpreserved nanofauna, and it thus avoids the problems of shrinkage (see above) and destruction due to fixation, preservation, and other chemical treatments. Alongi (1986) compared this method to other procedures and found recovery rates of about 85% in the tests with cultured ciliates and flagellates added to muds, and with even higher efficiencies for sandy sediments.

For density gradient centrifugation subsamples of 1.3 cm^3 of sediment were extruded from syringes into 30 ml centrifugation tubes, already containing 5 ml of a silica gel mixture. This material was vortexed for 1-2 minutes and centrifuged for 20 minutes with 490 g after 1 hour of particle settling. The supernatant was poured into a dish, and the procedure was repeated two times. Counting was achieved in a Petri dish with a 1 cm^3 grid marked on the bottom. Organisms were counted in randomly chosen squares and total numbers per area or volume were calculated.

To retard the motion of organisms in the counting tray, a 50% (weight to volume) $MgCl_2$ solution was added dropwise.

The silica gel used, a mixture of Percoll and sorbitol, was prepared according to Price et al. (1978) and Schwinghamer (1981). 91.1 g sorbitol, 2.64 g. Fris-HCl and 4.03 g. Tris base are dissolved in 0.5l Percoll. To this mixture 1.43 g $MgCl_2 \cdot 6H_2$ is dissolved in 100 ml filtered seawater are added. The volume is filled up to 1000 ml with Percoll. This Percoll-Sorbitol mixture is to be filtered and stored refrigerated to avoid growth of bacteria and diatoms. The density should be 1.15 g/cm^3 and must be checked regularly.

The silica gel Ludox TM was also tested, but it has to be detoxified by dialysis in running water for 12-14 hours, and it must be reliquefied with sterile water. But in the presence of salts Ludox TM still tends to gel. Percoll has the advantage of not gelling in seawater, but it is very expensive.

Subsamples from bathyal sites (Alongi, 1987) were

kept in ice before processing, and centrifugation occurred at ambient temperatures. The larger protozoans (>20 μm) were counted immediately, after 0.5-1 hour, and the smaller specimens were preserved in Lugol's solution for later evaluation.

References

Alongi, D.M.
1986. Quantitative Estimates of Benthos Protozoa in Tropical Marine Systems Using Silica Gel: A Comparison of Methods. *Estuarine, Coastal and Shelf Science*, 23:443-450.
1987. The Distribution and Composition of Deep-sea Microbenthos in a Bathyal Region of the Western Coral Sea. *Deep-sea Research* 34:1245-1254.

Burnett, B.R.
1973. Observation of the Microfauna of the Deep-sea Benthos Using Light and Scanning Electron Microscopy. *Deep-Sea Research*, 20:413-417.
1977. Quantitative Sampling of Microbiota of the Deep-sea Benthos. - I. Sampling Techniques and Some Data From the Abyssal Central North Pacific. *Deep-Sea Research*, 24:781-789.
1979. Quantitative Sampling of Microbiota of the Deep-sea Benthos. - II. Evaluation of Technique and Introduction to the Biota of the San Diego Trough. *Transactions of the American Microscopical Society*, 98:233-242.
1981. Quantitative Sampling of Nanobiota (Microbiota) of the Deep-sea Benthos. - III. The Bathyal San Diego Trough. *Deep-Sea Research*, 28:649-663.

Fenchel, T.
1967. The Ecology of Marine Microbenthos. I. The Quantitative Importance of Ciliates as Compared with Metazoans in Various Types of Sediments. *Ophelia*, 4:121-137.
1968. The Ecology of Marine Microbenthos. II. The Food of Marine Benthic Ciliates. *Ophelia*, 5:73-121.
1969. The Ecology of Marine Microbenthos. IV. Structure and Function of the Benthic Ecosystem, Its Chemical and Physical Factors and the Microfauna Communities with Special Reference to the Ciliated Protozoa. *Ophelia*, 6:1-182.

Humason, G.L.
1967. *Animal Tissue Techniques.* 386 pages. San Francisco: W.H Freeman.

Price, C.A., E.M. Reardon, and R.R.L. Guillard
1978. Collection of Dinoflagellates and Other Marine Microalgae by Centrifugation in Density Gradients of a Modified Silica gel. *Limnology and Oceanography*, 23:548-553.

Schwinghamer, P.
1981. Extraction of Living Meiofauna from Marine Sediments by Centrifugation in a Silica Sol - Sorbitol Mixture. *Canadian Journal of Fisheries and Aquatic Sciences*, 38:476-478.

Snider, L.J., B.R. Burnett, and R.R. Hessler
1984. The Composition and Distribution of Meiofauna and Nanobiota in a Central North Pacific Deep-sea Area. *Deep-Sea Research*, 31:1225-1249.

Stefanini, M.C., C. DeMartino, and L. Zamboni
1967. Observation of Ejaculated Spermatazoa by Electron Microscopy. *Nature*, 216:173-174.

Thiel, H.
1983. Meiobenthos and Nanobenthos of the Deep Sea. Pages 167-230 in G.T. Rowe, editor, *The Sea.* Volume 8. New York: John Wiley and Sons.

Uhlig, G.
1964. Eine einfache Methode zur Extraktion der vagilen, mesopsammalen Mikrofauna. *Helgoländer wissenschaftliche Meeresuntersuchungen*, 11:178-185.
1973. Preliminary Studies on Meiobenthic Ciliates from the Deep Sea. *Progress in Protozoology*, Abstract, page 417.

Venrick, E.
1971. The Statistics of Subsampling. *Limnology and Oceanography*, 16:811-818.

16. Taxonomic and Curatorial Considerations

Robert P. Higgins

The purpose of this essay is not to provide an extensive discourse on the subject of taxonomy or curation of meiofaunal collections. However, it seems appropriate to preface this section on meiofaunal taxa with a commentary on selected methodology and other subjects that relate to both the preceding chapters and the chapters to follow.

Contemporary meiofauna research, like research in many fields of biological science, finds itself dependent on taxonomic expertise that is rapidly diminishing. The field of systematics is facing a serious shortage of taxonomic specialists and a lack of financial support at the same time other disciplines are more and more relying on this expertise. In order for meiofaunal research to maintain its current rate of advancement, its taxonomic foundation must continue to grow as well. Much of this growth has been and will continue to be through the development of taxonomic expertise by investigators whose primary interest may be in a discipline other than taxonomy. Formal training in systematics is no longer included in the average university curriculum and future taxonomists may have to rely on a much less structured learning process. Perhaps because of this there are growing indications that some fundamental taxonomic and collection management procedures are not fully understood.

For the most part, there is a reasonable reference literature addressing the practice of taxonomy. A classic taxonomy text, *Taxonomy,* by R. Blackwelder (1967), remains very useful, despite both original and latent shortcomings, especially when used with more contemporary texts such as *Principles of Systematic Zoology,* by E. Mayr (1969), and *Biological Systematics,* by H. Ross (1974). Most of the fundamental processes in taxonomy are treated in these publications. All refer extensively to the primary resource literature such as the *International Code of Zoological Nomenclature* and the *Zoological Record* and its use.

Similarly, contemporary systematics and its relationship to the study of the pattern and processes of evolution is the subject of an extensive literature that began with Hennig (1950) and is now more recognizable under the textbook titles of *Phylogenetics* (Wiley, 1981) or *Phylogenetic Systematics* (Hennig, 1966). Taxonomy has profited from this more objective approach. In general, there is a sufficient and responsible taxonomic literature which, if consulted, should provide the guidelines for adequate and accurate descriptions of new taxa and nomenclatorial stability.

The subject of curation of specimens or collections of specimens is treated in the taxonomic texts cited above. In addition, one of the more valuable documents on this subject is *Guidelines for Acquisition and Management of Biological Specimens* (Lee et al., 1982).

Collections of Meiofauna

Collections of any faunistic or ecological assemblage constitute an important scientific resource. In general, these collections are the result of taxonomic studies as well as other basic research projects. Although such materials may find their way into permanently archived collections, far too often material from significant ecological (or other) studies is discarded or allowed to remain uncurated, and specimens that could be of considerable importance to future research in taxonomy, ecology, and zoogeography are lost forever. Voucher specimens from these collections are the only objective archival documentation of previous research, which often was accomplished at great expense, and should be cared for accordingly. "A voucher specimen is one which: 1. physically and permanently documents data in an archival report by verifying the identity of the organism(s) used in the study; and, 2. by so doing, ensures that a study, which otherwise could not be repeated, can be accurately reviewed or reassessed." (Lee et al., 1982).

There are three categories of voucher specimens: (1) type specimens upon which scientific names are based (eg., holotypes and paratypes), (2) taxonomic support specimens -- those not directly associated with nomenclatural perspectives, but which are important in range extensions and morphological variation, and (3) biological documentation specimens -- representatives of taxa referred to in ecological, faunistic, or other non-taxonomic studies.

In order to be of any value, voucher specimens

must be correctly preserved, properly stored, and thoroughly documented with collecting data. This may sound trite, but taxonomists and curators of museum collections constantly contend with the receipt of material that not only does not meet these fundamentally important requirements, but may be improperly "bottled" or carelessly packed for shipment.

Collection Data

Specimens are usually of no value without data. Data written on note paper or anything other than museum-quality label paper will eventually deteriorate. A high-rag-content, solvent-resistant bond paper should be used. This is especially important when bulk samples of preserved sediment are transported over rough seas or terrain, or over a long period of time. Many meiofaunists have had the unwelcome experience of finding that the label's data have been abraded beyond recognition; moreover, there is nothing more frustrating than to receive a vial of transparent, very small meiofaunal specimens that have been in the company of a label written on notebook paper. Sorting such specimens from the fibrous mass of deteriorated cheap paper is a chore to be avoided. Similarly, essential data may be lost forever if the ink with which it was written dissolves in the preservative. One should use a water-proof, chemical resistant ink, or medium soft pencil lead on labels.

It is important to have certain categories of data accompany all specimens. Based on the recommendations of a study by the Association of Systematics Collections, the following specimen data "are basic to, but exclusive of, documents such as scientific papers, summary reports, impact statements, etc., which summarize purpose, findings and conclusions of the original project."

1. A unique designation must be assigned to each sample collected during the collection process.
2. The position of a sample collection site shall be described so that the specific site and habitat can be revisited. Recommended elements of locality description as might apply to meiofaunal collections are:

   a. Continent or ocean
   b. Country (if within the recognized 200 mile limit) or ocean region
   c. State, province, county, or other political subdivision
   d. Sea, major island group, river system, lake
   e. Local place names
   f. Latitute and longitude or Universal Transverse Mercator grid designation
   g. Depth and altitude (lake or river)
   h. Habitat including such indications as substratum, bioassociations, tidal phase, etc.

Specimen Containers

In general, the size of meiofauna makes this problem relatively simple, or does it? Although meiofaunal specimens will fit in very small containers, perhaps several hundred thousand specimens in a 1-ml vial, there are several problems that must be addressed.

Small vials as well as large containers must have adequate closures to prevent evaporation or leakage of fluid contents. Screw-caps are unsatisfactory, unless taped tightly with pressure-sensitive tape, or dipped in a sealant compound. The practice of using cotton or pith plugs in shell vials is not a problem for specimens approaching the upper limits of meiofaunal dimensions, but those specimens near the lower size limits easily may be lost in the cotton fibers, or the cavernous spaces in any other kind of porous plug. Such containers must always be sealed to prevent evaporation, or (best) immersed in a second, larger, container filled with the same fluid and fitted with a proper closure, itself sealed to prevent leakage.

This may sound obvious to the more conscientious researcher, but too often samples are received in which the specimens are lost or damaged unnecessarily. Because meiofaunal specimens are very small we must recognize their potential loss when they come into contact with a closure. This includes closures such as plastic plugs that are beveled at their lower end, thereby leaving a slight gap between the vial wall and the plug. Minute specimens can be lodged in this space and damaged or destroyed when the plug is removed. Plugs that do not depend on contact with the internal surface of the vial must fit tightly at the vial's opening, or specimens can become lodged outside the lip of the vial and may desiccate or otherwise be damaged or lost.

Small vials have certain disadvantages. One problem arises when one tries to empty a small diameter tube of formalin. The surface tension of formalin may not allow the fluid to escape unless it is disrupted by shaking the vial or by the insertion of a small instrument (which risks the adherence of the specimen to the instrument). Also, small specimens can adhere to the inner surface of a container. In any event, the recovery of a single, small organism from the fluid contents of a container requires careful procedures. One way of more safely disrupting the surface tension to allow the contents of a vial to be emptied is to add a drop of ethanol or a drop of wetting agent. The contents of a small, narrow-mouth (<10 mm in diameter) tube or vial flow out much more easily into an examination dish if either of the above are added beforehand. Several rinses into examination dishes is a good habit to develop.

Surface tension in vials can create an additional problem. When transferring specimens either by Irwin loop, needle, bamboo splint, or pipette, one counts on the specimen floating off into the fluid in the vial. The surface tension of formalin may cause transferred specimens to be retained on the surface film rather than sink to the bottom of the vial. Again, a drop or two of ethanol or a wetting agent will facilitate the successful transfer of the specimen. If specimens accumulate on the surface film, the sorter may pick up more specimens than he or she tries to deposit, thereby causing an error in enumeration.

Although some meiofaunal taxonomists object to the presence of a label inside the container with the organism(s), this is usually a good practice. Regardless of the kind of label material used, there is always some potential of an organism adhering to it. Therefore, the label should be retrieved with fine forceps and washed carefully to remove any adherent organisms into an examination dish. The inner label need only be large enough to bear some form of identification number or code. A single label placed on the outer surface of a container is subject to damage and the contents may never be resolved with associated data. One of the best procedures is to place the organisms in a vial, then place the vial(s) into a larger container with the same preservative. A good quality label should be in the outer bottle.

Not all meiofauna are preserved in small containers. Some collections of meiofauna, usually in the form of unprocessed sediment, are stored in larger containers, usually with buffered formalin as a preservative and often with rose bengal stain introduced. In general, sediment samples stored in this manner have three potential problems: (1) too much sediment and/or organic matter relevant to the volume of preservative, (2) a minimum strength (4%) formalin solution may be added without considering its dilution by fluid already in the sediment, and (3) the potential loss of specimens through abrasion caused by agitation of the sediment.

In the first instance, the volume of sample should not exceed 50 percent of the volume of preservative within the container. Additionally, the preservative should be monitored for change in pH at least on an annual basis. For long term preservation, ethanol may be used in place of formalin. However, the substitution of ethanol should be considered in terms of the potential use of the sample. Although ethanol will not deteriorate calcium carbonate exoskeletons, as is the potential in the use of formalin, it will cause the collapse of the body wall in more delicate taxa such as the Tardigrada, Loricifera, Gastrotricha, and others. The strength of the preservative should take into consideration the relative amount of moisture already in the sediment. In the long run, it is better to have too much formalin than not enough.

In the case of too much agitation of the sediment, only common sense can prevail; the less the material is handled, the better. Agitation of such samples not only abrades specimens within the sediment, but the label as well. Thus there is considerable merit in placing in the container a rigid plastic bag with the sample's identification code indelibly written on it.

Seeking Assistance in Taxonomic Identification

One of the most common questions asked of a taxonomist or curator is "How do I go about obtaining expert help with the identification of specimens?" Keys to taxa are often unreliable, particularly when they are outdated or deal with taxa known to be poorly described or which are known to have many undescribed species. The literature on a given taxon may be overwhelming; if not overwhelming by the volume of references available to the investigator, then frustrating by its lack of availability. Sooner or later, we all turn to an authority on the taxonomy of a given group of organisms if, indeed, such an authority exists.

Information on the protocol of seeking assistance with the identification of specimens is not easily found, probably because it is such a difficult subject to discuss.

After determining that a competent authority exists, it is always necessary to correspond with the taxonomist to inquire whether or not he or she is willing to help in the identification process, whether or not this will be done as a professional courtesy or only under a contractural arrangement with payment involved, and whether or not there are any special requirements such as sending only specimens that have been properly relaxed, mounted in a particular manner, unmounted, etc. Assuming that one is seeking identifications as a professional courtesy, the requester should recognize that the consent to help may depend on certain conditions. In most cases, a taxonomist, especially one who holds the position as a museum curator, will welcome valuable additions to the collections and, assuming there are no storage space problems, will at least see to it that the material is properly archived, cared for, and available for study, if requested. But whether or not the taxonomist will be willing to allot the time and effort for identification, especially within a prescribed time limit, is the primary question that must be asked. Before asking, a requester must recognize that the response from the taxonomist is dependent on such variables as: (1) the number of specimens to be identified, (2) the condition of the specimens, (3) the locality from which the specimens were obtained, (4) the relative importance of the specimens to current interests or projects, or to possible future interest, (5) the degree of preparation of the specimens before identification can be achieved, (6) the perceived usefulness of the identification relevant to the project involved, (7) the amount of time available to the taxonomist for such identification, and (8) whether or not, after all the effort, the taxonomist will be able to describe any new material and/or retain material for his collections or museum collections.

In the initial communication with the taxonomist, all items, except for item 5, should be clearly addressed by the prospective requester. The last item, the extent of the taxonomist's prerogative to describe new taxa and retain certain voucher specimens for his muesum's collections is one of the more delicate subjects of this essay.

The taxonomist may expect to have the right to describe new material if it is sent to him for identification. The requester should make it clear that he is prepared to acknowledge these rights or negotiate some alternative arrangement. Most taxonomists who receive specimens for identification are willing to share authorship with the party requesting identification if that party demonstrates a genuine interest in the taxonomic problem, a willingness to participate in the work to whatever extent is practical, and communicates a desire to learn more about the taxonomic identification through the proposed collaboration.

At one end of the spectrum in this matter is the requester who asks (or worse, expects or even demands) that his or her specimens be identified promptly and returned so that any new species can then be described by the requester only. In this case the requester should not be surprised at a negative reply. On the other hand, a requester who offers to send specimens, along with copies of his or her preliminary illustrations of the species, an analysis of the taxonomic characters, and the apparent relationships between the specimen(s) and described taxa -- demonstrates a serious effort in independently attempting to identify the specimen(s) -- generally can expect most taxonomists to offer much more assistance. In most cases, the expert will confirm the sender's identification, correct the sender's identification, or declare the species undescribed. In the latter case, the expert may suggest that a coauthorship be arranged. In either case, the expert may request that some duplicate specimens be deposited in his museum upon the completion of the research.

A curator may also invite the requester to visit his museum and use the literature and reference collections to help the requester do his own identifications. Donors of specimens to any museum have the right to expect that a representative of each taxon, for which there are duplicates, be returned to a reputable museum in the donor's country, and/or in the country (applicable) from which the specimens were collected, and occasionally to the donor personally if sufficient material is available. Many museums archive such specimens in a manner so as to transfer representative specimens at a later date, often at such time museum facilities are available, if requested to do so. Collectors and/or donors of specimens also have the right to expect the courtesy of acknowledgment in any scientific publication(s) resulting from research on the specimens, as well as a copy of any publication(s). It would be well if journals had a policy of not publishing taxonomic or ecological studies without the author's inclusion of a statement on the location of voucher material.

Access to Voucher Specimens

Most museums have a clear policy of making specimens available for scientific study. Usually, the curator of the collection in question makes such decisions. Many museums will make the specimens available through loans, assuming the request is made by a responsible individual or institution. Some specimens such as holotypes may not be made available for loan, but an investigator may examine such material at the museum. Voucher material mentioned in publication should be clearly identifiable by a museum catalog or accession code or number in order to facilitate the retrieval of specimens for future studies.

Shipping of Specimens

Only a few comments need be made about the shipping of specimens. Clearly, the regulations concerning noxious materials must be addressed by the shipper. Fortunately, no marine meiofauna are on the "Endangered Species List," they have "no commercial value," and it is difficult to imagine that the shipping of such material would be for any other reason than "for scientific study." These three items should be noted on shipping labels and invoices. In addition, the shipper should be certain that all customs regulations are met when sending materials to another country.

Although one should not have to be reminded of the fragile nature of vials and other containers, far too many shipments of specimens, sorted and in small vials or in bulk samples of sediment, are not well protected against leakage, breakage, or complete annihilation by the postal system. Despite these comments, I have received, intact, one 0.5 ml glass screw-cap vial with a specimen in ethanol, shipped clad only in an airmail envelope. Considering the costs of making collections, it is folly not to be very careful in the packaging and shipping of materials.

References

Blackwelder, R.E.
1967. *Taxonomy.* xiv + 698 pages. New York: John Wiley and Sons.

Hennig, W.
1950. *Grundzüge einer Theorie der phylogenetischen Systematik.* 370 pages. Berlin: Deutscher Zentralverlag.

Lee, W.L., B.M. Bell, and J.F. Sutton
1982. *Guidelines for Aquisition and Management of Biological Specimens.* vii + 42 pages. Lawrence, Kansas: Association of Systematics Collections.

Mayr, E.
1969. *Principles of Systematic Zoology.* xi + 428 pages. New York: McGraw Hill, Inc.

NERC Working Party on the Role of Taxonomy in Ecological Research
1976. *The Role of Taxonomy in Ecological Research: Report of the NERC Working Party on the Role of Taxonomy in Ecological Research, Series B(14).* iv + 48 pages. London: Natural Environment Research Council.

Ride, W.D.L., C.W. Sabrosky, G. Bernardi, and R.V. Melville, editors
1985. *International Code of Zoological Nomenclature.* xx + 338 pages. Berkeley: University of California Press.

Ross, H.H.
1974. *Biological Systematics.* 345 pages. Reading, Massachusetts: Addison-Wesley Publishing Company, Inc.

Simpson, G.G.
1945. The Principles of Classification and a Classification of Mammals. *Bulletin of the American Museum of Natural History,* 85:1-350.

Wiley, E.O.
1981. *Phylogenetics: The Theory and Practice of Phylogenetic Systematics.* xv + 439 pages. New York: John Wiley and Sons.

17. Sarcomastigophora

Andrew J. Gooday

The phylum Sarcomastigophora includes an unwieldy, and probably polyphyletic assortment of protozoa, united by the possession of pseudopodia (the amoebae), one or more flagella (the flagellates), or both these kinds of locomotory organelle (Levine et al., 1980). Cilia are developed only in the parasitic opalinids. With the exception of some foraminifera, members of the phylum have only one kind of nucleus, a feature which distinguishes them from the ciliates. Reproduction is usually by binary fission, less commonly by multiple fission or budding. Sexual reproduction, where known, normally involves the fusion of gametes from different individuals to form a zygote (syngamy). Most sarcomastigophorans are free-living but a few are parasitic. Some mastigophorans (flagellates) form colonies. The sarcodinans (amoebae) may be naked or possess a secreted or agglutinated shell (test). In their size, members of the phylum span five orders of magnitude from tiny flagellates and amoebae a few microns long to the giant xenophyophores, some species of which construct tests up to 25 cm across.

Classification

The classification of the phylum, based on the scheme of Levine et al. (1980), is summarised in Table 17.1. The two principal subdivisions are distinguished mainly by the type of locomotory organelle present: flagella in the subphylum Mastigophora, and pseudopodia in the subphylum Sarcodina. However, this is probably artificial since many taxa possess both kinds of organelle.

The mastigophorans are abundant in marine (Lighthart, 1969; Sieburth, 1979) as well as in freshwater benthic habitats. The ecology of flagellates living in marine sediments has recently been reviewed by Patterson, Larson and Corliss (in press). Most flagellates are very small and belong with the nanobiota. The amoebae are subdivided according to whether they have lobose (class Lobosea), filiform (class Filosea), or branching (class Acarpomyxea) pseudopodia. The Lobosea embraces both naked amoebae (subclass Gymnamoebia) and many of the testate amoebae (subclass Testacealobosia). Naked amoebae are abundant and widespread in freshwater and in damp, terrestrial habitats such as soils and leaf litter (Page, 1976). Their importance in benthic marine environments is becoming appreciated to an increasing extent. Many species have been isolated from the intertidal zone and from shallow water (Bovee and Sawyer, 1979; Page, 1983) but they also occur in off-shore sands (Sawyer, 1980) and deep-sea muds (Burnett, 1981, and earlier papers). They consume a wide range of microorganisms, including bacteria, algae and other protozoa, and may well play an important role in marine ecology. Some amoebae are large enough to fall within the meiofaunal size range. However, even these forms are seldom seen in normal meiofaunal samples, and techniques involving isolation and culture are essential for their evaluation and identification. Page (1976, 1983) gives excellent introductions to the taxonomy of naked amoebae and to the methods involved in culturing and observing them. A key to marine representatives of the group is given by Bovee and Sawyer (1979).

Most testate amoebae have been described from terrestrial and freshwater habitats, for example, soils, *Sphagnum* mosses, shallow lakes and ponds (Ogden and Hedley, 1980). They may live also in brackish water, however, and sometimes are discharged into near-shore marine environments by rivers (Hamen, 1982), particularly after floods or storms (Bovee and Sawyer, 1979). Currently believed to be related to the testate amoebae, but sufficiently distinct to occupy a separate order, is the curious marine rhizopod *Trichosphaerium*, of which both spicule bearing (Sheehan and Banner, 1973) and naked (Page, 1983) variants occur.

The class Filosea includes a variety of testate amoebae, as well as a few naked forms. Like the testacealobosians, which they may resemble, most of these amoebae live in freshwater or damp terrestrial habitats (Ogden and Hedley, 1980). However, members of one family, the Psammonobiotidae, are common in supralittoral marine environments in many parts of the world. They are described in numerous papers by Golemansky (for example, 1974b, 1979, 1982; Couteaux and Golemansky, 1984). Sometimes occurring with psammonobiotids are tiny (<100 μm) testate rhizopods of the genus

Lagenidiopsis (Golemansky, 1974a). These are particularly intriguing because they resemble the foraminifera *Lagena* in overall shape but have a shell which is organic rather than calcareous. The affinities of *Lagenidiopsis* are unclear although Lee, Hutner, and Bovee (1985) place it in the filosean family Gromiidae. A larger and more familiar representative of the same family is *Gromia*, which includes *G. oviformis*, a well known inhabitant of marine tide pools.

In marine environments, the most conspicuous and easily studied sarcomastigophorans are the foraminifera. These are classified as an order within the class Granuloreticulosea because they possess delicate, filiform, granular pseudopodia which branch and anastomose. All foraminifera are testate, although some primitive forms may be able to leave their tests and live as naked amoebae recognizable as foraminifera only by their granuloreticulate pseudopodia (Marszaleck, Wright, and Hay, 1969). A characteristic feature of the group is the alternation of a haploid generation, which reproduces sexually, with a diploid generation which reproduces asexually. This haplodiplodont life-cycle is found in higher plants but is unique among protists and unknown in the animal kingdom (Grell, 1973).

The major subdivisions of the order Foraminiferida are based on the structure and composition of the test wall (suborders) and the major features of the test and its constituent

Table 17.1.—Classification of the phylum Sarcomastigophora (after Levine et al., 1980) and the major habitats in which the constituent taxa are normally found.

	Parasitic or Symbiotic	Terrestrial[1]	Freshwater	Marine
Phylum Sarcomastigophora				
Subphylum Mastigophora				
Class Phytomastigophorea	*	*	*	*
Class Zoomastigophorea	*	*	*	*
Subphylum Opalinata	*	-	-	-
Subphylum Sarcodina				
Superclass Rhizopoda				
Class Lobosea				
Subclass Gymnamoebia	*	*	*	*
Subclass Testacealobosia	-	*	*	*
Class Acarpomyxea	-	*	*	*
Class Acrasea	-	*	-	-
Class Eumycetozoea	-	*	-	-
Class Plasmodiophorea	*	-	-	-
Class Filosea	-	*	*	*
Class Granuloreticulosea	-	*	*	*
Class Xenophyophorea	-	-	-	*
Superclass Actinopoda				
Class Acantharea	-	-	-	*
Class Polycystinea	-	-	-	*
Class Phaeodarea	-	-	-	*
Class Heliozoea	-	*	*	*

[1]Soils, leaf litter, mosses, etc.

chambers (superfamilies and families) (Loeblich and Tappan, 1964, 1984; Haynes, 1981; Lee, Hutner, and Bovée, 1985). Most genera are defined on the basis of overall chamber arrangement and the nature and location of the aperture, while species are distinguished mainly by the number, size, and shape of the chambers. Lewis (1970) provides an illustrated glossary of the morphological terms applied to the test. Little is known about the taxonomic importance of cytological features although one genus has been established on the basis of nuclear structure.

The Allogromiidae, a family in which the test wall is predominately organic (proteinaceous), deserve particular attention. Allogromiids are sometimes common in meiofaunal samples (Ellison, 1984; Gooday, 1986a, 1986b) but, since many are small and delicate, they have often been overlooked or ignored. Some representative deep-sea specimens are illustrated in Figure 17.1a-f.

Because the tests of many species fossilize readily and are important in biostratigraphy and palaeoecology, foraminifera have been studied intensively by geologists who have generated much of the vast taxonomic literature. The textbooks of Phleger (1960), Murray (1973), Boltovskoy and Wright (1976), and Haynes (1981) are valuable sources of information on living forms. Foraminiferal biology, ecology and classification are reviewed by Hedley (1964), Lee (1980), Lipps (1983) and in the volume edited by Broadhead (1982). Important taxonomic works are those of Brady (1884), Cushman (1948), and Loeblich and Tappan (1964, 1984). Lewis (1970) and Lee, Hutner, and Bovee (1985) give keys for Recent taxa and Murray (1971, 1979), and Todd and Low (1981) provide descriptions and keys for important species in British waters and off the coast of the northeastern United States respectively. A valuable (but not widely available) aid for identifying species is the Ellis and Messina (1940), *Catalogue of Foraminifera*, a compilation of original descriptions which is constantly updated.

Two other free-living, aquatic sarcomastigophoran taxa require a brief mention. The Xenophyophorea are a class of testate rhizopods and include the largest of all protists (Tendal, 1972). They are restricted to the deep-sea. The Heliozoea include the only benthic representatives of the superclass Actinopoda (radiolarians and their allies). Some genera are attached by a stalk to hard substrates. Heliozoans have been described mainly from freshwater but also occur in shallow marine habitats (Mare, 1942; Jepps, 1956; Golemansky, 1976).

Most of the remainder of this chapter is concerned with the Foraminiferida. The xenophyophores are omitted because they are too large to be considered among the meiofauna. Many flagellates and naked amoebae, on the other hand, belong with the nanobiota. Moveover, even the larger species are not usually encountered during meiofaunal investigations and their study requires special techniques. The meiofauna of freshwater lakes and ponds may include some testate amoebae but since these rhizopods are more typical of damp, terrestrial habitats, they are not considered further. However, marine testate amoebae, which to some extent resemble foraminifera, are briefly mentioned and a few species illustrated (Figure 17.2a-d,g).

General accounts of living foraminifera are given by Jepps (1956), Loeblich and Tappan (1964), Murray (1973), Arnold (1974), Boltovskoy and Wright (1976) and Lipps (1983) and descriptions of particular species by authors such as Jepps (1942) and Arnold (1964, 1979). Most observations on live foraminifera have been made in culture. Active specimens are surrounded by a reticulate network of pseudopodia (Sheehan and Banner, 1972) which is in constant streaming motion. This network extends out for a distance of several test diameters and is used in locomotion and food collection. Some species, for example *Elphidium crispum*, collect food particles (mainly diatoms) into a feeding cyst which completely envelopes the test (Jepps, 1942). Several cysts may be formed and discarded during a single day. Other forms, for example, miliolids, accumulate a bolus of food around the aperature (Arnold, 1964). Debris may also adhere more generally to the test surface. Most shallow-water benthic foraminifera have protoplasm which is colored green, brown, purple, orange, yellow, or red during life by the presence of symbiotic algae or pigmented food particles (Vénec-Peyré and Le Calvez, 1986). The protoplasm of living foraminifera does not necessarily occupy the entire test and indeed may be withdrawn from the outer chambers (Marszalek, Wright, and Hay, 1969).

Pseudopodia are not usually visible in preserved specimens. These also lose much of their bright coloration, although some protoplasmic pigmentation is often present. The tests of preserved specimens may also retain adherent debris as well as food accumulations around the aperatures.

Foraminifera can occasionally be confused with other organisms. *Marenda nematoides,* a rhizopod described by Nyholm (1951) as a foraminifera but classified by Loeblich and Tappan (1964) as a testate amoeba, resembles a nematode. The tiny agglutinated species *Turritellella shoneana* lives enveloped in a delicate cocoon attached to an algal frond or other substrate (Arnold, 1979). These cocoons can be mistaken for hydroids. Some transparent allogromiids have protoplasm filled with brown debris and look like fecal pellets (Nyholm and Gertz, 1973).

Habitats and Ecological Notes

Occurrence.--Foraminifera occur in all marine environments and a few species penetrate into brackish or fresh water. They are normally most common in fine-grained sediments. High standing crops (>1000 live specimens per 10 cm^2) occur in areas such as intertidal mudflats (Ellison, 1984), coastal lagoons and associated marine marshes (Phleger, 1960, 1976; Matera and Lee, 1972) and regions of fine sediment on the continental shelf and upper slope where the nutrient supply is enhanced through upwelling (Murray, 1973; Phleger, 1976; Sen Gupta, Lee, and May, 1981). In some of these habitats, foraminifera are among the most important members of the meiofauna (Gerlach, 1971; Nyholm and Olsson, 1973; Olsson, 1975; Wefer and Lutze, 1976; Ellison, 1984). They are also of considerable importance in the deep-sea (Thiel, 1983; Snider, Burnett, and Hessler, 1984; Gooday, 1986b).

In general, foraminifera are less abundant in coarse sands without some silt and clay than in muds, presumably because food is scarcer. Species occurring in quartz sands may be attached to individual grains (Rhumbler, 1938). Usually, however, they cannot live in sands exposed to heavy surf action because their tests are unable to withstand the grinding action of the grains (Arnold, 1974; Phleger, 1976; Lipps and DeLaca, 1980).

Sediment Penetration.--Many sediment-dwelling foraminifera live on or near the sediment surface where the nutrients are concentrated and the porewater well aerated. Some species occur infaunally, usually in the upper few centimeters of sediment but sometimes deeper, while others may be both infaunal and epifaunal. Vertical stratification of species is known to occur in deep-sea sediments (Corliss, 1985; Gooday, 1986b). Epifaunal and infaunal forms are able to move on and within the sediment (Severin et al., 1982) and subsurface reproduction and feeding have been documented (Frankel, 1972, 1975).

Attachment to Substrates.--This is a common mode of life among foraminifera (Lipps, 1983). Attachment is achieved either by permanent cementation of the test or by the organisms clinging on with their pseudopodia, in which case they can move when necessary. Plants are also used as substrates. For example, species living in salt marshes may be attached to marine grasses (Lee et al., 1969) while coralline algae growing in intertidal rock pools frequently yield a rich harvest of foraminifera (Myers, 1940; Hedley, Hurdle, and Burdett, 1967). Lipps and DeLaca (1980) give a detailed account of the relationship between foraminifera and macroalgae around the Antarctic Peninsula. Only finely branched algal species associated with high concentrations of dissolved and particulate organic carbon and bacteria support foraminiferal populations. Foraminifera also live on other organisms, both sessile (Dobson and Haynes, 1973) and mobile (Mullineaux and DeLaca, 1984; Alexander and DeLaca, 1987), as well as on shell fragments, rock surfaces, pebbles, manganese nodules, and the dead tests of other foraminifera.

Spatial Variability.--In general, foraminifera are distributed unevenly in shallow-water, sedimentary environments, in terms of the standing crop of both the total assemblage and individual species. Richter (1961), for example, found that two samples (40 cm^3 of sediment) taken only a few meters apart in Jade Bay (North Sea) yielded 14 and 293 living specimens while Buzas (1970) recorded 1, 67, 294, 46 and 31 specimens of *Ammonia beccarii* per 10 cm^2 area in replicates from a station in Rehoboth Bay (Delaware). Such patches may result from asexual reproduction (Buzas, 1968; Murray, 1973), from the foraminifera clustering around food sources, or from local disturbances. Thus Lee et al. (1969) reported that areas of decaying *Enteromorpha* supported the highest foraminiferal standing crops in a salt marsh community. Murray (1976) concluded that patchiness generally increases with the variability of the habitat and, therefore, is particularly characteristic of marginal marine environments. However, it also exists in the deep-sea as demonstrated by Bernstein et al. (1978) who found both kilometer and centimeter scale patchiness among the total (live plus dead) foraminiferal fauna at abyssal depths in the central North Pacific. Populations of attached forms may display considerable small-scale variability arising, for example, from differences in the attractiveness of algal substrates or the need to occupy rock crevices to escape predation (Lipps and DeLaca, 1980).

Temporal Variability.--Foraminiferal populations can vary in time as well as space. In many nearshore habitats, species reproduce at particular times of the year and this leads to seasonal fluctuations in their abundance (Boltovskoy and Wright, 1976:33). Some species undergo rapid population increases in response to blooms in their algal food sources (Lee et al., 1969). Seasonal variations in total live foraminiferal abundances have been recorded in deeper water by Faubel et al. (1983) who discovered that, at a depth of 134 m in the North Sea, foraminifera were more abundant during August 1975 and July 1976 than between December 1975 and June 1976. They speculate that the highest reproductive activity may occur during June to August as a somewhat belated response to the spring bloom.

Methods of Collection.

Supralittoral Environments.—Testate amoebae often occur above the high-tide line in the interstitial water of sandy sediments. They can be collected by digging a shaft about a meter deep and then pipetting or siphoning off the water which seeps into the bottom, care being taken not to contaminate the sample with too much sand (C.G. Ogden, personal communication; Chappuis, 1942). A similar technique involving a trench rather than a shaft is described briefly by Gerlach (1972:121). Seepage into such shafts is very slow and hence the sample volume is only about 50 ml, although this obviously can be increased if the water is allowed to collect over an extended period of time. The testate amoebae are very small (15–70 μm) and therefore must be extracted from the groundwater in the laboratory using a Millipore filter with a porosity of 8 μm (Couteaux and Golemansky, 1984:266).

Intertidal and Near-shore Environments.—Epiphytic foraminifera are easily obtained by shaking algae or marine grasses vigorously in a container of sea water and sieving the resultant residue (Myers, 1940; Lee et al., 1969). Attention should be paid to algal holdfasts within which some species congregate. The holdfasts can be detached from their substrate by means of a hunting knife (Arnold, 1974). Another simple tool, made by recurving the leading edge of a putty knife (Arnold, 1974), is useful for scraping attached specimens from rock surfaces. Intertidal sediments can be sampled quantitatively simply by scooping the upper centimeter from an area of 100 cm^2 or more (Murray, 1979), or by inserting a hand-held coring tube. As waves lap up onto the beach they often carry with them light, empty foraminiferal tests which become concentrated and stranded at the highest point reached by each individual wave. Careful sampling of the sediment surface at such points often yields a good collection of dead specimens.

Beyond the maximum wading depth, qualitative samples may be taken by means of a shorebased pipe dredge or home-made epibenthic sledge (Arnold, 1974; Boltovskoy and Wright, 1976). A simple apparatus for obtaining quantitative material in shallow water (up to 6 meters) is the Lankford sampler, a small coring device driven into the sediment on the end of a long pole (Boltovskoy and Wright, 1976).

Arnold (1974) has described a sieving system for concentrating specimens and a microscope for examining them in the field. He emphasizes the value of thus being able to assess intertidal and near-shore samples at the time of their collection.

Deeper Water.—Foraminifera are usually sufficiently abundant in off-shore environments to be caught by any shipboard sampler that retains fine-grained sediments (Phleger, 1960; Murray, 1973; Arnold, 1974; Boltovskoy and Wright, 1976; Haynes, 1981; Douglas and Woodruff, 1982). Towed gears, such as the epibenthic sledge, Agassiz trawl and anchor dredge, are useful for obtaining a qualitative or semi-quantitative impression of the foraminiferal fauna. In the deep-sea, such samples may be a rich source of foraminiferal material, invaluable for taxonomic studies, but always more or less biased by the washing out of finer sediment particles and the resulting concentration of larger specimens. A further difficulty is that the volume of sediment from which the catch is concentrated is unknown, or at best subject to considerable uncertainty. In the case of the anchor dredge, the problem of sediment washing is minimized by the use of a closely woven canvas bag to retain the catch.

Quantitative foraminiferal samples from deeper waters have been obtained by a variety of grabs and corers which take sediment from a known area of sea floor. It is of crucial importance that the sediment-water interface is recovered in as undisturbed a condition as possible. Often, the superficial layer contains numerous delicate foraminifera. These are vulnerable to displacement by the bow-wave which precedes many such devices or are lost subsequently by washing and winnowing of the surface sediments as the gear returns through the water column. A vivid illustration of the impact of sample quality on foraminiferal standing crops is given by Douglas et al. (in Douglas and Woodruff, 1982, figure 3). They rated box cores on a scale of 1 to 5 (from poor to excellent quality). Samples considered to be poor yielded more than an order of magnitude fewer live foraminiferans than those rated excellent.

A corer which is widely used for the quantitative collection of live foraminifera has been described by Phleger (1960). It is able to recover fine-grained sediments from depths down to about 1000 meters and, with modifications, in even deeper water. A shorter and wider version of the Phleger corer is described by Lewis (1970:79). A gear with similar capabilities is the Lundquist corer, used by Höglund (1947) in his classic paper on the foraminiferal fauna of the Skagerrak and Gullmar Fjord. More recently, the Barnett-Watson Multiple Corer has proved ideal for collecting fine-grained sediments in deep-water (Barnett, Watson, and Connelly, 1984). One of the main virtues of this corer is its ability to obtain the sediment-water interface in a virtually undisturbed condition. It is useful also for studies of the small-scale distribution of foraminiferal populations. The upper centimeter or so of fine-grained, deep-sea sediment cores is usually very soft and must be removed with great care in order not to lose the vital flocculent surface layer. Phleger (1960) has illustrated a cutting device for slicing off the top 1 cm of small cores and for removing further sediment layers if necessary. Box-corers have been used with considerable success in a number of deep-sea foraminiferal studies (e.g., Tendal and Hessler, 1977, Bernstein et al., 1978; Douglas and Woodruff, 1982; Snider et al., 1984). They obtain a large block of sediment which can be subsampled by means of plastic or metal tubes, "meiostecher," or syringes.

Geologists have employed a variety of grabs to collect quantitative samples for foraminifera. Among those most commonly used are the Van Veen, orange peel, and Shipek grabs (Murray, 1973). However, many of the samples obtained by grabs are disturbed by bow-wave effects and sample washing and hence any numerical data they yield on living foraminifera must be treated with caution. Murray (1969) used a special grab which collects a relatively undisturbed slice of sea-floor 100 cm^2 in area.

Sample Size.--This factor must always be considered when foraminifera are to be collected quantitatively (Douglas, 1979, figure 1). To evaluate these protozoa as part of the meiofauna, only small samples are necessary, at least in fine-grained sediments. For example, Gooday (1986b) found that syringe subsamples (3.46 cm^2 surface area, 5 cm sediment penetration) from 1320 m in the NE Atlantic yielded large foraminiferal assemblages (434-759 live specimens). However, many specimens (particularly those retained on the 63 μm and 42 μm sieves) were juveniles with relatively few species being represented by sufficient adult specimens to document their importance adequately. Geologists study much larger sediment volumes but this is only practical if the fine residues are not examined. In many cases a standard sampling area of 10 cm^2 has been used although Murray (1976) argued that this is too small. He suggested that an area of 30 cm^2, should yield a suitable standing crop of 100-500 live individuals. Douglas (1979) and Douglas and Woodruff (1982) came to similar conclusions. Thus, there are difficulties involved in using one sample size to quantify both the larger and smaller foraminifera. The only way to evaluate the entire population at a particular locality is to use a range of sample sizes, with smaller samples being sieved on progressively finer meshes.

Methods of Extraction

Methods for isolating living foraminiferans from sediment or plant samples are dealt with exhaustively by Arnold (1974; see also Murray, 1979). Recently Schwinghamer (1981) has described a technique for quantitatively extracting delicate, living specimens of "soft" meiofauna. He found that the procedure, which involves centrifugation of marine sediments in a silica sol-sorbitol mixture, worked well with allogromiids but was less efficient for "testate" (presumably hard-shelled) foraminifera.

Preserved samples must be sieved with great care so as not to damage or destroy fragile forms. Tendal (1979) recommends a gentle elutriation method. The sieved residues are stained with Rose Bengal and the foraminifera picked out by hand. This process is best carried out in a Petri dish with the residue immersed in water or very dilute formalin. Larger and more robust specimens can be picked out from the coarser residues using a pair of fine forceps. However, most meiofaunal foraminifera are too small and delicate to be handled in this way and must be extracted by means of a fine pipette or an Irwin Loop.

Hand sorting, abandoned for most other meiofaunal taxa because it is so time-consuming, remains essential for the proper evaluation of preserved foraminifera which were living when captured. In part, this is because the group is so variable, both morphologically and in the composition and structure of the test wall, making it impossible for all the different types to be extracted using elutriation or flotation techniques. Another difficulty is that some species live in protected habitats, for example the empty tests of other rhizopods, or are attached to hard substrates such as mollusc shells. Foraminifera are often particularly difficult to extract from deep-sea samples. They may shelter inside empty globigerinacean shells or be hidden within mudballs or cocoons of fine sediment. According to Thiel et al. (1975), ultrasonic treatment of samples before sorting helps to reveal the more elusive specimens. They found that up to 30% of meiofaunal foraminfera in samples from the San Diego Trough were counted only after the application of ultrasound. However, a disadvantage of this method is that it tends to break up the delicate agglutinated forms which are often common in deep-sea material.

Distinction between Live and Dead Specimens.--This is central to the accurate quantification of foraminifera. The only truly reliable method for determining whether specimens are live or dead is to watch them carefully for signs of life (Arnold, 1974). However, with preserved material the standard procedure is to stain the protoplasm with Rose Bengal. As is well known (Boltovskoy and Wright, 1976; Vénec-Peyré and Le Calvez, 1986), this method has limitations. A cytological examination of foraminifera from low oxygen environments on the

Southern California Borderland revealed that 30% of stained specimens were not alive when collected (Douglas et al., 1980). The stained contents should be examined carefully to ensure that they constitute a fresh, protoplasmic mass. In the case of small foraminifera, the nature of the contents can often be determined if the test is immersed in glycerol and examined under a high-powered microscope. Larger specimens with opaque walls, however, must be crushed in order to determine whether they contain protoplasm.

Walker, Linton, and Schafer (1974) advocated the use of Sudan Black B as a superior stain to Rose Bengal for the recognition of live foraminifera. However, it is more complicated to use than Rose Bengal and has not been adopted widely. Tendal and Hessler (1977) considered Rose Bengal to be a poor stain for komokiaceans and recommended a mixture of Eosin B and Biebrich Scarlet diluted 1:1000 with either 11% formalin or distilled water. Members of this superfamily usually contain small amounts of protoplasm and live specimens can be recognised reliably only by sectioning (Tendal and Hessler, 1977).

Vénec-Peyré and Le Calvez (1986) recommended observation of the natural cytoplasmic coloration as a good method for recognising live foraminifera, even when preserved. Additional clues in preserved material are the fresh appearance of the test, the accumulation of a bolus of debris outside the aperture, and the development of a cocoon of fine sediment around the test. In the case of certain single chambered proteinaceous and agglutinated forms, much of the test volume in live specimens is occupied by stercomata (waste pellets) which disintegrate rather rapidly after death into a fine, dark grey powder.

Preparation of Specimens for Taxonomic Study

Fixation.--For routine purposes foraminifera, or sediments containing them, should be fixed in buffered 10% formalin. Detailed information about the fixative properties of formalin are given in Steedman (1976). The most suitable buffer for use with calcareous organisms such as foraminifera is borax (Bé and Anderson, 1976) which will maintain pH at around 8.2. Hexamine may also be used. In order to fix samples adequately, it is necessary to use a sufficient volume of formalin. A maximum sample to fixative volume ratio of 1:9 is suggested by Steedman (1976) for 5% formalin. Rather higher proportions are allowable if 10% formalin is used.

Following fixation, foraminifera are usually stored in 80% alcohol which is less noxious than formalin and non-acidic. However, alcohol causes soft-shelled forms, particularly allogromiids, to shrink through dessication and sometimes it may be necessary, therefore, to use buffered formalin as a storage fluid. If samples are stored in formalin for lengthy periods, it is essential to check the pH, preferably with a pH meter, several times a year and add additional borax if necessary. Steedman (1976) recommends 0.2 g of borax per 300 ml fixative to correct declining pH levels. Precipitation of borax occurs if this amount is exceeded.

A variety of other fixatives (Bouins, Carnoy, Clarke, Serra, and Zenker) has been used to fix foraminifera for special cytological purposes (Boltovskoy and Wright, 1976). Bé and Anderson (1976) described a procedure for fixing planktonic foraminifera in glutaraldehyde and post-fixing in osmium tetroxide. This gives excellent results if specimens are to be examined using transmission electron microscopy and should be applicable to benthic species.

Storage.--The most common storage method for hard-shelled foraminifera (mainly calcareous species and multichambered agglutinated species) is dry-mounting on cardboard or plastic slides (Boltovskoy and Wright, 1976:315; Haynes, 1981:15). Specimens are stuck down using a water-based glue such as gum tragacanth (made up with dilute formalin to prevent fungal growth). Proteinaceous and agglutinated species with flexible tests must not be dried. Often, they can be simply transferred to a cavity slide filled with glycerol. The slide should be placed in a desiccator or oven (35-40° C) for several days to render it anhydrous (Platt and Warwick, 1983), after which the cavity can be protected with a coverglass. Numerous small, fragile foraminifera, for example, all those from a particular residue, can be stored for months in this way.

Unfortunately, the hydroscopic properties of glycerol cause shrinkage of species with thin, proteinaceous test walls (many allogromiids). Larger specimens can be stored in a small vial with 4% formalin (not alcohol which also causes shrinkage). Smaller specimens, which could be lost in a vial, should be transferred gradually into anhydrous glycerol. The foraminifera are first placed in a glass cavity block containing 3% glycerol in distilled water. The water is then evaporated off by placing the cavity block in an oven for several days (35-40° C), leaving the specimens in pure glycerol. Even with this careful procedure, however, some delicate allogromiids may shrink.

Impregnation.--A promising method for examining foraminifera *in situ* is described by Frankel (1970). He fixed and stained cores and then impregnated them with epoxy resin (Epon 812). The hardened cores were sectioned and observations on foraminifera and other organisms in the sections made under a high-power microscope. This technique has yielded information on reproduction and feeding among infaunal foraminifera (Frankel, 1972, 1975). However, it can be used only with sediments which are sufficiently coarse to allow the resin to infiltrate.

Geological methods.--Much of the basic biological research on foraminifera carried out over the last forty years has been done by geologists whose methods for extracting live specimens from sediments often differ from those described above in two important respects. First, the finest size fraction examined is usually 63 μm (240 mesh) or 76 μm (200 mesh) while in some deep-sea studies the lower size limit adopted is 125 μm, 150 μm or even 250 μm (Schnitker, 1980; Douglas and Woodruff, 1982). The use of such relatively coarse meshes results in the loss of the smallest specimens in the meiofaunal size range. These "microforaminifera" (Boltovskoy and Wright, 1976:210) are often juveniles of species which are common in the coarser residues (Gooday, 1986b). However, they may constitute a significant proportion of the foraminiferal standing crop and therefore should be included in meiofaunal studies. Secondly, it is usual for samples to be sorted after being dried, either in air or in an oven at 60° C. Drying causes many delicate, soft-shelled foraminifera to become unrecognizable through shrinkage or to fragment and disappear from the sample. Therefore, species lists published by geologists are dominated by forms with rigid and, in most cases, multichambered tests. Wet samples on the other hand sometimes yield a diversity of fragile, single chambered species. A further disadvantage of drying samples is that tests often become less transparent and the protoplasm within them contracts, in some cases into a small, barely visible, lump. As a result, some foraminifera which contain obvious protoplasm when wet, display little trace of it when dry (Vénec-Peyré and Le Calvez, 1986).

Geologists sometimes use flotation methods to concentrate foraminifera. A heavy liquid such as tetrachloromethane (specific gravity 1.60) or bromoform (specific gravity 2.89), is poured onto the dried residue and the lighter, air filled tests float to the surface (Gibson and Walker, 1967; Lewis, 1970:80; Boltovskoy and Wright, 1976). The operation should be carried out in a fume chamber since these liquids give off highly toxic vapors. Floatation methods can be quantitative for hard- shelled species and are almost essential with large samples (Gibson and Walker, 1967). However, these methods are not really appropriate for meiofaunal studies because they involve drying and hence the loss or non-recognition of many delicate foraminfera.

Light Microscopy.--Preliminary observations can be made on specimens stored in anhydrous glycerol in a cavity slide. For more detailed examination of the test wall and cytoplasm, the methods used by nematode specialists have proved useful (Platt and Warwick, 1983:19). Specimens are transferred to another drop of anhydrous glycerol on a flat microscope slide and covered with a coverglass supported by suitably sized glass beads ("Ballotini" -- 100-125 μm diameter is appropriate for many meiofaunal foraminifera). The beads should be placed well clear of the specimens so as not to obscure them. A crystal of phenol is added to the glycerol to prevent the growth of molds. Once the coverglass is in place, the glycerol can be trapped by means of a suitable sealing agent (Platt and Warwick, 1983:20). I have found that the moutant D.P.X. provides an effective seal. Specimens can also be mounted permanently in dilute (10%) formalin if the coverglass is sealed in two stages using "Clearseal" and "Bioseal" (Northern Biological Supplies, Ipswich, Suffolk, U.K.). This may be essential for very delicate allogromiids which are damaged by immersion in glycerol.

A variety of stains has been used with foraminifera (Boltovskoy and Wright, 1976:311) but only rarely are they necessary for routine taxonomic purposes. However, it is sometimes worth staining with hematoxylin in order to bring out the nucleus and other cytoplasmic structures which occasionally may be of taxonomic importance. Soft-shelled forms should be stained in water since some species shrink when exposed to alcohol. An easy method is to pipette the specimens into a drop of aqueous hematoxylin in a cavity slide. Staining is usually rapid (less than 30 seconds) and to avoid overstaining it should be monitored under a stereomicroscope. The hematoxylin is then carefully pipetted off and replaced with tap water which, being slightly acidic, turns the stained cytoplasm blue. Finally, the specimen is transferred to glycerol in another cavity slide, dessicated and mounted as described previously.

Hard-shelled foraminifera can be examined in transmitted light if the test wall is not too thick. Clarifying agents such as glycerol (refractive index 1.473 when pure), xylene (refractive index 1.494) and immersion oil (refractive index 1.510) can be used to make the test appear more transparent. Specimens are best studied in a cavity slide or under a supported coverglass. Thick-walled specimens may have to be sectioned in order to study internal structures, although normally this lengthy procedure is not necessary for routine identifications. Details of the techniques involved are given by Boltovskoy and Wright (1976:327) and Haynes (1981:17).

Foraminifera mounted on glass slides can be examined under a high-powered microscope, a low power objective being adequate for general observations on the form of the cytoplasmic mass and its relationship to the test. To examine the test and cytoplasmic structures in detail, however, a 100x oil immersion lens is essential. Interference-contrast microscopy can be used for observing cytoplasmic structures. Phase-contrast is also useful and generally yields rather better photographic results because it involves a higher degree of contrast. Sieburth (1979) gives an account of these and other microscopic techniques as applied to marine protozoa.

Scanning Electron Microscopy.--In the case of many smaller foraminifera, the morphological features on which generic and specific identifications are based can be seen easily only in the SEM. Access to such an instrument is therefore almost essential for serious taxonomic work. A drawback of this technique, however, is that only external surfaces are visualized and so the foraminifera have a different appearance than when viewed under an ordinary light microscope. This, of course, applies particularly to small, transparent tests.

To prepare hard-shelled foraminifera for examination in the SEM, they must be rinsed carefully in 80% alcohol to remove any adhering debris, air dried, and then mounted on a specimen stub using Kodaflat glue, double-sided sticky tape or some other suitable adhesive before being coated with gold or aluminium in the usual way. Forms which collapse when dried, for example allogromiids and komokiaceans, are difficult to study by SEM although good results can be obtained if specimens are critical point dried before being mounted on stubs (Tendal and Hessler, 1977).

Internal structures can be studied by impregnating the test with a medium such as Canada Balsam or Lakeside 70 and then decalcifying in dilute acetic or hydrochloric acid (Boltovskoy and Wright, 1976:326; Haynes, 1981:17). When examined in the SEM, the resulting casts may reveal fine details of features such as pores and internal canals (for example, Alexander and Banner, 1984). To investigate wall structure by SEM, it is necessary first to treat fractured or sectioned surfaces with an etching agent, usually an aqueous EDTA solution (Haynes, 1981:18; Weston, 1984).

Type Collections

There are two main depositories for the types of Recent species:

British Museum (Natural History) houses specimens described by Brady (including the "Challenger" material), Heron-Allen and Earland (including the "Discovery," "Terra Nova," "Goldseeker" and Clare Island material) and, among others, by Brönnimann, Carpenter, Haynes, Norman, Parker and Jones, Sidebottom and Williamson (Adams, Harrison, and Hodgkinson, 1980).

National Museum of Natural History, Smithsonian Institution, Washington, D.C., houses many of Cushman's types (including species described in his North Pacific, Tropical Pacific, Philippine and Atlantic monographs) and also material of Flint, Parker, Todd, Loeblich and Tappan, Phleger, Bandy, and many others.

Other notable collections of Recent material include the following:

Royal Albert Memorial Museum, Exeter, contains the W.B. Carpenter collection which includes some important figured specimens (Murray and Taplin, 1984).

National Museum of Ireland, Dublin, houses material collected by Joseph Wright in Irish waters and includes a number of types.

Museum National d'Histoire Naturelle, Paris, houses the material of d'Orbigny.

Rijksmuseum van Natuurlijke Historie, Leiden, houses material described by Hofker (mainly during the Snellius 1, Luymes Saba Bank and Luymes Guyana Shelf expeditions).

Naturhistoriska Riksmuseet, Stockholm, houses material described by Höglund and Göes.

Naturhistorisches Museum, Vienna, houses the collection of Fitchel and Moll (Rögl and Hansen, 1984) which, although mainly fossil, includes several important species which are still extant.

Acknowledgments

I thank M.V. Angel, J.R. Haynes, J.W. Murray, F.C. Page, D.T. Patterson, and M.H. Thurston for critically reading the manuscript; C. Darter for drawing the plates from my rough sketches; V. Golemansky for permission to reproduce his published illustrations (Figures 17.2a,d-g); J.R. Haynes for providing foraminiferal material from Cardigan Bay; C.G. Ogden for information on testate amoebae, and J.E. de Hartog, R. Oleröd, C. O'Riorden, F. Rögl, L. Wallin, and J.E. Whittaker for information about museum collections.

References

Adams, C.G., C.A. Harrison, and R.L. Hodgkinson
1980. Some Primary Type Specimens of Foraminifera in the British Museum (Natural History). *Micropaleontology*, 26:1-16.

Alexander, S.P., and F.T. Banner
1984. The Functional Relationship Between Skeleton and Cytoplasm in *Haynesina germanica* (Ehrenberg). *Journal of Foraminiferal Research*, 14:159-170.

Alexander, S.P., and T.E. Delaca
1987. Feeding Adaptations of the Foraminiferan *Cibicides refulgens* Living Epizoically and Parasitically on the Antarctic Scallop. *Adamussium colbecki. Biological Bulletin*, 173:136-159.

Arnold, Z.M.
1964. Biological Observations on the Foraminifera *Spiroloculina hyalina* Schulze. *University of California Publications in Zoology*, 72:1-78, plates 1-7.

1974. Field and Laboratory Techniques for the Study of Living Foraminifera. Pages 153-206 in R.H. Hedley and C.G. Adams, editors, *Foraminifera*. London: Academic Press.

1979. A Cocoon-Building *Turritellella* (foraminifer) from California. *The Compass of Sigma Gamma Epsilon, Oklahoma*, 56:83-95.

Barnett, P.R.O., J. Watson, and D. Connelly
1984. A Multiple Corer for Taking Virtually Undisturbed Samples from Shelf, Bathyal and Abyssal Sediments. *Oceanologica Acta*, 7:399-408.

Bé, A.W.H., and O.R. Anderson
1976. Preservation of Planktonic Foraminifera and Other Calcareous Plankton. Pages 250-258 in H.F. Steedman, editor, *Zooplankton Fixation and Preservation*. Paris: The Unesco Press.

Bernstein, B.B., R.R. Hessler, R. Smith, and P.A. Jumars
1978. Spatial Dispersion of Benthic Foraminifera in the Abyssal North Pacific. *Limnology and Oceanography*, 23:401-416.

Boltovskoy, E., and R. Wright
1976. *Recent Foraminifera*. 515 pages. The Hague: W. Junk

Bovee, E.C., and T.K. Sawyer
1979. *Marine Flora and Fauna of the North-eastern United States. Protozoa: Amoebae*. NOAA Technical Report NMFS Circular 419:1-56. U.S. Department of Commerce.

Brady, H.B.
1884. Report on the Foraminifera Dredged by H.M.S. *Challenger* During the Years 1873-1876. *Reports of the Scientific Results of the Voyage of H.M.S. Challenger*, 9 (Zoology):1-814, plates 1-116.

Broadhead, T.M., editor
1982. Foraminifera. Notes for a Short Course. Organized by M.A. Buzas and B.K. Sen Gupta. *University of Tennessee, Department of Geological Sciences, Studies in Geology*, 6:1-219.

Burnett, B.R.
1981. Quantitative Sampling of Nanobiota (Microbiota) of the Deep-Sea Benthos - III. The Bathyal San Diego Trough. *Deep-Sea Research*, 28:649-663.

Buzas, M.A.
1968. On the Spatial Distribution of Foraminifera. *Contributions from the Cushman Foundation for Foraminiferal Research*, 19:1-11.

1970. Spatial Homogeneity: Statistical Analyses of Unispecies and Multispecies Populations of Foraminifera. *Ecology*, 51:874-879.

Chappuis, P.A.
1942. Eine neue Methode zur Untersuchung der Grundwasserfauna. *Acta Scientifica Mathematica et Naturalia Universitatis Francois-Joseph*, 6:1-7.

Corliss, B.H.
1985. Microhabitats of Benthic Foraminifera within Deep-Sea Sediments. *Nature*, 314:435-438.

Couteaux, M.-M., and V. Golemansky
1984. Premières observations sur les Thécamoebiens interstitiels du supralittoral marin des Antilles françaises. *Bulletin Muséum National d'Histoire Naturelle, Paris*, 4th series, 6, section A, number 2:263-277.

Cushman, J.A.
1948. *Foraminifera. Their Classification and Economic Use*. 605 pages. Cambridge, Massachusetts: Harvard University Press.

Dobson, M., and J. Haynes
1973. Association of Foraminifera with Hydroids on the Deep Shelf. *Micropaleontology*, 19:78-90, plate 1.

Douglas, R.C.
1979. Benthic Foraminiferal Ecology and Paleoecology: a Review of Concepts and Methods. Pages 21-53 in *Foraminiferal Ecology and Paleoecology*. SEMP Short Course No. 6. Houston: Society of Economic Paleontologists and Mineralogists.

Douglas, R.G., J. Liestman, C. Walch, G. Blake, and M.L. Cotton
1980. The Transition from Live to Sediment Assemblage in Benthic Foraminifera from the Southern California Borderland. Pages 257-280 in M.E. Field, A.H. Bouma, I.P.Colburn, R.E. Douglas, and J.C. Ingle, editors, *Quaternary Depositional Environments of the Pacific Coast*, Pacific Coast Paleography Symposium 4. Los Angeles: Pacific Section, Society of Economic Paleontologists and Mineralogists.

Douglas, R.C., and F. Woodruff
1982. Deep-Sea Benthic Foraminifera. Pages 1233-1327 in *The Sea*, volume 7, *The Oceanic Lithosphere*. New York: John Wiley and Sons.

Ellis, B.F., and A.R. Messina
1940. *Catalogue of Foraminifera*. New York: American Museum of Natural History.

Ellison, R.L.
1984. Foraminifera and Meiofauna on an Intertidal Mudflat, Cornwall, England: Populations; Respiration and Secondary Production; and Energy Budget. *Hydrobiologia*, 109:131-148.

Faubel, A., E. Hartwig, and H. Thiel
1983. On the Ecology of the Benthos of Sublittoral Sediments, Fladen Ground, North Sea. 1. Meiofauna Standing Stock and Estimation of Production. *"Meteor" Forschungsergebnisse, Serie D*, 36:35-48.

Frankel, L.
1970. A Technique for Investigating Microorganism Associations. *Journal of Paleontology*, 44:575-577.

1972. Subsurface Reproduction in Foraminifera. *Journal of Paleontology*, 46:62-65.

1975. Subsurface Feeding in Foraminifera. *Journal of Paleontology*, 49:563-565.

Gerlach, S.A.
1971. On the Importance of Marine Meiofauna for Benthos Communities. *Oecologia* (Berlin), 6:176-190.

1972. Meiobenthos. Pages 117-128 in C. Schlieper, editor, *Research Methods in Marine Biology*. London: Sedgwick and Jackson.

Gibson, T.G., and W.M. Walker
1967. Flotation Methods for Obtaining Foraminifera from Sediment Samples. *Journal of Paleontology*, 41:1294-1297.

Golemansky, V.
1974a. *Lagenidiopsis valkanovi* gen. n., sp. n. - un nouveau thécamoebien (Rhizopoda: Testacea) du psammal supralittoral des mers. *Acta Protozoologica*, 13:1-4, plate 1.

1974b. *Psammonobiotidae* fam. nov. - une nouvelle famille de thécamoebiens (Rhizopoda, Testacea) du psammal supralittoral des mers. *Acta Protozoologica*, 13:137-141.

1976. Contribution a l'étude des Rhizopodes et des Heliozoaires du psammal supralittoral de la Mediterranée. *Acta Protozoologica*, 15:34-45.

1979. Thécamoebiens psammobiontes du supralittoral vietnamien de la Mer Chinoise et description de *Cryptodifflugia brevicolla* sp. n. (Rhizopoda: Arcellinida). *Acta Protozoologica*, 18:285-292.

1982. Revision du genre *Ogdeniella* nom. n. (= *Amphorellopsis* Golemansky, 1970) (Rhizopoda, Gromida) avec considérations sur son origine et évolution dans le milieu interstitiel. *Acta Zoologica Bulgarica*, 19:3-12.

Gooday, A.J.
1986a. Soft-Shelled Foraminifera in Meiofaunal Samples from the Bathyal Northeast Atlantic. *Sarsia*, 71:275-287.

1986b. Meiofaunal Foraminiferans from the Bathyal Porcupine Seabight (Northeast Atlantic): Size Structure, Standing Stock, Taxonomic Composition, Species Diversity and Vertical Distribution in the Sediment. *Deep-Sea Research*, 33:1345-1373.

Grell, K.G.
1973. *Protozoology*. Third edition. 554 pages. Berlin, Heidelberg, New York: Springer Verlag.

Haman, D.
1982. Modern Thecamoebinids (Arcellinida) from the

Balize Delta, Louisiana. *Transactions of the 32nd Annual Meeting of the Gulf Coast Association of Geological Societies* (Houston, Texas), 353-376.

Haynes, J.R.
1981. *Foraminifera.* 433 pages. London and Basingstoke: Macmillan Publishers Ltd.

Hedley, R.H.
1964. The Biology of Foraminifera. Pages 1-45 in W.J.L. Felts and R.J. Harrison, editors. *International Review of General and Experimental Zoology, Volume 1.* New York: Academic Press.

Hedley, R.H., C.M. Hurdle, and I.D.J. Burdett
1967. The Marine Fauna of New Zealand. Intertidal Foraminifera of the *Corallina officinalis* Zone. *New Zealand Department of Science and Industry Research Bulletin,* 163:9-48.

Höglund, H.
1947. Foraminifera in the Gullmar Fjord and the Skagerak. *Zoologiska Bidrag fran Uppsala,* 26:1-328, plates 1-32.

Jepps, M.W.
1942. Studies on *Polystomella* Lamark (Foraminifera). *Journal of the Marine Biological Association of the United Kingdom,* 25:607-666.
1956. *The Protozoa, Sarcodina.* 183 pages. Edinburgh, London: Oliver and Boyd.

Lee, J.J.
1980. Nutrition and Physiology of the Foraminifera. Pages 43-66 in M. Levandowsky and S.H. Hutner, editors, *Biochemistry and Physiology of Protozoa,* volume 3. London: Academic Press.

Lee, J.J., W.A. Muller, R.J. Stone, M.E. McEnery, and W. Zucker
1969. Standing Crop of Foraminifera in Sublittoral Epiphytic Communities of a Long Island Salt Marsh. *Marine Biology,* 4:44-61.

Lee, J.J., S.H. Hutner, and E.C. Bovee
1985. *An Illustrated Guide to the Protozoa.* 629 pages. Lawrence, Kansas: Society of Protozoologists.

Levine, N.D., J.O. Corliss, F.E.G. Cox, G. Deroux, J. Grain, B.M. Honigberg, G.F. Leedale, A.R. Loeblich III, J. Lom, D. Lynn, E.G. Merinfeld, F.G. Page, G. Poljansky, V. Sprague, J. Vavra, and F.G. Wallace
1980. A Newly Revised Classification of the Protozoa. *Journal of Protozoology,* 27:37-58.

Lewis, K.B.
1970. A Key to the Recent Genera of the Foraminiferida. *Bulletin of the New Zealand Department of Scientific and Industrial Research,* 196:7-88.

Lighthart, B.
1969. Planktonic and Benthic Bacteriovorous Protozoa at Eleven Stations in Puget Sound and adjacent Pacific Ocean. *Journal of the Fisheries Research Board of Canada,* 26:299-304.

Lipps, J.H.
1983. Biotic Interactions in Benthic Foraminifera. Pages 331-376 in M.J.S. Tevesz and P.L. McCall, editors, *Biotic Interactions in Recent and Fossil Benthic Communities.* Plenum Publishing Corporation.

Lipps, J.H., and T.E. DeLaca
1980. Shallow Water Foraminiferal Ecology, Pacific Ocean. Pages 325-340 in M.E. Field, A.H. Bouma, I.P. Colburn, R.E. Douglas, and J.C. Ingle, editors, *Quaternary Depositional Environments of the Pacific Coast,* Pacific Coast Paleogeography Symposium 4. Los Angeles: Pacific Section, Society of Economic Paleontologists and Mineralogists.

Loeblich, A.R., and H. Tappan
1964. Protista 2. Sarcodina. Chiefly "Thecamoebians" and Foraminiferida. Pages 1-900 in R.C. Moore, editor, *Treatise on Invertebrate Paleontology,* Part C. Lawrence, Kansas: University of Kansas Press and the Geological Society of America.
1984. Suprageneric Classification of the Foraminiferida (Protozoa). *Micropaleontology,* 30:1-70.

Mare, M.F.
1942. A Study of a Marine Benthic Community with Special Reference to the Micro-Organisms. *Journal of the Marine Biological Association of the United Kingdom,* 25:517-554.

Marszalek, D.S., R.C. Wright and W.W. Hay
1969. Function of the Test in Foraminifera. *Transactions of the Gulf Coast Association of Geological Societies,* 29:341-352.

Matera, N.J., and J.J. Lee
1972. Environmental Factors Affecting the Standing Crop of Foraminifera in Sublittoral and Psammolittoral Communities of a Long Island Salt Marsh. *Marine Biology,* 14:89-103.

Mullineaux, L.S., and T.E. DeLaca
1984. Distribution of Antarctic Benthic Foraminifera Settling on the Pecten *Adamussium colbecki. Polar Biology,* 3:185-189.

Murray, J.W.
1969. Recent Foraminifers from the Atlantic Continental Margin of the United States. *Micropaleontology,* 15:401-409.
1971. *An Atlas of British Recent Foraminiferida.* 244 pages. London: Heinemann Educational Books Ltd.
1973. *Distribution and Ecology of Living Benthic Foraminiferida.* 274 pages. London: Heinemann Educational Books Ltd.
1976. Comparative Studies of Living and Dead Benthic Foraminiferal Distributions. Pages 45-109 in R.H. Hedley and C.G. Adams, editors, *Foraminifera,* volume 2. London: Academic Press.
1979. British Nearshore Foraminiferids. 68 pages in D.M. Kermack and R.S.K. Barnes, editors, *Synopses of the British Fauna,* no. 16. London, New York and San Francisco: Academic Press.

Murray, J.W., and C.M. Taplin
1984. The W.B. Carpenter Collection of Foraminifera: A Catalogue. *Journal of Micropaleontology,* 3:55-58.

Myers, E.H.
1940. Observations on the Origins and Fate of Flagellated Gametes in Multiple Tests of *Discorbis* (Foraminifera). *Journal of the Marine Biological Association of the United Kingdom,* 24:201-226, plates 1 and 2.

Nyholm, K.-G.
1951. A Monothalamous Foraminifer, *Marenda nematoides* n. gen. n. sp. *Contributions from the Cushman Foundation for Foraminiferal Research,* 2:91-95, plate 11.

Nyholm, K.-G., and I. Gertz
1973. To the Biology of the Monothalamous Foraminifer *Allogromia marina. Zoon,* 1:89-93.

Nyholm, K.-G., and I. Olsson
1973. Seasonal Fluctuations of the Meiobenthos in an Estuary on the Swedish West Coast. *Zoon,* 1:69-76.

Ogden, C.G., and R.H. Hedley
1980. *An Atlas of Freshwater Testate Amoebae.* 222 pages. London and Oxford: British Museum (Natural History).

Olsson, I.
1975. On Methods Concerning Marine Benthic Meiofauna. *Zoon,* 3:49-60.

Page, F.C.
1976. An Illustrated Key to Freshwater and Soil Amoebae. *Freshwater Biological Association, Special Publication,* no. 34:155 pages.
1983. *Marine Gymnamoebae.* 54 pages. Cambridge: Institute of Terrestrial Ecology, Natural Environment Research Council.

Patterson, D.J., J. Larsen, and J.O. Corliss
In press. The Ecology of Heterotrophic Flagellates and Ciliates Living in Marine Sediments. *Progress in Protistology,* 3.

Phleger, F.B.
1960. *Ecology and Distribution of Recent Foraminifera.* 297 pages. Baltimore: The Johns Hopkins Press.
1976. Benthic Foraminiferids as Indicators of Organic Production in Marginal Marine Areas. Pages 107-117 in C.T. Schafer and B.R. Pellietier, editors, *First*

International Symposium on Benthic Foraminifera of Continental Margins. Maritime Sediments Special Publication no. 1.

Platt, H.M., and R.W. Warwick
1983. Free-living Marine Nematodes. Part 1. British Enoplids. 307 pages in D.M. Kermack and R.S.K. Barnes, editors, *Synopsis of British Fauna* no. 28. Cambridge: Cambridge University Press.

Rhumbler, L.
1938. Foraminiferen aus dem Meeressand von Helgoland, gesammelt von A. Remane (Kiel). *Kieler Meeresforschungen*, 2:157-222.

Richter, G.
1961. Beobachtungen zur Ökologie einiger Foraminiferen des Jade-Gebietes. *Natur und Museum.* 91:163-170.

Rögl, F., and H.J. Hansen
1984. Foraminifera Described by Fichtel and Moll in 1798. A Revision of Testacea Miroscopica. *Neue Denkschriften des Naturhistorischen Museums in Wien*, 3:9-142.

Sawyer, T.K.
1980. Marine Amoebae from Clean and Stressed Bottom Sediments of the Atlantic Ocean and Gulf of Mexico. *Journal of Protozoology*, 27:13-32.

Schnitker, D.
1980. Quaternary Deep Sea Foraminifers and Bottom Water Masses. *Annual Review of Earth and Planetary Sciences*, 8:343-370.

Schwinghamer, P.
1981. Extraction of Living Meiofauna from Marine Sediments by Centrifugation in a Silica Sol-Sorbitol Mixture. *Canadian Journal of Fisheries and Aquatic Science*, 38:476-478.

Sen Gupta, B.K., R.F. Lee, and M.S. May
1981. Upwelling and an Unusual Assemblage of Benthic Foramanifera on the Northern Florida Continental Slope. *Journal of Paleontology*, 55:853-857.

Severin, K.P., S.J. Culver, and C. Blanpied
1982. Burrows and Trails Produced by *Quinquiloculina impressa* Reuss, a Benthic Foraminifer, in Fine-Grained Sediment. *Sedimentology*, 29:897-901.

Sheehan, R., and F.T. Banner
1972. The Pseudopodia of *Elphidium incertum*. *Revista Española de Micropaleontologia*, 4:31-63.

1973. *Trichosphaerium* - an Extraordinary Testate Rhizopod from Coastal Waters. *Estuarine and Coastal Marine Science*, 1:245-260.

Sieburth, J. McN.
1979. *Sea Microbes.* 491 pages. New York: Oxford University Press.

Snider, L.J., B.R. Burnett, and R.R. Hessler
1984. The Composition and Distribution of Meiofauna and Nanobiota in a Central North Pacific Deep-Sea Area. *Deep-Sea Research*, 31:1225-1249.

Steedman, H.F.
1976. General and Applied Data on Formaldehyde Fixation and Preservation of Marine Zooplankton. Pages 103-154 in H.F. Steedman, editor, *Zooplankton Fixation and Preservation.* Paris: The UNESCO Press.

Tendal, O.S.
1972. A Monograph of the Xenophyophoria (Rhizopodea, Protozoa). *Galathea Report*, 12:7-103, plates 1-17.

1979. Aspects of the Biology of Komokiacea and Xenophyophoria. *Sarsia*, 64:13-17.

Tendal, O.S., and R.R. Hessler
1977. An Introduction to the Biology and Systematics of Komokiacea (Textulariina, Foraminiferida). *Galathea Report*, 14:165-194, plates 10-26.

Thiel, H.
1983. Meiobenthos and Nanobenthos of the Deep-Sea. Pages 167-230 in G.T. Rowe, editor, *The Sea*, volume 8. New York: John Wiley and Sons.

Thiel, H., D. Thistle, and G.D. Wilson
1975. Ultrasonic Treatment of Sediment Samples for More Efficient Sorting of Meiofauna. *Limnology and Oceanography*, 20:472-473.

Todd, R., and D. Low
1981. *Marine Flora and Fauna of the Northeastern United States. Protozoa: Sarcodina: Benthic Foraminifera.* NOAA Technical Report NMFS circular 439:1-50. U.S. Department of Commerce.

Vénec-Peyré, M.-T., and Y. Le Calvez
1986. Foraminifères benthiques et phénomènes de transfert: importance des études comparatives de la biocénose et de la thanatocénose. *Bulletin du Museum National d'Histoire Naturelle*, Section C, Series 4, 8:171-184.

Walker, D.A., A.E. Linton, and C.T. Schafer
1974. Sudan Black B: A Superior Stain to Rose Bengal for Distinguishing Living from Non-Living Foraminifera. *Journal of Foraminiferal Research*, 4:205-215.

Wefer, G., and G.F. Lutze
1976. Benthic Foraminifera Biomass Production in the Western Baltic. *Kieler Meeresforschungen, Sonderheft*, 3:76-81.

Weston, J.F.
1984. Wall Structure of the Agglutinated Foraminifera *Eggerella bradyi* (Cushman) and *Karreriella bradyi* (Cushman). *Journal of Micropalaenotology*, 3:29-31.

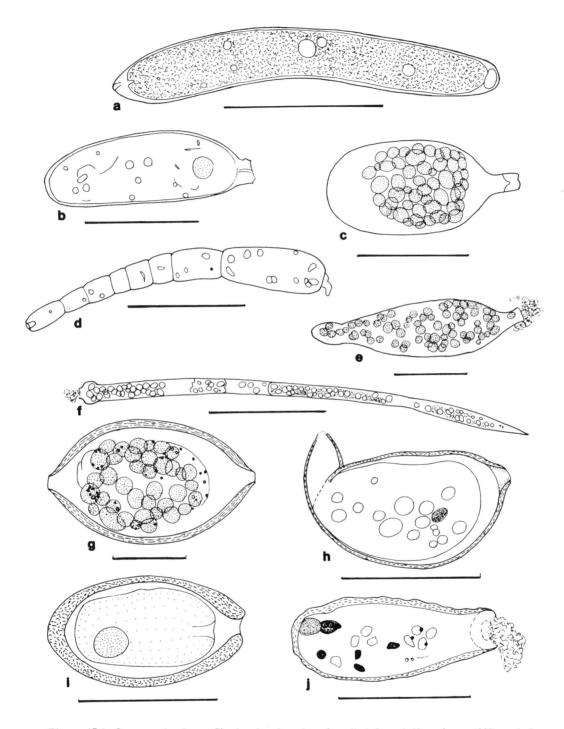

Figure 17.1.--Sarcomastigophora: Single chambered, soft-walled foraminifera from 4000 m (**a,j**) and 1320 m (**b-i**) in the NE Atlantic. **a-f** have proteinaceous tests (family Allogromiidae); **g-j** have proteinaceous/agglutinated tests (suborder Textulariina, Family Saccamminidae); all specimens are drawn as optical sections in transmitted light. **d**, *Nodellum membranacea* (Brady, 1879). **e,f**, two species resembling the genus *Nodellum*. **i**, *Ovammina* sp. Remainder are undescribed. (Scales = 50 μm (**e,g**), 100 μm (**b,c,f,h-j**), 200 μm (**a,d**).)

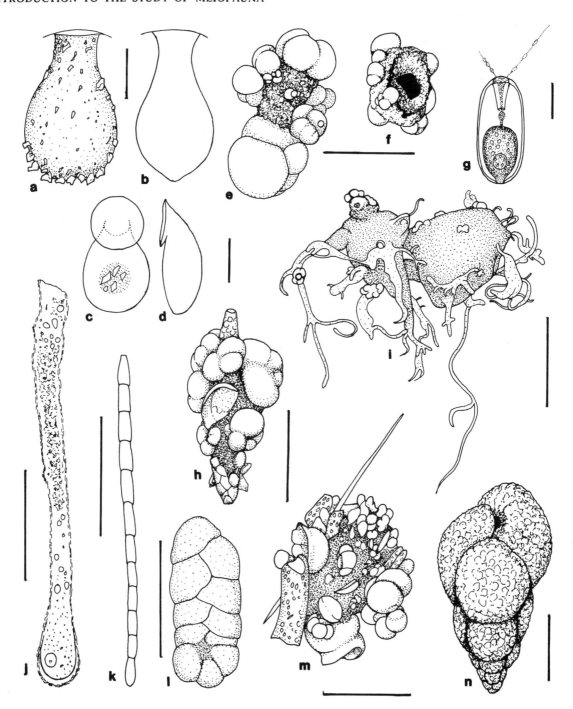

Figure 17.2.—Sarcomastigophora: a-d,g are testate amoebae from various supralittoral marine environments. e,f, h-n are agglutinated foraminifera (suborder Textulariina) from the NE Atlantic. a,b, *Ogdeniella maxima* Golemansky, 1982, front and side views (after Golemansky, 1982, fig. 17.1 c,d). c,d, *Psammonobiotus communis* Golemansky, 1967, front and side views (after Golemansky, 1974b fig.17.2d). e, *Crithionina* sp.; from 1320 m. f, same species with large *Globigerina* shell removed to show thick test wall and central protoplasmic mass. g, *Lagenidiopsis elegans* (Gruber) (after Golemansky, 1976, fig.5a). h, *Reophax* sp. composed mainly of *Globigerina* shells; from 1320 m. i, indeterminate two chambered agglutinated foraminifera with long filamentous appendages; from 1320 m. j, *Hyperammina* sp., a tiny species composed of mica flakes; from 4000 m. k, *Leptohalysis sp*; from 1320 m. l, *Spiroplectammina biformis* (Parker and Jones, 1865); from 1320 m. m, large, spherical, single-chambered foraminiferan; from 1320 m. n, *Eggerelloides scabrum* (Williamson, 1858); from 18 m in Cardigan Bay (Wales). (Scales = 20 μm (a-d), 50 μm (g), 100 μm (j-l), 200 μm (n), 500 μm (e,f,h,i,m).)

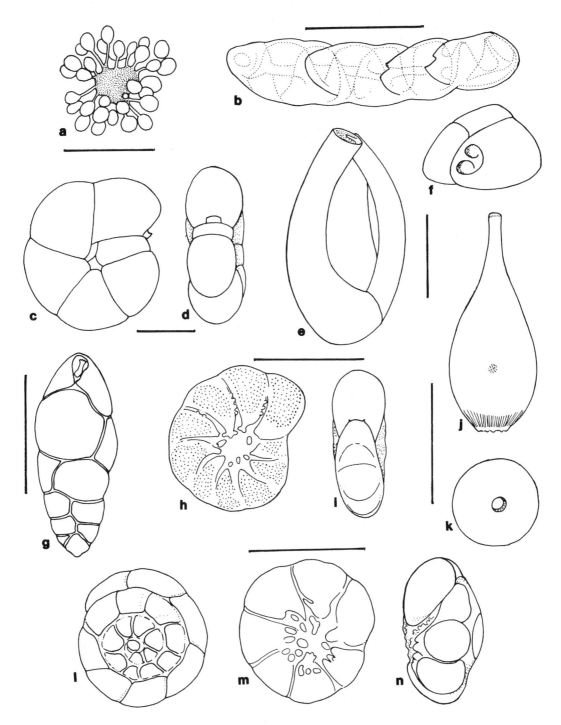

Figure 17.3.—**Sarcomastigophora:** Agglutinated foraminifera (suborder Textulariina) from 1320 m in the NE Atlantic (**a-d**) and calcareous foraminifera from 18 m in Cardigan Bay (Wales) (**e-n**). **a**, *Normanina* sp., a member of the newly described superfamily Komokiacea. **b**, *"Turritellella" laevigata* Earland, 1933. **c,d**, *Haplophragmoides bradyi* (Robertson, 1891), general and apertural views. **e,f**, *Quinquiloculina seminulum* (Linnaeus, 1767) (suborder Miliolina), general and apertural views. **g**, *Bulimina elongata* d'Orbigny, 1846 (suborder Rotaliina). **h,i**, *Elphidium selseyensis* (Heron-Allen and Earland, 1911), (suborder Rotaliina), general and apertural views. **j,k**, *Lagena semistriata* (Williamson, 1848) (suborder Lagenina), general and apertural views. **l-n**, *Ammonia batavus* (Hofker, 1951) (suborder Rotaliina), from left to right, spiral side, umbilical side, and lateral view. (Scales = 100 μm (**b-d**), 200 μm (**e-n**), 500 μm (**a**).)

18. Ciliophora

John O. Corliss, Eike Hartwig, and Susan E. Lenk

The ciliates, some 8,000 described species, comprise the phylum Ciliophora Doflein, 1901, of the Kingdom Protista Haeckel, 1866, and are cosmopolitan in distribution. These heterotrophic protists are characterized by two very distinctive features, nuclear dualism - diploid micronuclei and polyploid macronuclei - and a unique infraciliature or kinetidal system, consisting of single or paired kinetosomes with closely associated microtubules, microfibrils, and various specialized organelles. Most individuals are ciliated over at least a portion of the body during some stage in the life cycle. Possession of pellicular alveoli is another feature characteristic of most ciliates. Feeding structures in these protists vary considerably between groups, with some forms lacking any while others possess very elaborate oral apparatuses. Replication is usually via binary fission of a homothetogenic nature, essentially transverse or perkinetal (across the kineties or rows of cilia). Sexuality is represented by conjugation, not syngamy, with the exchange of haploid gametic nuclei. The cristae of the mitochondria are tubular. Most species are phagotrophic, holozoic, feeding on diverse substrates. Endosymbiotic forms have been described as well as other saprozoic species that exhibit osmotrophy. Other species show a type of "secondary phototrophy," serving as hosts for pigmented algal symbionts. Ciliates have been described from a wide range of habitats, both marine and freshwater as well as terrestrial; and they may occur as free-swimming or sessile or sedentary forms, while still others are ecto- or endosymbionts of a diversity of hosts.

Classification

Numerous other specialized characteristics within/among the ciliate taxa enable recognition of 10-26 classes and subclasses and 60-75 orders and suborders (Corliss, 1979; Small and Lynn, 1985). Nonetheless, the ciliates form a taxonomically and phylogenetically united group. Evolutionary pathways within the Ciliophora are still being debated (see Corliss, 1979, 1984; Corliss and Hartwig, 1977; Small, 1984; and references therein). But their "pre-ciliate" ancestry is even more highly conjectural: while no group among the 45 or so contemporary protistan phyla seems ideal as an origin, some of the striking similarities between dinoflagellates and ciliates have been the subject of interesting recent discussion (see Corliss, 1988).

Comprehensive keys to higher taxa within the Ciliophora are available in the monographic work of Kahl (1930-35; with an English translation of some of them by Patterson, 1978) and, most recently, in the treatment by Small and Lynn (1985). Simpler keys and ones often limited to special groups or habitats are represented in works such as those published by Bick (1972), Borror (1973), Curds (1969, 1982), Curds et al. (1983), Dragesco and Dragesco-Kernéïs (1986), Jahn et al. (1979), and Wright (1983).

Representatives of many taxa of free-living ciliates are found in marine and freshwater sediments (Figure 18.1). Perhaps the most exemplary of the interstitial forms are the karyorelictid ciliates, a group uniquely characterized by their diploid, non-dividing macronuclei (see Raikov, 1982, 1985, for review). This group exhibits many of the characteristics common to psammophilous ciliates and invertebrates - elongate, ribbon-like bodies, often highly contractile, adhesive, sometimes very fragile. Achieving lengths >3000 μm, these species may demonstrate traits representative of some primitive ciliate ancestor (Corliss, 1979; Lenk et al., 1984; Small, 1984; Raikov, 1982, 1985). Fauré-Fremiet (1951) termed these and other ciliates of similar appearance "microporals" because they generally inhabit sediments with grain sizes of about 250 μm in diameter. Typical genera of the microporal type include the karyorelictid genera (*Tracheloraphis, Trachelocerca, Trachelonema, Loxodes, Geleia, Remanella,* and *Kentrophoros*), and a number of non-karyorelictid genera such as *Condylostoma, Spirostomum, Helicoprorodon, Blepharisma, Gruberia, Lacrymaria, Litonotus, Dileptus,* and *Euplotes.*

The mesoporal ciliates (Fauré-Fremiet, 1951) are forms that inhabit much coarser sediments. These species tend to be considerably smaller in size and are somewhat flattened, swimming with a rather jerky motion rather than the gliding motion of the microporals. Typical mesoporal genera include: *Aspidisca, Dysteria, Cyclidium, Oxytricha, Pleuronema, Prorodon, Strombidium, Placus, Mesodinium, Disco-*

cephalus, and *Uronema*.

Finally, euryporal ciliates are those forms inhabiting both fine and coarse sediments: included are species of some of the above-listed genera. If sediments become very coarse, such as rocky beaches, or the pores fill in with silt and clay, such as in muddy areas, very few ciliates are present (see Elliott and Bamforth, 1975; Fenchel, 1969).

Many of the same genera can be found in freshwater habitats. Here, the karyorelictid ciliates are not the dominant forms: in fact, only *Loxodes* is represented in such non-saline sites. Lake and river bottoms are frequently covered with detritus and leaf-litter, offering a different habitat from that found in lakes with sandy bottoms. The populations reflect these differences (see Wilbert, 1986; Foissner, 1980, 1982; Foissner et al., 1982; and the ecology section of the present chapter).

Ciliates have been collected and described from sediments in the deep sea, where their numbers are small (Fenchel, 1987); but this may be due to the difficulties in collection (Burnett, 1973, 1977, 1979, 1981). Small and Gross (1985) found representatives of several ciliate classes in samples taken from deep-sea hydrothermal vents.

Habitats and Ecological Notes

Ciliates, as a group, are ubiquitous in their distribution. Habitats for free-living forms range from all types of freshwater, marine, and estuarine environments (including even ice slush and thermal springs) to soils, intertidal sands, muds, deserts, forest litter, and the like. Commensal and symbiotic forms may be found on or within a number of algal and other protists, and plants, and numerous animals, both vertebrate and invertebrate. Some species, notably those inhabiting the body cavity, digestive tract, or gills of (mainly invertebrate) hosts, can cause considerable harm as endoparasites (for a summary, see pages 452-455 in Corliss, 1979).

Most important for this volume are the meiobenthic forms from both marine and freshwater environs; to date, over 1,000 such species have been described. The bulk of these are marine, a complete list of which may be found in Patterson et al. (1988). The biotic and abiotic factors affecting the ciliate populations are essentially the same as for other members of the meiofaunal community (see Chapters 4 and 5).

The effects of grain size on the ciliate populations, as mentioned in the previous section, can be profound. Fenchel (1978) notes that true interstitial forms are absent in sediments in which the grain size falls below 100 µm. In medium sands (100-300 µm) that are well-mixed, lacking large percentages of silt and clay in the interstices, the ciliate populations can outnumber all other fauna. In coarse sands, >300 µm, the metazoan community becomes more predominant. Ciliates play a less substantial role in sediments with a high percentage of silt and clay. These non-capillary sediments are dominated by metazoa, whereas well mixed or capillary sands are dominated by ciliates. With respect to biomass, in well mixed sands Fenchel (1978) estimates that ciliates are present in numbers between 10^6 and 5×10^7 m^{-2}, with numbers in estuarine sediments being somewhat higher. These numbers correspond to a wet weight of 0.1-2.0 g m^{-2} (see Fenchel, 1978, for additional biomass estimates).

Ciliates play an important role in the food web, serving as prey for many metazoa and as predators (grazers) of/on bacteria. Ciliates act as nutrient recyclers for their bacterial prey. Many species are bactivorous and are found in the highest numbers in regions of high bacterial density, such as the redox transition layer where the aerobic and anaerobic sediments meet. Sulfur bacteria are a major food item for ciliates and are very active in this region. Carnivorous forms are subsequently found in this region as prey become available. Migration between different layers, vertical zonation, is very common.

This vertical zonation can be correlated with chemical zonation (Fenchel, 1978). Studies of spatial and temporal distribution of ciliates may be found in Fenchel (1969), Sorensen et al. (1979), and Fenchel and Staarup (1971). Although most ciliates are found in the top few centimeters of sediment, generally in the aerobic zone, there are species that live below the transition layer in the anaerobic zone. Some of these ciliates evolved with bacterial endosymbionts that function as mitochondria under anaerobic conditions; other species not harboring such endosymbionts have evolved other mechanisms for survival under low redox potentials (Fenchel et al., 1977; Lee, Soldo et al., 1985).

Many taxa are exclusively marine or freshwater, extremely sensitive to variations in salinity. Other species, especially those inhabiting estuarine or brackish water habitats, are more tolerant of salinity fluctuations. The contractile vacuole is a common means of osmoregulation in many freshwater forms, but it is much less prevalent among marine species.

Cyst formation is a mechanism that may have evolved as a means of survival under variable conditions. Many ciliates feed on very specific prey, such as diatoms or flagellates, that may appear in bloom populations themselves, encysting later when prey are less common. Encystment is also a means of survival against desiccation, which is more of a problem for freshwater forms than for marine species (Corliss and Esser, 1974). Many terrestrial forms excyst with rain or dew and encyst as this temporary water evaporates.

Papers on the distribution, vertical as well as geographic, and the ecology of ciliates are quite numerous; and they are increasing as the importance of ciliates in ecosystems becomes more apparent. The following books and papers may be consulted for more information: Borror, 1963, 1965; Burnett, 1973, 1977, 1979, 1981; Carey and Maeda, 1985; Corliss and Hartwig, 1977; Curds, 1982; Curds et al., 1983; Dragesco, 1960, 1963, 1965; Dragesco and Dragesco-Kernéïs, 1986; Dragesco et al., 1974; Elliott and Bamforth, 1975; Fauré-Fremiet, 1950, 1951; Fenchel, 1967, 1968a,b, 1969, 1975, 1978, 1987; Fenchel et al., 1977; Fenchel and Staarup, 1971; Finlay, 1978, 1980, 1981, 1982; Finlay et al., 1979; Finlay and Berninger, 1984; Finlay and Fenchel, 1986; Finlay and Ochsenbein-Gattlen, 1982; Foissner, 1980; Foissner et al., 1982; Goulder, 1974; Hartwig, 1973a,b, 1980a,b, 1986; Patterson et al., 1988; Raikov and Kovaleva, 1968; Sieburth, 1979; Small and Gross, 1985; Wilbert, 1986; Wright, 1982, 1983, 1984. The most comprehensive review of the reactions of meiobenthic ciliates to many physico-chemical factors of their habitats is the essay, currently in press, by Patterson et al. (1988).

Methods of Collection

Most meiobenthic ciliates are fragile and therefore easily damaged during collection from/with sediment material. In fact, some will be destroyed immediately on contact with the air/water interface. Nearly all species must be collected fresh, studied alive, and be specially prepared for identification and further study. Only a few species are readily identifiable under low magnification; usually a bright field or phase contrast microscope should be used. For identification, it is best to utilize stained material (see section on cytology) that facilitates revelation of many important structures not visible in living cells (such as the kinetosomes, microtubular/microfibrillar structures, nuclei, oral architecture, and extrusomes).

Because of their relatively small size and great abundance, ciliates can be collected using small-diameter core samples. For sublittoral sampling, the Reineck box corer, USNEL box corer, and their contemporary equivalents have been employed very successfully (Chapter 7). The Thiel "Meiostecher" (Thiel, 1966; Gerlach, 1968) is an alternative (see Hulings and Gray, 1971). For economic and rapid sampling of medium and fine sediments, a length of acetate tubing is very effective. This method may also be used to subsample a box corer.

Methods of Extraction

Perhaps the simplest means of collecting larger volumes of sediments for laboratory study is to place the sediments in a *plastic* bucket (avoid using metal, especially in sea water) to a depth of 5-10 cm. Air is bubbled over the surface of the sand through a large air stone in an effort to maintain a small current. Ciliates can be maintained under these conditions in the laboratory for several months, provided the buckets are kept at ambient temperatures. Plastic film may be used to cover the containers to avoid evaporation. Aquaria and glass jars have also been used successfully. Grow lights may be used to help support diatoms and other photosynthetic organisms that serve as food. Lights should be timed to mimic the natural photoperiod. Additional water may be added to keep the level constant. Carey (1986) devised a more elaborate method for long-term maintenance of psammophilic ciliates.

In order to avoid changes in the encased community, it is best to extract the ciliates within the first 24 hours following collection. Naturally, the fresher the sample, the more representative it will be of the natural population.

To produce samples of manageable size for studies of vertical zonation and the like, slicing a sediment core is recommended (see Boynton and Small, 1984). A piece of acetate tubing (ca. 20 cm in length) is plunged into the sediment. Before withdrawing the tube, a rubber stopper is inserted into the open top end; and, before lifting the tube entirely out of the water, a second stopper is placed in the open bottom end to avoid loss or disturbance of the sample.

The coring tube is clamped onto a ring stand. The top stopper is carefully removed and the water withdrawn with a pipette. Replace the stopper with a one-holed stopper fitted with a piece of clamped latex tubing. Gently remove the bottom stopper. By manipulation of the clamp, the sediment core will slide out and can be cut to the desired thickness: generally 2-cm slices are best.

In contrast to the above quantitative method, the ciliates can be extracted qualitatively from the sediment by the deterioration technique described in Chapter 9. This excellent method brings about a concentration of the fauna due to deoxygenation.

Another means of separation is to use the decantation method described by Boynton and Small (1984) (see Chapter 9). This technique is effective for samples of known volume as well as for samples of unknown volume. It is useful for a variety of sediments, including sands, silts, and clays. This method is not useful for removal of very adhesive species, which, therefore, may have to be observed directly.

For quantitative studies of living material, the Uhlig sea-water ice treatment is generally considered to be the most efficient (Uhlig et al., 1973). In this method, the infauna of a sample is influenced by two main ecological parameters: the change of the melting sea ice from high to low salinity and the low perfusion and flow of water through the sediment. The technique is most efficient with sandy sediments (capillary sediments); in the case of muddy sediments, treatment of prepartitioned sediment fractions is possible, with efficiency raised by repeated extraction.

The separation device is made simply by cutting the bottoms from two plastic cups. A piece of nytex nylon screening is held firmly in place between the cups across the bottom of the top cup. Sediments are placed in the cups followed by a layer of absorbent material. Ice made from water at the site is put on top of the absorbent material and allowed to melt. The water percolates through the sediments and is collected in a Petri dish below. It is wise to have an overflow dish to avoid floods! The ciliates are then isolated from the smaller Petri dish. This technique may be used with fresh water sediments using ice made from distilled water or spring water (avoid using tap water).

Although the Uhlig sea-water ice method works quite well for a wide range of specimens, the cold temperatures caused by the ice may be damaging to many cells, especially those collected during warmer months or from tropical or sub-tropical climates. Therefore, one of us (S.E.L., unpublished) has devised a modification of the Uhlig technique (Uhlig, 1968) that utilizes ambient water from the site rather than ice. This method is also good for field studies when ice may not be available.

Using the basic Uhlig setup, sediments are placed in the apparatus. The absorbent material is not used. Water from the site is repeatedly poured over the sediment and allowed to filter through several times. The organisms are collected from a Petri dish placed under the apparatus. By altering the mesh of the screen used, one can effectively limit larger invertebrates. This technique works extremely well with geotactic forms and may be used over a period of several hours to several days.

Helpful literature on collecting and extracting includes the following: Boynton and Small, 1984; Fenchel, 1969; Holme and McIntyre, 1984; Hulings and Gray, 1971; Lee, Small et al., 1985; Uhlig, 1964, 1968; Uhlig et al. 1973.

Methods of Preparation for Cytological and Taxonomic Studies

As is true for most minute meiofauna comprised of only soft parts, there is no suitable or feasible technique available for preserving ciliates directly in the sediments. After extraction of living specimens there are a few further approaches that can be taken, alone or in combination, as described below.

Ciliates isolated in a drop of water may be studied directly under a coverslip, with the disadvantages that one may not retrieve, manipulate, or maintain the sample. For freshwater samples, 2% aqueous methyl cellulose (sold commercially as Protoslo or Methocel) is effective as a means of slowing down rapid swimmers. Magnesium chloride used at a concentration of 7.5% for marine samples or 2.5% for freshwater ciliates is widely used. The concentration may have to be adjusted for the sample. Maugel et al. (1980) and Steedman (1976) describe several useful relaxants.

Hanging drops, observable with an inverted microscope (see Chapter 10) are useful in that they allow for less compact viewing. Several different types of microchambers and rotocompressors have been designed for long term and microcinematographical studies (see Uhlig and Heimberg, 1981; and Chapter 10).

One method to extend the life span of some ciliates when using the conventional coverslip is to place a ring of silicon grease between the cover glass and the slide. This is non-toxic and allows for oxygen exchange. Specimens may be kept alive for several hours to several days in this manner.

Assuming the specimens are abundant enough in the sample, one can fix some, apart from those used in making the observations described above, for subsequent staining and preservation as permanent preparations to be studied as time permits.

There are many sources of information on fixing and staining of ciliates, the most recent ones include Dragesco and Dragesco-Kernéïs, 1986; Lee, Small et al., 1985; Maugel et al., 1980; and references therein. Common fixatives include Schaudinn's fluid, Bouin's, glutaraldehyde, and osmium tetroxide, preceding appropriate stains or impregnation by silver. Often used for the nuclear apparatus is methyl green or the Feulgen nuclear reaction (see Chapter 10 for more fixatives). Unlike the case of many invertebrates, alcohol is not a good fixative for ciliates because the preservation is poor in their case.

Simple or compound ciliary structures (better, the infraciliature) are most reliably revealed by the use of one or the other of several techniques of silver impregnation. Citations or methodologies can be found in various of the above publications as well as in Corliss (1979). The best methods for ciliates are the Bodian Protargol technique and the Chatton-Lwoff technique.

Type Collections

Many ciliates have been described as new species in recent years; however, few type-specimens have been deposited in recognized repositories for type-material.' Partly this is due to the lack of enduringly fixed and stained materials. Protistologists should be aware of the International Collection of Ciliate Type-Specimens established at the Smithsonian Institution (Corliss, 1972) and feel free to address inquiries to the Curator, Dr. Klaus Ruetzler, Department of Invertebrate Zoology, National Museum of Natural History, Smithsonian Institution, Washington, D.C. 20560, USA.

References

Bick, H.
1972. *Ciliated Protozoa: An Illustrated Guide to the Species Used as Biological Indicators in Freshwater Biology.* 198 pages. World Health Organization, Geneva.

Borror, A.C.
1963. Morphology and Ecology of the Benthic Ciliated Protozoa of Alligator Harbor, Florida. *Archiv für Protistenkunde*, 106:465-534.
1965. New and Little-known Tidal Marsh Ciliates. *Transactions of the American Microscopical Society*, 84:550-565.
1973. Marine Flora and Fauna of the Northeastern United States. Protozoa: Ciliophora. *NOAA Technical Report NMFS Circular*, Number 378. 62 pages.

Boynton, J.E., and E.B. Small
1984. Ciliates by the Slice. *The Science Teacher*, 51:34-38.

Burnett, B.R.
1973. Observations of the Microfauna of the Deep-sea Benthos Using Light and Scanning Electron Microscopy. *Deep-Sea Research*, 20:413-417.
1977. Quantitative Sampling of Microbiota of the Deep-sea Benthos. I. Sampling Techniques and Some Data from the Abyssal Central North Pacific. *Deep-Sea Research*, 24:781-789.
1979. Quantitative Sampling of Microbiota of the Deep-sea Benthos. II. Evaluation of Technique and Introduction to the Biota of the San Diego Trough. *Transactions of the American Microscopical Society*, 98:233-242.
1981. Quantitative Sampling of Nanobiota of the Deep-sea Benthos. III. The Bathyal San Diego Trough. *Deep-Sea Research*, 28:649-663.

Carey, P.G.
1986. A Method for Long Term Laboratory Maintenance of Marine Psammophilic Ciliates. *Journal of Protozoology*, 33:442-443.

Carey, P.G., and M. Maeda
1985. Horizontal Distribution of Psammophilic Ciliates in Fine Sediments of the Chichester Harbour Area. *Journal of Natural History*, 19:555-574.

Corliss, J.O.
1972. Current Status of the International Collection of Ciliate Type-Specimens and Guidelines for Future Contributors. *Transactions of the American Microscopical Society*, 91:221-235.
1979. *The Ciliated Protozoa: Characterization, Classification, and Guide to the Literature.* Second Edition. 455 pages. New York: Pergamon Press.
1984. The Kingdom Protista and Its 45 Phyla. *BioSystems*, 17:87-126.
1988. The Quest for the Ancestor of the Ciliophora: A Brief Review of the Continuing Problem. *BioSystems*. (in press).

Corliss, J.O., and S.C. Esser
1974. Comments on the Role of the Cyst in the Life Cycle and Survival of Free-living Protozoa. *Transactions of the American Microscopical Society*, 93:578-593.

Corliss, J.O., and E. Hartwig
1977. The "Primitive" Interstitial Ciliates: Their Ecology, Nuclear Uniquenesses, and Postulated Place in the Evolution and Systematics of the Phylum Ciliophora. *Mikrofauna des Meeresbodens*, 61:65-88.

Curds, C.R.
1969. *An Illustrated Key to the British Freshwater Ciliated Protozoa Commonly Found in Activated Sludge.* 90 pages. Ministry of Technology, H.M.S.O., London.
1982. *British and Other Freshwater Ciliated Protozoa. Part 1. Ciliophora: Kinetofragminophora -- Keys and Notes for the Identification of the Free-Living Genera.* 387 pages. Cambridge: Cambridge University Press.

Curds, C.R., M.A. Gates, and D. McL. Roberts
1983. *British and Other Freshwater Ciliated Protozoa. Part II. Ciliophora: Oligohymenophora and Polyhymenophora -- Keys and Notes for the Identification of the Free-Living Genera.* 474 pages. Cambridge: Cambridge University Press.

Dragesco, J.
1960. Ciliés mesopsammiques littoraux. Systématique, morphologie, écologie. *Travaux de la Station Biologique de Roscoff*, 12:1-356.
1963. Compléments à la connaissance des ciliés mésopsammiques de Roscoff. I. Holotriches. *Cahiers de Biologie Marine*, 4:91-119; II. Heterotriches. III. Hypotriches. *Cahiers de Biologie Marine*, 4:251-275.
1965. Ciliés mésopsammiques d'Afrique Noire. *Cahiers de Biologie Marine*, 6:357-399.
1966. Observations sur quelques ciliés libres. *Archiv für Protistenkunde*, 109:155-206.
1970. Ciliés libres du Cameroun. *Annales de la Faculté des Sciences de Yaoundé.* 141 pages. Yaoundé.

Dragesco, J., and A. Dragesco-Kernéïs
1986. *Ciliés Libres de l'Afrique Intertropicale: Introduction à la Connaissance et à l'Etude des Ciliés.* 559 pages. Editions de l'ORSTOM, Paris.

Dragesco, J., F. Iftode, and G. Fryd-Versavel
1974. Contribution à la connaissance de quelques ciliés holotriches rhabdophores; I. Prostomiens. *Protistologica*, 10:59-76.

Dragesco, J., and T. Njiné
1971. Compléments à la connaissance des ciliés libres du Cameroun. *Annales de la Faculté des Sciences du Cameroun*, 7-8:97-140.

Elliott, P.B., and S.S. Bamforth
1975. Interstitial Protozoa and Algae of Louisiana Salt Marshes. *Journal of Protozoology*, 22:514-519.

Fauré-Fremiet, E.
1950 Ecologie des ciliés psammophiles littoraux. *Bulletin Biologique de la France et de la Belgique*, 84:35-75.
1951. The Marine Sand-dwelling Ciliates of Cape Cod. *Biological Bulletin*, 100:59-70.

Fenchel, T.
1967. The Ecology of Marine Microbenthos. I. The Quantitative Importance of Ciliates as Compared with Metazoans in Various Types of Sediments. *Ophelia*, 4:121-137.
1968a. The Ecology of Marine Microbenthos. II. The Food of Marine Benthic Ciliates. *Ophelia*, 5:73-121.
1968b. The Ecology of Marine Microbenthos. III. The Reproductive Potential of Ciliates. *Opehlia*, 5:123-136.
1969. The Ecology of Marine Microbenthos. IV. Structure and Function of the Benthic Ecosystem, Its Chemical and Physical Factors and the Microfauna Communities with Special Reference to the Ciliated Protozoa. *Ophelia*, 6:1-182.
1975. The Quantitative Importance of the Benthic Microfauna of an Arctic Tundra Pond. *Hydrobiologia*, 46:445-464.
1978. The Ecology of Micro- and Meiobenthos. *Annual Review of Ecology and Systematics*, 9:99-121.
1987. *Ecology of Protozoa: The Biology of Free-living Phagotrophic Protists.* 197 pages. Berlin: Springer-Verlag.

Fenchel, T., T. Perry, and A. Thane
1977. Anaerobiosis and Symbiosis with Bacteria in Free-living Ciliates. *Journal of Protozoology*, 24:154-163.

Fenchel, T., and B.J. Staarup
1971. Vertical Distribution of Photosynthetic Pigments and the Penetration of Light in Marine Sediments. *Oikos*, 22:172-182.

Finlay, B.J.
1978. Community Production and Respiration by Ciliated Protozoa in the Benthos of a Small Eutrophic Loch. *Freshwater Biology*, 8:327-341.
1980. Temporal and Vertical Distribution of Ciliophoran Communities in the Benthos of a Small Eutrophic Loch with Particular Reference to the Redox Profile. *Freshwater Biology*, 10:15-34.
1981. Oxygen Availability and Seasonal Migrations of Ciliated Protozoa in a Freshwater Lake. *Journal of General Microbiology*, 123:173-178.
1982. Effects of Seasonal Anoxia on the Community of Benthic Ciliated Protozoa in a Productive Lake. *Archiv für Protistenkunde*, 125:215-222.

Finlay, B.J., P. Bannister, and J. Stewart
1979. Temporal Variation in Benthic Ciliates and the Application of Association Analysis. *Freshwater Biology*, 15:333-346.

Finlay, B.J., and U.-G. Berninger
1984. Coexistence of Congeneric Ciliates (Karyorelictida: *Loxodes*) in Relation to Food Resources in Two Freshwater Lakes. *Journal of Animal Ecology*, 53:929-943.

Finlay, B.J., and T. Fenchel
1986. Physiological Ecology of the Ciliated Protozoon *Loxodes*. *Freshwater Biological Association Annual Report*, 54:73-96.

Finlay, B.J., and C. Ochsenbein-Gattlen
1982. *Ecology of Free-Living Protozoa: A Bibliography of Published Research Concerning Freshwater and Terrestrial Forms 1910-1981.* (Freshwater Biological Association Occasional Publication No. 17), 167 pages. Freshwater Biological Association, Ambleside, U.K.

Foissner, W.
1979. Taxonomische Studien über die Ciliaten des Grossglocknergebietes (Hohe Tauern, Österreich). Familien Microthoracidae, Chilodonellidae und Furgasoniidae. *Österreichische Akademie der Wissenschaften Mathematisch-Naturwissenschaftliche Klasse*, 188(1):27-43.
1980. Taxonomische Studien über die Ciliaten des Grossglocknergebietes (Hohe Tauern, Österreich). IX. Ordnungen Heterotrichida und Hypotrichida. *Berichte des naturwissenschaftlich-medizinischen Vereins in Salzburg*, 5:71-117.
1982. Ökologie und Taxonomie der Hypotrichida (Protozoa: Ciliophora) einiger österreichischer Böden. *Archiv für Protistenkunde*, 126:19-143.

Foissner, W., H. Adam, and I. Foissner
1982. Daten zur Autökologie der Ciliaten stagnierender Kleingewässer im Grossglocknergebiet (Hohe Tauern, Österreich). *Berichte des naturwissenschaftlich-medizinischen Vereins in Salzburg*, 6:81-101.

Fryd-Versavel, G., F. Iftode, and J. Dragesco
1975. Contribution à la connaissance de quelques ciliés gymnostomes. II. Prostomiens, pleurostomiens: morphologie, stomatogenèse. *Protistologica*, 11:509-530.

Gerlach, S.A.
1968. Meiobenthos. In C. Schlieper, editor, *Methoden der meeresbiologischen Forschung*, 322 pages. Jena: G. Fischer.

Goulder, R.
1974. The Seasonal and Spatial Distribution of Some Benthic Ciliated Protozoa in Esthwaite Water. *Freshwater Biology*, 4:127-147.

Grolière, C.-A.
1977. Contribution à l'étude des ciliés des sphaignes et des étendues d'eau acides. I. Description de quelques espèces de gymnostomes, hypostomes, hyménostomes et hétérotriches. *Annales de la Station Biologique de Besse-en-Chandesse*, Number 10 (year 1975-1976):265-297.

Guilcher, Y.
1951. Contribution à l'étude des ciliés gemmipares, chonotriches et tentaculifères. *Annales des Sciences Naturelles Zoologie*, 13(series 11):33-132.

Hartwig, E.
1973a. Die Ciliaten des Gezeiten-Sandstrandes der Nordseeinsel Sylt. I. Systematik. *Mikrofauna des Meeresbodens*, 18:387-453.
1973b. Die Ciliaten des Gezeiten-Sandstrandes der Nordseeinsel Sylt. II. Ökologie. *Mikrofauna des Meeresbodens*, 21:3-171.
1980a. A Bibliography of the Interstitial Ciliates (Protozoa): 1926-1979. *Archiv für Protistenkunde*, 123:422-438.
1980b. The Marine Interstitial Ciliates of Bermuda with Notes on their Geographical Distribution and Habitats. *Cahiers de Biologie Marine*, 21:409-441.
1986. Ciliophora. Page 10 in W. Sterrer, editor, *Marine Fauna and Flora of Bermuda*. New York: John Wiley and Sons.

Hill, B.F.
1980. *Classification and Phylogeny in the Suborder Euplotina (Ciliophora, Hypotrichida)*. Ph.D. Dissertation, University of New Hampshire, Durham.

Holme, N.A., and A.D. McIntyre, editors
1984. *Methods for the Study of Marine Benthos*. 387 pages. Oxford: Blackwell.

Hulings, N.C., and J.S. Gray, editors
1971. A Manual for the Study of Meiofauna. *Smithsonian Contributions to Zoology*, 78: 84 pages.

Jahn, T.L., E.C. Bovee, and F.F. Jahn
1979. *How to Know the Protozoa*. Second Edition. 279 pages. Dubuque, Iowa: William C. Brown.

Kahl, A.
1930-1935. Urtiere oder Protozoa. I: Wimpertiere oder Ciliata (Infusoria), eine Bearbeitung der freilebenden und ektokommensalen Infusorien der Erde, unter Ausschluss der marinen Tintinnidae. Pages 1-886, Parts 18 (Year 1930), 21 (1931), 25 (1932), 30 (1935) in F. Dahl, editor, *Die Tierwelt Deutschlands*. Jena: G. Fischer.

Lee, J.J., E.B. Small, D.H. Lynn, and E.C. Bovee
1985. Some Techniques for Collecting, Cultivating and Observing Protozoa. Pages 1-7 in J.J. Lee, S.H. Hutner, and E.C. Bovee, editors, *An Illustrated Guide to the Protozoa*. Lawrence, Kansas: Society of Protozoologists.

Lee, J.J., A.T. Soldo, W. Reisser, M.J. Lee, K.W. Jeon, and H.-D. Görtz
1985. The Extent of Algal and Bacterial Endosymbioses in Protozoa. *Journal of Protozoology*, 32:391-403.

Lenk, S.E., E.B. Small, and J. Gunderson
1984. Preliminary Observations of Feeding in the Psammobiotic Ciliate *Tracheloraphis*. *Origins of Life*, 13:229-234.

Maugel, T.K., D.B. Bonar, W.J. Creegan, and E.B. Small
1980. Specimen Preparation Techniques for Aquatic Organisms. *Scanning Electron Microscopy/1980/II*, pages 57-58, SEM Inc., AMF O'Hare, Illinois.

Patterson, D.J.
1978. *Kahl's Keys to the Ciliates*. 82 pages. England: University of Bristol.

Patterson, D.J., J. Larsen, and J.O. Corliss
1988. The Ecology of Heterotrophic Flagellates and Ciliates Living in Marine Sediments. *Progress in Protistology*, 3. (in press).

Puytorac, P. de, and T. Njiné
1971. Sur l'ultrastructure des *Loxodes* (ciliés holotriches). *Protistologica*, 6(year 1970):427-444.

Raikov, I.B.
1982. *The Protozoan Nucleus: Morphology and Evolution*. (Cell Biology Monographs, Volume 9). 474 pages. New York: Springer-Verlag.

1985. Primitive Never-dividing Macronuclei of Some Lower Ciliates. *International Review of Cytology*, 95:267-325.

Raikov, I.B., and V.G. Kovaleva
1968. Complements to the Fauna of the Psammobiotic Ciliates of the Japan Sea (Posjet Gulf). *Acta Protozoologica*, 6:309-333.

Sieburth, J. McN.
1979. *Sea Microbes*. New York: Oxford University Press. 491 pages.

Small, E.B.
1984. An Essay on the Evolution of Ciliophoran Oral Cytoarchitecture Based on Descent from within a Karyorelictean Ancestry. *Origins of Life*, 13:217-228.

Small, E.B., and M.E. Gross
1985. Preliminary Observations of Protistan Organisms, Especially Ciliates, from the 21° N Hydrothermal Vent Site. *Biological Society of Washington Bulletin*, 6:401-410.

Small, E.B., and D.H. Lynn
1985. Phylum Ciliophora Doflein, 1901. Pages 393-575 in J.J. Lee, S.H. Hutner, and E.C. Bovee, editors, *An Illustrated Guide to the Protozoa*. Lawrence, Kansas: Society of Protozoologists.

Sørensen, J., B.B. Jørgensen, and N.P. Revsbech
1979. A Comparison of Oxygen, Nitrate, and Sulfate Respiration in Coastal Marine Sediments. *Microbial Ecology*, 5:105-115.

Steedman, H.F.
1976. Narcotizing Agents and Methods. Pages 87-94 in H.F. Steedman, editor, *Zooplankton Fixation and Preservation*. London: UNESCO Press.

Thiel, H.
1966. Quantitative Untersuchungen über die Meiofauna von Tiefseeböden. *Veröffentlichungen des Instituts für Meeresforschung in Bremerhaven*, 2:131-147.

Uhlig, G.
1964. Eine einfache Methode zur Extraktion der vagilen, mesopsammalen Mikrofauna. *Helgoländer wissenschaftliche Meeresuntersuchungen*, 11:178-185.

1968. Quantitative Methods in the Study of Interstitial Fauna. *Transactions of the American Microscopical Society*, 87:226-232.

Uhlig, G., and S.H.H. Heimberg
1981. A New Versatile Compression Chamber for Examination of Living Microorganisms. *Helgoländer wissenschaftliche Meeresuntersuchungen*, 34:251-256.

Uhlig, G., H. Thiel, and J.S. Gray
1973. The Quantitative Separation of Meiofauna. *Helgoländer wissenschaftliche Meeresuntersuchungen*, 25:173-195.

Wicklow, B.J.
1982. The Discocephalina (n. subord.): Ultrastructure, Morphogenesis and Evolutionary Implications of a Group of Endemic Marine Interstitial Hypotrichs (Ciliophora, Protozoa). *Protistologica*, 18:299-330.

Wilbert, N.
1972. Morphologie und Ökologie einiger neuer Ciliaten (Holotricha, Cyrtophorina) des Aufwuchses. *Protistologica*, 7(year 1971):357-363.

1986. Ciliaten aus dem Interstitial des Ontario Sees. *Acta Protozoologica*, 25:379-396.

Wright, J.M.
1982. Some Sand Dwelling Ciliates of South Wales. *Cahiers de Biologie Marine*, 23:275-285.

1983. Sand Dwelling Ciliates of South Wales. *Cahiers de Biologie*, 24:187-214.

1984. The Ecology of Psammobiotic Ciliates of South Wales. *Cahiers de Biologie Marine*, 25:217-239.

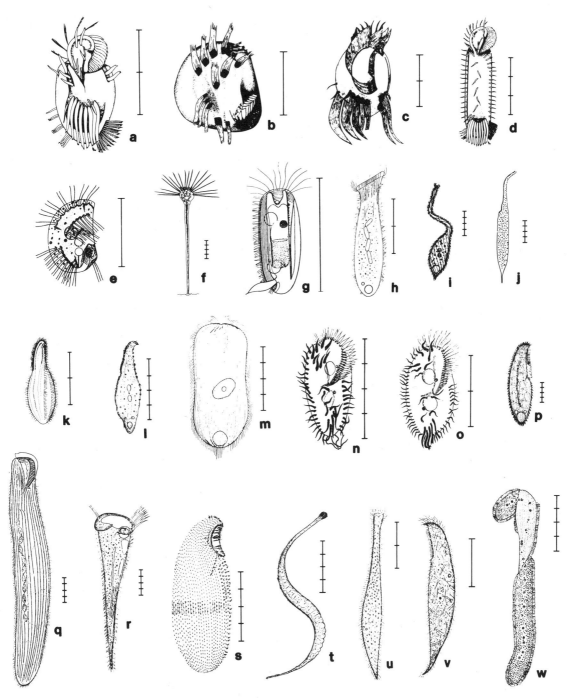

Figure 18.1--Ciliophora. a, *Discocephalus ehrenbergi* (from Wicklow, 1982); **b,** *Aspidisca costata* (from Hill, 1980); **c,** *Uronychia transfuga* (from Hill, 1980); **d,** *Psammocephalus faurei* (from Wicklow, 1982); **e,** *Microthorax pusillus* (from Foissner, 1979); **f,** *Ephelota gemmipara* (from Guilcher, 1951); **g,** *Dysteria scutellum* (from Wilbert, 1972); **h,** *Spathidium musicola* (from Dragesco, 1970); **i,** *Lacrymaria olor* (from Dragesco, 1966); **j,** *Dileptus anser* (from Dragesco, 1963); **k,** *Litonotus lamella* (from Fryd-Versavel et al., 1975); **l,** *Loxophyllum helus* (from Dragesco, 1966); **m,** *Prorodon teres* (from Grolière, 1977); **n,** *Stylonychia putrina* (from Dragesco and Njiné, 1971); **o,** *Oxytricha elliptica* (from Dragesco, 1970); **p,** *Blepharisma japonicum* (from Dragesco, 1970); **q,** *Condylostoma patulum* (from Dragesco, 1963); **r,** *Stentor roeseli* (from Dragesco, 1970); **s,** *Loxodes magnus* (from de Puytorac and Njiné, 1971); **t,** *Tracheloraphis phoenicopterus* (from Raikov and Kovaleva, 1968); **u,** *Trachelocerca tenuicolis* (from Dragesco, 1960); **v,** *Remanella margaritifera* (from Dragesco, 1960); **w,** *Kentrophoros flavum* (from Raikov and Kovaleva, 1968). Figures are all from Small and Lynn (1985); original sources credited in parentheses. (Smallest unit of each scale bar = 25 μm.)

19. Cnidaria

Hjalmar Thiel

The phylum Cnidaria is characterized by only two layers of their body wall, the ectoderm and the endoderm, separated by a generally thin acellular jelly layer, the mesogloea. This layer may become extensive in the medusae. All cnidarians are equipped with cnidocysts (nematocysts) of different types, which mainly have the function to catch prey but may be used for defense and in fixing vagile specimens to substrates. The species may be solitary or colonial. Many of them secrete an outer skeleton, which may be the chitinous periderm in Scyphozoa, Cubozoa, and Hydrozoa or the calcareous exoskeleton in Anthozoa and some Hydrozoa. In Scyphozoa, Cubozoa, and Hydrozoa metagenesis is developed, the alternating appearance of asexually propagating polyp and sexually reproducing medusa generations. In many cases the medusa generation is secondarily reduced to gonophores fixed to the polyp generation. In other species direct medusae development occurs through the loss of the polyp stages.

Classification

All four classes of the Cnidaria are represented in the meiofauna: two species in the Scyphozoa, four in the Cubozoa, one in the Anthozoa, and possibly 28 species in 15 genera in the Hydrozoa. The class Cubozoa, separated from the Scyphozoa only recently (Werner, 1973, 1975), have not been previously regarded as having meiofaunal representatives, but their polyps reach only about 1 mm in some species.

Meiofaunal size brings along a number of reductions. Most species are solitary, and only the scleractinian *Sphenotrochus* has its carbonate skeleton. Except in the Cubozoa, in which the polyp totally metamorphoses into a medusa, the gonophores are reduced generally providing the polyps hydra-like organization; mostly single eggs and sperm masses are produced in the epidermis.

The small number of meiofaunal species in the Cnidaria reveals a rather simple set of taxonomic characters for species determinations: general form, number of tentacles and their arrangement, presence of statoliths, and type of asexual reproduction. The only difficulties for species determination are inherent with the species of the Halammohydrina with their variable shapes and the existence of an apical adhesive groove.

The taxonomy of the Cnidaria is aided by the inventory of cnidocyst types, but with the small number of species involved most can be determined by morphological characters. However, species descriptions must include the cnidom and the researcher is referred to Werner (1965) and to the original papers of species descriptions. The taxonomy in this chapter follows Werner (1984) and to some extent Clausen and Salvini-Plawen (1986).

Scyphozoa.--*Stylocoronella riedli* (Figure 19.1a) was described by Salvini-Plawen (1966) as a meiofaunal scyphozoan polyp with a typical structure for this group. Its size is up to 400 µm. The adult animal, capable of asexual reproduction, has 16-24 tentacles with their ends mostly bent inwards and arranged in one circle. Young polyps were observed with only 7 tentacles, surrounding the protruding four-cornered proboscis. The cup-shaped body is followed by a narrow stalk and finally by a pedal disc for adhesion to the substrate. Sexual reproduction was not observed, but asexual budding occurs from stolonial protrusions, which appear interradially beyond the tentacles on the calyx. The distal ends of these stolons swell and are separated into planuloids (about 55-160 µm).

The gastric cavity is narrowed by four septa in which muscles extend to the pedal disc, but no funnel-like depressions between mouth and tentacles and no taeniols were observed. On the base of each tentacle a reddish brown spot is interpreted as an ocellus, and shadow reflexes were observed in the animals. The characteristics described are typical for scyphozoan polyps. The high number of tentacles would place them with the Semaeostomae, the ocelli would suggest the Stauromedusida.

The polyps live normally settled on sand grains but they may move with the aid of their haptic tentacle ends and their pedal disc. They colonize coarse sand with shell gravel in 4-7 m depth, where they direct their tentacle crowns against the current.

Material for the description of this species was collected near Rovigne (Adriatic Sea), but the same species was observed in sands from Marseille and Banyuls-sur-Mer (Monniot, 1962) and from Roscoff.

The second scyphozoan species, *Stylocoronella variabilis,* Salvini-Plawen, 1987 (first noted by Clausen and Salvini-Plawen in 1986), is 0.5-0.8 mm long and is distinguished by a smooth transition from calyx to stalk. The number of tentacles ranges from 16-28, with 12-19 straight and fairly contractible, and 5-12 more slender and coiling ones, which insert irregularly between and somewhat lower than the former tentacles. On the base of some of the straight tentacles between 8 and 11 ocelli were detected. No planuloid budding was observed.

This species was found in coarse sand in 11 and 13 m depth off Plymouth. Other specimens collected along French coasts may belong to this species (Salvini-Plawen 1987).

Cubozoa.--This class was separated from the Scyphozoa by Werner (1973). While the medusae show some superficial similarities to scyphomedusae, the polyps are clearly different and they metamorphose totally into the medusa generation. The polyps are only known from laboratory cultures, but they are of meiofaunal size. In most species the polyps are unknown.

Tripedalia. The polyp of *T. cystophora* (Figure 19.1b) is about 1 mm in height, drop-shaped with a conical hypostome, surrounded by 7-9 capitate tentacles with 20-40 nematocysts at their ends. It sits in a small periderm cup. Budding of daughter polyps occurs just above this cup (Werner, 1975). The young polyp has three tentacles, one stretched out in crawling. Together with the tentacles stretched out, the polyp may reach a length of 1.6 mm. Young specimens crawl on and between substrate particles and may well be found interstitially. However, they are known only from laboratory cultures (Werner, 1975).

Chironex. The "sea wasp," *C. fleckeri,* one of the most dangerous marine animals, is represented in the meiofauna by creeping forms (Hartwick, 1987) as in *T. cystophora*. The sedentary polyps develop 40-60 tentacles and reach a diameter of 0.8 mm; the tentacles may be stretched to a length of 2.4 mm.

Carybdea. The polyp of *C. marsupialis* (Figure 19.1c) is 2 mm high, and the rounded hypostome is surrounded by about 20 capitate tentacles carrying only one nematocyst. The lower part of the body is narrow and stalk-like. Daughter polyps may bud from the side. The young polyps of this species have a circle of 8 tentacles, may grow up to 1.2 mm in the laboratory, and they may crawl about as the above species, but none of the tentacles stretched out (Cutress and Studebaker, 1973). The polyp generation of *Carybdea alata* is very similar to that of *C. marsupialis.* It has up to 16 tentacles (Arneson and Cutress, 1976).

Hydrozoa.--Meiofaunal species of this taxon can be grouped into two categories of lebensformtypen by their specific life styles. One group consists of tiny typical polyps belonging in the suborder Hydrina. As a result of their simple structure some of the genera have the suffix "hydra" in their names. The other part is comprised of medusoids with totally reduced umbrella and rather long ciliated tentacles and manubrium. They constitute the suborder Halammohydrina and their descriptions are beyond the scope of this book. Further taxonomic details are discussed in several papers cited.

The suborder Hydrina is used in deviation from the taxonomic scheme of Clausen and Salvini-Plawen (1986) with arguments from Salvini-Plawen (1987) cited under the year 1984 in Clausen and Salvini-Plawen (1986).

Protohydra. This genus includes polyps without tentacles. *Protohydra leuckarti* was described by Greeff in 1868. Specimens are 1-3 mm long, appear worm-like with an uneven surface from the cnidocysts in the ectoderm. Reproduction is mainly by transverse fission and sometimes budding. The development of one egg, eventually two, and sperm masses in the epidermis was observed in some rare cases. *Protohydra leuckarti* lives in lagoons and in very shallow waters down to several meter depths from muddy to sandy habitats. This species seems to be world-wide in distribution. Dawidoff (1930) described *P. caulleryi* from Indo-China and Omer-Cooper (1957) reported on *P. psamathe* from South Africa. The three described species are distinguished by minor differences and should be compared based on more material for a taxonomic revision.

Psammohydra. This genus is monotypic. *Psammohydra nanna* (Figure 19.1d) was described by Schulz in 1950. It is distinguished by 4 short tentacles (in rare cases 5 or as few as 3) inserted at about 2/3 of the height which is about 0.4 mm. Contraction and expansion may vary the size from less than half to more than double the average. Reproduction occurs by transverse fission or by single egg production (Swedmark 1959). *Psammohydra nanna* lives in fine sand and was first found at a depth of 2 m in the Western Baltic. Later it was discovered by Swedmark (1956) in fine sand near Roscoff and by Salvini-Plawen (1966), near Rovigne, at depths of 4-8 m in medium fine to medium-coarse sand. The individuals adhere to sand grains with their pedal disc for most of the time, but they may move along by creeping with adhesion to grains by their pedal disc and their proboscis alternatively.

Boreohydra: Boreohydra simplex (Westblad, 1937; Figure 19.1e) bears 3 or 4 tentacles, similar to *Psammohydra nanna,* but these insert close to the small proboscis and are very short. The polyps reach a length of up to 1.8 mm. The lower part is a hydrocaulus covered by a detritus agglutinated sheath, which may be prolonged by rhizoids. Asexual reproduction is by transverse fission. Westblad (1947) described the development of medusoid gonophores below the middle of the hydranth; however, he found no germ cells. Egg cells were discovered by Nyholm (1951), but in the ectoderm above the hydrocaulus. *Boreohydra simplex* was found along the Scandinavian west coast between Northern Norway and Gullmar Fjord, north of Iceland and on the west and south coasts of Great Britain, always in mud at depths between 6 and 100 m (Westblad, 1953) and to 630 m (Rees 1938).

Siphonohydra. The most structured hydroid dwarf is *S. adriatica* (Salvini-Plawen, 1966), having a size of 0.75–1.2 mm, the lower 1/3 being the hydrocaulus. The polyp bears 2 circles of 4 tentacles each, small ones around the peristome, the other ones above the hydrocaulus. Buds with tentacle-like extrusions are found above the lower circle of tentacles. Below these but above the hydrocaulus gonophores bud beyond protective protrusions. Only 2 specimens were found living in coarse sand at 6 m depth near Rovigne (Adriatic Sea).

Meiorhopalon. Meiorhopalon arenicolum Salvini-Plawen (1987) (Figure 19.1f) was first noted by Clausen and Salvini-Plawen, 1986. The hydranth and hydrocaulus of this species are well separated; the tentacles are arranged in two circles of four, the shorter oral ones are capitate, the aboral ones filiform which may bud a young polyp. One of the four specimens found had four additional protrusions between the aboral tentacles which might have been buds or tentacles. At its lower end the hydranth bears four broad and rounded papillae, possibly static organs. The stalk is tapering, some thin filaments protrude from its lower part and a mucus sheath covers this hydrocaulus. Its total size is about 1 mm. Specimens have been found in coarse sand at about 11 m depth in the Plymouth Sound.

Acauloides. Acauloides ammisatum Bouillon, 1971 (Figure 19.1g) is a solitary polyp, 0.6 mm and rarely up to 2 mm in size, with 10–25 capitate tentacles spread irregularly over the head except that the upper 6–8 tentacles are arranged in a circle around the hypostome. The hydrocaulus is very short and covered by a cup-shaped mucous periderm. This species was found in coarse sands near Roscoff and Banyuls-sur-Mer. The polyps may fix themselves to the grains by secreting a mucus substance.

Eugymnanthea. Its former genus name *Anthohydra* Salvini-Plawen and Rao, 1973, suggests affinities with both the Anthozoa and the Hydrozoa. *Eugymnanthea psammobionta* Salvini-Plawen, 1987 (Figure 19.1h) resembles anthozoan polyps because of having 24 tentacles in a single circle surrounding the oral disc with a mouth cone. The distal end is a well developed, circular adhesive pedal disc. In contrast the solid tentacles and the gastric cavity, undivided by septa, point to the Hydrozoa, but morphological simplification might have occurred in these tiny organisms, which reach a height of only up to 1.5 mm. The specimens were found in medium to coarse sand in very shallow waters of the Little Andaman Islands (Eastern Indian Ocean).

The suborder Halammohydrina is composed of 2 genera and 12 species. All the species live in the interstitial environment of coarse sands. They are known as medusoid forms gliding through the grains with the aid of cilia, covering all the body. In their outer appearance and in developmental stages the species show similarities with the actinula larva of the Tubulariidae, and therefore they were combined into the order of Actinulida by Swedmark and Teissier (1959). However, most species are equipped with several statocysts resembling those of medusae. Remane (1927) placed them close to the Narcomedusae. Swedmark and Teissier argue that the statocysts developed in an actinula-like species inhabiting coarse sand, analogous to many other groups of interstitial fauna. This reasoning might have been valid for one statocyst, but a large number of these organs, arranged in a circle, point to their medusoid origin. This is especially apparent in the family Armorhydridae with strong medusoid characters.

Halammohydra. This genus (Figure 19.1i,j) is composed of 10 species which reach 0.3–1.3 mm in length and are characterized by a long manubrium or gastric tube and a small aboral cone, the strongly reduced umbrella without a gastric cavity. This bears two whorls of tentacles alternatively directed orally and aborally. Statocysts insert between the orally oriented tentacles. The aboral pole is deeply grooved by an adhesive organ. All the body is covered with cilia, which are used for locomotion. The gonads of both sexes are situated between ento- and ectoderm of the manubrium. The eggs develop directly into an actinula-like larva with 4 tentacles in each circle and 4 statocysts. Species are distinguished by body size, number of tentacles and statocysts, length and form of tentacles, shape and size of aboral cone and adhesive organ, and by the cnidome. Variation in species is discussed by Swedmark (1956) and by Clausen (1967). The latter author splits *H. coronata* into two subspecies (Clausen and Salvini-Plawen 1986). Members of this interstitial genus are found from the intertidal zone to depths of 60 m, preferring medium and coarse sands, and shell gravel. Species of this genus are known from Europe, India, North America and Japan.

Otohydra. This genus includes only two species. *Othydra vagans* (Swedmark and Teissier, 1958a) (Figure 19.1k), about 0.8 mm in length, has a more polypoid structure with a larger ovoid body with gastric cavity; with its 12–14 partly adhesive tentacles arranged near the short and distensible mouth, and a maximum of 12 lithostyles inserted between the tentacles. As in *Halammohydra,* the epidermis is ciliated, but there is no adhesive organ. The eggs of this hermaphroditic species develop in an inner brood chamber between ento- and ectoderm. Fertilization and early development are internal, the actinula-like larva is released with 8 tentacles. This species lives in coarse sand and shell gravel and was discovered near Roscoff and later found as well in the Mediterranean (Salvini-Plawen, 1966). The second species, *O. tremulans,* mentioned by Clausen and Salvini-Plawen (1986), is separated from *O. vagans* by its cnidome (Lacassagne, 1968a), and lacks a brood chamber (Salvini-Plawen, 1987).

The suborder Limnohydrina is known from continental shelf regions, brackish and freshwater. Because most of the polyps are very small they must be included with the meiofauna.

Armorhydra. While the outer appearance of the only species in this genus, *Armorhydra janowiczi* (Swedmark and Teissier, 1958b) (Figure 19.1l) is rather similar to *Otohydra,* the internal morphology clearly demonstrates the medusoid character. A large manubrium that carries the gonads is protected by an umbrella and a velum with the tentacles inserting in between. The species measures 1-2 mm and may stretch to 4 mm. Two types of 8-30 tentacles are arranged in one circle: filiform ones with cnidocysts alternate with capitate ones having adhesive papillae. Statocysts are absent. This species was found in sublittoral coarse sands, together with other species of *Halammohydrina* and other interstitial fauna, near Roscoff (Swedmark and Teissier, 1958b), off Rovigne (Salvini-Plawen, 1966), near Ischia in the Mediterranean (Clausen, 1971), and in the Plymouth Sound (Salvini-Plawen, 1987). Although a close resemblance exists in lebensformtyp and lifestyle to the Halammohydrina, Lacassagne (1968b) classifies *Armorhydra* with the Limnohydrina because of its cnidome. In the same paper the author mentions a polyp of 1 mm height, discovered on sand grains, producing tentacular frustules and podocysts. Because of its most similar cnidome it is suggested that it might be the polyp generation of *Armorhydra janowiczi.*

Gonionemus. The polyp of *G. vertens* (Figure 19.1m,n) may reach a height of 1 mm. Four or 6 tentacles insert around the conical proboscis, its base is covered by periderm. Asexual propagation may occur by elongated buds or by medusa buds which reach the size of the polyp itself. *Gonionemus vertens* was found in marine and brackish waters between sand grains along the coasts of the U.S.A. Together with seeding oysters the species was introduced to European waters early this century. Werner (1950) discovered the polyps fixed to the leaves of seagrass.

Craspedacusta. C. sowerbyi (Figure 19.1o,p) is the well-known and wide-spread freshwater medusa. The polyps reach a maximum of 2 mm in height. They have no tentacles but a cushion of nematocysts surrounding the mouth. The proximal end is covered by a thin periderm cup. The polyps may form colonies of a few individuals and they bud frustules and medusae. The benthic stages occur on plants and on the sediment surface (Reisinger, 1957).

Limnocnida. Both *L. indica* and *L. tanganyicae* are Indian and African freshwater medusae. They have polyp stages which are very similar to those of *Craspedacusta sowerbyi* according to Bouillon, 1957.

Microhydrula. Microhydrula pontica (Valkanov, 1965) was discovered in aquaria in Bulgaria, France, and the U.S.A. The rather spherical and sedentary polyps attain a size of 80-240 µm. Tentacles and permanent mouth are absent, a gastrovascular cavity is present. Reproduction apparently is only by the budding of elongated (about 40 x 240 µm) crawling frustels (Spoon and Blanquet, 1978).

Rhaptapagis. Rhaptapagis cantacuzeni (Bouillon and Deroux, 1967) was also found in aquaria in France. It is very similar to *Microhydrula pontica,* but differs in its cnidome.

Anthozoa.--*Sphenotrochus andrewianus* (Figure 19.1q) is a scleractinian of the family Caryophyllidae originally described by Milne-Edwards and Haime (1848). Specimens may grow to a height of 10 mm, but young ones showing asexual reproduction by transverse fission are 1-2 mm, therefore included within the definition of meiofauna. Bipolarity may occur during the fission process. The number of tentacles may reach 24, corresponding to 24 calcareous septa (Zibrowius, 1980). *Sphenotrochus andrewianus* was found in the western Mediterranean, around the British Isles and Ireland, to the south off Portugal and N.W.-Africa to the Cape Verde region and near the Azores. It always occurs in coarse sand or shell gravel, and off Marseille, young specimens were regularly recorded in *Amphioxus*-sand. The depth of occurrence ranges up to 170 m, although the deepest records may have been just skeletons (Zibrowius, 1980; Manuel, 1981).

Individual Size and Habitat

Meiofaunal Cnidaria exhibit the problem of size and meiofauna definition (Chapter 1). The reason for this is that interstitial organisms may fall out of the size range of meiofauna. When Remane 60 years ago studied the organisms inhabiting coarse sands, and when the terms "mesopsammon" and "interstitial fauna" were coined for those organisms moving in the interstitial space between grains (Remane 1927), the term "meiobenthos" (Mare, 1942) had not yet been introduced (Chapter 2).

Interstitial organisms may well be longer than 1 mm and may or may not pass a screen with 1 x 1 mm meshes. Therefore, relatively large cnidarians – as well as species from other taxa – belonging to the interstitial fauna may be evaluated with the meiofauna. As pertinent for other classes, in which meiofaunal organisms constitute the lower size range, there is a size group of organisms between meio- and macrofauna, not well defined, introducing difficulties to the inclusion of some species into this book. Solitary cnidarians were assigned to the interstitial fauna just because they were found in coarse sand, but their size would not allow to count them with the meiofauna. *Euphysa ruthae* (Norenburg and Morse, 1983) can be taken as an example of being regarded as interstitial but not meiofaunal. The species is found in coarse sand, it may survive burial for some time, but normally it lives fixed with its hydrocaulus to sand grains in or close to the sediment surface, extending its hydranth and part of the hydrocaulus into the open water. This maximally 40-50 mm high solitary hydropolyp finds its habitat for larval fixation in sands with well developed interstitial space, but its main life activities, feeding and development of reproductive organs, occur in the benthic boundary layer above the sediment surface. If this species is regarded as interstitial (Clausen and Salvini-Plawen, 1986) than a high number of other species must be included, e.g., *Euphysa aurata* (Rees, 1938), which gains a length of only 4-5 mm. But, because of living on muddy grounds, it never occurred to anyone to include this species with the meiofauna.

The same arguments could be brought forward in considering the affiliation of *Spenotrochus andrewianus* but this scleractinian is reproductive (albeit asexually by budding) at lengths of 1 mm. Many species, of course, fall in the temporary meiofauna, but they are not considered.

Actually, most meiofaunal cnidarians live in coarse sand and shell gravel, more or less vigorously moving around in the interstitial space. However, some species occur in muddy sediments or in detritus rich surface layers. Studying the meiofauna of continental slope and deep-sea areas with fine grained sediments, solitary undescribed Hydrozoa of very small size were discovered in different geographic regions (personal observations).

Methods of Collection, Extraction and Sorting

No specific methods are described for specific handling of meiofaunal Cnidaria. The best method for their collection is the inspection of unpreserved samples. The organisms begin to explore their environment for better conditions or for food and thus move around or stretch out, some time after they are poured into a dish together with a thin layer of sediment. Their movements help very much to discover them between the substrate. All species not permanently fixed to sand grains may be washed out by decantation, possibly after short freshwater shock or by elutriation, and some still fixed to a particle may successfully be transferred with these methods. Cnidarians can be kept alive for long periods of time together with some substrate, preferably in temperatures somewhat below ambient. They can be fed with ciliates; some larger cnidarians may be fed on small harpacticoids, or with tiny parts of mussel midgut-gland content.

In preserved samples, Cnidaria are often shrunk to indiscernible organic matter. Narcotization with $MgCl_2$ or MS-222 may extend specimens. Fixation and preservation may be done in 10% formalin. Shock fixation of narcotized specimens with hot, or 20%, or even higher concentrations of formalin may result in well expanded individuals. In Steedmans solution (Chapter 9), cnidarians can be stored for long periods. Descriptions of species from preserved and contracted materials must be done with reservation, and should be noted whether sizes are based on living or dead specimens.

References

Arneson, A.C., and C.E. Cutress
1976. Life History of *Carybdea alata* Reynaud, 1830 (Cubomedusae). Pages 227-236 in G.O. Mackie, editor, *Coelenterate Ecology and Behavior*. New York: Plenum Press.

Bouillon, J.
1957. Etude monographique du genre Limnocnida (Limnomedusa). *Annales de la Société Royale Zoologique de Belgique*, 87:253-500.
1971. Sur quelques hydroides de Roscoff. *Cahiers de Biologie Marine*, 12:323-364.

Bouillon, J., and G. Deroux
1967. Remarque sur les Cnidaires du type de *Microhydrula pontica* Valkanov, 1965 trouvés à Roscoff. *Cahiers de Biologie Marine*, 8:253-272.

Clausen, C.
1967. Morphological Studies of *Halammohydra* Remane (Hydrozoa). *Sarsia*, 29:349-370.
1971. Interstitial Cnidaria: Present Status of Their Systematics and Ecology. In N.C. Hulings, editor, Proceedings of the First International Conference on Meiofauna. *Smithsonian Contributions to Zoology*, 76:1-9.

Clausen, C., and L. v. Salvini-Plawen
1986. Cnidaria. In L. Botosaneanu, editor, *Stygofauna Mundi*. 740 pages. Leiden: E.J. Brill/Dr. W. Backhuys.

Cutress, C.E., and J.P. Studebaker
1973. Development of Cubomedusae, *Carybdea marsupialis*. *Proceedings of the Association of Island Marine Laboratories of the Caribbean*, 9:25.

Dawidoff, C.N.
1930. *Protohydra caulleryi* n.sp. des eaux indochinoises. *Archives de Zoologie Expérimentale et Génerale, Notes et Revue*, 70:55-57.

Greeff, R.
1870. *Protohydra leuckarti*. Eine marine Stammform der Coelenterata. *Zeitschrift für Wissenschaftliche Zoologie*,

Hartwick, B.
1987. The Box-jellyfish. Pages 99-105 in J. Covacevich, P. Davie, and J. Pearn, editors, *Toxic Plants and Animals, a guide for Australia*. Brisbane: Queensland Museum.

Lacassagne, M.M.
1968a. Deux nouveaux types de nematocystes astomacnides chez les Actinulides (Hydrozoaires): les euryteloides, spiroteles et aspiroteles. *Compte Rendu de la Académie des Sciences de Paris*, 266:892-894.
1968b. Anatomie et histologie de l'Hydroméduse benthique *Armorhydra janowiczi* Swedmark et Teissier, 1958. *Cahiers de Biologie Marine*, 9:187-200.

Manuel, R.L.
1981. British Anthozoa. In D.M. Kermack and R.S.K. Barnes, editors, *Synopsis of the British Fauna, New Series*, 18:1-241. London: Academic Press.

Mare, M.F.
1942. A Study of a Marine Benthic Community with Special Reference to the Micro-organisms. *Journal of the Marine Biological Association of the United Kingdom*, 25:517-554.

Milne-Edwards, H., and J. Haime
1848. Recherches sur les polypiers: 2. Mém. Monographie des Turbinolides. *Annales des Sciences Naturelles, 3. Série, Zoologie*, 9:211-343.

Monniot, F.
1962. Recherches sur le graviers à Amphioxus de la région de Banyuls-sur-Mer. *Vie et Milieu*, 13:231-322.

Norenberg, J.L., and M.P. Morse
1983. Systematic Implications of *Euphysa ruthae* n. sp. (Athecata: Corymorphidae), a Psammophilic Solitary Hydroid with Unusual Morphogenesis. *Transactions of the American Microscopical Society*, 102:1-17.

Nyholm, K.-G.
1951. Egg Cells in the Ectoderm of *Boreohydra simplex*. *Arkiv för Zoologie*, 2:531-534.

Omer-Cooper, F.L.S.
1957. On *Protohydra psamathe* n. sp. from South Africa. *Journal of the Linnean Society (Zoology)*, 45:145-150.

Rao, G.C.
1975. *Halammohydra chauhani* n. sp. (Hydrozoa) from Andamans, India. *Dr. B.S. Chauhan Commemorative Volume*: 299-303.
1978. On a New Species of *Halammohydra* (Actinulida, Hydrozoa) from Andamans, India. *Bulletin of the Zoological Survey of India*, 1:147-149.

Rao, G.C., and A. Misra
1980. On a New Species of *Halammohydra* (Actinulida, Hydrozoa) from Sagar Island, India. *Bulletin of the Zoological Survey of India*, 3:113-114.

Rees, W.J.
1938. Observations on British and Norwegian Hydroids and Their Medusae. *Journal of the Marine Biological Association of the United Kingdom*, 23:1-42.

Reisinger, E.
1957. Zur Entwicklungsgeschichte und Entwicklungsmechanik von *Craspadacusta* (Hydrozoa, Limnotrachilina). *Zeitschrift für Morphologie und Ökologie der Tiere*, 45:656-698.

Remane, A.
1927. *Halammohydra*, ein eigenartiges Hydrozoon der Nord- und Ostsee. *Zeitschrift für Morphologie und Ökologie der Tiere*, 7:643-677.

Salvini-Plawen, L. v.
1966. Zur Kenntnis der Cnidaria des nordadriatischen Mesopsammon. *Veröffentlichungen des Instituts für Meeresforschung in Bremerhaven*, Sonderband 2:165-186.
1987. On Mesopsammic Cnidaria from Plymouth. *Journal of the Marine Biological Association of the United Kingdom*, 67:623-637.

Salvini-Plawen, L. v., and G.C. Rao
1973. On Three New Mesopsammobiotic Representatives from the Bay of Bengal: Species of *Anthohydra* gen. nov. (Hydrozoa) and *Pseudovermis* (Gastropoda). *Zeitschrift für Morphologie der Tiere*, 74:231-240.

Schulz, E.
1950. *Psammohydra nanna*, ein neues solitäres Hydrozoon in der westlichen Beltsee. *Kieler Meeresforschungen*, 7:122-137.

Spoon, D.M., and R.S. Blanquet
1978. Life Cycle and Ecology of the Minute Hydrozoon *Microhydrula*. *Transactions of the American Microscopical Society*, 97:208-216.

Swedmark, B.
1956. Etude de la microfaune des sables marins de la région de Marseille. *Archive de Zoologie Expérimentale et Génerale, Notes et Revue*, 93:70-95.
1957. Variation morphologique des différentes populations régionales d'*Halammohydra*. *Anné Biologique*, 44:183-189.
1959. On the Biology of Sexual Reproduction of the Interstitial Fauna of Marine Sands. *Proceedings of the 15th International Congress on Zoology, London*, pages 327-329.

Swedmark, B., and G. Teissier
1957. *Halammohydra vermiformes* n. sp. et la famille des Halammohydridae Remane. *Bulletin de la Société Zoologique de France*, 82:38-49.
1958a. *Otohydra vagans* n. g., n. sp., hydrozoaire des sables, appasenté aux Halammohydridées. *Compte Rendu de la Académie des Sciences de Paris*, 247:238-240.
1958b. *Armorhydra janowiczi*, n.g., n.sp., hydroméduse benthique. *Compte Rendu de la Académie des Sciences de Paris*, 247:133-135.
1959. *Halammohydra* et *Othohydra*, hydrozoaires de la microfaune des sables et l'ordre des actinulides. *Proceedings of the 15th International Congress of Zoology, London*, pages 330-331.
1967. Structure et adaptation d'*Halammohydra adherens*. *Cahiers de Biologie Marine*, 7:63-74.

Valkanov, A.
1965. *Microhydrula pontica* n. g., n. sp. - Ein neuer solitärer Vertreter der Hydrozoen. *Zoologischer Anzeiger*, 174:134-147.

Werner, B.
1950. Weitere Beobachtungen über das Auftreten der Meduse *Gonionemus murbachi* Mayer im Sylter Wattenmeer und ihre Entwicklungsgeschichte. *Zoologischer Anzeiger*, 14 (Supplement, Verhandlungen der Deutschen Zoologischen Gesellschaft, Mainz 1949): 138-151.
1965. Die Nesselkapseln der Cnidaria, mit besonderer Berücksichtigung der Hydroida. I. Klassifikation und Bedeutung für die Systematik und Evolution. *Helgoländer wissenschaftlichte Meeresuntersuchungen*, 12:1-39.
1973. New Investigations on Systematics and Evolution of the Class Scyphozoa and the Phylum Cnidaria. *Publications of the Seto Marine Biological Laboratory*, 20:35-61.
1975. Bau und Lebensgeschichte des Polypen von *Tripedalia cystophora* (Cubozoa, class nov., Carybdeidae) und seine Bedeutung für die Evolution der Cnidaria. *Helgoländer wissenschaftlichte Meeresuntersuchungen*, 27:461-504.
1984. Stamm Cnidaria, Nesseltiere. In A. Kaestner, *Lehrbuch der speziellen Zoologie*, Band 1, Teil 2, 335 pages. Jena: Gustav Fischer Verlag.

Westblad, E.
1937. *Boreohydra simplex* n. gen., n. sp., ein Solitärpolyp von der norwegischen Küste. *Arkiv för Zoologie*, 29(B):1-6.
1947. Notes on Hydroids. *Arkiv för Zoologie*, 39(A):1-23.
1953. *Boreohydra simplex* Westblad, a "Bipolar" Hydroid. *Arkiv för Zoologie*, 4:351-354.

Zibrowius, H.
1980. Les scléractiniaires de la méditerranée et de l'Atlantique nord-oriental. *Memoires de l'Institut Oceanographique, Monaco* 11:1-227

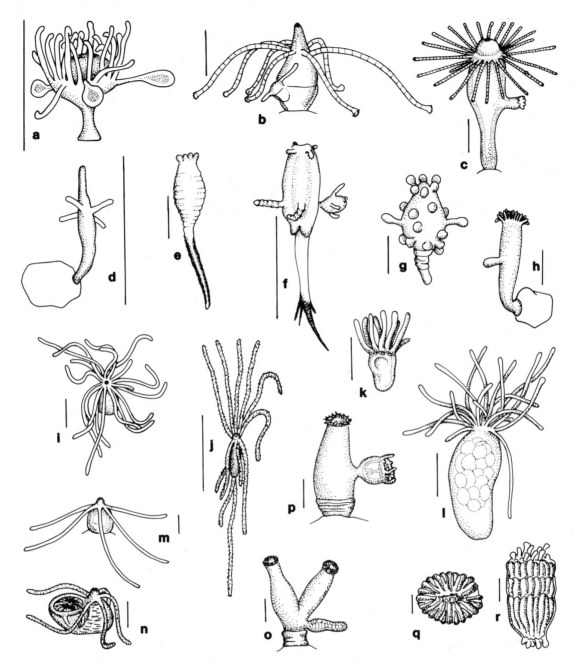

Figure 19.1.--Cnidaria: a, *Stylocoronella riedli* with 3 buds (from Salvini-Plawen, 1966); **b,** *Tripedalia cystophora* with periderm cup and bud (from Werner, 1984); **c,** *Carybdea marsupialis* with bud (from Werner, 1984); **d,** *Psammohydra nanna* on sand grain (from Clausen and Salvini-Plawen, 1986); **e,** *Boreohydra simplex* (from Westblad, 1937); **f,** *Meiorhopalon arenicolum* with 4 oral and 2 aboral tentacles and a bud (from Salvini-Plawen, 1987); **g,** *Acauloides ammisatum* (from Clausen and Salvini-Plawen, 1986); **h,** *Eugymnanthea psammobionta* with bud (from Salvini-Plawen and Rao, 1973); **i,** *Halammohydra schulzei* (from Clausen and Salvini-Plawen, 1986); **j,** *Halammohydra octopodides* (from Remane, 1927); **k,** *Otohydra vagans* with 3 statocysts and embryo in brood pouch (from Swedmark and Teissier, 1958a); **l,** *Armorhydra janowiczi*, interior shows gonad around gastric cavity (from Swedmark and Teissier, 1958a; and Clausen and Salvini-Plawen, 1986); **m,** *Gonionemus vertens* (from Werner, 1984); **n,** *Gonionemus vertens* with medusa bud (from Werner, 1984); **o,** *Craspedacusta sowerbyi*, double polyp with budding frustale (from Werner, 1984); **p,** *Craspedacusta sowerbyi* with medusa bud (from Werner, 1984); **q,** *Sphenotrochus andrewianus*, skeleton, top view (from Zibrowius, 1980); **r,** *Sphenotrochus andrewianus*, exceptional bipolar specimen, short tentacles extruded on both sides (from Rossi, 1961). (Scale = 500 μm.) (Plate redrawn by Thomas Soltwedel.)

20. Turbellaria

Lester R. G. Cannon and Anno Faubel

Von Graff gave exhaustive accounts of the "Turbellaria" in Bronn's *Klassen und Ordnungen des Tierreichs*, published in 1904-08 and 1912-17. A more recent overview, also in German, was published by Bresslau (1933) and another in French by de Beauchamp (1961). The only generalized account in English is by Hyman (1951) who should be consulted for historical details and references. Crezée (1982) presented a synopsis of families, but Cannon (1986) has provided a pictorial guide to families and lists of genera. Because of their position among the simplest of metazoans, the "Turbellaria" have always received considerable attention concerning their comparative morphology and systematics and the inter-relationship of these with their phylogeny or the evolution of the metazoa (see Dougherty, 1963; Boaden, 1975). More traditional studies have given way to work on ultrastructure (see Rieger, 1981, for a review) and the study of cytogenetics and karyotypes (e.g., papers in Schockaert and Ball, 1981; Tyler, 1986). Phylogenetic studies have been boosted by the ideas of Hennig (1966) which have been applied to several groups (Karling, 1974a - orders, Ball, 1977 - triclads, Ax, 1971a and Ehlers, 1972 - rhabdocoels). Recently Ehlers (1985) has questioned the validity of the taxon Turbellaria. He states "A monophylum "Turbellaria" does not exist in reality" (Ehlers, 1986). However, "turbellarian" may still represent a useful facies of a predominantly freeliving flatworm with a ciliated epidermis distinguishable from the similar Gnathostomulida by, among other things, the lack of a pair of jaws and base plate in the mouth.

Classification

Turbellarians are acoelomate bilateria without a definitive anus. They are largely freeliving, with a cellular, usually ciliated, epidermis. The worms are often dorsoventrally flattened with anterior sensory and glandular regions. Larger forms (especially polyclads and triclads) are often pigmented. With very few exceptions locomotion is by beating of the epidermal cilia which are often in restricted tracts in the larger or more specialized species. Many turbellarians have epidermal rhabdoids (rod-like bodies secreted by the epidermis). Sub-epithelial longitudinal and circular muscles combine with larger muscles through the parenchyma (the cellular body substance in which the organ systems are embedded) to provide an extremely flexible body. There is usually an anterior nerve ganglion and main anterior-posterior trunks with cross fibers and sensory tracts forming an elaborate network. The digestive system usually consists of a sac-like gut (often ramified in larger forms) with a mouth and a muscular pharynx for food capture. As there is no anus, wastes are regurgitated. There is also no circulatory system; there is, however, in many (notably the freshwater species) a protonephridial system with terminal flame cells.

There are 3 types of pharynges of importance for turbellarian taxonomy. The *pharynx simplex* which is characteristic of those worms in which yolk is incorporated into the oocyte. It is a mostly short, unfolded tube with totally ciliated inner walls, though within the Acoela it is often indistinct. The *pharynx plicatus* (characteristic of Polycladida, Seriata, and some Prolecithophora) exhibits a protrusible more or less tubular fold housed in the pharyngial cavity. The *pharynx bulbosus* is characteristic of some Prolecithophora and Rhabdocoela. There are 3 main types: *ph. variabilis* (Prolecithophora), *ph. rosulatus* (Typhloplanoida including the Kalyptorhynchia), and *ph. doliiformis* (Dalyellioida). The *ph. bulbosus* represents a bulb separated from the surrounding tissue by a muscular septum. The *ph. rosulatus* is mostly globular with a more or less vertical axis. The *ph. doliiformis* is barrel-shaped, anteriorly situated with a more or less horizontal axis and hardly protrusible.

Most turbellarians are hermaphrodites with cross fertilization following copulation; some have asexual stages and some are parthenogenetic, but most lay egg capsules from which one or more ciliated young emerge, grow, and differentiate directly without metamorphosis to the adult. Reproductive systems are often very complex - in the male system one or several testes connect to one or paired sperm ducts, united these form an ejaculatory duct leading to a seminal vesicle; there is often prostatic tissue and then a copulatory apparatus frequently armed with hard, sclerotic processes. The female systems may be equally complex and the degree to which the yolk or vitelline material is separated from the oocytes forms the fundamental basis for distinguishing the higher orders of the group. In the more primitive forms the yolk is incorporated into the oocytes and there may or may not be discrete ovaries and oviducts. In the more highly evolved groups oocytes develop in separate regions or discrete organs (ovaries) and the yolk develops in vitelline tissue or glands. A seminal receptacle is often present; so too is a region in which the egg capsule is formed. This is commonly called the uterus, which may have one or more secondary storage regions for capsules. The female gonopore, when present, is often, but not always, united with the male gonopore to form a common genital opening. The complexity and diversity of the reproductive systems form the main means of classifying turbellarians.

Order Catenulida.--Catenulids (Figure 20.1a-d) are the simplest of the flatworms. Chaining (asexual reproduction) is common. Catenulidae (Figure 20.1a) and Chordariidae (Figure 20.1b) have an unlobed brain which in the latter is far anterior to the mouth. Stenostomidae (Figure 20.1c) and Retronectidae (Figure 20.1d) have a lobed/divided brain. Of the 4 families, all are limnic except for the aberrant marine Retronectidae which lack protonephridia (and sometimes a gut). Important references: Luther, 1960; Marcus, 1945; Meixner, 1938; Sterrer and Rieger, 1974.

Order Nemertodermatida.--A small group consisting of a single family of rather specialized worms resembling acoels (Figure 20.1e). Important reference: Faubel and Dörjes, 1978.

Order Acoela.--A large group of worms often found in oxidized sediments rich in organic matter. They are difficult to identify below family level (Figure 20.1f-h). Important references: Dörjes, 1968; Ehlers and Dörjes, 1979; Faubel, 1976.

Order Haplopharyngida.--A small group (1 genus) formerly considered in the Macrostomida (Figure 20.1i). Important reference: Ax, 1971b.

Order Macrostomida.--Macrostomids (Figures 20.1j-m) include 3 families. They are common in both freshwater and saltwater (lacking protonephridia in saltwater). A pre-oral gut occurs in Microstomidae (Figure 20.1j) and chaining is common: the male stylet is simple. Macrostomidae (Figure 20.1k) and Dolichomacrostomidae lack a pre-oral gut and in the latter there is a common gonopore and two subfamilies: Dolichomacrostominae (Figure 20.1l) with complex stylet and female accessory pieces and Karlingiinae (Figure 20.1m) with a simple stylet as in the Macrostomidae. Important references: Faubel, 1974; Rieger, 1971a, b.

Order Polycladida.--Polyclads, consisting of 30 families, are usually large leaf-like worms commonly divided into two large groups - Acotylea (lacking a posterior adhesive organ (sucker) behind the female gonopore) and Cotylea (possessing such a structure). The animals are not usually considered part of the meiofauna. Important references: Faubel, 1983, 1984; Prudhoe, 1985.

Order Lecithoepitheliata.--A small group with some moderately large freshwater or terrestrial worms (Prorhynchidae) and some small marine ones (Gnosonesimidae) (Figure 20.1n). Important reference: Karling, 1968.

Order Prolecithophora.--Mostly small turbellarians (Figure 20.2a-d) (10 families) with either a *pharynx variabilis* or *plicatus* found in fresh or saltwater as well as several symbiotic groups. The Separata (Figure 20.2a) have mouth and gonopore separate. Worms with mouth and gonopore combined (Combinata) fall into two groups - Proporata (oral-genital pore in forebody) (Figure 20.2b,c) and Opisthoporata (oral-genital pore in hindbody) (Figure 20.2d) Important references: Karling, 1963, 1974b; Luther, 1960; Meixner, 1938.

Order Tricladida.--Usually large worms not considered part of the meiofauna. The 9 families can be organized into 3 groups - Terricola (terrestrial), Paludicola (freshwater), and Maricola (marine).

Order Proseriata.--A large group (7 families) of mainly interstitial worms (Figure 20.2e-k). The Bothrioplanidae (1 pair of ovaries) (Figure 20.2e) and the Nematoplanidae (2 pairs) (Figure 20.2f) both lack a statocyst which is found in the other groups. Monocelididae (ovaries anterior to pharynx) (Figure 20.2g) and Otomesostomidae (ovaries posterior to pharynx) (Figure 20.2h) both have separate male and female gonopores. [The Monotoplanidae resemble the Monocelididae but have multiple male organs.] With a common gonopore are the Otoplanidae (with prominent anterior bristles): worms in the subfamily Archotoplaninae (Figure 20.2i) are more or less evenly ciliated but those in other subfamilies, e.g., Otoplaninae (Figure 20.2j) have only a ciliated sole. Also with a common gonopore are the Coelogynoporidae (elongate, evenly ciliated and with feeble anterior bristles) (Figure 20.2k). Important references: Ax, 1956a; Ax and Ax, 1974a,b, 1977; Karling, 1974b; Sopott, 1972.

Order Rhabdocoela.—A very large group (35 families) of mainly small worms separated by the nature of the pharynx.

The Dalyellioida (Figure 20.2l–n) contains 11 families characterized by a barrel-like pharynx; many are symbiotic, but among free-living worms the Provorticidae (2 ovaries) (Figure 20.2l) and the Dalyelliidae (1 ovary) (Figure 20.2m) have a posterior gonopore whereas the Graffillidae (Figure 20.2n) have the gonopore in the forebody. Important references: Luther, 1955, 1962.

The Typhloplanoida have a rosulate pharynx and may be divided into those without an eversible proboscis (Typhloplanida) and those with an eversible proboscis (Kalyptorhynchia).

The suborder Typhloplanida consists of 8 families (Figure 20.3a–i). An elongate pharynx is found in the Solenopharyngidae (Figure 20.3a) and a ciliated one in the Ciliopharyngiellidae (Figure 20.3b), the Kytorhynchidae (Figure 20.2c) have a permanent anterior glandular pocket. Trigonostomidae (Figure 20.2d) have two connections between ovary and gonopore, but Promesostomidae (paired ovaries) and Typhloplanidae (single ovary) have only one. All three of these families have several subfamilies. Among the Promesostomidae, the Promesostominae (Figure 20.3e) have a long stylet and the Brinkmaniellinae (Figure 20.3f) a short one. Subfamilies of the Typhloplanidae are Olisthanellinae (testes above vitellaria, excretory pore in gonopore) (Figure 20.3g), Mesostominae (testes above vitellaria, excretory pore in mouth) (Figure 20.3h) and Typhloplaninae (testes below vitellaria) (Figure 20.3i). Important references: Ax, 1952a,b, 1959, 1971a; Ehlers, 1972; Karling, 1974b; Luther, 1962, 1963.

The suborder Kalyptorhynchia consists of 16 families (Figure 20.3j–o). A sub-group Schizorhynchia have a cleft proboscis and the presence of jaws (Karkinorhynchidae) (Figure 20.3j) or their absence (Schizorhynchidae) (Figure 20.3k) is used to distinguish them. Those without a cleft proboscis (Eukalyptorhynchia) may have hooks on the proboscis (Gnathorhynchidae) (Figure 20.3l), elaborations of the proboscis (e.g., Psammorhynchidae (Figure 20.3m)) or the body wall; without those features separation may concern the prostate tissue either separate from the ejaculatory duct (Polycystididae) (Figure 20.2n) or surrounding it (Koinocystidae) (Figure 20.3o). Important references: Karling, 1974b; Noldt and Hoxhold, 1984; Schilke, 1970; Schockaert, 1971, 1982.

Habitats and Ecological Notes

Turbellarians are ubiquitous forms in freshwater and marine habitats; they are common in moist terrestrial habitats and very many associate with other organisms to various degrees (some are parasitic). About 2/3 are small (1–2 mm long or less), but some, the marine polyclads and terrestrial triclads in particular, can grow to many centimeters in length.

Since turbellarians usually have to be sexually mature before they can be identified as to species, samples may have to be taken at different seasons of the year. It is also important that turbellarians are extracted alive because those taken from fixed samples are extremely hard to identify. In most cases fixed turbellarians exhibit little to distinguish them from other wormlike taxa and may be confused with large ciliates, gnathostomulids, some taxa of oligochaetes, archiannelids and other polychaetes and especially nemerteans. Nemerteans when fixed can break up into smaller pieces which can be virtually indistinguishable from turbellarians. For ecological work, determination of species extracted from fixed sediment samples can only be done with great care and experience of the fauna.

There are three main turbellarian life cycle types: univoltine, bivoltine, and polyvoltine. Among free-living forms, polyvoltine life cycles are only represented in the order Acoela which have either continuous or discontinuous sexual maturity throughout the year. Univoltine and bivoltine life cycles are predominantly found in all the other taxa.

Turbellarians are very common on and in various benthic substrates (soft bottom, sandy bottom, calcareous bottom, gravel, mussel associations, vegetation) both intertidally and subtidally. Some turbellarian species are pelagic but until now only a few have been reported in deep seas. Some turbellarians live in ground waters and caves (Riedl, 1959) in the stygobios.

In benthic systems turbellarians are generally represented in oxygenated substrates. In wave-protected areas with little mixing of the sediments such as sand or mud flats only a few surface millimeters and deeper in the deposit, the halo of oxic sand surrounding permanent burrows of large infaunal animals remains oxygenated. The high-oxygen end of the gradient, the chemocline, offers a poor line of demarcation, which is useful to differentiate between oxybiota and thiobiota (Meyers et al., 1987). Even in this oxygen-sulfide gradient of marine deposits, distinctive turbellarians are represented (Boaden, 1975, Crezée, 1976, Scherer, 1985).

The main factors which control the life and distribution of turbellarians in eulittoral and sublittoral zones (marine, brackish, limnic) are temperature, salinity, organic content, and grain size. These factors either operate directly or influence such factors as pore size and drainage. Also important are water content and the availability of oxygen and food. The importance of turbellarians in the marine meiobenthos was reviewed by Martens and Schockaert (1986).

Methods of Collection

Turbellarians are for the most part fragile. They have soft bodies and are easily damaged either when collecting or when preserving, so special care is often needed. Larger forms from marine, freshwater, or terrestrial habitats are best handled with moist brushes with which they may be coaxed into a vessel for later examination or fixation. Freshwater turbellarians, for example, may be collected in small nets by shaking water plants or stones over the nets while these are held under water. A fine paint brush can be used to transfer worms from moist surfaces to a vial. In water a wide bore pipette can be effective for transfer.

Methods of Extraction

Extracting microturbellarians from soil, mud, detritus, vegetation, or coarse sediments may be achieved in a variety of ways (Martens, 1984; Noldt and Wehrenberg, 1984 (see Chapter 9). Most turbellarians live in regions of high oxygen, i.e., on the surface or in the top few centimeters of sediments. By collecting large volumes, e.g., 1-10 l and allowing it to "decompose" for a few days, one can induce worms to migrate to the surface layers. If the substrate is very fine, e.g., silt or mud, the worms may be induced to migrate into coarse, clean (washed and sterilized) sand (from which they are more easily removed) by placing 2-5 cm of this above the decomposing habitat, i.e., the Übersand technique. When extracting microturbellarians from mud, one can also induce worms to migrate into the overlying water by placing 2 cm of this above the substrate and then keeping the dish at room temperature. After about 12-24 hr the overlying water can be decanted and worms counted out. To remove worms from sand three procedures are often used:

1. Surface material is mixed with 5-10 volumes of an anaesthetic (e.g., 7.5% $MgCl_2$ made up in water from the habitat). After 10 min vigorous stirring the supernatant is quickly decanted through suitable graded sieves (e.g., 140 μm). This may be repeated several times and then the sieve is backwashed with clean (filtered) water from the habitat.

2. The process of decanting may also be used without anaesthetic and for most turbellarians seems to work well. It is important in this case to use only fairly small volumes of sediment, e.g., 20 cm^3. Several washings (5-10) remove most of the worms without potential harm from the anaesthetic.

3. Sediment 10-30 cm^3 is placed on a suitable sieve (e.g., 140 μm or 250 μm) in a tube suspended over a Petri dish filled with clean (filtered) water from the habitat. Ice blocks made from the clean (filtered) water from the habitat are used to fill the tube above the sediment. Provided the water in the Petri dish is touching the bottom of the sieve, i.e., a meniscus forms, worms will migrate away from the cold water (from the melting ice), through the sieve and collect in the Petri dish. Excess fluid is caught in an outer, much larger, dish. This method is the Uhlig technique (see Chapter 7).

Preparation of Specimens for Taxonomic Study

Smaller forms, especially the free-living microturbellaria, may be adequately transferred with a Pasteur pipette while taking care to prevent them from attaching to the glass sides. As with larger forms they must be studied alive, particularly with the aid of a compound microscope. They may be mounted under a coverglass in a small drop of their natural medium and with the aid of tissue or blotting paper fluid can be removed until they are slowed sufficiently to observe. Remove too much fluid and they become squashed! (Methocel can prove useful in slowing worms, but they are less easy to recover subsequently.) A great deal can be learned from the live animal and at all times detailed notes and pictures (drawings or photographs) are an essential adjunct to subsequent identification. Color, for example, is most unstable, but often colors and patterns are characteristic.

To fix larger forms, the worms should be allowed to crawl out in a clean vessel (beaker, jar, vial, or petri dish) and when extended, flooded with a fast acting fixative. If large worms are merely immersed in fixative, they will curl and become difficult to study subsequently. They have a tendency to simply break up if they are not cool and healthy, i.e., keep them cool and fix them fresh. Anaesthetic ($MgCl_2$ or $MgSO_4$) may be used with discretion.

To fix small forms, the worms should be pipetted into a small petri dish, anaethetized by $MgCl_2$ or $MgSO_4$ if appropriate and then the water sucked off until the animal is restricted to only a small drop of water. When extended the worms should be flooded with a fast acting fixative, e.g., Bouin's fixative for light microscopy or glutaraldehyde (2.5%) for electron microscopy.

Fixation for light microscopy is achieved with most standard fixatives. For large worms formalin (10% by volume or somewhat less) is suitable. For better anatomical detail Bouin's fluid or fixatives containing mercuric chloride are recommended. Subsequently the worms may be embedded in wax or resin. Since serial reconstruction is necessary wax is the easier medium; however, resolution may be better with resin sections. Aqueous staining may be achieved after resin embedment by first removing the resin by soaking the slides in saturated sodium hydroxide in absolute ethanol for 5 min prior to washing and staining normally.

Serial sections can be stained in Heidenhain's iron hematoxylin, using orange G, eosin Y or Alcian Blue 8GX as a counterstain; brilliant staining may be achieved more easily by following the original procedure of Pasini (Romeis, 1968).

For electron microscopy it is important that all solutions should be cold and isotonic initially. Glutaraldehyde, 2.5% in cacodylate or phosphate buffer, for 1-2 hr at 4° C is recommended.

For all turbellarians it is necessary to have a combination of drawings and notes from life, a whole mount where possible, and sections for anatomical reconstruction. Whole mounts are often difficult to make and/or interpret. Turbellarians do not always seem to stain well as whole mounts. Clearing and mounting without staining will enable most larger structures to be seen and their relative positions noted. For most microturbellaria there is little point in making whole mounts except to preserve and reveal hard, i.e., "sclerotic" structures for fixation often renders most anatomical details unobservable. Hard structures can be observed by flattening worms under a coverglass and flooding by replacement with glycerin or CMCP, i.e., a water based clearing and mounting medium, drawn under the coverglass. The mount must then be ringed and sealed.

Distribution

Turbellarians are undoubtedly best known from Europe. Ecological studies on turbellarians have been hindered to some extent by problems of identification; however, notable contributions on freshwater forms have been made in Britain (Young, 1970), in Germany (see Schwank, 1982), and on marine forms from the sands of the island of Sylt in northern Germany by Ax and others (summarized by Ax, 1977).

The Acoela, Macrostomida, Proseriata, and Rhabdocoela have been studied in the North Sea (Ax, 1956a; Dörjes, 1968; Ehlers, 1972; Faubel, 1974, 1976; Martens, 1983; Schilke, 1970; Sopott, 1972; Westblad, 1948). These and other turbellarians from both marine and freshwater have been extensively studied in the Baltic (see contributions of Luther (1955-1963) and summary provided by Karling (1974b)). Marine forms have been studied in Britain (Boaden, 1963) and France (Ax, 1956b; Brunet, 1973). The freshwater fauna of Europe was reviewed by Lanfranchi and Papi (1978).

In North America Ruebush (1941) gave a key to freshwater turbellarians and the general accounts of Hyman and Jones (1959) and Pennak (1978) have augmented the knowledge. Marine turbellarians of the Atlantic coasts have been reported upon by Bush (1981) and by Rieger (e.g., 1974) and along the Pacific coasts by Karling (e.g., 1963, 1966).

Marcus, in a series of papers from 1945-1954, has given considerable information on the turbellarians of South America, notably from aquatic habitats, both freshwater and marine. Furthermore, the studies of Ax and colleagues in the Galapagos archipelago (Ax and Ax, 1974a, 1977; Ehlers and Ax, 1974; Ehlers and Dörjes, 1979) on mainly interstitial forms has greatly increased knowledge from the western Pacific.

Asia (outside the Palaearctic which has an essentially European fauna – see Evdonin, 1977: Kalyptorhynchia) has a poorly known fauna; however, the proseriates of Japan are currently receiving attention from Tajika (1982). Young (1976) gave an account of the freshwater fauna of Africa and Schockaert (1971, 1982) reported some Kalyptorhynchia from that continent.

Very few reports of freeliving microturbellaria exist from either fresh or marine habitats of the Australian region, though triclads and polyclads have received some attention as have symbiotic forms.

Type Collections

Collections of types are to be found in many museums, but among the more important collections are those of the American Museum of Natural History, New York; the National Museum of Natural History, Smithsonian Institution, Washington, D.C., USA; British Museum (Natural History), London, UK; Naturhistoriska Riksmuseet, Stockholm, Sweden; Naturmuseum und Forschungsinstitut Senckenberg, Frankfurt; Zoologisches Institut und Zoologisches Museum, Universität Hamburg, Hamburg; and II. Zoologisches Institut, Universität Göttingen, Göttingen; all in the Federal Republic of Germany.

Acknowledgments

We thank the Queensland Museum for permission to reproduce material, notably figures, from "Turbellaria of the World."

References

Ax, P.
1952a. Turbellarien der Gattung *Promesostoma* von den deutschen Küsten. *Kieler Meeresforschungen*, 8:218-226.
1952b. *Ciliopharyngiella intermedia* nov. gen. nov. spec., Repräsentant einer neuen Turbellarien-Familie des marinen Mesopsammon. *Zoologische Jahrbücher, Abteilung für Systematik, Ökologie und Geographie der*

Tiere, 81:286-312.
1956a. Monographie der Otoplanidae (Turbellaria). *Akademie der Wissenschaften und der Literatur Mainz, Mathematische-wissenschaftliche Klasse, Abhandlungen,* (1955): 499-796.
1956b. Les turbellariés des étangs côtiers du littoral méditerranéen de la France méridionale. *Vie et Milieu, Supplement,* 5:1-215.
1959. Zur Systematik, Ökologie und Tiergeographie der Turbellarienfauna in den ponto-kaspischen Brackwassermeeren. *Zoologische Jahrbücher, Abteilung für Systematik, Ökologie und Geographie der Tiere,* 87:43-184.
1971a. Zur Systematik und Phylogenie der Trigonostominae (Turbellaria, Neorhabdocoela). *Mikrofauna des Meeresbodens,* 4:1-84.
1971b. Neue interstitielle Macrostomida (Turbellaria) der Gattung *Acanthomacrostomum* und *Haplopharynx*. *Mikrofauna des Meeresbodens,* 8:295-308.
1977. Life Cycles of Interstitial Turbellaria from the Eulittoral of the North Sea. *Acta Zoologica Fennica,* 154:11-20.

Ax, P., and R. Ax
1974a. Interstitielle Fauna von Galapagos. V. Otoplanidae (Turbellaria, Proseriata). *Mikrofauna des Meeresbodens,* 27:571-598.
1974b. Interstitielle Fauna von Galapagos. VII. Nematoplanidae, Polystyliphoridae, Coelogynoporidae (Turbellaria, Proseriata). *Mikrofauna des Meeresbodens,* 29:611-638.
1977. Interstitielle Fauna von Galapagos. XIX. Monocelididae (Turbellaria, Proseriata). *Mikrofauna des Meeresbodens,* 64:395-437.

Ball, I.R.
1977. On the Phylogenetic Classification of Aquatic Planarians. *Acta Zoologica Fennica,* 154:21-35.

Beauchamp, P.M. de
1961. Classe de Turbellariés. Pages 35-212 in P.P. Grassé, editor, *Traité de Zoologie, Anatomie, Systématique, Biologie, IV, Plathelminties, Mésozoaires, Acanthocéphales, Némertiens.* Paris: Masson et Cie.

Boaden, P.J.S.
1963. The Interstitial Fauna of Some North Wales Beaches. *Journal of the Marine Biological Association of the United Kingdom,* 43:79-96.
1975. Anaerobiosis, Meiofauna and Early Metazoan Evolution. *Zoologica Scripta,* 4:21-24.

Bresslau, E.
1933. Turbellaria. Pages 52-320 in W. Kükenthal and T. Krumbach, editors, *Handbuch der Zoologie.* Berlin and Leipzig: Walter de Gruyter and Co.

Brunet, M.
1973. La famille des Cicerinidae (Turbellaria, Kalyptorhynchia). *Zoologica Scripta,* 2:17-31.

Bush, L.
1981. Marine Flora and Fauna of the Northeastern United States. Turbellaria: Acoela and Nemertodermatida. *NOAA Technical Report NMFS,* circular 440, 71 pages.

Cannon, L.R.G.
1986. *Turbellaria of the World: A Guide to Families and Genera.* 136 pages. Brisbane: Queensland Museum.

Crezée, M.
1976. Solenofilomorphidae (Acoela), Major Component of a New Turbellarian Association in the Sulphide System. *Internationale Revue der Gesamten Hydrobiologie,* 61:105-129.
1982. Turbellaria. Volume 1, pages 218-240 in S. P. Parker, editor, *Synopsis and Classification of Living Organisms.* New York: McGraw-Hill.

Dörjes, J.
1968. Die Acoela (Turbellaria) der deutschen Nordseeküste und ein neues System der Ordnung. *Zeitschrift für Zoologische Systematik und Evolutionsforschung,* 6:56-452.

Dougherty, E.C.
1963. *The Lower Metazoa — Comparative Biology and Phylogeny.* 478 pages. Berkeley: University of California Press.

Ehlers, U.
1972. Systematisch-phylogenetische Untersuchungen an der Familie Solenopharyngidae (Turbellaria, Neorhabdocoela). *Mikrofauna des Meeresbodens,* 11:1-77.
1985. *Das phylogenetische System der Plathelminthes.* 317 pages. Stuttgart: G. Fischer Verlag.
1986. Comments on a Phylogenetic System of the Platyhelminthes. *Hydrobiologia,* 132:1-12.

Ehlers, U., and P. Ax
1974. Interstitielle Fauna von Galapagos. VIII. Trigonostominae (Turbellaria, Typhloplanoida). *Mikrofauna des Meeresbodens,* 30:639-671.

Ehlers, U., and J. Dörjes
1979. Interstitielle Fauna von Galapagos. XXIII. Acoela (Turbellaria). *Mikrofauna des Meeresbodens,* 72:65-139.

Evdonin, L.A.
1977. [Turbellaria Kalyptorhynchia in the Fauna of the USSR and Adjacent Areas.] *Fauna USSR,* 115:1-400 [in Russian].

Faubel, A.
1974. Macrostomida (Turbellaria) von einem Sandstrand der Nordseeinsel Sylt. *Mikrofauna des Meeresbodens,* 45:339-370.
1976. Interstitielle Acoela (Turbellaria) aus dem Litoral der Nordfriesischen Inseln Sylt und Amrum (Nordsee). *Mitteilungen aus dem Hamburgischen Zoologischen Museum und Institut,* 73:17-56.
1983. The Polycladida, Turbellaria. Proposal and Establishment of a New System. Part I. The Acotylea. *Mitteilungen aus dem Hamburgischen Zoologischen Museum und Institut,* 80:17-121.
1984. The Polycladida, Turbellaria. Proposal and Establishment of a New system. Part II. The Cotylea. *Mitteilungen aus dem Hamburgischen Zoologischen Museum und Institut,* 81:189-259.

Faubel, A., and J. Dörjes
1978. *Flagellophora apelti* gen. n. sp. n.: A Remarkable Representative of the Order Nemertodermatida (Turbellaria: Archoophora). *Senckenbergiana maritima,* 10:1-13.

Hennig, W.
1966. *Phylogenetic Systematics.* 263 pages. Urbana: University of Illinois.

Hyman, L.H.
1951. *The Invertebrates: Plathyhelminthes and Rhynchocoela.* 550 pages. New York: McGraw-Hill.

Hyman, L.H., and E.R. Jones
1959. Turbellaria. Pages 323-365 in W.T. Edmondson, editor, *Freshwater Biology.* New York: John Wiley and Sons.

Karling, T.G.
1963. Marine Turbellaria from the Pacific Coast of North America. I. Plagiostomidae. *Arkiv för Zoologi,* 15:113-141.
1966. Marine Turbellaria from the Pacific Coast of North America. IV. Coelogynoporidae and Monocelididae. *Arkiv för Zoologi,* 18:493-528.
1968. On the Genus *Gnosonesima* Reisinger (Turbellaria). *Sarsia,* 33:81-108.
1974a. On the Anatomy and Affinities of the Turbellarian Orders. Pages 1-16 in N.W. Riser and M.P. Morse, editors, *Biology of the Turbellaria.* New York: McGraw-Hill.
1974b. Turbellarian Fauna of the Baltic Proper. *Fauna Fennica,* 27:1-101.

Lanfranchi, A., and F. Papi
1978. Turbellaria (excl. Tricladida). Pages 5-15 in J. Illies, editor, *Limnofauna Europaea.* Stuttgart: G. Fischer Verlag.

Luther, A.
1955. Die Dalyelliiden (Turbellaria Neorhabdocoela). *Acta Zoologica Fennica,* 87:1-337.
1960. Die Turbellarien Ostfennoskandiens. I. Acoela, Catenulida, Macrostomida, Lecithoepitheliata,

Prolecithophora, und Proseriata. *Fauna Fennica*, 7:1-155.
1962. Die Turbellarien Ostfennoskandiens. III. Neorhabdocoela 1. Dalyellioida, Typhloplanoida: Byrsophlebidae, Trigonostomidae. *Fauna Fennica*, 12:1-71.
1963. Die Turbellarien Ostfennoskandiens. IV. Neorhabdocoela 2. Typhloplanoida Typhloplanidae, Solenopharyngidae, Carcharodopharyngidae. *Fauna Fennica*, 16:1-163.

Marcus, E.
1945. Sôbre Catenulida brasileiros. *Boletins da Faculdade de Filosofia, Ciencias e Letras, Universidade de São Paulo. (Zoologia)*, 10:3-133.
1954. Turbellaria Brasileiros (11). *Papeis Avulsos do Departomento de Zoologia, Secretaria da Agricultura*, 11:419-489.

Martens, P.M.
1983. Three New Species of Minoninae (Turbellaria, Proseriata, Monocelididae) from the North Sea, with Remarks on the Taxonomy of the Subfamily. *Zoologica Scripta*, 12:153-160.
1984. Comparison of Three Different Extraction Methods for Turbellaria. *Marine Ecology Progress Series*, 14:229-234.

Martens, P., and E. Schockaert
1986. The Importance of Turbellarians in the Marine Meiobenthos. *Hydrobiologia*, 132:295-303.

Meixner, J.
1938. Turbellaria (Strudelwürmer) I. Pages 1-146 in G. Grimpe, E. Wagler, and A. Remane, editors, *Die Tierwelt der Nord- und Ostsee*, IVb.

Meyers, M.B., H. Fossing, and E.N. Powell
1987. Micro-distribution of Interstitial Meiofauna, Oxygen and Sulfide Gradients, and the Tubes of Macro-infauna. *Marine Ecology Progress Series*, 35:223-241.

Noldt, U., and S. Hoxhold
1984. Interstitielle Fauna von Galapagos. XXXIV. Schizorhynchia (Plathelminthes, Kalyptorhynchia). *Microfauna Marina*, 1:199-256.

Noldt, U., and C. Wehrenberg
1984. Quantitative Extraction of Living Plathelminthes from Marine Sands. *Marine Ecology Progress Series*, 20:193-201.

Pennak, R.W.
1978. *Freshwater Invertebrates of the United States*. 803 pages. New York: John Wiley and Sons.

Prudhoe, S.
1985. *A Monograph of the Polyclad Turbellaria*. 259 pages. London: Trustees of the British Museum (Natural History).

Riedl, R.
1959. Turbellarien aus submarinen Höhlen. 1. Archoophora. *Pubblicazioni della Statzione Zoologica di Napoli, Supplimento*, 30:178-208.

Rieger, R.M.
1971a. Die Turbellarienfamilie Dolichomacrostomidae nov. fam. (Macrostomida). I. Teil. Vorbemerkungen und Karlingiinae nov. subfam. *Zoologische Jahrbücher, Abteilung für Systematik, Ökologie und Geographie der Tiere*, 98:236-314.
1971b. Die Turbellarienfamilie Dolichomacrostomidae Rieger. II. Teil. Dolichomacrostominae 1. *Zoologische Jahrbücher. Abteilung für Systematik, Ökologie und Geographie der Tiere*, 98:569-703.

1974. A New Group of Turbellaria - Typhloplanoida with a Proboscis and Its Relationship to the Kalyptorhynchia. Pages 23-62 in N.W. Riser and P.M. Morse, editors, *Biology of the Turbellaria*. New York: McGraw-Hill.
1981. Morphology of the Turbellaria at the Ultrastructural Level. *Hydrobiologia*, 84:213-229.

Romeis, B.
1968. *Mikroskopische Technik*. 757 pages. München: R. Oldenbourg Verlag.

Ruebush, T.K.
1941. A Key to the American Freshwater Turbellarian Genera, Exclusive of the Tricladida. *Transactions of the American Microscopical Society*, 60:29-40.

Scherer, B.
1985. Annual Dynamics of a Meiofauna Community from the "Sulfide Layer" of a North Sea Flat. *Microfauna Marina*, 2:117-161.

Schilke, K.
1970. Kalyptorhynchia (Turbellaria) aus dem Eulitoral der deutschen Nordseeküste. *Helgoländer Wissenschaftliche Meeresuntersuchungen*, 21:143-265.

Schockaert, E.R.
1971. Turbellaria from Somalia. I. Kalyptorhynchia (part 1). *Monitore Zoologico Italiano*, supplemento, IV:101-122.
1982. Turbellaria from Somalia II. Kalyptorhynchia (part 2). *Monitore Zoologico Italiano*, supplemento, XVII:81-96.

Schockaert, E.R., and I.R. Ball
1981. *The Biology of the Turbellaria*. 300 pages. The Hague: Dr. W. Junk.

Schwank, P.
1982. Turbellarien, Oligochaeten und Archianneliden des Breitenbachs und anderer oberhessischer Mittelgebirgsbäche. IV. Allgemeine Grundlagen der Verbreitung von Turbellarien und Oligochaeten in Fliessgewässern. *Archiv für Hydrobiologie, Supplement*, 62:254-290.

Sopott, B.
1972. Systematik und Ökologie von Proseriaten (Turbellaria) der deutschen Nordseeküste. *Mikrofauna des Meeresbodens*, 13:165-236.

Sterrer, W., and R.M. Rieger
1974. Retronectidae -- a New Cosmopolitan Marine Family of Catenulida (Turbellaria). Pages 63-92 in N.W. Riser and P.M. Morse, editors, *Biology of the Turbellaria*. New York: McGraw-Hill.

Tajika, K.I.
1982. Morphologisch-phylogenetische Untersuchungen an der Familie Coelogynoporidae (Turbellaria, Proseriata). *The Journal of the Faculty of Science, Hokkaido University*, Series VI, 23:13-62.

Tyler, S.
1986. *Advances in the Biology of Turbellarians and Related Platyhelminthes*. 387 pages. The Hague: Dr. W. Junk.

Westblad, E.
1948. Studien über skandinavische Turbellarien. Acoela V. *Arkiv för Zoologi*, 41:1-82.

Young, J.O.
1970. British and Irish Microturbellaria. Historical Records, New Records and a Key to Their Identification. *Archiv für Hydrobiologie*, 67:210-241.
1976. The Freshwater Turbellaria of the African Continent. *Zoologischer Anzeiger*, 197:419-429.

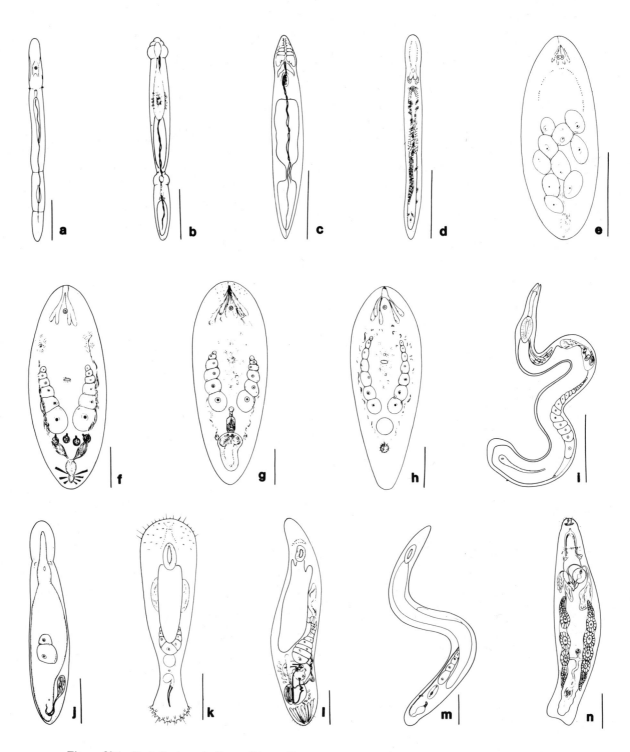

Figure 20.1.--Turbellaria: a-d, Catenulida, e, Nemertodermatida; f-h, Acoela; i, Haplopharyngida; j-m, Macrostomida; n, Lecithoepitheliata. (Scale: a-h and j-n = 200 μm; i = 1mm.)

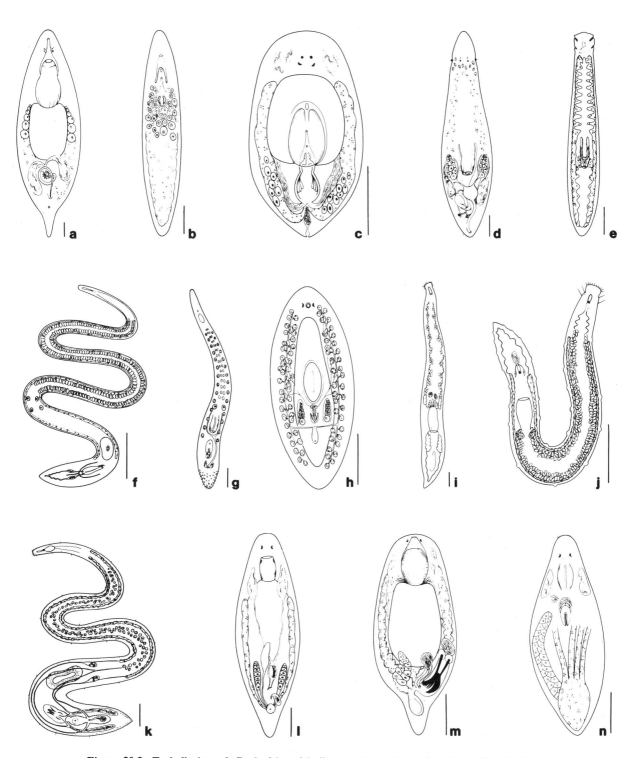

Figure 20.2.--Turbellaria: a-d, Prolecithoepitheliata; **e,** Proseriata, Bothrioplanidae; **f,** Proseriata, Nematoplanidae; **g,** Proseriata, Monocelididae; **h,** Proseriata, Otomesostomidae; **i,** Proseriata, Otoplanidae, Archotoplaninae; **j,** Proseriata, Otoplanidae, Otoplaninae; **k,** Proseriata, Coelogynoporidae; **l,** Rhabdocoela, Provorticidae; **m,** Rhabdocoela, Dalyelliidae; **n,** Rhabdocoela, Graffillidae. (Scale: a-d,g,i, and k-n = 200 μm; e,f,h, and j = 1 mm.)

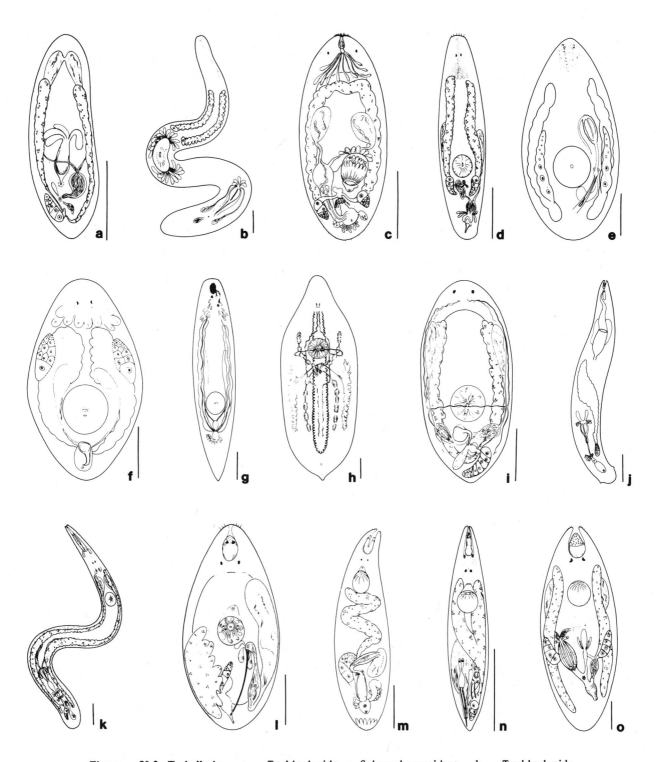

Figure 20.3.--Turbellaria: **a**, Typhloplanida, Solenopharyngidae; **b**, Typhloplanida, Ciliopharyngiellidae; **c**, Typhloplanida, Kytorhynchidae; **d**, Typhloplanida, Trigonostomidae; **e**, Typhloplanida, Promesostomidae, Promesostominae; **f**, Typhloplanida, Promesostomidae, Brinkmaniellinae; **g**, Typhloplanida, Typhloplanidae, Olisthanellinae; **h**, Typhloplanida, Typhloplanidae, Mesostominae; **i**, Typhloplanida, Typhloplaninae; **j-o**, Kalyptorhynchia. (Scale: a-g and i-o = 200 μm; h = 1 mm.)

21. Gnathostomulida

Wolfgang Sterrer and Richard A. Farris

Since their description (Ax, 1956) and recognition as a separate phylum (Riedl, 1969; Sterrer, 1972) Gnathostomulida have been encountered worldwide as an often abundant constituent of marine meiofauna. Most members of this phylum are of meiofaunal size and live in the interstitial spaces between sediment particles. At present, there are 18 genera with over 80 known species, but many more can be expected. The phylum is unique in having a body epithelium in which each cell carries only one cilium, and a bilaterally symmetrical muscular pharynx equipped with cuticular mouth parts. Although recently linked with Platyhelminthes into a new phylum, Platyhelminthomorpha (Ax, 1985), their systematic position continues to be dubious as they also share characters with the aschelminth complex (Sterrer et al., 1985).

Gnathostomulida are worm-shaped, cylindrical or with a slight dorsal-ventral flattening, and range from 300 to 3000 µm in length and 50 to 100 µm in diameter; however, the most frequently observed genera have a body length between 550 to 890 µm and a body width of 50 to 70 µm. The body is divided into a head which is commonly expanded and well delimited, a long trunk which constitutes 90% of the body, and a relatively short tail. Most are colorless, although a few species of the elongated Filospermoidea may be bright red. All move slowly by ciliary gliding. In many species one may observe occasional stopping often followed by simultaneous constriction of the body from both the posterior and anterior ends. Some species swim backward by reversing their ciliary beat, a diagnostic characteristic shared by only a few interstitial organisms such as the ciliates and some Catenulida (Turbellaria). The anterior end (rostrum) is usually beset with paired sensory cilia, and/or bristle-like compound cilia, and there may be "spiral ciliary organs" of unknown function embedded in the skin (Lammert, 1984), although the latter are not easily seen with a light microscope. The mouth is subterminal, and there is no permanent anus (a tissue connection that may function as a facultative anus has been observed in a few species).

The buccal cavity is followed by a muscular pharynx that contains, in all but one species (*Agnathiella beckeri* Sterrer), cuticular mouth parts usually consisting of an unpaired basal plate situated in the lower lip area, and paired forward-projecting jaws. In some species the jaws can be readily seen in an "unsqueezed" wet mount preparation under low power with a light microscope; others require a small amount of water to be removed from under the coverslip in order to compress or "squeeze" the specimen before the jaws become apparent. Further squeezing will enhance the appearance of the jaws, as well as other cuticularized structures, but at the sake of distorting the body morphology and soft tissue organs.

All gnathostomulids are hermaphrodites. The ovary is always unpaired and situated dorsally at about mid-body region, with eggs maturing caudally. Some taxa have a bursa for sperm storage behind the ovary which, if cuticular, appears as a rigid semitransparent structure in the middle of the body. One dorsolateral testis or pair of lateral testes occupy much of the posterior portion of the trunk; they discharge ventrocaudally to caudally into a male copulatory organ that may be a simple pore, or muscular penis, the latter may be provided with a cuticular stylet. In species possessing a cuticularized bursa and penis stylet, these structures, like the jaws, will become more obvious in squeezed preparations. Sperm is morphologically quite diverse within the phylum, ranging from filiform to droplet- to cone-shaped, the two latter types lacking a flagellum. Although Figure 21.1 shows the wide range of morphology present in the Gnathostomulida, the most commonly seen gnathostomulids belong to the genus *Gnathostomula* as represented in Figure 21.1h-j (*Gnathostomula tuckeri* Farris).

Classification

The phylum is divided into two orders, the elongated Filospermoidea (with filiform sperm; without paired sensory organs, and without a bursa, vagina, or penis stylet), and the usually more compact Bursovaginoidea (with sperm not filiform; and usually with paired sensory organs and a bursa). The latter order has two suborders, Scleroperalia (with a cuticular bursa and male stylet) and Conophoralia (without cuticular formations in bursa or penis; sperm cone-shaped). Families, genera, and species are characterized primarily by differences in the cuticular mouth parts such as size and structure, and the number and arrangement of teeth.

Habitats and Ecological Notes

Representatives of the phylum have been encountered in all the world's oceans, from the supratidal brackish groundwater (genus *Gnathostomaria*) to the intertidal, and the subtidal to 800 m. As a rule they occur in fine to coarse sand with a high admixture of organic detritus as typically found in sheltered bays, intracoastal waterways, between coral reefs, or wherever slightly reducing conditions may exist. Recently, they have been discovered abundantly on high energy (lotic) beaches in sediments trapped by the root systems of surf grasses which, due to a large input of organic matter, creates a reducing environment (Farris, in preparation). Only rarely do they venture into clean sand, mud, or other substrates. With the exception of some species of *Gnathostomula*, which frequent superficial sand layers, gnathostomulids reach their highest species diversity and population density a few centimeters to 15 cm below the sand surface, in the latter case in the proximity of polychaete burrows (Reise, 1981). The majority, however, tend to be in the surface 2-10 cm.

Their association with the "sulfide system" (Fenchel and Riedl, 1970; or "thiobios," Boaden, 1977), including the sediments underlying brine seeps (Powell and Bright, 1981), suggests a preference for boundary layers with steep oxygen-rich to hydrogen sulfide-rich chemoclines, probably in conjunction with a high physiological tolerance for near anoxic conditions and the use of sulfur bacteria for food. Since species distributions appear to be patchy, collecting small amounts of sand (300-500 cm^3) from a number of different locations on a beach is preferable to a large amount of sediment from one area.

Methods of Collection

For qualitative sampling, any device is appropriate that reaches into the subsurface sand layers, such as a spade in the intertidal, and a fine mesh but heavy epibenthic sled for the subtidal. Quantitative samples, usually 1-10 cm^3, are either taken directly in the field, or by subsampling from a box corer (such as the Reineck type, Farris and Crezée, 1976), by means of a vertically inserted small diameter (1-2 cm) corer; the resulting sediment core is then slid out and cut into individual samples. Sediments should be placed in a closed container or plastic bag and kept moist with seawater while in transit.

Methods of Extraction

For qualitative purposes, large sediment samples (10 liters or more) placed in a bucket and covered by only a few cm of seawater can be stored at a temperature roughly equivalent to or slightly less than the temperature of the seawater where they were collected. Allowing the sediment to sit for at least a day before extraction tends to concentrate specimens in the superficial sample layer. Specimens can now be collected over a period of weeks to months, by scraping off the surface centimeter of sand and extracting using the magnesium chloride method (Sterrer, 1971). About 100 cm^3 of sediment from the primary sample are gently mixed with about 1000 cm^3 of isotonic magnesium chloride, then left for 7-10 minutes during which the specimens become anesthetized. The sample is then shaken more or less vigorously and, after a brief rest to allow the sediment to settle, the supernatant is filtered through a 63 μm nylon sieve mounted on a plexiglas ring 2 cm high, 8-10 cm in diameter. This procedure is repeated twice, the last time using filtered seawater instead of magnesium chloride. The final sample retained in the sieve is then gently but thoroughly rinsed in running seawater and placed *as is* (not inverted!) in a Petri dish with just enough seawater to submerge it. Only minutes later, gnathostomulids (along with many other meiofauna taxa) will begin to crawl through the meshes of the sieve and seek the bottom of the dish from where they can easily be pipetted, almost free of sediment, after transfer of the sieve into an alternate Petri dish. In order to avoid evaporation, it is advisable to keep the dishes covered between observations. A second qualitative method that, judging from comparative results obtained with Turbellaria (Martens, 1984), may be equally effective is the seawater-ice technique (Uhlig, 1968).

For quantitative studies, small sediment samples (1-2 cm^3) of freshly collected sediments, are shaken with seawater in a beaker, and the supernatant then poured into a petri dish. This process is repeated several times. Gnathostomulids "usually showed up in the first 1 to 4 Petri dishes" (Reise, 1981).

Preparation of Specimens for Taxonomic Study

Gnathostomulida are best studied while in a living condition, anesthetized with magnesium chloride, and gently squeezed under a coverslip with tiny wax feet at its corners. For the study of delicate and minute details such as sensory bristles, cuticular parts, and sperm, 100x oil immersion phase contrast or Nomarski optics are indispensable; a drawing tube or other form of producing scale drawings is highly recommended, and videomicroscopy has recently been found to be an effective tool for specimen analysis (Farris and O'Leary, 1985). After study of live features, the specimen can either be made into a permanent mount (by adding formalin-glycerol), and sealing the edges of the coverslip with Eukitt or rubber cement, or revived in seawater and then fixed in Bouin's fluid for light microscopy. For electron microscopy, unsqueezed specimens are relaxed for 10 minutes in magnesium chloride, chilled to 4° C, double fixed with glutaraldehyde (3-3.5%, 3-5 h) and osmium tetroxide (1-2%, 1-2 h), both fixatives buffered with 0.1M Sorensen phosphate (10% sucrose).

References

Ax, P.
1956. Die Gnathostomulida, eine rätselhafte Wurmgruppe aus dem Meeressand. *Abhandlungen der Akademie der Wissenschaften und Literatur Mainz, mathematisch-naturwissenschaftliche Klasse*, 8:1-32.
1985. The Position of the Gnathostomulida and Platyhelminthes in the Phylogenetic System of the Bilateria. Pages 168-180 in S. Conway Morris et. al., *The Origins and Relationships of Lower Invertebrates*. Oxford: Clarendon Press.

Boaden, P.J.S.
1977. Thiobiotic Facts and Fancies (Aspects of the Distribution and Evolution of Anaerobic Meiofauna). *Mikrofauna des Meeresbodens*, 61:45-63.

Farris, R., and M. Crezée
1976. An Improved Reineck Box for Sampling Coarse Sand. *Internationale Revue der gesamten Hydrobiologie*, 61:703-705.

Farris, R., and D. O'Leary
1985. Application of Videomicroscopy to the Study of Interstitial Fauna. *Internationale Revue der gesamten Hydrobiologie*, 70:891-895.

Fenchel, T.M., and J.M. Riedl
1970. The Sulfide System: A New Biotic Community Underneath the Oxidized Layer of Marine Sand Bottoms. *Marine Biology*, 7:255-263.

Lammert, V.
1984. The Fine Structure of Spiral Ciliary Receptors in Gnathostomulida. *Zoomorphology*, 104:360-364.

Martens, P.M.
1984. Comparison of Three Different Extraction Methods for Turbellaria. *Marine Ecology Progress Series*, 14:229-234.

Powell, E.N., and T.J. Bright
1981. A Thiobios Does Exist Gnathostomulid Domination of the Canyon Community at the East Flower Garden Brine Seep. *Internationale Revue der Gesamten Hydrobiologie*, 66:675-683.

Reise, K.
1981. Gnathostomulida Abundant Alongside Polychaete Burrows. *Marine Ecology Progress Series*, 6:329-333.

Riedl R.J.
1969. Gnathostomulida from America. *Science*, 163:445-452.

Sterrer, W.
1971. Gnathostomulida: Problems and Procedures. Pages 9-15 in N.C. Hulings, editor, Proceedings of the First International Conference on Meiofauna. *Smithsonian Contributions to Zoology*, 76.
1972. Systematics and Evolution with the Gnathostomulida. *Systematic Zoology*, 21(2):151-173.
1982. Gnathostomulida. Pages 847-851 in S.P. Parker, editor, *Synopsis and Classification of Living Organisms*, Volume 1. New York: McGraw-Hill.

Sterrer, W., M. Mainitz, and R.M. Rieger
1985. Gnathostomulida: Enigmatic as Ever. Pages 181-199 in S. Conway Morris, editor, *The Origins and Relationships of Lower Invertebrates*. Oxford: Clarendon Press.

Uhlig, G.
1968. Quantitative Methods in the Study of Interstitial Fauna. *Transactions of the American Miscroscopical Society*, 87:226- 232.

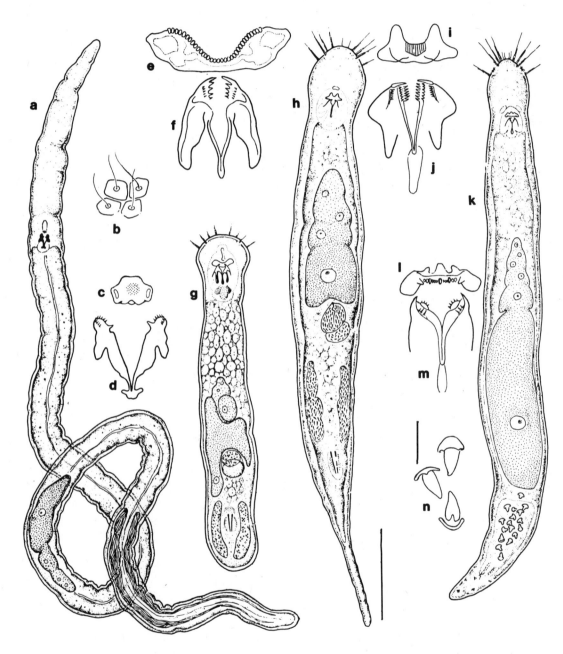

Figure 21.1.—**Gnathostomulida: a-f,** *Haplognathia* cf. *rosacea* Sterrer (Order Filospermoidea); **a**, habitus; **b**, monociliated epidermis cells; **c**, basal plate; **d**, jaws. **e-g,** *Problognathia minima* Sterrer and Farris (Order Bursovaginoidea, Suborder Scleroperalia); **e**, basal plate; **f**, jaws; **g**, habitus; **h-j,** *Gnathostomula tuckeri* Farris (Order Bursovaginoidea, Suborder Scleroperalia); **h**, habitus; **i**, basal plate; **j**, jaws; **k-n,** *Austrognathia microconulifera* Farris (Order Bursovaginoidea, Suborder Conophoralia); **k**, habitus; **l**, basal plate; **m**, jaws; **n**, sperm (conuli). (Scale = 100 μm for all habitus drawings; scale = 10 μm for all others.)

22. Nemertina

Jon L. Norenburg

The phylum Nemertina is comprised of non-segmented worms whose anatomy includes: the diagnostic rhynchocoel housing a protrusible proboscis, a closed blood-vascular system lined by endothelium, a regionated gut with anal pore, a fully ciliated and glandular epidermis, a nervous system with four cerebral ganglia and a pair of lateral nerve cords, and serial sacculate gonads. There are about 900 described species of nemertines. They are primarily marine and found in all major marine habitats, but they also are known from freshwater and terrestrial habitats. The size of adult worms ranges from 1 mm to many meters in length and 0.15 mm to a few centimeters in width.

Classification

The phylum consists of two classes, the Anopla and Enopla, the names reflecting respectively the absence or presence of specialized proboscis armature, the stylets. Anopla have a ventral mouth and all known meiofaunal Enopla have an anterior terminal pore that serves as mouth and proboscis pore. According to the classification of Iwata (1972) the Anopla consists of the orders Archi-, Palaeo- and Heteronemertina, whereas the Enopla includes only the order Hoplonemertina.

Enopla.--The hoplonemertean genus *Ototyphlonemertes* has a world-wide distribution, but there are major gaps in the geographic records (Kirsteuer, 1977). Members of the genus lack ocelli and are characterized by the presence of a pair of statocysts located dorsally on the posterior extensions of the ventral ganglia. Müller (1968) comments on 14 species known to that time and provides a key for them (see also Norenburg, 1988). Kirsteuer (1977) provides the most recent and extensive account on the principal features used in taxonomy of this group.

The remaining genera of interstitial hoplonemertines (Figure 22.1d-f) are each known from only one or a few species. We know little about which characteristics will prove to be significant to the taxonomy of these species or genera. *Annulonemertes minusculus,* and *Arenonemertes minutus,* both from northern Europe, have smooth stylets and are recognized by pseudosegmentation, annular constrictions, of the body wall in the intestinal region (Berg, 1985; Friedrich, 1949). Related species have been encountered in the Gulf of Maine and in Puget Sound (unpublished observations). These worms, as a group, are among the smallest nemertines known, adults often being less than 2 mm long and 150 μm in diameter. *Arenonemertes microps* is known only from near Kiel; it is 2-3 mm long, has four ocelli, the rhynchocoel and proboscis extend to the caudal terminus, and the stylets are smooth (Friedrich, 1933). A probable congener of *Prostomatella arenicola* has been encountered interstitially in shallow water off Florida (unpublished observations). *Otonemertes marcusi* is a small species (2 mm long) with a pair of ocelli, a pair of statocysts, and distinct mid-body pigmentation (Corrêa, 1958); little else is known about it.

Anopla.--Anoplans (Figure 22.1a-c) may be recognized by their non-regionated, fully protrusible proboscis that lacks discrete stylets. Interstitial archinemertines of the genera *Cephalothrix* (Figure 22.1a) and *Procephalothrix* may be recognized by the position of the mouth at the anterior end of the gut and far posterior to the cephalic tip (as much as 1 mm in a worm 7-10 mm long). The mouth of palaeonemertines, such as *Carinina arenaria* (Figure 22.1b), and heteronemertines is relatively more anterior, located just posterior to the cerebral ganglia. The archinemertines uniformly lack cerebral organs (glandular sensory pits, associated posteriorly with the cerebral ganglia of other anoplans), whereas they are well developed in *C. arenaria* and an undescribed, interstitial heteronemertine (Figure 22.1c) (unpublished observations). *Carinina arenaria,* in addition, is reported to possess statocysts posteriorly in the ventral ganglia (Hylbom, 1957). The species of interstitial *Cephalothrix* are characterized by a proboscis whose epithelium contains prominent cirri (long ciliary tufts), cells with a cluster of rhabdite-like inclusions, and cells with a pseudocnide (an urn with an eversible thread) (Gerner, 1969). All of these genera require the same detailed observations *in vivo* as do the hoplonemertines, but, because of the frequent lack of ocelli, discrete proboscis armature, and statocysts (except *Carinina arenaria*), differences between species, genera, and even orders of anoplans tend to be much more subtle and often require histological study.

Habitats and Ecological Notes

Macrofaunal species within each order have similar habitat preferences (Kirsteuer, 1971; Norenburg, 1985). The archinemertines are mostly of small diameter and live in soft mud or in interstitial spaces. Palaeo- and heteronemertines show tendencies to dramatic increases in size and in body-wall musculature and complexity, which correspond with increased burrowing ability compared to archinemertines (Norenburg, 1985). The hoplonemertines generally are small, relatively less muscular and tend to be epibenthic, thereby increasing the importance of cilia in locomotion (Norenburg, 1985). Similar segregation of these higher taxa is evident in the meiofauna.

Meiofaunal palaeonemertines, such as *Carinina coei*, *Hubrechtella dubia*, and *Tubulanus pellucidus*, live at or in the surface of muddy and silty marine bottoms (Coe, 1943; Hylbom, 1957; unpublished observations). The other three orders are represented in the mud fauna, but with lesser taxonomic diversity (Kirsteuer, 1963) and few would qualify as meiofauna except as juveniles (often abundant) (unpublished observations). Kirsteuer (1971) notes that the mud-dwelling nemertines "...have, with the exception of some statocyst-bearing palaeonemerteans, no developed morphological features which would make them quite distinct..." from their relatives in other marine habitats. In contrast, the nemertine fauna of the interstitial spaces of coarse marine sediments is morphologically distinct and hoplonemertines predominate in taxonomic diversity (Kirsteuer, 1971). The remainder of this review concerns those nemertines that may be considered interstitial.

Interstitial nemertines generally are less than 0.5 mm in diameter and may be up to 50 mm in length, but many are less than 0.3 mm in diameter and 10 mm in length. Kirsteuer (1971) points out that the marine interstitial nemertine fauna is relatively non-diverse systematically and shows clear morphological adaptations to the interstitial environment. About thirty species of interstitial nemertines have been described. Of the three orders, the hoplonemertines predominate in taxonomic diversity, with 23 species in five genera. They include, since the last review (Kirsteuer, 1971), a newly described genus, *Annulonemertes* Berg, 1985, and newly described species of *Ototyphlonemertes* (Mock and Schmidt, 1975). The literature references at least another 7 new, but undescribed, species of *Ototyphlonemertes* (Kirsteuer, 1977; Mock and Schmidt, 1975). In addition, a new hoplonemertine genus (Figure 22.1d), as well as new species in the genera *Annulonemertes* (Figure 22.1f), *Prostomatella*, and *Ototyphlonemertes* have been encountered in New England, Puget Sound, Florida, and Pacific Panama (Norenburg, 1988, unpublished observations). The described archinemertines are represented by two genera, *Cephalothrix* (Figure 22.1a) with 5 species and *Procephalothrix* with 1 species. *Carinina arenaria* (Figure 22.1b) is the only described interstitial palaeonemertine. Recently, the first truly interstitial heteronemertine (Figure 22.1c) was discovered in Puget Sound (Norenburg, 1988).

Information on the natural history of interstitial nemertines is scarce. They move principally by ciliary gliding, without displacing the substratum (Corrêa, 1949; Kirsteuer, 1971). Mock (1981a) was unable to substantiate Friedrich's (1935) observation that *Prostomatella arenicola* (Figure 22.1e) moves by nematode-like writhing. Interstitial nemertines are either carnivorous predators or scavengers (Corrêa, 1949; Mock, 1978, 1981a). Somewhat more is known of distribution and reproductive biology.

Most interstitial nemertines are recorded from intertidal and shallow subtidal clean, relatively coarse sand and shell hash, i.e., high-energy beaches or subtidal areas subject to considerable current action, where coarse sand and shell fragments accumulate. Interstitial nemertines are rarely found in areas with high loads of organic particulates or silt. Some anoplans tolerate greater amounts of silt than do hoplonemertines (personal observations). This is consistent with the generalized habit of their macrofaunal relatives (Kirsteuer, 1963; Norenburg, 1985). Kirsteuer (1986) notes that information concerning patterns of distribution of interstitial nemertines is sketchy and must be regarded with caution because surveys have often focused on a limited portion of the potential habitat of a species. Corrêa (1949) and Mock and Schmidt (1975) provide quantitative data for some *Ototyphlonemertes*. Corrêa (1949) indicates a distribution peak somewhat above mean low water level (unspecified species), while *O. fila* (sp. A of Kirsteuer, 1977), in studies by Mock and Schmidt (1975), appears to have a distribution peak just below mean low water. Other species have considerably broader distributions; e.g., *O. pallida* extends from the intertidal to depths of 10 m (Müller, 1968).

All interstitial nemertines described thus far are gonochoric. Gonads may be more or less regularly disposed among intestinal diverticula (when present) or they may be scattered about the intestine. Interstitial nemertines most commonly have only one ovum per ovary, less frequently 2-3 ova per ovary, evident. The number of mature ova per individual is a function of animal and ova size and ranges from less than 10 for the smallest species to more than 100 for the largest. Sperm morphology for representatives of each of the interstitial nemertine genera includes an elongate head and solitary flagellum. External fertilization is assumed, with the possible exception of *Cephalothrix germanica* (Gerner, 1969). Ova of interstitial hoplonemertines are relatively large, at 150-200 μm in diameter. They probably undergo direct development with lecithotrophy, as is common for other hoplonemertines and would be predicted with interstitial habit (Swedmark, 1964). Mock (1981b) has confirmed this for a species of *Ototyphlonemertes*. The mode of development for interstitial anoplans is of special interest, as macrofaunal cephalothricids usually have a simple, planktotrophic larval phase and macrofaunal heteronemertines generally have a specialized, often planktotrophic, larval phase.

Methods of Collection and Extraction

Most methods for the collection of other soft-bodied meiobenthic organisms are satisfactory for the collection of nemertines.

Meiofaunal nemertines must be sorted live. A simple method for extracting them in good condition from sand samples is to place sand in a pail to a depth of 5-8 cm and cover this with seawater to about 20 cm, stir the sand gently with a rotational lifting motion of the hand, about five times, so that less dense material (including the worms) separates from the sand as it falls back to the bottom. As the sand settles, but before the water stops spinning, quickly decant the water through a sieve (6-8 cm diameter) with a mesh size of 63-125 μm. Use a wash-bottle to quickly wash the contents of the sieve into a finger-bowl. Repeat the decanting procedure 3-4 times or until no more worms are obtained, then repeat the procedure with $MgCl_2$ solution (approximately 7.5%).

Other procedures (Chapter 9) may be preferable in some instances, e.g., where organic particulates or fine silt are a problem. Many nemertines may be kept for days to several months in their original substratum if stored in Whirlpak plastic bags in running seawater or stored in buckets over which a thin flow of seawater is maintained (there must be no macrofauna in these samples). Meiofaunal nemertines are best sorted live using a dissecting microscope at low magnification (10-15x). Most can be recognized by their long, relatively cylindrical shape as they glide smoothly through their surroundings (unlike the agitated movement of many otoplanid turbellarians). Hoplonemertines may often be recognized by an epidermal furrow that encircles the body in the vicinity of the cerebral ganglia.

Nemertodermatid turbellarians may be mistaken for the longer nemertines, even after long experience. Some nemertines (<10 mm) may be confused with kalyptorhynch and similar turbellarians, but with practice may be distinguished by their behavior and more cylindrical shape. Extraordinary efforts, such as wholemounts of individual worms, would be required to enable one to distinguish preserved meiofaunal nemertines from similar, preserved turbellarians.

Methods of Preparation for Taxonomic Study

Detailed observations of living specimens (preferably 10 or more) are essential for taxonomic work. Norenburg (1988) provides a key that is intended for use with living specimens and includes most of the described species and several undescribed species. With the dissecting microscope one records color of integument, gut, and cerebral ganglia, as well as information on size, shape and behavior of the worm while actively gliding, when "at rest," and when contracted. This information includes: shape of the cephalic region and its proportions relative to the cerebral ganglia; position and shape of the mouth for anoplans; shape of the caudal terminus and presence or absence of a caudal cirrus (tail-like extension of some anoplans) or adhesive plate (cannot always be detected in the living worm); and, when possible, relative extent of the rhynchocoel and proboscis. Some of these observations will require amplification with the compound light microscope.

Although nemertines in general have few external distinguishing features, interstitial nemertines have the advantage of small size and relative transparency, enabling one to make detailed observations of internal anatomy from living specimens (Figure 22.1). This is accomplished by placing a worm in a drop of seawater on a glass slide and floating or supporting, with bits of wax or clay, a coverslip on this drop so that the worm may still move (Kirsteuer, 1967). The coverslip is then carefully lowered, by drawing off water or pressing on it, so that the worm is somewhat flattened and locomotion is restricted. If one works rapidly and monitors evaporation the worm usually can be recovered intact.

Observations that need to be made with the compound microscope depend somewhat on the taxon in question and should be documented with sketches and with photomicrographs, if possible. In all cases it is desirable to obtain at least a good dorsal view to document the distribution of sensory cirri (also called sensory or ciliary bristles). Sometimes these bristles are numerous and readily evident, but in some species they may be lacking or difficult to observe until the worm rolls into the proper position. It is essential to determine whether cerebral organs are present, their form and position; size and shape of the cerebral ganglia; position and structure of ocelli and/or statocysts; extent of rhynchocoel; structure and length of proboscis and position of its posterior insertion; characteristics of stylet apparatus and other proboscideal elements such as rhabdite-like barbs and pseudocnides; and general construction of the gut. If at all possible, observations on an everted proboscis should be obtained. The proboscis may be everted voluntarily or under increased pressure from the coverslip. Sometimes dabbing the side of the coverslip with an irritant such as dilute acetic acid may provoke eversion of the proboscis. If an everted proboscis is obtained it should be carefully described and then preserved with the rest of the specimen.

Internal anatomy based on histology is an essential component of all nemertine taxonomy, but the subject is beyond the scope of this chapter.

Differences between any of the species of interstitial genera may be very subtle and, in general, considerable experience is required for taxonomic work at the specific level. All taxonomic descriptions of nemertines also require histological work. To facilitate the latter, specimens must be individually anesthetized and fixed. Kirsteuer (1967) emphasizes the importance of adequate narcotizing. Anesthetizing specimens in 7.5% $MgCl_2$ in distilled water is usually satisfactory. A few species react badly to this, evidenced by blistering or sloughing off of the epidermis or fragmenting of the worm. Other anesthetics that may be tried include urethane, chloral hydrate (both of which may be added to seawater as crystals or used in prepared solutions), propylene phenoxetol, or Lidocaine. As soon as the worm is suitably narcotized, within 1-3 minutes for most interstitial species (it may continue to glide by latent ciliary activity), it is pipetted with a minimum of fluid onto a glass slide so that it, or at least the anterior region, is relatively straight. Fixative is then immediately, but carefully, pipetted along the worm. When the fixative noticeably begins to take effect, gently flush the fluid back and forth so that the specimen will not adhere to the slide. After 1-2 minutes flush the worm into a vial of fixative. A good histological fixative, such as Hollande's cupri-picri-formol-acetic fluid, for light-microscopy, or a glutaraldehyde-based electron microscopy fixative (Chapter 10) should be used.

Acknowledgments

This paper constitutes Contribution Number 143 from the Marine Science Laboratory, Northeastern University, Nahant, Massachusetts, and Contribution Number 167 from the Smithsonian Marine Station at Link Port, Fort Pierce, Florida.

References

Berg, G.
1985. *Annulonemertes* gen. nov., A New Segmented Hoplonemertean. Pages 200-209 in S.C. Morris, J.D. George, R. Gibson, and H.M. Platt, editors, *The Origins and Relationships of Lower Invertebrates*. New York: Oxford University Press.

Boyden, C., and C. Little
1973. Faunal Distributions in Soft Sediments of the Severn Estuary. *Estuarine and Coastal Marine Science*, 1:203-223.

Bürger, O.
1895. Nemertinen. *Fauna und Flaura des Golfes von Neapel*, 22:1-743.

Coe, W.
1943. Biology of the Nemerteans of the Atlantic Coast of North America. *Transactions of the Connecticut Academy of Arts and Sciences*, 35:129-328.

Corrêa, D.
1948. *Ototyphlonemertes* from the Brazilian Coast. *Comunicaciones Zoologicas del Museo de Historia Natural de Montevideo*, 2:1-12.
1949. Ecological Study of Brazilian *Ototyphlonemertes*. *Comunicaciones Zoologicas del Museo de Historia Natural de Montevideo*, 3:1-7.
1950. Sôbre *Otothyphlonemertes* do Brasil. *Boletim de Faculdade de Filosofia, Ciências e Letras de Universidade de São Paulo, Zoologia*, 15:203-233.
1953. Sôbre a neurofisiologia locomotora de Hoplonemertinos e a taxonomia do *Ototyphlonemertes*. *Anais da Academia Brasileira de Ciências*, 25:545-555.
1958. Nemertinos do Litoral Brasileiro (VII). *Anais da Academia Brasileira de Ciências*, 29:441-455.
1961. Nemerteans from Florida and Virgin Islands. *Bulletin of Marine Science of the Gulf and Caribbean*, 11:1-44.

Friedrich, H.
1933. Morphologische Studien an Nemertinen der Kieler Bucht, I und II. *Zeitschrift für wissenschaftliche Zoologie*, 144:496-509.
1935. Studien zur Morphologie, Systematik und Ökologie der Nemertinen der Kieler Bucht. *Archiv für Naturgeschichte*, 4:293-375.
1949. Über zwei bemerkenswerte neue Nemertinen der Sandfauna. *Kieler Meeresforschungen*, 6:68-72.

Gerner, L.
1969. Nemertinen der Gattungen *Cephalothrix* und *Ototyphlonemertes* aus dem marinen Mesopsammal. *Helgoländer wissenschaftliche Meeresuntersuchungen*, 19:68-110.

Hylbom, R.
1957. Studies on Palaeonemerteans of the Gullmar Fiord Area (West Coast of Sweden). *Arkiv för Zoologi*, 10:539-582.

Iwata, F.
1972. Axial Changes in the Nemertean Egg and Embryo During Development and Its Phylogenetic Significance. *Journal of Zoology, London*, 168:521-526.

Kirsteuer, E.
1963. Zur Ökologie systematischer Einheiten bei Nemertinen. *Zoologischer Anzeiger*, 170:343-354.
1967. Marine, Benthonic Nemerteans: How to Collect and Preserve Them. *American Museum Novitates*, 2290:1-10.
1971. The Interstitial Nemertean Fauna of Marine Sand. Pages 17-19 in N.C. Hulings, editor, Proceedings of the First International Meiofauna Conference. *Smithsonian Contributions to Zoology*, 76.
1977. Remarks on Taxonomy and Geographic Distribution of the Genus *Ototyphlonemertes* Diesing (Nemertina, Monostilifera). *Mikrofauna des Meeresbodens*, 61:167-181.
1986. Nemertina. Pages 72-75 in L. Botosaneanu, editor, *Stygofauna Mundi. A Faunistic, Distributional, and Ecological Synthesis of the World Fauna Inhabiting*

subterranean Waters (Inluding the Marine Interstitial). Leiden: E.J. Brill.

Mock, H.
1978. *Ototyphlonemertes pallida* (Keferstein, 1862). *Mikrofauna des Meeresbodens*, 67:1-14.
1981a. Zur Morphologie von *Prostomatella arenicola* (Hoplonemertini, Monostilifera). *Helgoländer Meeresuntersuchungen*, 34:491-496.
1981b. Beobachtungen an einem Nemertinen-(Schnurwurm) Eigelege. *Microkosmos*, 4:102-104.

Mock, H., and P. Schmidt
1975. Interstitielle Fauna von Galapagos XIII. *Ototyphlonemertes* Diesing (Nemertini, Hoplonemertini). *Mikrofauna des Meeresbodens*, 51:1-40.

Müller, G.
1968. Betrachtungen über die Gattung *Ototyphlonemertes* Diesing, 1863 nebst Bestimmungsschlüssel der validen Arten (Hoplonemertini). *Senckenbergiana Biologica*, 49:461-468.

Norenburg, J.
1985. Structure of the Nemertine Integument With Consideration of Its Ecological and Phylogenetic Significance. *American Zoologist*, 25:37-51.
1988. Remarks On Marine Interstitial Nemertines and Key to the Species. *Hydrobiologia*, 156:87-92.

Swedmark, B.
1964. The Interstitial Fauna of Marine Sand. *Biological Reviews*, 39:1-42.

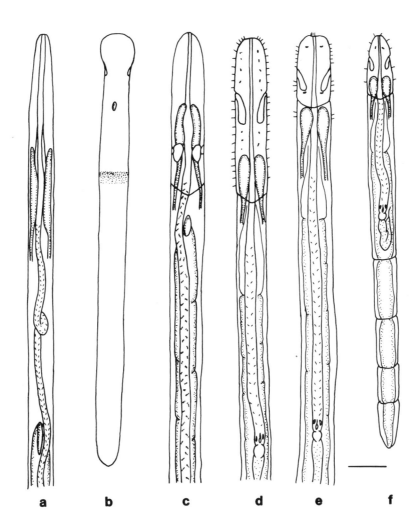

Figure 22.1.--Nemertina: Diagrams of representative interstitial nemertines, all except **b**, showing internal features observable in living specimens, cerebral organs (not present in **a**), cerebral ganglia, proboscis, and gut. **a**, Archinemertina: *Cephalothrix* sp.; **b**, Palaeonemertina: *Carinina arenaria* (after Hylbom, 1957); **c**, Heteronemertina sp. (Puget Sound specimen); **d-f**, Hoplonemertina, Monostilifera: **d**, Hoplonemertina sp. (New England specimen); **e**, *Prostomatella arenicola* (after Mock, 1981a); **f**, *Annulonemertes* sp. (Puget Sound specimen). (Scale = 250 μm.)

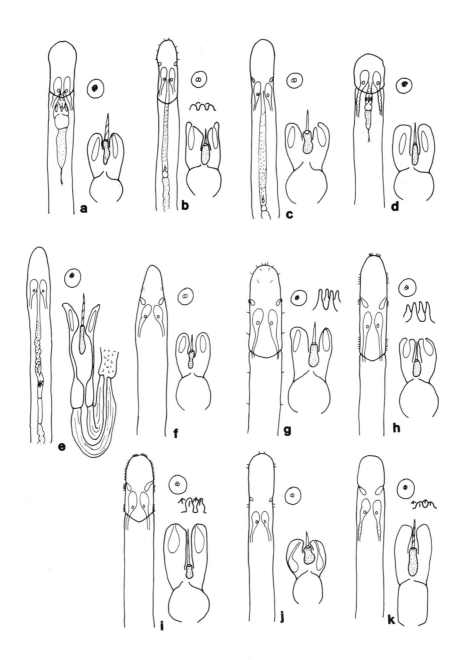

Figure 22.2.--Nemertina (continued): Diagrams depicting diversity and various characteristics observable in living specimens of *Ototyphlonemertes*. **a**, *O. americana* (after Gerner, 1969); **b**, *O. antipai* (after Müller, 1968); **c**, *O. aurantiaca* (after Gerner, 1969); **d**, *O. brevis* (after Corrêa, 1948); **e**, *O. macintoshi* (after Bürger, 1895); **f**, *O. brunnea (after Bürger, 1895);* **g**, *O. cirrula* (after Mock and Schmidt, 1975); **h**, *O. pallida* (after Mock, 1978); **i**, *O. erneba* (after Corrêa, 1950; Kirsteuer, 1977); **j**, *O. evelinae* (after Corrêa, 1958); **k**, *O. fila* (after Corrêa, 1953). (Diagrams not to scale.)

23. Nematoda

Franz Riemann

Nematodes can colonize virtually every moist habitat that can sustain metazoan life. In marine sediments they are the most abundant animals. Zoologists are usually familiar with the general construction plan of the nematode body and some aspects of the biology of parasitic nematodes. However, while realizing the numerical prominence of free-living nematodes, they have often developed an aversion to working with aquatic forms, anticipating particular taxonomic problems as an obstacle for ecological studies. Currently, however, some of these problems are considerably lessened. Briefly, the meiobenthologist working with nematodes is in the following fortunate situation: the worms can be preserved in formalin within the sediment in comparatively small samples at the collection site. The extractions can be conducted by simple mechanical methods (e.g., decantation and sieving). The isolated worms are transferred by an evaporation process into glycerin and are mounted in greater numbers in anhydrous glycerin on microscope slides. These easily fabricated permanent whole mounts are generally sufficient for species identification. Illustrated determination keys are available, as well as species checklists and general reviews (the marine biologists are a bit better supplied with the relevant information than the freshwater colleagues). I should like to encourage even amateur naturalists to work with aquatic nematodes. A well made glycerin-mounted preparation of such organisms presents a fascinating view and behavioral studies of living nematodes will be greatly appreciated by the nematological community.

Though the general construction plan of marine nematodes follows the textbook example of a parasite, there are notable peculiarities in the free-living marine forms. For instance, free-living nematodes have comparatively short and simple gonads, the number and position of which can be easily assessed in whole mounts. These characters have received high importance in the classification (Lorenzen, 1981). The cuticle may bear conspicuous bristles (setae), especially in nematodes from littoral sands, and is often very penetrable for fluids and stains. By secretions of the caudal glands and, possibly other glands too, the aquatic nematodes are often sticky. They can strongly adhere to sediment particles (and glass capillaries) and tend to hide within sediment agglutinations. Attached sediment particles and masses of symbiotic prokaryotes may camouflage the cuticle contours in certain species.

The length of marine nematodes is usually around 1–3 mm but sizes well over 10 mm may be attained, particularly in families associated with biotic structures such as kelp holdfasts or sponges (*Deontostoma timmerchioi* Hope, 1974, reaches 47 mm). The smallest adult nematodes (*Greeffiella* Cobb, 1922, a desmoscolecid) are shorter than 0.2 mm. Deep-sea sediments seem to contain smaller nematodes than shallow water regions. In the Iberian deep sea, the mean length of nematodes was found at 0.46 mm (Lavaleye, 1985, personal communication).

The following pertains mainly to the study of marine nematodes.

Classification

The class Nematoda consists of two subclasses, the Secernentea and the Adenophorea. The main diagnostic characters are the presence of caudal glands (secreting a sticky fluid), bristles, and conspicuous amphids (cephalic multifunction sense organs) in the majority of Adenophorea, being either absent or inconspicuous (amphids) in the Secernentea. Secernentean nematodes constitute the bulk of the parasites; the free-living representatives of this subclass are abundant in soils and organically polluted freshwaters. Many species may be considered as saprozoa. There are only two secernentean species found in abundance in marine habitats: *Rhabditis marina* Bastian, 1865, from rotting seaweeds at the shore and *Eudiplogaster pararmatus* (W. Schneider, 1938), from polluted estuarine mud flats. Both species occur in masses and are convenient laboratory animals. They are not typical marine nematodes, but should be regarded as halophilic freshwater worms.

Freshwater sediments and the phytal, when unpolluted, are dominated by members of the adenophorean families Tobrilidae, Tripylidae, Chromadoridae, Ironidae, Rhabdolaimidae, Leptolaimidae, Monhysteridae, and numerous representatives of the order Dorylaimida. Some limnetic sites, for instance the profundal of lakes, contain just a few species, which nevertheless may be very abundant.

Marine habitats exhibit an enormous variety of adenophorean taxa (450 genera, see Heip et al., 1982), and a handful of sublittoral sand contains more than 20 species. Fortunately, "contrary to popular opinion, marine nematodes do not 'all look the same' and it is time that this myth was finally put to rest" (Platt and Warwick, 1980, see Figure 23.2).

Habitat and Ecological Notes

Nematodes are usually strictly bound to a substrate. They can, however, be brought into suspension by water currents and are, therefore, often found in the water column together with seston particles. Indirect, positive evidence raised the discussion as to what extent nematodes can actively swim (Schneider, 1913; Jensen, 1981). In the Petri dish, members of a few genera can swim for a short distance.

Every type of sediment is colonized by nematodes, from almost dry dune sand to heavy surf beach sand, coarse shell sublittoral grounds, down to hadal trenches. Algal mats and bushes as well as phanerogams have their nematode populations, and other biotic habitats like crustacean gill chambers, mud enclosed by the pedal cavity of soft bottom dwelling actiniaria, worm tubes and fecal mounds also contain nematodes. Cold arctic waters as well as hot springs harbor these worms and, even in hypersalinic brines where salt is crystallizing, nematodes have been found.

While in muddy sediments usually around 90% of the nematodes are confined to the top 5 cm, a deeper penetration of the worms must be considered. Nematodes occur several dm deep in coarser sand (see Boaden, 1977; Heip et al., 1977; Munro et al., 1978), and, even in fine-sand flats, abundant species may have their population maximum in about 10 to 20 cm depth (e.g., *Theristus blandicor* Rachor, 1971). Many nematodes do not need a rich oxygen supply and may be regarded as facultative anaerobes. Certain taxa are regular inhabitants of sulfidic sediment layers, and one species (*Paramonohystera wieseri* Ott, 1977) has proved to be anoxybiotic. On the other hand, there are abundant species concentrated in the top cm of sediment (e.g., *Chromadora lorenzeni* Jensen, 1980) that might be pushed away in the course of careless sampling.

Methods of Collection

Any instrument that can be used to collect undisturbed sediment surfaces and has a sufficient depth of penetration is adequate for nematode collection. Phytal nematodes may be collected by introduction of the entire substrate into clear polyethylene bags. The assistance of SCUBA divers facilitates the collection of shallow-water nematodes.

Around 1000 nematodes can be expected under 10 cm^2 of any sediment surface or in plant habitats and, if we consider that about 300 specimens are sufficient to obtain a reasonable picture of the species composition in shallow waters (Boucher, 1980), then the employment of corers having about 20 mm diameter would presumably be satisfactory for faunistic studies. Under quantitative ecological aspects, however, such coring is inadequate because nematodes may occur in small patches (see Eskin and Coull, 1984). Small-scale physical inhomogeneities in the environment may be the cause for patchiness, as well as the attraction of nematodes to food items (Gerlach, 1977). Also behavioral reasons for the aggregation of nematodes in a homogeneous environment should be considered. The patchiness requires particular sampling strategies in quantitative studies. Tietjen (1980) while working in the sublittoral of the New York Bight found that three plastic core tubes (2.5 cm inside diameter) inserted into bottom samples taken with a grab showed a sample precision of ±50% of the mean nematode density at the station. In the range of ±22% are the deviations with four replicates taken with a 2.5 cm corer in intertidal fine sands (Blome, 1985, personal communication). Working with replicates involves some statistical exercises, but an alternative procedure based on the pooling of mini-cores to get an average value has rarely been practiced to my knowledge (see Rudnick et al., 1985).

Due to the high abundance of nematodes, subdivision of fine sediment samples brought into a homogeneous suspension, or an aliquot sample of the extracted preserved nematodes (see below) is often necessary to get a reasonably low number of specimens for the taxonomic work. In our laboratory, a Folsom plankton splitter is used for subdivision of a homogeneous suspension of nematodes in water (see Juario, 1975). Other methods of meiofauna sample splitting are described by Jensen (1982) and McIntyre and Warwick (1984), while Sherman et al. (1984) offer another approach for subsampling.

Methods of Extraction

Nematodes may be extracted either alive or from preserved sediment samples. In the past years, for taxonomic studies in our laboratory, we relied, whenever possible, on the extraction of live material but experience showed that nematodes may exhibit signs of poor condition, such as histolysis, after a short storage time even when they were still moving.

Aquatic nematodes are not as hardy as their inclusion among "hard meiofauna" would imply, and the condition of the specimens before fixation is as important as the fixation itself. Therefore, the bulk fixation of small sediment samples with 10% formalin used immediately upon retrieval should be generally considered as a basic, adequate method for nematode taxonomy. It is recommended that subsamples be treated this way whenever a collection is made.

Most of the meiofauna extraction methods listed by Uhlig et al. (1973) and McIntyre and Warwick (1984) are applicable for isolating living and preserved aquatic nematodes (see also Platt and Warwick, 1983).

For extracting living nematodes from clean sand, many workers apply the decantation and sieving method (see below) in the field. However, it is also useful to store a large volume of sand (e.g., 750 ml) in the laboratory in a high 1000 ml container filled with water from the sampling site. Under the influence of the deteriorating environmental conditions, the nematodes and other meiofauna concentrate near the sediment surface and can be pipetted off with some sediment into a Petri dish after about one or two days; the concentration factor is not high. The preliminary observation of living nematodes in their natural environment under the dissecting microscope and the rough assessment of the associated biota and of the sediment characteristics may be rewarding for the beginner.

After handling several Petri dishes containing a thin sand layer in this manner, the remaining sediment in the container may be subjected to the decantation and sieving method. Portions of the sediment are suspended by stirring and shaking with ample amounts of seawater in a bucket or stoppered measuring cylinder, then sedimentation is allowed to take place for 20-30 seconds so that the bulk of sediment particles can settle. The supernatant is decanted through a sieve with a mesh size of 42-63 μm and washed off the sieve into a petri dish from where the nematodes can be sorted out under a dissecting microscope. The procedure should be repeated about three times. This method does not extract all nematodes for quantitative evaluations; particularly nematodes from lotic habitats may effectively resist separation from the sediment by adhering firmly to heavier particles or sediment agglutinations the worms themselves produce. The extraction efficiency in this case can be increased by the relaxing effect of subsequent washings of the (marine) sediment with diluted seawater or freshwater (attention: tap water must be filtered to avoid the introduction of foreign meiofauna). If the sieve residue is instantly brought back into seawater, many nematodes will survive. The application of warm (40-50° C) seawater or freshwater will give a nearly complete extraction of nematodes from sand, but at the expense of an impaired preservation condition. Gentle heat is often used as a relaxant for nematodes, mostly for immobilizing single living worms on slides for the convenient observation under the microscope. However, the suitable period of heating and the best temperature is difficult to determine for nematode communities in sediments. Sometimes narcotization with magnesium chloride is recommended for nematodes, but I found that after the treatment specimens exhibit vacuolization of the body wall.

The Uhlig seawater ice method (Uhlig et al., 1973) delivers living nematodes in good condition, but sometimes at a low extraction efficiency, for instance, with epsilonematids. Though the Uhlig method is said to be most effective in sandy sediments, it can also be applied to muds after an elutriation of the finer particles. Thus, Uhlig (1971, personal communication) has demonstrated on-board ship the extraction of living deep-sea nematodes from muds.

A set-up similar to the Uhlig apparatus can be used without ice. If, after washing away the finest particles, the sieve is brought in contact with the surface of water; many nematodes from a thin layer of moist sediment on the sieve will crawl through the meshes into the collecting dish. A glass cover over the set-up is recommended to prevent excessive evaporation. Basically, this is a modification of the Baermann-funnel technique which is often used for the extraction of small soil invertebrates. Lorenzen (1969) used the method for the isolation of salt-marsh desmoscolecids, and I found it useful generally when working with detritus-rich estuarine intertidal muds. Living nematodes may also be extracted after washing away the finest particles, with the decanting and sieving method. Coarse detritus particles, however, conceal many worms and make a quantitative extraction by this simple method impossible.

For a quantitative extraction of nematodes, working with preserved material is preferred. For details, the reader should consult the review of McIntyre and Warwick (1984). Nematodes from sands are usually extracted by decanting and sieving methods as described for live material, or by a more automatic procedure employing elutriation with a constant, controlled water current that separates sand and nematodes in a separating funnel. For muds rich in detritus, centrifugation techniques have recently been developed which employ liquids with higher density than water, such as sugar solutions or silica sols (De Jonge and Bouwman, 1977). There is also a centrifugation method available for living meiofauna (Schwinghamer, 1981). The silica sol technique delivers the meiofauna after centrifugation free of sediment particles.

The recommended sieve mesh size for quantitative extraction of marine nematodes is 42 µm or even less (Bovée et al., 1974, Rudnick et al., 1985), though the 63 µm mesh size is also acceptable at times. For facilitating the sorting of nematodes in the presence of sediment particles, the staining of worms with Rose Bengal is often practiced. I found that this technique is indispensable when working with whitish deep-sea Globigerina oozes; in other cases, preserved nematodes are usually sufficiently in contrast against sediment particles under incident light. During the hand-sorting of preserved nematodes and other meiofauna, the worms may be transferred for short times into freshwater to avoid harmful effects of formaldehyde vapors; some laboratories use special sorting fluids (Steedman, 1976).

Because some of the manipulations involved in the quantitative extraction of nematodes may impair the preservation condition, it is advisable for the taxonomist to use more delicate semiquantitative extraction methods in order to get well preserved material for identification.

Most nematodes appear whitish under incident light in the petri dish, thus contrasting against a dark detritus background. Chromadorids are often iridescent, as are some sluggish oxystominids, which resemble a glass fiber.

The movements of nematodes are very different. Usually there is the undulatory propulsion, but exceptions exist like hopping, caterpillar- or leech-like crawling. Many nematodes normally appear motionless and may be neglected during the sorting of live samples.

The discrimination of dead aquatic nematodes from moribund or sluggish ones is not easy because, at least in cultures, the decay may proceed slowly. Tietjen (1967) observed the decomposition of monhysterids in the presence of bacteria alone and found intact nematode bodies even after two months. However, in the presence of euglenoids, the nematode remains disappeared entirely in less than 24 hours. There is no information to my knowledge about the degradation of dead nematodes in nature, but in population studies based on counts of preserved specimens at low magnification (e.g., in toxicity tests) a possible slow degradation of nematodes should be kept in mind.

Preparation of Specimens for Taxonomic Study

Only a few publications on the preparation techniques of marine nematodes exist in contrast to the wealth of instructions for the handling of soil nematodes. The "Laboratory Methods for Work with Plant and Soil Nematodes" (Southey, 1970) are recommended because many techniques described therein are also useful for work with aquatic nematodes. There is, to my knowledge, only one article comparing systematically the effects of various fixatives on the preservation of marine nematodes (Timm and Hackney, 1969). It seems that oral traditions concerning the processing of marine nematodes predominate in the different laboratories.

Personal experiences with different fixation techniques tested on nematodes from marine and brackish water communities (Riemann, 1979) have shown that it is not possible to obtain clear-cut results. There are not only differences in the reaction to fixatives in the various tissues (as is generally known), but also the various taxa show notable variations of preservation when subjected to fixatives. In addition, in every sample there are intraspecific differences in the preservation condition, which are beyond the examiner's control, and which make an evaluation of fixation tests problematical.

The best fixative for general routine work is still 10% formalin (=4% formaldehyde solution in water); 5% strength was found to be insufficient in some taxa, 7.5% may be considered as appropriate. The application of hot solutions often results in a better preservation of head structures. However, certain anatomical details may be obscured. Nevertheless, experiments with hot fixatives may be rewarding. Ice cooled fixatives were without advantage; therefore, fixation at 20° C is recommended. A fixation duration of 24 hours is sufficient for the subsequent preparation of permanent mounts in glycerin. We prefer a duration of three days and found a longer duration of no advantage. It is said that nematodes may be stored eternally in formalin without impairing the quality of preservation. Such is not my experience with unbuffered formalin; perhaps experiments with buffered formalin are recommendable. The fixative should be made up with distilled or freshwater. Seawater mixtures may result in an unfavorable strong light-refraction of certain cytoplasmic components concealing, for example, nuclei in gametocytes and fine fibers in the pharynx muscles. There is no osmosis phenomenon whatsoever that justifies the preference of fixatives made up with seawater.

The preservation and visibility of nuclei and cells can be greatly improved by the addition of acetic or propionic acid (final concentration 0.25 to 1%) to the formalin. Propionic acid has been recommended by Maggenti and Viglierchio (1965) and Netscher and Seinhorst (1969); it avoids excessive clearing of tissues seen sometimes in acetic acid preparations. The drawback of acetic or propionic acid/formalin mixtures as compared to pure formaldehyde solutions is unfortunately that nearly always head structures, which are so important for the determination, are obscured by adhering mucus and detritus particles, and the preservation of setae is poor. Moderate shrinking phenomena are sometimes evident in the course of further processing. The prominence of nuclei is better in fixatives made with distilled water than those made with seawater; however, in the latter case, the preservation condition of head structures may be more acceptable. As a compromise, the following procedure is proposed: prepare a fresh mixture of 10% formalin (=4% formaldehyde) and 0.5% propionic acid (1% is also recommendable) made with distilled water (starting with a 5% formaldehyde stock solution and a 5% propionic acid solution facilitates the preparation of the final mixtures). Collect the living nematodes in a 10 ml glass tube with 1 ml seawater. Add 1 ml of the fixative mixture and wait for about 30 minutes. Fill up the tube with 8 ml fixative and wait three days until processing further.

Hot fixatives containing acetic or propionic acid are sometimes advantageous but may give rise to brittleness of the nematodes. The bulk fixation of nematodes in marine sediments with acid fixatives is rarely successful because of the buffering capacity of calcareous components of the sediments.

The addition of ethanol to the fixatives improves the definition of glands and muscles, but the preservation of the cuticle ornamentation may be impaired by alcohol. The often stressed danger of shrinkage is inconsiderable in the majority of marine nematodes. A freshly prepared mixture consisting of 10% formalin (=4% formaldehyde), 1% propionic acid, and 30% ethanol made with distilled water (a modification of the old Ditlevsen's fixative) with a fixation duration of three days renders excellent specimens for anatomical studies.

Due to the effects of various fixation techniques, there may be an increase of body length amounting to more than 10% of the living nematode's length. Cold formaldehyde preserves the nematodes at almost the original length (Fagerholm, 1979).

Usually, the fixed nematodes are transferred through a slow evaporation procedure to pure, anhydrous glycerin and mounted in that medium as permanent whole mounts. This transfer, if not sufficiently gradual, may give rise to shrinking; however, the majority of marine nematodes are not prone to this problem. There are several transfer procedures to use (see Southey, 1970). We select the nematodes from the fixative by means of small hook-shaped needles or bristles stuck onto a handle, and transfer them into a cavity block (Boveri dish) to which has been added 5 ml of a mixture consisting of 35% pure, absolute ethanol, 4% glycerin and 61% distilled water. The fluid in the vessel is then subjected to evaporation in an oven at 35-40° C for about two days, whereby the evaporation time may be controlled by partially covering the vessel with a cover glass. Other workers conduct the evaporation at room temperature and prefer a mixture with 2% glycerin. After the evaporation, the nematodes are left in almost pure glycerin. Many workers then put the vessels for some days into a desiccator to remove the last traces of water, but we found no disadvantage in omitting this step. The worms are then mounted on a glass slide in a fresh drop of anhydrous glycerin. The stock amount of this glycerin should be kept permanently in a desiccator.

For another mode of transfer of nematodes from fixative to glycerin, a mixture consisting of 4% formaldehyde and 3% (or less) glycerin in distilled water may be subjected to the evaporation procedure, this being an excellent alternative (the initial fixation duration, for example, with the described formaldehyde-propionic acid mixtures, may be abbreviated) for the described procedure, provided the evaporation is conducted in a well aerated fume hood.

A coverglass support (glass rods, beads, or pieces of wire) having nearly the same diameter as the nematodes is necessary before sealing the slide with a xylene-resistant coverglass seal. As an alternative, particularly when many nematodes must be studied in ecological and taxonomical studies, we recommend the inclusion of the glycerin drop containing the nematodes in a paraffin ring. For this process, a mixture of 25% beeswax and 75% paraffin (melting point 42-44° C) is prepared; the final mixture has a melting point of about 60° C. The nematodes (after having passed the described evaporation procedure) are collected in a small glycerin drop on the slide then, by means of a heated spatulum, a flat rectangular frame of the paraffin mixture is arranged around the glycerin at a distance that matches the margins of the coverglass. When placing the coverglass, it must just touch the glycerin to avoid later inclusion of air bubbles or pieces of melting paraffin. Then the slide is heated on a thermostat plate at 70° C (careful heating on a flame may serve as a substitute) so that the paraffin will neatly flow around the glycerin. After cooling on a metal plate, remove the excess of paraffin protruding under the coverglass with a knife and repeat the heating, cooling, and so on; thus, size and thickness of the glycerin drop area will be controlled. The initial excess of paraffin is necessary to avoid trapping of air bubbles. Only practice can guide the preparator to estimate the appropriate size of the glycerin drop and the amount of paraffin necessary to get the best mounts. The inclusion of coverglass supports is dispensable in many cases. Sealing of the slides with a xylene-resistant cement or lac is useful. Slides processed in the described way in our laboratory two decades ago are still in good condition.

A good bright field microscope equipped with oil

immersion optics and a drawing apparatus is necessary for the taxonomic work with nematodes. Beginners may worry about the excessive transparency of glycerin-mounted nematodes; they will find semipermanent mounts containing a mixture of formaldehyde water and glycerin more convenient at times. Staining with cotton blue (achieved by adding a few grains of dye into the evaporation mixture, giving a light blue color to the solution) is often recommended for work with nematodes mounted in glycerin, but is problematical because some aquatic nematode taxa tend to overstain while others in the same sample remain unstained. Of enormous advantage is the availability of an interference contrast microscope (Nomarski type) that really adds new dimensions to studies of nematode whole mounts. Photography at high magnifications is particularly rewarding with this type of microscope.

There are no clear-cut rules as to which morphological data should be considered in the taxonomic practice. Valuable recommendations, including the terminology of structures, have been given by Coomans et al. (1979) and Platt and Warwick (1983).

Notes on the Literature.--A pictorial key to almost all marine nematode genera (worldwide) was given by Platt and Warwick (1983). The book of Goodey (1963) contains descriptions and illustrations of one representative of each genus encountered worldwide in soils and freshwater, with notes on bionomics. Andrássy (1976, see also 1984) made a checklist of all free-living non-marine nematodes genera. Descriptions and pictures of free-living non-marine nematodes of Europe were published by Meyl (1960). Gerlach and Riemann (1973/1974) gave references to almost all original and subsequent descriptions of aquatic adenophorean nematodes (excluding the Dorylaimida). Lorenzen's (1981) phylogenetic system of free-living nematodes found a wide acceptance, modifications were published by Inglis (1983). Heip et al. (1982, 1985) gave comprehensive reviews on the systematics and ecology of free-living marine nematodes. Nicholas (1984) gave a modern introduction into the biology of free-living nematodes. Important introductions into the morphology of nematodes are the books of Chitwood and Chitwood (1950) and Bird (1971).

References

Andrássy, I.
1976. *Evolution as a Basis for the Systematization of Nematodes.* 288 pages. London: Pitman Publishing Ltd.
1984. Klasse Nematoda (Ordnungen Monhysterida, Desmoscolecida, Araeolaimida, Chromadorida, Rhabditida). In H. Franz, editor, *Bestimmungsbücher zur Bodenfauna Europas.* 509 pages. Stuttgart: Gustav Fischer.

Bird, A.F.
1971. *The Structure of Nematodes.* 318 pages. New York: Academic Press.

Boaden, P.J.S.
1977. Thiobiotic Facts and Fancies (Aspects of the Distribution and Evolution of Anaerobic Meiofauna). *Mikrofauna des Meeresbodens,* 61:45-63.

Boucher, G.
1980. Impact of Amoco Cadiz Oil Spill on Intertidal and Sublittoral Meiofauna. *Marine Pollution Bulletin,* 11:95-101.

Bovée, F. de, J. Soyer, and Ph. Albert
1974. The Importance of the Mesh Size for the Extraction of the Muddy Bottom Meiofauna. *Limnology and Oceanography,* 19:350-354.

Chitwood, B.G., and M.B. Chitwood
1950. *An Introduction to Nematology.* Section 1, Anatomy, 2nd edition. 213 pages. Baltimore: Monumental Printing Company. (Reprint 1974, Baltimore: University Park Press. 334 pages.)

Coomans, A., L. De Coninck, and C. Heip
1979. Round Table Discussions and Addenda (First Workshop on the Systematics of Free-living Nematodes, Ghent, 13-16 August 1977). *Annales de la Société Royale Zoologique de Belgique,* 108:109-122.

De Jonge, V.N., and L.A. Bouwman
1977. A Simple Density Separation Technique for Quantitative Isolation of Meiobenthos Using the Colloidal Silica Ludox-Tm. *Marine Biology,* 42:143-148.

Eskin, R.A., and B.C. Coull
1984. A Priori Determination of Valid Control Sites: An Example Using Marine Meiobenthic Nematodes. *Marine Environmental Research,* 12:161-172.

Fagerholm, H.-P.
1979. Nematode Length and Preservatives, With a Method for Determining the Length of Live Specimens. *Journal of Parasitology,* 65:334-335.

Gerlach, S.A.
1977. Attraction to Decaying Organisms as a Possible Cause for Patchy Distribution of Nematodes in a Bermuda Beach. *Ophelia,* 16:151-165.

Gerlach, S.A., and F. Riemann
1973/ The Bremerhaven Checklist of Aquatic Nematodes.
1974. A Catalogue of Nematoda Adenophorea Excluding the Dorylaimida. *Veröffentlichungen des Instituts für Meeresforschung in Bremerhaven,* Supplement 4. 736 pages.

Goodey, T.
1963. *Soil and Freshwater Nematodes.* 2nd edition, revised by J.B. Goodey. 544 pages. London, New York: Methuen, Wiley.

Heip, C., M. Vincx, N. Smol, and G. Vranken
1982. The Systematics and Ecology of Free-living Marine Nematodes. *Helminthological Abstracts* (Series E), 51(1):1-31.

Heip, C., M. Vincx, and G. Vranken
1985. The Ecology of Marine Nematodes. *Oceanography and Marine Biology Annual Review,* 23:399-489.

Heip, C., K.A. Williams, and A. Goossens
1977. Vertical Distribution of Meiofauna and the Efficiency of the van Veen Grab on Sandy Bottoms in Lake Grevelingen (The Netherlands). *Hydrobiological Bulletin,* 11:35-45.

Inglis, W.G.
1983. An Outline Classification of the Phylum Nematoda. *Australian Journal of Zoology,* 31:243-255.

Jensen, P.
1980. Description of the Marine Free-living Nematode *Chromadora lorenzeni,* n.sp. with Notes on Its Microhabitats. *Zoologischer Anzeiger,* 205:213-218.
1981. Phyto-chemical Sensitivity and Swimming Behavior of the Free-living Marine Nematode *Chromadorita tenuis. Marine Ecology Progress Series,* 4:203-206.

1982. A New Meiofauna Sample Splitter. *Annales Zoologici Fennici*, 19:233-236.

Juario, J.V.
1975. Nematoda Species Composition and Seasonal Fluctuation of a Sublittoral Meiofauna Community in the German Bight. *Veröffentlichungen des Instituts für Meeresforschung in Bremerhaven*, 15:283-337.

Lorenzen, S.
1969. Desmoscoleciden (eine Gruppe freilebender Meeresnematoden) aus Küstensalzwiesen. *Veröffentlichungen des Instituts für Meeresforschung in Bremerhaven*, 12:231-265.
1981. Entwurf eines phylogenetischen Systems der freilebenden Nematoden. *Veröffentlichungen des Instituts für Meeresforschung in Bremerhaven*, supplement 7. 472 pages.

Maggenti, A.R., and D.R. Viglierchio
1965. Preparation of Nematodes for Microscopic Study: Perfusion by Vapor Phase in Killing and Fixing. *Hilgardia*, 36:435-463.

McIntyre, A.D., and R.M. Warwick
1984. Meiofauna Techniques. Pages 217-244. In N.A. Holm and A.D. McIntyre, editors, *Methods for the Study of Marine Benthos*, 2nd edition. Oxford: Blackwell.

Meyl, A.H.
1960. Die freilebenden Erd- und Süsswassernematoden. In P. Brohmer, P. Ehrmann, and G. Ulmer, editors, *Die Tierwelt Mitteleuropas*, 1(5a). Leipzig: Quelle und Meyer, 164 pages.

Munro, A.L.S., J.B.J. Wells, and A.D. McIntyre
1978. Energy Flow in the Flora and Meiofauna of Sandy Beaches. *Proceedings of the Royal Society Edinburgh*, 76B:297-315.

Netscher, C., and J.W. Seinhorst
1969. Propionic Acid Better than Acetic Acid for Killing Nematodes. *Nematologica*, 15:286.

Nicholas, W.L.
1984. *The Biology of Free-living Nematodes*. 2nd edition, 251 pages. Oxford: Clarendon Press.

Ott, J.A.
1977. New Freeliving Marine Nematodes from the West Atlantic. I. Four New Species from Bermuda with a Discussion of the Genera *Cytolaimium* and *Rhabdocoma* Cobb, 1920. *Zoologischer Anzeiger*, 198:120-138.

Platt, H.M., and R.M. Warwick
1980. The Significance of Free-living Nematodes to the Littoral Ecosystem. Pages 729-759 in J.H. Price, D.E.G. Irvine, and W.F. Farnham, editors, *The Shore Environment*, volume 2, Ecosystems. London: Academic Press.
1983. Free-living Marine Nematodes, part I, British Enoplids. In *Synopses of the British Fauna*, 28, 307 pages.

Rachor, E.
1971. *Theristus (Penzancia) blandicor* (Monhysterida) eine neue Nematodenart aus dem Eulitoral der Wesermündung. *Veröffentlichungen des Instituts für Meeresforschung in Bremerhaven*, 13:139-146.

Riemann, F.
1979. Nematoden aus dem Brackwasser des Weser-Ästuars und Beschreibung von drei Monhysteroidea. *Veröffentlichungen des Instituts für Meeresforschung in Bremerhaven*, 17:213-223.

Rudnick, D.T., R. Elmgren, and J.B. Frithsen
1985. Meiofauna Prominence and Benthic Seasonality in a Coastal Marine Ecosystem. *Oecologia (Berlin)*, 67:157-168.

Schneider, G.
1913. Nematoden als Fischnahrung. *Internationale Revue der gesamten Hydrobiologie und Hydrographie*, 6:489-490.

Schwinghamer, P.
1981. Extraction of Living Meiofauna from Marine Sediments by Centrifugation in a Silica Sol-sorbitol Mixture. *Canadian Journal of Fisheries and Aquatic Sciences*, 38:476-478.

Sherman, K.M., D.A. Meeter, and J.A. Reidenauer
1984. A Technique for Subsampling an Abundant Taxon While Completely Sorting Other Taxa. *Limnology and Oceanography*, 29:433-439.

Southey, J.F., editor
1970. Laboratory Methods for Work with Plant and Soil Nematodes. *Ministry of Agriculture Fisheries and Food, Technical Bulletin 2*. 148 pages. London: Her Majesty's Stationery Office.

Steedman, H.F.
1976. Miscellaneous Preservation Techniques. Examination, Sorting and Observation Fluids. Pages 175-183 in H.F. Steedman, editor, *Monographs on Oceanographic Methodology 4, Zooplankton Fixation and Preservation*. Paris: UNESCO Press.

Tietjen, J.H.
1967. Observations on the Ecology of the Marine Nematode, *Monhystera filicaudata* Allgén, 1929. *Transactions of the American Microscopical Society*, 86:304-306.
1980. Population Structure and Species Composition of the Free-living Nematodes Inhabiting Sands of the New York Bight Apex. *Estuarine and Coastal Marine Science*, 10:61-73.

Timm, R.W., and T. Hackney
1969. Effects of Fixation and Dehydration Procedures on Marine Nematodes. *Journal of Nematology*, 1:146-149.

Uhlig, G., H. Thiel, and J.S. Gray
1973. The Quantitative Separation of Meiofauna. A Comparison of Methods. *Helgoländer Wissenschaftliche Meeresuntersuchungen*, 25:173-195.

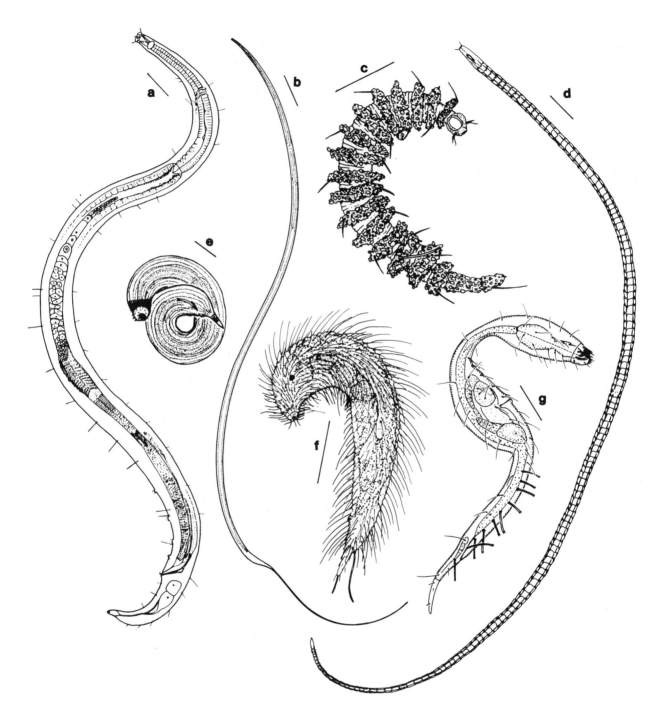

Figure 23.1.--Nematoda: a, *Echinotheristus,* (North sea, after von Thun and Riemann 1967, modified) example of typical nematode shape, contrasted with aberrant nematodes (b-g); **b,** *Halalaimus* (Antarctic deep-sea; original, Schrage); **c,** *Desmoscolex* (Iberian deep-sea; after Freudenhammer, 1975); **d,** *Pselionema* (North Sea; original, Schrage); **e,** *Richtersia* (Atlantic, USA; after Chitwood, 1936); **f,** *Greeffiella* (Skagerrak; original, Schrage); **g,** *Dracograllus* (Atlantic, USA; after Allen and Noffsinger, 1978). (Scale = 50 μm.)

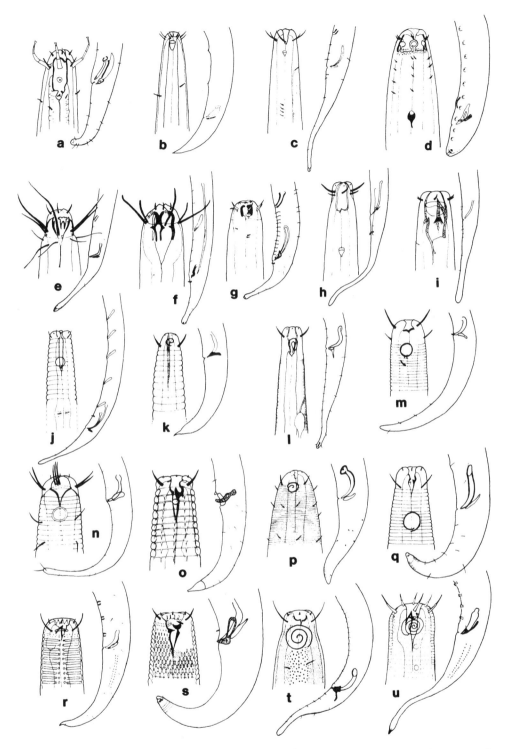

Figure 23.2.—Nematoda: The 21 most frequently recorded British free-living marine nematode genera. Note the different shapes of the buccal cavity corresponding to different feeding types, and the differences in the cuticular ornamentation and in the shape of cephalic sense organs: **a**, *Bathylaimus;* **b**, *Thalassoalaimus;* **c**, *Anticoma;* **d**, *Thoracostoma;* **e**, *Enoplolaimus;* **f**, *Enoploides;* **g**, *Enoplus;* **h**, *Anoplostoma;* **i**, *Viscosia;* **j**, *Leptolaimus;* **k**, *Camacolaimus;* **l**, *Axonolaimus;* **m**, *Metalinhomoeus;* **n**, *Daptonema;* **o**, *Monoposthia;* **p**, *Spirinia;* **q**, *Microlaimus;* **r**, *Neochromadora;* **s**, *Euchromadora;* **t**, *Sabatieria;* **u**, *Pomponema.* (After Platt and Warwick, 1983; courtesy of the Linnean Society of London.)

24. Gastrotricha

Edward E. Ruppert

Members of this phylum are small, strap or tenpin-shaped, acoelomate worms. Most adults are less than 1 mm in length, some species are less than 100 um, and others exceed 3 mm. The body is flattened ventrally and arched dorsally. It is divided into two regions, the head and trunk, which are often externally indistinct. The head bears the mouth, a tubular nematode-like pharynx, and the brain. The trunk contains a straight tubular gut and the reproductive organs, and bears the inconspicuous ventral anus. All gastrotrichs have ventral locomotory cilia and, with them, glide smoothly over the substratum. When disturbed they adhere with adhesive papillae, which are often numerous.

In the marine order Macrodasyida, the anterior adhesive papillae are ventral below the head whereas the posterior papillae are caudal behind the anus. Typically, there are one or more longitudinal rows of lateral papillae along the sides of the body, especially in the trunk region. Most members of the marine and freshwater order Chaetonotida have only the posterior adhesive papillae, which usually are reduced to a single pair. Each papilla is usually borne at the tip of a slender ramus. Together the two rami and their adhesive papillae constitute the furca, which resembles a two-tined fork. Unlike other chaetonotidans, species of *Neodasys* have lateral papillae like those of the macrodasyidans but, unlike them, lack anterior organs. Each adhesive papilla is enclosed in a cuticular sheath (as is the remainder of the body) which is usually cylindrical and stiff. Such adhesive papillae are termed adhesive tubes. A few freshwater chaetonotidans lack adhesive structures altogether.

All gastrotrichs have a cuticle which invests the body, adhesive papillae, and, microscopically, the cilia. In many species, the cuticle is thin, flexible, and featureless. In others, it is thicker and sculptured into small scales, spines, hooks, or combinations of these. Gastrotrichs with smooth flexible cuticles are themselves flexible and can elongate or shorten impressively. Those with sculptured cuticles often have stiffer bodies.

Gastrotrichs can be distinguished from other vermiform ciliated micrometazoans by a combination of features. Gastrotrichs glide smoothly over the substratum and do not typically writhe from side to side as is typical, for example, of flatworms. The gastrotrich mouth is anterior or, terminal, or slightly subterminal unlike that of most turbellarians and gnathostomulids in which it is ventral. The gastrotrich pharynx resembles that of nematodes, with which they are not easily confused, and is unlike those of turbellarians, annelids, and gnathostomulids. In the latter, the pharynx is often a compact muscular bulb, or, if tubular as in gastrotrichs, it is not often situated anteriorly in the longitudinal axis of the body. The occurrence of numerous, conspicuous, lateral, adhesive tubes is another hallmark of gastrotrichs. The minute body size and tenpin shape of many chaetonotidans readily distinguishes them from other animals, except for a few rotifers and archiannelids (e.g., *Diurodrilus*), both of which differ from gastrotrichs in internal organization. Species with sculptured cuticles and/or furca are readily identified as gastrotrichs.

The larger gastrotrichs, which include most macrodasyidans and a few chaetonotidans, are simultaneous hermaphrodites. Small species, which include most chaetonotidans, reproduce by parthenogenesis, although sometimes vestigial testes may be present. As far as is known, hermaphroditic gastrotrichs copulate and reciprocally cross-fertilize. Spermatophore formation has been documented in one genus (*Dactylopodola*).

Classification

The phylum Gastrotricha is divided into two orders, the marine Macrodasyida and the marine and freshwater Chaetonotida. The Macrodasyida includes six families and the Chaetonotida seven. Several of the families are difficult to define unambiguously, are probably unnatural taxa, and are in need of revision. As a result, the following family diagnoses are deliberately brief and hopefully emphasize useful characteristics. Because it is often easier to identify genera of gastrotrichs than to assign them to meaningful families, particular emphasis should be placed on the illustrations which show important generic characters. Additional taxonomic discussion is given by Hummon (1982).

Order Macrodasyida

The Macrodasyida are short to long worms. The body bears numerous adhesive tubes which are typically situated in bilaterally symmetrical anterior, lateral, and posterior groups. Posterior to the mouth, the pharyngeal lumen opens to the outside through a pair of pores (absent in *Lepidodasys* and the Chaetonotida). In transverse histological sections through the pharynx, the lumen is a triradiate inverted-Y. Macrodasyidans are hermaphrodites.

Family Dactylopodolidae.--Contains short-bodied (200-580 μm) animals each of which has a well-defined head and a stalked, biramous, posterior adhesive organ. Dactylopodolids are the only macrodasyidans having cross-striated body muscles and cilia borne on lumenal cells of the pharynx and intestine. The anterior testes and posterior ovaries are paired and lateral. The family contains three genera: *Xenodasys* (=*Chordodasys*) (3 spp., Figure 24.1a); *Dactylopodola* (5 spp., Figure 24.1b); and *Dendrodasys* (4 spp., Figure 24.1c).

Family Lepidodasyidae.--As presently constituted, this family is an unnatural taxon for which no satisfactory diagnosis can be written. It contains elongate worms (0.4-3.0 mm) which are distributed among seven genera: *Lepidodasys* (2 spp. Figure 24.2i); *Cephalodasys* (7 spp., Figure 24.1d); *Pleurodasys* (2 spp., Figure 24.1e); *Mesodasys* (4 spp., Figure 24.1f); *Megadasys* (=*Thiodasys*) (2 spp., Figure 24.1g); *Dolichodasys* (3 spp., Figure 24.2g); and *Paradasys* (5 spp., Figure 24.2h).

Family Macrodasyidae.--This family consists of soft-bodied, flexible, elongate worms which are 350-1600 μm long. The head grades imperceptibly into a parallel-sided trunk. The head often bears a pair of pestle organs. The trunk narrows terminally and forms a slender long or short tail. The body bears numerous anterior, lateral, and posterior adhesive tubes which are attached singly and never grouped into hand- or foot-like organs. The pharyngeal pores are often situated near the middle of the pharynx. Macrodasyids have well-developed copulatory organs and seminal receptacles. The muscular, tubular, copulatory organ bears an eversible tube (*Macrodasys*) or a rigid screw-like stylet (*Urodasys*) both of which are used in sperm transfer. The seminal receptacle opens ventrally (*Macrodasys*) or dorsally (some *Urodasys*) to the outside and anteriorly and internally to the ovum. The paired or unpaired sperm ducts do not internally join the copulatory organ. Sperm enter the copulatory organ when the body is flexed and the organ's pore is brought into contact with the gonopores. The Macrodasyidae includes two genera: *Macrodasys* (15 spp., Figure 24.1j) and *Urodasys* (8 spp., 24.1k).

Family Planodasyidae.--These are soft-bodied flexible worms which are similar to species of Macrodasyidae. In contrast to this latter family, planodasyid species have a biramous posterior end, not a uniramous tail, and each ramus bears adhesive tubes which are arranged fan-like around its margin. Planodasyids have numerous conspicuous and often large epidermal gland cells. The family includes two genera: *Planodasys* (1 sp., 1.5 mm, Figure 24.1h) and *Crasiella* (3 spp., 300-600 μm, Figure 24.1i).

Family Turbanellidae.--This family consists of soft-bodied, flexible, elongate animals (300 μm to 1.2 mm) each of which bears a pair of extensible, hand-like, anterior adhesive organs and a biramous posterior adhesive organ. Often, a swollen gland cell is situated in the angle between the two rami. Lateral and dorsolateral longitudinal rows of adhesive tubes may be numerous, few, or absent. Usually, a single pair of ventrolateral tubes, which is directed obliquely rearward not laterally, occurs in the midpharyngeal to anterior intestinal region. A pair of head sensory appendages, which may be simple pestle organs or conical or styliform derivatives of them, are common. The organization of the male reproductive system, except for *Paraturbanella*, is an important family characteristic. The paired anterior testes extend posteriorly as paired sperm ducts to about the midintestinal body region. Here each duct curves medially, then anteriorly, and extends to a common midventral gonopore in the anterior intestinal region. In *Paraturbanella*, the sperm ducts do not curve anteriorly but instead terminate at their posterior extremities in separate ventrolateral gonopores. The accessory reproductive organs of turbanellids (*Turbanella, Paraturbanella*) consists of three parts, all of which are situated middorsally above the gut and between the anus and ovum. The posterior blind, sac-like, copulatory organ opens ventrally near or in common with the anus whereas the anterior, thin-walled, sac-like, seminal receptacle communicates dorsally with the outside. Between the two organs, there is a compact cord of large featureless cells. The sperm are filiform. The Turbanellidae includes five genera: *Turbanella* (21 spp., Figure 24.2a,b); *Dinodasys* (1 sp., Figure 24.2d); *Pseudoturbanella* (1 sp., Figure 24.2c); *Paraturbanella* (11 spp., Figure 24.2e); and *Desmodasys* (1 sp., Figure 24.2f).

Family Thaumastodermatidae.--These constitute the most diversified natural taxon of marine Gastrotricha. Representatives of this family occur in nearly all sediments that contain gastrotrichs. In coarse offshore sediments, they achieve high species diversity, particularly in the genera *Acanthodasys* and *Tetranchyroderma*. Many of these species, unfortunately, are not yet described. The genus *Tetranchyroderma* is often diversified intertidally in medium to high energy beaches. All of the thaumastodermatids have a cuticle which is sculptured into plates, spines, or combinations of both. A few (*Platydasys*, *Ptychostomella*), however, have lost the cuticular sculpture. The only other macrodasyidans with sculptured cuticles are species of *Xenodasys* and *Lepidodasys*, which are members of the families Dactylopodolidae and Lepidodasyidae, respectively. Although some thaumastodermatids may be elongate and vermiform (up to 800 µm), for example some species of *Acanthodasys* and *Tetranchyroderma*, most are short-bodied and more or less dorsoventrally flattened. The mouth is often large and almost as wide as the head. Anterior, lateral, and posterior adhesive tubes are present. The anterior tubes arise singly and form a transverse row ventrally behind the lower lip. The lateral tubes are actually ventrolateral and often numerous and well developed, especially posteriorly. The posterior adhesive organ is usually biramous. Minimally, each ramus bears distally two tubes in the frontal plane and one (sensory?) tube which extends dorsally upward.

The Thaumastodermatidae includes the only gastrotrichs, in addition to unrelated species of *Mesodasys*, in which the sperm ducts *internally* join the copulatory organ. The family consists of eight genera: *Hemidasys* (not figured, see Remane, 1936); *Acanthodasys* (1 sp., Figure 24.2j-l); *Diplodasys* (4 spp., Figure 24.2m); *Platydasys* (10 spp., Figure 24.2n); *Tetranchyroderma* (25 spp., Figure 24.3a,b); *Pseudostomella* (5 spp., Figure 24.3c); *Ptychostomella* (4 spp., Figure 24.3d); and *Thaumastoderma* (6 spp., Figure 24.3e).

Order Chaetonotida

Except for *Neodasys* the Chaetonotida are readily distinguished from the Macrodasyida on the basis of their small size, tenpin body shape, caudal furca, each half of which bears a single adhesive tube, sculptured cuticle, and lack of anterior and lateral adhesive tubes. The mouth and buccal cavity are usually small but typically bear complex cuticular parts, the most common of which is a spiny basket in front of and surrounding the mouth. The triradiate pharynx lumen is Y-shaped in transverse section unlike the inverted Y-shape of macrodasyidans. It lacks pharyngeal pores. Most chaetonotidans, except for *Neodasys* and xenotrichulids, reproduce by parthenogenesis. Nevertheless, these often express vestigial testes and accessory reproductive organs ("X-organs," "X-bodies").

Family Neodasyidae.--The chaetonotidan suborder Multitubulatina comprises a single family, the Neodasyidae, and one genus, *Neodasys* (2 spp., Figure 24.3f). Species are elongate (400-800 µm), soft-bodied, flexible worms. The cuticle lacks sculpture and is smooth. The mouth is anterior and terminal and bears a well-developed, entire, cuticular funnel. The body lacks anterior adhesive organs (vestiges may be present) while the lateral structures are numerous but indistinct and papilliform not tubular. The posterior adhesive organ is biramous.

Family Xenotrichulidae.--The remaining six families are members of the suborder Paucitubulatina; all lack anterior and lateral adhesive structures and the caudal furca bears zero, usually two, or four adhesive tubes. The xenotrichulids are all tenpin-shaped, and range in length from 80-280 µm. Their ventral cilia are arranged in distinct tufts, called cirri, and each cirrus functions as a single locomotory unit. The furca bears two adhesive tubes. Most species occupy sediments which are disturbed by swift currents or breaking waves. Locomotion is often a rapid sequence of starts and stops reminiscent of that of hypotrichous ciliates or, on a larger body scale, otoplanid flatworms. Most species have a well-developed male reproductive system. There are three genera: *Draculiciteria* (1 sp., Figure 24.3j); *Xenotrichula* (13 spp., Figure 24.3k); and *Heteroxenotrichula* (7 spp., Figure 24.3l).

Family Chaetonotidae.--Members of this family lack cirri and instead have two unbroken, ventral, longitudinal rows of normal cilia. In a few genera, however, each continuous ciliary row breaks up into a series of isolated tufts but these always contain normal cilia not cirri. The body, which is 50-625 µm in length, is tenpin-shaped with each ramus of the caudal furca bearing a terminal adhesive tube. Most species reproduce by parthenogenesis. Although all chaetonotids share a similar overall body design, the cuticular sculpture differs conspicuously between species and genera. As a result, the taxonomy of this family depends almost entirely on an accurate analysis of the cuticle and its derivatives. The genera are: *Musellifer* (2 spp., Figure 24.3h); *Aspidiophorus* (12 spp., Figure 24.3m); *Chaetonotus* (121 spp., Figure 24.3n); *Halichaetonotus* (15 spp., Figure 24.3o); *Heterolepidoderma* (17 spp., Figure 24.3g,p); *Lepidodermella* (12 spp., Figure 24.3q); *Ichthydium* (26 spp., Figure 24.3r); and *Polymerurus* (16 spp.; Figure 24.3i,s).

Family Dichaeturidae.--This is a freshwater family containing small (100-200 µm) tenpin-shaped animals. The posterior end is biramous and each ramus bears two distal adhesive tubes. The continuous, ventral, ciliary stripes resemble those of the Chaetonotidae. The cuticle is smooth and lacks sculpture. Species reproduce by parthenogenesis. There are two genera: *Dichaetura* (2 spp., Figure 24.3w) and *Marinellina* (1 sp., Figure 24.3v).

Family Proichthydidae.--This freshwater family contains minute (70-150 µm) tenpin-shaped animals with a smooth, sculptureless cuticle. Each ramus of the caudal furca bears a distal adhesive tube. A band of cilia crosses the head dorsally. The locomotory cilia are normal. Reproduction is by parthenogenesis. The family contains two monotypic genera: *Proichthydium* (from Argentina, Figure 24.3u) and *Proichthydioides* (from Japan, Figure 24.3t).

Family Neogosseidae.--This freshwater family comprises tenpin-shaped species ranging from 100 to 270 µm in length. The head bears a pair of soft, fleshy, lateral palps, or tentacles, and well-developed swimming cilia. The posterior end is biramous but lacks adhesive tubes. Instead, it is either naked or bears long spines. The cuticle is sculptured into small spined scales. Ventrally, normal cilia are in tufts which are distributed in two longitudinal rows. Species reproduce by parthenogenesis. The family contains two genera: *Neogossea* (5 spp., Figure 24.3y) and *Kijanebalola* (1 sp., Figure 24.3x).

Family Dasydytidae.--This is a pelagic freshwater family that includes taxa without head palps and adhesive tubes. The cuticle bears long isolated spines or tufts of spines both of which can be moved actively. Head cilia are well-developed whereas the two ventral ciliated stripes are broken into several isolated tufts. Dasydytids range from 80-285 µm in length and reproduce by parthenogenesis. There are two genera: *Dasydytes* (15 spp., Figure 24.3z) and *Stylochaeta* (6 spp., Figure 24.3aa).

Habitats and Ecological Notes

Marine Gastrotricha are interstitial animals occupying the voids between sand grains. Because of the importance of interstitial free space, their occurrence is correlated with the distribution of intertidal and subtidal porous sands, although a few species occur in nonporous muds (*Musellifer*). In general, marine gastrotrichs are restricted to oxic sediments but some notable exceptions occur (*Dolichodasys, Megadasys*). Gastrotricha achieve high diversity in coarse, shelly, subtidal sands, and some families are particularly well represented in such areas (Dactylopodolidae, Macrodasyidae, and especially Thaumastodermatidae). Others may be well represented intertidally on exposed beaches (Chaetonotidae, Thaumastodermatidae), even in wave-pounded zones (Xenotrichulidae). In protected intertidal and shallow, subtidal sands, *Turbanella*, *Neodasys*, and *Macrodasys* often reach high densities. *Dolichodasys* and *Megadasys* can be abundant in fine, sulfide-blackened, subsurface sands.

Freshwater gastrotrichs are less common in sediments than their marine counterparts, perhaps because freshwater sediments are often poorly sorted and their interstices occluded. Instead, freshwater gastrotrichs are common in decaying benthic vegetation, organic muck, mats of floating plants (*Lemna*), mosses, submerged vegetation at the margins of lakes and ponds, and in tree holes and other small containers. Many freshwater gastrotrichs, unlike the majority of the marine species, are tolerant of micro-oxic and anoxic conditions.

Methods of Collection

Qualitative collections of sand containing marine Gastrotricha are made with a shovel and bucket, core tubes, plastic cups, or any convenient non-toxic container. Any of the readily available corers or dredges is useful in subtidal areas. Quantitative samples are generally taken in calibrated core tubes. Large and small qualitative samples can be stored for weeks and sometimes for months by keeping the covered buckets of sediment and cores in a cool place (12-15° C).

Freshwater gastrotrichs are collected using a hand-held jar, shovel and bucket, core tube, or dredge. Like their marine relatives, many species survive well in the laboratory, particularly if maintained in a wheat grain infusion or filtered suspension of yeast (Hummon, 1984).

Methods of Extraction

Living marine Gastrotricha are removed from sediments using either a relaxation-decantation technique (Rieger and Ruppert, 1978) or a seawater-ice technique (Ruppert, 1972). Both techniques are semi-quantitative. Preserved sediment samples will yield very little recognizable or taxonomically useful material unless a relaxant has been properly introduced before fixation. The fresh-water shock method (Kristensen and Higgins, personal communication), if done very carefully and rapidly, has produced good results but should be tested further.

Macrodasyida are removed effectively using the relaxation-decantation method. The following should be available: (1) a solution (7%) of $MgCl_2$ isotonic to seawater; (2) 500 cm^3 Erlenmeyer flasks; (3) a plastic ring (e.g., Plexiglas, 6-10 cm diameter, 2-3 cm height) with 63 µm mesh nylon netting glued tautly at one end to produce a sieve (the netting can also be made by placing the ring in a large Petri dish filled about 1 mm with chloroform for a few minutes, removing the ring, placing it upside-down on a flat surface, then placing the netting - stretched taut by a embroidery hoop - over the softened edge of the ring; after drying, the extra mesh can be trimmed with a scalpel); (4) some dishes (e.g., Petri dishes, Carolina culture dishes) to accommodate the sieve. Add 50-100 cm^3 of sediment to the flask and fill to the brim with $MgCl_2$ solution. Cover the opening with the palm of one hand and invert once gently to mix. Restore to the upright position, and let stand for 10 minutes. Invert the flask vigorously two or three times in succession to suspend the relaxed animals. Allow the sediment to settle and decant the supernatant through the sieve. After a brief rinse with fresh seawater, submerge the sieve in a dish of seawater to a few millimeters above the net, but not above the ring. Leave the dish overnight (a few hours minimum) in a cool place and then remove the sieve to a second dish of seawater. Inspect the first dish under transmitted light with a dissecting microscope. If gastrotrichs were present in the sample, they will have crept through the sieve. Then examine the second sieve in the second dish for any larger organisms that have been retained on its surface.

Marine Chaetonotida are removed effectively using seawater-ice. It is necessary to have available the following or their equivalents: (1) seawater-ice; (2) a short (e.g., 6 x 3 cm) plastic tube (e.g., Plexiglas); (3) 63 µm nylon mesh netting (see above); (4) a ringstand and clamp to hold the plastic tube; and (5) a culture dish to collect the extracted animals. Attach the netting securely to one end of the tube with elastic bands. Fill the tube 1/2-3/4 full with sediment. Clamp the tube vertically, sieve-end down, on the ringstand and place an empty dish a few cm below. Add one or more ice cubes to the upper end of the tube and allow them to melt at room temperature. Replace the collecting dishes as they fill with meltwater and inspect under transmitted light with a dissecting microscope. Continue adding ice until animals cease to appear in the collecting dish.

Freshwater Chaetonotida are usually separated from the substratum by inspecting small subsamples under a dissecting microscope. Gastrotrichs are then removed singly using a finely drawn Pasteur pipette. Gastrotrichs can be removed effectively from mosses by squeezing water from them into a dish (Kisielewski, 1981).

Preparation of Specimens for Taxonomic Study

Observations of wholemounts of living animals is a particularly useful first step in studying gastrotrichs but requires specialized, expensive, contrasting methods, such as differential interference contrast or phase contrast microscopy. In any case, it is important to have available a high-quality compound microscope. Wholemounts of preserved animals are also useful and modern critical techniques for preparing these have appeared recently (Smith and Tyler, 1984; Rieger and Ruppert, 1978; Cavey and Cloney, 1973). In most species of Chaetonotida, cuticular structures are the source of diagnostic characters. To study these, well prepared wholemounts are important for quality taxonomic work. On the other hand, most species of Macrodasyida and some of the Chaetonotida can only be well-studied by using serial sections to supplement the wholemounts. In most of these species, as in turbellarians, it is essential to study the reproductive anatomy in serial sections to properly identify species.

Most species of isolated marine and freshwater Gastrotricha react quickly to disturbances by adhering tenaciously with their adhesive organs to the culture dish. Handling them alive for observation and experimentation is complicated by this fact. Rapid transfers of individual specimens from the dish to microscope slides or experimental containers is executed best by pipetting them confidently and with dispatch. Micropipettes for this purpose can be made from a Pasteur pipette which has been drawn over a laboratory burner.

Fixation of marine gastrotrichs for light or electron microscopy should be preceded by complete relaxation in a solution of $MgCl_2$ prepared isotonic to seawater. Bouin's fluid is a good general fixative for light microscopy. A general fixative for electron microscopy (TEM) and critical light microscopy (LM) is: (1) primary fixation for 0.5 h at room temperature in 2.5% glutaraldehyde in 0.2 M Millonig's phosphate buffer at pH 7.5; (2) after a brief buffer rinse, postfix for 0.5 h at room temperature in 2.0% OsO_4 in the same buffer. Embed after complete dehydration in a resin such as Polybed 812.

Freshwater gastrotrichs can be relaxed in 1% $MgCl_2$ in distilled water, in water to which a drop of chloroform has been added, or in a drop of water which is inverted over phenol crystals. Critical fixation of freshwater gastrotrichs for TEM and LM requires experimentation. The following protocol may provide a good starting point: (1) primary fixation in 0.35% glutaraldehyde in 0.025 M cacodylate buffer for 2 h then (2) postfix; after a short buffer rinses, in 1% OsO_4 in 0.025 M cacodylate buffer for 1 h. Dehydrate and embed following the protocol described above. In general, the quality of fixation of gastrotrichs will very considerably among species.

Type Collections

Important collections of Gastrotricha are located at the National Museum of Natural History, Smithsonian Institution, Washington, D.C., USA; The American Museum of Natural History, New York, USA; Universitets Zoologisk Museum, Copenhagen, Denmark; Zoologisk Museum, Universitet I Bergen, Bergen, Norway; and the Musée National d'Histoire Naturelle, Paris, France. Other collections may be found at the Museo Civico di Storia Naturale, Verona, Italy, and the Zoological Survey of India, Calcutta, India.

Notes on Figures

Figures 24.1-3 are approximate scale drawings (scale = 200 µm) of species representing the genera of Gastrotricha. In order to include as much information as possible in each illustration, some liberties were taken with the conventional rules of illustration. Anterior batteries of adhesive tubes in Macrodasyida, although always on the ventral surface, are drawn as if on the dorsal side. The ventral, compound cilia ("cirri") in the Xenotrichulidae are drawn as if dorsally situated on one half of the body. Openings associated with accessory reproductive organs are dorsal or lateral if drawn as unbroken lines, ventral if indicated by dashed or dotted lines. Genera with sculptured cuticles are drawn as cutaways to simultaneously illustrate the cuticular pattern and internal anatomy. Detail drawings of cuticular spines and scales, indicated by arrows, illustrate their basic design (Chaetonotida) or give an impression of their diversity (Thaumastodermatidae). Variability in reproductive anatomy, particularly among species in genera with smooth cuticles, should be expected. An asterisk (*) preceding the figure designation indicates that information pertaining to that figure has been taken from the published literature on Gastrotricha.

Acknowledgments

My appreciation is extended to Professor William D. Hummon and Dr. Margaret R. Hummon, Department of Zoology and Biomedical Sciences, Ohio University, Athens, Ohio, USA, for their reviews of the manuscript.

References

Cavey, M.J., and R.A. Cloney
1973. Osmium-fixed and Epon-embedded Wholemounts of Delicate Specimens. *Transactions of the American Microscopical Society*, 92:148-151.

Colacino, J.M., and D.W. Kraus
1984. Hemoglobin-containing cells of *Neodasys* (Gastrotricha, Chaetonotida) - II. Respiratory Significance. *Comparative Biochemistry and Physiology*, 79A:363-369.

d'Hondt, J.-L.
1971. Gastrotricha. *Oceanography and Marine Biology Reviews*, 9:141-192.

Hagerman, G.M., and R.M. Rieger
1981. Dispersal of Benthic Meiofauna by Wave and Current Action in Bogue Sound, North Carolina USA. *Marine Ecology*, 2:245-270.

Hummon, M.R.
1984a. Reproduction and Sexual Development in a Freshwater Gastrotrich. 1. Oogenesis of Parthenogenic Eggs (Gastrotricha). *Zoomorphology*, 104:33-41.
1984b. Reproduction and Sexual Development in a Freshwater Gastrotrich. 2. Kinetics and Fine Structure of Postparthenogenic Sperm Formation. *Cell and Tissue Research*, 236:619-628.
1984c. Reproduction and Sexual Development in a Freshwater Gastrotrich. 3. Postparthenogenic Development of Primary Oocytes and the X-body. *Cell and Tissue Research*, 236:629-636.

Hummon, M.R., and W.D. Hummon
1983. Gastrotricha. Pages 195-205 in K.G. and R.G. Adiyodi, editors, *Reproductive Biology of Invertebrates*. Volume II. London: John Wiley and Sons, Ltd.

Hummon, W.D.
1971. Gastrotricha. Pages 485-506 in A.C. Giese, and J.S. Pearse, editors, *Reproduction of Marine Invertebrates*. Volume 1. New York: Academic Press, Inc.
1972. Dispersion of Gastrotricha in a Marine Beach of the San Juan Archipelago, Washington. *Marine Biology*, 16:349-355.
1974a. S$_{H'}$: A Similarity Index Based on Shared Species Diversity, Used to Assess Temporal and Spatial Relationships Among Intertidal Marine Gastrotricha. *Oecologia* (Berlin), 17:203-220.
1974b. Respiratory and Osmoregulatory Physiology of a Meiobenthic Marine Gastrotrich, *Turbanella ocellata* Hummon, 1974. *Cahiers de Biologie Marine*, 15:255-268.
1975. Seasonal Changes in Secondary Production, Faunal Similarity and Biological Accommodation, Related to Stability among the Gastrotricha of Two Semi-enclosed Scottish Beaches. In G. Persoone and E. Jaspers, editors, *Proceedings of the 10th European Symposium on Marine Biology*, 1:309-336.
1982. Gastrotricha. Pages 857-863 in S.P. Parker, editor, *Synopsis and Classification of Living Organisms*. Volume 1. New York: McGraw-Hill Book Company.

Hummon, W.D., and M.R. Hummon
1983. Gastrotricha. Pages 211-221 in K.G. and R.G. Adiyodi, editors, *Reproductive Biology of Invertebrates*. Volume I. London: John Wiley and Sons, Ltd.

Hyman, L.H.
1951. Gastrotricha. Pages 194-205 in *The Invertebrates*. Volume 3. New York: McGraw-Hill Book Company.

Kisielewski, J.
1981. Gastrotricha from Raised and Transitional Peat Bogs in Poland. *Monographie Fauny Polski*, 11:1-143.

Kraus, D.W., and J.M. Colacino
1984. The Oxygen Consumption Rates of Three Gastrotrichs. *Comparative Biochemistry and Physiology*, 79A:691-693.

Nixon, D.E.
1976. Dynamics and Spatial Pattern for the Gastrotrich *Tetranchyroderma bunti* in the Surface Sand of High Energy Beaches. *Internationale Revue der gesamten Hydrobiologie*, 61:211-248.

Powell, E.N., M.A. Crenshaw, and R.M. Rieger
1980. Adaptation to Sulphide in Sulphide-system Meiofauna. End Products of Sulphide Detoxification in 3 Turbellarians and a Gastrotrich. *Marine Ecology*, 2:169-177.

Remane, A.
1927. Beiträge zur Systematik der Süsswassergastrotrichen. *Zoologische Jahrbücher*, 53:268-320.
1936. Gastrotricha. In H.G. Bronn, editor, *Klassen und Ordnungen des Tierreichs*, 4:1-242.

Rieger, R.M.
1976. Monociliated Cells in Gastrotricha. Significance for Concepts of Early Metazoan Evolution. *Zeitschrift für zoologische Systematik und Evolutionsforschung*, 14:198-226.

Rieger, R.M., and M. Mainitz
1977. Comparative Fine Structure of the Body Wall in Gnathostomulida and Their Phylogenetic Position

Between Platyhelminthes and Aschelminthes. *Zeitschrift für zoologische Systematik und Evolutionsforschung*, 15:9-35.

Rieger, G.E., and R.M. Rieger
1977. Comparative Fine Structure Study of the Gastrotrich Cuticle and Aspects of the Cuticle Evolution within the Aschelminthes. *Zeitschrift für zoologische Systematik und Evolutionsforschung*, 15:81-124.
1980. Fine Structure and Formation of Eggshells in Marine Gastrotricha. *Zoomorphology*, 96:215-229.

Rieger, R.M., and E. E. Ruppert
1978. Resin Embedments of Quantitative Meioifauna Samples for Ecological and Structural Studies-Description and Application. *Marine Biology*, 46:223-235.

Rieger, R.M., E.E. Ruppert, and C. Schoepfer-Sterrer.
1974. On the Fine Structure of Gastrotrichs with Description of *Chordodasys antennatus* sp. n. *Zoologica Scripta*, 3:219-237.

Ruppert, E.E.
1972. An Efficient, Quantitative Method for Sampling the Meiobenthos. *Limnology and Oceanography*, 17:629-631.
1976. Zoogeography and Speciation in Marine Gastrotricha. *Mikrofauna des Meeresbodens*, 61:231-251.
1978a. The Reproductive System of Gastrotrichs. II. Insemination in *Macrodasys* a Unique Mode of Sperm Transfer in Metazoa. *Zoomorphologie*, 89:207-228.
1978b. The Reproductive System of Gastrotrichs III. Genital Organs of the Thaumastodermatinae subfam. n. and Diplodasyinae subfam. n. with Discussion of Reproduction in Macrodasyida. *Zoologica Scripta*, 7:93-114.
1982. Comparative Ultrastructure of the Gastrotrich Pharynx and the Evolution of Myoepithelial Foreguts in Aschelminthes. *Zoomorphology*, 99:181-220.

Ruppert, E.E., and K. Shaw
1977. The Reproductive System of Gastrotrichs. I. Introduction with Morphological Data for Two New *Dolichodasys* species. *Zoologica Scripta*, 6:185-195.

Ruppert, E.E., and P.B. Travis
1983. Hemoglobin-containing Cells of *Neodasys* (Gastrotricha, Chaetonotida). I. Morphology and Ultrastructure. *Journal of Morphology*, 175:57-64.

Schmidt, P., and G. Teuchert
1969. Quantitative Untersuchungen zur Ökologie der Gastrotrichen im Gezeiten-Sandstrand der Insel Sylt. *Marine Biology*, 4:4-23.

Smith, J.P.S., III, and S. Tyler
1984. Serial-sectioning of Resin-embedded Material for Light Microscopy: Recommended Techniques for Micro-metazoans. *Mikroskopie*, 41:259-270.

Teuchert, G.
1968. Zur Fortpflanzung und Entwicklung der Macrodasyoidea (Gastrotricha). *Zeitschrift für Morphologie und Ökologie der Tiere*, 63:343-418.
1972. Die Feinstruktur des Protonephridialsystems von *Turbanella cornuta* Remane, einem marinen Gastrotrich der Ordnung Macrodasyoidea. *Zeitschrift für Zellforschung*, 136:277-289.
1974. Aufbau und Feinstruktur der Muskelsysteme von *Turbanella cornuta* Remane (Gastrotricha, Macrodasyoidea). *Mikrofauna des Meeresbodens*, 39:223-246.
1975. Differenzierung von Spermien bei dem marinen Gastrotrich *Turbanella cornuta* Remane (Ordnung Macrodasyoidea. *Verhandlungen der anatomischen Gesellschaft*, 69:743-748.
1976a. Elektronenmikroskopische Untersuchung über die Spermatogenese und Spermatohistogenese von *Turbanella cornuta* Remane (Gastrotricha). *Journal of Ultrastructure Research*, 56:1-14.
1976b. Sinneseinrichtungen bei *Turbanella cornuta* Remane (Gastrotricha). *Zoomorphologie*, 83:193-207.
1977a. The Ultrastructure of the Marine Gastrotrich *Turbanella cornuta* Remane (Macrodasyoidea) and its Functional and Phylogenetical Importance. *Zoomorphologie*, 88:189-246.
1977b. Leibeshöhlenverhältnisse von dem marinen Gastrotrich *Turbanella cornuta* Remane (Ordnung Macrodasyoidea) und eine phylogenetische Bewertung. *Zoologische Jahrbücher, Abteilung für Anatomie*, 97:586-596.

Teuchert, G., and A. Lappe
1980. Zum sogenannten "Pseudocoel" der Nemathelminthes. - Ein Vergleich der Leibeshöhlen von mehreren Gastrotrichen. *Zoologische Jahrbücher, Abteilung für Anatomie*, 103:424-438.

Travis, P.B.
1983. Ultrastructural Study of Body Wall Organization and Y-cell Composition in the Gastrotricha. *Zeitschrift für zoologische Systematik und Evolutionsforschung*, 21:52-68.

Tyler, S., L. Melanson, and R.M. Rieger
1980. Adhesive Organs of the Gastrotricha. II. The Organs of *Neodasys*. *Zoomorphologie*, 95:17-26.

Tyler, S., and G.E. Rieger
1980. Adhesive Organs of the Gastrotricha. I. Duo-gland Organs. *Zoomorphologie*, 95:1-15.

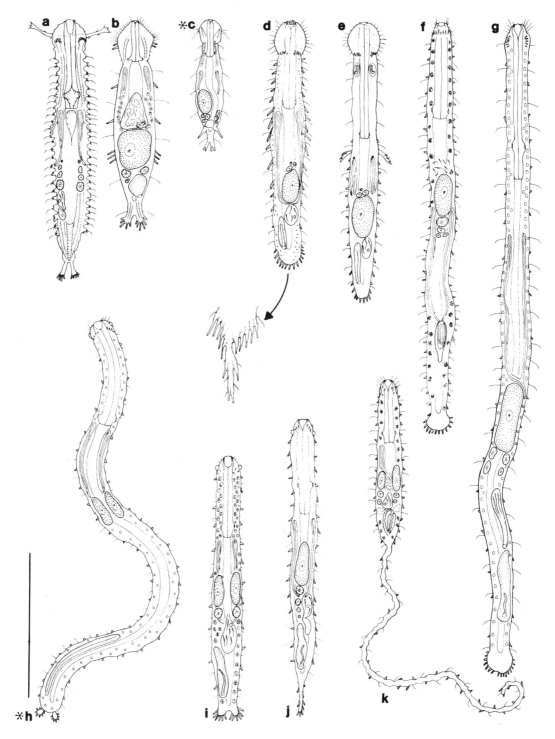

Figure 24.1.--Gastrotricha: a, *Xenodasys;* b, *Dactylopodola;* c, *Dendrodasys;* d, *Cephalodasys;* e, *Pleurodasys;* f, *Mesodasys;* g, *Megadasys;* h, *Planodasys;* i, *Crasiella;* j, *Macrodasys;* k, *Urodasys.* (See special note in text for further explanation of illustrations; scale = 200 μm.)

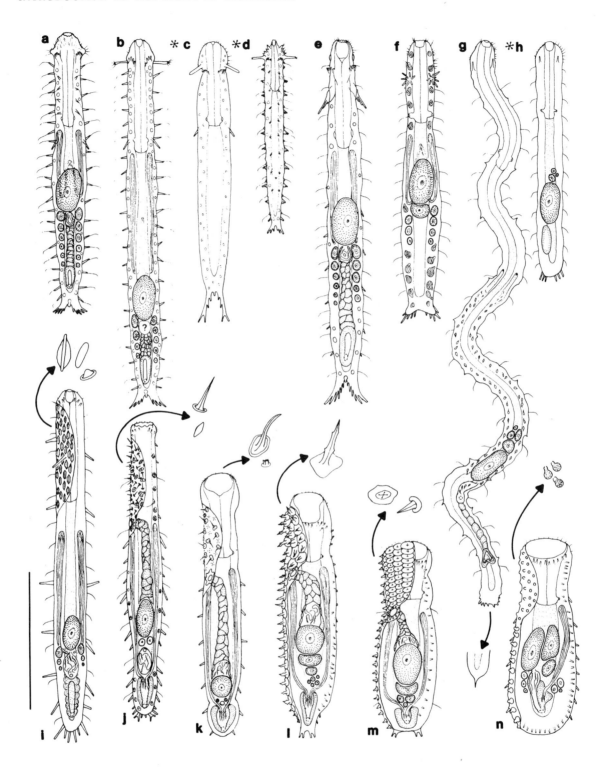

Figure 24.2.--Gastrotricha (continued): **a**, *Turbanella;* **b**, *Turbanella* **c**, *Pseudoturbanella;* **d**, *Dinodasys;* **e**, *Paraturbanella;* **f**, *Desmodasys;* **g**, *Dolichodasys;* **h**, *Paradasys;* **i**, *Lepidodasys;* *Acanthodasys;* **m**, *Diplodasys;* **n**, *Platydasys.* (See special note in text for further explanation of illustrations; scale = 200 μm.)

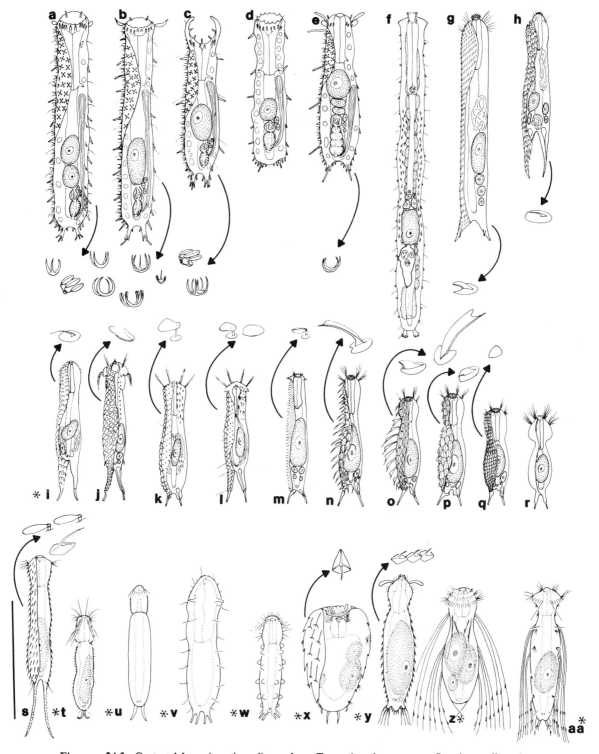

Figure 24.3.--Gastrotricha (continued): **a,b,** *Tetranchyroderma;* **c,** *Pseudostomella;* **d,** *Ptychostomella;* **e,** *Thaumastoderma;* **f,** *Neodasys;* **g,** undescribed genus? (*Heterolepidoderma?*); **h,** *Musellifer;* **i,** *Polymerurus ? delamarei* (*Musellifer?*); **j,** *Draculiciteria;* **k,** *Xenotrichula;* **l,** *Heteroxenotrichula;* **m,** *Aspidiophorus;* **n,** *Chaetonotus;* **o,** *Halichaetonotus;* **p,** *Heterolepidoderma;* **q,** *Lepidodermella;* **r,** *Ichthydium;* **s,** *Polymerurus;* **t,** *Proichthydioides;* **u,** *Proichthydium;* **v,** *Marinellina;* **w,** *Dichaetura;* **x,** *Kijanebalola;* **y,** *Neogossea;* **z,** *Dasydytes;* **aa,** *Stylochaeta.* (See special note in text for further explanation of illustrations; scale = 200 μm.)

25. Rotifera

Paul N. Turner

As the name Rotifera suggests, one of their most diagnostic features is the corona or "wheel"-like complex of cilia in the head region which functions in locomotion and feeding. They are commonly vermiform metazoans, sometimes with elaborate cuticular spines or ornamentation, but rarely exceeding 2 mm in length and more commonly less than 0.5 mm long. They are usually bilaterally symmetrical, have a body wall devoid of muscle, but responsible for the secretion of a cuticle which may be thin and flexible, rigidly thickened to form a lorica in the trunk region of the body, superficially annulate with the resulting regions somewhat telescopic, or in the form of a secreted tube. Caudally, there usually is a foot with two adhesive toes although toes may be absent or be present in numbers up to four.

Most rotifers are translucent and the internal anatomy may give clues that may assist in their recognition as a member of this taxon. The mouth may be anterior to ventral in the coronal region and leads to a distinctive pharynx with a bulbous complex of muscles and sclerotized jaws (trophi) collectively called the mastax. The remaining digestive tract consists of a short esophagus, usually with lobate glands, a straight stomach and intestine. The anus, or cloacal aperture, is usually dorsally displaced and opens near the base of the foot. Although sexes are separate, males are uncommon or unknown. The female rotifer usually has one or two ovaries, usually with only one egg developing at a time. The reproductive life history is usually complex. The nervous system consists of a large ganglionic mass dorsal to the mastax, various cephalic sensory structures, the most conspicuous of which is the middorsal "antenna," a complex of smaller ganglia in various regions of the body, all connected by nerve fibers. A flame-bulb protonephridial osmoregulatory system is usually present. For a more comprehensive contemporary review of rotifer biology one should consult Pennak (1978) or Nogrady (1982).

Classification

An estimated 2,000 species of rotifers have been described and divided into three classes, five orders up to 22 families and 123 genera and subgenera. Ten of the families have meiobenthic representatives, most are between 65 µm and 510 µm in length. Although the majority are known from freshwater, rotifers are also found in brackish water and in the sea. Clearly, rotifers have been very successful in invading all kinds of freshwater habitats from lakes to temporary ponds, damp soil to damp beach sand, from water collected in bromeliads to submerged aquatic plants, and in even mosses and liverworts. Although they are often planktonic, they are equally well adapted as benthic organisms. In the same ways but in far less abundance, rotifers are found in marine habitats; mostly in association with submerged vegetation or in the interstices of both intertidal and subtidal medium to coarse sands.

The following summary of rotifer classification is taken primarily from Koste (1978) and de Ridder (1986).

The Rotifera are composed of three classes. The class Pararotatoria is represented by a single genus, *Seison,* an exclusively marine taxon epizoic on the gills of the crustacean *Nebalia*. The remaining two classes include the Digononta and the Eurotatoria; the latter class contains most of the known species of the phylum.

The Digononta (two ovaries) contains the single order Bdelloidea, named after their leech-like creeping motion and vermiform appearance. Typically, bdelloid rotifers have retractable anterior coronal discs; when retracted a rostrum folds into an anterior position and is used in locomotion along with the adhesive toes. Males are unknown; ovaries have vitellaria, and reproduction is entirely parthenogenetic. They may be oviparous or viviparous. Almost all members of this class are found in freshwater or damp terrestrial habitats such as mosses and soil; some are found in marine sediments. This class contains 4 families: Habrotrichidae, Adinetidae, Philodinavidae (mostly confined to soil lichens and mosses), and Philodinidae (members of the latter family are commonly associated with benthic habitats).

Family Philodinidae.--Common representatives of

this family are found in the genus *Rotaria* (with three toes and eyespots on the rostrum) including *R. rotatoria* (Pallas) (Figure 25.1l) and *R. tardigrada* (Ehrenberg). *Philodina* has eyespots in the neck region and four toes; *Dissotrocha* has spiny dorsal projections. Live members of this family may be recognized on the basis of two distinct, somewhat elevated trochal disks.

The largest class of the Rotifera is the Eurotatoria (=Monogononta Plate, 1889). This class is divided into two subclasses: Pseudotrocha, with a single order Ploima, and Gnesiotrocha, with the orders Monimotrocha and Paedotrocha. The order Ploima is represented in the benthos by 10 of its 17 families; four of these families have only a single genus and three have only a single species each. The class Eurotatoria contains over 1,000 described species and most of its benthic species are found in seven of these families. The only benthic representatives of the subclass Gnesiotrocha are members of a single family of the order Paedotrocha.

Family Brachionidae.--This family has a few endemic psammic representatives in Lake Baikal; these include *Notholca kozhovi* Vasil'eva and Kutikova (see Arov, 1985:191) as well as *N. baikalensis* Jaschonov. *Notholca psammarina* Buchholz and Rümann is also a psammophile (Koste, 1978), but virtually all others are planktonic.

Family Collothecidae.--*Collotheca wiszniewskii* Varga (Figure 25.1k) is the only known psammobiotic sessile rotifer. *Collotheca campanulata* (Dobie), *C. ornata* (Ehrenberg) and *C. voigti* Berzins are all known from, but not restricted to, "the biogenous bottom" (Berzins, 1951).

Family Colurellidae.--These rotifers have a lorica and a telescopic three- or four-segmented foot with two toes protruding posteroventrally from an opening on the lorica. The laterally constricted, psammophilic genus *Colurella* includes *C. colurus* (Ehrenberg) (Figure 25.1a), a facultative anaerobe that feeds on bacteria; *C. obtusa* (Gosse); *C. dicentra* (Gosse); and, *C. grandiuscula* Kutikova and Arov which is endemic to Lake Baikal. Members of this genus possess a movable shield which covers the corona. Some dorsoventrally flattened members of this family are found in the genus *Lepadella*, including one saltwater psammobiont, *L. psammophila* Tzschaschel (Figure 25.1b). *Lepadella triptera* Ehrenberg is a psammophilic species associated with loose sediment.

Family Dicranophoridae.--Most of the members are omnivores or voracious carnivores. Of the 12 genera, all but 4 may be found in aquatic sands. The genus *Wigrella* (Figure 25.1o) is found almost exclusively in the psammon. *Encentrum lineatum* Tzschaschel (*Wierzejskiella elongata* (Glascott)) (Figure 25.1j), *E. villosum* Harring and Myers, and *E. marinum* (Dujardin) (Figure 25.1n) are three of 22 species that are present in marine sand; 5 others are found in freshwater sand. This latter genus can be recognized by its oblique corona, teeth only at the top of the trophi rami, and relatively short toes. *Dicranophorus* (Figure 25.1q) is nearly as well represented in the psammon although apparently restricted to freshwater. Members of this genus have a ventral corona, teeth on the inner margin of the trophi rami, and relatively long sometimes jointed toes. *Paradicranophorus* (Figure 25.1i) is found only in freshwater and marine benthic sediments. This genus has a pear-shaped body and laterally positioned small, pointed toes. *Aspelta* (Figure 25.1h) and *Inflatana* have only one freshwater species each. *Pedipartia gracilis* (Myers) is found only in northeastern North America. *Myersinella tetraglena* Harring and Myers is very similar to the genus *Erignatha*.

Family Euchlanidae.--This group has only one consistent representative in the freshwater psammon, *Euchlanis arenosa* Meyers (Figure 25.1e).

Family Lecanidae.--This is a loricate family consisting of a single genus and three subgenera, two of which are represented in the psammolittoral. These include *Lecane* with two toes, and *Monostyla* with a single toe. The former genus has no marine representatives, but 9 freshwater species are known including *L. (M.) psammophila* Myers (Figure 25.1d) and *L. (sensu strictu) mucronata* Harring and Myers (Figure 25.1c).

Family Lindiidae.--This family consists of a single genus and a single psammobiontic species, *Lindia janickii* Wiszniewski. *Lindia (Halolinda) tecusa* Harring and Myers may also be found in both freshwater and marine sand and detritus.

Family Notommatidae.--Up to 21 genera with many littoral representatives are found in this family. *Cephalodella* is a common representative, having many species, in the freshwater psammolittoral. *Cephalodella compacta* Wiszniewski (Figure 25.1f) and two other species may be considered freshwater psammobionts; *Notommata bennetchi* Myers is the only member of this genus in the freshwater hygropsammon.

Family Proalidae.--This family consists of 4 genera, 2 with representatives in the meiobenthos: *Proales* and *Bryceella*. *Proales* species can be found in freshwater littoral areas including periphyton, but

P. germanica Tzschaschel (Figure 25.1g) and *P. halophila* Remane are marine psammobionts. *Bryceella tenella* Bryce (Figure 25.1p) is found only in freshwater.

Family Trichocercidae.--Two of the 3 genera of this family may be represented in the meiobenthos. *Elosa worallii* Lord and four species of *Trichocera*, including *T. (Diurella) pygocera* Wiszniewski (Figure 25.1m), may be found in fresh or brackish waters.

Habitats and Ecological Notes

A conscientious look at rotifers in the psammolittoral began in the early 1930's. Wiszniewski (1934b) proposed three categories of benthic rotifers: (1) *psammobiotic*, rotifers almost exclusively found in sandy sediments; (2) *psammophilic*, rotifers preferential to sand, but also living in other regions of the littoral; and (3) *psammoxenic*, rotifers found occasionally in sand but generally living elsewhere and normally destined to perish in this habitat. These groupings are used in the following discussion with some qualifications; since comparatively few documented records exist for psammic rotifers, these terms, as applied to given taxa, may not be entirely accurate. The number of significant contributions to benthic rotifers, especially those found in coarse sediments, is very small. Even less is known about the presence of rotifers in muddy sediments. Most of the literature would suggest their absence from such habitats, but rotifers are present in freshwater mud, usually at or near the sediment-water interface; their presence in similar marine habitats may have been overlooked by the very nature of their small size and amorphous configuration when subjected to bulk formalin fixation.

Both marine and freshwater psammolittoral rotifers seem to be found mostly in a mixture of coarse and medium sand, yet sand grain size is not the apparent overriding factor determining whether or not rotifers will be present. Studies showing relationships between granulometry and the presence or absence of rotifers are indeed sparse; most notable are those studies by Arov (1985), Pennak (1940), and Ruttner-Kolisko (1953, and especially 1954). The number of rotifers found per unit area appears to be influenced as much by sand grain size as it is by the physical distance from the water's edge and the ability of the sand to hold water.

Marine and Brackish Water Habitats.--Marine and brackish-water psammolittoral studies occasionally include rotifers, but even fewer are devoted exclusively to this phylum. Those that do include papers by Althaus (1957), Fenchel and Jansson (1966), Remane (1949), Thane-Fenchel (1968a), and Tzschaschel (1979, 1980); all papers concern studies conducted only in European coastal waters. These papers account for most of the rotifers known from the thalassopsammon and illustrate the virtual absence of data for all other regions of the world. Tzschaschel (1979), in a single paper concerning the North Sea, increased the number of psammobiotic and psammophilic marine species by eight.

Most rotifers associated with sandy sediments are interstitial. Many are found intertidally in marine beaches where the interstices are kept moist by capillary action. Tzschaschel (1983) demonstrated that marine rotifers prefer the upper 20-cm horizon of sand, generally above the water-line, although they can be found as deep as 40 cm in the sand as long as no sulfide layer is present. Although he records no rotifers from the sublittoral sands of a semilotic beach included in his studies, Remane (1949), Althaus (1957), and Thane-Fenchel (1968a) all routinely encountered rotifers in this zone from the Baltic and Black Seas. Similarly, in studies currently being conducted on the meiofauna of subtidal coarse sands at depths between 15-20 m off the Atlantic coast of Florida, U.S.A., and at depths of between 250-350 meters off the coasts of North Carolina and South Carolina, U.S.A., bdelloid rotifers are found regularly, but not in large numbers (Higgins, personal communication). In both study areas, rotifers have been found along with the recently described species of Loricifera (Higgins and Kristensen, 1986); these two authors note that specimens of both phyla (Rotifera and Loricifera) are occasionally difficult to separate from one another during the sorting process (Higgins and Kristensen, personal communication).

Tzschaschel (1980) records *Proales halophila* (forms), *Colurella colurus*, and *Lepadella psammophila* all year round; *Proales syltensis* Tzschaschel is present only in summer months, and *Encentrum axi* Tzschaschel, *E. permutandum* Tzschaschel, and *Proales germanica* Tzschaschel only in colder seasons. Tzschaschel (1983) demonstrated that *Encentrum* (Dicranophoridae) is most abundant in cooler or cold seasons and absent from marine beaches in warmer summer months. Wulfert (1936) found an increase in freshwater species of this genus in autumn, winter, and spring.

Freshwater Habitats.--When found in freshwater sandy biotopes, rotifers are more commonly above the water-line in well-oxygenated areas. Pennak (1939, 1940), Wiszniewski (1934b, 1935, 1936, 1937), and Ruttner-Kolisko (1953, 1954) have demonstrated the presence of large numbers of rotifers in the hygropsammic areas of freshwater beaches, from the surface of the sand to horizons several cm deep.

Wiszniewski (1934b) and Pennak (1940) indicate that psammic rotifer populations appear in the spring (April) and disappear in winter (November). Wiszniewski (1934b) found population maxima in June and again in late September.

Evans (1982) found that the majority of rotifers occurred in the top 3 cm of sand, at or very near the water's edge in Lake Erie; the highest population count was 23.1 individuals per cm^2. Pennak (1940) found rotifers distributed from the water's edge to 250 cm landward; 90% of his rotifers were found in the upper 2 cm of sand. Ruttner-Kolisko (1954) also found the majority of rotifers in this same zone. Fenchel and Jansson (1966) have shown that only a few rotifers such as *Lecane (Monostyla) cornuta* (Müller) and *Rotaria citrina* (Ehrenberg) tolerate the sulfide layer of sediments. Thane-Fenchel (1968a) and Pennak (1951) both indicate that rotifers are more abundant in freshwater than in marine sands although the reason for this remains unclear.

Some species, *Colurella colurus*, *Lecane (M.) cornuta*, *Trichocerca (Diurella) taurocephala* (Hauer), *Encentrum marinum*, and *Proales halophila*, are known to occur in both freshwater and marine habitats. Of the psammic rotifers, only the genera *Pedipartia* and *Myersinella* are restricted to freshwater.

Additional information on the ecology of freshwater rotifers can be found in Chapter 4.

Methods of Collection and Extraction

Since most benthic rotifers have been found in coarse sediments, it follows that those methods especially suited for the sampling of this habitat are the most practical. Both coring devices and dredges designed to sample the uppermost strata are effective. In both freshwater and marine beaches, large sand samples will yield rotifers simply by placing the sediment in a bucket of water, mixing the sand into a momentary suspension, and decanting the water through a fine mesh net (42 µm is recommended). Core samples should be treated in a similar manner but on a smaller scale. In all instances, a preliminary step to narcotize the animals in the sediment will yield much better results relevant to both the quantity of rotifers obtained and the quality of the specimens. Since it is of paramount importance to have non-loricate rotifers fully relaxed for identification, this latter point cannot be overemphasized. A portion of the sample should also be fixed quickly to insure that the cohabiting loricate rotifers adequately contract for their specific identification.

The extraction of marine rotifers from sand by subjecting the sample to a freshwater shock (Kristensen and Higgins, 1984) produces some well-extended specimens; however, careful study of the results of this technique has not been undertaken to date. The more tested method of removal of (live) marine rotifers from sand involves the Uhlig seawater-ice technique (Uhlig, 1964, 1966) discussed in Chapter 9.

Freshwater rotifers may be effectively relaxed using neosynepherine hydrochloride, carbonated water, or asprin in various concentrations in acid waters; 2% butyn and 2% hydroxylamine hydrochloride (Pennak, 1978) are more effective in alkaline water. Live specimens of non-loricate rotifers are preferable to preserved specimens since fixation nearly always causes some distortion of the animals. Soft bodied ploimate and bdelloid rotifers may be killed with an equal portion of boiling water poured quickly and directly into the rotifer sample. The rotifers must be removed rapidly and washed because of the adhesive nature of the cilia after this treatment. A similar procedure involves the use of 70% alcohol heated to its boiling point. Some excellent results have been obtained by placing a freshwater core-sample into a beaker of boiling alcohol and then decanting off the fluid through a fine-mesh net. It is a technique that should be tested further.

Preparation for Taxonomic Study

Rotifers should normally be fixed in 5-10% formalin. Further processing often requires a careful examination of the trophi of non-loricate rotifers (except bdelloids) since it may be the only means of identification at the species level. Trophi clearance and isolation are facilitated by the judicious use of sodium hypochlorite (household bleach), or any suitable basic solution such as 5.0% potassium hydroxide. Techniques for trophi manipulation and extraction can be found in Myers (1937) and Russel (1961a).

The mounting of specimens for study usually requires that either formalin or alcohol fixed specimens be placed into either 70% ethanol containing 2-5% glycerin or 5-10% formalin with an equal amount of glycerin. This solution is allowed to evaporate slowly so as leave the specimens in glycerin. Specimens are then transferred to a suitable mounting medium such as polyvinyl lactophenol lactophenol, modified Hoyer's, or glycerin as described in Chapter 10.

References

Althaus, B.
1957. Neue Sandbodenrotatorien aus dem Schwarzen Meer. *Wissenschaftliche Zeitschrift der Martin Luther*

Universität Halle-Wittenberg, Mathematisch Naturwissenschaftliche Reihe, 6:445-458.

Arov, I.V.
1985. Psammolittoral Rotifers of Lake Baikal. Pages 189-198 in L.A. Kutikova, editor, *Proceedings of the Second All Union Symposium of Rotifers.* Leningrad: Nauka Publishing House. [In Russian.]

Berzins, B.
1951. Contribution to the Knowledge of Marine Rotatoria of Norway. *Universitetet i Bergen, Arbok, Naturvitenskapelig Rekke,* 6:1-11.

Czapik, A.
1952. Untersuchunge über die Infusorien und Rotatorien des Küstengrundwassers und des Sandbodens der Stalin-Bucht. *Arbeiten aus der Biologischen Meeresstation in Stalin.* 17:61-65. [In Bulgarian.]

Edmondson, W.T.
1948. Two New Species of Rotatoria from Sand Beaches, with a Note on *Collotheca wiszniewskii*. *Transactions of the American Microscopical Society,* 67:149-152.

Erben, R.
1987. Rotifer Fauna of the Periphyton of Karst Rivers in Croatia, Yugoslavia. *Hydrobiologia,* 147:103-105.

Evans, W.A.
1980. *Seasonal Abundances of Psammic Rotifers Along a Gradient of Acid Mine Pollution in a Southeastern Ohio Stream.* Ph.D Thesis, Ohio University, Athens, Ohio.
1982. Abundances of Micrometazoans in Three Sandy Beaches in the Island Area of Western Lake Erie. *Ohio Journal of Science,* 82:246-251.

Fenchel, T.M.
1978. The Ecology of Micro- and Meiobenthos. *Annual Review of Ecology and Systematics,* 9:99-121.

Fenchel, T., and B.-O. Jansson
1966. On the Vertical Distribution of the Microfauna in Sediments of a Brackish Water Beach. *Ophelia,* 3:161-177.

Higgins, R.P., and R.M. Kristensen
1986. New Loricifera from Southeastern United States Coastal Waters. *Smithsonian Contributions to Zoology,* 438:1-70.

Koste, W.
1961. *Paradicranophorus wockei* nov. spec., ein Rädertier aus dem Psammon eines norddeutschen Nierungsbaches. *Zoologischer Anzeiger,* 167:138-141.
1978. *Rotatoria. Die Rädertiere Mitteleuropas, Begründet von Max Voigt. Monogononta.* 673 pages. Berlin and Stuttgart: Gebrüder Borntraeger.

Kristensen, R.M., and R.P. Higgins
1984. A New Family of Arthrotardigrada (Tardigrada: Heterotardigrada) from the Atlantic Coast of Florida, U.S.A. *Transactions of the American Microscopical Society,* 103:295-311.

Kutikova, L.A.
1970. *The Rotifer Fauna of USSR. Fauna USSR.* Volume 104. 744 pages. Leningrad: Akademia Nauk SSSR.

Kutikova, L.A., and I.V. Arov
1985. New Species of Psammophilic Rotifers (Rotatoria) in Lake Baikal. Pages 50-82 in L.A. Kutikova, editor, *Proceedings of the Second All Union Symposium on Rotifers.* [In Russian.]

Moore, G.M.
1939. A Limnological Investigation of the Microscopic Benthic Fauna of Douglas Lake, Michigan. *Ecological Monographs,* 9:537-582.

Myers, F.J.
1936. Psammolittoral Rotifers of Lenape and Union Lakes, New Jersey. *American Museum Novitates,* 830:1-22.
1937. A Method of Mounting Rotifer Jaws for Study. *Transactions of the American Microscopical Society,* 56:256-257.
1942. The Rotatorian Fauna of the Pocono Plateau and Environs. *Proceedings of the Academy of Natural Sciences, Philadelphia,* 94:251-285.

Neel, J.K.
1948. A Limnological Investigation of the Psammon in Douglas Lake. *Transactions of the American Microscopical Society,* 67:1-53.

Neiswestnowa-Shadina, K.
1935. Zur Kenntnis des rheophilen Mikrobenthos. *Archiv für Hydrobiologie,* 28:555-582.

Nogrady, T.
1982. Rotifera. Pages 866-872 in S.P. Parker, editor, *Synopsis and Classification of Living Organisms.* New York: McGraw-Hill Book Company, Inc.

Pennak, R.W.
1939. A New Rotifer from the Psammolittoral of Wisconsin Lakes. *Transactions of the American Microscopical Society,* 58:222-223.
1940. Ecology of the Microscopic Metazoa Inhabiting Sandy Beaches of Some Wisconsin Lakes. *Ecological Monographs,* 12:537-615.
1951. Comparative Ecology of the Interstitial Fauna of Fresh-water and Marine Beaches. *Année Biologique,* 27:449-480.
1978. *Fresh-water Invertebrates of the United States.* Second Edition. 803 pages. New York: John Wiley and Sons.

Pejler, B.
1962. On the Taxonomy and Ecology of Benthic and Periphytic Rotatoria. Investigations in Northern Swedish Lapland. *Zoological Bidrag, Uppsala,* 33:328-422.

Remane, A.
1949. Die psammonbionten Rotatorien der Nord- und Ostsee. *Kieler Meeresforschungen,* 6:59-67.

Ricci, C.
1983. Rotifera or Rotatoria? *Hydrobiologia,* 104:1-2.
1987. Ecology of Bdelloids: How to be Successful. *Hydrobiologia,* 147:117-127.

Ridder, M. de
1986. *Annotated Checklist of Non-marine Rotifers from African Inland Waters.* 123 pages. Tervuren (Brussels): Museum of Central Africa.

Russell, C.R.
1961a. A Simple Method of Permanently Mounting Rotifer Trophi. *Journal of the Quekett Microscopical Club,* 5:377-378.
1961b. Simple Method of Mounting Rotifers and Other Small Animals in Glycerin Jelly. *Journal of the Quekett Microscopical Club,* 28:384-386.

Ruttner-Kolisko, A.
1953. Psammonstudien I. Das Psammon des Torneträsk in Schwedisch-Lappland. *Sitzungberichte der Österreichischen Akademie der Wissenschaften, mathematisch-naturwissensschaftliche Klasse,* 162:129-161
1954. Psammonstudien II. Das Psammon des Erken in Mittelschweden. *Sitzunberichte der Österreichischen Akademie der Wissenschaften, mathematisch-naturwissensschaftliche Klasse,* 163:301-324.
1956. Psammonstudien III. Das Psammon des Lago Maggiore in Oberitalien. *Memorie dell'Instituto Italiano di Idrobiologia Dott. Marco de Marchi,* 9:356-402.

Sassuchin, D.N., N.M. Kabanov, and K.S. Neiswestnova
1927. Über die mikroskopische Pflanzen- und Tierwelt der Sandfläche des Okaufers bei Murom. *Russische Hydrobiologische Zeitschrift,* 6:59-83. [In Russian with German summary.]

Svedmark, B.
1964. The Interstitial Fauna of Marine Sands. *Biological Reviews,* 39:1-42.

Thane-Fenchel, A.
1968a. Distribution and Ecology of Non-planktonic Brackish-water Rotifers from Scandinavian Waters. *Ophelia,* 5:273-297.
1968b. A Simple Key to the Genera of Marine and Brackish-water Rotifers. *Ophelia,* 5:299-311.

Tzschaschel, G.
1979. Marine Rotatoria aus dem Interstitial der Nordseeinsel Sylt. *Mikrofauna des Meeresbodens,*

71:1-64.
1980. Verteilung, Abundanzdynamik und Biologie mariner interstiteller Rotatoria. *Mikrofauna des Meeresbodens,* 81:1- 56.
1983. Seasonal Abundance of Psammon Rotifers. *Hydrobiologia,* 104:275-278.

Uhlig. G.
1964. Eine einfache Methode zur Extraktion der vagilen mesopsammalen Mikrofauna. *Helgoländer wissenschaftliche Meeresuntersuchungen,* 11:178-185.
1966. Untersuchungen zur Extraktion der vagilen Mikrofauna aus marinen Sedimenten. *Zoologischer Anzeiger* (Supplement), (2)9:151-157.

Varga, L.
1938. Vorläufige Untersuchungen über die mikroskopischen Tiere des Balaton-Psammons. *Arbeiten des Ungarischen Biologischen Forschungsinstitutes,* 10:101-138.
1957. Neuere Daten über die Mikrofauna des Balaton Psammons. *Magyar Tudomanyos Akademia Tihanyi Biologiai Kutato Intezetenek Evkonyve,* 271-282.

Wiszniewski, J.
1929. Zwei neue Rädertierarten: *Pedalia intermedia* n. sp. und *Paradicranophorus limosus* n. g. n. sp. *Bulletin academie polonaise science letters, Classe sciences natural,* Krakow, Serial B (II):137-153.
1934a. Les rotifères psammiques. *Annales Musei Zoologici Polonici,* 10:339-399.
1934b. Recherces écologiques sur le psammon. *Archives d'Hydrobiologie et d'Ichthyologie,* 8:149-271.
1935. Notes sur le psammon II. Rivière Czarna auz environs de Varsovie. *Archives d'Hydrobiologie et d'Ichthyologie,* 9:221-238.
1936. Notes sur le psammon. III. Deux tourbiéres aux environs de Varsovie. *Archives d'Hydrobiologie et d'Ichthyologie,* 10:173-187.
1937. Differenciation ecologique des Rotifers dans le psammon d'eaux douces. *Annales Musei Zoologici Polonici,* 13:1-13.

Wulfert, K.
1936. Beiträge zur Kenntnis der Rädertierfauna Deutschlands. II. Teil. *Archiv für Hydrobiologie,* 30:401-437.
1942. Über Meeres- und Brackwasser-Rotatorien in der Umgebung von Rovigno d'Istria. *Thalassia,* 4:3-26.

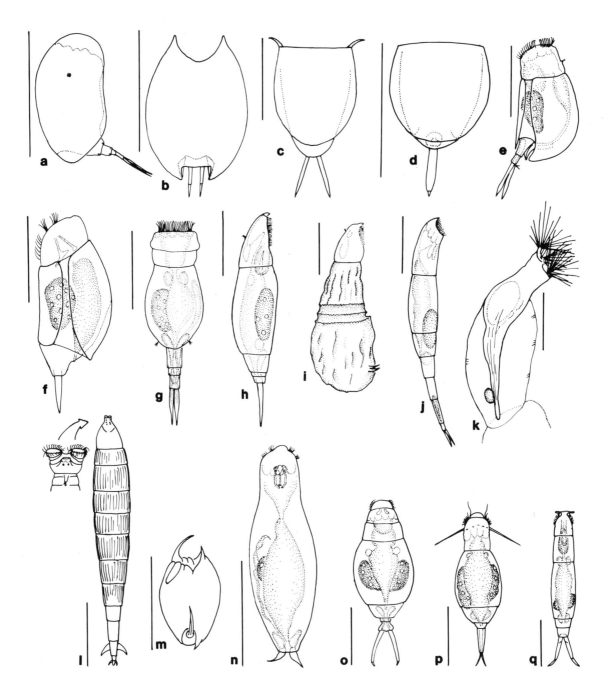

Figure 25.1.—**Rotifera. a,** *Colurella colurus,* lateral view; **b,** *Lepadella psammophila,* dorsal view (after Tzschaschel, 1979); **c,** *Lecane mucronata,* dorsal view; **d,** *Lecane (Monostyla) psammophila,* dorsal view; **e,** *Euchlanis arenosa,* right lateral view (after Myers, 1936); **f,** *Cephalodella compacta,* lateral view (left), dorsal view (right); **g,** *Proales germanica,* dorsal view (after Tzschaschel, 1979); **h,** *Aspelta egregia,* lateral view; **i,** *Paradicranophorus hudsoni,* lateral view (after Wiszniewski, 1929); **j,** *Encentrum lineatum (=Wierzejskiella elongata),* lateral view; **k,** *Collotheca wiszniewskii,* right lateral view, attached to sand grain (after Varga, 1938); **l,** *Rotaria rotatoria,* dorsal view, head (only) extended (left) and retracted into body with rostrum extended (right); **m,** *Trichocerca (Diurella) pygocera,* ventral view (after Wiszniewski, 1934a); **n,** *Encentrum marinum,* dorsal view; **o,** *Wigrella depressa,* dorsal view (after Wiszniewski, 1934a); **p,** *Bryceella tenella,* dorsal view (after Wiszniewski, 1934a); **q,** *Dicranophorus hercules capucinoides,* dorsal view (after Wiszniewski, 1934a). (Scale = 100 μm.)

26. Loricifera

Robert P. Higgins and Reinhardt M. Kristensen

The loriciferans are the most recent phylum described (Kristensen, 1983). The first known representative of this exclusively marine meiobenthic taxon was discovered in 1974 in medium sand at 400 m depth off the coast of North Carolina (Higgins and Kristensen, 1986). Since the description of the phylum nine species representing three genera and two families have been described, but over 23 other species are known currently.

Adult loriciferans are bilaterally symmetrical and between 115-383 µm long. The body is divided into a spherical, eversible head or "introvert," a neck and thorax, all of which are retractable into a loricate abdominal region. A mouth cone with 8-9 oral styles or 6-16 oral ridges marks the anteriormost portion of the body. Within the mouthcone is centered an extrusible cuticularized buccal canal. The head may be armed with up to nine variably distinct rings of spine-like appendages, or scalids. Sexual dimorphism may be apparent in the structure of the first row of eight clavoscalids. Two segments make up the thorax: the anterior portion or neck bears special appendages called trichoscalids; the appendageless posterior region joins with the abdomen. The abdomen is covered by a more heavily cuticularized lorica, consisting of six plates with hollow spines along the anterior margin or 22 or more folds or plica. Flosculi, when present, are located posteriorly on the lorica. Saccate gonads open terminally. One pair of protonephridia is present. Also present is a triradiate myoepithelial pharynx bulb which may have five rows of placoids (cuticularized rods as found also in the Tardigrada).

The immature stages in the life history of Loricifera, named "Higgins-larvae" by Kristensen (1983) are 80-385 µm long and have the same body regions as the adult. The larval mouth cone may have 6-12 oral stylets or none. A hexagonally arranged internal armature is present in some. The head always bears a ring of eight clavoscalids and usually six or sometimes seven additional rings of scalids are present. A single middorsal scalid, highly modified, may be present. A collar-like neck area may be set off from the remaining thoracic region. It may bear minute structures which could be the anlangen of the trichoscalids and are presumed to act as part of the closing apparatus for the withdrawn head. Several distinctive plates on the anteroventral region of the larval thorax may also have this function. The remaining thoracic region consists of 5-6 rows of plates formed from transverse and longitudinal folds. These, too, fold in over the withdrawn head.

The lorica consists of a series of longitudinal folds and plates. Some Higgins-larvae have three distinct abdominal regions formed by constrictions in the lorica. Two or three locomotory and/or sensory spines are present near the anteroventral margin of the lorica; two or three sensory setae are present more posteriorly, usually in dorsal, lateral, and caudal positions. Three flosculi may be present, two laterals and one middorsal close to the anus. A pair of protonephridia is located laterally. Each of the seven terminal cells has one cilium and opens near or into the rectum. Where toes are present in the larvae, there are a pair of large adhesive glands located basally in the abdomen; the adhesive gland pore is near the spiny tip of the toe. Development is by a series of molts of the larva. A postlarval stage may intervene between the larva and adult stages. When the postlarva is present, it appears similar to the adult.

Classification

The phylum Loricifera consists of a single order, the Nanaloricida Kristensen, 1983, and two described families. The Nanaloricidae consists of two genera each with three species (only one genus with a single species is currently described, *Nanaloricus mysticus*, Figure 26.1a,d). The adult of this species has sexually dimorphic differences in the first ring of head appendages or clavoscalids; all are similarly uniramus in the female, but in the male (Figure 26.1a), the middorsal, dorsolateral, and ventrolateral pairs of clavoscalids are divided into three branches. The ventral pair of clavoscalids is uniramus as are all clavoscalids of the female. The lorica consists of six plates forming a strongly coronate anterior margin. The Higgins-larva of this species (Figure 26.1d) has a long, unarmed cylindrical mouth cone, its five rows of thoracic plates are interrupted at the midventral margin by two slightly longer plates. The three midventral locomotory/sensory spines at the junction of the thorax and abdomen are fused basally. The larval lorica consists of four large elements, one dorsal, one ventral, and two lateral, all with longitudinal folds. The toes at the posterior end of the lorica extend laterally and are beset with mucros (leaf-like structures) and extend laterally. Eight to twelve striated muscles attach to the base of each toe.

The family Pciliciloricidae has two genera, *Pciliciloricus* (Figure 26.1b,e) and *Rugiloricus* (Figure 26.1c,f). Family characteristics of the adults include the presence of two kinds of unbranched clavoscalids, a "double organ" consisting of two variously fused and modified midventral second-ring head appendages, and a cuticle with 22-60 longitudinal folds or plicae; the pcilicilorid Higgins-larvae have only two pairs of ventral locomotory/sensory setae at the junction of the thorax and abdomen, and the toes tend to extend anteroventrally along the abdomen. *Pciliciloricus* has five species; adults (Figure 26.1b) have both single and double trichoscalids, Higgins-larvae (Figure 26.1e) have a collar-like closing apparatus at the anterior region of the thorax. *Rugiloricus* has three species; adults (Figure 26.1c) have only single trichoscalids, Higgins-larvae (Figure 26.1f) lack the collar and may or may not have toes.

Habitats and Ecological Notes

Loriciferans have been found at depths from intertidal to 8000 m; in coarse sandy-shell sediments as well as in very fine sediments. No special pattern has been determined regarding habitat preference. Specimens have been found within the Arctic Circle (Greenland), coastal northern Europe, the Mediterranean, the southeastern and Gulf coasts of the United States, deep sea mid-Pacific and shallow coral lagoons of the Chesterfield Reefs off the east coast of Australia. Our knowledge of this newest phylum currently is in its infancy; therefore, it is unwise to establish firm ecological perspectives.

Methods of Collection

It would appear that the loricifera occur in the uppermost (oxygenated) layers of sediment and the standard methods of collecting material from this zone should be sufficient. Highly efficient qualitative sampling devices, such as an anchor dredge, may insure sufficient material for taxonomic examination whereas core samples are less likely to provide more than one or two specimens.

Methods of Extraction

Loriciferans have been found using standard elutriation techniques. In a few cases, the bubbling technique (see Chapter 9) has produced a few specimens also. Kristensen (1983) and Higgins and Kristensen (1986) were especially successful in obtaining large numbers of loriciferans using the fresh-water osmotic shock technique (see Chapter 9), but this may or may not provide a demonstrable advantage other than for its tendency to expand specimens for more adequate examination.

Preparation of Specimens for Taxonomic Study

The most satisfactory fixation of specimens is with 2% glutaraldehyde, trialdehyde, or 6-8% buffered formalin. Sorted specimens, to be used primarily for the study of external structures, should be placed in 70% ethanol if prolonged storage is anticipated, but in doing so, it is best to begin with 10% ethanol and transfer specimens through a series of progressively higher concentrations of ethanol until the desired concentration is reached. There is some evidence that ethanol may not be good for the preservation of internal structures. Specimens are often very delicate; when fixed in 2% glutaraldehyde, the cuticle tends to remain more rigid, offering a distinct advantage if specimens are to be used for SEM study. In preparing specimens for light microscopy, specimens in formalin should be placed in a solution of 2% glycerin in distilled water and slowly evaporated to glycerin over a period of several days; specimens in 70% ethanol should be transferred first to distilled water and then to a solution of 2% glycerin in distilled water for evaporation.

The use of Cobb aluminum slide mounts or the new H-S slide mounts (see Chapter 10) is extremely useful since both sides of a mounted specimen must be studied. Mounting in glycerin affords the most satisfactory preparations, although if enough specimens are available for study, prior immersion in Hoyer's-50, or even mounting in this medium, offers some advantage by slightly clearing the specimens. As in all such preparations, the coverslip should be carefully sealed by a reliable medium.

Type Collections

Currently, the only type collections of the phylum Loricifera are found in the National Museum of Natural History, Smithsonian Institution, Washington, D.C., USA; Musée National d'Histoire Naturelle, Paris; and Universitets Zoologisk Museum, Copenhagen, Denmark.

References

Higgins, R.P., and R.M. Kristensen
1986. New Loricifera from Southeastern United States

Coastal Waters. *Smithsonian Contributions to Zoology*, 438:1-70.

Kristensen, R.M.
1983. Loricifera, a New Phylum with Aschelminthes Characters from the Meiobenthos. *Zeitschrift für Zoologische Systematik und Evolutionsforschung*, 21:163-180.

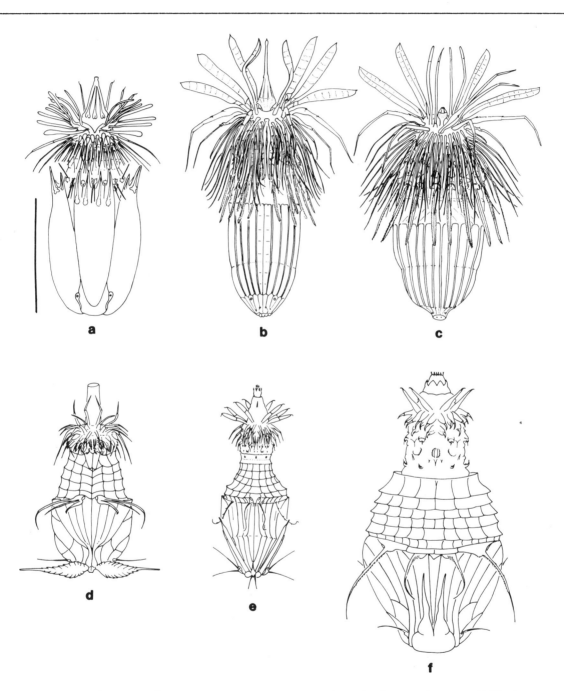

Figure 26.1.--Loricifera: **a**, *Nanaloricus mysticus*, adult male (from Kristensen, 1983); **b**, *Pliciloricus gracilis*, adult male (from Higgins and Kristensen, 1986); **c**, *Rugiloricus cauliculus*, adult male (from Higgins and Kristensen, 1986); **d**, *N. mysticus*, Higgins-larva (from Kristensen, 1983); **e**, *P. gracilis*, Higgins-larva (from Higgins and Kristensen, 1986); **f**, *R. carolinensis*, Higgins-larva (from Higgins and Kristensen, 1986). (Scale = 100 μm.)

27. Priapulida

C. Bradford Calloway

Prior to 1968, the phylum Priapulida was considered to be a cold-water, macrobenthic taxon, represented in the meiofauna only by larval stages. The 7 species and 4 genera were all placed in a single family. Beginning with the discovery of *Tubiluchus corallicola* from the shallow, warm waters of Curaçao by van der Land (1968), the number of priapulid taxa has steadily increased to 15 species, 7 genera, and 3 families. Most (7 species) or all (8 species) of these new taxa are meiofaunal. However, they are not merely miniatures of their larger macrofaunal relatives. They exhibit many new morphological and ecological characters demonstrating that the basic priapulid body plan is much more plastic than it was once thought to be.

The Priapulida are vermiform marine invertebrates which are bilaterally symmetrical with a secondary radial tendency, presumably a consequence of their infaunal burrowing habit. The body is unsegmented, cylindrical, and generally has three body regions: an introvert, abdomen, and tail or caudal appendage. The introvert is the anteriormost body region. It is invaginable into the abdomen, and bears the mouth centrally on its apical surface. Small conical projections of the introvert surface called scalids are arranged in circles and longitudinal rows. The structure, number, and disposition of scalids are important taxonomic characters, but scalid function is largely unknown. The cylindrical abdomen bears the anus on its posterolateral or posteroventral surface. Urogenital pores are located ventrolaterally anterior to the anus. The surface of the abdomen generally possesses a variety of protuberances of obscure function which are used in classification. The tail, considered to be a cylindrical post-anal extension of the abdomen by van der Land (1970), is not present in all species. A chitinous cuticle secreted by the single layer of epidermal cells of the body wall is molted periodically.

The alimentary canal is a straight or slightly curved tube lying essentially free within the body cavity. It consists of a buccal tube extending from the mouth to the pharynx, an esophagus, intestine, and a short rectum terminating at the anus. The muscular pharynx is lined with posteriorly directed cuticular teeth, the number, structure and arrangement of which are important taxonomic characters. The buccal tube and anterior pharynx may be evaginated through the mouth to form a cone-shaped structure protruding from the apical surface of the introvert. During evagination the anterior pharynx turns inside out and its teeth become directed anteriorly. Feeding generally involves a repeated evagination and invagination of this apparatus.

The large body cavity between the alimentary canal and the body wall extends into the tail and is not divided by septa. It serves as a circulatory system as well as a hydrostatic skeleton. Two urogenital complexes, each consisting of gonads and solenocytic protonephridia combined into an organ system, lie laterally in the posterior abdomen. Sexes are separate and fertilization is generally external.

Larval stages have an introvert and a neck region, both of which may be invaginated into an abdomen covered by a thick cuticle, the lorica. The cuticle is molted between larval stages. At metamorphosis the lorica is lost, and, depending on the species, a tail (or tails) may develop.

Classification

Fourteen of the 15 species of extant priapulids are included in three families: Chaetostephanidae, Priapulidae, and Tubiluchidae. The systematic placement of the recently described species *Meiopriapulus fijiensis* Morse, 1981 is controversial. All large, macrofaunal Priapulids are included in the family Priapulidae and range in length from approximately 0.5 to 20 cm although individuals in excess of 10 cm are rare. The smaller priapulid species range in length from approximately 0.5 to 3.7 mm excluding the tail and are, at the upper end of their size range, technically not meiofaunal. However, being elongate animals with a diameter generally less than 0.5 mm, they readily pass through a 1 mm sieve and are sampled with the meiofauna. Meiofaunal priapulids include all species of the families Chaetostephanidae and Tubiluchidae plus *Meiopriapulus fijiensis* Morse, 1981. In addition, two small, poorly described species occur in the family Priapulidae. *Priapulopsis* (?) *cnidephorus* Salvini-

Plawen, known only from a single specimen described as a young individual 1.5 mm long and 0.2–0.25 mm in diameter, is tentatively included with the meiofaunal priapulids. The single specimen of *Acanthopriapulus horridus* (Théel, 1911) for which measurements are recorded is a robust animal 3.9 mm in length exclusive of its tail and 1.5 mm in diameter. It will not pass through a 1 mm sieve and, consequently, is not considered meiofaunal. Priapulid larvae, being small in size (generally less than 2 mm in length even in large macrofaunal species) and infaunal, are included in the meiofauna. The larvae and possibly the earliest postlarvae of macrofaunal species that grow out of the meiofaunal size range during development are temporary meiofauna.

The Chaetostephanidae is a monotypic family represented by *Maccabeus cirratus* Malakhov, 1979, and *M. tentaculatus* Por, 1973 (Figure 27.1e). *Chaetostephanus praeposteriens* Salvini-Plawen, 1974, is congeneric and probably conspecific with *M. tentaculatus*. The body (introvert and abdomen) is less than 3 mm long in *M. tentaculatus* and 3.6 mm long in the single known specimen of *M. cirratus*. There is no tail. The introvert bears highly modified anterior scalids. Eight short, presumably sensory papillae ringing the mouth are surrounded by a circlet of 25 scalids, each of which is divided distally into 2 elongate tentacles. Posteriorly, 3 or 4 types of scalids are arranged in 4–8 circlets. The abdomen is covered with short, crenulate, circularly arranged tubercles which are interrupted near the posterior end by a band of longitudinal ridges bearing 1–3 circlets of conical spines and, at its posterior end, a circlet of hooks. The circlet of hooks delimits the posterior end of the abdomen which is rounded to cone-shaped and bears 2 large spines adjacent to the terminal anus. The pharyngeal teeth are cuspate and are arranged in circlets of 5. Only females are known, leading to speculation that *Maccabeus tentaculatus* is parthenogenetic. The larvae of *M. tentaculatus*, like those of the Priapulidae, have a bilaterally symmetrical, dorsoventrally compressed lorica that reaches a length of as much as 0.7 mm. This species lives in a tube of agglutinated particles at the sediment-water interface and feeds presumably by catching small organisms with the crown of modified scalids which it projects into the water column.

The Tubiluchidae is a monotypic family represented by 4 species: *Tubiluchus australensis* van der Land, 1985a; *T. corallicola* van der Land, 1968; *T. philippinensis* van der Land, 1985a; and *T. remanei* van der Land, 1982. The body consists of a bulbous introvert, an elongate abdomen and a long tail (Figure 27.1a,b). The body, exclusive of the tail, ranges from 0.4 mm to 2.5 mm long; large individuals in an uncontracted state may have a body diameter of 0.32 mm. The introvert, described only in *T. corallicola*, bears scalids arranged quincuncially in circlets and in longitudinal rows. Eight large scalids form the first circlet surrounding the mouth. These are followed by 25 longitudinal rows of scalids each consisting of 6–9 scalids progressively decreasing in size posteriorly. The abdomen is covered by small conical elevations (tumuli) arranged in circles and in longitudinal rows and by sparsely scattered long, thin tubes (tubuli) and flower-shaped sensillae (flosculi). It extends posteriorly as a long, thin cylindrical tail. Pharyngeal teeth are pectinate and arranged in diagonal and longitudinal rows. The polythyridium, an organ consisting of circlets of cuticular plates surrounded by circular muscle, lies in the alimentary canal at the entrance to the intestine.

In tubiluchids the sexes are separate and dimorphic. The male is identified by extensive ventral setation and by integumental structures associated with the urogenital pores (Figure 27.1a). These specializations, which are absent in the female (Figure 27.1b), serve to differentiate species. Larvae are known only in *Tubiluchus corallicola* and *T. remanei*. The lorica, in contrast to those of other priapulids, is circular in cross section, has 20 longitudinal cuticular ridges, and a complex closing apparatus formed by the cuticular folds of the neck (Figure 27.1c,d). In *T. corallicola* the lorica may reach a length of 0.46 mm whereas fully extended larvae may reach lengths of up to 0.75 mm.

Meiopriapulus fijiensis Morse, 1981, is known only from a single beach on Viti Levu, Fiji. The body consists of a bulbous introvert which is invaginable into an elongate abdomen and reaches lengths of up to 3 mm. There is no tail (Figure 27.1f). The introvert bears an anterior circlet of 8 elongate scalids surrounding the mouth. These are followed by a second and a third circlet of 8 elongate but progressively smaller scalids (Storch and Higgins, personal communication, 1987). Posteriorly, the remaining introvert surface is covered with about 150 longitudinal rows of minute scalids with 8–9 scalids in each row. Large triangular scalids are found on what appears to be the anterior end of the abdomen adjacent to the introvert. Posteriorly, the remaining abdominal surface is covered by small spherical tubercles and scattered tubuli-flosculi complexes (Storch and Higgins, personal communication, 1987). A circlet of recurved hooks delimits the rounded to cone-shaped posterior abdomen that bears the anus centrally on its posteriormost end. Pharyngeal teeth are pectinate and arrranged in approximately 16 rows with at least 6 teeth per row. A polythyridium-like organ is present in the alimentary canal. Sexes are reported to be separate. Larvae are unknown.

Priapulopsis (?) *cnidephorus* Salvini-Plawen, 1973, is based on a single young individual 1.5 mm long and 0.2–0.25 mm in diameter. The body consists of an introvert bearing scalids of unknown morphology and disposition, an elongate abdomen with wart-like hills containing unexploded nematocysts on its external surface, and two short, non-vesiculate tails. Pharyngeal teeth are cuspate. Larvae are unknown. This species is provisionally included with the meiofauna.

Larval Stages

Priapulid larvae have an introvert and neck, both of which are invaginable into an elongate abdomen that is covered by a thick cuticle, the lorica (Figure 27.1c,d,g,h). A pharynx with cuspate teeth is always present. The introvert bears the mouth centrally on its apical surface. It is provided with scalids and, in larval *Halicryptus* (Family Priapulidae), with adhesive tubes on its posterior end. As in the adult, the structure, number, and disposition of the scalids and pharyngeal teeth are important taxonomic characters. The neck is characterized by cuticular folds that in the Tubiluchidae form a closing apparatus for the lorica (Figure 27.1c,d). In *Priapulus, Priapulopsis* (Family Priapulidae), and *Maccabeus* the posterior neck (or anterior lorica) bears cuticular accessory plates that close the anterior end of the lorica when the introvert and neck are invaginated (Figure 27.1g,h). The morphology, number, and position of the plates, which appear to be absent in *Halicryptus*, are taxonomic characters.

The lorica is characterized by longitudinal ridges or folds that divide it into a series of longitudinally arranged fields plates, the number and arrangement of which are important taxonomic characters. The lorica of the Priapulidae and Chaetostephanidae is dorsoventrally flattened (Figure 27.1g,h) whereas that of the Tubiluchidae is circular in cross-section (Figure 27.1c,d). In all larvae, except those of *Halicryptus*, the lorica ridges bear tubuli. Adhesive glands are reported to be present at the posterior end of the larva of *Maccabeus tentaculatus*.

Habitats and Ecological Notes

Tubiluchus is the most widely distributed of the meiofaunal priapulids. It has been considered to have a disjunct circumtropical to subtropical distribution and to be confined to shallow, warm water carbonate sediments. However, the discoveries of unidentified tubiluchids from relatively deep, cool water (to 270 m) on the Blake Plateau off the North and South Carolina coasts in non-carbonate sediments (Higgins, personal communication, 1986) and from the coast of Sweden (Erséus, personal communication, 1987) demonstrate that tubiluchids are not confined to shallow, warm carbonate environments and may help to explain their presence in widely separated, often insular habitats.

Maccabeus tentaculatus has been reported only from the Mediterranean Sea at depths of 60–550 m in sandy to clayey muds. *M. cirratus* is known only from a single specimen collected from the Indian Ocean at a depth of 2520 m. A specimen similar to *M. tentaculatus* has been reported from the Mozambique Channel. *Meiopriapulus fijiensis* is known only from a medium to coarse sand beach on Viti Levu, Fiji. *Priapulopsis* (?) *cnidephorus* is known from one specimen collected in the Adriatic Sea at a depth of approximately 30 m in mud.

Meiofaunal priapulids are generally associated with an abundant and diverse meiofaunal assemblage. This is in marked contrast to the macrofaunal priapulids which appear to be most common in sediments with a poor fauna (Land, 1970). Though all meiofaunal priapulids, with the possible exception of the tube-dwelling *Maccabeus*, appear to be active burrowers like the larger macrofaunal species, they are characterized by a variety of anatomical specializations suggestive of a more sessile habit. Whereas macrofaunal priapulids are active predators feeding on slow-moving prey, it appears that *Meiopriapulus* and *Tubiluchus* are probably detritivores scraping, raking, or sieving small particles off of surfaces or out of the sediment with their pectinate pharyngeal teeth. Though a predator, the tube-dwelling *Maccabeus* is the only priapulid known to feed by entrapment. *Priapulopsis* (?) *cnidephorus* is the only priapulid reported to harbor nematocysts, suggesting that it feeds on Cnidaria.

Methods of Collection

All meiofaunal priapulids live within the upper few centimeters of the sediment and thus any means for obtaining surficial sediment is suitable for their collection. As priapulids are generally an uncommon component of meiofaunal assemblages, relatively large quantities of sediment must often be collected and processed to obtain sufficient specimens.

Methods of Extraction

Simple decantation and sieving will separate most meiofaunal priapulids from the sediment (See Chapter 9). The smallest larval stages may require the use of mesh sizes as small as 42 μm. Extraction may be facilitated by the freshwater shock technique or by narcotization with $MgCl_2$ (see Chapter 9). However, as these techniques may damage the specimens, they must be used with caution particularly when the specimens are destined for histological studies. Staining sediment samples with Rose Bengal greatly enhances the chances of finding the smaller larval stages. Adults are easily manipulated with an Irwin Loop, whereas smaller larval stages are best isolated with a transfer pipette.

Preparation of Specimens for Taxonomic Study

The taxonomy of priapulids is based almost entirely upon external morphology, including the surface features of the evaginated anterior pharynx. Adults are easily identified to genus since most of their taxonomic characters, including the overall shape and size of the introvert and abdomen, the gross external structure of the introvert, and the absence or presence, morphology, and number of tails are easily observed at low magnification. The determination of species as well as the identification of all larvae requires a more detailed analysis of the morphology, number, and placement of a variety of small external structures including scalids, pharyngeal teeth, tubuli, spines, the plates of the larval lorica, etc. These structures are routinely observed by examining whole mounts with a compound microscope equipped with phase contrast or differential phase interference optics. Surface specializations are particularly well visualized in whole mount squeeze-preparations.

Introverts of macrofaunal priapulids are normally extended following fixation, but meiofaunal priapulids as well as all larvae often withdraw the introvert upon fixation or other disturbance. The surface features of the introvert and the anterior alimentary canal are most easily studied when the introvert and pharynx are extended. This can be accomplished by (1) narcotizing live animals with menthol crystals in sea water or 7.5% $MgCl_2$ in distilled water, (2) exposing live animals to freshwater or to a hypotonic salt solution, or (3) squeezing the animals in whole mounts.

Specimens may be whole mounted live in sea water or, following fixation, in the fixative or in a variety of media including glycerin (see Chapter 26) and modifications of Hoyer's medium containing decreased amounts of chloral hydrate, e.g. Hoyer's-125 (Higgins, 1983). Whole mounts in glycerin or Hoyer's may be made permanent by sealing them with slide ringing compound.

Scanning electron microscopy (SEM) is an extremely valuable technique for studying the morphology of priapulid body surfaces. Successful SEM examination requires clean, well fixed specimens. Specimens sorted from routinely fixed meiofaunal sediment samples may be satisfactory for SEM examination, though introverts are often withdrawn. Best results are obtained using live sorted specimens that are narcotized and then fixed in a glutaraldehyde based fixative adjusted to be approximately isotonic with sea water.

Type Collections

The disposition of meiofaunal priapulid type material is noted in the original species descriptions. Material can be found in The Australian Museum, Sidney, Australia; Hebrew University of Jerusalem, Jerusalem, Israel; The Institute of Oceanology, USSR Academy of Sciences, Moscow, USSR; National Museum of Natural History, Smithsonian Institution, Washington, D.C., USA; Naturhistorisches Museum, Wien, Austria; and the Rijksmuseum van Natuurlijke Historie, Leiden, The Netherlands.

Macrofaunal priapulids, though rare, are much more common in museum collections than meiofaunal species. See van der Land (1970) for the location of type material and additional specimens of extant macrofaunal priapulids and their larvae and postlarvae, and Conway Morris (1977) for the location of collections of fossil priapulids.

References

Alberti, G., and V. Storch
1983. Fine Structure of Developing and Mature Spermatozoa in *Tubiluchus* (Priapulida, Tubiluchidae). *Zoomorphology*, 103(3):219-227.
1986. Zur Ultrastruktur der Protonephridien von *Tubiluchus philippinensis* (Tubiluchidae, Priapulida). *Zoologischer Anzeiger*, 217(3/4):259-271. [In German with English abstract.]

Calloway, C.B.
1975. Morphology of the Introvert and Associated Structures of the Priapulid *Tubiluchus corallicola* from Bermuda. *Marine Biology*, 34(2):161-174.
1982. Priapulida. Pages 941-944, plates 86-87 in S.P. Parker, editor, *Synopsis and Classification of Living Organisms, volume 1*. New York: McGraw-Hill.

Conway Morris, S.
1977. Fossil Priapulid Worms. *Special Papers in Palaeontology*, 20: iv + 95 pages, 30 plates. London: The Palaeontological Association.

Higgins, R.P.
1983. The Atlantic Barrier Reef Ecosystem at Carrie Bow Cay, Belize, II: Kinorhyncha. *Smithsonian Contributions to the Marine Sciences*, 18:1-131.

Higgins, R.P., and R.M. Kristensen
1986. New Loricifera from Southeastern United States Coastal Waters. *Smithsonian Contributions to Zoology*, 438:1-70.

Hulings, N.C., and J.S. Gray, editors
1971. A Manual for the Study of Meiofauna. *Smithsonian Contributions to Zoology*, 78:1-83.

Kirsteuer, E.
1976. Notes on Adult Morphology and Larval Development of *Tubiluchus corallicola* (Priapulida), Based on *in vivo* and Scanning Electron Microscopic Examinations of Specimens from Bermuda. *Zoologica Scripta*, 5(6):239-255.

Kirsteuer, E., and J. van der Land
1970. Some Notes on *Tubiluchus corallicola* (Priapulida) from Barbados, West Indies. *Marine Biology*, 7(3):230-238.

Kirsteuer, E., and K. Rutzler
1973. Additional Notes on *Tubiluchus corallicola* (Priapulida), Based on Scanning Electron

Microscope Observations. *Marine Biology*, 20(1):78-87. Evolutionsforschung, 21:163-180.

Land, J. van der
1968. A New Aschelminth, Probably Related to the Priapulida. *Zoologische Mededelingen*, 42:237-250.
1970. Systematics, Zoogeography and Ecology of the Priapulida. *Zoologische Verhandelingen*, 112:1-118, plates 1-5.
1982. A New Species of *Tubiluchus* (Priapulida) from the Red Sea. *Netherlands Journal of Zoology*, 32(3):324-335.
1985a. Two New Species of *Tubiluchus* (Priapulida) from the Pacific Ocean. *Proceedings of the Koninklijke Nederlandse Akademie van Wetenschappen*, ser. C., 88(3):371-377.
1985b. Abyssal *Priapulus* (Vermes, Priapulida). Pages 379-383 in L. Laubier and Cl. Monniot, editors, *Peuplements profonds du golfe de Gascogne*. Brest: IFREMER, Service Documentation Publication.

Land, J. van der, and A. Nørrevang
1985. Affinities and Intraphyletic Relationships of the Priapulida. Pages 261-273 in S. Conway Morris, J.D. George, R. Gibson, and H.M. Platt, editors, *The Origins and Relationships of Lower Invertebrates*. The Sytematics Association Special Volume no. 28. Oxford: Clarendon Press.

Lang, K.
1953. Die Entwicklung des Eies von *Priapulus caudatus* Lam. und die systematische Stellung der Priapuliden. *Arkiv för Zoologi*, ser. 2, 5:321-348.

Malakhov, V.V.
1979. A New Representative of Sedentary Priapulids *Chaetostephanus cirratus* sp.n. *Zoologichesky Zhurnal*, 58(9):1410-1412. [In Russian with English summary.]

McLean, N.
1984. Ameobocytes in the Lining of the Body Cavity and Mesenteries of *Priapulus caudatus* (Priapulida). *Acta Zoologica*, 65(2):75-78.

Morse, M.P.
1981. *Meiopriapulus fijiensis* n.gen., n.sp.: An Interstitial Priapulid from Coarse Sand in Fiji. *Transactions of the American Microscopical Society*, 100(3):239-252.

Por, F.D.
1973. Priapulida from Deep Bottoms near Cyprus. *Israel Journal of Zoology*, 21:525-528.
1983 Class Seticoronaria and Phylogeny of the Phylum [1984]. Priapulida. *Zoologica Scripta*, 12(4):267-272.

Por, F.D., and H.J. Bromley
1974. Morphology and Anatomy of *Maccabeus tentaculatus* (Priapulida: Seticoronaria). *Journal of Zoology*, 173(2):173-197.

Salvini-Plawen, L. von
1973. Ein Priapulide mit Kleptocniden aus dem Adriatischen Meer. *Marine Biology*, 20(2):165-169. [In German with English abstract.]
1974. Zur Morphologie und Systematik der Priapulida: *Chaetostephanus praeposteriens*, der Vertreter einer neuen Ordnung Seticoronaria. *Zeitschrift für zoologische Systematik und Evolutionsforschung*, 12:31-54. [In German with English summary.]
1977. Caudofoveata (Mollusca), Priapulida und apode Holothurien (*Labidoplax, Myriotrochus*) bei Banyuls und im Mittelmeer allgemein. *Vie et Milieu*, ser. A, 27(1):55-81. [In German with English abstract.]

Storch, V., and G. Alberti
1985a. Zur Ultrastruktur des Darmtraktes von *Tubiluchus philippinensis* (Tubiluchidae, Priapulida). *Zoologischer Anzeiger*, 214(5/6):262-272. [In German with English abstract.]
1985b. Ultrastructural Investigation of the Integument of *Tubiluchus philippinensis* (Priapulida, Tubiluchidae). *Zoologica Scripta*, 14(4):265-272.

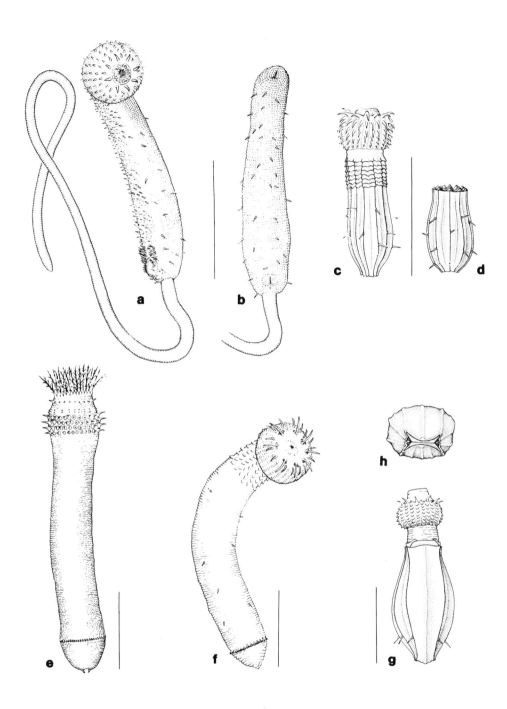

Figure 27.1.—**Priapulida: a**, *Tubiluchus corallicola*, adult male; **b**, same, adult female with introvert retracted; **c**, same, larva; **d**, same, with introvert and neck retracted; **e**, *Maccabeus tentaculatus*, adult; **f**, *Meiopriapulus fijiensis*, adult; **g**, *Priapulopsis* sp., larva; **h**, same, with introvert retracted. (Scale bar = 0.5 mm.) (Illustrations by Laszlo Meszoly.)

28. Kinorhyncha

Robert P. Higgins

The Phylum Kinorhyncha is sometimes considered a class of the Aschelminthes or Nemathelminthes and may also be referred to as the Echinoderida, especially in the older literature. This phylum of marine pseudocoelomates is exclusively meiobenthic in marine and estuarine sediments, normally less than 1 mm in length, bilaterally symmetrical, and distinctly segmented. The first of 13 segments consists of an eversible, spherical head with up to 7 rows of recurved spines or scalids, and a terminal protrusible mouth cone surrounded by 9 oral styles. By everting and withdrawing the head, the animal is able to move through sediment as well as feed. The second segment consists of a neck, usually equipped with a series of plates which function as a closing apparatus for the head when it is withdrawn into the anterior one-third of the next series of 11 segments which constitute the trunk. The trunk segments may be variously subdivided longitudinally into a series of separate plates. The number and arrangement of neck and trunk plates in addition to the number and arrangement of spines and adhesive tubes constitute the basis for further classification of the 100 species whose descriptions are based on adult specimens; nearly 60 additional species have been described from juvenile stages which likely will be unrecognizable with an adult.

Classification

The Kinorhyncha are divided into 2 orders, the Cyclorhagida and the Homalorhagida. The cyclorhagid mouth cone is usually equipped with jointed oral styles, the neck usually consists of a series of 14-16 plates (placids) or patches of denticles, and the trunk is usually round to oval in cross-section. Three suborders are based upon the closing apparatus; the Cyclorhagae have a series of 14-16 placids articulating with an undivided cylindrical trunk segment; placids of the Conchorhagae articulate with a bilaterally divided cylindrical trunk segment which also may function in closing off the withdrawn head; and the Cryptorhagae possess a series of 14 patches of denticles which are not articulated with the incompletely divided cylindrical first trunk segment.

There are 2 families in the Cyclorhagae. The Echinoderidae are represented currently by a single genus *Echinoderes* (Figure 28.1a,b,j,k) (42 species), the most commonly encountered kinorhynch genus and the one with the most species. Adults of this genus lack a midterminal spine, and have their first 2 trunk segments undivided.

The family Centroderidae consists of 3 genera, each with only a few species, all with midterminal spines throughout their life history, and with only the first trunk segment undivided. The genus *Condyloderes* (Figure 28.1d) (2 species) lacks lateral terminal accessory spines and ventral adhesive tubes; the genus *Campyloderes* (Figure 28.1c) (3 species) has a midterminal spine shorter than the lateral terminal spines in addition to a pair of ventral adhesive tubes and long ventral spines on the first trunk segment; the genus *Centroderes* (Figure 28.1e) (2 species) has a midterminal spine longer than the lateral terminal spines, lacks ventral adhesive tubes on the first trunk segment, but has a pair of long ventral spines on this segment.

The Conchorhagae are represented by a single family, Semnoderidae, with 2 genera. The genus *Semnoderes* (Figure 28.1f) (3 species) has a first trunk segment consisting of two single lateral halves; the genus *Sphenoderes* (Figure 28.1g) (1 species) has a middorsal and midventral plate interposed between the posterior portions of the two lateral halves.

The only remaining suborder, the Cryptorhagae, consists of a single family, the Cateriidae, and a single genus, *Cateria* (Figure 28.1h) (3 species).

The Homalorhagida consists of a single suborder, the Homalorhagae, The second segment consists of a series of fewer than 8 placids, (dorsally and ventrally oriented). The placids articulate with the first trunk segment which is distinctly triangular in cross-section and longitudinally separated into at least 2 plates; the ventral plate may be partially or completely divided into three separate plates.

The more plesiomorphic of the two homalorhagid families is the Neocentrophyidae. In this family, the oral styles are segmented and the midsternal plate of the first trunk segment is not completely defined. This family consists of 2 genera: *Paracentrophyes* (Figure 28.1p) (2 species) and *Neocentrophyes* (Figure 28.1q) (1 species). The former genus has lateral terminal spines in addition to a midterminal spine, the latter genus has only a small midterminal spine.

The family Pycnophyidae consists of two genera, both relatively common and with many species. The genus *Pycnophyes* (Figure 28.1l-n) (23 species) has lateral terminal spines in the adult and late juvenile stages; the genus *Kinorhynchus* (Figure 28.1o) (17 species) does not.

Habitats and Ecological Notes

There are three kinds of habitats in which kinorhynchs are found. Most are found in the upper 0-3 cm of organically enriched mud to muddy sand in estuaries or the sea, from intertidal to abyssal depths. Some are found in association with algae or other invertebrates. Algal holdfasts, whether in large algae as *Fucus* or in clumps of *Enteromorpha*, often accumulate sediment which makes it difficult to classify this habitat as strictly algal. In the case of algae such as *Gracillaria* some kinorhynchs may be found in association with the detrital floc on the surface of the plant.

Occasionally, kinorhynchs have been found in association with other invertebrates. Oftentimes, as in the case of the algal habitats, kinorhynchs are actually in the sediment that is collected with the invertebrate. The first record of this phylum is from the detritus in jars containing oysters from the northern coast of France (Dujardin, 1851). Other invertebrates harboring kinorhynchs have included polychaetes (Southern, 1914), bryozoans (Higgins, 1977), and sponges (Higgins, 1978).

The third habitat in which kinorhynchs may be found is the interstices of medium to coarse marine sand or gravel; either intertidal sand of high energy beaches or subtidal medium to coarse sand, "shell-gravel," "shell-hash," "*Amphioxus*-sand", "*Dentalium*-sand," or even "*Halimeda*-sand."

Methods of Collection

Any method which collects the uppermost (oxygenated) layers of sediment with a minimum loss of fine materials is likely to produce kinorhynchs. The most productive method is to use a meiobenthic dredge (see Chapter 7), specifically designed to collect only the uppermost few centimeters of substrate in a fine mesh net protected by an abrasion and tear resistant cover. Quantitative samplers considered adequate for the collection of undisturbed sediment are often very useful in the collection of kinorhynchs, but have the shortcoming of not always collecting sufficient material to provide adequate numbers of specimens for taxonomic or life-history studies.

In the case of coarse sediments, kinorhynchs have been found as deep as 1 m in high energy marine beaches. The problems of sampling gear penetration prevents us from knowing within-habitat depth limits of kinorhynchs and other meiofauna in subtidal coarse sediment habitats. In any event, relatively large samples of coarse sediment are preferable because of the generally reduced numbers of individuals.

Methods of Extraction

The simplest technique for removing kinorhynchs from muddy sediments is by the bubbling technique. The unpreserved mud sample should be diluted to about twice its volume with sea water so as to allow stirring the mud into a temporary suspension. Immediately after suspension, a stream of fine air-bubbles should be introduced for about one minute. The most effective apparatus for this procedure is a simple hand-operated bicycle tire pump with an aquarium airstone attached by a short rigid plastic tube. The air bubbles bring the kinorhynchs (and a selection of other meiofauna) to the surface film where they are trapped by the surface tension. After bubbling, the suspension should be allowed to settle for a few minutes. Next, a piece of moderately absorbent paper is lightly placed on the surface film and immediately removed. The "float" of invertebrates will adhere to the paper and may be washed directly into a container for observation or for preservation. For either purpose, it is best to repeat the bubbling-blotting-washing process many times using a 63 µm mesh net or sieve to concentrate the specimens. A brief (1 minute) freshwater rinse of the specimens in the net or sieve will protrude the head of many of the specimens through an osmotic reaction; this allows a more adequate observation of the external anatomy but is not advisable if histological studies are contemplated.

The bubbling technique is non-quantitative; best estimates, however, suggest that about 90-95% of the kinorhynchs can be removed by this method. Variations of this technique are also effective; they include any means of "frothing" the suspension such as pouring the suspension from one bucket to another, holding the full bucket high enough above the other bucket in the pouring process so as to create a maximum disturbance of the suspension as it is poured.

Certain species found only in intertidal mud apparently do not have the same hydrophobic cuticle as subtidal kinorhynchs and may not respond to the bubbling technique. If a relatively small sediment sample is to be processed for specimens, centrifugation techniques are useful. Elutriation techniques are also satisfactory. However, regardless of the technique or fluid in which the kinorhynchs may be placed, they have a tendency to accumulate on the surface film.

Extraction from algae or other invertebrates is simply a matter of dislodging the kinorhynchs by some form of agitation in sea water and subsequent elutriation, sieving and sorting. The same is true for extracting kinorhynchs from coarse sediments. Fine sand is usually unproductive; medium to coarse sand or gravel must be rinsed in sea water, a relaxing solution, or freshwater. In the latter instance, only a brief contact with the freshwater is advisable (see Chapter 8).

Preparation of Specimens for Taxonomic Study

The most satisfactory fixation of specimens is with 6-10% buffered formalin. Sorted specimens should be placed in 70% ethanol if prolonged storage is anticipated. The use of the Irwin Loop (see Chapter 10) is almost essential in the manipulation of kinorhynchs. The identification of kinorhynchs demands that specimens be mounted in a perfectly dorsoventral aspect; specimens mounted otherwise are virtually worthless. In addition, specimens must be moderately cleared in order to observe the details of external anatomy – details upon which classification is based. A modified Hoyer's solution (use 62.5% of the amount of chloral hydrate; i.e., 125 gms instead of 200 gms) or "Hoyer's-125" is the most satisfactory mounting medium despite its potential long-term problems of crystalization or over-clearing. Reduction of the amount of chloral hydrate in the standard formula for Hoyer's medium, and careful sealing of the coverslip are extremely important in producing a permanent mount.

Cobb aluminum slide mounts or the new H-S slide mounts (see Chapter 10) are the most satisfactory because in the mounting process, despite great care, specimens may orient themselves either dorsally or ventrally and both surfaces must be observable. The specimen is placed in a small drop of Hoyer's-125 on a coverslip, oriented by the Irwin Loop or a needle, the second (no. 0, 12 mm circle) coverslip is placed on the surface of the medium and, as the coverslip settles and the medium spreads beneath it, the coverslip must be pressed in such a way as to cause the specimen to orient itself in the dorsoventral position without crushing it or breaking off spines. Hoyer's-125 is the only medium which can render the cuticle pliable enough to allow this absolutely necessary procedure.

Preparation of specimens for SEM observation requires no special technique. Preparation of specimens for sectioning requires procedures that will overcome the almost impermeable cuticle. Cutting a specimen or puncturing the cuticle during the fixation process appears to be the best approach despite the shortcomings of dealing with varying extents of structural damage.

Type Collections

The most extensive collections of Kinorhyncha are located at the National Museum of Natural History, Smithsonian Institution, Washington, D.C., USA. Other museums having several types include: British Museum (Natural History), London, UK; Naturhistoriska Riksmuseet, Stockholm, Sweden; and the National Museum of Ireland, Dublin, Ireland.

References

Claparède, E.
1863. *Beobachtungen über Anatomie und Entwicklungsgeschichte wirbelloser Tiere an der Küste der Normandie angestellt.* 120 pages, 18 plates. Leipzig: Wilhelm Engelmann.

Dujardin, F.
1851. Sur un petit animal marin, l'Echinodère, formant un type intermédiaire entre les Crustacés et les Vers. *Annales des Sciences Naturelles, Zoologie,* series 3, 15:158-160.

Gerlach, S.A.
1956. Über einen aberranten Vertreter der Kinorhynchen aus dem Küstengrundwasser. *Kieler Meeresforschungen,* 12:120-124.

Higgins, R.P.
1968. Taxonomy and Postembryonic Development of the Cryptorhagae, a New Suborder for the Mesopsammic Kinorhynch Genus *Cateria. Transactions of the American Microscopical Society,* 87:21-39.
1969a. Indian Ocean Kinorhyncha: 1. *Condyloderes* and *Sphenoderes,* New Cyclorhagid Genera. *Smithsonian Contributions to Zoology,* 14:1-13.
1969b. Indian Ocean Kinorhyncha. 2. Neocentrophyidae, a New Homalorhagid Family. *Proceedings of the Biological Society of Washington,* 82:113-128.
1977. Two New Species of *Echinoderes* (Kinorhyncha) from South Carlina. *Transactions of the American Microscopical Society,* 96(3):340-354.
1978. *Echinoderes gerardi* n. sp. and *E. riedli* (Kinorhyncha) from the Gulf of Tunis. *Transactions of the American Microscopical Society,* 97(2):171-180.
1981. Kinorhyncha. Pages 873-877, 4 plates in S.P. Parker, editor, *Synopsis and Classification of Living Organisms. Volume 1.* New York: McGraw-Hill.
1983. Atlantic Barrier Reef Ecosystem at Carrie Bow Cay, Belize, II. Kinorhyncha. *Smithsonian Contributions to the Marine Sciences,* 18:1-131.
1986. Kinorhyncha. Pages 110-118 in L. Botosaneanu, editor, *Stygofauna Mundi,* Leiden: E.J. Brill.

Higgins, R.P., and J.W. Fleeger
1980. Seasonal Changes in the Population Structure of *Echinoderes coulli* (Kinorhyncha). *Estuarine and Coastal Marine Science,* 10:495-505.

Southern, R.
1914. Nemathelmia, Kinorhyncha and Chaetognatha. In Clare Island Survey, Part 54. *Proceedings of the Royal Irish Academy,* 31:1-80.

Zelinka, C.
1928. *Monographie der Echinodera.* iv + 396 pp. Leipzig: Wilhelm Engelmann.

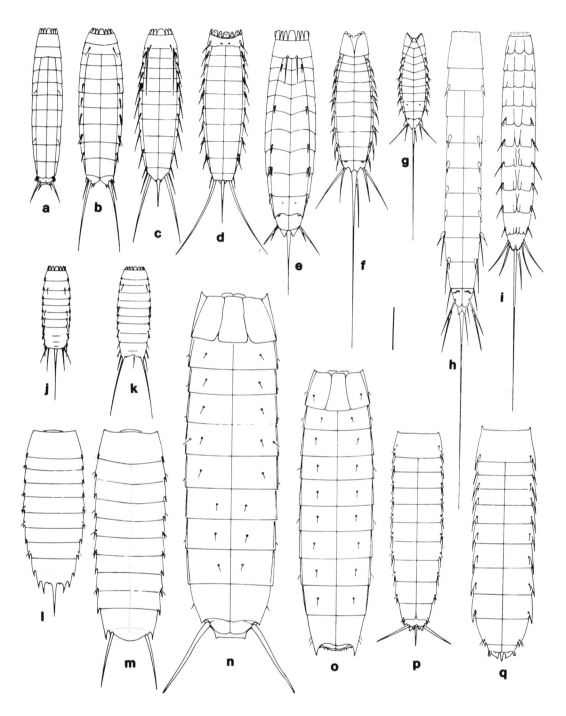

Figure 28.1.--Kinorhyncha: a, *Echinoderes coulli*, adult female; b, *Echinoderes dujardinii*, adult female; c, *Campyloderes macquariae*, adult female; d, *Condyloderes multispinosus*, adult female; e, *Centroderes spinosus*, adult female; f, *Semnoderes armiger*, adult female; g, *Sphenoderes indicus*, adult female; h, *Cateria submersa*, adult female; i, *Cateria gerlachi*, adult female; j, *Echinoderes dujardinii*, juvenile stage 3; k, *Echinoderes dujardinii*, juvenile stage 4; l, *Pycnophyes greenlandicus*, juvenile stage 3; m, *Pycnophyes greenlandicus*, juvenile stage 4; n, *Pycnophyes greenlandicus*, adult female; o, *Kinorhynchus mainensis*, adult female; p, *Paracentrophyes praedictus*, adult female; q, *Neocentrophyes satyai*, adult female; a,c,d-g,i,n,o,q after Higgins, 1986, (courtesy E.J. Brill) (Scale = 100 μm.)

29. Polychaeta

Wilfried Westheide

Polychaeta are relatively primitive annelids, apparently very close to the stem species of the Articulata. Their monophyly, however, is not proven – derived characters (=synapomorphies) for all Polychaeta are as yet unknown. A general definition of the Polychaeta does not exist, except that they are "worm-like," soft-bodied invertebrates with a metamerically organized body. For morphological details see textbooks of invertebrate zoology. Glossaries on the terminology generally used in systematic polychaete literature: Fauchald (1977); Mikkelsen and Virnstein (1982).

Classification

The more than 10,000 species of polychaetes can be arranged into about 80 mostly well-defined families, which, however, often show little similarity. Pettibone (1982) gives an up-to-date synopsis of all suprageneric polychaete taxa. Differences from Clitellata, especially from Oligochaeta, to which several polychaete taxa are externally very close, are usually large number of fertile segments, lack of clitellum and cocoons, and a dioecious organization. Consecutive and simultaneous hermaphrodites, however, are also known in Polychaeta.

The meiofaunal polychaetes do not represent a single taxonomic group. They evolved in several independent lines from larger species which may have lived in or on top of the sediment. Among them are many highly derived, secondarily reduced or distinctly paedomorphic (=progenetic) species. Segmental parapodia or cirri as well as head appendages which are so characteristic for most polychaete species may be lacking in these forms. Even chaetae may be completely absent.

Length of polychaete species range between 3 m to less than 1 mm; the smallest species are those of the genus *Diurodrilus* (about 300 µm); the smallest adult individual is the male of *Dinophilus gyrociliatus*, with a length of 50 µm. Defining meiofaunal polychaetes and distinguishing them from large benthic forms are as problematic as the general definition of meiofauna. Some species generally considered to be meiofaunal polychaetes do not pass through a sieve of 1 mm mesh size. Although many inhabit the interstices of sediments, others are infaunal, burrowing in soft sediments by pushing aside sediment particles to make space in which to move. Many of these species exhibit typical meiofaunal characteristics: e.g., specific adhesive organs, reproductive organs that allow direct transmission of sperm, and low number of oocytes (Westheide, 1971, 1984).

Another problem of distinction concerns the many small polychaete species from non-sedimentary habitats. Numerous species live as sessile or mobile forms, e.g., on algae and rock surfaces, and a considerable number of species inhabit crusts of colonial sedentary animals such as bryozoans or hide, e.g., in coral reefs. Though some of these species can be identified to family or even genus level by the plates (Figures 19.1–19.4), these non-sedimentary polychaetes of small body dimensions are not dealt with in this chapter. The numerous small juvenile sand-inhabiting individuals of large-dimensioned species forming the so-called temporary meiofauna are not included either. All small parasitic species are also excluded.

Thus, the selection made is commensurate with the fundamental difficulties of defining meiofauna in the first place, and may be considered controversial. On the whole only a small percentage of all known species of Polychaeta belong to the group of sediment-inhabiting meiofaunal polychaetes, most of them being part of the "interstitial" fauna inhabiting pore spaces between sediment grains. Almost all meiofaunal polychaetes inhabit exclusively marine sediments in intertidal as well as in subtidal regions. Limnic groundwater sediments are the habitat of *Troglochaetus beranecki* (Nerillidae). *Hesionides riegerorum* lives in river sand banks. A terrestrial polychaete species from forest soils is *Parergodrilus heideri* (Parergodrilidae), and the recently described *Hrabeiella periglandulata* (Pizl and Chalupsky, 1984), inhabiting forest soil, most likely also belongs to the Polychaeta.

Whereas most of the meiofaunal polychaete species are part of suprageneric taxa with a majority of macrofaunal species, there are several families which are exclusively meiofaunal: Dinophilidae, Diurodrilidae, Nerillidae, Protodrilidae, Protodriloidae, Saccocirridae, Polygordiidae, Parergodrilidae, Psammodrilidae; nearly all the Pisionidae and many Syllidae are meiobenthic.

Family Phyllodocidae.--The very elongate, generally medium sized or large species of this large taxon (about 30 genera) have a distinct prostomium with 4 frontal antennae, in some taxa also a median appendage and 2 to 4 pairs of tentacular cirri. The numerous homonomous uniramous parapodia possess usually leaf-like dorsal and ventral cirri. There is a long eversible proboscis. The approximately 10 species of the true interstitial genus *Hesionura* Hartmann-Schröder, 1958, are thread-like, vividly moving and often colored. The elongate prostomium is without a median appendage. There are 3 pairs of slender tentacular cirri on 2 segments; the tentacular segments lack chaetae; dorsal cirri are absent on segment 3. Parapodial cirri are spindle-shaped. The genus *Mystides* Théel, 1879, also contains small species. Important taxonomic features for *Hesionura* (separation of species unsatisfactory) include presence of eyes, color, chaetation, especially details of compound chaetae. (Figure 29.3b.) References: Hartmann-Schröder (1963), Laubier (1967b).

Family Goniadidae.--Of the generally medium-sized, very slender, multisegmented species only two of the genus *Goniadides* Hartmann-Schröder, 1960, may be considered to be mesopsammic forms. Important taxonomic features include pharyngeal organs, parapodia, and branchiae. (Figure 29.3c.) Reference: Hartmann-Schröder (1962).

Family Sphaerodoridae.--The family comprises slender multisegmented usually medium-sized species (up to 50 mm long) and short plump species with few segments (length down to 1.2 mm). The dorsum has large and small spherical protuberances, often in transverse rows; cylindrical papillae may be present on any part of the body. The head region, not distinctly separated from the rest of the body, possesses 2 to 6 prostomial appendages and 1 pair of tentacular cirri. The parapodia are uniramous. Important taxonomic features are presence and arrangement of tubercles and papillae, and chaetation. (Figure 29.1k.) Reference: Fauchald (1974).

Family Hesionidae.--The small and medium-sized species with a moderate number of homonomous segments have sub-biramous or biramous parapodia. There are several prostomial appendages and a variable number of segments with tentacular cirri. The pharynx is cylindrical and partly eversible. Out of the 30 genera, 3 are considered typically meiobenthic: the hermaphroditic genus *Microphthalmus* Mecznikow, 1865 (25 species), with 5 prostomial appendages, 6 pairs of tentacular cirri, and a single anal adhesive lobe, *Hesionides* Friedrich, 1937 (11 species), with 5 prostomial appendages, 4 pairs of tentacular cirri, and 1 or 2 anal adhesive lobes, and *Heteropodarke* Hartmann-Schröder, 1962 (3 species), with 5 prostomial appendages of which the unpaired one inserts frontally, and a variable number of tentacular cirri (up to 8 pairs). Important taxonomic features are details of chaetae, shape of anal lobes, and structure of reproductive organs, particularly in the males. (Figure 29.1l-p.) References: Westheide (1967a, 1977a), Westheide and Rao (1977).

Family Syllidae.--This large taxon, to which a large part of the meiobenthic species belongs, comprises very small to medium-sized usually slender species with few to numerous segments. The rounded prostomium has 3 dorsal and 2 usually short, thick, often fused anteroventral appendages (=palps). One segment bears 1 or 2 pairs of tentacular cirri. The parapodia are uniramous. The foregut is tripartite: a smooth pharyngeal tube with a single tooth, a distal ring of teeth or unarmed, a muscular proventriculus, and an esophagus. Various kinds of brooding and fission are characteristic. The approximately 80 genera with about 800 species are especially numerous in the shallow subtidal region between sponges, hydroids, algae, in crevices of corals and other hard substrates. Common species in sandy and muddy sediments, some of them worldwide distributed, belong to the following selection of genera: *Typosyllis* Langerhans, 1879; *Eusyllis* Malmgren, 1867; *Petitia* Siewing, 1955; *Pionosyllis* Malmgren, 1867; *Streptosyllis* Webster and Benedict, 1884; *Syllides* Oersted, 1845; *Brania* Quatrefages, 1865; *Exogone* Oersted, 1845; *Plakosyllis* Hartmann-Schröder, 1956; *Sphaerosyllis* Claparède, 1863. Important taxonomic features include number and shape of head appendages, presence and shape of parapodial cirri, shape and armament of foregut structures, chaetation, and types of reproduction. Revisions are needed for a number of genera (Figure 29.1a-j). References: Imajima (1966a-e), Westheide (1974b), Perkins (1981), San Martin (1984).

Family Sigalionidae.--This taxon of usually long and narrow, elytra-bearing scaleworms with numerous segments comprises about 160 species in about 21 genera. Small species especially occur in the genera *Pholoe* Johnston, 1839, *Pholoides* Pruvot, 1895, and *Metaxypsamma* Wolf, 1986. Important taxonomic features are structure of anterior end, including surface ornamentation of appendages, shape and ornamentation of elytra and parapodia, and chaetation. (Figure 29.3a.) References: Laubier (1975),

Pettibone (1982), Wolf (1986a).

Family Pisionidae.--The small to medium-sized slender species of this small taxon have numerous homonomous segments. Their surface is smooth and iridescent. The prostomium is very small, indistinct or completely reduced, with 1 pair of prominent slender appendages surrounded by and fused with the tentacular segment which bears a pair of anteriorly directed slender tentacular cirri and usually a pair of short, flask-shaped ventral cirri. There is usually a pair of stout acicular chaetae obliquely in front of the mouth. The dorsal and ventral flask-shaped cirri are missing on segment 2, which has slender ventral cirri, or on segment 3, which may have long dorsal cirri. The uniramous parapodia are equipped with complex copulatory appendages in the males in a species specific number of middle segments. The prominent muscular pharynx may have paired jaws. Species of the 4 genera especially occur in coarse sand sediments: *Pisione* Grube, 1857 (about 16 species) (Figure 29.3d); *Pisionella* Hartmann, 1939 (1 species); *Pisionidens* Aiyar and Alikunki, 1943 (3 species), without chaetae in adults; *Anoplopisione* Laubier, 1967 (1 species), jaws are lacking. Important taxonomic features include chaetation, number of fertile segments, number and details of male copulatory organs. References: Laubier (1967b,c), Stecher (1968), Yamanishi (1976).

Family Dorvilleidae.--The species range from very small forms with few segments to medium-sized multi-segmented animals with high morphological variety. The taxa can be arranged in a morphological sequence with increasing degree of paedomorphic organization, strongly suggesting their progenetic origin. The prostomium has 2 pairs of appendages which may be reduced to only 1 or none at all and followed by 2 achaetous rings. The pygidium possesses 2 or 3 appendages. There are sub-biramous, uniramous or no parapodia. Chaetae are simple and compound, only simple, or absent. Smaller forms possess ciliary rings and a distinct pattern of sensoria, especially on the prostomium. The refractive epidermal glands are obvious. All but 2 genera show cuticular mouthparts of high variety and complexity. The 70 species in about 16 genera live in intertidal, mostly subtidal, sand sediment: *Dorvillea* Parfitt, 1866; *Ophryotrocha* Claparède and Mecznikow, 1869 (often in marine aquaria); *Protodorvillea* Pettibone, 1961; *Parapodrilus* Westheide, 1965; *Parophryotrocha* Hartmann-Schröder, 1971; *Exallopus* Jumars, 1974; *Meiodorvillea* Jumars, 1974; *Pettiboneia* Orensanz, 1973; *Ikosipodus* Westheide, 1982; *Gymnodorvillea* Wainwright and Perkins, 1982; *Apodotrocha* Westheide and Riser, 1983; *Coralliotrocha* Westheide and von Nordheim, 1985; *Arenotrocha* Westheide and von Nordheim, 1985; *Pusillotrocha* Westheide and von Nordheim, 1985; *Eliberidens* Wolf, 1986; *Diaphorosoma* Wolf, 1986; *Ougia* Wolf, 1986; *Parougia* Wolf, 1986; *Westheideia* Wolf, 1986; *Petrocha* von Nordheim, 1987. (Figure 29.2c-h.)

Important taxonomic features include degree of development of head appendages and notopodial cirri, structure of cuticular jaw elements, and types of chaetae. References: Jumars (1974), Westheide (1982), Westheide and von Nordheim (1985), Wolf (1986b,c).

Family Dinophilidae.--The short species with few segments have a length up to 2 mm. There are no parapodia, chaetae, or appendages except for a short unpaired anal appendage that is more or less visible in species of *Dinophilus*. The prostomium is characterized by distinct, stiff sensoria. Ciliary rings on prostomium and segments are obvious. The animals move with a ciliary band on their ventral side. The monomorphic or dimorphic - dwarf males! - species of *Dinophilus* O. Schmidt, 1848 (taxonomic validity of several of them doubtful), are often found in marine aquaria. *Trilobodrilus* Remane, 1925 (4 species), live in clean intertidal and subtidal sands. The taxonomic position of *Apharyngtus* Westheide, 1971 (1 species), is uncertain. Important taxonomic features include number of segments, epidermal glands, ciliation, and pattern of sensoria. (Figure 29.2j,l.) References: Westheide (1967b), Rao (1973), Donworth (1986).

Family Nerillidae.--All species of the 12 genera are short, with few segments, and usually move by gliding with the ventral ciliary band. The prostomium carries at least 2 appendages. The parapodia are reduced with bundles of fine simple or compound chaetae. *Nerilla* O. Schmidt, 1848; *Troglochaetus* Delachaux, 1921; *Nerillidium* Remane, 1925; *Mesonerilla* Remane, 1949; *Thalassochaetus* Ax, 1954; *Meganerilla* Boaden, 1961; *Paranerilla* Jouin and Swedmark, 1965; *Psammoriedlia* Kirsteuer, 1966; *Nerillidopsis* Jouin, 1966; *Afronerilla* Faubel, 1978; *Bathychaetus* Faubel, 1978; *Bathynerilla* Faubel, 1978. Important taxonomic features include number and shape of prostomial appendages, number of segments, presence of parapodial cirri, number and shape of chaetae, number of intracellular pharyngeal hard structures, and reproductive organs. (Figure 29.4o-s.) References: Jouin (1967), Swedmark (1959), Schmidt and Westheide (1977).

Family Orbiniidae.--The genus *Schroederella* Laubier, 1962 (2 species), with long and pointed prostomium and branchiae on the posterior segments may belong to the meiofauna. Important taxonomic features are shape of prostomium, chaetation and anal cirri (Figure 29.3e). Reference: Laubier (1962).

Family Spionidae.--This large family comprises more than 300 species of which many are smaller than 6 mm. They commonly inhabit muddy and sandy sediments. These elongate animals with numerous similar segments are characterized by a pair of conspicuous extensile, mobile head tentacles (=palps) with a ciliated groove for gathering food. Reference: Light (1978), Blake and Kudenov (1978).

Family Protodriloidae.--This recently erected family comprises two species only (genus *Protodriloides*, Jouin, 1966) which resemble the *Protodrilus* species; their anterior appendages, however, are solid without internal canals and close together basally. *Protodriloides chaetifer* with chaetae. (Figure 29.4e,f.) References: Jouin (1966a), Purschke and Jouin (in press).

Family Protodrilidae.--Very slender, generally multisegmented species reveal one pair of prostomial appendages. The posterior end is a bilobed or trilobed adhesive appendage. The animals move by gliding with a ciliary band on the ventral body side. Two genera: *Protodrilus* Hatschek, 1880 (30 species) (basis of head appendages distinctly separated), and *Astomus* Jouin, 1979 (1 species) (gutless). Important taxonomic features include number of segments, epidermal glands, position of salivary glands, ciliated lateral organs and spermducts in the males, and position of fertile segments. Immature individuals are difficult to identify. Discrimination of several species is possible by ultrastructure of, e.g., sperm. (Figure 29.4a-d.) References: Jouin (1970b, 1971), von Nordheim (1983, 1987).

Family Saccocirridae.--The family consists of one genus only (*Saccocirrus* Bobretzky, 1872, with about 17 species). The slender, multi-segmented animals possess one pair of long, motile prostomial appendages. The short parapodial stumps have highly retractile chaetae of different types, often chisel-shaped. Usually anal appendages are bilobed and adhesive. The species are highly active, particularly common in coarse intertidal and shallow subtidal sand throughout warm and tropical regions. Important taxonomic features include chaetae, position of gonads and copulatory organs, anal appendages, and the presence of a pharyngeal bulb. (Figure 29.4g-k.) References: Jouin (1971), Brown (1981), Jouin and Rao (in press).

Family Psammodrilidae.--The more or less slender almost completely ciliated species belong to 2 genera: *Psammodrilus* Swedmark, 1952 (2 species), and *Psammodriloides* Swedmark, 1958 (1 species). There are no prostomial or anal appendages. The parapodia are reduced except for 6 thoracic dorsal cirri, supported with aciculae. (Figure 29.3j,k.) References: Swedmark (1955, 1958), Kristensen and Nørrevang (1982).

Family Paraonidae.--This family includes mostly medium-sized elongated species of which a few small forms may be considered belonging to the meiofauna. The prostomium is oval or cone-shaped with an unpaired median dorsal antenna. The antenna may be absent. The parapodia are biramous and have simple chaetae. Several anterior segments possess simple branchiae. There are 2 to 3 anal appendages. Important taxonomic features include presence or absence of antenna, number and distribution of branchiae and chaetation. (Figure 29.4t.) References: Strelzow (1973), Hartley (1981).

Family Questidae.--The 2 genera *Questa* Hartman, 1966 (3 species), and *Novaquesta* Hobson, 1970 (1 species) have slender bodies. The prostomium is without appendages. Cirri and several pairs of branchiae may occur on posterior segments. There are different types of simple chaetae. Important taxonomic features are details of chaetae, posterior appendages, and reproductive organs (Figure 29.3g,h). References: Giere and Riser (1981), Jamieson and Webb (1984).

Family Ctenodrilidae.--The short (15 segments) to slender (40 segments) species are often found in marine aquaria. The pigmented inclusions (often dark purple or red brown) resemble conditions in *Aeolosoma*. The species are without any appendages or with filamentous branchiae on few or several anterior segments. There are no parapodia but simple chaetae. Asexual reproduction by means of fission is common. Four genera: *Ctenodrilus* Claparède, 1863 (3 species); *Rhaphidrilus* Monticelli, 1910 (1 species); *Raricirrus* Hartman, 1961 (1(?) species); *Aphropharynx* Wilfert, 1974 (1 species). Important taxonomic features include number of segments, presence of appendages, details of chaetae, beginning of the midgut region, number of zooids, and position of fission zone. (Figure 29.3m.) References: Wilfert (1973, 1974), George and Petersen (in press).

Family Parergodrilidae.--The two enchytraeid-like species have a short cylindrical body without any appendages. They possess simple chaetae in up to 15 segments. *Stygocapitella subterranea* (Figure 29.3f) lives in the upper beach slope. *Parergodrilus heideri* is a terrestrial species in forest soils. References: Karling (1958), Purschke (1987).

Family Acrocirridae.--The medium-sized to small

slender species have numerous or a moderate number of segments. True meiobenthic species belong to the genus *Macrochaeta* Grube, 1850 (7 species), characterized by 2 ventral prostomial appendages and up to 6 pairs of relatively long filiform branchiae on the anterior most segments, which are delicate and usually partly or completely lost in fixed specimens. The body surface is rugose; the epidermis has yellow refractive glands. There are prominent papillae all over the body. Distinct parapodial lobes do not exist, but a few long chaetae can be found widely protruding in each segment. There are simple, pseudocompound, or compound hooded chaetae. Important taxonomic features include number and shape of papillae, number of branchiae, and chaetation. (Figure 29.3o.) References: Banse (1969), Westheide (1981).

Family Fauveliopsidae.--The genus *Fauveliopsis* McIntosh, 1922 (about 11 species), occurring mostly in deep sea regions, consists of elongate, subcylindrical species of medium and small size, which are without any appendages and parapodial lobes. The head region is usually not visible. (Figure 29.3n.) References: Katzmann and Laubier (1974), Riser (1987).

Family Sabellidae.--This large taxon of mostly macrofaunal and sessile species has also a number of smaller species that are able to move around outside their soft membranous tubes, if any exist. Their body consists of a few chaetigers only. *Euchone* Malmgren, 1866; *Fabricia* Blainville, 1828; *Fabriciola* Friedrich, 1939; *Manayunkia* Leidy, 1859, (brackish water species). Important taxonomic features include shape, arrangement of tentacles and other details of tentacular crown, pattern of epidermal gland cells (stained with methyl green), and chaetation. (Figure 29.3l.) References: Fauchald (1977), Banse (1970), Hobson and Banse (1981).

Family Diurodrilidae.--Minute, short species with 5 trunk segments of the genus *Diurodrilus* Remane, 1925 (6 species), about 400 µm. They can easily be mistaken for chaetonoid gastrotrichs. There are no head appendages, parapodia, or chaetae, but sensoria are present especially on the prostomium. The posterior end has toe-like adhesive organs, the ventral side is characterized by a distinct pattern of cells with numerous cilia. Alterations between fast jerky locomotion and position of rest is characteristic. Important taxonomic features are pattern of ciliation on the ventral side, pattern of stiff, adjoined cilia (=sensoria) and shape of posterior toes. (Figure 29.2m.) References: Ax (1967), Mock (1981), Kristensen and Niilonen (1982).

Family Polygordiidae.--The approximately 15 species (revision necessary) of the single genus *Polygordius* Schneider, 1868, are slender, thread-like animals whose many segments are externally not visible. The surface is iridescent and completely smooth. The prostomium bears 2 short stiff stout appendages, with bases close together. The posterior end of the body is slightly expanded, and has adhesive glands; some species carry paired anal appendages. Chaetae are lacking. Important taxonomic features are the circulatory system and the posterior end of the body. (Figure 29.4l-n.) References: Jouin (1970a), Jouin and Rao (in press).

Family Lobatocerebridae.--The turbellariomorph animals of the single monotypic genus *Lobatocerebrum* are slender, completely ciliated, without any segmentation, but with a male genital system, and an anus-hindgut system that support provisional classification as Annelida. Species are probably worldwide distributed, in intertidal to deep subtidal sandy and muddy sediments. (Figure 29.2a,b.) References: Rieger (1980).

Methods of Collection and Extraction

Polychaetes can be collected by nearly all of the methods described for other meiofaunal taxa. And, as in most instances, it is important to consider the habitat involved (see notations in "Classification"). The extraction of marine meiofaunal polychaetes from sediment is facilitated by narcotization of the sample with an isotonic solution of magnesium chloride; the seawater ice method is also highly recommended (Chapter 9). Extraction must be conducted with recognition that many species are very fragile. thus, the removal from the sediment in a living condition and subsequent relaxation and fixation (ca. 10% buffered formalin) is preferable.

Preparation of Specimens for Taxonomic Study

For determination and description of small polychaete species, investigation of living individuals is highly recommended, and for achaetous taxa it is a must. Handling of living animals should follow the instructions laid down in Chapter 10.

Most polychaetes can be narcotized with 4-8% $MgCl_2$ solution. By slowly sucking off the water under the coverglass, many polychaete species can be heavily squashed before being eventually destroyed. Thus important structural details are easy to observe in living animals under the compound microscope, e.g., epidermal glands, sensory structures, ciliation, pharynx, jaw elements, chaetae. The highest magnification (100x objective, oil immersion), however, has to be used, preferably with interference contrast optics. Interference contrast microscopy also allows recognition of structures like ciliary tufts and bands even in fixed specimens when they are heavily squashed. In studying exterior structures such as cilia bands, nuchal organs or lateral organs in alcohol preserved specimens, short-term staining (less than one minute) with Mayer's muchaematein (water solution) was recommended by light-microscopists. The best information on ciliation results from scanning microscopy.

For descriptions of species, the following standard drawings are recommended: anterior end that shows all appendages and prominent sensory organs, including tentacular segments and first chaetigerous segments; posterior end with pygidium and anal cirri; parapodia with fascicles of chaetae, preferably with the anterior side facing the observer; single chaetae of all types observed; extended pharynx - if possible - is also useful. The use of a camera lucida is absolutely necessary, at least for the contours of the chaetae. For drawings of parapodia a dissection of single segments with a pair of parapodia is best obtained from fixed specimens. The standard fixing agent is 10% formalin in sea water. Bouin's fluid is better for preserving the body shape of small specimens. Ethanol should never be used for fixation because it especially ruins the chaetae after a certain time in storage. After at least 24 hours, however, fixed specimens should be transferred from the fixative to 70-80% alcohol; salt should be washed out before transfer by rinsing the fixed specimens in distilled water. If only a few individuals are available, narcotized animals which have been carefully used for live observation may also be fixed by placing drops of Bouin's fluid next to the coverslip.

Tiny needles (insect pins glued to wooden kabob skewers with fingernail polish), iris scissors or small scalpels made by pieces of razor blades glued to wooden skewers are needed for the preparation of parapodia and distinct parts of the polychaete body. These parts are mounted on a flat slide in glycerin-alcohol, glycerin-distilled water or mounting media like Zeiss W 15 that mix easily with water. The latter medium can be used also for a permanent mount, but phase or interference contrast is necessary for observation.

A variety of lactophenol mounting and clearing media is used by polychaete systematists (for recipes see Hobson and Banse, 1981), but usually are not recommended for small species. Tiny fixed specimens may be prepared, dehydrated, cleaned and mounted on glass slides under continuous observation with a dissecting microscope, e.g., with increasing concentrations of alcohol, subsequently successive additions of toluol or xylol, followed by a mounting medium like DEPEX or Technicon Mounting Medium, and then covered with a cover glass. For larger species this treatment should be done in depression slides or small dishes, and from here specimens have to be removed to flat slides.

Mounting of the entire animal is also recommended for deposition of tiny species, especially for type material. Specimens, including type material, are usually best preserved in 70-80% ethanol, which allows preparations to be made if necessary.

For detailed additional references of useful sample treatment, curation, and techniques used for identification and description of especially larger polychaete species see Fauchald (1977). Keys treating only meiobenthic polychaete species do not exist; so-called stygobiontic Polychaeta comprising also the marine interstitial species were recently listed by Hartmann-Schröder (1986). A most valuable worldwide key to supraspecific polychaete taxa is that of Fauchald (1977). Polychaete faunas with keys and diagnoses for specific geographic areas which also address the small species are, e.g., Hartmann-Schröder (1971) for the German coasts and neighboring regions, Banse and Hobson (1974), Hobson and Banse (1981) for the northern Pacific coast of North America, and Uebelacker and Johnson (1984) for the Gulf of Mexico. For questions of nomenclature the "Catalogue of the Polychaetous Annelids of the World" by Hartman (1959, 1965) is still indispensable.

Type Collections

Large collections of meiofauna polychaete species are in the Zoologisches Institut und Zoologisches Museum, Universität Hamburg, Hamburg, FRG, and in the National Museum of Natural History, Smithsonian Institution, Washington, D.C., USA.

References

Aiyar, R.G., and K.M. Alikunhi
1944. On Some Archiannelids from the Sandy Beach, Madras. *Proceedings of the National Institute of Sciences of India*, 10:113-140.

Alikunhi, K.H.
1948. On Some Archiannelids of the Krusadai Island. *Proceedings of the National Institute of Sciences of India*, 14:373-383.

Ax, P.
1967. *Diurodrilus ankeli* nov. spec. (Archiannelida) von der nordamerikanischen Pazifikküste. Ein Beitrag zur Morphologie, Systematik und Verbreitung der Gattung *Diurodrilus*. *Zeitschrift für Morphologie und Ökologie der Tiere*, 60:5-16.

Banse, K.
1969. Acrocirridae, n. fam. (Polychaeta Sedentaria). *Journal of Fisheries Research Board of Canada*, 26:2595-2620.

1970. The Small Species of *Euchone* Malmgren (Sabellidae, Polychaeta). *Proceedings of the Biological Society of Washington*, 83:387-408.

Banse, K., and K.D. Hobson
1974. Benthic Errantiate Polychaetes from British Columbia and Washington. *Bulletin of the Fisheries Research Board of Canada*, 185:1-111.

Blake, J.A., and J.D. Kudenov
1978. The Spionidae (Polychaeta) from Southeastern Australia and Adjacent Areas with a Revision of the Genera. *Memoirs of the National Museum Victoria*, 39:171-280.

Boaden, P.J.S.
1961. *Meganerilla swedmarki*, nov. gen., nov. spec., An Archiannelid of the Family Nerillidae. *Arkiv för Zoologie*, series 2, 2:553-559.

Brown, R.
1981. Saccocirridae (Annelida: Archiannelida) from the Central Coast of New South Wales. *Australian Journal of Marine and Freshwater Research*, 32:439-456.

Dohle, W.
1967. Zur Morphologie und Lebensweise von *Ophryotrocha gracilis* Huth, 1934 (Polychaeta, Eunicidae). *Kieler Meeresforschungen*, 23:68-74.

Donworth, P.J.
1986. A Reappraisal and Validation of the Species *Dinophilus taeniatus* Harmer, 1889 and of Taxonomically Significant Features in Monomorphic Dinophilids (Annelida: Polychaeta). *Zoologischer Anzeiger*, 6:1-19.

Fauchald, K.
1974. Sphaerodoridae (Polychaeta: Errantia) from World-Wide Areas. *Journal of Natural History*, 8:257-289.
1977. The Polychaete Worms. Definitions and Keys to the Orders, Families and Genera. *Natural History Museum of Los Angeles County, Science Series*, 28:1-190.

George, J.D., and M.E. Petersen
1988. The Validity of the Genus *Zeppelina* Vaillant (Polychaeta: Ctenodrilidae). (in press).

Gerlach, S.A.
1953. Zur Kenntnis der Archianneliden des Mittelmeeres. *Kieler Meeresforschungen*, 9:248-251.

Giere, O., and N.W. Riser
1981. Questidae - Polychaetes with Oligochaetoid Morphology and Development. *Zoologica Scripta*, 10:95-103.

Hartley, J.P.
1981. The Family Paraonidae (Polychaeta) in British Waters: A New Species and New Records with a Key to Species. *Journal of the Marine Biological Association of the United Kingdom*, 61:133-149.

Hartman, O.
1959. Catalogue of the Polychaetous Annelids of the World. *Allan Hancock Foundation Occasional Paper*, 23, 629 pages.
1965. Catalogue of the Polychaetous Annelids of the World. *Allan Hancock Foundation Occasional Paper*, Supplement and Index, 197 pages.

Hartmann-Schröder, G.
1956. Polychaeten-Studien I. *Zoologischer Anzeiger*, 157:87-91.
1962. Zweiter Beitrag zur Polychaetenfauna von Peru. *Kieler Meeresforschungen*, 18:109-147.
1963. Revision der Gattung *Mystides* Theel (Phyllodocidae; Polychaeta, Errantia). Mit Bemerkungen zur Systematik der Gattungen *Eteonides* Hartmann-Schröder und *Protomystides* Czerniavsky und mit Beschreibungen zweier neuer Arten aus dem Mittelmeer und einer neuen Art aus Chile. *Zoologischer Anzeiger*, 171:204-243.
1971. Annelida, Borstenwürmer, Polychaeta. *Die Tierwelt Deutschlands und der angrenzenden Meeresgebiete*, 58:1-594.
1977. Die Polychaeten der Kubanisch-Rumänischen Biospeologischen Expedition nach Kuba 1973. *Résultats des Expéditions Biospéologiques Cubano-Roumaines à Cuba*, 2:51-63.
1986. Polychaeta (incl. Archiannelida). Pages 210-233 in L. Botosaneanu, editor, *Stygofauna Mundi*. Leiden: Brill/Backhuys.

Hobson, K.D.
1970. *Novaquesta trifurcata*, A New Genus and Species of the Family Questidae (Annelida, Polychaeta) from Cape Cod Bay, Massachusetts. *Proceedings of the Biological Society of Washington*, 83:191-194.

Hobson, K.D., and K. Banse
1981. Sedentariate and Arachiannelid Polychaetes of British Columbia and Washington. *Canadian Bulletin of Fisheries and Aquatic Science*, Number 209:1-132.

Jamieson, B.G.M., and R.I. Webb
1984. The Morphology, Spermatozoal Ultrastucture and Phylogenetic Affinities of a New Species of Questid (Polychaeta: Annelida). Pages 21-34 in P.A. Hutchings, editor, *Proceedings of the First International Polychaete Conference, Sydney*. New South Wales: The Linnean Society of New South Wales.

Jmajima, M.
1966a. The Syllidae (Polychaetous Annelids) from Japan. I. Exogoninae. *Publications of the Seto Marine Biology Laboratory*, 13:385-404.
1966b. The Syllidae (Polychaetous Annelids) from Japan. II. Autolytinae. *Publications of the Seto Marine Biology Laboratory*, 14:27-83.
1966c. The Syllidae (Polychaetous Annelids) from Japan. III. Eusyllinae. *Publications of the Seto Marine Biology Laboratory*, 14:85-111.
1966d. The Syllidae (Polychaetous Annelids) from Japan. IV. Syllidae 1. *Publications of the Seto Marine Biology Laboratory*, 14:219-252.
1966e. The Syllidae (Polychaetous Annelids) from Japan. V. Syllidae 2. *Publications of the Seto Marine Biology Laboratory*, 14:253-294.

Jouin, C.
1966a. Morphologie et anatomie comparée de *Protodrilus chaetifer* Remane et *Protodrilus symbioticus* Giard; création du nouveau genre *Protodriloides* (Archiannélides). *Cahiers de Biologie Marine*, 7:139-155.
1966b. Hermaphrodisme chez *Nerillidopsis hyalina* n.g. n.spec. et chez *Nerillidium* Remane, Archiannélides Nerillidae. *Comptes Rendues de l'Académie des Sciences Paris*, 263:412-415.
1967. Etude morphologique et anatomique de *Nerillidopsis hyalina* Jouin et de quelques *Nerillidium* Remane (Archiannélides, Nerillidae). *Archives de Zoologie Experimentale et Générale, Paris*, 108:97-110.
1970a. Archiannélides interstitielles de Nouvelle-Calédonie. Pages 149-167 in *Expedition Française sur Recifs Coralliens de la Nouvelle Caledonie*, Volume 4. Paris: Edition de Fondation Singer-Polignac.
1970b. Recherches sur les Protodrilidae (Archiannélides: I. Etude morphologique et systématique du genre *Protodrilus*. *Cahiers de Biologie Marine*, 11:367-434.
1971. Status of the Knowledge of the Systematics and Ecology of Archiannelida. Pages 47-56 in N.C. Hulings, editor, *Proceedings of the First International Conference on Meiofauna*. *Smithsonian Contributions to Zoology*, 76.

Jouin, C., and G.C. Rao
1988. Morphological Studies on Some Polygordiidae and Saccocirridae (Polychaeta) from the Indian Ocean. *Cahiers de Biologie Marine, (in press)*.

Jumars, P.
1974. A Generic Revision of the Dorvilleidae (Polychaeta) with Six New Species from the Deep North Pacific. *Journal of the Linnean Society of London, Zoology*, 54:101-135.

Karling, T.G.
1958. Zur Kenntnis von *Stygocapitella subterranea* Knöllner und *Parergodrilus heideri* Reisinger (Annelida). *Arkiv för Zoologie*, 11:307-342.

Katzmann, W.
1973. Zwei neue Sphaerodoridae (Polychaeta/Meiofauna) aus der Adria. *Annalen des Naturhistorischen Museums Wien*, 77:283-286.

Katzmann, W., and L. Laubier
1974. Le genre *Fauveliopsis* (Polychète sédentaire) en

Mediterranée. *Mikrofauna des Meeresbodens,* 50:1-16.

Kristensen, R.M., and T. Niilonen
1982. Structural Studies on *Diurodrilus* Remane (Diurodrilidae fam. nov.) with Description of *Diurodrilus westheidei* sp. n. from the Arctic Interstitial Meiobenthos, W. Greenland. *Zoologica Scripta,* 11:1-12.

Kristensen, R.M., and A. Nørrevang
1982. Description of *Psammodrilus aedificator* sp. n. (Polychaeta), With Notes on the Arctic Interstitial Fauna of Disko Island, W. Greenland. *Zoologica Scripta,* 11:265-279.

Laubier, L.
1962. *Schroederella pauliani* gen. nov., sp. nov., un nouvel orbiniide (polychètes sédentaires) de la faune interstitielle d'Afrique. *Annals of the Transvaal Museum,* 24:231-238.
1967a. Annélides Polychètes interstitielles de Nouvelle-Calédonie. Pages 91-101 in *Expedition Française sur Recifs Coralliens de la Nouvelle Calédonie.* Paris: Edition de Fondation Singer-Polignac.
1967b. Quelques Annélides Polychètes interstitielles d'une plage de Côte d'Ivoire. *Vie et Milieu,* 18(Serie A):573-594.
1967c. Présence d'une Annélide Polychète de la famille des Pisionidae appartenant a un genre nouveau dans les eaux interstitielles littorales de Côte d'Ivoire. *Comptes Rendues de l'Académie des Sciences, Paris,* 264:1431-1433.
1967d. Sur quelques *Aricidea* (Polychètes, Paraonidae) de Banyuls-sur-Mer. *Vie et Milieu, Série A,* 18:99-132.
1975. Adaptations morphologiques et biologiques chez un aphroditien interstitiel: *Pholoe swedmarki* sp. n. *Cahiers de Biologie Marine,* 16:671-683.

Light, W.J.
1978. Spionidae (Polychaeta, Annelida). Pages 1-211 in W.L. Lee, editor, *Invertebrates of the San Francisco Bay Estuary system.* Pacific Grove, California: Boxwood Press.

Mikkelsen, P.S., and R.W. Virnstein
1982. An Illustrated Glossary of Polychaete Terms. *Harbor Branch Foundation, Inc., Technical Report,* Number 46:1-92.

Mock, H.
1981. Zur Kenntnis von *Diurodrilus subterraneus* (Polychaeta, Dinophilidae) aus dem Sandhang der Nordseeinsel Sylt. *Helgoländer Meeresuntersuchungen,* 34:329-335.

Perkins, T.H.
1981. Syllidae (Polychaeta) Principally from Florida, with Descriptions of a New Genus and Twenty-one New Species. *Proceedings of the Biological Society of Washington,* 93:1080-1172.

Pettibone, M.H.
1982. Annelida. In S.P. Parker, editor, *Synopsis and Classification of Living Organisms.* Volume 2. New York: McGraw-Hill Book Company.

Purschke, G.
1987. Anatomy and Ultrastructure of Ventral Pharyngeal Organs and Their Phylogenetic Importance in Polychaeta (Annelida). III. The Pharynx of the Parergodrilidae. *Zoologische Jahrbücher, Anatomie,* 115:331-362.

Purschke, G., and C. Jouin
1988. Anatomy and Ultrastructure of the Ventral Pharyngeal Organs of *Saccocirrus* and *Protodrilus* with Remarks on Their Phylogenetic Relationships within the Protodrilida (Annelida, Polychaeta). *Journal of Zoology* (London). In Press.

Rao, G.C.
1973. *Trilobodrilus indicus* n. sp. (Dinophilidae, Archiannelida) from Andhra Coast. *Proceedings of the Indian Academy of Sciences, Section B,* 77:101-108.

Rieger, R.M.
1980. A New Group of Interstitial Worms, Lobatocerebridae nov. fam. (Annelida) and Its Significance for Metazoan Phylogeny. *Zoomorphologie,* 95:41-84.

Riser, N.W.
1987. A New Interstitial Polychaete (Family Fauveliopsidae) from the Shallow Subtidal of New Zealand with Observations on Related Species. *Bulletin of the Biological Society of Washington,* Number 7, 1987:211-216.

San Martin, G.
1984. Estudio biogeográfico, faunístico y sistemático de los poliquetos de la familia Silidos (Syllidae: Polychaeta) en Baleares. Tesis Doctoral. Editorial de la Universidad Complutense de Madrid.

Schmidt, P., and W. Westheide
1977. Interstitielle Fauna von Galapagos. XVII. Polygordiidae, Saccocirridae, Protodrilidae, Nerillidae, Dinophilidae (Polychaeta). *Mikrofauna des Meeresbodens,* 62:1-38.

Stecher, H.-J.
1968. Zur Organisation und Fortpflanzung von *Pisione remota* (Southern) (Polychaeta, Pisionidae). *Zeitschrift für Morphologie der Tiere,* 61:347-410.

Strelzow, V.
1973. Polychaetous Annelids of the Family Paraonidae Cerruti, 1909 (Polychaeta, Sedentaria) [in Russian]. *Akadademiya Nauk, USSR, Leningrad,* 170 pages.

Swedmark, B.
1955. Recherches sur la morphologie, le développement et la biologie de *Psammodrilus balanoglossoides,* Polychète Sédentaire de la microfauna des sables. *Archives de Zoologie Expérimentale et Génerale,* 92:141-220.
1958. *Psammodriloides fauveli* n. gen., n. sp. et la famille Psammodrilidae (Polychaeta Sedentaria). *Arkiv för Zoologie,* series 2, 12:55-64.
1959. Archiannélides Nerillidae des côtes du Finistère. *Archives de Zoologie Expérimentale et Génerale,* 98:26-42.

Uebelacker, J.M.
1984. Family Syllidae Grube, 1850. Chapter 30, pages 1-150 in J.M. Uebelacker and P.G. Johnson, editors, *Taxonomic Guide to the Polychaetes of the Northern Gulf of Mexico. Final Report to the Minerals Management Service, Contract 14-12-001-29091.* 7 volumes. Mobile, Alabama: Barry A. Vittor and Associates.

Uebelacker, J.M., and P.G. Johnson, editors
1984. *Taxonomic Guide to the Polychaetes of the Northern Gulf of Mexico. Final Report to the Minerals Management Service, Contract 14-12-001-29091.* 7 volumes. Mobile, Alabama: Barry A. Vittor and Associates.

von Nordheim, H.
1983. Systematics and Ecology of *Protodrilus helgolandicus* sp. n., an Interstitial Polychaete (Protodrilidae) from Subtidal Sands of Helgoland, German Bight. *Zoologica Scripta,* 12:171-177.
1987. *Anatomie, Ultrastruktur und Systematik der Gattung Protodrilus (Annelida, Polychaeta).* 298 pages. Dissertation, Universität Osnabrück.

Westheide, W.
1965. *Parapodrilus psammophilus* nov. gen. nov. spec. eine neue Polychaeten-Gattung aus dem Mesopsammal der Nordsee. *Helgoländer wissenschaftliche Meeresuntersuchungen,* 12:207-213.
1967a. Monographie der Gattungen *Hesionides* Friedrich und *Microphthalmus* Mecznikow (Polychaeta, Hesionidae). Ein Beitrag zur Organisation und Biologie psammobionter Polychaeten. *Zeitschrift für Morphologie der Tiere,* 61:1-159.
1967b. Die Gattung *Trilobodrilus* (Archiannelida, Polychaeta) von der deutschen Nordseeküste. *Helgoländer wissenschaftliche Meeresuntersuchungen,* 16:207-215.
1970. Zur Organisation, Biologie und Ökologie des interstitiellen Polychaeten *Hesionides gohari* Hartmann-Schröder (Hesionidae). *Mikrofauna des Meeresbodens,* 3:1-37.
1971. *Apharyngtus punicus* nov. gen. nov. spec., ein aberranter Archiannelide aus dem Mesopsammal der tunesischen Mittelmeerküste. *Mikrofauna des Meeresbodens,* 6:1-19.

1974a. Interstitielle Polychaeten aus brasilianischen Sandstränden. *Mikrofauna des Meeresbodens*, 31:1-16.
1974b. Interstitielle Fauna von Galapagos. XI. Pisionidae, Hesionidae, Pilargidae, Syllidae, (Polychaeta). *Mikrofauna des Meeresbodens*, 44:1-146.
1977a. Phylogenetic Systematics of the Genus *Microphthalmus* (Hesionidae) Together With a Description of *M. hartmanae* nov. spec. Pages 103-113 in D.J. Reish and K. Fauchald, editors, *Essays on Polychaetous Annelids in Memory of Dr. Olga Hartman*. Allan Hancock Foundation Special Publication.
1977b. Interstitielle Fauna von Galapagos XVIII. Nereidae, Eunicidae, Dorvilleidae (Polychaeta). *Mikrofauna des Meeresbodens*, 63:1-39.
1981. Interstitielle Fauna von Galapagos. XXVI. Questidae, Cirratulidae, Acrocirridae, Ctenodrilidae (Polychaeta). *Mikrofauna des Meeresbodens*, 82:1-23.
1982. *Ikosipodus carolensis* gen. et sp. n., An Interstitial Neotenic Polychaete from North Carolina, U.S.A., and Its Phylogenetic Relationships Within Dorvilleidae. *Zoologica Scripta*, 11:117-126.
1984. The Concept of Reproduction in Polychaetes with Small Body Size: Adaptations in Interstitial Species. In A. Fischer and H.-D. Pfannenstiel, editors, *Polychaete Reproduction. Fortschritte der Zoologie*, 29:265-287.

Westheide, W., and G.C. Rao
1977. On Some Species of the Genus *Hesionides* (Hesionidae, Polychaeta) from Indian Sandy Beaches. *Cahiers de Biologie Marine*, 18:275-287.

Westheide, W., and N.W. Riser
1983. Morphology and Phylogenetic Relationships of the Neotenic Interstitial Polychaete *Apodotrocha progenerans* n. gen., n. sp. (Annelida). *Zoomorphology*, 103:67-87.

Westheide, W., and H. von Nordheim
1985. Interstitial Dorvilleidae (Polychaeta) from Europe, New Zealand and Australia. *Zoologica Scripta*, 14:183-199.

Wilfert, M.
1973. Ein Beitrag zur Morphologie, Biologie und systematischen Stellung des Polychaeten *Ctenodrilus serratus*. *Helgoländer wissenschaftliche Meeresuntersuchungen*, 25:332-346.
1974. *Aphropharynx heterochaeta* nov. gen. nov. spec., ein neuer Polychaet aus der Familie Ctenodrilidae Kennel 1882. *Cahiers de Biologie Marine*, 15:495-504.

Wolf, P.S.
1986a. A New Genus and Species of Interstitial Sigalionidae and a Report on the Presence of Venom Glands in Some Scale-worm Families (Annelida: Polychaeta). *Proceedings of the Biological Society of Washington*, 99:79-83.
1986b. Four New Genera of Dorvilleidae (Annelida: Polychaeta) from the Gulf of Mexico. *Proceedings of the Biological Society of Washington*, 99:616-626.
1986c. Three New Species of Dorvilleidae (Annelida: Polychaeta) from Puerto Rico and Florida and a New Genus for Dorvilleids from Scandinavia and North America. *Proceedings of the Biological Society of Washington*, 99:627-638.

Yamanishi, R.
1976. Interstitial Polychaetes of Japan I. Three New Pisionid Worms from Western Japan. *Publications of the Seto Marine Biological Laboratory*, 23:371-385.

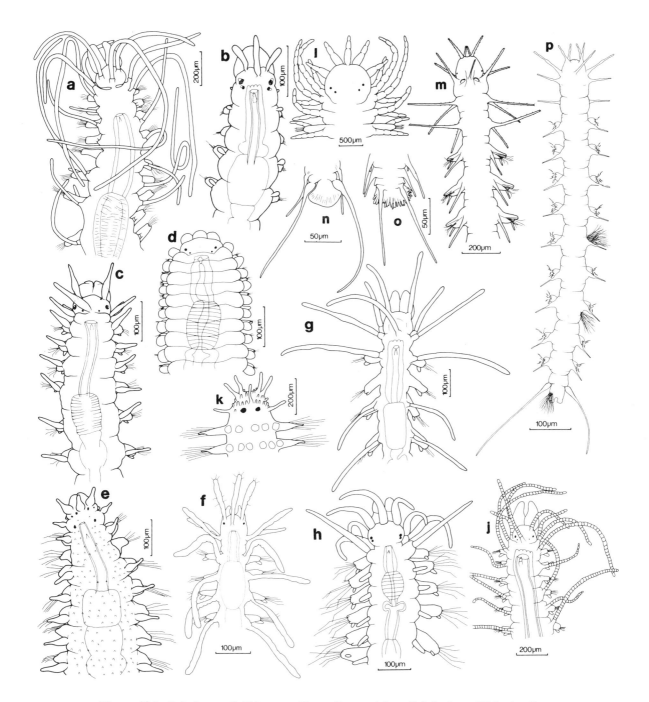

Figure 29.1.--**Polychaeta**: Syllidae: **a**, *Pionosyllis* sp. (after Uebelacker, 1984); **b**, *Exogone naidinoides* Westheide (after Westheide, 1974b); **c**, *Brania oculata* (Hartmann-Schröder) (after Westheide, 1974a); **d**, *Plakosyllis brevipes* Hartmann-Schröder, anterior end (after Hartmann-Schröder, 1956); **e**, *Sphaerosyllis centroamericana* Hartmann-Schröder (after Westheide, 1974b); **f**, *Syllides* sp. (after Westheide, 1974b); **g**, *Eusyllis homocirrata* Hartmann-Schröder (after Westheide, 1974b); **h**, *Petitia amphophthalma* Siewing (after Hartmann-Schröder, 1977); **j**, *Typosyllis glarearia* Westheide (after Westheide, 1974b); Sphaerodoridae: **k**, *Clavodorum adriaticum* Katzmann (after Katzmann, 1973); Hesionidae: **l**, *Heteropodarke heteromorpha* Hartmann-Schröder, anterior end (after Laubier, 1967a); **m**, *Microphthalmus sczelkowii* Mecznikow, anterior end (after Westheide, 1967a); **n**, *Microphthalmus listensis* Westheide, posterior end (after Westheide, 1967a); **o**, *Microphthalmus similis* Bobretzky, posterior end (after Westheide, 1967a); **p**, *Hesionides gohari* Hartmann-Schröder (after Westheide, 1970).

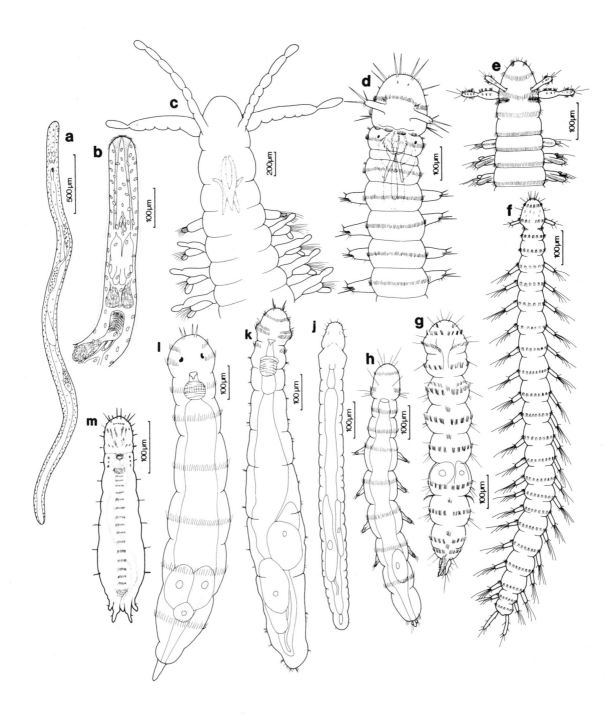

Figure 29.2.—Polychaeta (continued): Lobatocerebridae: **a**, *Lobatocerebrum psammicola* Rieger (after Rieger, 1980); **b**, *L. psammicola* anterior end; Dorvilleidae: **c**, *Dorvillea pacifica* (Westheide) (after Westheide, 1977b); **d**, *Ophryotrocha gracilis* Huth (after Dohle, 1967); **e**, *Pettiboneia australiensis* Westheide and von Nordheim (after Westheide and von Nordheim, 1985); **f**, *Pusillotrocha akessoni* Westheide and von Nordheim (after Westheide and von Nordheim, 1985); **g**, *Apodotrocha progenerans* Westheide and Riser (after Westheide and Riser, 1983); **h**, *Parapodrilus psammophilus* Westheide (after Westheide, 1965); Dinophilidae: **j**, *Apharyngtus punicus* Westheide (after Westheide, 1971); **k**, *Trilobodrilus axi* Westheide (after Westheide, 1967b); **l**, *Dinophilus gyrociliatus* O. Schmidt, female; Diurodrilidae: **m**, *Diurodrilus westheidei* Kristensen and Niilonen, ventral side (after Kristensen and Niilonen, 1982).

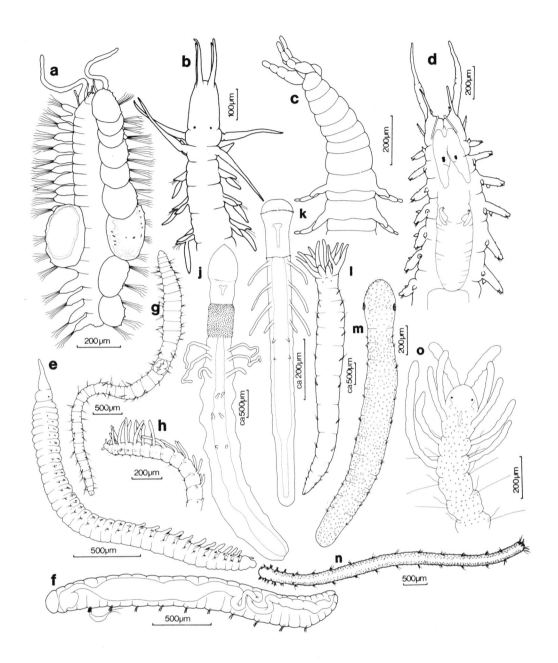

Figure 29.3.--**Polychaeta** (continued): Sigalionidae: **a**, *Pholoe swedmarki* Laubier (after Laubier, 1975); **b**, *Hesionura laubieri* (Hartmann-Schröder) (after Westheide, 1974a); Goniadidae: **c**, *Goniadides falcigera* Hartmann-Schröder, anterior end (after Hartmann-Schröder, 1962); Pisionidae: **d**, *Pisione galapagoensis* Westheide (after Westheide, 1974b); Orbiniidae: **e**, *Schroederella pauliani* Laubier (after Laubier, 1962); Parergodrilidae: **f**, *Stygocapitella subterranea* Knöllner (Courtesy of G. Purschke); Questidae: **g**, *Novaquesta trifurcata* Hobson (after Hobson, 1970); **h**, *Questa media* Westheide, posterior end (after Westheide, 1981); Psammodrilidae: **j**, *Psammodrilus balanoglossoides* Swedmark (after Swedmark, 1955); **k**, *Psammodriloides fauveli* Swedmark (after Swedmark, 1958); Sabellidae: **l**, *Manayunkia aestuarina* (Bourne) (after Wesenberg-Lund from Hartmann-Schröder, 1971); Ctenodrilidae: **m**, *Ctenodrilus serratus* (O. Schmidt) (after Wilfert, 1973); Fauveliopsidae: **n**, *Fauveliopsis brevis* (Hartman) (after Katzmann and Laubier, 1974); Acrocirridae: **o**, *Macrochaeta multipapillata* Westheide (after Westheide, 1981).

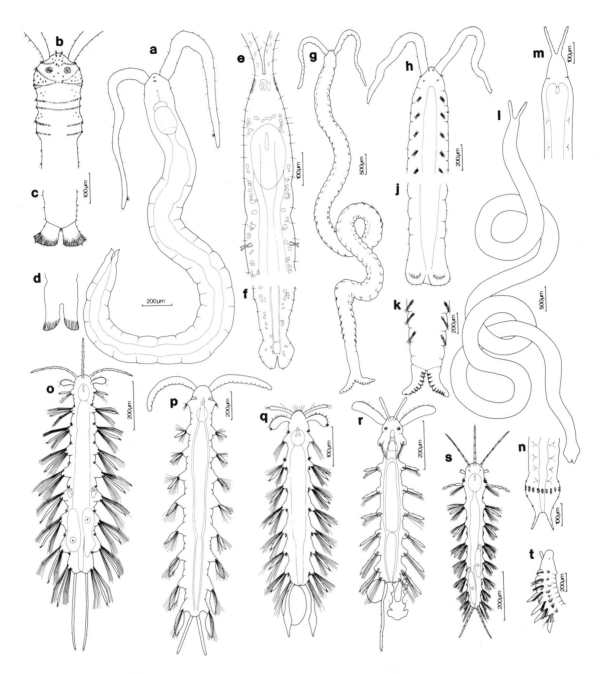

Figure 29.4.--Polychaeta (continued): Protodrilidae: **a**, *Protodrilus brevis* Jouin (after Jouin, 1970); **b**, *Protodrilus helgolandicus* von Nordheim, anterior end; **c**, *P. helgolandicus*, posterior end (after von Nordheim, 1983), **d**, *P. adhaerens* Jägersten, posterior end (after Gerlach, 1953); Protodriloidae: **e**, *Protodriloides chaetifer* (Remane), anterior end (after Jouin, 1966a); **f**, *P. chaetifer*, posterior end (after Jouin, 1966a); Saccocirridae: **g**, *Saccocirrus sp.* (after Jouin, 1971); **h**, *Saccocirrus minor* Aiyar and Alikunhi, anterior end, (after Aiyar and Alikunhi, 1944); **j**, *S. minor*, posterior end; **k**, *Saccocirrus krusadensis* Alikunhi, posterior end (after Alikunhi, 1948); Polygordiidae: **l**, *Polygordius sp.* (after Jouin, 1971); **m**, *P. madrasensis* Aiyar and Alikunhi (after Aiyar and Alikunhi, 1944), anterior end; **n**, *P. madrasensis*, posterior end; Nerillidae: **o**, *Mesonerilla ecuadoriensis* Schmidt and Westheide (after Schmidt and Westheide, 1977); **p**, *Meganerilla swedmarki* Boaden (after Boaden, 1961); **q**, *Nerillidium lothari* Schmidt and Westheide, with brooded egg (after Schmidt and Westheide, 1977); **r**, *Nerillidopsis hyalina* Jouin, with brooded egg and embryo (after Jouin, 1967); **s**, *Nerilla parva* Schmidt and Westheide (after Schmidt and Westheide, 1977); Paraonidae: **t**, *Aricidea (Acesta) cerrutii* Laubier, anterior end, (after Laubier, 1967d).

30. Aeolosomatidae and Potamodrilidae

Dieter Bunke

The Aeolosomatidae and Potamodrilidae are represented by minute annelid worms living in freshwater habitats. Both taxa are made up of only a few species, the Potamodrilidae consists of only a single species, *Potamodrilus fluviatilis*. Conventionally, they are grouped within the Oligochaeta/Clitellata, but recent ultrastructural findings indicate that close affinities with oligochaete taxa do not exist.

The Aeolosomatidae (Figure 30.1a-c) are about 0.3-10 mm long and 0.04-0.06 mm in diameter. In front, a large lobe-like prostomium is present, the ventral side of which is ciliated and serves as a locomotory device. Usually a ciliary groove occurs on either side. The trunk consists of a series of setal segments that, in general, represent a chain of zoids produced by paratomy. The setae, which are absent in the first segment or peristomium, are arranged in four bundles per segment, two of these are ventrolateral and two are dorsolateral. In most species only hair setae are present, varying in length (0.05-0.6 mm) and number (1-12). Additionally, one to three sigmoid setae, about 0.04-0.07 mm long, may occur, especially in the ventrolateral bundles of the posterior body region or, alternatively, sigmoid setae are exclusively present in all bundles.

The colored epidermal glands are a striking feature found all over the body surface. Each gland cell consists of a large vacuole, rounded or lobate in shape and filled with red, green, blue-green, or yellow liquid, depending on the species. Sometimes the contents are colorless or the gland cells may be lacking. A satellite cell may be attached to the gland cell. The chemical nature of the liquid and the function of the gland cells are still unknown. In most species the mesenteries and septa (except for the first) are reduced. Correspondingly, the vascular system is simple, restricted to the wall of the gut. The brain is located in the prostomium and the nerve cord is attached to the ventral epithelium. Paired metanephridia are serially arranged but usually are not in all segments. The alimentary canal consists of an esophagus which extends through the anterior segments, a dilated midgut in the main region of the trunk, and a narrow posterior part that terminates as an anus on the pygidium.

Most species reproduce exclusively by paratomy. Sexual reproduction generally seems to be suppressed. The animals usually represent a chain of 2-8 zoids. The number of setal segments of the first zoid is species-specific and varies from 5-17. Genital organs are only temporarily developed and have been observed only in a few species. Aeolosomatidae are hermaphrodites. Paired female gonads are produced from the ventral coelomic epithelium in the midregion of the body. In general, only one ovary reaches maturity. The ventral epithelium beneath the mature ovary becomes glandular and forms a female pore. Paired male gonads develop from the ventral coelomic epithelium in both anterior and posterior segments. Spermatozoa are discharged through the metanephridia. The spermatozoa seem to be transferred directly into the receptacula of the partner. These receptacula are tiny epidermal sacs located ventrally in the anterior body region. Oviposition and full development of eggs have been observed only in one species.

Potamodrilus fluviatilis (Figure 30.1d) is about 1.3 mm long and up to 0.11 mm in diameter. It always has seven segments clearly marked by deep furrows and each slightly subdivided into three subunits. The large, lobate prostomium has a tripartite locomotory ciliary field on the ventral surface. The animals move quickly with short interruptions; no body wall musculature is used in locomotion. The peristomium has a broad mouth opening on the ventral side and lacks setae. Each of the six trunk segments has two ventrolateral and two dorsolateral bundles of setae. Each bundle of setae consists of a long (up to 0.15 mm) and a short (0.07 mm) hair seta with a slightly serrated shaft. At the posterior end of the body there is a short tail appendage about 0.015 mm long. The animal uses this appendage to adhere temporarily to sand grains.

No colored skin glands are present in contrast with the Aeolosomatidae. The body wall musculature is weakly developed with the exception of two longitudinal strands on the ventral side which allow immediate curling. The mesenteries and the septa, except for one, as well as the blood vascular system are either absent or simplified; the peritoneum that covers the hind gut forms a thick layer. Only two metanephridia are present. A special feature is the muscular tongue-like bulbus of the ventral pharynx wall that fills the mouth cavity. It is protrusible and serves in trapping food particles from the substratum. The middle part of the alimentary canal is dilated; the anus is located ventrally at the beginning of the caudal appendage.

Potamodrilus is a hermaphrodite; only sexual reproduction occurs. Paired ovaries develop in the fifth segment at the ventral coelomic epithelium. Eggs are shed through a ventromedian pore located in the intersegmental furrow between the sixth and seventh segment. It is surrounded by large gland cells and immediately in front of these is the opening of an unpaired receptaculum seminis. Two pairs of male gonads are found ventrally in the fourth and fifth segment. Spermatogenesis takes place mainly in the coelomic fluid. Spermatozoa are discharged through two coiled vasa deferentia each of which opens into a median superficial furrow on the sixth segment. The mode of development is not known. Newly hatched juveniles possess three segments.

Classification

According to most textbooks, the Aeolosomatidae comprise four genera: *Aeolosoma*, *Rheomorpha*, *Hystricosoma*, and *Potamodrilus*. Histological and recent ultrastructural investigations on *Potamodrilus* revealed that this form has no synapomorphies with the *Aeolosoma* species, and a new monotypic family, Potamodrilidae, has been established for it (Bunke, 1967, 1985). On the other hand, there is no doubt that the genera *Rheomorpha* and *Hystricosoma* form sister-groups with one of the *Aeolosoma* species or species-groups and that these three genera constitute the monophyletic taxon Aeolosomatidae.

Ultrastructural investigations on the spermatozoa of the Potamodrilidae and Aeolosomatidae corroborate the view that neither taxon is closely allied with any of the oligochaete or even clitellate taxa, and, at present, classification within the Oligochaeta/Clitellata has no empirical support. In addition, there is no clear synapomorphic correspondence between either taxon and any of the non-clitellate annelid groups (Bunke, 1985, 1986).

The genus *Aeolosoma* (Figure 30.1a) consists of about 26 species. In the older literature a number of additional poorly described forms can be found (species inquirendae: van der Land, 1971). The phylogenetic relationships of the species are obscure. Diagnostic characteristics include body dimensions, number of setal segments of the first zoid, color and other details of the skin glands, features of ciliation at the prostomium, type, length, and distribution of the setae, and the position of the dilated portion of the alimentary canal. It is possible that the two main species groups with either red or greenish epidermal glands represent monophyletic subgroups, assuming that development of sigmoid setae and disappearance of skin glands are convergent phenomena.

The genus *Rheomorpha* is monotypic, represented only by *R. neizvestnovae* (Figure 30.1b). Diagnostic features include the absence of setae, occurrence of adhesive glands in the epidermis, and a bilobed pygidium. The skin glands are faintly green or colorless, with a satellite cell attached; sometimes the glands are lacking. The thickened ventral wall of the pharynx has a deep furrow resulting in a secondary posterior lip at the mouth.

The genus *Hystricosoma* is made up of three species, *H. chappuisi*, *H. insularum*, and *H. pictum*. The last two species are inadequately known and it is even doubtful if they belong to this genus. Michaelsen (1926) described *H. chappuisi* and noted some external features based on preserved specimens only. The original observations have been corrected and accomplished in a redescription by Pop (1975). *Hystricosoma chappuisi* (Figure 30.1c) is epizoic on crayfish (*Astacus astacus*). Contrary to *Aeolosoma*, it does not move with the cilia of the prostomium but by means of body contractions and dilations. Anatomical characters include: a first zoid of about 0.6-0.86 mm long, consisting of 10 or 11 segments, setal bundles exclusively with rigid sigmoid setae serrated at the convex side of the tip and about 0.04-0.09 mm long, setae of the dorsal bundles arranged in two transverse rows which are parallel and with setae running in opposite directions, and skin glands with orange-red vacuoles, present only on the dorsal side and accumulated at the bases of the setal bundles.

Habitats and Ecological Notes

Most *Aeolosoma* species are found in phytal regions and on the detritus-rich bottoms of freshwater habitats, especially in ponds and lakes, but also brooks and streams. In addition, some inhabit the humid litter of forest soils (O- and A-horizons), where they may be found in an encysted state. Another habitat is the littoral mesopsammal region of streams and brooks where animals may be found between 10-70 cm deep, both in the water-covered bottom and in the hygropsammic or hydropsammic regions of sand beaches. A few of these latter species have been found in brackish waters of estuaries or marine beaches (up to 0.5% salinity). Only one species is known (*A. maritimum*) exclusively from a marine habitat (2.93-3.46% salinity), in humid sand and groundwater of a beach in Tunisia (Westheide and Bunke, 1970). The *Rheomorpha* species is known only from the limnetic mesopsammal of streams and lakes. *Potamodrilus* is found exclusively in the interstices of coarse and medium-fine sands from the bottoms of streams.

A number of *Aeolosoma* species have a worldwide distribution while others appear to be restricted to only certain countries. *Rheomorpha neizvestnovae, Hystricosoma chappuisi,* and *Potamodrilus fluviatilis* have been found only in a few European countries.

Methods of Collection and Extraction

Plant material from the littoral zone or organic detritus from the bottom of freshwater habitats may be washed in small portions into a Petri dish in order to obtain sufficient numbers of specimens for a survey of the *Aeolosoma* species present. A suction pump tube which can extract just the surface material from submersed plants or detrital layer is often effective. Psammobiontic forms are collected by stirring a sample of sand in a water-filled bucket to separate the animals from the sediment and then decanting the suspended material through a fine mesh sieve as the sediment settles. A quantative method for determining the distribution of species and specimens in a sandy beach has been described by Schmidt (1968) using an extraction device designed by Uhlig (Westheide and Schmidt, 1969).

Preparation of Specimens for Taxonomic Study

A reliable determination of *Aeolosoma* species is possible only with living specimens because, after fixation, body dimensions and the color of skin glands are indeterminable. Also, because of individual variations, determination often requires multiple specimens; if only one animal is available, culturing should be attempted in order to obtain sufficient material for a proper determination. A number of species are easily cultured using an Erdschreiber solution with green algae as food, e.g., *Chlorogonium elongatum*.

For light microscopical histological study, specimens should be fixed with Bouin's fluid and paraffin sections should be stained with hematoxylin and eosin. For electron microscopy, conventional fixation with glutaraldehyde or a mixture of glutaraldehyde and paraformaldehyde and osmium tetroxide are recommended. These methods are not as satisfactory in the case of *Potamodrilus*, but can be improved by further en-block-staining with uranyl acetate.

Type Collections

No major type collections exist for most of the species of the two families. This is partially because of the poor information content of preserved material. Types of a few species of *Aeolosoma* are found in the Eveline Marcus Oligochaeta Collection at the University of São Paulo, Brasil. The types of *Aeolosoma maritimum* and *Hystricosoma insularum* are deposited in the Zoologisches Museum der Universität Hamburg, FRG.

References

Brinkhurst, R.O.
1971. Phylogeny and Classification. Part I in R.O. Brinkhurst and B.G.M. Jamieson, editors, *Aquatic Oligochaeta of the World.* Edinburgh: Oliver and Boyd.
1982. Evolution in the Annelida. *Canadian Journal of Zoology,* 60:1043-1059.

Bunke, D.
1967. Zur Morphologie und Systematik der Aeolosomatidae Beddard 1895 und Potamodrilidae nov. fam. (Oligochaeta). *Zoologische Jahrbücher Abteilung für Systematik, Ökologie und Geographie der Tiere,* 94:187-368.
1985. Ultrastructure of the Spermatozoon and Spermiogenesis in the Interstitial Annelid *Potamodrilus fluviatilis. Journal of Morphology,* 185:203-216.
1986. Ultrastructural Investigations on the Spermatozoon and Its Genesis in *Aeolosoma litorale* with Considerations on the Phylogenetic Implications for the Aeolosomatidae (Annelida). *Journal of Ultrastructure and Molecular Structure Research,* 95:113-130.

Land, van der, J.
1971. Family Aeolosomatidae. Pages 665-706 in R.O. Brinkhurst and B.G.M. Jamieson, editors, *Aquatic Oligochaeta of the World.* Edinburgh: Oliver and Boyd.

Michaelsen, W.
1926. Schmarotzende Oligochäten nebst Erörterungen über verwandtschaftliche Beziehungen der Archioligochäten. *Mitteilungen aus dem Zoologischen Staatsinstitut und Zoologischen Museum Hamburg,* 42:91-103.

Parker, S.P., editor
1982. *Synopsis and Classification of Living Organisms.* Volume 2. 1232 pages. New York: McGraw-Hill Book Company.

Pop, V.
1975. Was ist *Hystricosoma chappuisi* Michaelsen (Aeolosomatidae, Oligochaeta)? *Mitteilungen aus dem Hamburgischen Zoologischen Museum und Institut,* 72:75-78.

Reynolds, J.W., and D.G. Cook
1976. *Nomenclatura Oligochaetologica.* 217 pages. Fredericton, New Brunswick: University of New Brunswick.

Schmidt, P.
1968. Die quantitative Verteilung und Populationsdynamik des Mesopsammons am Gezeiten-Sandstrand der Nordseeinsel Sylt. I. Faktorengefüge und biologische Gliederung des Lebensraumes. *Internationale Revue der gesamten Hydrobiologie,* 53:723-779.

Westheide, W., and D. Bunke
1970. *Aeolosoma maritimum* nov. spec., die erste Salzwasserart aus der Familie Aeolosomatidae (Annelida: Oligochaeta). *Helgoländer wissenschaftliche Meeresuntersuchungen.* 21:134-142.

Westheide, W., and P. Schmidt
1969. Von der Kleintierwelt im Meeresstrand. I. Fang und Untersuchung. *Mikrokosmos,* 58:257-262.

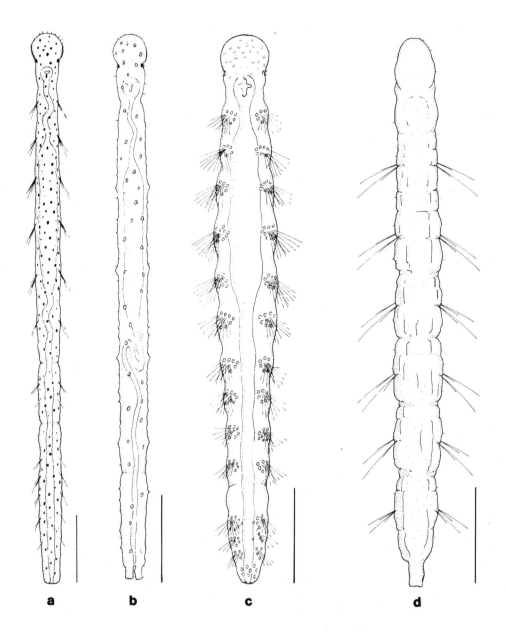

Figure 30.1.--Aeolosomatidae and Potamodrilidae: **a-c**, Aeolosomatidae: **a**, *Aeolosoma hemprichi*, three zoids (from Bunke, 1967); **b**, *Rheomorpha neizvestnovae*, two zoids (from Bunke, 1967); **c**, *Hystricosoma chappuisi*, two zoids (redrawn from Pop, 1975); **d**, Potamodrilidae: *Potamodrilus fluviatilis* (from Bunke, 1967). (Scale = 0.2 mm.)

31. Oligochaeta

Christer Erséus

The Oligochaeta are considered a class, sometimes a subclass (within the class Clitellata), within the Annelida. When sexually reproducing (asexual forms occur), oligochaetes are hermaphroditic, with direct development of the young in cocoons secreted by the clitellum (eggs laid singly in at least some gutless Tubificidae). This characterization would also include leeches and related groups (Hirudinoidea), which in fact genealogically may be oligochaetes too, but which traditionally have been regarded as clitellate groups of the same rank as the Oligochaeta. The latter *(sensu stricto)* comprise about 25 different families, the majority of which are large terrestrial forms commonly known as earthworms. Some of the families are, however, aquatic and they are often referred to as microdriles in recognition of their small size in relation to the earthworms (=megadriles). Still, many microdriles are macrofaunal rather than meiofaunal. In this chapter the smallest oligochaetes are treated. They escape through sieves of a 0.5-1 mm mesh size and thus can be regarded as members of the meiofauna, although some of them (very slender, generally interstitial species) may reach lengths of 10 mm or more.

The aquatic oligochaetes are similar to earthworms in general appearance, although the clitellum (when developed) is only one cell thick and the eggs are few and large; in earthworms, the clitellum is multilayered and the eggs are generally small. Oligochaetes are distinguished from most polychaetes by their lack of anterior appendages and parapodia. However, some polychaetes (including what used to be called "archiannelids" and the family Capitellidae) also lack such structures; these oligochaetoid forms may be difficult to separate from true oligochaetes unless one is able to study their genital systems. In the case of capitellids, which probably are those most often confused with oligochaetes, parapodia are generally at least rudimentary in some segments, and/or the two bundles of setae on one side of the worm are more close together (reflecting the setal arrangement on a parapodium) than any of them are to a bundle on the other side. In addition, in capitellids hair (capillary) setae are always present in both dorsal and ventral bundles of a least a few anterior segments, and the other setae (uncini) are hooded, whereas in oligochaetes — with a single known exception (*Capilloventer*) — hairs, if present at all, are restricted to dorsal bundles, and the other setae (crotchets) are never hooded. The stiff, filiform and transparent species of *Grania* (Enchytraeidae) may superficially resemble nematodes, but they have a clear internal segmentation and their (stout) setae are generally visible at the higher magnification of a good dissecting microscope.

A detailed account of the external and internal morphology of aquatic oligochaetes has been given by Cook (1971). The only worldwide review of their taxonomy is the extensive and still useful monograph by Brinkhurst and Jamieson (1971), updated with regard to the freshwater forms by Brinkhurst and Wetzel (1984), but in recent years also a high number of new marine forms have been described, most of which are meiofaunal. A tentative review of the taxonomic criteria used for marine oligochaetes was published by Erséus (1980a), a first key (largely British) to marine and estuarine oligochaetes by Brinkhurst (1982a). The latter work also includes useful chapters on general morphology, ecology, collection and preservation. Still, however, most of the information on meiobenthic marine oligochaetes must be searched for in the original taxonomic literature (e.g., Brinkhurst, 1985, 1986; Brinkhurst and Coates, 1985; Coates and Erséus, 1985; Erséus, 1979, 1980b, 1981, 1982, 1983, 1984; Erséus and Strehlow, 1986; Harman and Loden, 1984; Righi and Varela, 1983). The biology and ecology of marine Oligochaeta were reviewed by Giere and Pfannkuche (1982).

Families comprising at least some meiofaunal forms are treated below. The families Aeolosomatidae and Potamodrilidae are no longer regarded as oligochaetes (Brinkhurst, 1982b).

General Remarks on Morphology

Microdrile oligochaetes are generally slender and flexible, and they often exhibit jerking movements. Several species are reddish, brownish or orange due to the coloration of the blood or the chloragogen tissue covering the gut. Some species are colorless and transparent, others conspicuously white. The segmentation is often indistinct as many septa are either thin or incompletely developed.

The setae of aquatic oligochaetes are generally bifid or single-pointed crotchets, sometimes accompanied by hair setae. If present, hairs occur in dorsal bundles only; an exception being those of *Capilloventer atlanticus* (Capilloventridae) which has both dorsal and ventral hair setae. Generally, the setae are few, inconspicuous and arranged in four bundles per segment (sometimes a single seta replaces the bundle). The bifid (or trifid) setae of many of the marine species of Tubificidae possess a "subdental ligament," which connects the lower tip of the seta with the setal shaft, but the tips are never hooded as in capitellid polychaetes.

Two aberrant genera of Tubificidae, *Inanidrilus* and *Olavius*, invariably lack a normal alimentary system and are dependent upon epidermal uptake of nutrients, a mechanism involving the symbiotic bacteria found in high numbers in their cuticle/epidermis interface. All other oligochaetes have a tubular gut running through the whole body and opening to the exterior at both ends. The pharyngeal pad is always dorsal, a feature separating the group from several polychaetes, including the oligochaetoid (but dioecious) family Questidae (cf. Giere and Riser, 1981).

All sexually reproducing oligochaetes are hermaphrodites. The genital system consists of testes, ovaries, male and female efferent ducts, and spermathecae. The latter are ampullar invaginations of the body wall that receive the sperm during copulation. The segmental arrangement of the sexual organs constitutes the most important feature used for the distinction of different families of the Oligochaeta. The principal morphology of the genitalia (male ducts, spermathecae) provides the fundamental generic characters, whereas particular details in the genitalia (proportions, thickness of muscle layers, penis shapes, etc.) and setal characteristics are used to distinguish species.

Classification

The classification and phylogeny of the aquatic Oligochaeta have been subject to some recent discussion. Oligochaete families with meiobenthic species are only found within the order Tubificida as defined by Brinkhurst (1982b), and in the recently discovered Narapidae, Capilloventridae, and Randiellidae.

Family Tubificidae.--No eyes; male pores usually in segment XI, spermathecae in X; setae of varying shapes, mostly bifid, but also single-pointed, trifid or pectinate setae may be present. Hairs, if present (rare in meiofaunal species), in dorsal bundles only. Modified genital setae (penial or spermathecal setae) particularly common in marine forms; they replace the normal ventral setae in the penial and/or spermathecal segments, or sometimes in an adjacent segment. The Tubificidae comprise the largest of the aquatic oligochaete families with over 400 species described, a majority of which are marine. The bulk of marine species are interstitial and can be regarded as members of the meiofauna, whereas many freshwater forms are macrofaunal and burrowing rather than interstitial. Five subfamilies: Phallodrilinae, Limnodriloidinae, Rhyacodrilinae, Tubificinae and Telmatodrilinae, the first two exclusively marine; all except the last-mentioned with at least some meiobenthic species.

Some of the largest genera are mentioned here: *Aktedrilus* (Phallodrilinae) (Figure 31.1d), with about 25 species, all small and inhabiting marine or brackish water littoral habitats; with spermatheca unpaired, mid-dorsal in segment X; penes present, sometimes cuticularized. *Phallodrilus* (Phallodrilinae), a large and heterogeneous genus with more than 50 marine littoral, subtidal, deep-sea, and also some limnic species; atria each with two discrete prostate glands; penes generally poorly developed; penial setae generally present. *Inanidrilus* and *Olavius* (Figure 31.1e) (both Phallodrilinae), with over 40 described species, all devoid of an alimentary canal; occur mostly in coral sands in tropical and subtropical reefs; worms conspicuously white due to occurrence of symbiotic bacteria in body wall; genitalia and penial setae similar to those of *Phallodrilus*, but often more elaborate. *Heterodrilus* (Rhyacodrilinae) (Figure 31.1f), an homogeneous group of 17 marine species, also associated with coral reefs; very characteristic, rather stout worms, with trifid setae in preclitellar segments, and generally with penial setae. *Limnodriloides* (Figure 31.1a) (with over 40 species) and other genera of the Limnodriloidinae; common in muddy marine sediments; characterized by modified part of esophagus in segment IX (esophagus either widened and glandular or bearing a pair of anterior diverticula in that segment); male genital ducts sometimes elaborate; penial setae rare, but spermathecal setae present in several species of *Limnodriloides*. *Tubificoides* (Tubificinae), with over 30 species, mostly temperate and Arctic, generally confined to muddy sediments; with or without hair setae; sometimes with fine papillation over whole or parts of body; cuticularized penes present.

Family Naididae.--Simple eyes sometimes present. Male pores in segments V, VI, VII or VIII; spermathecal pores in segment immediately anterior to that bearing male pores, but asexual reproduction by budding or fragmentation predominant. Setae of varying shape, generally of more than one type:

bifids and/or needles in combination with hair setae; dorsal needles usually different from ventral setae; dorsal setae sometimes absent from a few anterior segments. Penial setae sometimes present. Mostly freshwater and brackish water species, all more or less cosmopolitan.

Totally about 20 genera, virtually all meiofaunal. *Nais* (about 15 species), a freshwater genus with several common species; normally with eyes and anterior segments pigmented; dorsal setae beginning in segment VI, comprising hairs usually together with bifid or single-pointed, rarely palmate or pectinate, needles; penial setae present (but worms seldom sexually mature). *Paranais* (about 10 species), the only primarily marine and brackish water genus of Naididae; lacking hair setae; dorsal (bifid) setae beginning in segment V; ventral setae of II often somewhat longer than rest and generally more lateral in position; penial setae present. *Chaetogaster* (7 species), small freshwater and brackish water species, some of which are commensal with, or parasitic upon, molluscs; lacking dorsal setae; prostomium weakly developed.

Family Narapidae.--No eyes. Setae absent. Male genital ducts in segment VI. Penes present. Spermathecae in VII. This family was recently erected to accommodate the only species, *Narapa bonettoi* Righi and Varela, from the Paraná River in Argentina.

Family Enchytraeidae.--No eyes. Setae single-pointed, straight or sigmoid. Hairs absent. Setae totally absent in a few species. Cuticle thick, rendering worms stiffer than most other aquatic oligochaetes. Male pores in segment XII, spermathecae in V. Penial setae absent. Pharyngeal glands (often called septal glands) as discrete bodies in segments IV-VI (or -VII). Most species are terrestrial or semiterrestrial, but about 150 aquatic species (majority of which are marine) are known.

Marionina, an heterogeneous group in need of revision (actual number of species difficult to estimate); mostly very small and interstitial, marine (intertidal) or limnic forms; setae small and generally few, sometimes absent; male ducts and spermathecae simple. *Grania* (Figure 31.1b), exclusively marine, with about 20 slender, transparent, almost nematode-like species occurring in various habitats ranging from the intertidal to the deep sea; setae, if present, only as unisetal "bundles," always with a strong tendency towards complete reduction in anterior segments; sperm as small rings in spermathecal wall.

Family Capilloventridae.--Monotypic, marine, comprising the poorly known *Capilloventer atlanticus* Harman and Loden from Brazil. No eyes. Dorsal and ventral setae 2 (1 hair and 1 bifid crotchet) per bundle. Penial setae hair-like and coiled. Male pores in segment XII; spermathecae in VII.

Family Randiellidae.--Exclusively interstitial marine oligochaetes. No eyes. Somatic setae very small, single-pointed, often more than 2 per bundle anteriorly. Penial setae very long, generally hair-like, but otherwise hair setae absent. One or two pairs of sperm funnels present posteriorly in segment X, or in X-XI (if 2 pairs), but male ducts and male pores have not been observed. Spermathecae, multiple pairs, in VII, VII-VIII, or VIII.

Randiella Erséus and Strehlow (Figure 31.1c), the only genus, with four species, three of which intertidal, one subtidal.

Habitats and Ecological Notes

Meiobenthic oligochaetes occur in a great variety of aquatic habitats. In freshwater, they are largely confined to the coarse sands and gravels of lakes and river beds (tubificids, naidids, narapids and enchytraeids) or associated with vegetation (naidids); the mud-dwelling species are generally more or less macrobenthic. The estuarine and marine environments harbor a diverse fauna of small oligochaetes. Virtually any kind of marine sediment contains at least one oligochaete species, although not always in very high densities.

In littoral sands most oligochaetes are meiobenthic and show adaptations characteristic of interstitial animals; slender body shape and possession of adhesive glands. The genera *Aktedrilus* and *Marionina* are particularly important.

In any given area, the sublittoral oligochaetes can generally be divided into two assemblages, one characteristic of muddy bottoms (e.g., *Tubificoides, Limnodriloides*) and one, often richer in species, in coarse sands and gravel. The latter assemblage consists exclusively of interstitial forms. Some of these (species of *Grania, Heterodrilus, Bathydrilus, Phallodrilus, Duridrilus*) have attained a rather stiff body shape and have setae that are reduced in number but often large in size. In stable (often poorly oxygenated) sands in tropical and subtropical coral reefs, high numbers of the gutless tubificids (*Inanidrilus, Olavius*) occur.

Tubificids and members of the enchytraeid genus *Grania* are present also in the deep sea, particularly in bathyal depths on the continental slopes, although generally in low numbers. One species, *Bathydrilus hadalis,* is known from 7,298 m depth in the Aleutian Trench (North Pacific).

The ecological role of marine meiobenthic oligochaetes have been discussed by Giere (1975) and

Giere and Pfannkuche (1982). Most of the species seem to represent final links of short food chains, with microorganisms as their main source of nutrition. Some gutless tubificids appear to benefit not only from the metabolic products excreted by their symbiotic, chemoautotrophic bacteria, but also from direct digestion of the cell material of the latter (Giere and Langheld, 1987).

Methods of Collection and Extraction

Meiobenthic oligochaetes can be collected for quantitative work by taking grab or core samples of the sediment. In some coarse bottoms, these animals appear to penetrate deep into the sand, and therefore the best results are obtained when the gear is operated by a diver. If qualitative samples are sufficient, e.g., for taxonomic studies, a variety of methods including scooping by hand and dredging from a boat can be used.

Extraction of the worms from the sediment generally has to be performed by means of elutriation as they are smaller than a great proportion of the actual sediment particles. Most oligochaetes except some juveniles and cocoons will be retained on a 250 μm screen; this mesh size is therefore recommended for most cases. The adhesive behavior of some marine littoral species (e.g., some *Aktedrilus, Marionina*) make them more difficult to extract than others. For this, anaesthesia (e.g., $MgCl_2$) may be needed, but also a freshwater shock can be used for the extraction of such worms from the sand. When freshwater is added to a marine littoral sample, most interstitial oligochaetes tend to have a weaker ability to adhere to the sand grains and are more easily washed out. As soon as the material has been collected on the screen it should, however, be transferred back into sea water if one intends to study the specimens alive.

For sorting the oligochaetes, small portions of the material collected in the sieve are inspected in a Petri dish under a dissection microscope.

Preparation of Specimens for Taxonomic Study

Identification of aquatic oligochaetes is, in most cases, possible only if sexually mature specimens are available; most of the taxonomically important features are restricted to the genitalia and a few other internal structures. This, as well as the fact that the setae, which have great taxonomic relevance at the species level, are indeed minute in meiobenthic oligochaetes, puts demands on the quality of the microscope optics. For studies immediately upon extraction from the sediment, a worm can be temporarily mounted alive in habitat water on a slide. The coverslip should be subjected to mild pressure, which will hinder the worm's movements and flatten it to facilitate observations of the internal features.

Conventional chemicals can be used for fixation. Ten percent formalin is adequate if the material will be used for straight forward identification of known species, but for more advanced taxonomic work, histological techniques must be employed. Fixation in Bouin's solution gives excellent results for sections as well as whole mounts, provided the worms are transferred into alcohol (70–80%) after a maximum of a few days (minimum 2–3 hours); longer periods in Bouin's make the animals hard and brittle.

For detailed studies of the internal organs of a particular species, a few specimens should preferably be sectioned, but as the meiobenthic species are not very thick, whole mounts cleared in toluene or xylene (or similar) are generally adequate even for internal studies; the genitalia are reasonably visible in such preparations to make identification possible. If an ordinary light microscope is used, it is preferable to stain the cell nuclei of the specimens prior to dehydration, clearing and mounting. This can be done, for example, in haematoxylin, borax carmine or alcoholic paracarmine. If an interference contrast microscope is used, however, the tissues should not be stained. For permanent mounts, Canada balsam or artificial balsam can be used as mounting medium. For temporary mounts of fixed worms, glycerin is an excellent medium as, while it has a clearing effect, it does not harden the specimen. The standard method of using Amman's lactophenol for clearing/mounting should be restricted to studies in which (1) one does not mind the material deteriorating after some time (months), and (2) the species studied have enough setal characteristics or cuticular penis sheaths, to enable secure identification (provided the taxonomy of the species studied has already been well established). Soft internal structures are not always clearly visible in lactophenol.

Type Collections

The most extensive collections of aquatic, meiobenthic Oligochaeta are located at the National Museum of Natural History, Smithsonian Institution, Washington, D.C., USA; the British Museum (Natural History), London, UK; and the Naturhistoriska Riksmuseet, Stockholm, Sweden.

References

Brinkhurst, R.O.
1982a. *British and Other Marine and Estuarine Oligochaetes,* in D.M. Kermack and R.S.K. Barnes, editors, *Synopsis of the British Fauna.* 127 pages. Cambridge: Cambridge University Press.
1982b. Evolution in the Annelida. *Canadian Journal of*

Zoology, 60:1043-1059.
- 1985. A Further Contribution to the Taxonomy of the Genus *Tubificoides* Lastockin (Oligochaeta: Tubificidae). *Canadian Journal of Zoology*, 63:400-410.
- 1986. Taxonomy of the Genus *Tubificoides* Lastockin (Oligochaeta, Tubificidae): Species with Bifid Setae. *Canadian Journal of Zoology*, 64:1270-1279.

Brinkhurst, R.O., and K.A. Coates
- 1985. The Genus *Paranais* (Oligochaeta: Naididae) in North America. *Proceedings of the Biological Society of Washington*, 98:303-313.

Brinkhurst, R.O., and B.G.M. Jamieson
- 1971. *Aquatic Oligochaeta of the World*. 860 pages. Edinburgh: Oliver and Boyd.

Brinkhurst, R.O., and M.J. Wetzel
- 1984. Aquatic Oligochaeta of the World: Supplement. A Catalog of New Freshwater Species, Descriptions, and Revisions. *Canadian Technical Report of Hydrography and Ocean Sciences*, 44:1-101.

Coates, K.A., and C. Erséus
- 1985. Marine Enchytraeids (Oligochaeta) of the Coastal Northwest Atlantic (Northern and Mid U.S.A.). *Zoologica Scripta*, 14:103-116.

Cook, D.G.
- 1971. Anatomy. Microdriles. Pages 8-41 in R.O. Brinkhurst and B.G.M. Jamieson, editors, *Aquatic Oligochaeta of the World*. Edinburgh: Oliver and Boyd.

Erséus, C.
- 1979. Taxonomic Revision of the Marine Genus *Phallodrilus* Pierantoni (Oligochaeta, Tubificidae), with Descriptions of Thirteen New Species. *Zoologica Scripta*, 8:187-208.
- 1980a. Specific and Generic Criteria in Marine Oligochaeta, with Special Emphasis on Tubificidae. Pages 9-24 in R.O. Brinkhurst and D.G. Cook, editors, *Aquatic Oligochaete Biology*. New York: Plenum Publishing Corporation.
- 1980b. Taxonomic Studies on the Marine Genera *Aktedrilus* Knöllner and *Bacescuella* Hrabe (Oligochaeta, Tubificidae), with Descriptions of Seven New Species. *Zoologica Scripta*, 9:97-111.
- 1981. Taxonomic Revision of the Marine Genus *Heterodrilus* Pierantoni (Oligochaeta, Tubificidae). *Zoologica Scripta*, 10:111-132.
- 1982. Taxonomic Revision of the Marine Genus *Limnodriloides* (Oligochaeta: Tubificidae). *Verhandlungen des naturwissenschaftlichen Vereins in Hamburg* (Neue Folge), 25:207-277.
- 1983. Deep-sea *Phallodrilus* and *Bathydrilus* (Oligocheata, Tubificidae) from the Atlantic Ocean with Descriptions of Ten New Species. *Cahiers de Biologie Marine*, 25:125-146.
- 1984. Taxonomy and Phylogeny of the Gutless Phallodrilinae (Oligochaeta, Tubificidae), with Descriptions of One New Genus and Twenty-two New Species. *Zoologica Scripta*, 13:239-272.

Erséus C., and D.R. Strehlow
- 1986. Four New Interstitial Species of Marine Oligochaeta Representing a New Family. *Zoologica Scripta*, 15:53-60.

Giere, O.
- 1975. Populations Structure, Food Relations and Ecological Role of Marine Oligochaetes, with Special Reference to Meiobenthic Species. *Marine Biology*, 31:139-156.

Giere, O., and C. Langheld
- 1987. Structural Organization, Transfer and Biological Fate Of Endosymbiotic Bacteria in Gutless Oligochaetes. *Marine Biology*, 93:641-650.

Giere, O., and O. Pfannkuche
- 1982. Biology and Ecology of Marine Oligochaeta. A Review. *Oceanography and Marine Biology Annual Review*, 20:173-308.

Giere, O., and N.W. Riser
- 1981. Questidae-Polychaetes with Oligochaetoid Morphology and Development. *Zoologica Scripta*, 10:95-103.

Harman, W.J., and M.S. Loden
- 1984. *Capilloventer atlanticus* gen. and sp.n., a Member of a New Family of Marine Oligochaeta from Brazil. *Hydrobiologia*, 115:51-54.

Lasserre, P.
- 1971. Oligochaeta from the Marine Meiobenthos: Taxonomy and Ecology. *Smithsonian Contributions to Zoology*, 76:71-86.

Righi, G., and M.E. Varela
- 1983. *Narapa bonettoi*, gen.nov., sp.nov. (Oligochaeta, Narapidae, fam.nov.) de agua doce da Argentina. *Revista de la Asociacion de Ciencias Naturales del Litoral*, 14:7-15.

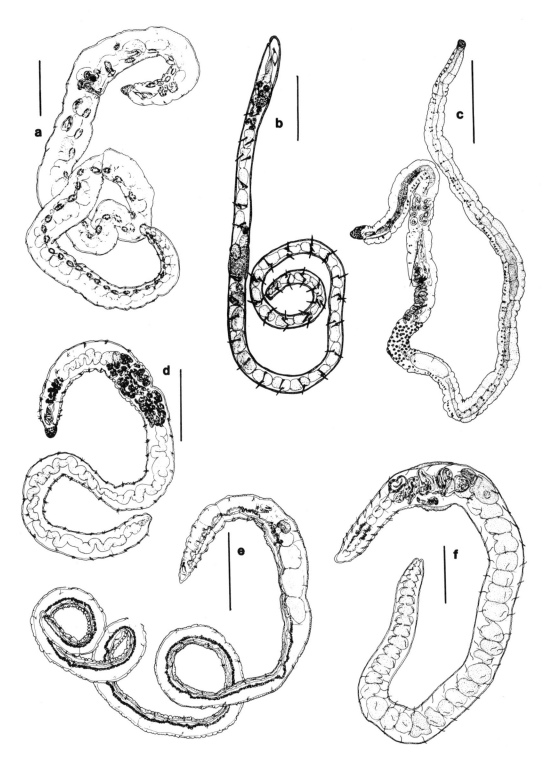

Figure 31.1.--Oligochaeta: a-f, Drawings of whole-mounted (fixed, contracted and somewhat compressed) specimens of oligochaetes. **a**, *Limnodriloides barnardi* Cook; **b**, *Grania pusilla* Erséus; **c**, *Randiella multitheca* Erséus and Strehlow; **d**, *Aktedrilus monospermathecus* (Knöllner); **e**, *Olavius geniculatus* (Erséus); **f**, *Heterodrilus jamiesoni* Erséus. (Scale = 500 μm.)

32. Sipuncula

Mary E. Rice

Sipunculans are recognized as a phylum of unsegmented, coelomate marine worms, numbering from 200 to 300 species (Stephen and Edmonds, 1972; Gibbs and Cutler, 1987). The group is characterized by a unique combination of characters, namely, division of the body into a thickened posterior trunk and a more narrow anterior introvert which can be retracted into the trunk; a recurved gut with elongate esophagus extending from the mouth at the tip of the introvert to the intestine, usually coiled with descending and ascending portions, leading to a dorsal anus on the anterior trunk; one to four long retractor muscles traversing the coelom from anterior introvert to attach on the body wall of the trunk; a median, unpaired ventral nerve cord, circum-esophageal connectives and an anterior supra-esophageal brain; one or two metanephridia; a gonad situated at the base(s) of the ventral retractor muscle(s); a spacious, undivided coelomic cavity filled with coelomocytes and often developing gametes. Although the vast majority of sipunculans are macrofaunal, some meet the criteria for meiofauna. The body diameter may have a maximum dimension of less than 1 mm even though the length of the organism is greater than 1 mm. If sieved alive, sipunculans several mm in length may easily squeeze through 1 mm mesh openings, or pass through in a contracted state.

Classification

Two recent classifications of the phylum are those of Stephen and Edmonds (1972) and Gibbs and Cutler (1987). The former authors recognize 4 families, whereas the latter divide the group into 6 families, 4 orders and 2 classes. Important characters for the distinction of higher taxa are: the structure of the body wall including the presence or absence of banding of the circular and longitudinal muscle layers and presence or absence of coelomic channels; structure of the tentacular crown and its position relative to the mouth; the presence of hardened structures ("shields") at either end of the trunk. Species-specific characters include, but are not limited to, overall body size, relative length of introvert and trunk, presence and structure of hooks on the introvert, and form and distribution of papillae on the body. Internal features which may be used in distinguishing species are number of longitudinal muscle bands, number and position of attachment of retractor muscles, attachment of nephridia to body wall and the location of small muscles attaching the gut to the body wall.

Until recently sipunculans were thought to be only temporary components of meiofauna, represented by numerous, usually unidentifiable, juveniles. Studies of sediments on the continental shelf off the central east coast of Florida have now demonstrated the presence of an undescribed, small, interstitial adult sipunculan, of the genus *Phascolion*. This species averages 4 mm in length, the largest specimens having a maximum width of 0.5 mm. The ratio of introvert to trunk is approximately 2 to 1 (Figure 32.1a). The anus opens on the introvert at a point located about one-fifth the total length of the introvert from its base, where the body broadens into the thicker trunk. The tentacular crown at the end of the introvert surrounds the mouth and is comprised of 4 to 8 filiform tentacles.

Posterior to the tentacles a bulbous expansion bears 3 to 4 rows of curved hooks. Prominent, rounded papillae are scattered over the surface of the trunk. The convex curvature of the body is dorsal rather than the more usual ventral curvature of sipunculans. Internally a single retractor muscle extends the length of the introvert and in the anterior third of the trunk divides into a thickened dorsal retractor and a narrow, thin ventral retractor, both of which attach at a level near the posterior end of the trunk. The structure and arrangement of the retractor muscles are typical of the subgenus *Phascolion* as defined by Gibbs (1985). The esophagus remains attached to the ventral retractor for a short distance before continuing into several loosely formed intestinal loops. The rectum is long and straight, opening at the anus on the posterior introvert. As in all *Phascolion* there is a single nephridium to the right of the nerve cord opening to the exterior ventrally in the anterior trunk. Despite the small size of the specimens, they have been demonstrated to be adults rather than juveniles by the presence of mature gametes in the coelomic cavity which have been observed in the laboratory to be spawned and to develop into swimming larvae.

Another sipunculan, reported as part of the meiofaunal assemblage on marine beaches, is *Aspidosiphon exiguus* Edmonds, 1974 (Edmonds, 1982). Although the specimens have been presumed to be adults, no gametes have been observed in any of the animals examined. The trunk of the specimens is about 3 mm and the introvert 3 to 5 times the length of the trunk (Figure 32.1b). The shields at the anterior and posterior trunk, thickened areas of skin characteristic of all members of this genus, are pale yellow, composed of small polygonal plates, and lack furrows. The posterior shield is often not well developed. At the distal end of the introvert there are 10 to 30 rows of small, clear, double-pointed hooks. Internally the longitudinal body wall musculature is smooth, without bands, and the single retractor muscle divides into two short branches near the posterior shield. A long coiled gut opens at a dorsal anus just below the anterior shield. Openings of the two nephridia are at the level of the anus or immediately posterior to it.

Habitats and Ecological Notes

Sipunculans are widely distributed in the oceans of the world from intertidal waters to abyssal depths, occupying such habitats as sand and mud, burrows in calcareous substrata, interstices of rocks and algal holdfasts. Meiofaunal species have been found in sandy sediments on the continental shelf in subtropical waters, as well as in coarse sand on marine beaches. *Phascolion* sp. (Figure 32.1a) occurs at depths of 8 to 27 m in sediments characterized as medium to coarse sand and shell hash. Unlike other known species of this genus, it does not occupy discarded gastropod shells or tubes of other animals, but is instead found dwelling among sand grains. *Aspidosiphon exiguus*, on the other hand, is reported from coarse coral sand in the intertidal zone.

Methods of Collection

Subtidal meiofaunal sipunculans are collected by dredging sediment which is then sieved through a screen with a mesh opening of 1 mm or less; sieved samples are examined under a dissecting microscope and sipunculans sorted from the samples. Species inhabiting marine beaches may be collected from holes dug in the sand from 0.5 m to 1 m above the line of wave action. The water filling these holes is passed through a net of 42 µm mesh, which retains the interstitial specimens.

Preparation of Specimens for Taxonomic Study

For taxonomic studies it is critical that specimens be carefully relaxed with tentacles extended prior to fixation. This may be accomplished by placing animals in 10% ethanol in sea water or in 7.5% magnesium chloride until introvert and tentacular crown are extruded. For reference specimens, a good fixative is 10% formalin in 70% ethanol, from which specimens are transferred after one to two days to 70% ethanol for storage.

Acknowledgments.

Contribution Number 221 of Smithsonian Marine Station at Link Port.

References

Edmonds, S.J.
1974. A New Species of Sipuncula (*Aspidosiphon exiguus* n. sp.), Belonging to the Interstitial Fauna of Marine Beaches Collected by Mr. L. Botosaneanu During the Second Cuban-Rumanian Biospeleological Expedition to Cuba 1973. *International Journal of Speleology*, 6:187-192.
1982. A Sipunculan Reported to be "Interstitial" From the Netherland Antilles. *Bijdragen tot de Dierkunde*, 52(2):228-230.

Gibbs, P.E.
1985. On the Genus *Phascolion* (Sipuncula) With Particular Reference to the North-East Atlantic Species. *Journal of the Marine Biological Association of the United Kingdom*, 65:311-323.

Gibbs, P.E., and E.B. Cutler
1987. A Classification of the Phylum Sipuncula. *Bulletin of the British Museum Natural History (Zoology)*, 52(1):43-58.

Stephen, A.C., and S.J. Edmonds
1972. *The Phyla Sipuncula and Echiura.* 128 pages. London: British Museum (Natural History).

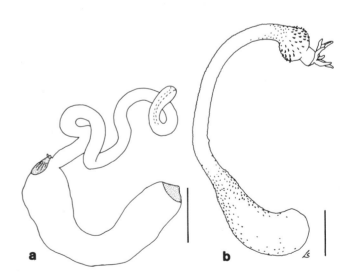

Figure 32.1.--Sipuncula: a, *Phascolion* sp.; b, *Aspidosiphon exiguus*, introvert not fully extended, thus tentacles are not shown (after Edmonds, 1982). (Scale = 1 mm.)

33. Tardigrada

J. Renaud-Mornant

The Tardigrada are a phylum of uncertain phylogenetic position, but with closest affinities to the Arthropoda. Commonly called the "water-bears," they are bilaterally symmetrical invertebrates, rarely exceeding 1 mm in length – most marine species are less than 0.5 mm in length. The body is divided into five variably distinct body segments including a cephalic segment, three trunk segments each bearing a pair of anteroventrally directed legs, and a terminal segment with a pair of posteroventrally directed legs; the four pairs of legs terminate in claws or toes of variable number and structure.

The cuticle, which may be chitinous, may be smooth or with sculpturing including elaborate cuticular projections; in some, especially the "terrestrial" members of the otherwise marine class Heterotardigrada, the cuticle reaches its greatest complexity, taking the form of distinctly armored plates.

The nervous system consists of a relatively large six-lobed brain which innervates the cephalic sensory appendages. A circumoesophagal ring connects the brain to a double-ventral nerve chord connecting at each leg-pair by a ganglion.

The digestive tract (foregut, midgut, and hindgut) is usually colored by the ingested material, mostly algal. The buccal apparatus consists of a buccal tube opening into a muscular pharyngeal bulb variously provided with cuticularized elements; it is flanked by stylets and salivary glands; stylets supports may be present. The bulb is followed by a short esophagus leading to a large intestine which is diverticulate in the primarily marine class Heterotardigrada. In this latter class, both anus and genital openings exist; in the predominantly freshwater and "terrestrial" class Eutardigrada, the genital duct and excretory system open into a cloaca. In many species the coelomic fluid of the body cavity contains numerous coelomocytes.

Reproductive organs are located dorsally, above the intestine. The sexes in the Heterotardigrada are always separated in the marine heterotardigrades although the primarily "terrestrial" family Echiniscidae may be parthenogenetic and many eutardigrade species are considered hermaphroditic. From the sac-shaped gonads two spermiducts in males and a single oviduct in females emerge. These ducts pass around the intestine and rectum and exit ventrally. Sexual dimorphism exists at least in the case of many marine heterotardigrades. This dimorphism is evidenced by the locations of the gonopores; in the female, a rosette of plates is located anterior to the anus. When seminal receptacles are present, the shape and location within the female are taxonomically important. In the male, a rounded or tubular opening is present very close to the anus. No gonopore dimorphism is present in *Renaudarctus* and in *Halechiniscus* dimorphism is apparent only in the ratio of the length of clavae to the length of the lateral cirri.

The Heterotardigrada lack excretory glands, but three and sometimes four such glands are present in the Eutardigrada.

Many Tardigrada develop directly from eggs deposited into the recently shed cuticle of the female. In some tardigrades, the eggs are deposited singly. Larvae resemble miniature adults with a reduced number of claws and digits and underdeveloped reproductive structures. Full development of trunk appendages is acquired through periodic molting. During a molt, egress occurs through an opening in the head region and the old buccal apparatus is lost as a new one is being formed. The pedal glands play an important role in the production of new feet. Apparently, only during the intermolt period can defecation, oviposition, and sperm penetration occur.

The muscles, composed of striated fibers, form dorsal and longitudinal bands dorsally and ventrally. Circular muscles are absent. There are pedal retractor muscles, and radiating pharyngeal muscles control the buccal apparatus.

Circulatory and respiratory systems are lacking.

Classification

The Phylum Tardigrada, despite some similarities with the aschelminth phyla clearly is part of the Annelid-Arthropod line of evolution. Some characteristics of the Rotifera and Nematoda include the absence of circular muscles and presence of a tripartite foregut. Annelid-like characters are represented in the morphology of the ventral nervous system and the organization of the leg, (they appear to be derived from annelid parapodia). Characters suggesting a relationship with Arthropoda are many and include a chitinous procuticle, lack of ciliated epithelium, large haemocoel in the adult, and metameric arrangement of legs and muscular system.

The pattern of development, with irregular cleavage of coelomic pouches, of which only the gonocoel is retained in the adult, is unlike any other group, and approaches that of deuterostomous invertebrates.

The phylum is divided into 2 large classes: the Heterotardigrada, including 65 marine species in 26 genera from the order Arthrotardigrada, plus 170 terrestrial species in 8 genera from the order Echiniscoidea; and the Eutardigrada, including 300 terrestrial and freshwater species plus 6 marine species in 19 genera from the order Parachela, plus 2 terrestrial species in 2 genera, from the order Apochela.

An intermediate class Mesotardigrada has been erected to include a monotypic hot-spring species seen only once.

The most prominent discriminative characters of the Heterotardigrada include: presence or absence of metameric division into plates; presence or absence of digits and/or claws; arrangement, number, and size of cephalic and somatic appendages; morphology of buccal apparatus, genital ducts, seminal receptacle ducts, oviduct, spermiducts, and gonopores.

Family characters are differentiated by presence or absence of digits with or without claws, presence or absence of plates, and possession of unpaired median cirrus. Details of the buccal apparatus, currently used to distinguish terrestrial taxa, are not heavily relied upon in differentiating marine species since they are often very difficult to observe. Somatic cirrus E, leg 4 papilla, and caudal apparatus morphology are the most important specific characters, together with size and shape of cephalic appendages.

The Eutardigrada lack the distinctive division of the body into segments, the cephalic appendages, and rarely have any cuticular appendages. The buccal apparatus of the eutardigrades usually is more complex, three and sometimes four excretory glands may be present, the gonad opens into the hindgut thereby creating a cloaca. Claws are usually present and are different in structure than those of the Heterotardigrada.

Subordinate taxa, consisting of 6 families, are based primarily on the morphology of the buccal apparatus and claws.

Phylogenetic Remarks

The fact that over 26 genera of marine Tardigrada have been erected to include 65 species only, demonstrates the wide morphological diversity of the group and indicates its very ancient origin.

Because of their occurrence in a conservative biotope, mostly in sedimentary lacunar systems, their cosmopolitan generic distribution and high morphological diversity, Arthrotardigrada are considered as archaic forms that originated from the early paleocontinent littoral sediments before the fractioning of the pangean world. Different evolutionary developments could have followed. Some taxa with high genetic stability have undergone little evolution and remained in the sand as relict forms; others colonized different marine habitats such as submarine caves, higher tidal zones (phytal species) or abyssal oozes. The less genetically stable taxa could have progressively acquired characters specialized toward life in freshwater and soil (order Echiniscoidea and class Eutardigrada). Among the latter, some species were secondarily readapted to marine life (*Halobiotus, Hypsibius*). In comparison, the diversification of cuticular structures and sensorial organs within the Arthrotardigrada, and the relative uniformity in appearance and simplicity in morphology of Eutardigrada are striking. Simplification and disparateness of external structures, together with the presence of additional internal organs (Malpighian tubules) in this class (Eutardigrada) attest to its apomorphic condition.

Habitat and Ecological Notes

Sandy biotopes are the most common residence of most of the marine tardigrade species. Tardigrades occur interstitially both in intertidal and subtidal habitats. Some species of *Stygarctus* and *Renaudarctus* are found as deep as 180 and 100 cm in the sand.

In low energy beaches and subtidal flats of fine or muddy sands, where silt and debris are more abundant and may clog the sediment lacunar system, Tardigrada are less diversified and live on or above the sediment interface (e.g., *Styraconyx* and *Tholoarctus*). The most diversified community in a medium energy beach is the area located between mid- and low-water tide marks.

Most ecological studies of intertidal Tardigrada have shown that species with "migrating" possibilities follow the oscillations of the shore line, according to beach water drainage and temperature – both factors linked with sediment texture and seasonal fluctuations. For species living in high energy beaches or subjected to severe weather conditions, especially in winter, sediment transport by storm waves easily modifies their habitat and may completely rearrange their distribution, thus making quantitative sampling somewhat unreliable.

Reproduction occurs throughout the entire year in the case of temperate interstitial species, but peaks in density occur in both spring and autumn, a phenomenon probably related to favorable trophic conditions during these periods.

The amount of available organic (probably phytal) matter is important in the distribution and reproduction pattern of sand dwelling Tardigrada although its role has not been fully elucidated. According to stomach coloration, algae constitutes the most common food of littoral forms whose buccal morphology is well developed. Deep-sea species with reduced pharyngeal structure are supposed to be mostly detritus feeders.

In oligotrophic sandy biotopes, chiefly coralline sands, pouches of symbiotic bacteria are frequent in tardigrades, and it seems likely that the hosts may utilize some substance synthetized by their symbionts.

Marine biotopes include seaweeds such as *Enteromorpha* and even lichens of the spray zone. In the latter case, this association is specific. Other marine tardigrades are reported to live ectocommensally on various other marine invertebrates including Echinodermata and various Crustacea (barnacles) and have been found more or less restricted to these habitats. Some marine taxa may also be found in freshwater habitats; e.g., *Styraconyx hallasi* was reported from a warm spring in Greenland (Kristensen, 1977). Conversely, some predominantly freshwater taxa may also be found in seawater; e.g., *Hypsibius itoi* was described by Tsurusaki (1980) from intertidal marine waters of Japan.

Marine Tardigrada are also reported from deeper water habitats. Endemic forms, or species from common littoral genera may live in sublittoral caves or in the bathyal zone off the continental shelf. At abyssal depths, a small number of specialized families of tardigrades occur. Considering the collecting difficulties encountered at great depths, a relatively high diversity has been recorded. Marine Tardigrada are not restricted to the temperate zone. They occur in higher latitudes (70° N in Greenland) as well as in equatorial (Galapagos Islands) and tropical areas (Caribbean, Polynesia).

Freshwater tardigrades, mostly in the class Eutardigrada and members of the heterotardigrade order Echiniscoidea, are found in sediments (including damp soil), but the majority of the species have been found in association with plants, mostly mosses and lichens. Studies of the hydrophyllous tardigrade fauna of freshwater sediments are few contrasted with studies of the hygrophylous species of mosses, lichens, and damp soil.

One of the earliest studies on freshwater tardigrades of beaches was that of Pennak (1940). Since this author has an extensive review of the ecology of freshwater tardigrades in Chapter 4 of this book, only a few additional remarks are necessary here. Freshwater tardigrades, although most commonly reported from mesopsammal habitats of freshwater beaches (Pennak, 1940; Schuster et al., 1978), littoral sandy biotopes (Nalepa and Quigley, 1983) or even at depths of 150 m in lakes (Forel, 1901), are also found in similar habitats in freshwater streams (Sudzuki, 1975; Whitman and Clark, 1984). The most common families found in such biotopes include the Macrobiotidae, Hypsibiidae, and less commonly the Milnesiidae. The genera involved usually are *Macrobiotus, Dactylobiotus, Hypsibius, Isohypsibius, Thulinia, Itaquascon,* and *Milnesium*. The genus *Carphania*, reported from freshwater habitats (Binda, 1978) is a representative of the class Heterotardigrada.

Methods of Collection

Collecting methods are the same as those general methods described for most meiofaunal organisms. In general, those methods used to collect the uppermost strata of sediments are the most effective, especially if larger numbers of specimens are desired. Core samplers used separately or as subsamplers of material taken by other means (such as a box corer) are mandatory if quantitative data are needed. In the case of collecting in sandy beach habitats, one can either dig a hole to the water table and sieve the collected water through a fine-mesh net (42-63 µm), or process sand by placing it in a pail of freshwater or seawater (depending on the habitat), stirring it to suspend the tardigrades, and decanting through a net. The use of various methods to relax or otherwise cause the animals to release from the sand grains is very important.

In sublittoral areas sediment can also be collected by diving and using various techniques of skimming of the sediment surface. This is an especially useful technique in areas where submerged obstacles prevent the use of other devices. For deep water sediments, a USNEL corer is the most reliable gear both for quantitative and qualitative studies.

Methods of Extraction

As Tardigrada cling strongly to sand grains, are often covered with debris, or entangled in other material, their extraction requires vigorous procedures including powerful water currents combined with proper anesthetization techniques. In the case of marine tardigrades, the $MgCl_2$ method is not effective for all species; soaking the sand in ten times its volume of 3.5% ethanol is better. Recently, Kristensen obtained some of the rarest species by subjecting large quantities of marine sand to a freshwater rinse for a very brief period (usually less than a minute). The resultant osmotic shock causes the animals to release their hold on the sediment particles. By pouring this water and the suspended meiofauna through a fine-mesh net and immediately re-introducing this material into seawater, the procedure has little effect on the animals. It should be noted that 42-63 µm mesh net is satisfactory for most adult tardigrades but a 23 µm mesh net is necessary if the smallest larval stages are to be collected.

Methods of Preparation for Taxonomic Study

Tardigrada should be preserved in 3-7% formalin buffered with Borax, or killed by the addition of polyvinyl-lactophenol introduced under the coverglass of a microslide preparation after living examination is completed.

For mounting long lasting microscopic preparations, the Kristensen and Higgins (1984) technique is as follows: the animals "were transferred with a drop of 2% formalin to microslides and covered with coverslips. The formalin preparation was infused with a 10% solution of glycerin in 96% ethyl alcohol and allowed to evaporate to glycerin over a period of several days; the resulting whole-mount preparation was sealed by Murrayite."

Phase and interference-contrast (Normaski) optics are necessary for systematic fine observations. SEM and TEM investigation techniques are useful to solve fine morphological and cytological problems. These techniques are found in Kristensen (1976).

Type Collections

The most significant collections of the marine Tardigrada are located at: Musée National d'Histoire Naturelle, Paris, France; Universitets Zoologisk Museum, Copenhagen, Denmark; National Museum of Natural History, Smithsonian Institution, Washington, D.C., USA; Istituto di Zoologia, Universita di Bari, Italy; Department of Invertebrate Zoology, University of California at Davis, Davis, California, USA.

References

Binda, M.G.
1978. Risistemazione di alcuni Tardigradi con l'istituzione di un nuovo genere di Oreeleidae e della nuova famiglia Archechiniscidae. *Animalia*, Catania, 5:307-314.

Crisp, M., and R.M. Kristensen
1983. A New Marine Interstitial Eutardigrade From East Greenland With Comments on Habitat and Biology. *Videnskabelige Meddeleser fra Dansk Naturhistorisk Forening i Kobenhavn*, 144:99-114.

Forel, F.A.C.
1901. Le Léman. *Monographie Limnologique. Tome III. Biologie, Histoire, Navigation, Peche.* 715 pages. Lausanne: F. Rouge.

Grimaldi de Zio, S., M. D'Addabbo Gallo, M.R. Morone de Lucia, and P. Grimaldi
1980. Ulterior i Dati sui Tardigradi del Mesopsammon di Alcune Spiagge Pugliesi. *Thalassia Salentina*, 10:45-65.

Grimaldi de Zio, S., M. D'addabbo Gallo, R. Morone de Lucia, R. Vaccarella, and P. Grimaldi
1982a. *Neostygarctus acanthophorus* n. gen., n. sp., Nuovo Tardigrado Marino del Mediterraneo. *Cahiers de Biologie Marine*, 23:319-373.
1982b. Quatro Nuove Specie di Halechiniscidae Rinvenute in due Grotte Sottomarine dell'Italia Meridionale. (Tardigrada: Heterotardigrada). *Cahiers de Biologie Marine*, 23:415-426.

Kristensen, R.M.
1976. On the Fine Structure of *Batillipes noerrevangi* Kristensen 1976. 1. Tegument and Moulting Cycle. *Zoologischer Anzeiger*, 197:129-150.
1977. On the Marine Genus *Styraconyx* (Tardigrada, Heterotardigrada, Halechiniscidae), with Description of a New Species from a Warm Spring on Diskö Island, West Greenland. *Astarte*, 10:87-91.
1978. Notes on Marine Heterotardigrades 1. Description of Two New *Batillipes* Species, Using the Electron Microscope. *Zoologischer Anzeiger*, 200:1-17.

Kristensen, R.M., and T.E. Hallas
1980. The Tidal genus *Echiniscoides* and Its Variability, With Erection of Echiniscoididae Fam. N. *Zoologica Scripta*, 9:113-127.

Kristensen, R.M., and R.P. Higgins
1984a. Revision of *Styraconyx* (Tardigrada: Halechiniscidae), With Descriptions of Two New Species from Diskö Bay, West Greenland. *Smithsonian Contributions to Zoology*, 361:1-40.
1984b. A New Family of Arthrotardigrada (Tardigrada: Heterotardigrada) From the Atlantic West Coast of Florida, U.S.A. *Transactions of the American Microscopical Society*, 103:295-311.

Lindgren, E.W.
1971. Psammolittoral Marine Tardigrades from North Carolina and Their Conformity to Worldwide Zonation Patterns. *Cahiers de Biologie Marine*, 12:481-496.

McKirdy, D., P. Schmidt, and M. McGinty-Bayly
1976. Interstitielle Fauna von Galapagos. XVI. Tardigrada. *Mikrofauna der Meeresbodens*, 58:409-449.

Marcus, E.
1936. Tardigrada. *Das Tierreich*, 66:1-340.

Nalepa, T.F. and M.A. Quigley
1983. Abundance and Biomass of the Meiobenthos in Nearshore Lake Michigan with Comparisons to the Macrobenthos. *Journal of Great Lakes Research*, 9(4):530-547.

Pennak, R.W.
1940. Ecology of the Microscopic Metazoa Inhabiting the Sandy Beaches of Some Wisconsin Lakes. *Ecological Monographs*, 10:537-615.

Pollock, L.W.
1970. Distribution and Dynamics of Interstitial Tardigrada at Woods Hole, Massachusetts, U.S.A. *Ophelia*, 7:145-165.
1976. Marine Flora and Fauna of the Northeastern United States, Tardigrada. *NOAA Technical Report NMFS Circular*, 394:1-25.

Ramazzotti, G., and W. Maucci
1983. Il Phylum Tardigrada. *Memorie Dell'Istituto Italiano di Idrobiologia Dott. Marco di Marchi*, 41:1-1012.

Renaud-Debyser, J.
1956. Répartition de deux Tardigrades *Batillipes mirus* Richters et *Stygarctus bradypus* Schulz dans un

Segment de Plage du Bassin d'Arcachon. *Compte-rendus des Séances de l'Académie des Sciences, Paris,* 243:1365-1369.
1963. Recherches Écologiques sur la Faune Interstitielle des Sables (Bassin d'Arcachon; Île de Bimini, Bahamas). *Vie et Milieu,* Supplement, 15:1-157.

Renaud-Mornant, J.
1967. Tardigrades de la Baie Saint Vincent, Nouvelle Calédonie. Pages 103-118 in *Expédition Francaise sur Récifs Coralliens de la Nouvelle Calédonie.* Paris: Editions de la Fondation Singer-Polignac.
1975. Deep-sea Tardigrada from the "Meteor" Indian Ocean Expedition. *"Meteor" Forschungergebnisse.* Series D, 21:54-61.
1979a. Tardigrades Marins de Madagascar, I: Halechiniscidae et Batillipedidae. *Bulletin du Muséum National d'Histoire Naturelle* (Paris), 4th Series, 1 (Section A):257-277.
1979b. Tardigrades Marins de Madagascar. II. Stygarctidae et Oreellidae. III. Considérations Écologiques Générales. *Bulletin du Muséum National d'Histoire Naturelle, Paris.* 4th Series, 1 (Section A):339-351.
1982. Species Diversity in Marine Tardigrada. Pages 149-178, 11 figures, in D. Nelson, editor, *Proceedings of the Third International Symposium on the Tardigrada, August 3-6, 1980. Johnson City, Tennessee, USA.* Johnson City: East Tennessee State University Press.
1983. Halechiniscidae (Heterotardigrada) de la Campagne "Benthedi" canal du Mozambique. *Bulletin du Muséum National d'Histoire Naturelle (Paris),* 4th Series, 6 (Section A):67-88.

Schmidt, P.
1969. Die quantitative Verteilung und Populationsdynamik des Mesopsammons am Gezeiten-Sandstrand der Nordsee-Insel Sylt. II. Quantitative Verteilung und Populationsdynamik einzelner Arten. *Internationale Revue der gesamten Hydrobiologie,* 54:95-174.

Schuster, R.O., E.C. Toftner, and A.A. Grigarick
1978. Tardigrada of Pope Beach, Lake Tahoe, California. *The Wasmann Journal of Biology,* 35(1):115-136.

Sudzuki, M.
1975. Lotic Tardigrada from the Tama River with Special Reference to Water Saprobity, pages 377-391 in R.P. Higgins, editor, International Symposium on Tardigrades. *Memorie Dell'Istituto Italiano Di Idrobiologia,* 32(Supplement):1-469.

Tsurusaki, N.
1980. A New Species of Marine Interstitial Tardigrada of the Genus *Hypsibius* from Hokkaido, Northern Japan. *Annotationes Zoologicae Japonenses,* 533(4):280-284.

Whitman, R.L., and W.J. Clark
1984. Ecological Studies of the Sand-Dwelling Community of an East Texas Stream. *Freshwater Invertebrate Biology,* 3(2):59-79.

Zio, S. de
1965. Distribuzione del Mesopsammon in Rapporto alla Distanza della Linea di Riva e alla Distanza da un Corso di Acqua Salmastra. *Bolletino di Zoologia,* 32:525-537.

Zio, S. de, and P. Grimaldi
1964. Ricerche Sulla Distribuzione ed Ecologie di *Batillipes pennaki* Marcus in una Spiaggia Pugliese (Heterotardigrada). *Cahiers de Biologie Marine,* 5:271-285.
1966. Ecological Aspects of Tardigrada Distribution in South Adriatic Beaches. *Veröffentlichungen des Instituts für Meeresforschung in Bremerhaven,* 2:87-94.

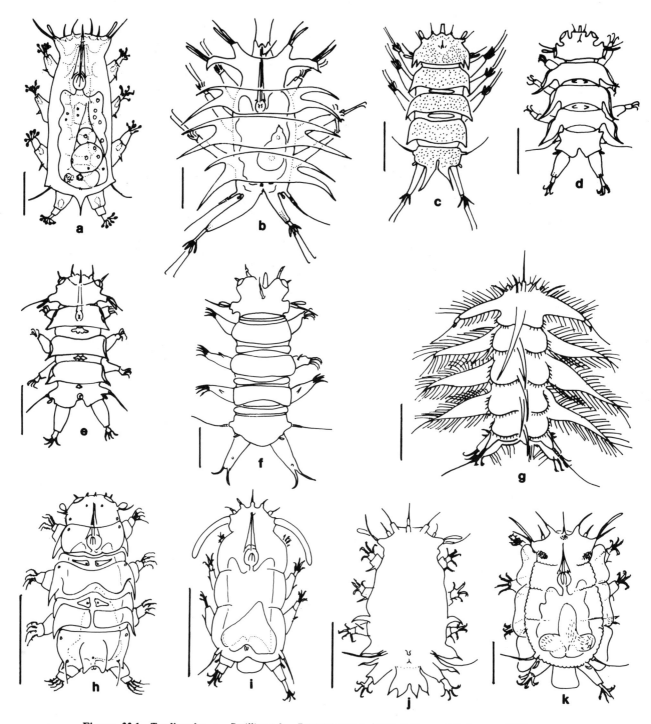

Figure 33.1.—Tardigrada: a, *Batillipes;* **b,** *Parastygarctus higginsi* Renaud-Debyser, 1965; **c,** *Stygarctus granulatus* Pollock, 1970; **d,** *Pseudostygarctus triungulatus* McKirdy et al., 1976; **e,** *Mesostygarctus intermedius* Renaud-Mornant, 1979; **f,** *Megastygarctides orbiculatus* McKirdy et al., 1976; **g,** *Neostygarctus acanthophorus* Grimaldi de Zio et al, 1982; **h,** *Renaudarctus psammocryptus* Kristensen and Higgins, 1984; **i,** *Ligiarctus eastwardi* Renaud-Mornant, 1982; **j,** *Halechiniscus flabellatus* Grimaldi de Zio et al., 1982; **k,** *Wingstrandarctus corallinus* Kristensen, 1984. (Scale = 50 μm.)

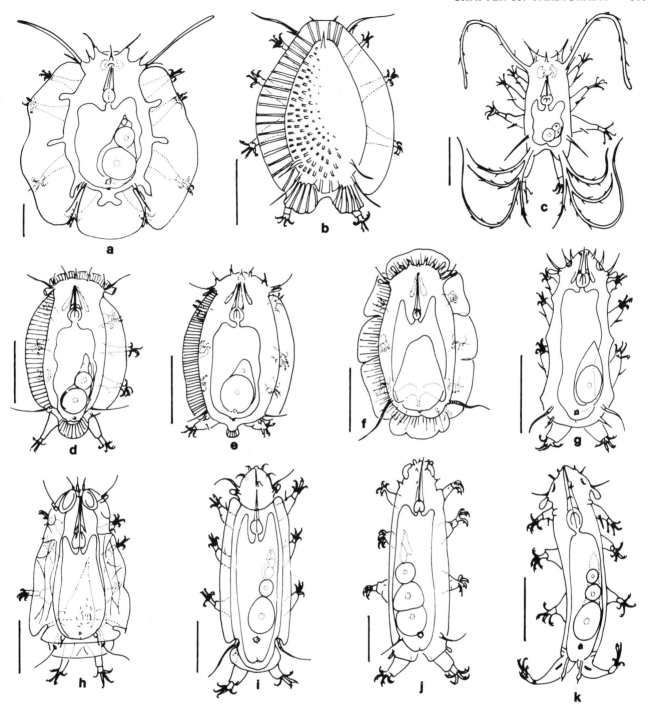

Figure 33.2.--Tardigrada (continued): **a**, *Florarctus salvati* Delamare Deboutteville and Renaud-Mornant, 1965; **b**, *Actinarctus doryphorus* Schulz, 1935; **c**, *Tanarctus arborspinosus* Lindgren, 1971; **d**, *Raiarctus colurus* Renaud-Mornant, 1981; **e**, *Rhomboarctus thomassini* Renaud-Mornant, 1984; **f**, *Parmursa fimbriata* Renaud-Mornant, 1984; **g**, *Styraconyx kristenseni* Renaud-Mornant, 1981; **h**, *Tholoarctus natans* Kristensen and Renaud-Mornant, 1983 (a semi-benthic species); **i**, *Lepoarctus coniferus* (Renaud-Mornant, 1983); **j**, *Clavarctus falculus* Renaud-Mornant, 1983; **k**, *Angursa lanceolata* Renaud-Mornant, 1981. (Scale = 50 μm.)

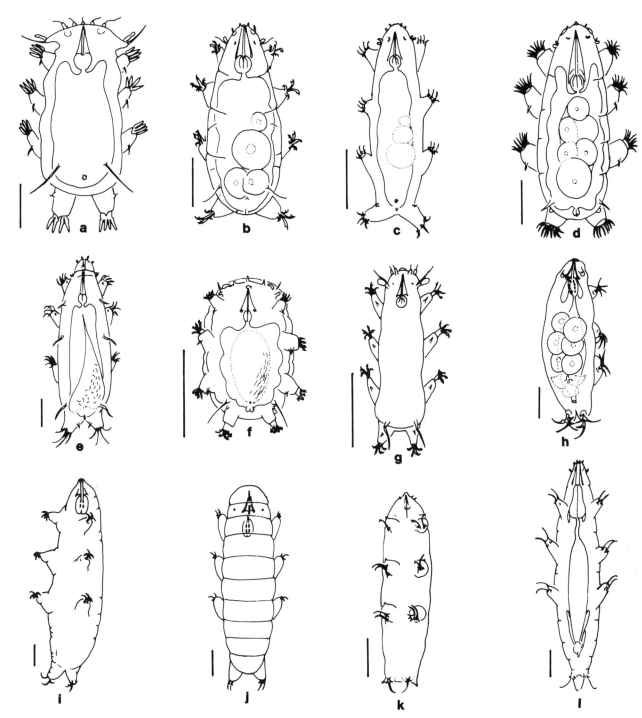

Figure 33.3.--Tardigrada (continued): **a**, *Orzeliscus belopus* Bois-Reymond-Marcus, 1952; **b**, *Archechiniscus marci* Schulz, 1953; **c**, *Anisonyches diakidius* Pollock, 1975; **d**, *Echiniscoides heopneri* Kristensen and Hallas, 1980 (a parasitic species on barnacles); **e**, *Coronarctus tenellus* Renaud-Mornant, 1974; **f**, *Tetrakentron synaptae* Cuénot, 1892 (a parasitic species on a sea cucumber); **g**, *Pleocola limnoriae* Cantacuzéne, 1951 (a parasitic species on Isopoda); **h**, *Halobiotus arcturulius* Crisp and Kristensen, 1983 (a marine eutardigrade); **i**, *Thulinia ruffoi* Bertolani, 1981 (freshwater eutardigrade); **j**, *Hypsibius convergens* (Urbanowicz, 1925); *Carphania fluviatilis* Binda, 1978; **k**, *Milnesium tardigradum* Doyère, 1840. (Scale = 50 μm.)

34. Cladocera

David G. Frey

The Cladocera constitute one of the four orders of the crustacean class Branchipoda, the other three being the Notostraca, Conchostraca, and the Anostraca. The Cladocera are rather primitive crustaceans, having a body that is completely covered by a shell except in some predacious species, and having 9 or 10 pairs of appendages – antennules, antennae, maxillae, and mandibles attached to the head and five or six trunk limbs attached to the abdomen, which are used in food getting and respiration (Fryer, 1963, 1968, 1974). The abdomen ends in a characteristic structure – the postabdomen – with a series of denticles along the dorsal margin and clusters of setae laterally, and with a terminal pair of postabdominal claws having characteristic basal spines and rows of setules. Trunk limbs I and II are adapted to the particular nutrition of the species.

All the meiobenthic species have an unhinged shell completely covering the body and attached to the headshield anteriorly. The headshield-shell junction constitutes the ecdysial line, which spreads apart at molting and allows the next-instar individual to escape from the old exoskeleton. On the headshield of the chydorids is a series of median pores and two minor pores, the configuration of which is characteristic of the subfamilies and their genera (Frey, 1959, 1966). The macrothricids tend to have a large median structure resembling a pore – the nuchal organ – but which is blocked by an imperforate chitinous membrane. No other pores are known in the macrothricids except a minute one at the extreme front end between the bases of the antennules in *Guernella*. Some of the sidids also have a nuchal organ comparable to that in the macrothricids. *Sida crystallina* can become temporarily attached to the substrate by the suction exerted by this structure.

The space enclosed between the shell and the body is the brood chamber, in which the parthenogenetic eggs undergo development to independent first-instar individuals, which are then released. When bisexual reproduction occurs, the fertilized egg becomes a resting egg, which becomes enclosed by a modified part of the shell (ephippium) on molting. Such resting eggs survive extreme conditions of drought, temperature, and salinity for many years. They are the chief means whereby a species spreads to new regions. One resting egg is sufficient to establish a new population.

The shell and head can be sculptured in various ways, even to having a deep honeycomb-like structure. Such details, plus the pattern of the headpores, the structure of the postabdomen and terminal claws, features of the head appendages and the first two trunk limbs, as well as the special morphology of the three male instars and of the ephippium provide most of the characters used to distinguish the species of chydorids. Some of these characters are important in the other two families as well, in addition to some other characters not present in the chydorids at all.

Size.--Species in the chydorid subfamilies Chydorinae and Aloninae range roughly from 0.3 to 1.3 mm in length, with a strong peak at 0.5 mm. Sixty percent of the species are 0.3 to 0.5 mm long. The eight species in the subfamilies Sayciinae and Eurycercinae are much longer, up to 6.1 mm. The family Macrothricidae has fewer species and their size ranges from 0.3 to 2.2 mm, with a weak peak about 0.7 mm; 43% of these species measure 0.3 to 0.7 mm. Ninety-one percent of adult chydorids and 65% of adult macrothricida are less than 1.0 mm long. Meiobenthic members of the Sididae are larger, ranging as adults from about 1 mm to 4 mm, but maximum size is a poor descriptor of cladoceran populations. All species mature through a series of juvenile stages, gradually becoming larger at each molt. Because mortality of all individuals is high, the major peak in population size-frequency generally comes among the two pre- reproductive instars, followed by a rapid trailing off into the mature instars. Accordingly, an arbitrary upper size limit cannot be used effectively to separate the benthic Cladocera into macrofauna and meiofauna.

Classification

In the cladocera there are about 400 to 500 described species. The order is divided into two suborders, the Haplopoda with a single family and single genus, and the Eucladocera with three superfamilies and 10 families. All but a few families of both superorders are mostly or totally planktonic, living in the water column totally independent of contact with any benthic substrate. Almost all members of the eucladoceran families Chydoridae and Macrothricidae and some members of the Sididae, on the other hand, are substrate oriented. They occur on and in various types of substrate and swim only short distances from one activity place to another. Although not included among the meiobenthos previously, they properly belong here because of their small size and habitat.

There are about 200 species of Chydoridae, 60 species of Macrothricidae, and a much smaller number of meiobenthic Sididae.

In addition to references mentioned in the text are included the major taxonomic works from various countries of the World that will help in identifying species found in those countries. For precise taxonomy, though, one must realize that most species, at least of the Chydoridae, are not so widely distributed as currently believed.

Habitats and Ecological Notes

Cladocera as a group occur primarily in fresh water. The only truly marine taxa are a few planktonic species in the family Podonidae and in the genus *Penilia* of the family Sididae. The other members of the Podonidae and some members of the Cercopagidae occur in the saline inland waters of the Ponto-Aralo-Caspian Basin. In arid and semi-arid regions, e.g., Australia, where the electrolyte concentration of the waters increases from vigorous evaporation, the diversity of Cladocera present declines drastically as the salinity increases. Although few species can tolerate high concentrations of electrolytes, there are some species, such as in the genera *Daphniopsis* and *Moina* especially, that can thrive at salt concentrations at least twice as high as in the oceans. In weakly brackish water, as in the Baltic, some littoral Cladocera can occur at salinities up to 5 and even 10 parts per thousand.

Meiobenthic Cladocera inhabit all kinds of substrate in shallow water. The number of species diminishes rapidly with increasing depth, although some species persist to 100 m or more if dissolved oxygen is continuously present. Diversity of substrate increases the diversity of niches and hence the diversity of Cladocera. Bare mud and sand have relatively few chydorids and macrothricids, and these are specifically adapted to these particular materials. The highest diversity occurs in beds of aquatic macrophytes, which obviously provide the highest diversity of habitat of any type of substrate.

Methods of Collection

Meiobenthic Cladocera can be collected simply with a plankton net (120-150 μm mesh) with a 5-mm mesh screen across the mouth to keep out leaves, macrophytes, snails, and other large materials. Mounted on a pole, such a net can be used effectively to sample all kinds of vegetation and other substrates. This method is qualitative or at best only semi-quantitative. Other small dredges designed to remove only the upper horizons of sediment are also useful, as are bottom grabs.

Quantitative samples can be obtained in a number of ways. Smirnov (1971) sampled the beds of macrophytes with a But weed sampler and a grab for the bottom sediments. The unit reported is number per "combined" meter, which represents 1 m^2 of bottom plus the overlying 1 m^3 of water with its macrophytes. Whiteside et al. (1978) used an array of funnels with bottles at the top to catch the animals as they migrate upward at night. One run of the funnels over-night is claimed to catch more than 90% of the total meiobenthic population. Goulden (1971) used a small sediment corer (1.64 cm diameter) to study the chydorids in a lake in Pennsylvania. Several corers were operated simultaneously at each station to provide some statistical estimates of reliability of the results.

Such quantitative methods reveal the tremendous abundance of these animals. In the Volga River, Smirnov (1971) obtained biomasses of 1 g to more than 9 g of Cladocera per combined meter, which, considering the small sizes of these animals, represent very large numbers of individuals. Whiteside consistently obtained more than 1 million animals per m^2, even under the ice. Goulden, working in a more oligotrophic environment with less habitat diversity obtained a maximum of about 300,000 animals per m^2. Note that this number represents animals from the sediments only. Cladocera constitute one of the most important converters of algae, bacteria, and detritus in the lentic freshwater environments. Any study of meiobenthos in lakes and ponds possibly down to considerable depths, in endorheic waters that are not too saline, and in moderately brackish marine environments is certain to yield a diversity and probably a considerable number of these organisms.

Methods of Extraction and Preparation of Specimens for Taxonomic Study

Extraction of Cladocera from sediment can usually be accomplished by a series of sieves or by differential gravity centrifugation. Specimens are prepared for taxonomic study by methods generally used for all crustacean taxa. Specimens should be fixed in 4% to 5% formalin containing 40 g of sucrose per liter, then stored in this mixture if permitted, otherwise in 70% ethanol.

References

Alonso, M.
1985. Las Lagunas de la España Peninsular: Taxonomía, Ecología y Distribución de los Cladoceros. 795 pages. Ph.D. dissertation, Universidad de Barcelona.

Amoros, C.
1984. Crustacés Cladocères. *Bulletin Mensuel de la Société Linnéenne de Lyon*, 43:72-145.

Birge, E.A.
1918. The Water Fleas (Cladocera). Pages 676-740 in H.B. Ward and G.C.Whipple, editors, *Fresh-water Biology*. New York: John Wiley and Son.

Bowman, T.E., and L.G. Abele
1983. Classification of the Recent Crustacea. Pages 1-27 in L.G. Abele, editor, *Biology of Crustacea. Volume 1. Systematics, the Fossil Record, and Biogeography*. New York: Academic Press.

Brooks, J.L.
1959. Cladocera. Pages 587-656 in W.T. Edmondson, editor, *Fresh-water Biology, 2nd Edition*. New York: John Wiley and Son.

Chiang, S.-c., and N.-s. Du.
1979. Fauna Sinica, Crustacea, Freshwater Cladocera. vi + 297 pages. Peking: Science Press, Academia Sinica.

Daday, E.
1898. Mikroskopische Süsswasserthiere aus Ceylon. *Természetrajzi Füzetek, Anhangsheft* 21:1-123.
1905. Untersuchungen über die Süsswasser-Mikrofauna Paraguays. *Zoologica, Stuttgart*, 18(44):1-374.

Fernando, C.H.
1974. Guide to the Freshwater Fauna of Ceylon (Sri Lanka), Supplement 4. *Bulletin Fisheries Research Station, Sri Lanka (Ceylon)*, 25:27-81.

Flössner, D.
1972. Kiemen- und Blattfüsser, Branchiopoda Fischläuse, Branchiura. *Die Tierwelt Deutschlands*, 60:1-501. Jena: Gustav Fischer.

Frey, D.
1959. The Taxonomic and Phylogenetic Significance of the Head Pores of the Chydoridae (Cladocera). *Internationale Revue der gesamten Hydrobiologie*, 44:27-50.
1966. Phylogenetic Relationships in the Family Chydoridae. *Marine Biological Association of India, Proceedings of the Symposium on Crustacea*, 1:29-37.
1982. Questions Concerning Cosmopolitanism in Cladocera. *Archiv für Hydrobiologie*, 93:484-502.
1986. The Non-Cosmopolitanism of Chydorid Cladocera: Implications for Biogeography and Evolution. Pages 237-256 in K.L. Heck and R.H. Gore, editors, *Crustacean Issues, Vol. 4, Crustacean Biogeography*. Rotterdam: A.A. Balkema.
1987. The Taxonomy and Biogeography of the Cladocera. In L. Forró and D.G. Frey, editors, *Cladocera: Proceeding of the Cladocera Symposium, Budapest 1985. Hydrobiologia*, 145:5-17.

Fryer, G.
1963. The Functional Morphology and Feeding Mechanism of the Chydorid Cladoceran "*Eurycercus lamellatus*" (O.F. Müller). *Transactions of the Royal Society of Edinburgh*, 65:334-381.
1968. Evolution and Adaptive Radiation in the Chydoridae (Crustacea: Cladocera): A Study in Comparative Functional Morphology and Ecology. *Philosophical Transactions of the Royal Society of London, Series B*, 254:221-385

1974. Evolution and Adaptive Radiation in the Macrothricidae (Crustacea: Cladocera): a Study in Comparative Functional Morphology and Ecology. *Philosophical Transactions of the Royal Society of London, Series B*, 269:137-274.

Goulden C.E.
1971. Environmental Control of the Abundance and Distribution of the Chydorid Cladocera. *Limnology and Oceanography*, 16:320-331.

Idris, B.A.G.
1983. *Freshwater Zooplankton of Malaysia (Crustacea: Cladocera)*. vii + 148 pages. Serdang, Selangor: Penerbit Universiti Pertanian Malaysia.

Lilljeborg, W.
1901. Cladocera Sueciae, oder Beiträge zur Kenntniss der in Schweden lebenden Krebsthiere von der Ordnung der Branchiopoden und der Unterordnung der Cladoceren. *Nova Acta Regiae Societatis Scientiarum Upsaliensis*, 3:1-701.

Mamaril, A.C., Sr., and C.H. Fernando
1978. Freshwater Zooplankton of the Philippines (Rotifera, Cladocera, and Copepoda). *Natural and Applied Science Bulletin, University of the Philippines*, 30:105-221.

Margalef, R.
1953. *Los Crustaceos de las Aguas Continentales Ibericas. Biología de las Aguas Continentales. X*. 243 pages. Madrid: Instituto Forestal de Investigaciones y Experiencias.

Margaritora, F.G.
1985. Cladocera. 399 pages in *Fauna d'Italia*. Bologna: Edizoni Calderini.

Michael, R.G., and B.K. Sharma
in press Indian Cladocera. *Fauna of India Series*. Calcutta: Zoological Survey of India.

Müller, O.F.
1776. *Zoologiae Daniae Prodromus seu Animalium Daniae et Norvegiae indigenarum characteres, nomina, et synonyma imprimis popularium*. Havniae.
1785. *Entomostraca seu Insecta Testacea, quae in aquis Daniae et Norvegiae reperit, descripsit et iconibus illustravit*. Lipsiae et Havniae.

Negrea, Ş.
1983. Cladocera. *Fauna Republicii Socialiste România, Crustacea*. Volume 4, Fascicle 12, 399 pages.

Notenboom-Ram, E.
1981. Verspreiding en ecologie van Branchiopoda in Nederland. *Rijksinstituut voor Natuurbeheer, Rapport*, 81/14:1-95.

Rey, J., and L. Saint-Jean
1980. Branchiopodes (Cladocères). Pages 307-332 in J.R. Durand and C. Lévêque, editors, *Flore et Faune Aquatiques de l'Afrique Sahelo-Soudanienne*, Volume 1. Office de la Recherche Scientifique et Technique Outre-Mer, Collection Initiations, Documentations Techniques Number 44.

Sars, G.O.
1865. Norges Ferskvandskrebsdyr. Første Afsnit Branchiopoda. I. Cladocera Ctenopoda (Fam. Sididae & Holopedidae). *Christiania*, vii + 71 pages.
1885. On Some Australian Cladocera, Raised from Dried Mud. *Christiania Videnskabs-Selskabs Forhandlinger*, 1885(8):1-46.
1888. Additional Notes on Australian Cladocera, Raised from Dried Mud. *Christiania Videnskabs-Selskabs Forhandlinger*, 1888(7):1-74.
1894. Contributions to the Knowledge of the Fresh-Water Entomostraca of New Zealand as Shown by Artificial Hatching from Dried Mud. *Videnskabs-Selskabets Skrifter, I. Mathematisk-Naturvidenskab Klasse*, 1894(5):1-62.
1895. On Some South-African Entomostraca Raised from Dried Mud. *Videnskabs-Selskabets Skrifter, I. Mathematisk-Naturvidenskab Klasse*, 1895(8):1-56.
1896. On Fresh-water Entomostraca from the Neighbourhood of Sydney, Partly Raised from

Dried Mud. *Archiv for Mathematik og Naturvidenskab, Kristiania,* 18(3):1-81.
1901. Contributions to the Knowledge of the Fresh-Water Entomostraca of South America, as Shown by Artificial Hatching from Dried Material. Part 1. Cladocera. *Archiv Mathematik og Naturvidenskab, Kristiania,* 23(3):1-101.
1916. The Fresh-Water Entomostraca of Cape Province (Union of South Africa). Part I: Cladocera. *Annals of the South Africa Museum,* 15:303-351.

Scourfield, D.J., and J.P. Harding
1966. A Key to the British Species of Freshwater Cladocera, 3rd Edition. *British Freshwater Biological Association Scientific Publication 5,* pages 1-55.

Smirnov, N.N.
1971. Chydoridae Fauny Mira. *Fauna SSSR, New Series,* Number 101. Rakoobraznyye, T. 1, vyp. 2, 531 pages. [Also available as an English translation by A. Mercado for Israel Program for Scientific Translations, Keter Publishing House, Jerusalem. 1974.]
1976. Macrothricidae i Moinidae Fauny Mira. *Fauna SSSR,* Rakoobraznyye, T. 1, vyp. 3, 237 pages.

Smirnov, N.N., and V.B. Timms
1983. A Revision of the Australian Cladocera (Crustacea). *Records of the Australian Museum,* Supplement 1:1-132.

Uéno, M.
1927. The Freshwater Brachiopoda of Japan I. *Memoirs of the College of Science, Kyoto Imperial University,* Series B. 2(5) (article 12):259-311.

Whiteside, M.K., J.B. Williams, and C.P. White
1978. Seasonal Abundance and Pattern of Chydorid Cladocera in Mud and Vegetative Habitats. *Ecology,* 59:1177-1188.

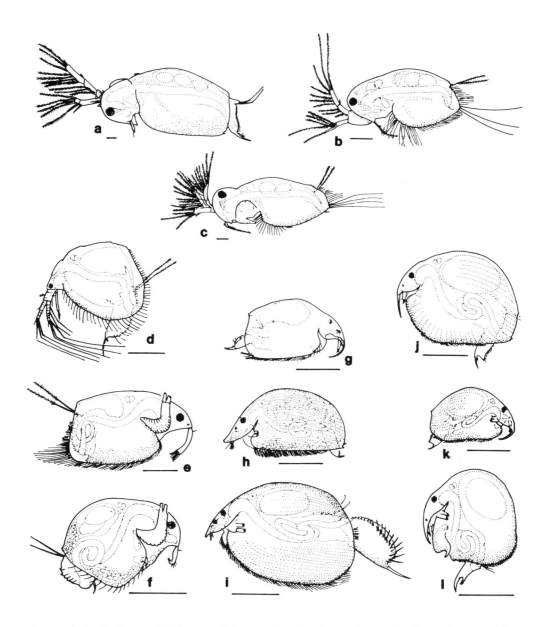

Figure 34.1.--Cladocera: Sididae: **a**, *Sida crystallina*, **b**, *Latonopsis occidentalis*, **c**, *Latona setifera;* Macrothricidae: **d**, *Ilyocryptus spinifer*, **e**, *Acantholeberis curvirostris*, **f**, *Streblocerus serricaudatus;* Chydoridae, Aloninae: **g**, *Rhynchotalona falcata*, **h**, *Graptoleberus testudinaria*, **i**, *Leydigia acanthocercoides;* Chydoridae, Chydorinae: **j**, *Pleuroxux denticulatus*, **k**, *Disparalona dadayi*, l, *Anchistropus minor*. (From Birge, 1918). (Note the antennae in Figures 34.1e-l are not shown completely, and in Figures 34.1g and k are not shown at all) (Scale = 200 μm.)

35. Ostracoda

Dietmar Keyser

Ostracods are small crustaceans ranging in length from 0.08 to 32 mm. Their entire body is encased in a bivalved, calcified carapace which can be smooth to variously ornamented. The two valves are joined by a dorsal hinge opposed by closing muscle.

The body is unsegmented and has a reduced number of limbs. The head is larger than both the thorax and abdomen combined. It bears five paired appendages: first and second antennae, mandibles, and the first and second maxillae. Commonly, two additional thoracopods are also present. The second maxillae and the two thoracopods are often used as walking or cleaning legs. The abdomen terminates in a pair of furcae. Between the last thoracopods and the furcae there may be a pair of large male copulatory organs.

The gut is divided into an atrium, esophagus, midgut (intestine), and rectum which terminates either in front of or behind the furcae. A pair of hepatopancreatic tubes have their origin on the midgut.

Most Ostracoda possess nauplier eyes. Only the Myodocopa have compound eyes as well. The nervous system can be straight or compressed.

The circulatory system normally is reduced; a heart with two ostii and a short aorta may be present, but only in the Myodocopa. Gill-like structures are also found only in some myodocopids. Vibratory plates of the limbs are often used to furnish the thin cuticular walls of the inner side of the carapace with a flow of water for respiration and excretion.

Antennal and maxillary excretory glands are present. Most Ostracoda have paired reproductive organs. Sometimes they are very large, extending into the carapace folds. The spermatozoa of some species represent the longest in the animal kingdom, sometimes reaching 1.0 mm, as much as ten times longer than the animal itself. The gonopores open in front of the furcae. As already mentioned the paired male copulatory organ can be huge and take up half of the carapace space. The eggs emerge either directly into the water or into the carapace where they pass into the brood chamber situated dorsal to the furcae. They remain within the carapace up to the third larval stage or may be cemented to some suitable substrate.

The nauplius emerges with its carapace already developed and undergoes from five to eight molts before attaining adulthood where further molting does not occur.

Ostracoda can be bisexual or parthenogenetic. Parthenogenetic forms are mainly found within the limnetic Cypridoidea. Sexual dimorphism is often made obvious by the presence of a very large penis or distinct brood chamber.

Ostracods live in almost all aquatic environments. From marine pelagic (mainly Myodocopa) to benthic (mainly Podocopa). They are common in both brackish and freshwater environments, freshwater ostracods may even live in temporary ponds, small puddles, in the water held by the leaves of bromeliads, in mosses and salt ponds. Groundwater and springs contain ostracods, and specialized ostracods are found as part of the mesopsammon of sandy biotopes. One group of Ostracoda (Entocytheridae) are commensals, living on or within other animals.

Ostracods are able to swim, although most of them crawl. They have the capability of producing a special thread to secure themselves to a substrate or to crawl on it. Sometimes they use this sticky thread for trapping algae as a source of food. Most obtain their food by filter-feeding, grazing, sucking, or as predators of scavengers.

These small crustaceans are easily confused with some larvae of Cirripedia, the so called Cypris-larvae, and sometimes with very small bivalves and even with some Cladocera. They may be differentiated from barnacle Cypris-larvae on the basis of the position of the eye spot. In the Ostracoda the nauplier eye is always at the anterodorsal edge of the carapace; in the Cypris-larvae, the eyes are separated and situated at the anteromedial part of the lateral shell. Another feature of the Cypris-larvae is the presence of a ventral connection between the valves. Bivalve larvae do not possess appendages and are concentric with a dorsal muscle. Cladocera always have a head and antennae which are not included in the carapace.

Classification

The Ostracoda are one of the few groups which have been easily fossilized because of their calcified valves. They have a long fossil record; there are more than 13,000 post-Paleozoic species, which includes about 5,000 recent species (Hartmann, 1975). Ostracods are found back to early Cambrian times and furnish an excellent tool for stratigraphy and paleoecology.

Living ostracods are divided into two main groups: the Myodocopa and the Podocopa.

The Myodocopa are best characterized as fairly large (1-29 mm) animals with a mostly lightly calcified shell. This shell has a convex marginal area and can be smooth or ornamented at the lateral sides. A rostral incisure is often present. The exopod of the second antenna has 7-9 joints. A heart and a compound eye as well as naupliar eye is often present. These animals are found only in marine or slightly brackish waters. Most of them live pelagic.

The Podocopa are small (0.08-2.5 mm) and bear an elongated shell with a concave ventral margin. They do not possess a rostral incisure. The exopod of the second antenna never has more than two joints. They do not have a heart or compound eyes. Naupliar eyes are frequently developed. They occur in nearly every aquatic biotope and are mostly benthic animals.

The nearly 600 species of Myodocopa are divided into three orders: The Myodocopida are characterized by a rounded or elongated shell with a rostral incisure. The body has seven appendages, where the seventh is vermiform, reduced or absent. A heart is present and so is the naupliar eye. These animals are usually larger than 1 mm.

The Halocyprida possess a shell with an almost straight dorsal hinge. In this case a rostral incisure is present. In the family Thaumatocyprididae the shell is rounded, but without rostral incisure. The halocyprid body has seven pairs of appendages but the seventh limb is short with two terminal bristles. Naupliar eyes are sometimes present, compound eyes are absent.

The Cladocopida are very small Myodocopids (smaller than 1 mm). They carry a rounded shell without a rostral incisure. The body has only five pairs of appendages. No eye or heart is present.

In the Myodocopa most of the adults do not qualify as meiofauna, only members of the Cladocopida have many species small enough to qualify as meiofauna.

The two orders of the Podocopa comprise about 13,000 species, of which the Platycopida have only one living family with about 70 species. They are characterized by an elongated rounded shell with a slight concave ventral margin. The shell is smooth or slightly ornamented. Parallel to the margin there is a groove which ensures a tight seal during closure. The muscle scars form a rosette of several imprints.

All other ostracod species belong to the order of Podocopida. Members of the Superfamily Darwinuloidea are characterized by a cuneiform carapace, pointed in front and showing also a rosette-like muscle-scar field.

The shell of the Bairdioidea is mostly rhombic. The second maxilla is a walking leg and the exopod of the second antenna is a scale with three bristles.

The exopod of the second antenna in the Cytheroidea is normally a spinning bristle. The second maxilla is a walking leg, so are also the one or two pairs of the thoracic limbs. The furcae are reduced to little bristles.

In the Cypridoidea a pair of furcae is normally present. The second maxilla is either a prehensile joint or a lamelliform joint. The second thoracic limb is an inward turned limb with no walking function. In males a huge spined ductus ejaculatorius is present.

Habitats and Ecological Notes

As already mentioned ostracods are found in nearly all aquatic environments. Most groups are marine, however. The Myodocopa are often suspended among the plankton and hence collected as plankton. The benthic community typically has many representatives of the Podocopa. These animals crawl on the grains of sediment and on the algae. They adhere often to their substrate (rocks or plants) when taken out of the water. Also the mesopsammal forms can glue themselves to the sand grains. Several forms swim just above the bottom. Especially in fresh water this is widespread. Puddles and temporary pools often contain a rapidly growing population of parthenogenetic Cypridoidea. Their eggs are often very sturdy and resistant to desiccation, freezing, and even the passage through the gut of fishes and birds. They can easily be recovered alive from a sample of dried mud, even after some years. But this is true only for some freshwater ostracods.

Some ostracods occur year-round, others may have one or two population peaks each year, either in winter or summer. They endure the unfavorable season either in a special larval stage or as eggs or, sometimes, as adults. They live from 14 days up to several years.

Methods of Collection and Extraction

Collecting of Ostracoda is mainly done in a conventional way. Benthic taxa are collected best by using a plankton net with a rugged rim so as to withstand the abrasion caused by dragging it over the sediment. A meiobenthic dredge, equipped with a plankton net and designed to sample only the uppermost few centimeters of the sediment, is also effective. The net should have a mesh size of about 150-200 μm. Ostracods are most often found in the upper 2 cm of sediment, but swimming forms are also taken by this method. Any method which maximizes the collection of the uppermost sediment layers is likely to produce ample numbers of ostracods for study.

The sediment should be transferred into a bucket and stirred briskly in a rotating manner to suspend the sediment and animals it contains. Simple slow decantation of this suspension through a 42 or 63 μm mesh net will concentrate the ostracods.

Coarser sediment, including pebbles or small rocks, or algae should be transferred to a bucket of water to which a few drops of full-strength formalin have been added. The formalin will cause the ostracods to close their shells and remain unattached to any substrate. The pebbles or small rocks, or algae can be removed and discarded to facilitate subsequent decantation in the manner described above. The remaining sediment should have additional water added to it if animals are to be studied alive.

Collecting intertidal mesopsammal ostracods can be either by digging the sand down to the water level and passing this water through a fine-mesh net (42-63 μm), or by placing the sediment in a bucket, adding seawater, stirring, and decanting the seawater with its suspended ostracods through a fine-mesh net. Sampling in springs and wells is usually by some sort of plankton net device.

In the deep-sea either box corer (if quantitative data is desirable) or meiobenthic dredge (if abundance of specimens is more desirable) is a suitable collecting device. In the lab the ostracods are picked one by one under a stereo-microscope and either stored in vials with alcohol or very low concentrations of formalin (0.5% buffered). When storing in alcohol care has to be taken: an intermediate change in freshwater is necessary for marine or brackish water animals, because alcohol gives a precipitate with salt water.

Preparation of Specimens for Taxonomic Study

In some instances, one needs only the calcified shell of an ostracod to identify it to a given level. In this instance, simple air drying of the material is sufficient; exceptions occur in the case of the Myodocopida.

For more general preservation of ostracods, especially if the appendages must be studied, 10% buffered formalin should be used. Buffering with borax (pH 8.0) is satisfactory. One can use as little as 0.5% formalin in seawater to maintain the integrity of the calcification of the shell.

In freshwater fixation in 70% ethanol is best. If special studies such as those using electron microscopy are desirable, one should consult special literature on this subject.

For dissection of ostracods we use steel-microneedles ("Minutienstifte"), which are used for pinning minute insects in collection boxes. These needles are secured in a wax-filled glass tube, so as to allow for a quick change of needles by warming the tip of the tube. For very critical work we use tungsten wire chemically tapered by dipping it into molten $NaNO_2$ and slowly withdrawing it. This is repeated several times to avoid overheating of the wire.

The shell of the ostracod is separated from the body by running the needle along the inner side of the carapace, thus tearing off the closing muscle from the shell. Pinching the body with the second needle to the glass slide, the other shell is torn away in the same way. After this the calcified shells are kept dry in a covered slide. Care has to be taken against electrostatic charge of the slide material, which causes the valves to stick to the coverglass. Special plastic slides or paper-board slides can be used for the shells.

The body is transferred into a drop of polyvinyl-lactophenol (PVL) stained with a small amount of Orange-G. Separation of the appendages is then achieved in two steps. First the body is torn into a front and back part at the border of the head and thorax. Care has to be taken to see where the second maxilla is located during this tearing process. Next, the parts can be turned dorsally and one appendage after the other can be separated. They are arranged in the PVL drop and covered with a coverglass. After curing them for two or more hours at 60° C they can be stored for several years in the dark.

Type Collections

The largest collections of Ostracoda are found at the National Museum of Natural History, Smithsonian Institution, Washington, D.C., USA; the British Museum (Natural History), London, UK; the Zoologisches Institut und Zoologisches Museum Hamburg, Hamburg, FRG; the Geological Institute, University of Tokyo, Tokyo, Japan; and the Institute of Marine Biology, Vladivostok, USSR.

References

Bonaduce, G., G. Ciampo, and M. Masoli
1975. Distribution of Ostracoda in the Adriatic Sea.

Publicacione Stazione Zoologica Napoli, 40, Supplement: 1-304.

de Deckker, F., and P.J. Jones
1978. Check List of Ostracoda Recorded from Australia and Papua New Guinea 1845-1973. *Australian Government Publishing Service Report,* 195:1-184.

Gottwald, J.
1983. Interstitielle Fauna von Galápagos XXX. Podocopida I (Ostracoda). *Mikrofauna des Meeresbodens,* 90:1-187.

Hanai, T.
1982. Studies on Japanese Ostracoda. *University Museum of Tokyo Bulletin,* 20:1-172.

Hartmann, G.
1964a. Zur Kenntnis der Ostracoden des Roten Meeres. *Kieler Meeresforschungen,* 20:35-127.
1964b. Asiatische Ostracoden. *Internationale Revue der gesamten Hydrobiologie und Hydrographie,* Systematik Beiheft, 3:1-154.
1966-1975. Ostracoda. In Bronn's *Klassen und Ordnungen des Tierreichs,* V. Band, 1. Abteilung, 2. Buch, IV. Teil, 1-4. Lieferung: 1-786. Leipzig: Akademische Verlagsgesellschaft Geest and Portig, K.-G. (1966, 67, 68) and Jena: VEB Gustav Fischer Verlag (1975).
1974. Zur Kenntnis des Eulitorals der afrikanischen Westküste zwischen Angola und Kap der guten Hoffnung und der afrikanischen Ostküste von Südafrika und Mocambique unter besonderer Berücksichtigung der Polychaeten und Ostracoden. Teil III. Die Ostracoden des Untersuchungsgebietes. *Mitteilungen aus dem Hamburgischen Zoologischen Museum und Institut,* Supplement 69:229-520.
1979. Die Ostracoden der Ordnung Podocopia G.W. Müller, 1894 der warm-temperierten (antiborealen) West- und Südwestküste Australiens (zwischen Perth im Norden und Eucla im Süden). *Mitteilungen aus dem Hamburgischen Zoologischen Museum und Institut,* 76:219-301.
1980. Die Ostracoden der Ordnung Podocopida G.W. Müller 1894 der warmtemperierten und subtropischen Küstenabschnitte der Süd- und Südostküste Australiens (zwischen Ceduna im Westen und Lakes Entrance im Osten). *Mitteilungen aus dem Hamburgischen Zoologischen Museum und Institut,* 77:111-204.
1981. Die Ostracoden der Ordnung Podocopia G.W. Müller, 1894 der subtropisch-tropischen Ostküste Australiens (zwischen Eden im Süden und Heron-Island im Norden). *Mitteilungen aus dem Hamburgischen Zoologischen Museum und Institut,* 78:97-149.
1982. Beitrag zur Ostracodenfauna Neuseelands (mit einem Nachtrag zur Ostracodenfauna der Westküste Australiens). *Mitteilungen aus dem Hamburgischen Zoologischen Museum und Institut,* 79:119-150.
1984. Zur Kenntnis der Ostracoden der polynesischen Inseln Huahiné (Gesellschaftsinseln) und Rangiroa (Tuamotu-Inseln). *Mitteilungen aus dem Hamburgischen Zoologischen Museum und Institut,* 81:117-169.
1985. Ostracoden aus der Tiefsee des Indischen Ozeans und der Iberischen See sowie von ostatlantischen sublitoralen Plateaus und Kuppen. *Senckenbergiana Maritima,* 17(1/3):89-146.

Hartman, G., and H.S. Puri
1974. Summary of Neontological and Paleontological Classification of Ostracoda. *Mitteilungen des Hamburgischen Zoologischen Museums und Institutes,* 70:7-73.

Hartmann-Schröder, G., and G. Hartmann
1978. Zur Kenntnis des Eulitorals der australischen Küsten unter besonderer Berücksichtigung der Polychaeten und Ostracoden. *Mitteilungen aus dem Hamburgischen Zoologischen Museum und Institut,* 75:63-219.

Hazel, J.E.
1967. Classification and Distribution of the Recent Hemicytheridae and Trachyleberididae (Ostracoda) off Northeastern North America. *Geological Survey Professional Paper,* 546.

Hornibrook, N. De B.
1952. Tertiary and Recent Marine Ostracoda of New Zealand. Their Origin, Affinities and Distribution. *Paleontology Bulletin Wellington,* 18:1-82.

Keij, A.J.
1964. Neogene to Recent Species of Cytherelloidea from Northwestern Borneo. *Micropaleontology,* 10, No. 4.

Klie, W.
1938. Ostracoda, Muschelkrebse. Pages 1-230 in F. Dahl, editor, *Die Tierwelt Deutschlands und der angrenzenden Meeresteile.* Volume 34.

Kornicker, L.
1975. Antarctic Ostracoda (Myodocopina), Parts 1 and 2. *Smithsonian Contributions to Zoology,* 163:1-720.

Kornicker, L.S., and M.V. Angel
1975. Morphology and Ontogeny of *Bathyconchoecia septemspinosa* Angel, 1970 (Ostracoda: Halocypridae). *Smithsonian Contributions to Zoology,* 195:1-21.

Kornicker, L.S., and I.G. Sohn
1976. Phylogeny, Ontogeny and Morphology of Living and Fossil Thaumatocypridavea (Myodocopa: Ostracoda). *Smithsonian Contributions to Zoology,* 219:1-124.

Kornicker, L.S., and F.P.C.M. Van Morkhoven
1976. *Metapolycope,* A New Genus of Bathyal Ostracoda from the Atlantic (Suborder Cladocopina). *Smithsonian Contributions to Zoology,* 225:1-29.

Liebau, A.
1975. Comment on Suprageneric Taxa of the Trachyleberididae s.l. (Ostracoda, Cytheracea). *Neues Jahrbuch für Minerologie, Geologie und Paläontology Abhandlungen,* 148:353-379.

Löffler, H.
1961. Zur Systematik und Ökologie den Chilenischen Süsswasser-entomostraken. *Beiträge zur Neotropischen Fauna,* 2:143-222.
1963. Zur Ostracoden- und Copepodenfauna Ekuadors. *Archiv für Hydrobiologie,* 59:196-243.

Maddocks, R.F.
1969a. Revision of Recent Bairdiidae (Ostracoda). *United States National Museum Bulletin,* 295:1-126.
1969b. Recent Ostracodes of the Family Pontocyprididae Chiefly from the Indian Ocean. *Smithsonian Contributions to Zoology,* 7:1-56.

Moore, R.C., and C.W. Pitrat, editors
1961. *Treatise on Invertebrate Paleontology, Part Q, Arthropoda 3, Crustacea, Ostracoda.* Lawrence: University of Kansas Press.

Morkhoven, F.P.C.M. Van
1962. *Post Paleozoic Ostracoda. Their Morphology, Taxonomy and Economic Use.* 1:1-204 (1962) and 2:1-478 (1963). New York: Elsevier Publication Company.

Müller, G.W.
1894. Die Ostracoden des Golfes von Neapel und der angrenzenden Meeresabschnitte. *Fauna und Flora des Golfes von Neapel,* 21(1-8):1-404.

Neale, J.W., editor
1969. *The Taxonomy, Morphology and Ecology of Recent Ostracoda.* Edinburg: Oliver and Boyd.

Okubo, I.
1980. Taxonomic Studies on Recent Marine Podocopid Ostracoda from the Inland Sea of Seto. *Publication Seto Marine Biology Laboratory,* 30(5/6):389-443.

Pokorny, V.
1958. *Grundzüge der zoologischen Mikropaläontologie, Ostracoda-Muschelkrebse.* 2(15):66-323. Berlin: VEB Deutscher Verlag der Wissenschaften.

Poulsen, E.
1962. Ostracoda-Myodocopa. Part I: Cypridiniformes-Cypridinidae. *DANA-Report,* 57:1-414.
1965. Ostracoda-Myodocopa, Part II: Cypridiniformes-Rudidermatidae, Sarsiellidae and Asteropidae. *DANA-Report,* 65:1-483.

Sandberg, P.A.
1964. The Ostracod Genus Cyprideis in the Americas. *Stockholm Contribution to Geology,* 12:1-178.

Sars, G.O.
1928. An Account of the Crustacea of Norway. *Ostracoda,* 9:1-277.

Skogsberg, T.
1920. Studies on Marine Ostracods, Part I (Cypridinids, Halocyprids and Polycopids). *Zoologica Bidrag, Supplement,* 1:1-784.

Sylvester-Bradley, P.C., and D.J. Siveter, editors
1973. *A Stereo-Atlas of Ostracode Shells.* Leicester: Department of Geology, University of Leicester.

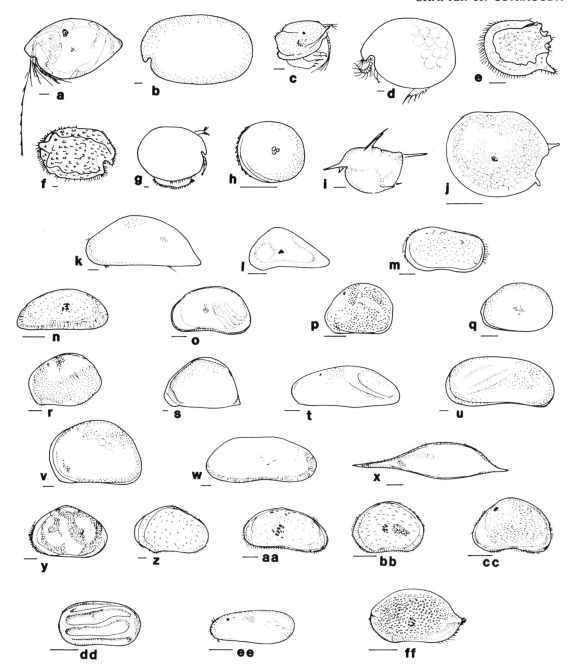

Figure 35.1--Ostracoda: **a**, *Cypridina sinosa* (after Poulsen, 1962); **b**, *Diasterope schmitti* (after Kornicker, 1975); **c**, *Rutiderma rostrata* (after Poulsen, 1965); **d**, *Philomedes globosus* (after Sars, 1928); **e**, *Eusarsiella cornuta* (after Poulsen, 1965); **f**, *Asteropteron skogsbergi* (after Poulsen, 1965); **g**, *Cycloleberis galatheae* (after Poulsen, 1965); **h**, *Polycopsis serrata* (after G.W. Müller, 1894); **i**, *Bathyconchoecia septemspinosa* (after Kornicker and Angel, 1975); **j**, *Danielopolina orghidani* (after Kornicker and Sohn, 1976); **k**, *Macrocypris minna* (after Sars, 1928); **l**, *Propontocypris cedunaensis* (after Hartmann, 1980); **m**, *Ilyocypris gibba* (after Sars, 1928); **n**, *Phlyctenophora* aff. *zealandica* (from Van Morkhoven, 1963); **o**, *Candona candida* (after Hiller, 1977); **p**, *Cypria ophthalmica* (after Sars, 1928); **q**, *Cyclocypris globosa* (after Sars, 1928); **r**, *Notodromas monacha* (from Klie, 1938); **s**, *Clamydotheca rudolphi* (after Triebel, 1939); **t**, *Dolerocypris fasciata* (after Sars, 1928); **u**, *Herpetocypris reptans* (after Sars, 1928); **v**, *Cyprois marginata* (after Sars, 1928); **w**, *Stenocypris major* (after Triebel, 1953); **x**, *Strandesia bicornuta* (after Hartman, 1964); **y**, *Heterocypris salina* (from Elie, 1938); **z**, *Cypris maculosa* (after Hartmann, 1964); **aa**, *Eucypris afghanistanensis* (after Hartmann, 1964); **bb**, *Cypretta foveata* (after Hartmann, 1964); **cc**, *Cypridopsis newtoni* (from Klie, 1938); **dd**, *Cytherelloidea damericacensis* (from Van Morkhoven, 1963); **ee**, *Darwinula stevensoni* (after Sars, 1928); **ff**, *Bairdioppilata* sp. (after Maddocks, 1975); (Scale = 200 μm.)

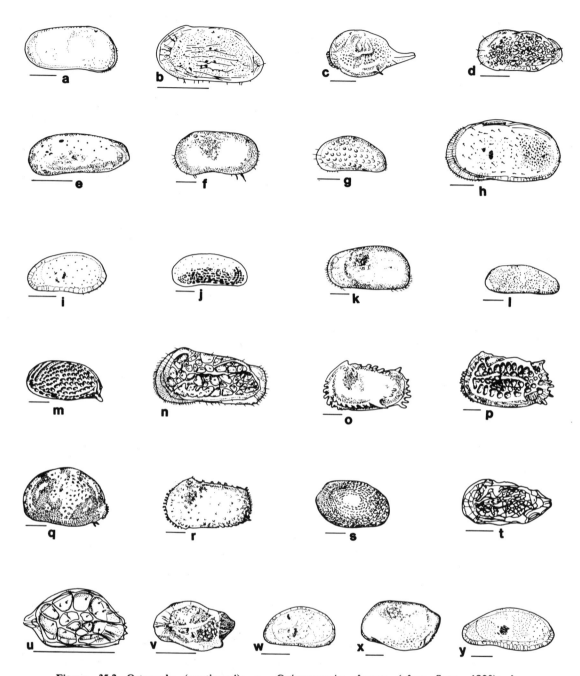

Figure 35.2.--Ostracoda (continued): a, *Cytheromorpha fuscata* (after Sars, 1928); b, *Microcytherura nigrescens* (after G.W. Müller, 1894); c, *Paijenborchella cymbula* (after Ruggieri, 1950); d, *Perissocytheridea meyerabichi* (after Hartmann, 1953); e, *Leptocythere levis* (after G.W. Müller, 1894); f, *Limnocythere sanctipatricii* (after Sars, 1928); g, *Eucythere argus* (after Sars, 1928); h, *Cyprideis lengae* (after Hartmann, 1961); i, *Paracytheroma sudaustralis* (after Hartmann, 1980); j, *Cushmanidea seminuda* (after Cushman, 1906), k, *Krithe bartonensis* (after Sars, 1928); l, *Neocytherideis senescens* (after Ruggieri, 1952); m, *Leguminocythereis oertlii* (after Keij, 1958); n, *Patagonacythere tricostata* (after Hartmann, 1962); o, *Pterygocythereis jonesi* (after Sars, 1928); p, *Bradleya? dyction* (after Hornibrook, 1952); q, *Aurila convexa* (after G.W. Müller, 1894); r, *Echinocythereis dunelmensis* (after Sars, 1928); s, *Loxoconcha propunctata* (after Hornibrook, 1952); t, *Paracytheridea luandensis* (after Hartmann, 1974); u, *Hemicytherura videns* (after Hartmann, 1964); v, *Cytheropteron abyssicolum* (after G.W. Müller, 1894); w, *Xestoleberis baja* (after Hartmann, 1974); x, *Bythocythere turgida* (after Sars, 1928); y, *Sclerochilus semivitrens* (after Schornikov, 1981). (Scale = 200 μm.)

36. Mystacocarida

Robert R. Hessler

The Mystacocarida is a distinct taxon of the Class Crustacea, co-equal to the Copepoda, Cirripedia, Branchiura, Tantulocarida, and Ostracoda. Most recent classifications unite these taxa (with the possible exception of the Ostracoda) under the Subclass Maxillopoda, which would be equal to the Malacostraca, Branchiopoda, and so on. Many aspects of mystacocarid external morphology are probably a result of paedomorphosis, as is the case for the Maxillopoda in general.

As is so common among interstitial meiofauna, mystacocarids are markedly elongate. The body is divided into a cephalon with the usual five pairs of appendages, and a trunk with appendages anteriorly and on the telson. A unique feature is the presence of paired lateral or dorsolateral, toothed furrows at the posterior end of the cephalon and on each trunk segment except the telson. Glands are associated with the furrows, whose function is not known.

The cephalon is constricted behind the attachment of the first antenna. There are no eyes. An unusually long ventral labrum covers the mouth and extends posteriorly well beyond it. The first antenna is an unbranched, elongate, sensory structure that stretches anteriorly. The second antenna is biramous, with a flagellar exopod and four-segment endopod. The mandible is very similar to the second antenna, but has a gnathic process on the coxa. Both second antenna and mandible extend laterally. The ventrolaterally-directed first and second maxillae are unbranched. They bear strong enditic setae on all podomeres.

There are 11 trunk segments, including the telson. The first is slightly reduced and bears a maxilliped with strong enditic setae, but which may or may not have an endopod and exopod. Segments 2-5 have paired, unsegmented, ventrolaterally-directed limbs that are little larger than buds. The limb of segment 4 bears the gonopores. The last of these bud-like limbs is armed with coupling hooks in the male. Segments 6-10 lack limbs, but the telson bears a pair of large, claw-like caudal rami that are covered with tiny combs.

The only other meiofaunal taxon with which mystacocarids could be confused is the Harpacticoida. The long first antenna, the large, similar, laterally-directed second antenna and mandible, the wide spacing of the head appendages, and the minute size of the anterior trunk appendages on the mystacocarids allow reliable field discrimination. Also, females do not carry an egg sac. When not surrounded with sediment, mystacocarids are not capable of effective locomotion; this is also a good field character if living animals are being observed.

Sexes are separate. Females produce only a few large eggs. We know from the absence of brooding individuals that the eggs are released into the environment, but a free egg has never been identified.

Development is direct. The first free-living instar is a metanauplius (Figure 36.1b) which, in addition to the nauplial appendages, carries a rudimentary first maxilla; this larva looks much like an adult except for the abbreviated number of limbs and trunk segments. It is less than half an adult's length. The 2 species that have been studied in detail both have 9 instars: 7 larval, 1 juvenile, and 2 adult. Development is gradual, except for the sudden loss of the naupliar process (a feeding endite arising from the coxa of the second antenna) after the sixth instar.

All stages live interstitially as does the adult. There is no evidence for a planktonic dispersal stage.

Classification

The Mystacocarida contains only 12 species divided among 2 genera: *Derocheilocaris* (Figure 36.1a) and *Ctenocheilocaris*. This group is extremely conservative. Differences between species, and even between genera, are very small; for the most part, even setal counts are constant between species. In short, superficially, all mystacocarids look alike.

Those structures which discriminate the species include: (1) antennular portion of cephalon: size and shape of processes and notches; presence of tubercles and spines; (2) labrum: shape of distal end; (3) shape of lateral tooth furrows; distinctness of dentition; (4) first antenna: relative size; setation of basal podomere; (5) first maxilla: number of setae on protopodal endites and distal podomere; (6) second maxilla: number of setae on distal podomere; (7)

maxilliped: presence or absence of endopod and exopod (this feature is of generic value); number of setae on endopod, when present; (8) trunk limbs 3-5: setation; (9) combs on ventral surface of telson: shape and dentition; (10) supra-anal process: spine- or tuft-like apex; number and size of secondary spines; size of terminal seta; (11) caudal ramus: size, number and position of setae and spines; (12) body size (0.3-1.0 mm) long.

Habitats and Ecological Notes

Mystacocarids are only known to live in the interstitial spaces of sandy substrates. They are most frequently encountered in intertidal sandy beaches. Here, they display unexplained patchy distributions. They have also been collected subtidally (at depths of over 20 m), but nothing is known about the details of subtidal existence. Interestingly, the same species may occur both intertidally and subtidally.

These crustaceans are found most frequently in beaches of clean and well-sorted medium-sized sands, but they are also known from coarse and fine sands. Usually the sand is silica, but calcium carbonate localities are also known. Mystacocarids occupy both tidal beaches, such as on the Atlantic coast, or virtually atidal beaches, as in the Mediterranean. In tidal beaches, they are most likely to occur near the high-tide level, but have been collected down to low-tide level as well. The animals may migrate vertically as tidal level fluctuates, to avoid unsaturated pore space. Because of this, individuals are not likely to be at the exposed sand surface at low tide. In atidal beaches, peak mystacocarid abundance may occur as much as 20 m away from the shore; ground-water salinity and temperature are important in determining the landward limit of their distribution.

Locomotion requires that the animal be able to contact solid surfaces dorsally as well as ventrally; only in this way can its limbs gain sufficient purchase. The second antenna and mandible are the principle locomotory organs, but the maxillae and trunk limbs participate under some circumstances. The long, flexible trunk, coupled with the claw-like caudal rami implement the ability to turn around in tight spaces.

These animals feed on small particles which are scraped toward the mouth by the many medially directed setae on the maxillae and maxilliped. The first antenna is a sensory structure. Combs on the caudal rami are used for grooming.

Mystacocarids have been sought in many places, but their known global distribution is still restricted. They have been found along the Atlantic and Gulf of Mexico coasts of the United States, the Atlantic coast of southern Europe, the Mediterranean Sea, the Atlantic coast of Africa continuing slightly into the Indian Ocean, and the Atlantic and Pacific coasts of middle South America. The South American localities are all of the genus *Ctenocheilocaris*; all the other localities are occupied by *Derocheilocaris*. It is likely that other localities will be found. Two localities are known where two species of *Derocheilocaris* co-exist.

Methods of Collection and Extraction

Any method of collecting sandy sediments, intertidal or subtidal, will produce mystacocarids if they are present. Because mystacocarids do not adhere to sand grains, simple elutriation techniques are adequate for quantitative sampling. First-stage larvae will be retained on a 63 μm mesh sieve.

Preparation of Specimens for Taxonomic Study

No special fixation technique is required; however, if body length is to be measured, lactic acid is the recommended killing agent because all individuals die in the same, relaxed posture. Fine cuticular details are rendered visible with a lactic acid solution of methyl blue, with subsequent inspection in clear lactic acid. If animals are to be kept for more than a few days, they should be transferred to ethanol.

References

Hall, J.R.
1972. Aspects of the Biology of *Derocheilocaris typica* (Crustacea: Mystacocarida). II. Distribution. *Marine Biology*, 12:42-52.

Hessler, R.R.
1972. New Species of Mystacocarida from Africa. *Crustaceana*, 22:259-273.

Hessler, R.R., and H.L. Sanders
1966. *Derocheilocaris typicus* Pennak & Zinn (Mystacocarida) Revisited. *Crustaceana*, 11:141-155.

Jansson, B.-O.
1966. On the Ecology of *Derocheilocaris remanei* Delamare and Chappuis (Crustacea, Mystacocarida). *Vie et Mileu*, 17(A):143-186.

Lombardi, J., and E.E. Ruppert
1982. Functional Morphology of Locomotion in *Derocheilocaris typica* (Crustacea, Mystacocarida). *Zoomorphology*, 100:1-10.

McLachlan, A.
1977. The Larval Development and Population Dynamics of *Derocheilocaris algoensis* (Crustacea, Mystacocarida). *Zoologica Africana*, 12:1-14.

Renaud-Mornant, J.
1978-
1979. Mystacocarides du Bresil. Description de deux espèces nouvelles du genre *Ctenocheilocaris* Renaud-Mornant, 1976 (Crustacea). *Vie et Milieu*, 28-29(AB):393-408.

Renaud-Mornant, J., and C. Delamare Deboutteville
1976. L'originalité de la sous-classe de Mystacocarides

(Crustacea) et le probleme de leur repartition. *Annales de Speleologie,* 31:75-83.

Zinn, D.S., B.W. Found, and M.G. Kraus
1982. A Bibliography of the Mystacocarida. *Crustaceana,* 43:270-274.

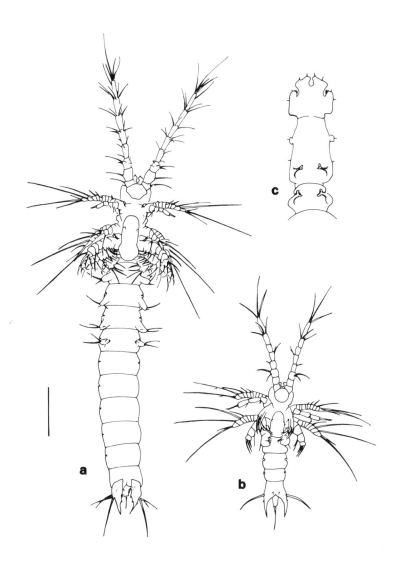

Figure 36.1.--Mystacocarida: a, *Derocheilocaris typica* Pennak and Zinn, 1943, adult male, ventral view; b, first instar (metanauplius), ventral view; c, cephalon, and first trunk segment, adult male, dorsal view. (Scale = 100 μm.)

37. Copepoda

J.B.J. Wells

In terms of meiobenthos abundance the Copepoda are second only to the Nematoda in sediments and probably exceed them in the phytal. In terms of biomass they are often the most important taxon. There is considerable species diversity and copepods inhabit all available benthic habitats in the sea, freshwater, and inland saline waters. They are common in damp terrestrial habitats, e.g., among mosses and bromeliads, particularly in warm climatic regions.

The vast majority of meiobenthic copepod species belong to the Order Harpacticoida. Among the Order Cyclopoida, species of the Family Cyclopidae are commonly taken in freshwater sediment and phytal samples (though it may be doubted that all of these are truly benthic in habit) and the genus *Halicyclops* often is abundant in estuarine sediments. Species of the Family Cyclopinidae occur regularly, albeit usually in very small numbers, in sandy marine substrata. Occasional specimens of "semi-parasitic" cyclopoid species, whose hosts are benthic macrofauna, occur in marine sediment and phytal samples. A few species of the mainly planktonic Order Calanoida (in the families Pseudocyclopidae, Ridgewayiidae, and Epacteriscidae) are epibenthic in habit.

This chapter concentrates on the Harpacticoida, though the techniques are equally applicable to all copepod taxa. Karl Lang's *Monographie der Harpacticiden* (Lang, 1948) remains the standard monograph. It is indispensable but as it contains less than half of the fauna now known it must be supplemented by Bodin's (1979) catalogue of new species and Wells' (1976-1985) identification keys.

The copepod body consists of three tagmata: the cephalosome, the metasome, and the urosome. The cephalosome consists of the head and first thoracic segment. The tergites and pleurites of these segments are fused together to give a continuous head shield. The cephalosome bears the head appendages - antennule (=1st antenna), anntenna (=2nd antenna), mandible, maxillule (=1st maxilla), maxilla (=2nd maxilla) - and the maxilliped of the first thoracic segment. In most harpacticoids the second thoracic segment has become fused with the cephalosome to form a cephalothorax (cf. Figure 37.2a-c with Figure 37.2d,e).

The metasome primitively consists of the thorax, except for the first segment, and bears the first five pairs of "swimming legs," or pereiopods, the P.1 to P.5. The cephalosome and metasome together are termed the prosome.

The urosome is usually demarcated from the metasome by an obvious movable articulation, at a point which primitively is the narrowest part of the body. These features are not obvious in those interstitial harpacticoids that have evolved the truly cylindrical or vermiform body shape (e.g., Figure 37.3a-g). Primitively it consists of the abdomen only and terminates in the paired caudal rami. In harpacticoids the urosome also contains the last thoracic segment. The first abdominal segment contains the genital apertures and may bear a sixth pair of pereiopods (P.6), which are always reduced in form and often represented by setae only.

Harpacticoids may be distinguished from cyclopoids and calanoids by their particular combination of three characteristics: (1) the prosome-urosome articulation is between the fifth and sixth postcephalosome segments in harpacticoids and cyclopoids, but between segments six and seven in calanoids; (2) the antennule is generally longest in calanoids (usually >22 segments), shortest in harpacticoids (4-10 segments) and intermediate in cyclopoids (usually 10-22 segments); and (3) the antenna is biramous in calanoids but uniramous (i.e., lacking an exopod) in cyclopoids. It is usually biramous in harpacticoids, but the exopod may be greatly reduced (e.g., *Heterolaophonte*, some Cletodidae, most Darcythompsoniidae), or even absent (e.g., Metidae, Ancorabolidae).

Additionally, the maxilla and maxilliped are often of much simpler construction in harpacticoids, while the P.1-P.4 are often reduced in setation and show much greater variability in design than in calanoids or cyclopoids.

The P.5 differs markedly from the P.1-P.4. Usually it is fused and flattened and in the female serves to hold the extruded ovisac against the urosome; occasionally the pair of P.5 serve as a true brood chamber (Figure 37.2k-m).

Figures 37.1, 37.2 and 37.3 illustrate the diversity

of body form. In harpacticoids body form reflects ecology more than phylogeny, with the result that similar forms have evolved independently many times and primitive families do not necessarily have the primitive body shape.

According to Noodt (1971) the primitive body form is semi-cyclopoid. This is represented in several families and in all habitats – epibenthic, phytal, and interstitial (Figure 37.1a-e). From this form have arisen, repeatedly and independently, a number of common modifications.

1. A trend to dorsoventral compression is common in the phytal habitat (Figure 37.1f-j).

2. Another development is for the prosome to deepen in the vertical plane, leading to a pyriform shape or to extreme lateral compression (Figure 37.1k-o).

3. The commonest trend is to increasing linearity, with the body shape seen in *Canuella* (Figure 37.2a-c) and *Harpacticus* (Figure 37.2d,e) being a generalized one appearing in many families and being equally common in the epibenthic and phytal habitats. The extremes of this latter trend are either an elongate cylindrical form (e.g., Figure 37.3b) or a short, stout, somewhat dorsally compressed body (e.g., Figure 37.2j). The former is typical of, but not exclusive to interstitial species, while the latter occurs in all habitats, including the interstitial (Figure 37.2f-j, Figure 37.3a-g).

4. Some bizarre forms have arisen through the production of spinous excrescences. Most of these species are mud dwellers, often from deep water (Figure 37.3h-k).

5. In a few genera the enlargement in the female of the pair of P.5 to form a brood pouch makes them distinctive (Figure 37.2k-m).

The life cycle (Figure 37.4) typically includes six nauplius larval stages and always includes five sub-adult copepodid stages during which there is a progressive addition of prosome and urosome segments and development of their appendages. The sexes can be distinguished from the fourth copepodid stage. Little is known of the ecology of the nauplii for most benthic species. It had been assumed that all life history stages were spent in or on the substratum, but it is now certain that this is not always the case, although it is probable that the copepodids of most species are holobenthic.

All species are sexually dimorphic but the only universal morphological difference is the structure of the first two abdominal segments (cf. Figure 37.2b,c). In the female these are fused into a genital somite and the P.6 is reduced to one or two setae flanking the single, median genital aperture. The demarcation between the two segments often is marked by a suture line dorsally and/or laterally, but never ventrally. In the male the two segments remain separate and the P.6 is far more elaborate than the female. Females are usually larger than males. Other almost universal sexual differences occur in the male antennule, which usually is prehensile to a greater or lesser degree, and P.5 which usually is smaller and less elaborate than the female. Sexual dimorphism may also be apparent in P.1-P.4 but there is great variability in the form of such differences between, and even within, families. The most consistent of such male characters is in the endopod of P.2 or P.3, which may aid in spermatophore transfer during copulation. In a few cases (mainly in Family Diosaccidae) species, or even genera, are distinguished on male characters only, females being morphologically indistinguishable.

Field sampling generally shows that females outnumber males, but there is no evidence that the genetic sex ratio departs from 1:1. This imbalance, which can be constant or temporary, cyclic or acyclic, may be caused by one or more of several factors, e.g., phenotypic expression of sex, inbreeding, sex selective predation, crowding under high population densities, shorter male longevity cf. females. Males are unknown in many of the rarer species. Parthenogenesis occurs in some freshwater species but has not been proved in marine species.

Classification

At present the Order Harpacticoida contains about 3300 species distributed among 398 genera in 34 families. Most authorities acknowledge that some families are not natural units. The suprafamily classification proposed by Lang is still in use but in practical terms is not of value in the construction of keys for identification and need concern us no further here. Undoubtedly many species await discovery and it is most probable that as data accumulates the number of genera will increase considerably and taxonomic revision will erect additional families. The greater part of the freshwater and terrestrial environment and large areas of the world oceans, especially the deep sea, remain unexplored for harpacticoids.

Harpacticoids range in size from 0.2 to 2.5 mm and thus can all be classed as meiofauna. However, a few marine species are holoplanktonic or parasitic while about 700 species (including four complete families and about 30 complete genera) are confined to freshwater and terrestrial habitats and thus the marine meiobenthos contains about 2400 species in 340 genera and 26 families.

Taxonomically important characters are all external and to a large extent families and genera can be distinguished on readily observable characteristics of the body shape and of the antennule, antenna, maxilliped, P.1-P.5, and caudal rami. These structures also often serve to distinguish between species. The mouthparts (mandible, maxillule, maxilla) are more difficult to study and with their general functional similarity across the order they mainly offer differences in fine detail that are most useful at the intrageneric level. Similarly, the patterns formed by the body surface ornamentation of sensilla, spines, and setae often can be species specific. Wells (1976) and Coull (1977) give illustrated accounts of the taxonomic characters used for identification purposes.

Habitats and Ecological Notes

Meiobenthic copepods occur in most aquatic habitats. In the sea they range from supralittoral pools to the abyssal zone. In freshwater they occur wherever currents are slow enough to allow them to establish, including underground waters and within riverine sands. They are associated with water held by terrestrial vegetation. They are found in all salinity regimes from brackish water to hypersaline pools and inland seas, and in all temperatures from sub-zero polar waters to moderately hot springs. They are associated with all forms of soft substrata. In sediments they are restricted to the oxygenated regions, so that they tend to be found on or just beneath the surface of muds but extend deep within sands and gravels to the level of the permanent water table. In the sea they are associated with sessile epibenthic macrofauna and are especially abundant and diverse on macrophytes, where they form a large part of the phytal meiobenthos.

In the marine environment most harpacticoid species are present throughout the year, though their abundance often shows wide seasonal fluctuations, usually associated with peaks of reproductive activity. Several freshwater species, and one marine species, are known to encyst in winter or when temporary pools dry out.

Hicks and Coull (1983) review the ecology of marine meiobenthic harpacticoids and Noodt (1971) reviews the relationship between habitat and copepod body form.

Methods of Collection

Quantitative samples in littoral sediments should be collected by coring to the depth required. In muds or mud/sand mixtures this is generally to the depth of the anoxic layer which is seldom more than 10 cm and may be as little as 0.5 cm. Quantitative sub-littoral sampling should be made by cores taken directly by SCUBA or remotely by subsampling from a box corer. There is much evidence that the sampling action of most grabs and all gravity corers prevents their use in quantitative meiobenthos sampling, and may seriously bias qualitative, faunistic sampling.

In marine littoral clean sand beaches, the interstitial fauna may be present from the surface to the depth of the permanent water table, which at the high tide level may be 1 m or more below the surface. There is a similar distribution in riverine sands, where an interstitial fauna can be present in the water table even when the stream may no longer be visible on the surface. Sampling to these depths by coring from the surface is possible but corers of this length are cumbersome and usually have to be driven into the sand with a hammer. One alternative is to plunge a corer of about 30 cm length into the sand, excavate down to the level of the bottom of the corer and remove and repeat as often as is required. Data on vertical distribution of this deep fauna obtained with these techniques must be treated with caution as the disturbance factor they introduce may affect the natural vertical distribution of the animals, though this aspect has not been adequately tested.

Core diameter and the number of replicates must express a balance between convenience and the accuracy required for the estimate of the population variable being investigated. There is no substitute for trial sampling to determine this number, but as a rough guide to corer size, small diameters (10-20 mm) may be appropriate for muds and larger diameters (25-35 mm) may be more suitable for sands.

Quantitative sampling of the phytal fauna presents problems of a different order. While it is not difficult to remove either single plants or the flora of an entire known area and to count the extracted fauna, it is more difficult to relate this to the surface area. The methods adopted by Hicks (1977) are recommended for their simplicity and ease of use, though there are indications that they may not be universally applicable.

Methods of Extraction

Compared to some other taxa, harpacticoids are relatively easy to remove from the substratum. For live extraction of both sediment and phytal fauna, simple agitation (shaking or stirring) of the sample in water of the correct salinity will release most of the animals into the water which then is decanted through a fine mesh sieve. Most adults can be collected on a 120-150 µm sieve; most copepodids and the smaller or more vermiform species will certainly be retained by a 63 µm sieve; but the smallest nauplii may require 42 µm mesh. The animals are then hand picked from the sieve contents under a stereomicroscope. Side lighting may help to concentrate some phototactic species. Repeating this procedure three times generally will collect 90-95%

of the animals in sediments, but may be less effective with the phytal fauna. Narcotization, by shaking in isotonic $MgCl_2$ for a few minutes before sieving (Chapter 9) can increase extraction efficiency to greater than 95% for all substrata. Total extraction of live animals is difficult and involves hand picking the final 5% or so from the residue of the sample.

This "wash-and-decant" method is equally suitable for preserved sediment samples and since harpacticoids can be identified to species from dead specimens, it is often more convenient to preserve the whole sample in 4% buffered formalin immediately upon collection. One advantage here is that the addition of a little Rose Bengal stain before, or with, the dilute formalin will stain most of the animals red and make subsequent sorting under the microscope easier (Chapter 10). However, the stain necessarily obscures natural color patterns and can make it more difficult to see fine structural detail. Since both of these features can be important for species identification and for assessment of inter- and intra-population variability staining is not recommended for preliminary surveys nor until species of the community can be recognized without recourse to such detailed examination.

It is not advisable to preserve macrophyte samples in this way as it causes many phytal harpacticoids to cling more tightly to the plant and thus makes extraction of the preserved animals more difficult. It is best to wash and decant phytal material and preserve the sieve contents.

These methods result in sieve contents that contain variable quantities of sand grains, fecal pellets, detritus, plant debris and mucilage, which has to be searched for the contained animals. This final stage of extraction can introduce errors of greater magnitude than those caused by the relatively crude wash and decant extraction technique. However, such errors may still be acceptable within the framework of a properly constituted replicate sampling program. For extraction of live animals these errors probably must be tolerated but techniques are available to reduce them when sorting preserved samples. The Ludox flotation and centrifugation technique (Chapter 10), by reducing the quantity of detritus and debris present in the final aliquot, can dramatically reduce this error as well as speed up the extraction process. While it can cause damage to the animals, subjecting the sample to a short burst of ultrasound (ca. 10 seconds) can also help efficiency by breaking up aggregated material and fecal pellets.

In general the more automated extraction methods, such as Boisseau elutriation and Ludox centrifugation (Chapter 10), can be very effective on preserved samples but often are less efficient, or not usable, with live material. They may be advantageous over wash-and-decant techniques for quantitative analyses, especially in samples with a high detritus content.

Copepods should be fixed in 4-5% buffered formalin. They may also be stored in this medium but traditionally have been transferred to 70% alcohol (usually ethanol or isopropyl) for long term storage.

Methods of Preparation for Taxonomic Study

Initial identification to species almost always involves examination of P.1-P.5 structure and setation and may require observations of the head appendages and body surface spinulation patterns, particularly of the abdomen. Such details are seen most clearly in dissected specimens mounted on slides. Thankfully, in most communities once species have been identified they can be recognized subsequently on gross structures visible with a stereomicroscope.

For permanent mounts, whole animals, or dissected parts, should be mounted in a gum arabic based medium, such as Hoyer's or Reyne's, or in fluid mountants, such as glycerol or lactic acid. Except for Reyne's these mountants must be ringed. Lactic acid is a powerful clearing agent and is inadvisable for long term storage, but this problem is substantially reduced if it is mixed with glycerol (at about 1 part of lactic acid in 4 of glycerol). Polyvinyl lactophenol has been widely used, but tends to shrink with time and there is little doubt that the gum arabic mountants give the best long term results.

It is possible to stain the material (before dissection) with, for example, Rose Bengal, Lignin Pink or Chlorazol Black. Staining may give some benefit under bright field microscopy, but viewing unstained material with a combination of bright and dark field, positive phase contrast and Nomarski Interference Contrast microscopy is a much preferred way of studying all the required detail. Nomarski is particularly valuable in distinguishing surface detail.

A surprisingly large amount of structural detail is visible on temporarily mounted whole animals. Place the specimen in a generous quantity of glycerol (or a glycerol/lactic acid mixture) on a microscope slide and gently lower a cover slip on to it. With care and practice the animal can be contained in a layer of mountant thick enough for it to be rolled over without damage as the cover slip is pushed, but thin enough for it to be viewed with a high power microscope. In this way the animal may be viewed in a variety of orientations and much detail observed. Using a depression or cavity slide makes this technique easier but a standard slide gives greater control over the rolling process. Relatively high magnifications can be used with this preparation, including oil immersion. Phase contrast microscopy is possible, but good results are difficult to obtain with Nomarski.

If dissection is required I recommend the following technique, for which the rapid drying property of Reyne's mountant is ideal. Dissecting needles are made from thin tungsten wire (diameter ca. 0.2 mm) sharpened by dipping into molten sodium nitrite. It will be necessary to experiment with a variety of holders to find the correct weight and balance for you, e.g., pin vise, glass capillary tube, sticks of balsa wood.

1. Place the animal in a drop of water at the end of a clean microscope slide.

2. Place a streak of Reyne's mountant transversely across the center of the slide.

3. By cutting between the somites with a pair of needles divide the body into the cephalothorax, the individual thoracic segments, and the entire abdomen.

4. Transfer the parts as they are dissected to the streak of mountant, being careful to arrange them in the sequence they are found in the animal; it is very important to get the often very similar P.2–P.4 in the correct order. The abdomen should be placed dorsal surface uppermost so that details of the anal operculum can be seen. While this must obscure the ventral structures to some extent, in practice details of the genitalia and of ventral somite ornamentation are usually adequately visible in such a preparation.

5. The head appendages should then be separated from the cephalic shield. In large specimens it may be possible to separate all the mouthparts and arrange them in sequence. In small animals this may not be possible but it should always be possible to separate at least the antennule and antenna.

6. To prevent movement of the parts in the next stage allow the mountant to become tacky. If it becomes tacky during the dissecting and transfer stages, add a drop of water with the dissecting needle to the part of the streak being worked with.

7. Place a drop of mountant on top of the streak and gently lower the cover slip. My own preference is to use 18 mm diameter circular cover slips.

These preparations take a day or two in air to harden but may be examined immediately provided oil immersion is not used.

Several specimens may be arranged beneath a single cover slip, thus making comparisons easier, but in that case be very careful to allow adequate drying time before putting on the cover slip to prevent movement and mixing of adjacent sets of parts.

These techniques are modified from Hamond (1969) who gives a more detailed evaluation of techniques for preserving, examining and drawing copepods.

Type Collections

Extensive type collections and reference collections are located in the British Museum (Natural History), London, UK; National Museum of Natural History, Smithsonian Institution, Washington, D.C., USA; Naturhistoriska Riksmuseet, Stockholm, Sweden; Zoologisk Museum, Universitet I Oslo, Oslo, Norway.

References

Bodin, P.
1979. *Catalogue des Nouveaux Copépodes Harpacticoides Marins. (nouvelle édition).* 228 pages. Brest: Université de Bretagne Occidentale, Laboratoire d'Océanographie Biologique.

Coull, B.C.
1977. Marine Flora and Fauna of the Northeastern United States. Copepoda: Harpacticoida. Pages 1-47 in *NOAA Technical Report. NMFS Circular* 399.

Hamond, R.
1969. Methods of Studying the Copepods. *Microscopy: Journal of the Quekett Microscopical Club*, 31:137-149.

Hicks, G.R.F.
1977. Species Composition and Zoogeography of Marine Phytal Harpacticoid Copepods from Cook Strait, and Their Contribution to Total Phytal Meiofauna. *New Zealand Journal of Marine and Freshwater Research*, 11:441-469.

Hicks, G.R.F., and B.C. Coull
1983. The Ecology of Marine Meiobenthic Harpacticoid Copepods. *Oceanography and Marine Biology Annual Review*, 21:67-175.

Lang, K.
1948. *Monographie der Harpacticiden.* 2 volumes. 1682 pages. Lund: Håkan Ohlsson.

Noodt, W.
1971. Ecology of the Copepoda. Pages 97-102 in N.C. Hulings, editor, Proceedings of the First International Conference on Meiofauna. *Smithsonian Contributions to Zoology*, 76.

Wells, J.B.J.
1976. *Keys to Aid in the Identification of Marine Harpacticoid Copepods.* 215 pages. Aberdeen: University of Aberdeen, Department of Zoology. [Note: These keys are available from the author. To supplement and update them Wells has published five Amendment Bulletins that can be accessed in the following citations].

1978-1985. Keys to Aid in the Identification of Marine Harpacticoid Copepods. Amendment Bulletin Numbers 1-5. *Zoology Publications from Victoria University of Wellington*, 70:1-11 (Number 1, 1978); 73:1-8 (Number 2, 1979); 75:1-13 (Number 3, 1981); 77: 1-9 (Number 4, 1983); 80: 1-19 (Number 5, 1985). [Note: further Bulletins are expected and a revision of the keys is under consideration].

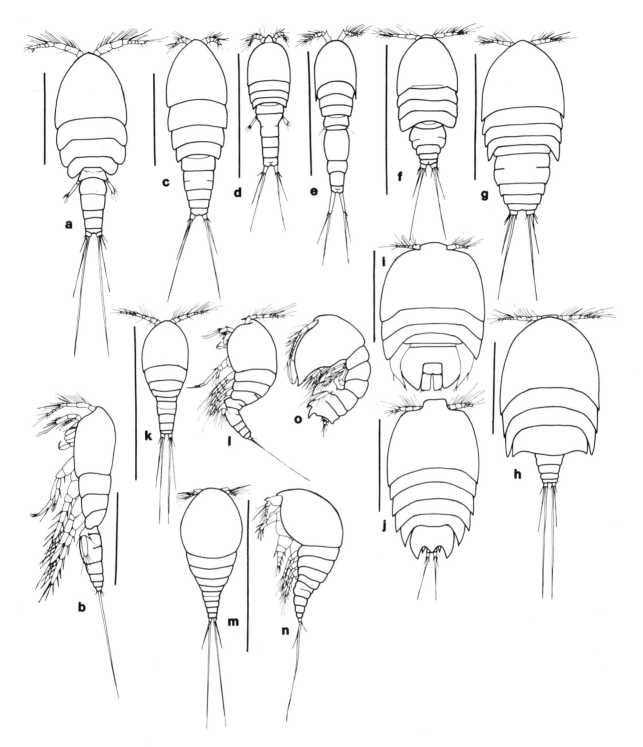

Figure 37.1.--Copepoda: a,b, *Tisbe furcata* (Tisbidae), dorsal and left lateral views; **c,** *Tachidius discipes* (Tachidiidae); **d,** *Stenhelia normani* (Diosaccidae); **e,** *Paramesochra dubia* (Paramesochridae); **f,** *Zaus spinatus* (Harpacticidae); **g,** *Amenophia peltata* (Thalestridae); **h,** *Sacodiscus littoralis* (Tisbidae); **i,** *Porcellidium viride* (Porcellidiidae); **j,** *Peltidium purpureum* (Peltidiidae); **k,l,** *Diarthrodes major* (Thalestridae), dorsal and left lateral views; **m,n,** *Metis ignea* (Metidae), dorsal and left lateral views; **o,** *Tegastes falcatus* (Tegastidae), left lateral view. (Scale = 500 μm.)

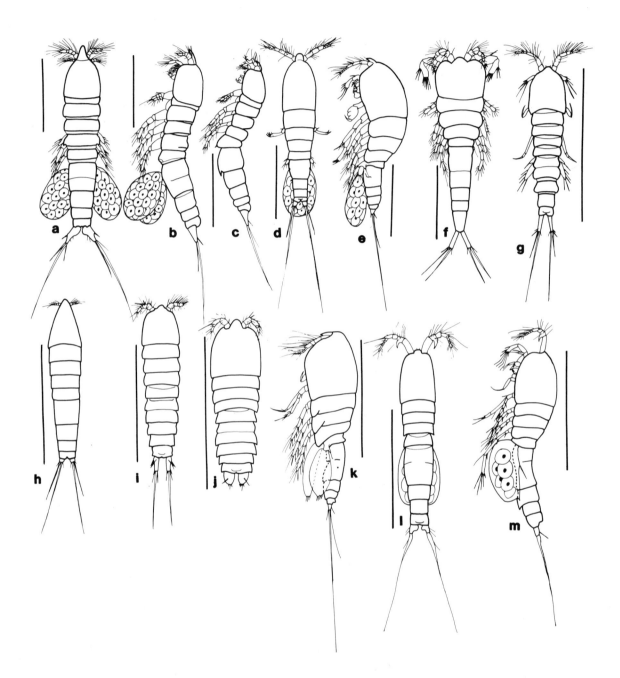

Figure 37.2.—Copepoda (continued): **a-c**, *Canuella perplexa* (Canuellidae), female dorsal and left lateral, male left lateral views; **d,e**, *Harpacticus chelifer* (Harpacticidae), dorsal and left lateral views; **f**, *Cervinia bradyi* (Cerviniidae); **g**, *Laophonte thoracica* (Laophontidae); **h**, *Ectinosoma melaniceps* (Ectinosomatidae); **i**, *Enhydrosoma buchholtzi* (Cletodidae); **j**, *Asellopsis hispida* (Laophontidae), **k**, *Phyllothalestris mysis* (Thalestridae), left lateral view; **l,m**, *Phyllopodopsyllus bradyi* (Tetragonicipitidae), dorsal and left lateral views. (Scale = 500 μm.)

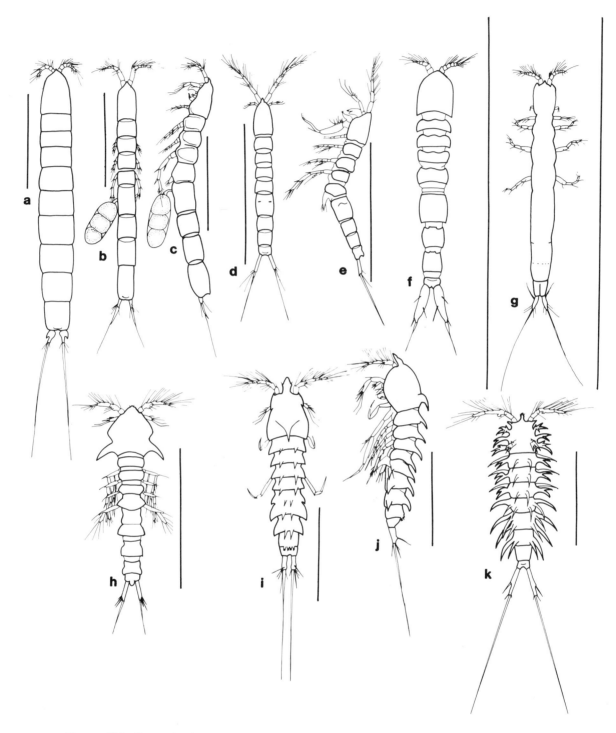

Figure 37.3.--Copepoda (continued): a, *Darcythompsonia fairliensis* (Darcythompsoniidae); b-c, *Cylindropsyllus laevis* (Cylindropsyllidae), dorsal and left lateral views; d-e, *Leptastacus macronyx* (Cylindropsyllidae), dorsal and left lateral views; f, *Scottopsyllus pararobertsoni* (Paramesochridae); g, *Apodopsyllus vermiculiformis* (Paramesochridae); h, *Laophontodes bicornis* (Ancorabolidae); i-j, *Echinolaophonte horrida* (Laophontidae), dorsal and left lateral views; k, *Ancorabolus mirabilis* (Ancorabolidae). All figures are of females in dorsal view unless otherwise indicated. Figures 37.3f,g after Karl Lang; others after G.O. Sars. (Scale = 500 µm.)

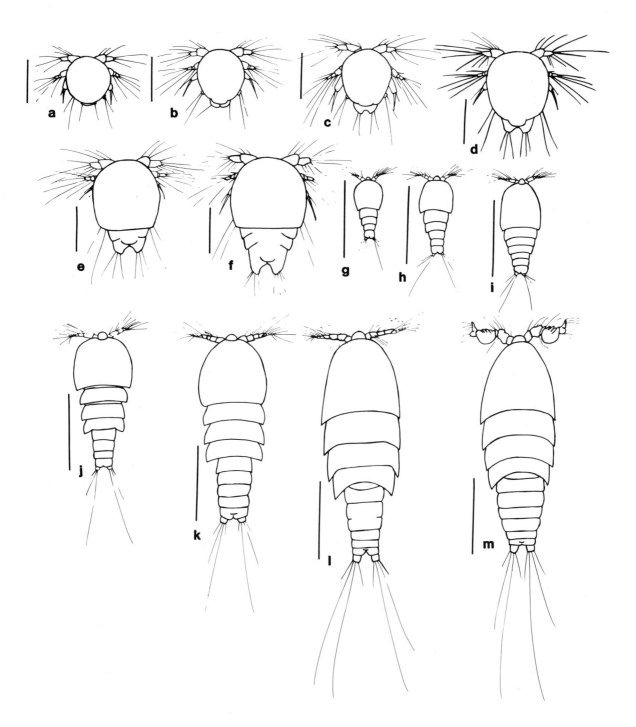

Figure 37.4.--Copepoda (continued); Semi-diagrammatic drawings of the life history stages of a typical harpacticoid (in dorsal view); **a-f**, nauplius stages I-VI (scale = 100 μm); **g-k**, copepodid stages I-V; **l,m**, adult female and male; a-k modified from various authors; l,m original. (Scale = 300 μm.)

38. Syncarida

Jürgen Sieg

Within the eumalacostracan superorder Syncarida currently there are two orders, the Anaspidacea and the Bathynellacea. From the total of about 150 known species only 16 belong to the first order while the overwhelming majority is assigned to the second order. Syncarids have been found in both surface and subterranean freshwater, but a few species of the genus *Hexabathynella* occur also in oligo- to polyhaline coastal waters.

Of the two orders, the Anaspidacea represent the most plesiomorphic taxon; much of the morphology of the Bathynellacea are simplified due to reduction of body size which adapt them to interstitial habitats.

The first thoracic segment is free (Bathynellacea) or fused with the head (Anaspidacea). Pedunculate eyes are present or absent and a carapace is lacking. The first antenna has a 3-segmented peduncle and two flagella, its basal segment may have a statocyst which does not have a statolith. The second antenna has a 2-segmented peduncle. The mandibles are asymmetrical, without a lacinia mobilis but with a row of serrate setae; the first maxilla consists of two endites, without any palp. Thoracopods I-VIII are biramous, without oostegites; the first thoracopod is very similar to the remaining ones, only in the family Stygocarididae is the first thoracopod specialized as a maxilliped, lacking epi- and exopodite. Thoracopods II-VII typically bear a bilobed epipodite; thoracopod VIII is simplified in Anaspidacea and transformed in male Bathynellacea to a copulatory organ and is variably reduced in the female. In the Anaspidacea the pleon consists of six free pleomeres and a telson, which in the Stygocarididae has the form of a truncate cone in contrast to the five pleonites and pleotelson in the Bathynellacea. Pleopods, if present are multiarticulated; the most distal pair is transformed to uropods which may form a tail fan with the telson. The Anaspidacea have the endopodites of the second pleopod transformed to a petasma.

The minute bathynellaceans lack eyes; the inner flagellum of the first antenna is reduced, and the statocyst has been lost, but there are still two of the sensory setae. Instead of the second pair of pleopods, the thoracopod VIII is transformed to a copulatory organ. Pleopods are typically absent, but occasionally there persist remnants (uniramous, 1- or 2- segmented) of the first two pairs. Finally, the uropods are styliform.

Development

Embryology and postembryonic development has recently been summarized and discussed (Schminke, 1981). It is suggested that the ancestors of the Recent Syncarida had a development similar to that of the decapodan superfamily Penaeoidea.

Within the Anaspidacea the larval stages are condensed thereby resulting in a nearly direct development. Schminke suggests (1981:606, figure 19) that an "egg nauplius," an "egg protozoea," and an "egg zoea" can be recognized. Anaspidaceans then hatch as postlarvae which molt into the adult stage.

In the Bathynellacea, the mode of development is characterized by neotony (Schminke, 1981). A "nauplioid" phase within the egg membrane still exists, but members of this order hatch as a "parazoea" (three stages) which is followed by the bathynellid phase (four or five stages). These stages correspond to the zoea in the decapod superfamily Penaeoidea. Thus, development in bathynellaceans ceases before metamorphosis and they become sexually mature at a time when Penaeoidea are still in the larval phase.

Habitat and Ecological Notes

Recent Syncarida never have been found in the marine habitats and only the anaspidacean family Stygocarididae and the order Bathynellacea are represented in the meiofauna.

Anaspididae appear to be restricted to the Southern Hemisphere. Members of Anaspididae are endemic to Tasmania and typically occur in small upland streams, moorland pools, and lakes. Occasionally they are also collected in crayfish burrows which seem to be the typical habitat for the family Koonungidae. Members of the family Psammaspidae are subterranean and are found in the interstitial spaces of coarse sediments. The minute members of the family Stygocarididae have been found in interstitial habitats of southern South America, New Zealand, and Australia.

The Bathynellacea are confined mainly to

freshwater interstitial habitats except for some species of *Hexabathynella* which occur in oligo- to polyhaline waters. They have been recorded from all continents except Antarctica, and those areas in the Northern Hemisphere effected by the latest glaciation. The intercontinental relationship of the various bathynellid genera has been analyzed by Schminke (1974, 1975) who suggests that the order most probably has evolved in eastern Asia. From there they have invaded Europe, Africa, Australia, and New Zealand as well as across Beringia North America. South America has been reached twice by this group; one track probably came from Australia via Antarctica to the southern part of the subcontinent while a second group had its origin in Africa.

Very little is known about the biology of the Syncarida. The larger anaspidaceans are omnivorous, feeding on algae, detritus, and small animals such as worms, chironomid larvae, and tadpoles. It is assumed that bathynellids feed on bacteria, fungi, and detritus.

Classification

The Syncarida consist of the fossil Palaeocaridacea, which were distributed in shallow marine waters of the Northern Hemisphere throughout Carboniferous and Permian times, and the Recent Anaspidacea and Bathynellacea.

The Anaspidacea include four families, the most plesiomorphic are members of the family Anaspididae which are medium to large in size. This family contains 5 species in 3 genera. The sister-group of these two taxa is represented by the families Psammaspididae and Stygocarididae both of which exhibit neotenic features. The Psammaspididae consist of 2 species in 2 genera. The family Stygocarididae originally were established as a separate order, but the discovery of stygocarids in Australia and New Zealand (Schminke, 1978, 1980) and the subsequent reassessment of their morphological characters have resulted in their reassignment to family status within Anaspidacea.

The Bathynellacea are subdivided into two closely related families: Bathynellidae and Parabathynellidae. The Bathynellidae among other things, are characterized by a many plesiomorphic characters. At present, the family contains about 60 species in 13 genera. Members of the Parabathynellidae appear to be more advanced; currently, this family consists of about 75 species in 25 genera.

Methods of Collection and Extraction

The methods described in previous chapters on interstitial meiofauna are equally applicable in regard to the Syncarida. Of particular note is the Bou-Rouch technique (Bou and Rouch, 1967) which has accounted for a large number of successful collections of small crustaceans, especially by scientists working in the area of the Mediterranean Sea, and more recently in the West Indies (Stock, 1976). This technique and certain ecological ramifications of its use are reviewed by Pennak in Chapter 4. The second useful collecting technique is that described by Cvetkov (1968). Whereas the Bou-Rouch method is designed to sample hyporheic and phreatic habitats, e.g., the interstitial waters flowing beneath and adjacent to river beds, the Cvetkov method is intended for the study of the small invertebrates of wells. Basically, any method used to sieve will extract syncarids if they are present. Specimens, sieved through fine mesh net are further processed by sorting under a stereomicroscope. Although staining with Rose Bengal is often helpful, it is less necessary since these techniques generally do not involve much extraneous material such as sediment and detritus.

Preparation of Specimens for Taxonomic Study

Specimens should be fixed in 6-10% formalin and transferred to 70% ethanol for storage. When a more intensive examination is conducted, the specimens should be placed in a 2% glycerin-in-alcohol solution which is allowed to evaporate very slowly, thereby dehydrating the specimens, leaving them in glycerin. One should be careful to note any shrinkage of the specimens by measuring them before and after this process; normally, very little shrinkage is experienced. Glycerin, being more viscous than alcohol, makes dissection easier. Microdissection needles are used for the dissection of the appendages of taxonomic importance. The dissected parts should be transferred to microslides for permanent mounts.

As in all other crustacean groups, especially those represented by small to minute animals, detailed description of all appendages including full setation is required. Chiefly within Bathynellacea, which are characterized by a lack of morphological structure, identification is dependent on the position, number, and type of setae. Of considerable taxonomic importance is the transformed male thoracopod VIII.

References

Bartok, P.
1944. Die morphologische Entwicklung von *Bathynella chappuisi*. *Acta Scientiarum Mathematicarum e Naturalium Kolozsvar*, 21:1-26.

Birstein, J.A., and S.I. Ljovuschkin
1964. Occurrence of Bathynellacea (Crustacea, Syncarida) in Central Asia. *Zoologiske Zhurnal*, 43(1):17-27.

1968. Biospeologica Sovietica XXXVIII. The Order Bathynellacea (Crustacea, Malacostraca) in the U.S.S.R. 2. The Parabathynellidae Family and a Zoogeographical Review. *Byulleten Moskovskogo Obshchestva Ispytatelei Prirody, Otdel Biologicheskii,* 73(6):55-64.

Bou, C.
1975. Les methodes de recolte dans les eaux souterraines interstitielles. *Annales de Speleologie,* 29(4):611-619.

Bou, C., and R. Rouch
1967. Un nouveau champ de recherches sur la faune souterraine. *Comptes Rendus de l'Academie des Sciences (Paris),* 265:369-370.

Brooks, H.K.
1962. On the Fossil Anaspidacea, With a Revision of the Classification of the Syncarida. *Crustaceana,* 4(3):229-242.

Cannon, H.G., and S.M. Manton
1929. On the Feeding-Mechanism of the Syncarid Crustacea. *Transaction of the Royal Society of Edinburgh,* 56:175-189.

Chappius, P.A.
1927. Anaspidacea. In Kükenthal, W. and T. Krumbach, editors, *Handbuch der Zoologie,* 3(1):593-606.

1948. Le developpement larvaire de *Bathynella. Bulletin de la Societé des Sciences de Cluj,* 10:305-309.

Cvetkov, L.
1968. Un filet phreatobiologique. *Bulletin d'Institut de Zoologie et Musée (Academie des Sciences de Bulgare),* 27:215-218.

Delamare Deboutteville, C.
1960. *Biologie des eaux souterraines littorales.* 740 pages. Paris: Hermann.

Delamare Deboutteville, C., and P.A. Chappuis
1953. Les bathynelles de France et d'Espagne avec diagnoses d'especes et de formes nouvelles. *Vie et Millieu,* 4(1):114-115.

1954a. Recherches sur les Crustaces souterrains. V. Les *Bathynella* de France et d'Espagne. *Archives de Zoologie Experimentale et Generale,* 91(1):51-73.

1954b. Recherches sur les Crustaces souterrains. Remarques sur le developpement des bathynelles. *Archives de Zoologie Experimentale et Generale,* 91(1):74-82.

1954c. Recherches sur les Crustaces souterrains. VI. Revision des genres *Parabathynella* Chappuis et *Thermobathynella* Capart. *Archives de Zoologie Expeirmentale et Generale,* 91(1):83-102.

Gordon, I.
1964. On the Mandible of the Stygocaridae (Anaspidacea) and Some Other Eumalacostraca, with Special Reference to the lacinia mobilis. *Crustaceana,* 7(2):150-157.

Hickman, V.V.
1937. The Embryology of the Syncarid Crustacean *Anaspides tasmaniae. Papers of the Royal Society of Tasmania,* 1936:1-36

Jakobi, H.
1954. Biologie, Entwicklungsgeschichte und Systematik von *Bathynella natans* Vejd. *Zoologische Jahrbücher (Abteilung für Systematik),* 83(1-2):162.

Knott, B., and P.S. Lake
1980. *Eucrenonaspides oinotheka* gen. sp. n. (Psammaspididae) from Tasmania, and a New Taxonomic Scheme for Anaspidacea (Crustacea, Syncarida). *Zoologica Scripta,* 9:25-33.

Manton, S.M.
1930. Notes of the Habits and Feeding Mechanisms of *Anaspides* and *Paranaspides* (Crustacea, Syncarida). *Proceedings of the Zoological Society of London,* 1930(3):791-800.

Miura, Y., and Y. Morimoto
1953. Larval Development of *Bathynella morimotoi* Ueno. *Annotationes Zoologicae Japonenses,* 26(4):238-245.

Noodt, W.
1963. Anaspidacea (Crustacea, Syncarida) in der südlichen Neotropis. *Verhandlungen der Deutschen Zoologischen Gesellschaft,* 1962:568-578.

1964. Natürliches System und Biogeographie der Syncarida (Crustacea, Malacostraca). *Gewässer und Abwässer,* 37-38:77-186.

1968. Deuten die Verbreitungsbilder reliktärer Grundwasser-Crustaceen alte Kontinentalzusammen-hänge an? *Naturwissenschaftliche Rundschau,* 21(11):470-476.

Pennak, R.W., and J.V. Ward
1985. Bathynellacea (Crustacea: Syncarida) in the United States, and a New Species from the Phreatic Zone of a Colorado Mountain Stream. *Transactions of the American Microscopical Society,* 104:209-215.

Schminke, H.K.
1974. Mesozoic Intercontinental Relationship as Evidenced by Bathynellid Crustacea (Syncarida: Malacostraca). *Systematic Zoology,* 23:157-164.

1975. Phylogenie und Verbreitungsgeschichte der Syncarida (Crustacea, Malacostraca). *Verhandlungen der Deutschen Zoologischen Gesellschaft,* 1974:384-388.

1978. Die phylogenetische Stellung der Stygocarididae (Crustacea, Syncarida) unter besonderer Berücksichtigung morphologischer Ähnlichkeit mit Larvenformen der Eucarida. *Zeitschrift für Zoologische Systematik und Evolutionsforschung,* 16(3):225-239.

1980. Zur Systematik der Stygocarididae (Crustacea, Syncarida) und Beschreibung zweier neuer Arten (*Stygocarella pleotelson* gen. n. sp. n. und *Stygocaris giselae* sp. n.). *Beaufortia,* 30(6):139-154.

1981. Adaption of Bathynellacea (Crustacea, Syncarida) to Life in the Interstitial ("Zoea Theory"). *Internationale Revue der gesamten Hydrobiologie,* 66(4):575-637.

Siewing, R.
1959. Syncarida. In Schellenberg, A., et al. editors, *Bronns Klassen und Ordnungen des Tierreichs* (2. Auflage), Band V, Abteilung I, Heft 4, Teil 2:1-121.

Stock, J.H.
1976. A New Genus and Two New Species of the Crustacean Order Thermosbaenacea from the West Indies. *Bijdragen tot de Dirkunde,* 46(1):47-70.

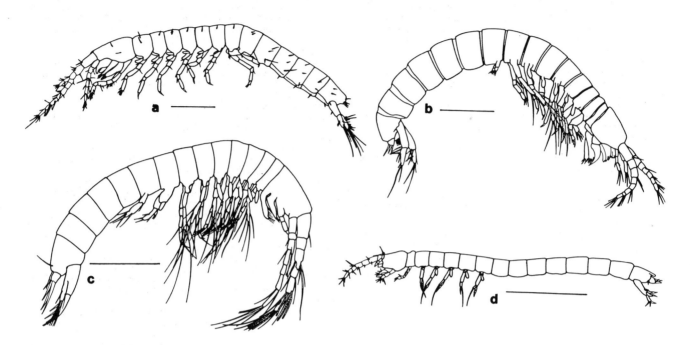

Figure 38.1.--Syncarida: **a**, Anaspidacea, Stygocarididae, *Stygocarella pleotelson*, Schminke, 1980, (modified after Schminke, 1978); **b**, Bathynellidae, *Bathynella paranatans* Serban, 1971 (after Serban, 1972); **c**, Parabathynellidae, *Parabathynella motasi* Dancan and Serban, 1963 (after Serban, 1972); **d**, Bathynellidae, *Hexabathynella halophila* Schminke, 1972 (after Schminke, 1972). (Scale = 200 μm.)

39. Thermosbaenacea

Jürgen Sieg

This order of small crustaceans is placed in the superorder Pancarida which is considered the sister group of the Peracarida. The order presently contains about 10 species in 4 genera. They seem to be more abundant in nearcoastal habitats. Recently a member has been caught in a cave connected to the open sea.

The body is more or less stout and only the first thoracic segment is fused to the cephalon (cephalothorax). Eyes are typically lacking, except for *Theosbaena cambodjiana* Cals and Boutin, 1985, which has ocular lobes without visual elements. The carapace is large, covering up to the third or fourth free thoracic segment. Especially the anterior lateral part is ventilated by the backwards bent maxillipedal epipodite. In the female the carapace is enlarged and swollen forming a dorsal brood-chamber (Figure 39.1a,e). The remaining seven thoracic segments each bear one pair of pereopods (except *Thermosbaena* in which the two last pairs of thoracopods are wanting; Figure 39.1d). Genital openings are found on the same segments as in other malacostracans. Testes open separately on two large penes originating from the coxae of the last pair of pereopods. Ovary openings are situated on the coxal plates of the fifth pair of pereopods (=sixth thoracomere; Siewing, 1958 Abb. 1 B) and not "on the seventh somite" as stated by Bousfield (1982:241). The abdomen consists of six free pleonites and a telson (except *Thermosbaena* in which the last pleonite is fused to the telson – Figure 39.1c).

Antenna 1 is biramous with a 3-segmented peduncle and antenna 2 is uniramous consisting of a 5-segmented peduncle and a multisegmented flagellum (up to 26 in *Theosbaena*). Mouthparts are similar to those of the Peracarida. The mandible has a lacinia mobilis and a 3-segmented palpus. In maxilla 1 as well as in maxilla 2 a 3-segmented palpus is found. There is no epipodite in maxilla 2. The first thoracopod is developed as a maxilliped which shows similarities to the eucarids in having a coxal endite. In some cases (e.g., *Monodella sanctaecrucis* Stock, 1976; *M. atlantomaroccana* Boutin and Cals, 1985) this pair of legs shows sexual dimorphism. Females have only a reduced 2-segmented endopodite while in these species it is 5-segmented (ischium-dactylus) and enlarged in the male (Figure 39.1g). The remaining thoracopods are typically biramous (except *Theosbaena* which lacks an expodite on the first pair of legs) and may be called pereopods. Oostegites or epipodites are missing. In some cases the expodite is reduced (e.g., *Theosbaena cambodjiana, Monodella atlantomaroccana*), lacking setation. Pleopods are only developed on the first two pleonites and reduced to a uniramous, 1-segmented appendage. Uropods are biramous with the endopodite and the expodite 1-segmented and 2-segmented, respectively.

Classification

The order is subdivided in the monogeneric family Thermosbaenidae (*Thermosbaena mirabilis*) and Monodellidae (*Halosbaena, Limnosbaena, Monodella, Theosbaena*).

Determination of thermosbaenaceans, as in all small or minute crustaceans, requires a detailed and very carefull study of all appendages. Generic discriminants (characters used to distinguish genera) include the number of pereopods or pleonites present, shape of maxillipedal endopodite, morphology of expodite or first as well as seventh pereopod, and armament of maxilla 2. Characters frequently used for species identification consist of the mandibular palp, shape and number of joints in the endopodite of maxilla 2, shape of pleotelson, and exopodite articles in the uropods.

Habitats and Ecological Notes

Thermosbaenaceans are typical interstitial forms, only *Thermosbaena mirabilis* lives in hot springs in North Africa. Originally all records were from coastal habitats, e.g., brackish waters in caves close to the sea, mesopsammal beach communities, or near desiccated salty inlets. More recently members from purely limnic habitats have also been recorded (i.e., *Monodella texana* Maguire, *Limnosbaena finki* (Mestrov and Latinger-Penko)) which support the idea of a penetration of the continental subterranean waters in Miocene-Pliocene times (Freyer, 1965; Stock, 1976, 1977b). Bowman and Iliffe (1986) recently described a new species in the Grand Canaries collected in a cave which is connected to the open sea. They believe that this discovery supports Stock's and Freyer's hypothesis for a marine origin of this order. As pointed out, all known present-day localities lie on or near the Oligocene shore-lines. During Miocene regressions of the sea level they, as well as the Microparasellidae "became adapted to the brackish "Küstengrundwasser," and evolved finally into limnic mesopsammal forms" (Stock, 1976a:90).

Methods of Collection and Extraction

The method most commonly cited in the collection and extraction of the Thermosbaenacea is the Bou-Rouch pump method (Chapter 7). In reality, almost any method of collecting and sieving water from the proper habitat will yield these rare crustaceans if they are present.

Methods of Fixation and Preparation for Taxonomic Study

Specimens should be fixed in 4-6% formalin and transferred to 70% ethanol for storage. The methods common in the preparation and taxonomic study of other small crustaceans are applicable to the Thermosbaenacea.

References

Barker, D.
1959. The Distribution and Systematic Position of the Thermosbaenacea. *Hydrobiologia*, 13:209-235.
1962. A Study of *Thermosbaena mirabilis* (Malacostraca, Peracarida) and Its Reproduction. *Quarterly Journal of Microscopical Science*, 103(2):261-286.

Botosaneanu, L., and C. Delamare Deboutteville
1967. Fossiles vivants des eaux soutereraines. *Sciences (Paris)*, 52:17-22.

Bou, C.
1975a. Les methodes de récolte dans les eaux souterraines interstielles. *Annals de Spéléologie*, 29(4):611-619.
1975b. Récherches sur la faune des eaux souterraines de Grece. *Biologia gallo-hellenica*, 6(1):101-115.

Bou, C., and R. Rouch
1967. Un nouveau champ de récherches sur la faune souterraine. *Comptes réndus de l'Academie des Sciences (Paris)*, 265:369-370.

Bousfield, E.L.
1982. Thermosbaenacea. Page 241 in S.P. Parker, editor, *Synopsis and Classification of Living Organisms, Volume 2*. New York: McGraw-Hill.

Boutin, C., and P. Cals
1985. Importance en biogeographie évolutive de la découverte d'un Crustace phréatobie, *Monodella atlantomaroccana*, n.sp. (Thermosbaenacea) dans la plaine alluviale de Marrakech (Maroc atlantique). *Comptes réndus de l'Academie des Sciences (Paris)*, (3)300(7):267-270.

Bowman, T.E., and T.M. Iliffe
1986. *Halosbaena fortunata*, a New Thermosbaenacean Crustacean from the Jameos del Agua Marine Lava Cave, Lanzarote, Canary Islands. *Stygologia* 2(1/2):84-89.

Bruun, A.F.
1939. Observations on *Thermosbaena mirabilis* Monod from the Hot Springs of El-Hamma, Tunesia. *Videnskabelige Meddelelser fra Dansk naturhistorisk Forening i Kjobenhavn*, 103:492-501.

Cals, P., and C. Boutin
1985. Découverte au Cambodge, domaine ancien de la Tethys orientale, d'un nouveau "fossile vivant" *Theosbaena cambodjiana*, n.g., n.sp. (Crustacea, Thermosbaenacea). *Comptes réndus de l'Academie des Sciences (Paris)*, (3)300(8):337-340.

Chappuis, P.A.
1942. Eine Methode zur Untersuchung der Grundwasserfauna. *Acta Scientiarum Mathematicarum e Naturalium Kolozsvar*, 6:3-7.

Chelazzi, L., and G. Messana
1982. *Monodella somala*, n.sp. (Crustacea, Thermosbaenacea) from the Somali Democratic Republic. *Monitore Zoologico Italiano*, Supplement 16(7):161-172.

Cvetkov, L.
1968. Un filet phréatobiologique. *Bulletin de l'Institut de Zoologie et Musée (Academie des Sciences de Bulgare)*, 27:215-218.

Delamare Deboutteville, C.
1960. *Biologie des eaux souterraines littorales*. 740 pages. Paris: Hermann.

Freyer, G.
1965. Studies on the Functional Morphology and Feeding Mechanism of *Monodella argentarii* Stella (Crustacea, Thermosbaenacea). *Transactions of the Royal Society of Edinburgh*, 66(4):49-90.

Hessler, R.R.
1969. Order Thermosbaenacea. In R.C. Moore, editor, *Treatise on Invertebrate Paleontology, Part R (Arthropoda)*, volume 4(1):R366-R367.

Karaman, S.
1953. Über einen Vertreter der Ordnung Thermosbaenacea (Crustacea, Peracarida) aus Jugoslavien, *Monodella halophila*, n.sp. *Acta adriatica*, 5(3):1-22.
1954. Über unsere unterirdische Fauna. *Acta Musei Macedonici Scientiarum Naturalium*, 1(9):195-216.

Maguire, B.
1965. *Monodella texana*, n.sp., an Extension of the Range of the Crustacean Order Thermosbaenacea to the Western Hemisphere. *Crustaceana*, 9(2):149-154.

Mestrov, M., and R. Lattinger-Penko
1969. Sur la présence de Thermosbaenacés (Crustacea, Peracarida) dans des eaux interstielles continentales de la Yougoslavie (*Monodella finki*, n.sp.). *Annales de Spéléologie*, 24(1):111-123.

Monod, T.
1924. Sur un type nouveau de Malacostracé: *Thermosbaena mirabilis*, n.gen., n.sp. *Bulletin de la Société zoologique France*, 49(2):58-68.
1927a. Nouvelles observations sur la morphologie de *Thermosbaena mirabilis*. *Bulletin de la Société zoologique France*, 52(3):196-200.
1927b. *Thermosbaena mirabilis* Monod - Remarques sur sa morphologie et sa position systématique. *Faune des Colonies françaises*, 1(2):29-51.
1940. Thermosbaenacea. *Bronns Klassen und Ordnungen des Tierreichs*. Bd. 5, Abt. 1, Buch 4, Teil IV:1-24.

Por, F.D.
1962. Un nouveau Thermosbaenacé, *Monodella relicta*, n.sp. dans la dépression de la Mer Morte. *Crustaceana*, 3(4):304-310.

Rouch, R.
1965. Contribution a la connaissance du genre *Monodella* (Thermosbaenacés). *Annales de Spéléologie*, 19(4):717-727.

Ruffo, S.
1949a. *Monodella stygicola*, n.g., n.sp., nuovo Crostaceo Thermosbaenaceo delle acque sotterranee della penisola Salentina. *Archivio Zoologico Italiano*, 34:31-48.
1949b. Sur *Monodella stygicola* Ruffo des eaux souterraines de l'Italie méridionale, deuxième espece connue de

l'ordre des Thermosbaenacés (Malacostraca Peracarida). *Hydrobiologia*, 2:56-63.

Siewing, R.
1958. Anatomie und Histologie von *Thermosbaena mirabilis*. *Akademie der Wissenschaften und der Literatur in Mainz, Abhandlungen der Mathematisch- naturwissenschaftlichen Klasse*, 1957(7):195-270.

Stella, E.
1951a. *Monodella argentarii*, n.sp. di Thermosbaenacea (Crustacea, Peracarida) limnotroglobio di Monte Argentario. *Archivio Zoologico Italiano*, 36:1-15.
1951b. Notizie biologiche su *Monodella argentarii* Stella, Thermosbaenaceo delle acque di una grotta di Monte Argentario. *Bolletino Zoologia*, 18:227-233.
1953. Sur *Monodella argentarii* Stella, éspèce de Crustacé Thermosbaenacé des eaux d'une grotte di l'Italie centrale (Monte Argentario, Toscana). *Hydrobiologia*, 5:226-234.
1955. Behaviour and Development of *Monodella argentarii* Stella, a Thermosbaenacean from an Italian Cave. *Proceedings of the International Association of Theoretical and Applied Limnology*, 12:464-466.
1959. Ulteriori osservazioni sulla riproduzione e lo sviluppo di *Monodella argentarii* (Pancarida Thermosbaenacea). *Rivista Biologia*, 51:121-144.

Stock, J.H.
1976. A New Genus and Two New Species of the Crustacean Order Thermosbaenacea from the West Indies. *Bijdragen tot de Dierkunde*, 46(1):47-70.
1977a. Microparasellidae (Isopoda, Asellota) from Bonaire with Notes on the Origin of the Family. *Studies on the Fauna of Curaçao and Other Caribbean Islands*, 51(168):69-91.
1977b. The Zoogeography of the Crustacean Suborder Ingolfiellidea with Descriptions of New West Indian Taxa. *Studies on the Fauna of Curaçao and Other Caribbean Islands*, 55(178):131-146.

Taramelli, E.
1954. La posizione sistematica dei Thermosbaenacei quale risulta dallo studio anatomico di *Monodella argentarii* Stella. *Monitore Zoologico Italiano*, 62(1):9-24.

Vandel, A.
1964. *Biospéologie - La Biologie des animaux cavernicoles*, Paris: Gauthier-Villars.

Vigna-Taglianti, A., V. Cotarelli, and R. Argano
1969. Messa a punto di metodiche per la raccolta della faune interstiziale e freatica. *Archivio Botanico Biogeografico Italiano*, 45(4):375-381.

Zilch, R.
1972. Beitrag zur Verbreitung und Entwicklungsbiologie der Thermosbaenacea. *Internationale Revue der gesamten Hydrobiologie*, 57(1):75-107.
1974. Die Embryonalentwicklung von *Thermosbaena mirabilis* Monod. (Crustacea, Malacostraca, Pancarida). *Zoologische Jahrbücher (Anatomie)*, 93:462-576.

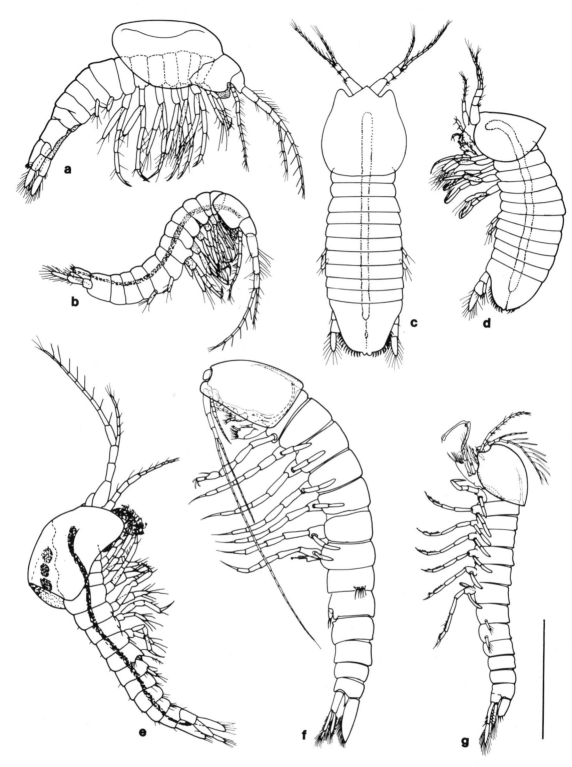

Figure 39.1 --Thermosbaenacea: a, *Monodella sanctaecrucis* Stock, female with empty brood pouch (after Stock, 1976); b, same, male (after Stock, 1976); c, *Thermosbaena mirabilis* Monod, dorsal view (after Monod, 1924); d, same, lateral view (after Monod, 1924); e, *Halosbaena acanthura* Stock (after Stock, 1976); f, *Theosbaena cambodjiana* Cals and Boutin (after Cals and Boutin, 1985); g, *Monodella atlantomaroccana* Boutin and Cals (after Boutin and Cals, 1985) (scale = 1 mm.)

40. Isopoda

Roberto Argano

The wide adaptive radiation of the Isopoda is one of the most interesting features of this order of Crustacea. These animals populate many different habitats of the marine benthic systems in the world; they live in the surface and subterranean fresh-water habitats; one entire suborder of isopods have become terrestrial, and they are also parasitic epibiotic on many marine organisms (particularly on fish and other crustaceans).

Many species, belonging to different suborders, take a dominant role in the meiobenthos. They exhibit a variety of adaptations relevant to their specific benthic habitat. Species living on the surface of soft bottoms are often flat in shape or, in case of elongated bodies, the legs are very long and they allow the animals to move over soft surfaces. Most of the known species live in fouling communities or associated with sediments of hard bottoms. Therefore, they have elongated bodies and comparatively shorter legs to exploit the smallest crevices or to attach to substratum. Often these animals are longer than 1 mm although very narrow, hence they can pass through a 1 mm mesh sieve. The interstitial species, living in medium to coarse sand are even smaller, blind, colorless, filiform, and have very short appendices.

Classification

The following brief description gives the diagnostic characters of the order and allows one to distinguish both meiobenthic and other isopods with the exception of such taxa as the Phreatocoidea and the parasitic taxa (Epicaridea and Gnathiidea) which are more or less greatly modified.

As crustaceans they have two pairs of antennae, as malacostracans they have an 8-segmented pereion (but the first maxillipedal segment is fused with cephalon), and as peracaridans the females are provided with a brood pouch. The diagnostic features of the Isopoda include a dorsoventrally flattened body, a complete absence of carapace, seven pairs of legs, seven thoracic free segments, respiratory pleopods, and sessile eyes.

The function of the first three pairs of pleopods (copulatory, natatory, opercular), the morphology of the anterior pereiopods (ambulatory or prehensile), and the number of the pleonal free segments (0-5) all are variable features depending on the suborders or minor taxa. The structure of the pleonal copulatory appendices is, in general, of a great importance for specific diagnosis.

The post-marsupial stage (manca), for all marine isopods, can be regarded as a mero-meiobenthonic condition.

There are ten different suborders in the Isopoda. These include: Asellota, Microcerberoidea, Anthuridea, Calabozoidea, Flabellifera, Oniscoidea (terrestrial), Valvifera, Phreatocoidea, Gnathiidea, and Epicaridea (parasites on other crustaceans). Excluding the orders Oniscoidea and Epicaridea, all suborders have meiobenthic species: the orders Microcerberoidea and Calabozoidea are exclusively meiobenthic, but the greater number of meiobenthic species are found in the suborder Asellota (more than 30 different families).

A highly depressed body, a wide pleotelson, one or two free reduced pleonal segments and the terminal position of the biramous (with subequal rami) uropods are general features that allow an easy identification of an Asellota without dissection of the specimen. The wide pleotelson forms a branchial chamber that includes at least three pairs of laminar respiratory pleopods. The chamber is closed by an operculum whose structure is of taxonomic relevance at superfamiliar rank.

The superfamily Janiroidea includes the majority of marine meiobenthic isopods species. In this taxon, the first male pleopods are fused in a narrow structure forming, with the laminar second pleopod, both the opercular and the copulatory system. The second female pleopods are fused in a single laminar operculum (the first pair is lacking).

The wide exopods of the third pleopods form the operculum in the superfamily Aselloidea. Pleopod I is lacking in females whereas it protects the copulatory pleopod II in males. Some tiny freshwater species, both phreatic and hyporheic, belong to this superfamily.

An analogous organization may be found in the Stenetroidea, a relatively small marine superfamily. Unlike the Aselloidea, the Stenetroidea have the basal portion of the first pleopods fused in the males, and in the female the second pleopod (formed by a single foliaceus structure) is not opercular. The operculum consist of pleopod III.

The superfamily Gnathostenetrioidea includes both marine and freshwater isopods having an opercular structure intermediate to the Stenetrioidea and the Janiroidea: the very large first male pleopods, medially fused in the basal portions (like in the Stenetrioidea), cover all other pleopods while the female operculum is very similar to that of Janiroidea.

The exclusively interstitial (both marine and freshwater) isopods of the suborder Microcerberoidea are easily identifiable by their subcylindrical body, often subchelate first pereiopods, absence of eyes, two large pleonal free segments, and terminal uropods with rudimentary endopod. Recent data suggest that the ancestry of Microcerberoidea could be investigated within the Asellota, while since their discovery they have been regarded as being related to the following suborder, the Anthuridea.

The suborder Anthuridea also have elongate subcylindrical bodies and subchelate pereiopods I (pereiopods II and III can be subchelate, too), but they are chiefly characterized by five short pleonites, either free or fused, and laminar uropods often folded over the pleotelson. They are part of the marine meiobenthos (often burrowing) of the coastal and open sea, of the infauna and of the stygobiont communities.

At present only one phreatic and hyporheic species from Venezuela is known for the suborder Calabozoidea. This taxon is distinguished by an oblong-oval body, the pereiopods are all ambulatory; it has three free pleonal segments, an enormous pleotelson, pleopods I and II are modified for copulatory purposes in males, and the lateral uropods extremely reduced.

The suborder Flabellifera includes species with a dorsoventrally depressed body, large pleotelson fused with a variable number of segments, natatory anterior pleopods and lateral flattened uropods forming a tail fan. This taxon is the largest in marine, brackish and hypogean species number, but only few of them pertain to meiobenthos.

The Valvifera are also poorly represented in meiobenthos. Their more distinctive feature is the seeming absence of the uropods: in reality they have a valve-like uropod structure forming a hard operculum enclosing five pleopods in the respiratory chamber.

The archaic gondwanian suborder of Phreatocoidea comprises species with smooth, elongate and laterally compressed body. The suborder is restricted to Australia, New Zealand, South Africa, and India.

Finally, sexual dimorphism is the major characteristic of Gnathiidea. A very wide cephalon with frontal forceps-like mandibles, flattened first pereiopods (pleopods) operculating the buccal cavity, only five walking legs, pleon abruptly narrower than pereion are unmistakable features of males. Females are characterized by small and narrow cephalon with conspicuous eyes, produced mouthparts, a very large pereion with an internal marsupium and a tiny pleon. Juveniles (praniza stage) are external parasites on fish.

Habitats and Ecological Notes

Because of the wide adaptive radiation of the order, meiobenthic isopods are present in a variety of habitats: (1) phreatic (wells); (2) hyporheic (beside streams and rivers) (refer to Chapter 4); (3) cavernicolous; (4) interstitial (both freshwater and marine); (5) within soft sediments (infauna); (6) or on top of sediments (epibenthic); (7) epibiontic on sessile organisms; and (8) under submerged stones.

Methods of Collection

The method of collection depends on the kind of habitat being investigated. Sampling the phreatic habitats typical of wells is best accomplished by a Cvetkov net (a plankton net with a valve preventing the water reflux from the terminal collecting container). Hyporheic habitats are best sampled by the Bou pump (a hand-operated pump which withdraws interstitial water and accompanying organisms from sandy river beds, including hyporheic areas adjacent to the river). Cavernicolous habitats require the use of smaller hand-operated nets or dredges equipped with plankton netting. Interstitial habitats (aside from those associated with the above habitats) can be sampled by numerous methods such as the addition of water to a sand sample with subsequent stirring and decantation (with or without narcotization techniques), or by the Karaman-Chappuis method (a hole dug in the sand to the water-table and the subsequent sieving of the interstitial water that accumulates).

Meiobenthic isopods from soft sediments (epibenthic or infaunal) are best collected with a dredge that samples just the uppermost strata of the sediment. Epibiontic species must be removed by judicious scraping techniques and, for those species living under stones, a small, fine-mesh hand-operated net may be used to sample the underlying sediment, and the stones should be carefully placed in a plastic bag with dilute formalin where they can subsequently be washed free of the detritus and any specimens.

Methods of Extraction

Most techniques applicable for the extraction of

crustaceans (as well as other organisms) are well-suited for the extraction of meiofaunal isopods from sediments. In general, these techniques involve placing the sediment in a plastic pail, adding water, stirring and decanting through a fine-mesh sieve or graded series of sieves. Like all meiofauna, the final sorting of material requires the use of a stereomicroscope. Various kinds of elaborate apparatus designed for the removal of meiofauna often are no better than one form or another of the above elutriating technique. As in all situations where quantitative analyses are required, no elutriation technique can be considered 100% effective.

Preparation of Specimens for Taxonomic Study

Specimens should be fixed in 4-6% formalin and later transferred to 70% ethanol. Preparation for the first *in toto* observation requires the specimen to be transferred to glycerin, preferably by first placing the specimen in a glycerin-alcohol solution and allowing the alcohol to evaporate. The specimen should be placed in glycerin on a microslide and a drawing of the animal should be made (preferably with a camera lucida); this is a very important step for subsequent analysis and is essential for proper description.

For a diagnosis at specific level a complete dissection is indispensable (18 couples of appendices = 36 pieces besides the body). It is advisable to execute the dissection with microneedles directly on a microslide in a drop of mounting medium (faure, polyvinylactophenol, euparal, etc.). Obviously, a routine dissection restricted to the appendages of diagnostic value (i.e., copulatory pleopods) is often enough.

References

Argano, R.
1979. *Isopodi (Crustacea Isopoda). Guide per il riconoscimento delle specie animali delle acque interne italiane. Progetto finalizzato "Promozione della qualita' dell'ambiente."* 63 pages. Rome: Consiglio Nazionale delle Ricerche [AQ/1/43].

Birstein, Y.A.
1963. *Deep Water Isopods (Crustacea, Isopoda) of the North-western Part of the Pacific Ocean.* [Translated from Russian (1973)]. Washington: Publication for the Smithsonian Institution and the National Science Foundation by the Indian National Scientific Document Centre, New Delhi.

Kensley, B.
1978. *Guide to the Marine Isopods of Southern Africa.* 173 pages. Cape Town: Trustees of the South African Museum.

Naylor, E.
1972. *British Marine Isopods.* 86 pages. London and New York: Academic Press.

Schultz, G.A.
1969. *The Marine Isopod Crustaceans. Handbook of North American Coasts.* 359 pages. Dubuque, Iowa: Wm. C. Brown Company.

Wolff, T.
1962. The Systematics and Biology of Bathyal and Abyssal Isopoda Asellota. *Galathea Reports,* 6:1-320.

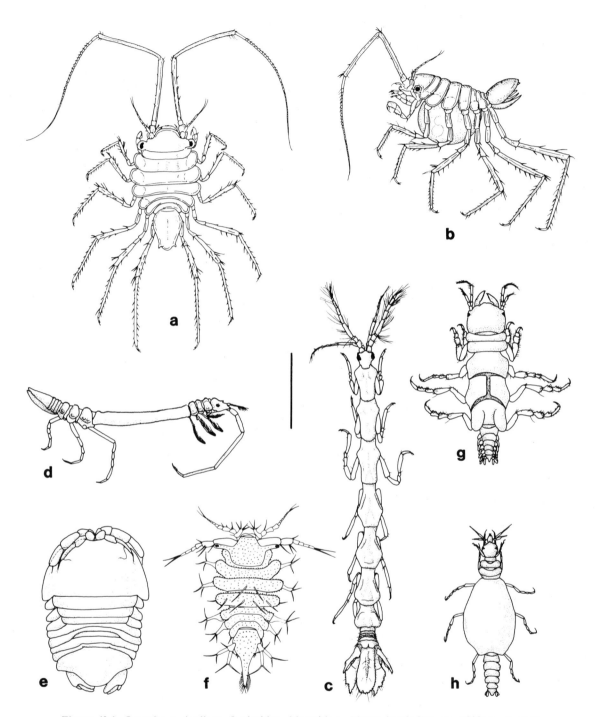

Figure 40.1.--Isopoda: a, Asellota, Janiroidea, Munnidae, *Munna boecki* Kroyer, 1839; **b,** Asellota, Janiroidea, Munnidae, *Munna boecki* Kroyer,1839, female with brood pouch, lateral view; **c,** Anthuridea, Hyssuridae, *Eisothistos antarcticus* Vanhöffen, 1914; **d,** Valvifera, Arcturidae, *Arcturopsis rudis* Koehler, 1911; **e,** Flabellifera, Serolidae, *Basserolis kimblae* Poore, 1985; **f,** Asellota, Janiroidea, Pleurocopidae, *Plurocope dasyura* Walker, 1901; **g,** Gnathiidea, Gnathiidae, *Paragnathia formica* (Hesse, 1864), male. **h,** Gnathiidea, Gnathiidae, *Paragnathia formica* (Hesse, 1864), female. (Scale = 500 μm.)

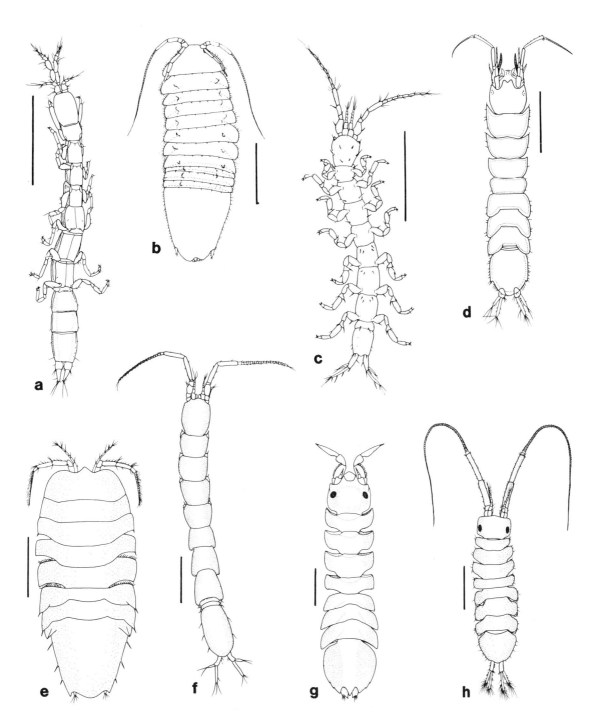

Figure 40.2.--**Isopoda** (continued): **a**, Microcerberoidea, Microcerberidae, *Microcerberus* sp.; **b**, Calabozoidea, Calabozoidae, *Calabozoa pellucida* Van Lieshout, 1983; **c**, Asellota, Janiroidea, Microparasellidae, *Angeliera phreaticola* Chappuis and Delamare Debouteville, 1952; **d**, Asellota, Gnathostenetroidea, Gnathostenetroidae, *Gnathostenetroides laodicense* Amar, 1957; **e**, Asellota, Haploniscidae, *Haploniscus laticephalus* Birstein, 1968; , **f**, Asellota, Gnathostenetroidea, Gnathostenetroidae, *Caecostenetroides ischitanum* Fresi and Schiecke, 1968; **g**, Asellota Jaeropsidae, *Jaeropsis beuroisi* Kensley, 1975; **h**, Asellota, Janiroidea, Janiridae, *Janaira gracilis* Moreira and Pires, 1977. (Scale = 500 μm.)

41. Tanaidacea

Jürgen Sieg

The Tanaidacea are an order of free-living and exclusively benthic peracaridean Malacostraca with three recent suborders and about 23 families. The body is more or less cylindrical (Figure 41.1f-k) or dorsoventrally depressed (Figure 41.1a-e) and ranges in length from 1-37 mm. Although primarily macrobenthic because the mean body length is about 2 mm, some tanaidaceans, certainly many juveniles, pass through a 1 mm-mesh sieve and are a common element in meiobenthic samples.

The first two thoracic segments are fused with the head, forming a cephalothorax. Within the Apseudoidea normally there remains free second thoracic sternite. There is only a relatively small carapace enclosing a respiratory chamber. Typically the last abdominal segment is fused with the telson, forming a pleotelson. Besides this, in several independent lines, additional segments are fused to the pleotelson or have been reduced.

There are two different types of mouthparts. The apseudomorphan type is characterized by a monocondyle mandible usually bearing a 3-segmented palp (reductions are very seldom observed) in addition to a well-developed spine row, a first maxilla consisting of two endites, one of which bears a palp with several long setae, a complex structured second maxilla (with a fused inner and outer lobe – probably movable – and a fixed endite), and a labium (=paragnaths) bearing a palp with two or three spine-like setae at the tip. The tanaidomorphan type generally is characterized by reduction. The mandible is dicondyl, lacks a palpus as well as the spine row. Within the different families many kinds of reductions can be observed in the pars molaris, lacinia mobilis, and the pars incisiva. The first maxilla consists of a single endite with a palp which normally bears only two long setae at the tip. The second maxilla always is reduced to a small oval plate. The labium with one exception (family Tanaidae) lacks the distal palp and consists of an inner and outer lobe.

The first pair of thoracic legs (the maxillipeds) have an epignath and are unfused or fused medially. The second pair of thoracic legs are developed as chelipeds which show a strong sexual dimorphism within several tanaidomorphan families. The following six pairs are called peraeopods. Only the first one may bear an exopodite. Within the apseudomorphans there exist at least three different leg-types (digging or fossorial, climbing, and walking) while within the tanaidomorphans the legs are more or less cylindrical. Testes open separately on one or two genital cones on the last thoracic segment and the female gonopores are located on the coxal plates of the fourth pair of peraeopods. Pleopods may be present or absent and are relatively uniform. They do not serve as respiratory organs. The uropods are filiform or styliform, and uni- or biramous. Within the different tanaidomorphan families uropodal reduction is a common phenomenon.

Development

The postmarsupial development of the order is only partly understood. In all cases, development is direct and there is no planctonic larval stage. After hatching, the broodpouch embryo gradually changes (without a molt) to the first postembryonic stage (manca-I), the last pair of peraeopods and all pleopods are missing. In the second stage (manca-II) these appendages are rudimentary. Besides this, both stages are easily recognized by a proportionally much smaller peraeonite compared to the adult. The next stage is called neuter (preparatory female/male) and is followed by the adult.

Apseudomorphans are probably all primarily gonochoristic (Sieg, 1984) and follow that scheme established for *Pargurapseudes largoensis* by Messing (1979, 1983). Sexual dimorphism in this group is only weakly developed, affecting at most the first antenna and the chelipeds.

Tanaidomorphans show a great variety of postmarsupial development types. For plesiomorphic taxa (e.g., Tanaidae) we also have to accept a gonochoristic type of development (Figure 41.2a). Reduction of the mouthparts in the male phase sometimes leads to a highly complicated postmarsupial development expressed by up to four different types of males (e.g. Leptocheliidae; Figure 41.2c). Mainly the secondary males may show strongly developed dimorphism resulting in huge and striking chelipeds as well as totally different first antenna. More apomorphic families (e.g., Anarthruridae, Pseudotanaidae, and others) are also gonochoristic but the sole primary male is adapted for the search of the female ("swimming male"; Figure 41.2c). However, for many taxa there never have been males reported and for these parthenogenesis might be considered (Figure 41.2e).

Protogyny in Neotanaidae (Figure 41.2b) is considered as a parallel event to that of the Tanaidomorpha by Sieg (1983b, 1984).

Classification

As recently shown (Sieg, 1983b, 1984), a given character can vary widely within a taxon. Based on various methods of tube construction, apseudomorphans are characterized by differentiation and specialization in the peraeopods while tanaidomorphans have relatively uniform legs but show a great variety in the structures of the mouth-parts (Sieg, 1983b, 1984).

Within the apseudomorphan superfamily Apseudoidea, three different leg-types are recognized in the P.1 (see above) of which at least the climbing type has evolved convergently twice (Sieg, 1984). The original climbing leg is found within the Metapseudidae. In contrast, the Cyclopoapseudidae have a climbing leg which is derived from the fossorial type by reduction. Besides the P.1, the P.2-P.3 can also be modified in the same way accompanied by a differentiation of the propodus in the P.4. If the P.6 is also specialized, we may find up to five different structured legs for a single species. The combination of these various leg types within the Apseudoidea are the basis for the grouping on family and genus level (Sieg, 1984).

There are several additional important taxonomic characters but these have been disregarded so far. Among these are the number of joints in antenna-1, the shape and presence or absence of the lacinia mobilis as well as that of the mandibular palp, and general structure of maxilliped. There is no evidence for using the number of oostegital plates for the gross-classification as recently proposed by Gutu (1982). Species separation mainly can be done by using the shape of the cephalothorax (presence or absence of ocular spines, shape of rostrum and ocular lobes), shape of the peraeonites (length-width-ratio, lateral spines, etc.), shape of pleonites (mainly shape of epipleura), shape of the pleotelson, number of articles in the antennular flagellum, setation of maxilla-2, shape of cheliped and pleopods.

Neotanaidae, the single family of the suborder Neotanaidomorpha mainly follows the same scheme.

Classification of higher taxa within the Tanaidomorpha is much more subtle. Tube-constructing by using silk produced by glands of the first three pairs of peraeopods, and the addition of feces, results in a very homogenous environment and does not cause leg differentiation. All six pairs of peraeopods, therefore, are very similar. Logically, they are only of subordinate value for familiar classification, but setation may play a certain role for establishing genera. Higher taxonomic categories are mainly characterized by the mode of reproduction because the various types are accompanied by very different shaped males. The most plesiomorphic family Tanaidae still is gonochoristic and males are only different in the shape of antenna-1 and the chelipeds. Leptocheliidae have the most complex postmarsupial development within Tanaidacea. Development is characterized by protogynous hermaphroditism producing one primary and up to three different types of secondary males. Sexual dimorphism is strongly pronounced, differences are found in antenna-1, shape of cephalothorax, chelipeds, and mouth parts (totally reduced in the male). Remaining tanaidomorphan families all appear to be gonochoristic, too. But the male is totally different in body shape; mainly the pleon is much more enlarged and the pleopods are much better developed. It might be called "swimming male" because it is well adapted to find the females (Sieg, 1984). Additionally, the mode of the articulation of the chelipeds (Lang, 1970; Sieg, 1984) and the structure of the maxilliped (presence or loss of coxa, basis fused or unfused medially) are taxonomically important.

With regard to the general shape of the males on the genus level, gross morphology of mandibles, setation of peraeopods and pleopods is important. Species determination is sometimes subtle and mainly is done by structures of the mandibles (lacinia mobilis, pars molaris) and length-width ratios of peraeonites. Additional characters are found on all appendages and may be recognized by a different setation or proportion of segments. Normally in all these details the males are differing and, therefore, only females can be used for correct species determination (Sieg, 1978).

Habitats and Ecological Notes

Tanaidaceans are typically marine animals and are distributed world-wide from the shore down to hadal depths (9000 m). Occasionally they occur in brackish waters and in hyperhaline or even fresh water beach pools. Subordinate taxa may show different distribution patterns. In general, all families are distributed worldwide, but depending on their southern origin many of them are not reaching the northern North Atlantic and the Polar Basin. Some families are restricted to shallow waters (e.g., Kalliapseudidae, Tanaidae, and others) while some occur only in deeper waters, showing an increase in species diversity with increasing depth (Neotanaidae, Gardiner, 1975; Leptognathiidae, Sieg, 1984). Comparison of the numerical abundance in the deep sea often shows that tanaidaceans are the second most abundant crustacean group besides the amphipods. Therefore, tanaidacean assemblages of the world oceans vary greatly depending on the region and depths.

Tanaidaceans are tube-dwellers which seldom leave their tubes. This is true of all stages, including juvenile animals (sexually immature). When building a new tube, juveniles bore through the wall of the "mother tube" and build their own nearby (Buckle-Ramirez, 1965). Therefore, tanaidaceans have patchy distribution with high population densities. Mainly in shallow water areas can very high population densities (10^4–$10^5/m^2$) be observed. Thus they are an important part of the food chain within marine ecosystems. Depending on species size and mode of life, tanaidaceans are part of the food of polychaetes, amphipods, decapods, fishes, and some water fowl.

Analyses of the stomach and gut contents are indicating that tanaidaceans are usually scavengers or detritivores, some may be raptorial carnivores. Their food normally consists of detritus or small algae (mainly diatoms). Occasionally they also may feed on nematodes and harpacticoids (Feller, 1978). Only kalliapseudids are filter feeders (Lang, 1956a) as indicated by the setae structure on the maxilliped and cheliped (Sieg, 1984).

All species are live within the substrate (tube-dwelling). Apseudomorphans and neotanaids are all probably tunnelers. "The tube is prevented from caving in by means such as the trampling of a dust floor, and the packing of fine sediment into the walls with the aid of the peraeopods." (Johnson, 1982:85). It might be that additional mucous from unicellular glands of the body wall is used for stabilizing sediment. Tanaidomorphans are exclusively tube-building forms. To construct their tubes within the sediment or between algal mats or filaments, they use spinning silk produced by glands having their outlet on the dactylar tip of the first to third peraeopod. During construction, feces and detritus may be added to the silk. The fine texture of these tubes constructed by all members of a population reduce erosion which leads to a more stable sandy bottoms (Richards, 1969).

Very little is known on habitat preference. Within the suborder Tanaidomorpha there are some indications that several families occur in more or less regularly distinct habitats. Tanaidae are common in algal mats. If there is also plenty of sediment between these algal filaments, members of the family Paratanaidae are also found in this habitat although they are more typical of sandy bottoms. If the sand is less coarse or if the percentage of mud increases, members of the Leptognathiidae and Pseudotanaidae would replace members of these other families.

Methods of Collection and Extraction

Most all standard meiobenthic collecting methods are suitable for tanaidaceans. Elutriation-sieving methods are satisfactory techniques for the removal of specimens from sediment. To some extent, the bubbling technique (see Chapter 9) may often be effective for living material. Specimens should be sorted using a dissecting microscope and fine forceps, needles, or Irwin Loops.

Preparation of Specimens for Taxonomic Study

Specimens should be fixed in 10% formalin but later transferred to 70% ethanol. When ready for further processing, specimens should be transferred to alcohol containing 5% glycerin and the alcohol allowed to evaporate slowly, thus leaving the animals in glycerin which makes their dissection easier. The smaller specimens must be dissected with extremely fine needles. The dissected parts should be transferred to specially prepared slides and mounted for further examination (Sieg, 1973).

For description of a taxon, detailed and very careful drawings of every appendage have to be done. Even in small species an exact knowledge of all mouthparts is required. Mainly within the suborder Tanaidomorpha, the complete setation also has to be drawn for insuring correct classification. A special terminology has been developed for describing the exact placement of setae and spines on appendages (Lang, 1968, Sieg, 1977, 1980a).

Type Collections

The most extensive tanaidacean collections are located at the British Museum (Natural History), London, U.K.; National Museum of Natural History, Smithsonian Institution, Washington, D.C., U.S.A; Swedish Museum of Natural History, Stockholm, Sweden; Musée d'Histoire naturelle "Grigore Antipa," Bucaresti, Rumania; and Universitets Zoologisk Museum, Copenhagen, Denmark.

References

Andersson, A., E. Hallberg, S.-B. Johnson
1978. The Fine Structure of the Compound Eye of *Tanais cavolinii* Milne-Edwards (Crustacea: Tanaidacea). *Acta Zoologica (Stockholm)*, 59(1):49-55.

Belyaev, G.M.
1966. *The Hadal Bottom Fauna of the World Ocean.* 199

pages. Academy of Sciences of the USSR, Institute of Oceanology, Moscow. [Israel Program of Scientific Translation, Jerusalem, 1972].

Bückle-Ramirez, L.F.
1965. Untersuchungen über die Biologie von *Heterotanais oerstedi* (Kroyer). *Zeitschrift für Morphologie und Ökologie der Tiere*, 55:711-782.

Claus, C.
1884. I. Über *Apseudes latreillii* M.-Edw. und die Tanaiden. *Arbeiten aus dem Zoologischen Institut der Universität Wien*, 5:319-332.
1888. II. Über *Apseudes latreillii* M.-Edw. und die Tanaiden. *Arbeiten aus dem Zoologischen Institut der Universität Wien*, 7:139-220.

Cotelli, F., and C. Lora-Lamia-Donin
1980. The Spermatozoon of Peracarida. II. The Spermatozoon of Tanaidacea. *Journal of Ultrastructure Research*, 73(3):263-268.

Dennel, R.
1937. On the Feeding Mechanism of *Apseudes talpa*, and the Evolution of the Peracaridean Feeding Mechanism. *Transactions of the Royal Society of Edinburgh*, 59:57-78.

Dohle, W.
1972. Über die Bildung und Differenzierung des postnauplialen Keimstreifs von *Leptochelia* sp. (Crustacea, Tanaidacea). *Zoologische Jahrbücher (Anatomie)*, 89:503-566.

Feller, R.J.
1978. Predation on Meiofauna Established with Immunological Methods (Abstract). *American Zoologist*, 18(3):662.

Gardiner, L.F.
1975. The Systematics Postmarsupial Development, and Ecology of the Deep-Sea Family Neotanaidae (Crustacea: Tanaidacea). *Smithsonian Contributions to Zoology*, 170:1-264.

Gutu, M.
1972. Phylogenetic and Systematic Considerations upon the Monokonophora (Crustacea, Tanaidacea) with the Suggestion of a New Family and Several New Subfamilies. *Revue Roumaine de Biologie (Zoologie)*, 17:297-305.
1982. A New Contribution to the Systematics and Phylogeny of the Suborder Monokonophora (Crustacea, Tanaidacea). *Travaux du Museum d'Histoire naturelle "Grigore Antipa,"* 23:81-108.

Hansen, H.J.
1913. Crustacea Malacostraca II. *Danish Ingolf-Expedition*, 3(3):1-145.

Hessler, R.R.
1982. The Structural Morphology and Walking Mechanisms in the Eumalacostracan Crustaceans. *Philosophical Transactions of the Royal Society of London*, 296(B):245-298.

Highsmith, R.C.
1982. Induced Settlement and Metamorphosis of Sand-Dollar (*Dendraster exentricus*) Larvae in Predator-free Sites: Adult Sand-Dollar Beds. *Ecology*, 63(2):329-337.
1983. Sex Reversal and Fighting Behaviour: Coevolved Phenomena in a Tanaid Crustacean. *Ecology*, 64(4):719-726.

Holdich, D.M., and J.A. Jones
1983a. Tanaids. Keys and Notes for the Identification of the Species. *Synopsis of the British Fauna (New Series)*, 27:1-98.
1983b. The Distribution and Ecology of British Shallow-water Tanaid Crustaceans (Peracarida, Tanaidacea). *Journal of Natural History*, 17:157-183.

Johnson, S.B.
1982. Functional Models, Life History, and Evolution of Tube-dwelling Tanaidacea (Crustacea). 113 pages. Doctoral Thesis, University of Lund.

Kaestner, A.
1970. *Invertebrate Zoology. Crustacea.* (Translated and adapted by H.W. Levi and L.R. Levi). New York: Interscience Publishers.

Kudinova-Pasternak, R.K.
1966. Tanaidacea Crustacea of the Pacific Ultra-abyssals. *Zoologiske Zhurnal (Moskva)*, 45:518-535.
1970. Tanaidacea Kurilo-Kamciatkogo jeloba. *Trudy Instituta Okeanologii Moskva*, 86:341-380.
1973. Tanaidacea (Crustacea, Malacostraca) Collected on the R/V *Vitjaz* in Regions of the Aleutian Trench and Alaska. *Trudy Instituta Okeanologii Moskva*, 91:141-168.
1975. Tanaidacea (Crustacea, Malacostraca) from the Atlantic Sector of Antarctic and Subantarctic. *Trudy Instituta Okeanologii Moskva*, 103:194-228.
1977. Tanaidacea (Crustacea, Malacostraca) from the Deep-Sea Trenches of the Western Part of the Pacific. *Trudy Instituta Okeanologii Moskva*, 108:115-135.

Kudinova-Pasternak, R.K., and F.A. Pasternak
1978. Deep-Sea Tanaidacea (Crustacea, Malacostraca) Collected in the Caribbean Sea and Puerto-Rico Trench During the 16th Cruise of R/V *Akademik Kurchatov* and the Resemblance Between the Fauna of the Deep-Sea Tanaidacea of the Caribbean Region and the Pacific. *Trudy Instituta Okeanologii Moskva*, 113:178-197.
1981. Tanaidacea Collected by the Soviet Antarctic Expedition. *Trudy Instituta Okeanologii Moskva*, 115:108-125.

Lang, K.
1956a. Kalliapseudidae, A New Family of Tanaidacea. Pages 205-225 in K.G. Wingstrand, editor, *Bertil Hanström. Zoological Papers in Honour of His 65th Birthday*.
1956b. Neotanaidae nov. fam., with Some Remarks on the Phylogeny of the Tanaidacea. *Arkiv för Zoologie*, 9(2):469-475.
1968. Deep-Sea Tanaidacea. *Galathea Report*, 9:23-209.
1970. Taxonomische und phylogenetische Untersuchungen über die Tanaidaceen. 4. Aufteilung der Apseudidae in vier Familien nebst Aufstellung von zwei Gattungen und einer neuen Art der Familie Leiopidae. *Arkiv för Zoologie*, 22(2):596-626.
1973. Taxonomische und phylogenetische Untersuchungen über die Tanaidaceen. 8. Die Gattungen *Leptochelia* Dana, *Paratanais* Dana, *Heterotanais* G.O. Sars und *Nototanais* Richardson. Dazu einige Bemerkungen über die Monokonophora und ein Nachtrag. *Zoologica Scripta*, 2:197-229.

Lauterbach, K.E.
1970. Der Cephalothorax von *Tanais cavolinii* Milne-Edwards. Ein Beitrag zur vergleichenden Anatomie und Phylogenie der Tanaidacea. *Zoologische Jahrbücher (Anatomie)*, 87:94-204.

Messing, C.G.
1979. *Pagurapseudes* (Crustacea: Tanaidacea) in Southeastern Florida: Functional Morphology, Postmarsupial Development, Ecology, and Shell Use. 242 pages. Doctoral Thesis, University of Miami.
1983. Postmarsupial Development and Growth of *Pagurapseudes largoensis* McSweeny (Crustacea, Tanaidacea). *Journal of Crustacean Biology*, 3(3):380-408.

Nierstrasz, H.F.
1913. Die Isopoden der Siboga-Expedition. I. Isopoda Chelifera. *Siboga Expedition*, 32(A):1-56.

Richards, L.
1969. Tanaidacea (Crustacea: Peracarida) of the San Juan Island. *Friday Harbor Laboratories Zoological Reports*, 533:1-18.

Richardson, H.
1905. A Monograph on the Isopods of North America. *Bulletin of the U.S. National Museum*, 54:1-727.

Sars, G.O.
1882. Rivision af Gruppen: Isopoda Chelifera. *Archiv for Mathematik og Naturvidenskab*, 7:1-54.
1886. Nye bidrag til kundskaben om middel havetsinvertebrat fauna. III. Middelhavets saxipoder

(Isopoda Chelifera). *Archiv for Mathemathik og Naturvidenskab*, 11:263-368.
1896. Isopoda. *An Account of the Crustacea of Norway*, 2:1-270.

Scholl, G.
1963. Embryologische Untersuchungen an Tanaidaceen (*Heterotanais oerstedi* Kroeyer). *Zoologische Jahrbücher (Anatomie)*, 80:500-554.

Sieg, J.
1973. Zum Problem der Herstellung von Dauerpräparaten von Klein-Crustaceen, insbesondere von Typus-exemplaren. *Crustaceana (Leiden)*, 25(2):222-224.
1977. Taxonomische Monographie der Familie Pseudotanaidae (Crustacea: Tanaidacea). *Mitteilungen aus dem Zoologischen Museum Berlin*, 53:3-109.
1978. Bermerkungen zur Möglichkeit der Bestimmung der Weibchen bei den Dikonophora und der Entwicklung der Tanaidaceen. *Zoologischer Anzeiger*, 200:233-241.
1980a. Taxonomische Monographie der Tanaidae Dana, 1849 (Crustacea: Tanaidacea). *Abhandlungen der senckenbergischen naturforschenden Gesellschaft*, 537:1-267.
1980b. Sind die Dikonophora eine polyphyletische Gruppe? *Zoologischer Anzeiger*, 205(5-6):401-416.
1982a. Katalog der Tanaidaceen-Literatur (1808-1982). (2. Auglage). *Tanaidacea News*, 14:1-65.
1982b. *Alphabetische Liste der Tanaidacea*. 35 pages. Vechta: Universität Osnabrück Abteilung Vechta.
1983a. Tanaidacea. Pages 1-552 in H.E. Gruner, and L.B. Holthuis, editors, *Crustaceorum Catalogus*. Part 6. The Hague: W. Junk Publishers.
1983b. Evolution of Tanaidacea. Pages 229-256 in F.R. Schram, editor, *Crustacean Phylogeny*.
1984. Neuere Erkenntnisse zum natürlichen System der Tanaidacea. Eine phylogenetische Studie. *Zoologica (Stuttgart)*, 136:1-132.

Sieg, J., and R.N. Winn
1978. Keys to Suborders and Families of Tanaidacea (Crustacea). *Proceedings of the Biological Society of Washington*, 91(4):840-846.
1981. The Tanaidea (Crustacea: Tanaidacea) of California with a Key to the World Genera. *Proceedings of the Biological Society of Washington*, 94(2):315-343.

Siewing, R.
1953. Morphologische Untersuchungen an Tanaidaceen und Lophogastriden. *Zeitschrift für wissenschaftliche Zoologie*, 157:333-426.

Wolff, T.
1956. Crustacea Tanaidacea From Depths Exceeding 6000 Meters. *Galathea Report*, 2:187-214.

Zimmer, C.
1927. Tanaidacea in Kükenthal and Krumbach, *Handbuch der Zoologie*, 3(l):683-696.

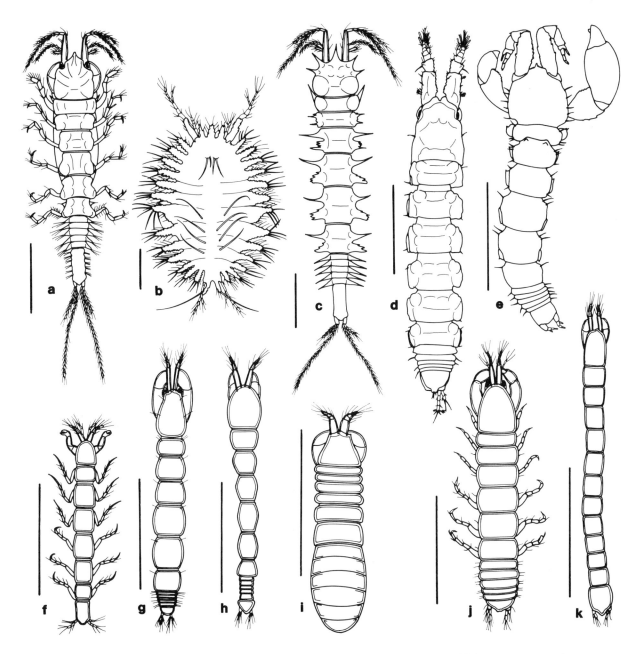

Figure 41.1.--Tanaidacea: a, *Apseudes spinosus* (M. Sars) (after G.O. Sars, 1896); b, *Tansanapseudes longiseta* Bacescu (after Bacescu, 1975); c, *Apseudes setosus* Lang (after Lang, 1968); d, *Synapseudes idios* Gardiner (after Gardiner, 1973); e, *Pagurapseudes largoensis* McSweeny (after McSweeny, 1982); f, *Anarthrura simplex* G.O. Sars (after G.O. Sars, 1896); g, *Pseudozeuxo belizensis* Sieg (after Sieg, 1982); h, *Agathotanais hanseni* Lang (after Lang, 1971); i, *Mirandotanais vorax* Kussakin and Tzareva (after Sieg, 1984); j, *Heterotanais oerstedi* (Kroyer) (after G.O. Sars, 1896); k, *Filitanais rebainsi* Kudinova-Pasternak (after Kudinova-Pasternak, 1975). (Scale = 1 mm.)

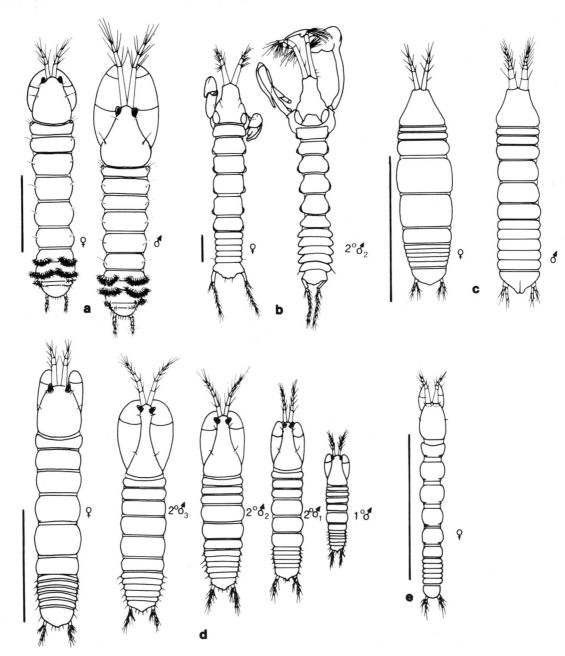

Figure 41.2.—Tanaidacea (continued): Sexual-dimorphism in body shape. **a**, "primary" gonochoristic type, superfamily Apseudoidea, Family Tanaidae; **b**, hermaphroditic type with 2 male types, family Neotanaidae; **c**, hermaphroditic type with 4 male types, family Leptocheliidae; **d**, "secondary" gonochoristic type, family Leptognathiidae, Pseudotanaidae, and others; **e**, parthenogenetic (?) type, family Leptognathiidae. (Scale = 1 mm.)

42. Amphipoda

Les Watling

Amphipods can be distinguished from all other eumalacostracan crustaceans by their possession of the following set of features: second antenna uniramous; eyes sessile; maxilliped without epipod; carapace absent; pereopods uniramous; 3 pairs of pleopods and 3 pairs of uropods on abdomen. In general, the presence of a laterally compressed body has also been used to diagnose the group; however, many meiofaunal-sized amphipods have bodies that are circular in cross-section.

The general form of a meiofaunal amphipod can be seen in Figure 42.1. The seven species shown represent much of the morphological diversity to be found in the group. Using these species as examples, the variability in amphipod morphology seen in meiofaunal forms will be described, beginning with the features on the head and progressing posteriorly.

The amphipod body is generally strongly laterally flattened, but in many smaller species may be cylindrical in cross-section. In the laterally compressed forms, the pereopodal coxae are expanded into plates, enhancing the flattened nature of the body. In the cylindrical forms the coxae are usually short. Some families, for example, the stenothoids and nihotungids, have one or more coxae that are greatly enlarged, covering several of the pereopods well beyond their bases.

The antennules are often short and robust. The transition from the three peduncular articles to the flagellar articles is usually not marked. There are usually few flagellar articles. An accessory flagellum is generally absent in these small forms. The antennae are often similar in length and morphology to the antennules. Aesthetascs (large sensory setae) are occasionally present on the antennules, but calceoli, which may be found on both antennal pairs, are generally not seen in meiofaunal amphipods.

The mouthparts, consisting of the upper lip, mandibles, lower lip, first and second maxillae, and maxillipeds, are located ventrally on the head in a buccal mass. While each of these appendages have their own distinguishing features, they are all modified together in response to the food resource of the species. Of all the mouthparts, the mandible and maxilliped show the greatest diversity of form, but it appears that the construction of the mandible determines the shape of the other buccal appendages. For example, in the genus *Colomastix*, all mouthpart appendages are reduced to vestiges of those seen in larger amphipods. In the genus *Nihotunga*, with the elongate, "styliform," development of the mandible, all mouthpart appendages are elongate, resulting in what is often termed a "conical" mouthpart bundle. The maxilliped may be narrow or laterally expanded, depending on its role in keeping food in the vicinity of the other mouthparts.

The first two pairs of pereopods are almost always modified and are termed gnathopods 1 and 2. The form of the gnathopods ranges from simple (where the leg looks like those immediately following), i.e., in *Wandelia*, to subchelate (where the dactyl is recurved along the propodus), i.e., in the male of *Colomastix*, to chelate (where the dactyl and a distal extension of the propodus form a claw), i.e., in *Seba*. The gnathopods, along with the following two pairs of pereopods are directed forwards. In pereopods 3 and 4 the dactyls are directed posteriorly.

Pereopods 5-7 are generally alike, directed posteriorly, with their dactyls pointing anteriad. The bases of these pereopods may be narrow or expanded, usually in response to the need for directing water currents generated by the pleopods (Dahl, 1977).

The abdomen is subdivided into pleosome (bearing the pleopods) and urosome (bearing the uropods). The pleopods are the primary impellers of water flow past the gills, their beating either being continuous or intermittent, depending on the rate of oxygen consumption from the water. Clearly, if the pleopods are much reduced, the animal must be living in a habitat with strong water movement. The first two pairs of uropods are usually alike and are biramous. The third uropods, however, may differ significantly from the first two pairs. They are often much shortened and are commonly uniramous. The telson shows few variations in meiofaunal amphipods. It is generally short and without any subdivisions (referred to as "entire").

Classification

The Amphipoda have generally been considered an order within the Superorder Peracarida. Recently, however, with the renewed debate on phylogenetic relationships within the Eumalacostraca, most investigators have concluded that Amphipoda stand apart from the other peracaridan orders (Hessler, 1983; Watling, 1981, 1983; Schram, 1986), possibly meriting the rank of Superorder as has been formally proposed by Watling (1983). Within the Superorder there are four Orders: Hyperiidea, Caprellidea, Ingolfiellidea, and Gammaridea. The latter two orders contain all known meiofaunal amphipods.

The ingolfiellids show the extreme adaptations towards the "Lebensformtyp Mesopsammon" (Remane, 1952), viz., an elongated, cylindrical body, loss of pigmentation and eyes, reduction of appendages (especially at the posterior of the body), and small body size (Stock, 1976). The Order is currently divided into two families, the Metaingolfiellidae (containing a single genus, *Metaingolfiella)* and the Ingolfiellidae (containing two genera, *Ingolfiella* and *Trogloleupia).* In the Metaingolfiellidae the somite bearing the first gnathopod is fused with the head, the second gnathopod is chelate, and the pleopods are biramous, whereas in the Ingolfiellidae the somite bearing the first gnathopod is not fused to the head, the second gnathopod is non-chelate, and the pleopods are rudimentary and uniramous.

Gammaridean amphipods are very diverse in body form and size, ranging from the very small cylindrical meiofaunal species to the extremely large (>20 cm) abyssal lysianassid species. The order currently has been divided into as many as 91 families, although debate on this issue may result in many fewer than that (Barnard and Barnard, 1983). Most of the families under dispute are from Eurasian freshwater and differ in small details. The marine gammaridean genera were admirably summarized by Barnard (1969) who also outlined the major features to be noted in determining the identity of species. Of primary importance are characters such as: whether an accessory flagellum is present on the first antenna; the morphology of the mandible, especially concerning details of the molar and incisor processes; the size of coxal plates 1-4; characteristics of the maxilliped; the relative sizes, shape and form of the gnathopods; shape of the basis of pereopods 5-7; size and form of uropod 3. There are many other details to be noted in the proper determination of any species. These are very nicely laid out in the appendix to Barnard's (1969) monograph.

Habitats and Ecological Notes

Meiofaunal amphipods can be found in three very different habitats: interstitial in sands, at the bases of marine algae, and in hypogean freshwater. While the meiofaunal-sized ingolfiellids are found almost exclusively in sands (in groundwater, in caves, and on coral flats, as well as in the deep-sea) (Stock, 1976), very few of the smallest gammarideans are found in the psammic habitats.

Two features of the distribution of meiofaunal gammarideans stand out. First, most (approximately 70%) have been found in the sediment and shell rubble that accumulates at the bases of algal clumps, both intertidally and subtidally. In its physical and chemical attributes this is essentially a psammic habitat, but it is rarely investigated. However, it differs from the typical sandy beach and subtidal sand in the amount of organic matter trapped in the sediments and perhaps in the degree of water flushing through the interstitial spaces. Nearly all of the records from the algal habitat are from the southern hemisphere. It is not clear whether this is simply a result of sampling emphasis. In any case, no meiofaunal-sized gammarideans have been found in the algae along the northeast coast of North America even though many larger amphipods are found in the same habitat (personal observation) and very few are found along the extensively sampled European coasts. Our knowledge of species from the algal habitat derives primarily from the work of Barnard (1972, 1974).

Second, many of the remaining meiofaunal gammarideans dwell in hypogean freshwater. These are primarily representatives of the Family Bogidiellidae and have been found living in environments ranging from cold mountain springs to warm mineral springs to infralittoral, marine conditions (Stock, 1981). Most of these specimens were taken from drilled or dug wells, or by using various types of pumps designed for extracting water from a meter or so beneath the surface of sand or gravel beds.

Methods of Extraction

When dealing with sandy habitats, common methods of extraction can be used. However, for the algal habitat, the best procedure is to sample the algae and its substrate together and immerse both in 5% formalin. Intertidally, this is best accomplished by scraping the surface of the rock on which the alga is growing and retaining on the scraper as much of the debris at the base of the alga as possible. Subtidally, the best approach is to envelop the alga and as much of the substratum as possible in a plastic bag before any scraping or movement of the substratum is attempted. Any organisms that flee the disturbance are then caught in the bag. After removal to the surface of the water, enough formalin is added to make a 5% solution. This is left for an hour or so (even overnight is all right), then the contents are washed over a fine mesh sieve. During the washing process, large pieces of the algae are discarded. In this way, large algal samples can be concentrated into relatively small vials (2-4 drams).

Preparation for Taxonomic Study

Amphipoda can be identified to family level, with experience, using a dissection microscope. However, correct determination of specific identity will almost always require the dissection of the specimen and the use of a compound microscope. Dissection can be carried out on a slide with the specimen in glycerin or a mixture of glycerin and alcohol. The appendages can then be teased off the body and spread out. In my own experience, it is easiest to begin by removing the head from the body and transferring it to a separate slide. The mouthparts can then be gradually teased off and arranged for good viewing on that slide. Returning to the body, only the appendages from one side of the body need to be removed. The others should be left in place so that any relative lengths or features of overlap can be determined.

References

Barnard, J.L.
1969. The Families and Genera of Marine Gammaridean Amphipoda. *Bulletin of the United States National Museum*, 271:1-535.
1970. Sublittoral Gammaridea (Amphipoda) of the Hawaiian Islands. *Smithsonian Contributions to Zoology*, 37:1-286.
1972. The Marine Fauna of New Zealand: Algae-living Littoral Gammaridea (Crustacea Amphipoda). *New Zealand Oceanographic Institute Memoir*, 62:1-216.
1974. Gammaridean Amphipoda of Australia, Part II. *Smithsonian Contributions to Zoology*, 139:1-148.

Barnard, J.L., and C.M. Barnard
1983. *Freshwater Amphipoda of the World*. Mt. Vernon, Virginia: Hayfield Associates.

Dahl, E.
1977. The Amphipod Functional Model and Its Bearing upon Systematics and Phylogeny. *Zoologica Scripta*, 6:163-166.

Hessler, R.R.
1983. A Defense of the Caridoid Facies: Wherein the Early Evolution of the Eumalacostraca is Discussed. Pages 145-164 in F.R. Schram, editor, *Crustacean Phylogeny*. Rotterdam: A.A. Balkema.

Lincoln, R.J.
1979. *British Marine Amphipoda: Gammaridea*. London: British Museum (Natural History).

Noodt, W.
1965. Interstitielle Amphipoden der konvergenten Gattungen *Ingolfiella* Hansen und *Pseudingolfiella n. gen.* aus Südamerika. *Crustaceana*, 9:17-30.

Remane, A.
1952. Die Besiedlung des Sandbodens im Meere und die Bedeutung der Lebensformtypen für die Ökologie. Verhandlungen *der Deutschen Zoologischen Gesellschaft, Wilhelmshaven*, 1951:327-359.

Schram, F.R.
1986. *Crustacea*. 606 pages. New York: Oxford University Press.

Stock, J.
1976. A New Member of the Crustacean Suborder Ingolfiellidea from Bonaire, with a Review of the Entire Suborder. *Studies of the Fauna of Curaçao and other Caribbean Islands*, 50:56-75.
1981. The Taxonomy and Zoogeography of the Family Bogidiellidae (Crustacea, Amphipoda), with Emphasis on the West Indian taxa. *Bijdragen Dierkunde*, 51:345-374.

Watling, L.
1981. An Alternative Phylogeny of Peracarid Crustaceans. *Journal of Crustacean Biology*, 1:201-210.
1983. Peracaridan Disunity and Its Bearing on Eumalacostracan Phylogeny with a Redefinition of Eumalacostracan Superorders. Pages 213-228 in F.R. Schram, editor, *Crustacean Phylogeny*. Rotterdam: A.A. Balkema.

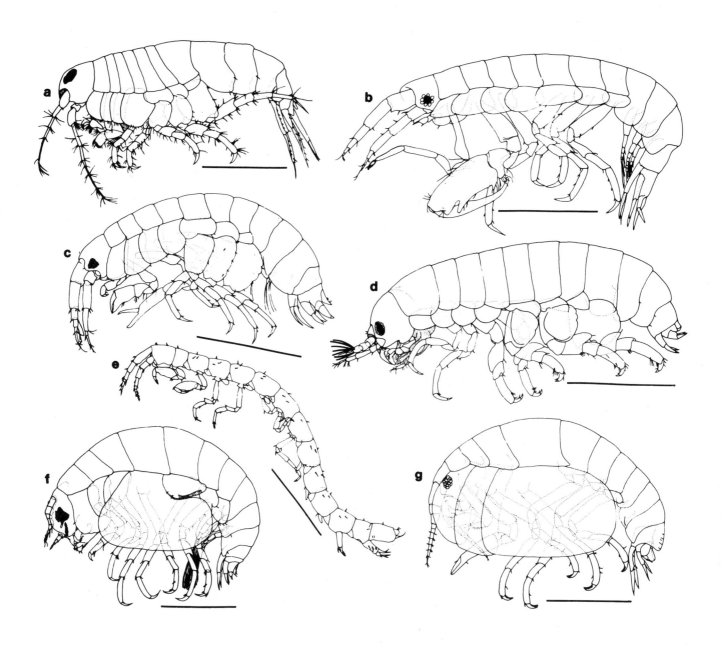

Figure 42.1.--Amphipoda: a, Oedicerotidae, *Kanaloa manoa*, J.L. Barnard, 1970; **b**, Colomastigidae, *Colomastix kapiolani* J.L. Barnard, 1970; **c**, Sebidae, *Seba ekepuu* J.L. Barnard, 1970; **d**, Eophliantidae, *Wandelia wairarapa* J.L. Barnard, 1972; **e**, Ingolfiellidea, *Ingolfiella putealis* Stock, 1976; **f**, Nihotungidae, *Nihotunga noa* J.L. Barnard, 1972; **g**, Stenothoidae, *Raumahara dertoo* J.L. Barnard, 1972. (Scale = 500 μm.)

43. Cumacea

Les Watling

While more or less shrimp-like in form, cumaceans are very distinct members of the eumalacostracan crustaceans. Their principal distinguishing features include: a carapace that covers at least three thoracic somites in forming a branchial cavity; uniramous antennae, which are generally minute in females; mandible without a palp; three anterior pairs of thoracic legs modified as maxillipeds while the remaining five pairs function as walking legs (pereopods); with one exception, pleopods are absent from the abdominal somites of females, while in males they may also be absent or as many as five pairs may be present; a free telson may be present or absent.

The general form of meiofaunal-sized cumaceans is shown in Figure 43.1. The males and females of the species illustrated represent the predominant morphologies seen.

The cumacean body consists of two distinct parts, the cephalothorax and the abdomen. The cephalothorax is composed of the head and anterior thoracic somites covered by the carapace, and the remaining thoracic somites. The abdomen consists of six somites that are always cylindrical in cross-section, and usually it is of a smaller diameter than the cephalothorax. The last abdominal somite may bear the anus posteriorly or a telson may be present with the anus located on its ventral surface.

The carapace is usually extended antero-dorsally into a short pseudorostrum, beneath which lies the siphon. The latter is an anterior extension of the first maxilliped and serves to direct the exhalant branchial current. In smaller forms the carapace is usually smooth and unarmed, but species living in carbonate sediments often have a robust and strongly ridged carapace, while those living in the deep-sea may have the carapace beset with long, delicate spines. An eye-lobe is usually present, but it may not always bear eyes.

The antennules are generally biramous, with both rami reduced in length. In contrast, the antennae are nearly completely reduced in females, but in males may be very long. In some cumaceans, especially those found in dynamic environments, the male antennae may be short and recurved and are clearly used for clinging to the dorsum of the female during amplexus. The remaining head appendages, the upper lip, mandibles, lower lip, maxillules and maxillae, are all hidden from view by the carapace and the three pairs of maxillipeds. With the exception of the mandibles and maxillae, the head appendages play only a minor role in the classification of the species in this group.

The first maxillipeds, which are modified first thoracic appendages, are one of the most complex of cumacean appendages. The coxa, or first leg segment, bears a complex branchial apparatus, part of which is directed backwards and may be highly convoluted into lobe-like gills, and part of which is directed forwards to form the branchial siphon. The remainder of the leg is reduced and functions to help direct food particles to the mouth.

The second maxilliped is much simpler in structure, consisting of a slightly elongated leg which, in females, also bears small brood plates. The third maxilliped is much more leg-like in structure, and in fact, often has the form of the first pereopod (walking leg). An exopod is often present on the third maxilliped and the first pereopod, but may be absent from the remaining legs. The remaining pereopods are much alike, becoming shorter in length posteriorly.

Abdominal appendages consist of pleopods, when present, and the uropods. With the singular exception of a deep-sea species, pleopods are never found in females. The number of pleopod pairs occuring on the male abdomen varies from zero to five, and is usually consistent within a family or subfamily. The uropods are always biramous and give cumaceans the impression of having a forked tail.

Classification

Cumaceans have generally been considered as an order within the Superorder Peracarida. Recently, however, Schram (1981) proposed the Superorder Brachycarida for eumalacostracan crustaceans bearing a short carapace used primarily as a branchial structure. Other orders within this group are the Spelaeogriphacea, Thermosbaenacea, and Tanaidacea (Watling, 1983). This arrangement has engendered considerable debate (e.g., Hessler, 1983), however, and is not yet generally accepted.

Within the Order, features of importance in

determining the proper placement of a species include: presence or absence of a free telson; shape of the mandible; number of pereopods with exopods; number of pleopods in males; structure of the branchial apparatus; and features of the maxilla. There are currently 8 families recognized, with one, the Bodotriidae, subdivided into three subfamilies (a fourth is currently under consideration, but has not yet been formally proposed). The Bodotriidae, Nannastacidae, and Leuconidae do not have free telsons, whereas all the others do. To date, only the families Bodotriidae, Nannastacidae, and Gynodiastylidae have been found to have meiofaunal members. Representatives of these families are shown in Figure 43.1.

Habitats and Ecological Notes

Cumaceans are dwellers of the soft benthos, including sediment that may be trapped in tropical algal turf. It is probable that most species inhabit sediments whose grains lie within a narrow size range. Off the Ivory Coast, Leloeuff and Intes (1972) noted that only 2 of 19 species were found to have wide grain size tolerances. Similar preferences were shown for species along the British coast (Pike and LeSueur, 1958). In the Delta area of the Netherlands, Vader and Wolff (1973) found that only the more stenohaline marine species were limited by sediment type, the truly estuarine species being able to deal with a wider range of grain size.

Nearly all cumaceans inhabit truly marine waters, however, there are brackish water representatives in most families. The latter may be found in the estuaries and inland seas of most continents and are probably independently derived. In the seas, cumaceans have been taken at all depths, but most occur on the continental shelves and upper slopes. Jones (1969) noted that in some genera, notably *Makrokylindrus*, there was an increase in body size with water depth. In general, there is not a pronounced miniaturization of cumaceans with depth in the ocean. All meiofaunal cumaceans have thus far been taken in very shallow coastal waters.

Cumaceans live buried in the sediment, usually with only the anterior part of the cephalothorax exposed (Foxon, 1936). A respiratory current is taken in under the mouthparts and is exhaled through the siphon. In the filter-feeding genus, *Gephyrocuma*, the first pereopods, which bear long filtratory setae distally, are also exposed above the bottom (Hale, 1943). The last three pairs of pereopods are used to dig the body into the bottom after the animal has settled from the water column.

In most shallow water environments, cumaceans can be taken in plankton tows at night. Anger and Valentin (1976) found that the maximum intensity of pelagic activity of *Diastylis rathkei* occurred from about 2200 to 2300 hrs in the Kiel Bight. There was a strong attraction to light after the animal left the bottom, which would explain the ease with which others have taken cumaceans with night lights (e.g., Hale, 1953). Many explanations have been given for this tendency of cumaceans to leave the bottom sediments at night, but Anger and Valentin (1976) state that *D. rathkei* migrated into the water solely for the purpose of molting. It is also likely that, at certain times of the year, mating follows molting in the water column. Meiofaunal cumaceans, particularly in tropical habitats, also seem to undergo nightly migrations although it is not possible at present to know what proportion of the population is involved on any given night.

Cumaceans are believed to be primarily deposit-feeders. In sands finer than 150 μm, *Cumella vulgaris* fed by ingesting deposited materials, but in coarser sediments it fed as an epistrate feeder (Wieser, 1956). In the latter mode, grains would be grasped and manipulated by the mouthparts in such a way that material could be scraped from the surface of the grain. This feeding mode has also been observed in *Cumopsis* (Dixon, 1944). Dennell (1934) suggested that *Diastylis bradyi* mouthparts showed features typical of a filter-feeding malacostracan, presumably removing particles from a current set up by the movements of the maxillae and maxillipeds. However, cumaceans living in finer sediments have been observed to have a series of plumose setae on the maxillipeds that acted as sieves to keep the gills from becoming fouled by fine particles (Dixon, 1944). It may be that this was, in fact, what Dennell was observing rather than a strict feeding current. Some cumaceans, such as *Gephyrocuma*, show a more likely model of the morphology of a filter-feeding cumacean. Some cumaceans, for example *Campylaspis*, have styliform mandibular molars and are thought to be predators, perhaps on foraminiferans and small crustaceans.

Methods of Collection and Extraction

Standard methods of coring and dredging are effective in the collection of cumaceans. In particular, dredges that concentrate the uppermost levels of the sediment are best.

Cumaceans can be easily removed from sediment by elutriation, formalin washing, or by using freshwater to drive living animals from the substratum. Its also possible to collect large numbers of cumaceans by shining a flashlight on the water at night and moving a small net around the periphery of the cone of light. Fixation is best accomplished in 5% formalin which should be followed, for long term preservation, by 70% alcohol.

Preparation of Specimens for Taxonomic Study

Since there are only eight families of cumaceans, identification to family level can be accomplished relatively readily. Details of some appendages will sometimes need to be checked using a compound microscope. Dissection is best accomplished in a small dish of alcohol with or without glycerin. Usually only the walking legs and the uropods will have to be removed. Very small specimens can be mounted whole under a coverslip and most details will be observable. Specimens generally do not need to be stained.

References

Anger, K., and C. Valentin
1976. In Situ Studies on the Diurnal Activity Pattern of *Diastylis rathkei* (Cumacea, Crustacea) and its Importance for the "Hyperbenthos." *Helgoländer wissenschaftliches Meeresuntersuchungen*, 28:138-144.

Day, J.
1980. South African Cumacea, Part 4, Families Gynodiastylidae and Diastylidae. *Annals of the South African Museum*, 82:187-292

Dennell, R.
1934. The Feeding Mechanism of the Cumacean Crustacean *Diastylis bradyi*. *Transactions of the Royal Society of Edinburgh*, 58:125-142.

Dixon, A.Y.
1944. Notes on Certain Aspects of the Biology of *Cumopsis goodsiri* (van Beneden) and Some Other Cumaceans in Relation to their Environment. *Journal of the Marine Biological Association of the United Kingdom*, 26:61-71.

Foxon, G.E.H.
1936. Notes on the Natural History of Certain Sand-dwelling Cumacea. *Annals and Magazine of Natural History*, series 10, 17:377-393.

Hale, H.M.
1943. Notes on Two Sand-dwelling Cumacea (*Gephyrocuma* and *Picrocuma*). *Records of the South Australian Museum*, 7:337-342.

1953. Notes on Distribution and Night Collecting with Artificial Light. *Transactions of the Royal Society of South Australia*, 76:70-77.

Hessler, R.R.
1983. A Defense of the Caridoid Facies; Wherein the Early Evolution of the Eumalacostraca is Discussed. Pages 145-164 in F.R. Schram, editor, *Crustacean Phylogeny*. Rotterdam: A.A. Balkema.

Jones, N.S.
1969. The Systematics and Distribution of Cumacea from Depths Exceeding 200 Meters. *Galathea Reports*, 10:99-180.

1976. British Cumaceans, Arthropoda: Crustacea. Keys and Notes for the Identification of the Species. *Synopses of the British Fauna*, No. 7. London and New York: Academic Press.

1984. The Family Nannastacidae (Crustacea: Cumacea) from the Deep Atlantic. *Bulletin of the British Museum (Natural History) Zoology Series*, 46:207-289.

LeLoeuff, P., and A. Intes
1972. Les Cumacés du Plateau Continental de Côte d'Ivoire. *Cahier de O.R.S.T.O.M., Serie Oceanographique*, 10:19-46.

Pike, R.B., and R.F. LeSueur
1958. The Shore Zonation of Some Jersey Cumacea. *Annals and Magazine of Natural History*, series 13, 1:515-523.

Schram, F.R.
1981. On the Classification of the Eumalacostraca. *Journal of Crustacean Biology*, 1:1-10.

Vader, W., and W.J. Wolff
1973. The Cumacea of the Estuarine Area of the Rivers Rhine, Meuse and Scheldt (Crustacea, Malacostraca). *Netherlands Journal of Sea Research*, 6:365-375.

Watling, L.
1982. Cumacea. Pages 243-245 in S.B. Parker, editor, *Synopsis and Classification of Living Organisms*. New York: McGraw-Hill Book Company.

1983. Peracaridan Disunity and its Bearing on Eumalacostracan Phylogeny with a Redefinition of Eumalacostracan Superorders. Pages 213-228 in F.R. Schram, editor, *Crustacean Phylogeny*. Rotterdam: A.A. Balkema.

Wieser, W.
1956. Factors Influencing the Choice of Substratum in *Cumella vulgaris* Hart (Crustacea, Cumacea). *Limnology and Oceanography*, 1:274-285.

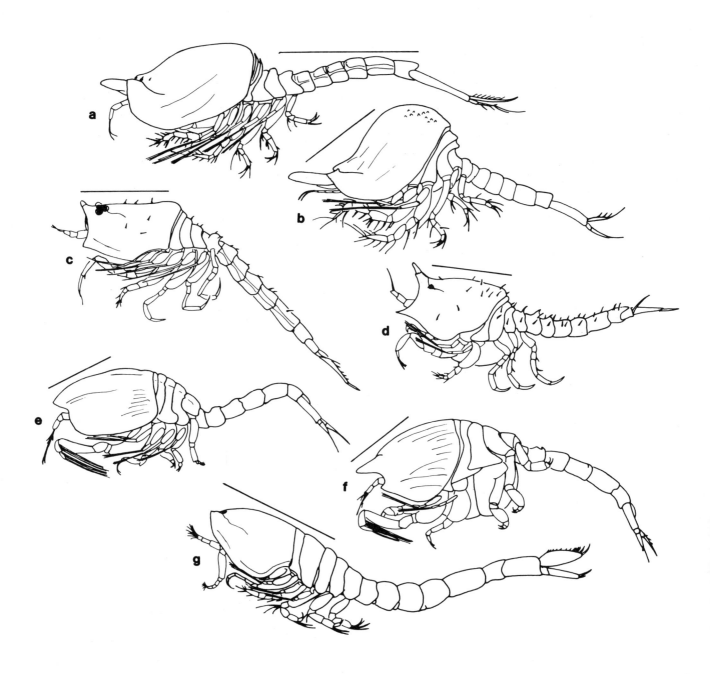

Figure 43.1.--Cumacea. a-d, Nannastacidae; a, *Campylaspis minor* Hale, male; b, female; c, *Nannastacus inflatus* Hale, male; d, female; e,f, Gynodiastylidae, *Gynodiastylis lata* Hale, male; f, female; g, Nannastacidae, *Picrocuma poecilata* Hale, male. (Scale = 500 μm.)

44. Halacaroidea

Ilse Bartsch

The Halacaroidea (suborder Prostigmata) are part of the marine, brackish and freshwater benthos. The halacarid body is divided into an anterior gnathosoma and a posterior idiosoma. The gnathosoma may be almost as long as the idiosoma, but usually it is smaller, sometimes concealed beneath the idiosoma. The gnathosoma includes the gnathosoma base, rostrum, chelicerae, and palps. The chelicerae are stout or stylet-like, the ventral digit usually is large and serrate, the dorsal digit reduced to a membranous flap. The palps have 2, 3, or 4 segments. They are either attached laterally or dorsally to the gnathosoma base. The idiosoma is 0.15-2.0 mm in length, it is rhombic or elliptical in outline, flattened in some genera, slender and almost subcylindrical in interstitial forms. The idiosoma is commonly covered with 4 sclerotized dorsal plates (anterodorsal, posterordorsal, right and left ocular plates) and 4 ventral plates (anterior epimeral, genitoanal, right and left posterior epimeral plates). Two or more of the dorsal, ventral, or both dorsal and ventral plates may be fused. On the other hand, the plates may be divided, the ocular and the dorsal plate may be minute or lacking.

Males and females are very similar in the general appearance, with marked differences in the genital region only. In the middle or in the posterior half of the genital plate there is the genital opening (genital foramen). In females, the genital foramen is surrounded by 2-15 pairs of perigenital setae; in males, four to almost 100 pairs of setae are present. Internal structures are: in females an ovipositor, in males a spermapositor, and in both sexes 3 pairs of genital acetabula. In the freshwater genera, internal genital acetabula are lacking, instead acetabula-like structures are found either on the genital sclerites or on the genital plate. The 4 pairs of legs are attached marginally or dorsomarginally. Legs I and II are directed forwards, legs III and IV are directed backwards. Leg IV is lacking in larvae. The legs of the adults are six segmented (trochanter, basifemur, telofemur, genu, tibia, tarsus). The tarsi terminate with one, two, or three claws. In phytophagous species, all legs are similar in outline, all are used when walking. In predatory genera, the first leg is longer and wider than the following legs, leg I is provided with spines or bristles, thus equipped to grasp and hold prey organisms. Taxonomic important characters are: outline and ornamentation of dorsal and ventral plates; outline of gnathosoma; length and outline of leg segments; cuticular ornamentations and lamellae on plates and legs; insertion of gland pores; number and insertion of setae.

Most halacarids are dioecious, probably all are oviparous. The life cycle runs from the eggs through 1 larval and 1-3 nymphal stages (proto-, deuto-, tritonymph). The larvae have 6 legs, with 5 segments only. The nymphs have 4 pairs of legs; leg IV of the protonymph is 5-segmented, all the other legs are 6-segmented. Though differing in outline of the plates and number of setae, the juveniles commonly can be attributed to the adults on base of insertion of setae and gland pores.

Classification

The superfamily Halacaroidea includes several subfamilies and genera. In all, almost 700 species are described, the majority lives in marine and brackish water habitats, 40 species are known from freshwater.

The Halacarinae (14 marine genera, 5 freshwater genera, more than 200 species) have the palps inserted lateral to the rostrum. The majority of the species are supposed to be carnivorous (predators, scavengers, parasites), though algivory and fungivory probably exist. Species of the genus *Halacarus* (50 species) often have a long frontal spine. Leg I usually is longer and stronger than the other legs, it is provided with slender spines; the genu on leg I is as long or longer than the telofemur or tibia. The palps are four segmented, the third segment (P-3) bears a spine. *Halacarellus* (=*Thalassarachna*) (65 species) includes a variety of forms. Leg I may be stouter than the other legs. The genu I is shorter than the telofemur or tibia. The palps are four segmented, P-3 bears a spine or bristle. *Agauopsis* (55 species) usually has a wide and flattened idiosoma. Leg I is stronger than the other legs. The gnathosoma is short, its base is quadrangular. The palps are four segmented, they hardly surpass the rostrum.

Within the genus *Agaue* (33 species), the majority of the species has elaborate cerotegumental lamellae on the idiosoma, often also on the legs. The rostrum and the 4-segmented palps are elongate. The legs never bear stout spines. The tarsi terminate with a very minute median claw and two large lateral claws. Species of *Bradyagaue* (10 species) are specialized for living on stolonaceous organisms, i.e., colonial hydrozoans. The tibiae and tarsi of the posterior legs are curved, the median claw is enlarged, thus adapted to climb and cling to stolons. The rostrum and the 4-segmented palps are elongate. The legs never bear stout spines.

Acarochelopodia (5 species) and *Anomalohalacarus* (14 species) are both exclusively arenicolous in habit. *Acarochelopodia* is characterized by the very large leg I with strong bipectinate ventral bristles on the tibia and the slender, finger-like tarsus. *Anomalohalacarus* has a spindle-shaped body, with the very slender legs inserting rather terminal. The dorsal and ventral plates are largely reduced, often divided. The dorsal and lateral setae on the idiosoma and the legs are long.

Copidognathus (Copidognathinae) (240 species) is the most successful genus, it contains more than one-third of all halacarid species described. *Copidognathus* is found in littoral and abyssal depths, in marine and limnic waters. Leg I is not markedly stronger than the other legs. The tibiae I and II each have 3 ventral bristles. The palps are 4-segmented; P-3 lacks a spine or bristle.

The Simognathinae (2 genera, 24 species) are heavily sclerotized forms with a short and conical gnathosoma and the palps attached dorsally. *Simognathus* and *Acaromantis* are separated by leg I. Tibia I is enlarged in *Acaromantis*, tarsus I is very short, not longer than high, and ends with one minute claw, whereas in *Simognathus* the tarsus I is distinctly longer than high and ends with one stout and two slender, scythe-shaped claws. The tibia and tarsus I form a subchela in *Simognathus*.

The Lohmannellinae (3 marine genera, 1 limnic genus, 40 species) have a wide, often flattened idiosoma. The palps are attached dorsal to the rostrum, i.e., the P-1 are separated by less than the width of P-1. *Lohmannella* (marine) and *Porolohmannella* (limnic) have a slender, often elongate rostrum and slender, 4-segmented palps. The palps and the rostrum form a pincer. *Scaptognathus* is characterized by its large gnathosoma, in some species it is almost as long as the idiosoma. The rostrum is spathula-shaped. The 2-segmented palps end with stout spines.

Soldanellonyx (Limnohalacarinae) is a limnic genus. Leg I may be larger than the other legs. The claws on tarsi I have umbrella-like arranged claw teeth.

The Rhombognathinae (4 genera, 70 species) are phytophagous, hence with green gut content. They are restricted to the photic zone. The gnathosoma is short, in *Isobactrus* species often hidden from dorsal view. The palps are 3- or 4-segmented, closely attached to the rostrum. Between the tip of the tarsi and the claws, there is a small sclerite (carpit) inserted.

Non-halacarid mites found in tidal marine habitats are the Oribatei, Hyadesiidae, Rhodacaridae, Uropodidae, Erythraeidae, and Pontarachnidae, in freshwater there are the "Hydrachnellae" and Oribatei. Species of the suborder Oribatei (beetle mites) are characterized by the usually dark-brown color due to the pigmentation of the cuticle, a commonly strongly sclerotized large dorsal shield and the position of the stigmata at the base of legs II and III. The Hyadesiidae (superfamily Acaroidea, suborder Astigmata) are distinguished by a commonly light brown color, bulbous idiosoma, and the tarsi I and II each ending with a long fleshy stalk and a small claw. The Rhodacaridae and Uropodidae (suborder Mesostigmata) have lateral stigmata, strongly sclerotized chelicerae and one or two dorsal shields. Most of the Rhodacaridae have an elongate idiosoma, yellow or whitish in color. The male chelicerae are modified for sperm transfer. The Uropodinae usually are flattened, yellow-brown in color; the male chelicerae are not modified. The Erythraeidae (superfamily Erythraeoidea, suborder Prostigmata) are soft-bodied, with a red or velvet color. The idiosoma is covered with a large number of minute bristles or setae. The Pontarachnidae ("Hydrachnella," suborder Prostigmata) are soft-bodied, the idiosoma is bulbous, the 4 legs insert adjacent. The Pontarachnidae live subtidally in warm water areas. The freshwater "Hydrachnellae" include several superfamilies (Hygrobatoidea, Hydrachnoidea, Hydryphantoidea, Hydrovolzioidea). Some of them are extremely soft-bodied, others sclerotized, most have a rather sac-like idiosoma.

Habitats and Ecological Notes

Halacarids are exclusively benthonic - planktonic forms are not known. Marine mites inhabit almost all parts of the oceans. Some species are adapted to survive in the spray zone, others to live in deep-sea trenches. The deepest record is from 6850 meters. Almost 50 species inhabit brackish or freshwaters. Halacarids live in almost all sorts of substrata, within tufty algae, on large fronds of algae, on colonies of hydrozoans and bryozoans, within coarse and fine sands, in rooted salt marsh sediment, and in the fine ooze in brackish water lagoons. They are usually lacking in silty or almost oxygen-free sediments. Halacarids are often found on large animals (crustaceans, gastropods) with a rough (hairy) surface or a rich epiflora and epifauna, and in gill chambers of crustaceans. Only a few true parasitic forms are known.

On algae in the upper tidal zone, halacarids often are the dominant meiofauna group, both in numbers and in biomass. The percentage of halacarids in numbers of meiofauna may surpass 90%. In other biotops, halacarids are of less importance (Bartsch, 1979a, 1982). In muddy sand most halacarids are found in the upper 0-3 cm, whereas in sandy beaches, halacarids penetrate to depths of 100 cm (Bartsch and Schmidt, 1978). Some species are known to invade along the ground water horizon into continental waters for meters or even kilometers (Petrova, 1972). In sorted, medium-grained sand, the halacarids usually make up less than 5% of the number of meiofauna specimens. In coarse and rather unsorted sediment and in brackish water sand flats, the percentage of halacarids may increase to 15% (Bartsch, 1979a, 1982; Bartsch and Schmidt, 1979).

Methods of Collection

In the intertidal zone, substrata (sediment, algae, animal colonies) are collected by hand at low tide; 100-500 cm^3 of the substratum may contain several hundred halacarids. Collections from deeper water can be made with grabs, trawls, and dredges. A meiobenthic dredge will give good results for inhabitants of sediments; halacarids adapted to living on larger animals or animal colonies will be found in collections taken with grabs or trawls.

Methods of Extraction

Halacarids, especially intertidally living species, are very resistant to unfavorable conditions; shock techniques with freshwater, anaesthetizing media, ethanol, etc., will only slightly affect them. The ice water technique is not very efficient. All sorts of mechanical disturbing (stirring, bubbling, strong jet of water) can be used to extract halacarids. Sediment samples are stirred vigorously in water, then the supernatant water is poured through a 63 µm mesh sieve. A minority of species has a hydrophobic cuticle, these mites are easily trapped at the water surface film. A substratum like algae or animal colonies is placed in a series of sieves, the upper one with a mesh size of 1-2 mm, the lower one with 63 µm, and washed with a strong jet of water. The halacarids are pushed through the upper and retained on the lower sieve.

Rose Bengal staining can be used when sorting preserved sediment samples.

Preparation of Specimens for Taxonomic Study

Ethanol (70%) should be used for fixation and storing. For closer examination the mites have to be cleared and mounted on slides. The mite is placed into a clearing medium - pepsin or warm lactic acid (50° C), the gnathosoma is torn off with help of a sharp needle. After some minutes (in older material after some hours) the gut and associated tissues are squashed out with help of slight pressure of a blunt needle. Specimens which have been in formalin cannot be treated with pepsin. Material preserved and stored in formalin for long periods (months or years) is more difficult to clear. If transferred into a medium containing ca. 2% glacial acetic acid (e.g., glycerin-acetic acid), and stored there for a couple of weeks or months, the contents of the idosoma can be dissolved in lactic acid.

For permanent mounts, Hyrax, modified Hoyer's, glycerin-jelly, or glycerin can be used. Hyrax gives good contrast even for fine details, but the mounting in Hyrax is time consuming (methods see Newell, 1947:7-9); remounting can be done by heating the slide; sealing is not required. Mounting in modified Hoyer's (50 ml distilled water, 30 g gum arabic, 125 g chloral hydrate, 20 ml glycerin), glycerin-jelly, or glycerin is easy; the mites are rinsed after clearing and then placed into the mounting medium; remounting is possible; sealing of the coverslip is required, e.g., with a standard microscopical sealing agent, good quality nail varnish, marine epoxy paint, or electrical insulating paint. Polyvinyl lactophenol, easy to handle as the mites can be placed from alcohol, water, etc., into the mounting medium, is excellent for short time mounts. It should not be used for permanent mounts; after several years the cuticles of the mites are over-cleared and fine details in setation are hardly or no longer distinguishable.

To allow viewing of both dorsal and ventral aspects, mounting between two coverslips is recommended. The mounts can be stored in aluminum slides or attached to a glass slide by a drop of glycerin.

Type Collections

Extensive collections of Halacaroidea are located at: Musée National d'Histoire Naturelle, Paris, France; National Musuem of Natural History, Smithsonian Institution, Washington, D.C., USA; Zoologisches Institut und Zoologisches Museum, Universität Hamburg, Hamburg, FRG. Other museums are: British Museum (Natural History), London, UK; Naturmuseum und Forschungsinstitut Senckenberg, Frankfurt/Main, FRG; University of California, Riverside, USA; Oregon State University,

Corvallis, USA; Department of Scientific and Industrial Research, Auckland, New Zealand.

References

André, M.
1946. Halacariens marins. *Faune de France,* Volume 46, 152 pages. Paris: P. Lechevalier.

Bartsch, I.
1972. Ein Beitrag zur Systematik, Biologie und Ökologie der Halacaridae (Acari) aus dem Litoral der Nord- und Ostsee. I. Systematik und Biologie. *Abhandlungen und Verhandlungen des naturwissenschaftlichen Vereins, Hamburg (NF),* 16:155-230.

1974. Ein Beitrag zur Systematik, Biologie und Ökologie der Halacaridae (Acari) aus dem Litoral der Nord- und Ostsee. II. Ökologische Analyse der Halacaridenfauna. *Abhandlungen und Verhandlungen des naturwissenschaftlichen Vereins, Hamburg (NF),* 17:9-53.

1977. Interstitielle Fauna von Galapagos. XX. Halacaridae (Acari). *Mikrofauna des Meeresbodens,* 65:1-108.

1978. Verbreitung der Halacaridae (Acari) im Gezeitenbereich der Bretagneküste, eine ökologische Analyse. I. Verbreitung der Halacariden. *Cahiers de Biologie Marine,* 19:363-383.

1979a. Verbreitung der Halacaridae (Acari) im Gezeitenbereich der Bretagneküste, eine ökologische Analyse. II. Quantitative Untersuchungen und eine Faunenanalyse. *Cahiers de Biologie Marine,* 20:1-28.

1979b. Halacaridae (Acari) von der Atlantikküste Nordamerikas. Beschreibung der Arten. *Mikrofauna des Meeresbodens,* 79:1-62.

1982. Halacaridae (Acari) von der Atlantikküste des borealen Nordamerikas. Ökologische und tiergeographische Faunenanalyse. *Helgoländer Meeresuntersuchungen,* 35:13-46.

Bartsch, I., and P. Schmidt
1978. Interstitielle Fauna von Galapagos. XXII. Zur Ökologie und Verbreitung der Halacaridae (Acari). *Mikrofauna des Meeresbodens,* 69:1-38.

1979. Zur Verbreitung und Ökologie einiger Halacaridae (Acari) in Sandstränden der Ostsee (Kieler Bucht), der Nordsee (Sylt) und des Europäischen Nordmeeres (Tromsö). *Mikrofauna des Meeresbodens,* 74:1-37.

Green, J., and M. Macquitty
1987. Halacarid Mites (Arachnida: Acari). Keys and Notes for the Identification of the Species. *Synopsis of the British Fauna,* 36, 178 pages. Leiden: E.J. Brill/Dr. W. Backhuys.

Newell, I.M.
1947. A Systematic and Ecological Study of the Halacaridae of Eastern North America. *Bulletin of the Bingham Oceanographic Collection,* 10:1-232.

1971. Halacaridae (Acari) Collected During Cruise 17 of the R/V Anton Bruun, in the Southeastern Pacific Ocean. *Anton Bruun Report,* 8:3-58.

1984. Antarctic Halacaroidea. *Antarctic Research Series,* 40:1-284.

Petrova, A.
1972. Sur la présence d'*Halacarellus subterraneus* Schulz, 1933 et *Halacarellus phreaticus* n. sp. (Halacaridae, Acari) en Bulgarie. *Acarologia,* 13:367-373.

Pugh, P.J.A., and P.E. King
1985. Vertical Distribution and Substrate Assocation of the British Halacaridae. *Journal of Natural History,* 19:961-968.

Viets, K.
1927a. Die Halacaridae der Nordsee. *Zeitschrift für Wissenschaftliche Zoologie,* 130:83-173.

1927b. Halacaridae. Volume XIc, pages 1-72 in G. Grimpe and E. Wagler, editors, *Die Tierwelt der Nord- und Ostsee.* Leipzig: Akademische verlagsgesellschaft.

1939/ Meeresmilben aus der Adria. *Archiv der*
1940. *Naturgeschichte N.F.,* 8:518-550, 9:1-135.

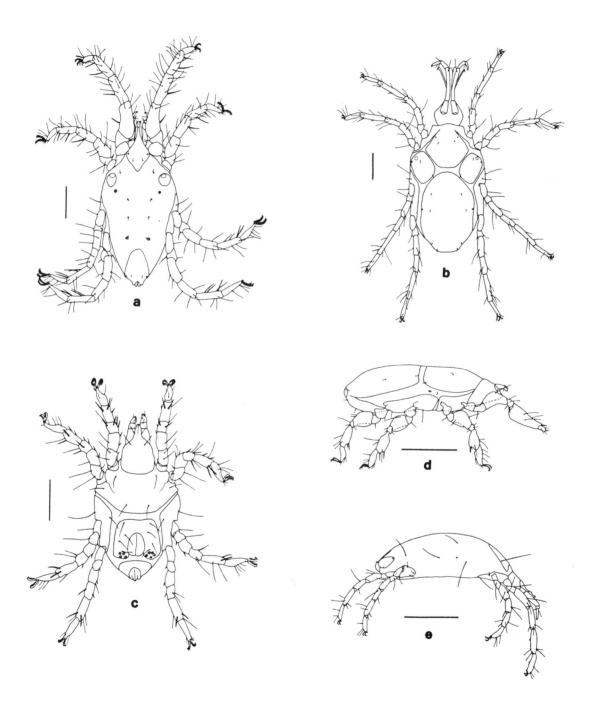

Figure 44.1.--**Halacaroidea**: **a**, *Halacarus rismondoi*, dorsal, female; **b**, *Porolohmannella violacea*, dorsal, female; **c**, *Soldanellonyx monardi*, ventral, female; **d**, *Acaromantis subasper*, lateral, female; **e**, *Anomalohalacarus minutus*, lateral, female. (Scale = 100 µm.)

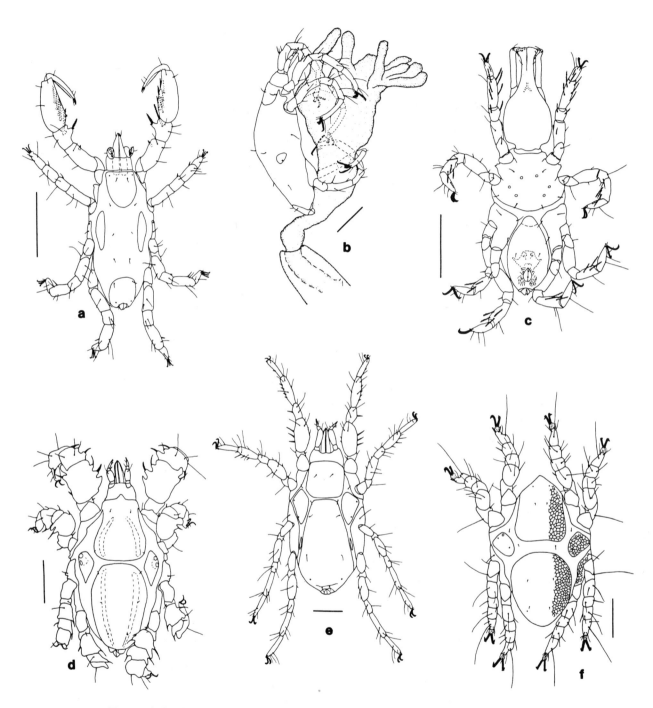

Figure 44.2.--Halacaroidea (continued): a, *Acarochelopodia cuneifera*, dorsal, female; b, *Bradyagaue drygalski*, lateral, larva; c, *Scaptognathus hallezi*, ventral, male; d, *Agauopsis pteropes*, dorsal, female; e, *Halacarellus subterraneus*, dorsal, female; f, *Isobactrus hartmanni*, dorsal, male. (Scale = 100 μm.)

45. Pycnogonida

C. Allan Child

The size of most of the approximately 900 known species of Pycnogonida, mostly attributable to their extremely long and fragile legs and body projections, makes their inclusion in the meiofauna questionable. Nonetheless, a few have been found in coarse sand - allegedly interstitial - and some, probably young adults, have been found among the meiofauna of sediment samples. Their feeding behavior, sucking out the tissue fluids of sessile organisms, and their extremely slow movements also reduce the number of likely candidates that might be considered meiofaunal. Thus, only a few species might qualify on the basis of size. Meiobenthologists should carefully evaluate the status of the pycnogonids if and when they are found in collections of meiofauna.

Pycnogonid literature, particularly that of the past 30 years, has included an increasing number of habitat descriptions as "under rocks," "under rubble," "in sand among rocks," "sandy substrate," and other habitats wherein meiofauna are present. It is difficult and often impossible to interpret this collecting data as to whether the pycnogonids were on top of or within the sediment.

Classification

The class Pycnogonida has been placed in the Arachinda or other chelicerate groups because of the functional chelae of several families and genera, or among the Crustacea because of their chitinous exoskeleton. That they are not closely related to either group is made clearer when the combination of eight walking legs, curved ovigers, cement glands in males, a suctorial, unsegmented proboscis without articulated mouth parts, and a highly reduced abdomen are considered.

There are eight or nine families containing 78 genera that make up the class. Orders are meaningless in that all Recent families are within the order Pantopoda. Most of the taxonomy of the class is based on morphology of the appendages, segments, presence or absence of ovigers in the female, chelifores, chelae, palpi, and auxiliary claws.

Of the eight or nine families, only three are known to have small enough species to be considered meiofauna, at least as small adults as well as earlier life history stages. These families include the Ammotheidae Leach, 1814; Phoxichilidiidae Sars, 1891, and Rhynochothoracidae Thompson, 1909.

Habitats and Ecological Notes

The genus *Rhynchothorax* Costa, containing about 14 or 15 known species, is probably more suited than most others for at least an interstitial life since these species have legs shorter and more robust than the usual pycnogonid appendages, a proboscis, trunk and short abdomen carried in a flat plane rather than at oblique angles to each other as in most pycnogonids, and are otherwise relatively small (a leg span of from about 2 to 5 mm). Members of this genus are commonly found in sandy habitats.

Hedgpeth (1951) described *Rhynchothorax philopsammum* (Figure 45.1a), the first of nine other members of this genus from sand. In the case of Hedgpeth's species, it was found five or six inches below the surface of coarse sand. Since it has a maximum leg span of only 2.4 mm and is otherwise unencumbered with protrusions which would slow or prevent movement between coarse sand grains, it has been assumed to be interstitial at least, and some stages of its life history are small enough to conform to the definition of meiofauna. Other littoral or shallow-water species of this genus have been recorded from sand or mixed sand habitats; some could be at least interstitial. Species from deeper water, taken with trawls or dredges, cannot be classified other than being associated with the sediment.

Nymphonella lambertensis Stock has been found in interstitial sand habitats, down as much as sixty cm into the sand at 33 m of water depth. (Arnaud, personal communication). A second of a total of three species in this genus, *N. lecalvezi* Guille and Soyer, 1967, has been described from fine sand at depths of 5-11 m.

There are three species of the prolific epifaunal genus *Anoplodactylus* which have been taken in sandy biotopes and should be mentioned. *Anoplodactylus arescus* du Bois-Reymond Marcus (Figure 45.1b), less than 3 mm in maximum dimension; *A. evelinae* Marcus, slightly over 2 mm in maximum dimension, found at least once in tropical beach sand (Child 1979); and *A. tarsalis* Stock, about 3 mm in maximum dimension.

It is probable that no pycnogonid is exclusively interstitial, but it is likely that the various species found in sand venture there in search of food. The size and sparse numbers of these tiny, robust animals make study of their ecology difficult, but some attempts should be made to examine a sufficient number of live phyllopsammous pycnogonids in order to better evaluate their status within the benthic community.

Methods of Collection, Extraction, and Preparation for Taxonomic Study

Pycnogonida found in sandy biotopes have been collected by various methods; most any technique used to sample the upper layers of sandy habitats is likely to be effective. Subtidally, dredges designed to collect a large amount of the uppermost layers of coarse sediment are judged to be the most effective way of finding specimens, considering their relatively low numbers per unit of surface area.

The sea spiders are easily extracted from sediments by sieving. Some care, however, should be used because of the fragile nature of their appendages. Some do pass through a 1 mm mesh net, they may be confused momentarily with the more common, and sometimes large mites. Fixation by 4-6% formalin and storage in 70% ethanol is standard. Whole-mount preparations or careful dissections of specimens is routine in their identification.

Type Collections

The most extensive type collections of Pycnogonida are located at the Zoological Museum, Amsterdam, The Netherlands, and at the National Museum of Natural History, Smithsonian Institution, Washington, D.C., USA. Other museums where lesser amounts of type material may be found include the British Museum (Natural History), London, UK; the Musée National d'Histoire Naturelle, Paris, France; Zoologisches Museum Berlin, Berlin, GDR; and the South African Museum, Cape Town, South Africa.

References

Child, C.A.
1979. Shallow-Water Pycnogonida of the Isthmus of Panama and the Coasts of Middle America. *Smithsonian Contributions to Zoology*, 293:1-86.

Costa, O.G.
1861. Microdoride mediterranea, o descrizione de'poco ben conosciuti od affatto ignoti viventi minuti e microscopici del Mediterraneo, etc., 1:i-xviii, 1-80.

Guille, A., and J. Soyer.
1967. Nouvelle signalisation du genre *Nymphonella* Ohshima à Banyuls-sur-Mer: *Nymphonella lecalvezi* n. sp. *Vie et Milieu (A)*, 18(2A):345-353.

Hedgpeth, J.W.
1951. Pycnogonids from Dillon Beach and Vicinity, California, with Descriptions of Two New Species. *The Wasmann Journal of Biology*, 9(1):105-117.

Marcus, E.
1940. Os Pantopoda brasileiros e os demais sul-americanos. *Boletin da Faculdade de Filosofia, Ciencias e Letras da Universidade de São Paulo*, 19(Zool. 4):3-179.

Marcus, E. du Bois-Reymond
1959. Ein neuer Pantopode aus Foraminiferensand. *Kieler Meeresforchungen*, 15(1):105-107.

Stock, J.H.
1966. Sur quelques Pycnogonides de la région de Banyuls. 3ème Note. *Vie et Milieu*, 17(IB):407-417.

1968. Pycnogonida Collected by the *Galathea* and *Anton Bruun* in the Indian and Pacific Oceans. *Videnskabelige Meddeleser fra Dansk Naturhistorisk Forening i Kjobenhavn*, 131:7-65.

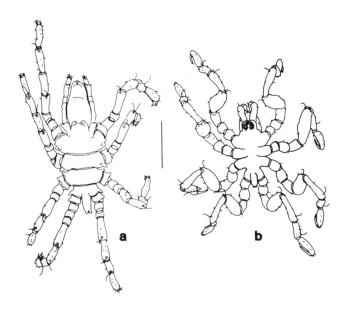

Figure 45.1.--Pycnogonida: a, *Rhynchothorax philopsammum* Hedgpeth; b, *Anoplodactylus arescus* Marcus. (Scale = 0.5 mm.)

46. Palpigradida

Bruno V. N. Condé

The Order Palpigradida, formerly called Microteliphonida, on account of an external similarity to minute whip-scorpions, is considered now the most primitive among the living Arachnida, in spite of the much better-known but highly specialized Scorpions, and despite the only unmistakable fossil, *Paleokoenenia mordax* Rowland and Sissom, 1980, from the middle or late Cenozoic of Arizona.

The body length is generally less than 2 mm for the adults (the giant being a cave-dwelling species from Mallorca, 2.8 mm long), and only 1 mm or less for the rare meiofaunal representatives. All are blind and whitish, with some parts translucent and others pale brownish-yellow. The appendages are more or less elongated, with the pedipalps and first legs directed forward. The plurisegmented flagellum is most characteristic of the group; unfortunately, it is very often broken. If so, it can be confused with some mites (e.g., *Rhagidia*). The "peduncle" between prosoma and opisthosoma is more or less conspicuous and even lacking in the genus *Leptokoenenia*.

Classification

There is only a single Recent family, Eukoeneniidae Petrunkewich, 1955. The extinct family Sternarthronidae Haase, 1890, is of uncertain position and only tentatively referred to the Palpigradida. Among the 6 genera of Eukoeneniidae, *Prokoenenia* Börner, 1901, has ventral sacs on opisthosoma ("lung-sacs") and is thereby very distinctive; three other genera (*Eukoenenia* Börner, 1901; *Allokoenenia* Silvestri, 1913; *Koeneniodes* Silvestri, 1913) require careful microscopical examination of the glandular complexes and other characters in order to distinguish them; *Paleokoenenia* Rowland and Sissom, 1980, is a fossil with very long pedipalps and strong chelicerae; *Leptokoenenia* (Condé, 1965) is a more slender and elongated genus with relatively shorter appendages. There are about 60 species in the order.

Habitats and Ecological Notes.

The tropical origin of the palpigrades is inferred from their recent distribution and thermal preferenda. They live primarily in moist soils, from the surface to more than 1 m deep. About one-third of the known species inhabit caves up to latitudes where, with a few remarkable exceptions, the endogean forms are no longer present. These exceptions are related to some very peculiar local survival conditions, e.g., nunatakker in the Alps (Kaiser-gebirge, northern Tirol, Mahnert and Janetschek, 1970) or xerothermal habitats (Lower Austria: Schauboden Heide, Warmeinsel, and Hochriess Lumperheide, Condé, 1984b).

The marine littoral origin of the phylum has been hypothesized several times. Savory (1974) has summarized the process when he wrote: "We may reasonably imagine a small marine "proto-arachnid" leaving the almost uninhabitable sea and seeking asylum under the organic debris that littered the primaeval shore. Here, "in dark and damp" it was able to survive, its outer surface always sufficently moist. Provided only that it could feed itself on whatever it may have been that its new home provided, continued life was reasonably assured." The small size of the palpigrades ensures a surface large enough to allow gaseous exchange by diffusion alone, as the surface is wet. In fact, no respiratory organ has evolved in the group, with the possible exception of the so-called "lung-sacs" of *Prokoenenia*, a genus supposed to be morphologically the most primitive, despite its occurrence in a dry area rather than along the shore.

There is an alternative proposal: since palpigrades are abundant in terrestrial habitats, there is a strong possibility that they may have migrated from mesic and wet lands to the seashore, rather than the reverse. A third suggestion is the possibility that beaches house both terrestrial species and true marine meiofaunal ones from a single genus; the few specimens collected up to now favors such an assumption.

Less than 20 specimens have been collected from six littoral locations: Madagascar (two, one lost), Red Sea (one), Western Africa (? ten to twelve), Cuba (one), the Philippines (one), and Thailand (one).

Remy (1960) was the first to investigate an interstitial palpigrade, collected by R. Paulian on the southern coast of Madagascar (Faux-Cap) in the phreatic water of the beach, less than 1 m from the sea, using the Chappuis method, in January, 1952. It is an adult female, 0.70 mm long, thought to belong to an undetermined species of *Eukoenenia*. As some details of the sternal chetotaxy were reminiscent of *Leptokoenenia* Condé, I have taken the opportunity of this contribution to re-examine the defective slide preserved in the Paris Muséum national d'Historie naturelle (Zoologie: Arthropodes). Even in its rather poor actual condition (flattened and excessively cleared) the specimen is obviously a member of the genus *Leptokoenenia*. Another specimen was obtained from an adjacent location (Cap Sainte-Marie), but remains unpublished and was not found in the Remy collection.

The genus *Leptokoenenia* Condé, 1965, is noticeable by a slender body and short appendages (legs two to four especially), as found in many other meiofaunal groups. The type specimen of the genus, a 0.70 mm larva of the first or "A"-stage, was discovered by Gerlach in November, 1964 on Sarad Sarso Island in the Farasan Archipelago (Red Sea), at about 25 cm depth in wet coral sand, near the higher limit of sea water, in association with Turbellaria, Nematoda, Polychaeta, Oligochaeta, Copepoda, and Halacaroidea.

At first, the true significance of *L. gerlachi* Condé was overlooked by the present author who was even doubtful about the specific habitat of the species. In fact, a number of small true terrestrial Arthropods (Pauropoda, Symphyla, Protura, Collembola, Diplura, and Acarina) are not rare in samples from interstitial fresh, brackish, or salt waters; some of them, living just above the water level, are swept along by running water. The discovery at Pointe-Noire (Congo) of a small population of *Leptokoenenia* by F. Monniot, in April, 1964, and her subsequent observations on live specimens, changed my opinion.

Leptokoenenia scurra Monniot, 1966, was collected from a large "puisard Chappuis," at a depth of 1.5 m in slightly diluted seawater. It was found in the sand of the beach between the tide-marks. No animal was seen floating on the surface and, apart from palpigrades, the sample consisted of Halacaroidea, Oligochaeta, Polychaeta, and Ciliata. The major observation was that the *Leptokoenenia* specimens extracted from the net and returned to salt water had a "normal" behavior, clutching at the grains of sand and swimming with their legs, without any disordered motions. This condition is very impressive, for it is well known that the cuticle of palpigrades is usually waterproof and that a water surface is the best trap for any endogean or cavernicolous species (Condé, 1984a:).

The description of *L. scurra*, based on a few male and female adult specimens using a phase contrast microscope, is of an unexcelled accuracy (even if the "double sensory setae" were proven to be an optical artifact, Monniot, 1970) and fully illustrated. It is a model for those who should be fortunate enough to find new specimens. Nevertheless, no additional reliable morphological character, associated with the meiofaunal environment, was found. The number of specimens investigated and the relative fraction of males, females, and immatures unfortunately are omitted; furthermore, no type specimen was selected.

As far as close comparison of the specimens is possible, owing to the scarcity and legibility of the material available, and with the help of Remy's short description of "*Eukoenenia*" sp., the Madagascarian specimen is slightly different from *Leptokoenenia scurra* and could be the adult of the Red Sea species, *Leptokoenenia gerlachi*.

More recently (Condé, 1981), I described *Koeneniodes deharvengi* from Mindoro (Philippines). It was an adult female, 0.92 mm long, obtained from beach sand through the elutriation method. The sand was collected 2.5 m from the sea, just above ground-water level, together with a terrestrial halophilous fauna. No adaptative characters were found, so the occurence in that particular biotope could be accidental. The same is true for an adult male *Eukoenenia* sp., badly damaged, collected on a Cuban beach (Bañes Oriente) by C. Delamare Deboutteville, and for an adult female *E. angusta* Hansen, collected in Thailand (Wanakhon beach, south of Prauchuap Khiri Khan) by P. Leclerc.

The lack of palpigrades in freshwater meiofauna is noticeable and could suggest that the terrestrial soil was directly colonized from the shore in passing from the marine interstitial environment.

Methods of Collection

The few specimens of *Leptokoenenia* known at this time were collected incidentally from holes bored in the beach sand by the so-called Chappuis method, and passing the water through a fine-mesh net since representatives of this genus do not float.

Preparation of Specimens for Taxonomic Study

The most satisfactory fixation of specimens is with 75-95% ethanol. Mounting on a slide (Cobb aluminum slide if available) in Marc André II or Hoyer's medium is a difficult manipulation when the specimen is not rectilinear, which is quite frequent. The best is to put a coverslip supported at one side by a small cylindrical piece of spun glass and wait a few days, without heating, for the total extension of the specimen. Then, it is easy to induce the rotation of the specimen by moving carefully the coverslip. It is often necessary to inspect the dorsal, lateral and ventral sides of the body. If it is desirable to replace the specimen in alcohol, the slide can be placed in tepid water into which alcohol is added until the desired concentration is reached.

Type Collections

The most extensive collections of Palpigradida are located at The Laboratorio di Entomologia Agraria "Filippo Silvestri," Portici, Italy; Musée National d'Histoire Naturelle, Remy collection, Paris, France; Muséum d'Histoire Naturelle, Geneve, Switzerland; and Musée de Zoologie, Nancy, France.

References

Condé, B.
1965. Présence de Palpigrades dans le Milieu Interstitiel Littoral. *Comptes Rendus de l'Académie des Sciences de Paris*, 261:1898-1900.
1981. Palpigrades des Canaries, de Papouasie et des Philippines. *Revue Suisse de Zoologie*, 88(4):941-951.
1984a. Les Palpigrades: Quelques Aspects Morphobiologiques. *Revue Arachnologique*, 5(4):133-143.
1984b. Palpigrades (Arachnida) d'Europe, des Antilles, du Paraguay et de Thaïlande. *Revue Suisse de Zoologie*, 91(2):369-391.

Mahnert, V., and H. Janetschek
1970. Bodenlebende Palpenlaüfer in den Alpen (Arach., Palpigradida). *Oecologia* (Berlin), 4:106-110.

Monniot, F.
1966. Un Palpigrade Interstitiel, *Leptokoenenia scurra* n. sp. *Revue d'Ecologie et de Biologie du Sol*, 3(1):41-64.
1970. Quelques Caractères Morphologiques de *Eukoenenia mirabilis* (Arachnide-Palpigrade) Observés au Stereoscan. *Revue d'Ecologie et de Biologie du Sol*, 7(4):559-562.

Remy, P.A.
1960. Palpigrades de Madagascar. II. *Mémoires de l'Institut Scientifique de Madagascar*, Série A, 13(1959):33-66.

Savory, T.
1974. On the Arachnid Order Palpigradi. *Journal of Arachnology*, 2:43-45.

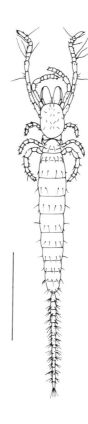

Figure 46.1.--Palpigradida: *Leptokoenenia scurra* Monniot, dorsal view (after Monniot, 1966). (Scale = 500 μm.)

47. Insecta

Alexander D. Huryn and Daniel G. Perlmutter

Insects are potentially significant components of freshwater meiofauna (e.g., Strayer and Likens, 1986). Although generally not as well represented among the marine meiofauna, insects, mostly larvae of non-biting midges (Diptera: Chironomidae) may be relatively common in some shallow near-shore marine habitats. Unlike most other groups, insects must be considered temporary meiofauna (McIntyre, 1969) in that only the earliest instars generally fall within the size range of organisms treated in this volume. As development proceeds, most taxa outgrow the meiofaunal size range and may change habitat preference (Williams, 1984) and feeding habit (Cummins and Merritt, 1984). Exceptions may be found among Coleoptera which contain taxa that are considered permanent meiofauna (e.g., Phreatodytidae, Matsumoto, 1976). The continuous presence of insects within meiofaunal assemblages may be maintained by multivoltine taxa, taxa with extended periods of recruitment, and by seasonal replacement of taxa. The role of the insects as components of the meiofauna is not well understood, due primarily to difficulties in taxonomy, lack of appropriate methodologies, as well as their temporary meiofaunal status. Regardless, available information indicates a prominent role of meiobenthic insects in secondary production and energy flow in both lentic (Strayer and Likens, 1986) and lotic (Huryn and Wallace, 1986) freshwater systems.

The taxonomy of insects that are potential components of meiofauna (body length <2 mm) presents major problems. Insects within this size range are primarily early instars; however, the taxonomy of the immature stages of the aquatic orders is based, almost without exception, on characters of late-instar larvae (i.e., those that are about to pupate or molt to reproductive stages). These characters are often not sufficiently developed in early instars, making adequate identification using current taxonomic sources impossible. However, through the knowledge of the taxonomy and life histories of the insect fauna inhabiting a given study area, unequivocal identification through association with later developmental stages is often possible. Although there are several adequate manuals dealing with the taxonomy of late-instar larvae of the aquatic insect orders (see References), there are no general treatments of early instars. The purpose of the present work is to provide a taxonomic base for the identification of the early instars of insect orders most likely to be encountered in samples of meiofauna.

Classification and Ecological Notes

Members of the class Insecta (sensu lato) may be distinguished from all other arthropods by the presence of 3 principal body regions: head, thorax and abdomen, coupled with the presence of 3 pairs of thoracic legs at some stage during the life history of the animal. However, these ground-plan characters are often not displayed by the immature stages of some orders (e.g., Diptera).

The diversity of the body-forms of larval insects is so remarkable that generalization at the class level is impractical. The majority of legless insect larvae can be distinguished from other morphologically similar invertebrate phyla (e.g., Annelida, Nematoda) by the presence of a sclerotized external head-capsule (Figure 47.2a-c) or internal cephalopharyngeal skeleton (Figure 47.2e). It should be noted that although larval forms of aquatic Acarina also may have 3 pairs of legs, the head and thorax are fused. The following is a diagnostic treatment of early instar larvae of insect orders that may be found in freshwater and marine sediments. First-instar larvae of the insect taxa considered herein range from <1 mm to approximately 2 mm in length. Orders with aquatic taxa that are generally associated with specific host plants (e.g., Lepidoptera) or animals (Neuroptera, Hymenoptera), or that are almost exclusive inhabitants of the water column or surface film (Hemiptera) are omitted. An exception is the Collembola which are often incidentally included in samples due to their omnipresence in shallow aquatic habitats. General biological information is given with each description. Assignments of feeding habits follow Cummins and Klug (1979) and generally apply to later instar larvae. The early instars of most taxa are thought to be primarily collector-gatherers of fine particulate detritus.

CHAPTER 47: INSECTA 429

Collembola.—Segment[...] present; colophore (or [...] extending ventrad fro[...] body form ranging f[...] globular. A furcula (ju[...] many collembolan taxa a[...] venter of the fourth abd[...] undergo simple metamorp[...] adults are generally indi[...] instars except by size.

The Collembola are ubi[...] shallow-water aquatic con[...] mountain streams to tidal [...] expected to be collected in [...] taken from shallow aquatic h[...] be particularly abundant on o[...] intertidal habitats where there [...] macroalgae. A few specialized [...] among sand grains of open be[...] be found to a depth of 50 cr[...] Higgins, personal communicati[...] are generally considered colle[...] taxa may be specifically adapted [...] matter associated with the air-w[...]

Ephemeroptera.—Segmented tho[...] present; 3 caudal filaments (Fig[...] commonly, 2 (Figure 47.1g); ta[...] antennae gradually tapering with >[...] form ranging from radically [...]essed (e.g., Heptageniidae, Figure 47.1d) to cylindrical (e.g., Baetidae, Figure 47.1g). The presence of three caudal filaments distinguishes larval ephemeropterans from all other aquatic orders of insects (but see Odonata: Zygoptera). Those taxa characterized by two caudal filaments (e.g., some Heptageniidae, some Baetidae) may be confused with the Plecoptera (see below); however, the Plecoptera have two tarsal claws (Figure 47.1e,f,i).

The Ephemeroptera occur in most permanent freshwater habitats but attain greatest diversity in streams. Early instars of robust forms (e.g., Ephemerellidae, Baetidae; Figure 47.1a,g) were found to depths of 30-70 cm below an Ontario (Canada) stream bed. More delicate forms (e.g., *Paraleptophlebia*, Figure 47.1c) were more common on the substrate surface (Williams, 1984). The Ephemeroptera includes collector-gathers and filterers, scrapers, shredders, and predators.

Odonata.—Segmented thoracic legs always present; labium hinged to form extensile grasping organ which covers oral region when retracted (Figure 47.1h). Two general body forms occur within the order. Members of the suborder Zygoptera are cylindrical and the terminus of the abdomen is occupied by three [...] (e.g., Grieve, 1937). Larval Zygoptera [...] associated with beds of submerged [...] probably will not be encountered [...] samples. The body forms of the [...]re 47.1h) range from spindle-shaped [...] they have internal respiratory [...] within a modified rectal chamber. [...]onata are found throughout a wide [...]nt and temporary lotic and lentic [...] diversity is reached in lentic [...]ata are thought to be predacious, [...] on other invertebrates.

[...]nted thoracic legs always present; [...] each with at least 3 segments; [...] 8 segments; tarsal claws paired; [...]ustly spindle-shaped (e.g., [...] ire 47.1e), dorsoventrally [...]loroperlidae, Figure 47.1i), to [...]tridae, Figure 47.1f). Larval [...]nfused with larvae of certain [...] taxa (see Ephemeroptera,

[...]vide range of permanent and [...] habitats; however, greatest [...]cool lotic situations. Early- [...] represented among the [...]reams (Williams, 1984). Feeding [...]nclude collector-gathering, shredding, scraping, and predation.

Megaloptera.—Segmented thoracic legs always present; abdomen with fleshy filaments laterally, terminating in either: (1) a single elongate filament (Sialidae), or (2) a pair of prolegs, each with a pair of hooks (Corydalidae, Baker and Neunzig, 1968); body form depressed-cylindrical.

Larval megalopterans occur in a wide range of temporary to permanent habitats including seeps, streams, rivers, ponds, lakes, and swamps. Although first instar larvae of the Sialidae fall within the size range treated here (<2 mm), many of the Corydalidae range considerably larger and will probably not be encountered among samples of meiofauna. Larvae are generally considered predators; however, early instars may feed upon detritus.

Trichoptera.—Segmented thoracic legs always present; terminus of abdomen with pair of sclerotized hooks laterally (Figure 47.2g,h); body forms generally cylindrical. Larvae free-living or inhabiting cases or fixed retreats and filtering nets constructed of silk (secreted from labial glands) and detritus and/or minerals (see Diptera: Chironomidae).

The Trichoptera are well represented in a wide range of aquatic environments ranging from cool lotic through lentic habitats (temporary and permanent), brackish marshes, and the marine intertidal (see Winterbourne and Anderson, 1980). The order attains greatest diversity in cool headwater streams; however, the Leptoceridae and Hydroptilidae (Figure 47.2h) are particularly well represented in ponds and the littoral zones of lakes.

Coleoptera.--Segmented thoracic legs almost always present but may be reduced; head strongly sclerotized to form distinct external head capsule; antennae 2-4 segmented except in Scirtidae (=Helodidae) (see below); tarsal claws either paired (suborder Adephaga) or single (suborder Polyphaga); body-form exceedingly diverse ranging from plate-like (e.g., Eubriidae, Figure 47.2i) through cylindrical (e.g., Elmidae, Figure 47.2f). Curculionidae are legless; prolegs are absent. Early instars of most aquatic curculionids generally feed within roots and leaves of aquatic macrophytes and will normally not be encountered in samples of benthic materials. Larvae of Coléoptera are likely to be confused with Trichoptera and Megaloptera but can be separated by: (1) absence of paired anal prolegs bearing sclerotized hooks (larvae of Gyrinidae have abdomens terminated by a single median lobe bearing four hooks (Tonapi, 1959), cf., Megaloptera, Trichoptera), or (2) absence of the combination of single anal filament and laterally projecting filaments along abdomen (see Megaloptera). Larvae of the Scirtidae are exceptional in the long multi-segmented antennae (ca. 100+ segments) and may be confused with the Plecoptera; however, their cerci are unapparent and they have single tarsal claws. *Donacia* spp. (Chrysomelidae) are characterized by reduced thoracic legs and large, sclerotized hooks (not to be confused with anal hooks of Trichoptera) formed from the spiracles of the eighth abdominal segment. Larvae of *Donacia* feed externally upon roots of various wetland macrophytes and obtain oxygen by tapping into the host plant's vascular system with the modified posterior spiracles (Houlihan, 1969) and may be encountered in collections of sediments taken in association with root systems of *Nuphar, Sparganium,* and other emergent macrophytes occurring in freshwater marshes.

Coleoptera occur in almost all freshwater environments as well as littoral zones of brackish habitats and may be found in the marine intertidal. The greatest diversity occurs in ponds and littoral zones of lakes. Matsumoto (1976) briefly discusses species of phreatic dytiscids and phreatodytids from western Japan. Larvae of Coleoptera exhibit many kinds of feeding habits, ranging from collector-gatherers, shredders (utilizing various dead and living plant materials), scrapers, and predators.

Diptera.--Segmented thoracic legs always absent although unsegmented thoracic and abdominal prolegs may be present (e.g., Figure 47.2b,c,e); head capsule either: (1) well sclerotized, external (Figure 47.2a-c) or internal (suborder Nematocera) or, (2) internal and reduced to a complex structure of rods and phragmata (suborder Brachycera) (Figure 47.2e); body forms diverse, ranging from depressed spindle-shaped with an integument armored by mineralized plates (e.g., Stratiomyidae) through extreme elongate-cylindrical (e.g., Ceratopogonidae, Figure 47.2a). Absence of legs will usually serve to distinguish larvae of Diptera from other taxa (but see Coleoptera).

Members of the Nematocera, notably Ceratopogonidae (Figure 47.2a) and Chironomidae (Figures 47.2b,c), will probably be the insect larvae most commonly encountered among the meiobenthos of any freshwater, brackish, or nearshore marine habitat. In marine habitats, chironomid larvae may be particularly common near mangrove areas and may be relatively common in nearshore waters to depths of 1-2 m (Higgins, personal communication). Chironomids are well represented among the hyporheic fauna of streams where they have been recovered from depths of 70 cm (Williams, 1984) (see also Pennak and Ward, 1986 for a review of hyporheic Chironomidae). The Chironomidae may be distinguished from most other meiobenthic Diptera by the external head capsule combined with presence of prolegs (usually paired and invested with sclerotized hooks) arising from the anal and first thoracic segments. Like the Trichoptera, many chironomid larvae produce silk which is used to construct portable and fixed retreats, often with associated filtering devices.

Most commonly encountered Ceratopogonidae (e.g., Ceratopogoninae) have an elongate external head capsule and a long narrow body that, except for constrictions between body segments, has a smooth almost featureless integument (Figure 47.2a). Larvae of marine ceratopogonids occur in habitats such as the intertidal and supratidal zones of sheltered beaches, mangrove swamps, salt marshes, and tidal flats. Apparently no taxa inhabit sand beaches exposed to heavy wave action (Cheng, 1976). Members of the Dolichopodidae, Empididae, Tabanidae (orthorrhaphous Brachycera, Figure 47.2e), or Ephydridae (cyclorrhaphous Brachycera) may commonly be encountered within sediments of freshwater, marine, and inland saline wetlands. *Notiphila* spp. (Ephydridae) which may be extremely common in freshwater and some brackish or alkaline wetland habitats, feeds upon sediments associated with root systems of emergent macrophytes and, like *Donacia* (Coleoptera), obtains oxygen by inserting spine-like posterior spiracles into the vascular tissue of the host plant (Bussaca and Foote, 1978). In general, larval Diptera can be expected to occur in almost any aquatic habitat ranging from thermal springs to the marine intertidal. Feeding habits are diverse, and include collector-gathering, collector-filtering, shredding, scraping, and predation.

Methods of Collection

Sampling for insects within the meiofaunal size range is incidental to sampling for other groups of freshwater meiofauna, but differs from procedures usually used to sample larger instars of aquatic insects. Samples may be taken from lentic environments with instruments analogous to marine benthic samplers. In streams, netted samplers (e.g., Surber and Hess samplers) are often used to collect insect fauna (Merrit et al., 1984). The operator of these instruments entrains sediments and associated organisms into the flowing water where they are retained by the collection net. Reduction of the mesh to a size appropriate for retention of meiofauna will increase the likelihood of clogging and backwash with the resultant loss of material (Resh, 1979).

Considering the limitations and restrictions of the above equipment, coring and suction devices appear to be the most appropriate instruments for sampling insect meiofauna. Corers may be used to sample units of small surface area but potentially to great depths and thus allow for estimates of both vertical and horizontal heterogeneity of meiofaunal distribution (e.g., Williams, 1984). This information is especially useful in streams wherein there may be a great diversity of benthic habitats over short reaches (e.g., Minshall, 1984).

In shallow-water lentic and lotic environments, transparent handcores can be inserted into the sediment to prescribed depths, plugged and removed. The integrity of the core can be maintained during removal with a slight suction provided by first withdrawing a sealed sleeve or flanged insert. Alternatively, shallow cores may be capped at their bottom ends to prevent loss of material. Substrate composed of coarse mineral particles may make operation of coring devices difficult. In such substrates, materials within may be removed by hand if the core is of sufficient size. An alternative is use of suction devices (Pearson et al., 1973; Gale and Thompson, 1975; Drake and Elliott, 1982; Boulton, 1985). It should be noted that suction samplers designed to collect macrofauna may need modification to be useful in obtaining samples of meiofauna. Modifications such as the reduction of the mesh size of collection nets may reduce the overall efficiency of these devices. Specimens of insects collected by suction sampling are variously reported as being either minimally damaged to badly mutilated. Frozen-core techniques may prove to be of particular value in the study of insect meiofauna (Shapiro, 1958; Efford, 1960; Stocker and Williams, 1972; Lotspeich and Ried, 1980; Pugsley and Hynes, 1983; Williams, 1984).

Funnel traps may compliment sediment removal techniques. These traps are constructed by fitting a funnel in the opening of a wide-mouth bottle. When the bottle is placed (funnel-opening down) on the sediment surface, organisms leaving the sediments may swim upward through the funnel and its stem to enter the collection bottle. This method has been used to selectively capture early instar dipterans and ephemeropterans, organisms difficult to separate from whole sediments (Whiteside and Lindegaard, 1980). For species-level determination of most insects, adults must be obtained. Collections of adult insects are probably most efficiently made using emergence traps (Merritt et al., 1984).

Leaf litter retained in mesh bags or as leaf packs attached to bricks has frequently been used to investigate invertebrate fauna associated with organic detritus in streams. The sample units are anchored to the stream bottom to be removed after a specified period of incubation. Immediately prior to removal, the leaf material can be enclosed with plexiglass sleeves or mesh or plastic bags to reduce loss of organisms.

Methods of Preservation, Sorting, and Extraction

Field samples preserved in 5-10% formalin store well with little evaporative loss compared to alcohol. Insects and other arthropods preserved in formalin become brittle with time and their appendages are more easily lost during sample sorting. The transfer of insects to 70-80% ethanol or isopropanol maintains the suppleness of specimens. However, in ecological studies where biomass is to be directly determined from sample materials, preservation in alcohol will result in substantial weight loss due to leaching of soluble materials (Leuven et al., 1985). Weight loss may be less severe when formalin is used as the preservative (Ross, 1982).

Many procedures used to extract meiofauna from sediment samples are applicable to insects. These include decantation, sugar density centrifugation, phase separation, bubbling, Ludox flotation, sonification, and elutriation, or a combination of these methods (Anderson, 1959; Heip et al., 1974; Thiel et al., 1975; Karlsson et al., 1976; Lawson and Merritt, 1979; Barmuta, 1984). When applying these procedures, it must be remembered that insects that build mineral cases (e.g., Trichoptera and Chironomidae) are not as efficiently removed by elutriation and density separation techniques. In addition, excessive agitation or harsh sonification of sediments may damage specimens as can the use of sieves with mesh sizes similar to the dimensions of the organisms of interest. Among insect orders, the Ephemeroptera are particularly fragile. Gentle agitation of detritus in a flow-through chamber immersed in an ultrasonic water bath may aid in separating insects from organic debris.

Although time consuming, a thorough hand-sorting of sediment samples will probably result in a more complete inventory of insect meiofauna (but see Karlsson et al., 1976; Barmuta, 1984). Large samples should be subsampled appropriately to sizes that can be processed most efficiently within imposed time restrictions and finances. For samples containing substantial quantities of detritus, processing of small subsamples may result in much greater yields of invertebrates compared to treatment of the whole samples (e.g., Wallace et al., 1986). In this regard, subsamplers following the design of Waters (1969) have proven to be of great utility. Sorting of insect meiofauna from sediment samples preserved in formalin is facilitated by staining with Rose Bengal or Phloxine B. These dyes are added to the formalin solution used to preserve field samples.

Preparation of Specimens for Taxonomic Study

In many cases no preparation of insect specimens for taxonomic study is necessary as characters may be readily observed with a dissecting microscope. However, in some cases placing the specimen in glycerin (liquid or solid) facilitates appropriate orientation of body and appendages. The Chironomidae must be mounted on microscope slides for examination of characters of the head capsule which are necessary for proper identification (see references for Diptera). The most convenient method of mounting is by transferring specimens directly from formalin or alcohol to a drop of Hoyer's or CMC-10 on a microscope slide. The head capsule is teased away from the body with a needle and positioned ventral side up. A coverslip is added and gentle pressure is applied to spread out mouthparts and antennae. As the mounting medium hardens, the specimens will become adequately cleared. The use of the above media results in high quality permanent mounts and takes less time compared to mounting in media requiring dehydration of the specimen. However, many workers prefer to clear specimens in 10% KOH and mount them in Euparol. Specimens can be transferred to Euparol directly from 90-95% ethanol. Although time-consuming, the clearing step undoubtedly results in higher-quality specimens and should be considered when preparing voucher specimens.

References

Anderson, R.O.
1959. A Modified Flotation Technique for Sorting Bottom Fauna. *Limnology and Oceanography*, 4:223-225.

Azam, K.M., and N.H. Anderson.
1969. Life History and Habits of *Sialis californica* Banks and *Sialis rotunda* Banks in Western Oregon. *Annals of the Entomological Society of America*, 62:549-558.

Baker, J.R., and H.H. Neunzig.
1968. The Egg Masses, Eggs, and First Instar Larvae of Eastern North American Corydalidae. *Annals of the Entomological Society of America*, 61:1181-1187.

Barmuta, L.A.
1984. A Method for Separating Benthic Arthropods from Detritus. *Hydrobiologia*, 112:105-107.

Bertrand, H.P.I.
1972. *Larves et nymphes des Coleopteres aquatiques du globe.* 804 pages. Paris: Centre National de la Recherche Scientifique.

Boulton, A.J.
1985. A Sampling Device that Quantitatively Collects Benthos in Flowing or Standing Waters. *Hydrobiologia*, 127:31-39.

Brigham, A.R., W.U. Brigham, and A. Gnilka, editors
1982. *Aquatic Insects and Oligochaetes of North and South Carolina.* 837 pages. Mahomet, Illinois: Midwest Aquatic Enterprises.

Brigham, W.U.
1982a. Megaloptera. Pages 7.1-7.12 in A.R. Brigham, W.U. Brigham, and A. Gnilka, editors, *Aquatic Insects and Oligochaetes of North and South Carolina.* 837 pages. Mahomet, Illinois: Midwest Aquatic Enterprises.

1982b. Coleoptera. Pages 10.1-10.136 in A.R. Brigham, W.U. Brigham, and A. Gnilka, editors, *Aquatic Insects and Oligochaetes of North and South Carolina.* 837 pages. Mahomet, Illinois: Midwest Aquatic Enterprises.

Brittain, J.E.
1982. Biology of Mayflies. *Annual Review of Entomology*, 27:119-147.

Bussacca, J.D., and B.A. Foote
1978. Biology and Immature Stages of Two Species of *Notiphila*, with Notes on Other Shore Flies Occurring in Cat-tail Marshes (Diptera: Ephydridae). *Annals of the Entomological Society of America*, 71:457-466.

Cheng, L., editor
1976. *Marine Insects.* 591 pages. North Holland, Amsterdam.

Christiansen, K.
1964. Bionomics of the Collembolla. *Annual Review of Entomology*, 9:147-148.

Christiansen, K., and R.J. Snider.
1984. Aquatic Collembolla. Pages 82-93 in R.W. Merritt and K.W. Cummins, editors, *An Introduction to the Aquatic Insects of North America.* 2nd edition. 722 pages. Iowa: Kendall/Hunt.

Coffman, W.P., and L.C. Ferrington, Jr.
1984. Chironomidae. Pages 551-652 in R.W. Merritt and K.W. Cummins, editors, *An Introduction to the Aquatic Insects of North America.* 2nd edition. 722 pages. Iowa: Kendall/Hunt.

Conrad, J.E.
1976. Sand Grain Angularity as a Factor Affecting Colonization by Marine Meiofauna. *Vie et Milieu*, 26:181-198.

Corbet, P.S.
1980. Biology of Odonata. *Annual Review of Entomology*, 25:189-217.

Corpus, L.D.
1986. Biological Notes and Descriptions of the Immature Stages of *Pelastoneurus vagans* Loew (Diptera: Dolichopodidae). *Proceedings of the Entomological Society of Washington*, 88:673-679.

CSIRO
1970. *The Insects of Australia.* 1029 pages. Division of Entomology, Commonwealth Scientific and Industrial Research Organization, Australia.

Cummins, K.W., and M.J. Klug
1979. Feeding Ecology of Stream Invertebrates. *Annual Review of Ecology and Systematics,* 10:147-172.

Cummins, K.W., and R.W. Merritt
1984. Ecology and Distribution of Aquatic Insects. Pages 59-66 in R.W. Merritt and K.W. Cummins, editors *An Introduction to the Aquatic Insects of North America.* 2nd edition, 722 pages. Iowa: Kendall/Hunt.

Davis, K.C.
1903. Sialidae of North and South America. Pages 442-486 in J.G. Needham, A.D. MacGillivray, O.A. Johannsen, and K.C. Davis, editors, Aquatic Insects in New York State. *New York State Museum Bulletin,* 68:199-517.

Deonier, D.L., and J.T. Regensberg
1978. Biology and Immature Stages of *Parydra quadrituberculata. Annals of the Entomological Society of America,* 71:341-353.

Deonier, D.L., W.N. Mathis, and J.T. Regensberg
1979. Natural History and Life-cycle Stages of *Notiphila carinata* (Diptera: Ephydridae). *Proceedings of the Biological Society of Washington,* 91:798-814.

Drake, C.M., and J.M. Elliott
1982. A Comparative Study of Three Air-lift Samplers Used for Sampling Benthic Macro-Invertebrates in Rivers. *Freshwater Biology,* 12:511-533.

Durand, J.R., and C. Lévêque, editors
1981. *Flore et Faune Aquatiques de L'Afrique Sahelo-Soudanienne.* Volume 2, 849 pages. Paris: Editions l'Office Recherche Scientifique Technique Outre-Mer Collection Initiations - Documentations Techniques, number 45.

Eastin, W.C., and B.A. Foote
1971. Biology and Immature Stages of *Dichaeta caudata* (Diptera: Ephydridae). *Annals of the Entomological Society of America,* 64:271-279.

Edmunds, G.F., Jr.
1984. Ephemeroptera. Pages 94-125 in R.W. Merritt and K.W. Cummins, editors, *An Introduction to the Aquatic Insects of North America,* 2nd edition, 722 pages. Iowa: Kendall/Hunt.

Edmunds, G.F., Jr., S.L. Jensen, and L. Berner
1976. *The Mayflies of North and Central America.* 330 pages. Minneapolis: University of Minnesota Press.

Efford, I.E.
1960. A Method of Studying the Vertical Distribution of the Bottom Fauna in Shallow Waters. *Hydrobiologia,* 16:288-292.

Evans, E.D., and H.H. Neunzig
1984. Megaloptera and Aquatic Neuroptera. Pages 261-270 in R.W. Merritt, and K.W. Cummins, editors, *An Introduction to the Aquatic Insects of North America.* 2nd edition, 722 pages. Iowa: Kendall/Hunt.

Foote, B.A., and W.C. Eastin
1975. Biology and Immature Stages of *Discocerina obscurella* (Diptera: Ephydridae). *Proceedings of the Entomological Society of Washington,* 76:401-408.

Gale, W.F., and J.D. Thompson
1975. A Suction Sampler for Quantitatively Sampling Benthos on Rocky Substrates in Rivers. *Transactions of the American Fisheries Society,* 2:398-405.

Gisin, H.
1960. *Collembolenfauna Europas.* 312 pages, Geneva.

Grieve, E.G.
1937. Studies on the Biology of the Damselfly *Ischnura verticalis* Say, with Notes on Certain Parasites. *Entomologica America,* 17:121-153.

Harper, P.P.
1979. Observations on the Early Instars of Stoneflies (Plecoptera). *Gewässer and Abwässer,* 64:18-28.

Harper, P.P., and K.W. Stewart
1984. Plecoptera. Pages 182-225 in R.W. Merritt and K.W. Cummins, editors, *An Introduction to the Aquatic Insects of North America.* 2nd edition. 722 pages. Iowa: Kendall/Hunt.

Heip, C., N. Smol, and W. Hautekiet
1974. A Rapid Method of Extracting Meiobenthic Nematodes and Copepods from Mud and Detritus. *Marine Biology,* 28:79-81.

Hickin, N.E.
1967. *Caddis Larvae.* 480 pages. London: Hutchinson.

Hitchcock, S.W.
1974. Guide to the Insects of Connecticut. Part IV. The Plecoptera or stoneflies of Connecticut. *Bulletin of the Connecticut State Geological and Natural History Survey,* 107:1-262.

Houlihan, D.F.
1969. Respiratory Physiology of the Larva of *Donacia simplex,* a Root-piercing Beetle. *Journal of Insect Physiology,* 15:1517-1536.

Huryn, A.D., and J.B. Wallace
1986. A Method for Obtaining *in situ* Growth Rates of Larval Chironomidae (Diptera) and Its Application to Studies of Secondary Production. *Limnology and Oceanography,* 31:216-222.

Huggins, D.G., and W.U. Brigham
1982. Odonata. Pages 4.1-4.100 in A.R. Brigham, W.U. Brigham, and A. Gnilka, editors, *Aquatic Insects and Oligochaetes of North and South Carolina.* 837 pages. Mahomet, Illinois: Midwest Aquatic Enterprises.

Hynes, H.B.N.
1976. The Biology of Plecoptera. *Annual Review of Entomology,* 21:135-153.

Illies, J.
1978. *Limnofauna Europaea.* 2nd edition, 532 pages. New York: Gustav. Fischer-Verlag.

James, H.G.
1970. Immature Stages of Five Diving Beetles (Coleoptera: Dytiscidae), Notes on Their Habits and Life History, and a Key to Aquatic Beetles of Vernal Woodland Pools in Southern Ontario. *Proceedings of the Entomological Society of Ontario,* 100:52-97.

Karlsson, M., T. Bohlin, and J. Stenson
1976. Core Sampling and Flotation: Two Methods to Reduce Costs of a Chironomid Population Study. *Oikos,* 27:336-338.

Kawai, T., editor
1985. *An Illustrated Book of Aquatic Insects of Japan.* 409 pages. Teokyeo: Teokai Daigaku Shuppankai.

Lawson, D.L., and R.W. Merritt
1979. A Modified Ladell Apparatus for the Extraction of Wetland Macroinvertebrates. *Canadian Entomologist,* 111:1389-1393.

Lepneva, S.G.
1964. *Fauna of the U.S.S.R.* Trichoptera. Larvae and Pupae of Annulipalpia. Zoological Institute of the Academy of Sciences of the USSR. New Series No. 88. 560 pages.

1966. *Fauna of the U.S.S.R.* Trichoptera. Larvae and Pupae of the Integripalpia. Zoological Institute of the Academy of Sciences of the USSR. New Series No. 95. 700 pages.

Leuven, R.S.E.W., T.C.M. Brock, and H.A.M. van Druten
1985. Effects of Preservation on Dry- and Ash-free Dry Weight Biomass of Some Common Aquatic Macroinvertebrates. *Hydrobiologia,* 127:151-159.

Lotspeich, F.B., and B.H. Reid
1980. Tri-tube Freeze-core Procedure for Sampling Stream Gravels. *Progressive Fish-Culture,* 42:96-99.

Mackay, R.J., and G.B. Wiggins
1979. Ecological Diversity in Trichoptera. *Annual Review of Entomology,* 24:185-208.

Marlier, G.
1962. Genera des Trichopteres de L'Afrique. Musée Royale de L'Afrique Amtral-Tervuran, Belgique Annales, seine 8°. *Sciences Zoologiques,* Number 109.

Matsumoto, K.
1976. An Introduction to the Japanese Groundwater Animals with Reference to Their Ecology and Hygienic Significance. *International Journal of Speleology,* 8:141-155.

Matsunae, T.

1962. *Aquatic Entomology.* 269 pages. Tokyo: Hokuryu-Kan Co., Ltd.

McAlpine, J.F., B.V. Peterson, G.E. Shewell, H.J. Teskey, J.R. Vockeroth, and D.M. Wood
1981. Manual of Nearctic Diptera, Volume 1. *Research Branch Agriculture Canada,* 27:1-674.
1987. Manual of Nearctic Diptera. Volume 2. *Research Branch Agriculture Canada,* 28:675-1332.

McIntyre, A.D.
1969. Ecology of Marine Meiobenthos. *Biological Review,* 44:245-290.

Merritt, R.W., and K.W. Cummins, editors
1984. *An Introduction to the Aquatic Insects of North America,* 2nd edition, 722 pages. Iowa: Kendall/Hunt.

Merritt, R.W., K.W. Cummins, and V.H. Resh
1984. Collecting, Sampling, and Rearing Methods for Aquatic Insects. Pages 11-26 in R.W. Merritt, and K.W. Cummins, editors, *An Introduction to the Aquatic Insects of North America,* 2nd edition. 722 pages. Iowa: Kendall/Hunt.

Morse, J.C., and R.W. Holzenthal
1984. Trichoptera Genera. Pages 312-347 in R.W. Merritt and K.W. Cummins, editors, *An Introduction to the Aquatic Insects of North America.* 2nd edition, 722 pages. Iowa: Kendall/Hunt.

Minshal, G.W.
1984. Aquatic Insect-substratum Relationships. Pages 358-400 in V.H. Resh and D.M. Rosenberg, editors, *The Ecology of Aquatic Insects.* 625 pages. New York: Praeger.

Needham, J.G., and M.J. Westfall, Jr.
1955. *A Manual of the Dragonflies of North America (Anisoptera) Including the Greater Antilles and the Provinces of the Mexican Border.* 615 pages. Berkeley: University of California Press.

Oliver, D.R.
1971. Life History of the Chironomidae. *Annual Review of Entomology,* 16:211-230.

Pennak, R.W., and J.V. Ward
1986. Interstitial Faunal Communities of the Hyporheic and Adjacent Groundwater Biotopes of a Colorado Mountain Stream. *Archiv für Hydrobiologie,* Supplement 74:356-396.

Pearson, R.G., M.R. Litterick, and N.V. Jones
1973. An Air-lift for Quantitative Sampling of the Benthos. *Freshwater Biology,* 3:309-315.

Pinder, L.C.V.
1986. Biology of freshwater Chironomidae. *Annual Review of Entomology,* 31:1-23.

Pugsley, C.W., and H.B.N. Hynes
1983. A Modified Freeze-core Technique to Quantify the Depth Distribution of Fauna in Stony Streambeds. *Canadian Journal of Fisheries and Aquatic Sciences,* 40:637-643.

Resh, V.H.
1979. Sampling Variability and Life History Features: Basic Considerations in the Design of Aquatic Insect Studies. *Journal of the Fisheries Research Board of Canada,* 36:290-311.

Resh, V.H., and D.M. Rosenberg, editors
1984. *The Ecology of Aquatic Insects.* 625 pages. New York: Praeger.

Ross, D.H.
1982. Production of Filter Feeding Caddisflies (Trichoptera) in a Southern Appalachian Stream System. Ph.D. Dissertation. 109 pages. University of Georgia, Athens.

Ross, H.H.
1944. The Caddis Flies, or Trichoptera, of Illinois. *Bulletin of the Illinois Natural History Survey,* 23:1-326.

Shapiro, J.
1958. The Core Freezer - a New Sampler for Lake sediments. *Ecology,* 39:758.

Smith, M.E.
1952. The Immature Stages of the Marine Fly *Hypocharassus pruinosus* Wh., with a Review of the Biology of the Immature Dolichopodidae. *American Midland Naturalist,* 48:421-432.

Stocker, Z.S.J., and D.D. Williams
1972. A Freezing Core Method for Describing the Vertical Distribution of Sediments in a Streambed. *Limnology and Oceanography,* 17:136-138.

Strayer, D., and G.E. Likens
1986. An Energy Budget for the Zoobenthos of Mirror Lake, New Hampshire. *Ecology,* 67:303-313.

Teskey, H.J.
1984. Larvae of Aquatic Diptera. Pages 448-465 in R.W. Merritt and K.W. Cummins, editors, *An Introduction to the Aquatic Insects of North America.* 2nd edition. 722 pages. Iowa: Kendall/Hunt.

Thier, R.W., and B.A. Foote
1980. Biology of Mud-shore Ephydridae (Diptera). *Proceedings of the Entomological Society of Washington,* 82:517-535.

Thiel, H., D. Thistle, and G.D. Wilson
1975. Ultrasonic Treatment of Sediment Samples for More Efficient Sorting of Meiofauna. *Limnology and Oceanography,* 20:472-473.

Thienemann, A.
1954. *Chironomus.* Leben, Verbreitung und wirtschaftliche Bedeutung der Chironomiden. *Die Binnengewässer,* 20:1-834.

Tonapi, G.T.
1959. A Note on the Eggs and Larva of *Dineutes indicus* Aub. (Coleoptera, Gyrinidae). *Current Science,* 4:158-159.

Uéno, S.
1957. Blind Aquatic Beetles of Japan, with Some Accounts of the Fauna of Japanese Subterranean Waters. *Archiv für Hydrobiologie,* 53:250-296.

Unzicker, J.D., and P.H. Carlson
1982. Ephemeroptera. Pages 3.1-3.97 in A.R. Brigham, W.U. Brigham, and A. Gnilka, editors, *Aquatic Insects and Oligochaetes of North and South Carolina.* 837 pages. Mahomet, Illinois: Midwest Aquatic Enterprises.

Unzicker, J.D., and V.H. McCaskill
1982. Plecoptera. Pages 5.1-5.50 in A.R. Brigham, W.U. Brigham, and A. Gnilka, editors, *Aquatic Insects and Oligochaetes of North and South Carolina.* 837 pages. Mahomet, Illinois: Midwest Aquatic Enterprises.

Unzicker, J.D., V.H. Resh, and J.C. Morse
1982. Trichoptera. Pages 9.1-9.138 in A.R. Brigham, W.U. Brigham, and A. Gnilka, editors, *Aquatic Insects and Oligochaetes of North and South Carolina.* 837 pages. Mahomet, Illinois: Midwest Aquatic Enterprises.

Walker, E.M.
1953. *The Odonata of Canada and Alaska.* Part I. General, Part II. The Zygoptera-damselflies. Volume 1, 292 pages. Toronto: University of Toronto Press.
1958. *The Odonata of Canada and Alaska.* Anisoptera. Volume 2, 318 pages. Toronto: University of Toronto Press.

Walker, E.M., and P.S. Corbet
1975. *The Odonata of Canada and Alaska.* Macromiidae, Corduliidae, Libellulidae. Volume 3, 307 pages. Toronto: University of Toronto Press.

Wallace, J.B., D.S. Vogel, and T.F. Cuffney
1986. Recovery of a Headwater Stream from an Insecticide-induced Community Disturbance. *Journal of the North American Benthological Society,* 5:115-126.

Waltz, R.D., and W.P. McCafferty
1979. Freshwater Springtails (Hexapoda: Collembola) of North America. *Agriculture Experimental Station Research Bulletin 960.* Lafayette, Indiana: Purdue University.

Waters, T.F.
1969. Subsampler for Dividing Large Samples of Stream Invertebrate Drift. *Limnology and Oceanography,* 14:813-815.

Westfall, M.J., Jr.
1984. Odonata. Pages 126-176 in R.W. Merritt and K.W. Cummins, editors, *An Introduction to the Aquatic Insects of North America.* 2nd edition, 722 pages.

Iowa: Kendall/Hunt.

White, D.S., W.U. Brigham, and J.T. Doyan
1984. Aquatic Coleoptera. Pages 361-437 in R.W. Merritt and K.W. Cummins, editors, *An Introduction to the Aquatic Insects of North America*. 2nd edition, 722 pages. Iowa: Kendall/Hunt.

Whiteside, M.C., and C. Lindegaard
1980. Complementary Procedures for Sampling Benthic Invertebrates. *Oikos*, 35:317-320.

Wiederholm, T., editor
1983. Chironomidae of the Holarctic Region, Keys and Diagnoses. Part 1. Larvae. *Entomological Scandinavica Supplement* 19:1-457.

Wiggins, G.B.
1977. *Larvae of the North American Caddisfly Genera*. 401 pages. Toronto: University of Toronto Press.
1984. Trichoptera. Pages 271-311 in R.W. Merritt and K.W. Cummins, editors, *An Introduction to the Aquatic Insects of North America*. 2nd edition, 722 pages. Iowa: Kendall/Hunt.

Williams, D.D.
1984. The Hyporheic Zone as a Habitat for Aquatic Insects and Associated Arthropods. Pages 430-456 in V.H. Resh and D.M. Rosenberg, editors, *The Ecology of Aquatic Insects*. 625 pages. New York: Praeger.

Winterbourne, M.J., and N.H. Anderson
1980. The Life History of *Philansius plebius* Walker (Trichoptera: Chathamiidae), a Caddisfly Whose Eggs Were Found in a Starfish. *Ecological Entomology*, 5:293-303.

Winterbourne, M.J., and K.L.D. Gregson
1981. Guide to the Insects of New Zealand. *Bulletin of the Entomological Society of New Zealand*, 5:1-80.

Zwick, P.
1980. *Handbuch der Zoologie. Band IV. 7. Pleocoptera (Steinflieger)*. 115 pages. Berlin: Walter M. Gruyten.

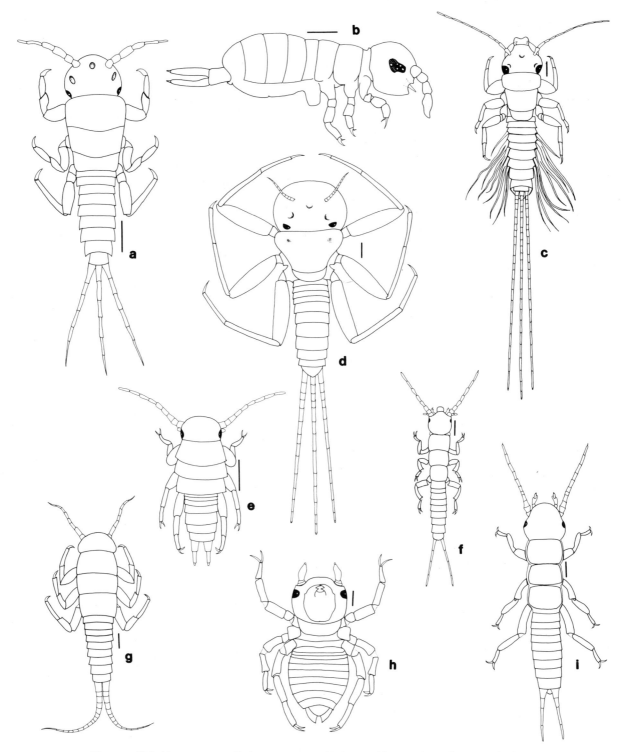

Figure 47.1.—Insecta: a, Ephemeroptera, Ephemerellidae (dorsal); **b,** Collembola, Entomobryidae (lateral); **c,** Ephemeroptera, Leptophlebiidae, *Paraleptophlebia* (dorsal); **d,** Ephemeroptera, Heptageniidae, *Stenacron* (dorsal); **e,** Plecoptera, Peltoperlidae, *Peltoperla* s.s. (dorsal); **f,** Plecoptera; Leuctridae, *Leuctra* (dorsal); **g,** Ephemeroptera, Baetidae, *Baetis* (dorsal); **h,** Odonata, Gomphidae, *Lanthus* (ventral); **i,** Plecoptera, Chloroperlidae, *Sweltsa* (dorsal). (Scale = 100 μm.)

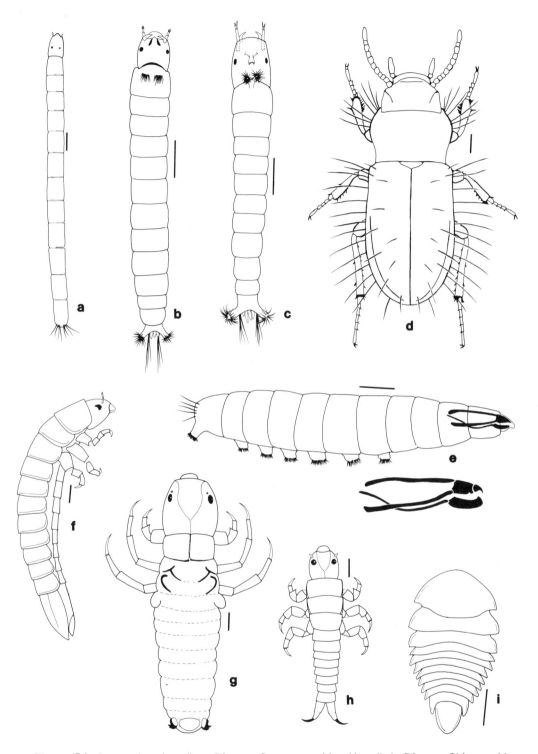

Figure 47.2.--Insecta (continued): **a**, Diptera, Ceratopogonidae (dorsal); **b**, Diptera, Chironomidae, Orthocladiinae, *Stilocladius* (ventral); **c**, Diptera, Chironomidae, Tanypoinae (ventral); **d**, Coleoptera, Phreatodytidae, *Phreatodytes relictus* (dorsal, redrawn from Figure 1 of Uéno, 1957); **e**, Diptera, Empididae (lateral with detail of internal cephalopharyngeal skeleton); **f**, Coleoptera, Elmidae, *Optioservus* (lateral); **g**, Trichoptera, Lepidostomatidae, *Lepidostoma* (dorsal); **h**, Trichoptera, Hydroptillidae, *Hydroptilla* (dorsal); **i**, Coleoptera, Eubriidae, *Ectopria* (dorsal); (Scale = 100 μm.)

48. Bryozoa

Patricia L. Cook

All members of the Phylum Bryozoa are colonial, and all colonies are formed by repeated budding of genetically (but not always morphologically) identical, physically connected, intercommunicating member zooids. Body walls are cuticular and flexible, but are variously strengthened by calcification in most species. Zooidal coeloms communicate by means of special cells which plug interconnecting pores in body walls, and nutrients and nervous impulses are transferred within colonies so that damage is repaired, non-feeding parts maintained and colony wide functions coordinated.

Feeding zooids (autozooids) possess, at some period or periods of their lives an evertible, usually circular, crown of ciliated tentacles with supporting structures (lophophore). The everted lophophore can be withdrawn by a retractor muscle through an orifice in the body wall. The orifice is closed by a sphincter and is sometimes protected by a flap of body wall (operculum). When the lophophore is everted, the tentacles filter food particles from the surrounding water and pass them through a central mouth into a U-shaped gut, the anus opening outside the lophophore. The lophophore and viscera, with some muscles, degenerate and regenerate cyclically during colony life. Polymorphic zooids include some without orifice and viscera (kenozooids). Kenozooids may have muscles and often have uncalcified body walls and function as anchoring rhizoids.

Other zooids, which usually lack a feeding capacity (avicularia), have variously enlarged and modified opercula (mandibles) which function in cleaning, stabilizing, supporting and defending colonies, and in a few species, sustain locomotion. Other non-feeding zooids are known to be males which are modified for sperm dissemination. Ova are usually brooded within diverticula of the body wall, and are often associated with calcified structures (ovicells) which may be partially or wholly kenozooidal. The maternal zooid, which produces the ovum, and sometimes the surrounding zooids, are often morphologically modified. Colony wide (extrazooidal) structures may include strengthening calcification, and partially, or wholly uncalcified rhizoid systems (Boardman et al., 1983).

All colonies are founded by metamorphosis of a motile larva (which is usually lecithotrophic), into a primary zooid (ancestrula) or ancestrular complex, which is almost invariably attached to a substratum. In meiofaunal species, this substratum is usually a minute fragment, and the attachment may be only temporary. Three kinds of colony morphology may develop. The first may consist of 4 or 5 zooids only and may breed without further growth. The second may grow beyond the substratum, sometimes incorporating it within a mass of basal calcification. These colonies have avicularia with long, bristle-like (setiform) mandibles, which stabilize and support, and sometimes sustain locomotion. The third kind of colony maintains position among the particles of sediment by means of adherent or adhesive, flexible rhizoids or kenozooidal processes. Attachment to or entanglement among particles may be temporary, and colonies may feed within the sediment or above its surface. Species in this third group tend to have colonies which consist of one relatively large feeding zooid, with several processes or rhizoids, but others have globular or conical colonies composed of small zooids, anchored by a single rhizoid.

Species of all three kinds have evolved convergently among a wide diversity of systematically unrelated groups. More than 150 species are now known, or may be inferred, to spend their early life, at least, as part of the interstitial meiofauna. The previous paucity of reported forms (Gray, 1971) is entirely the result of a lack of examination of sediments. Colonies are often difficult to recognize as bryozoans (d'Hondt, 1975; Berge et al., 1985). Some closely resemble foraminiferans (Cook, 1981) or ascidians (compare *Pachyzoon* Figure 48.1e with the interstitial solitary ascidian, *Heterostigma fagei*, illustrated by Monniot, 1971). The different kinds of colony forms tend to occur in parallel groups from very deep or from shallow water. This suggests that some of the species described here from abyssal depths may have as yet unknown, shallow-water analogues.

Classification

Two of the three Classes of Bryozoa are found in Recent seas; the Gymnolaemata, which includes the Orders Ctenostomata and Cheilostomata, and the Stenolaemata, which includes the Order Tubuliporata (or Cyclostomata). Although some interstitial cyclostomes have recently been described (Hakansson and Winston, 1985; Winston and Hakansson, 1986), the meiofaunal forms dealt with below include only ctenostomes and cheilostomes. The Ctenostomata have uncalcified body walls, and colonies are formed with various specific patterns of autozooids and kenozooids (Hayward, 1985). The Cheilostomata have partially calcified body walls, opercula, and a diversity of polymorphic zooids (Cheetham and Cook, 1983).

Family Monobryoontidae.--Within the Ctenostomata, the Family Monobryozoontidae includes a single genus, *Monobryozoon,* the most widely known meiofaunal bryozoan group. Colonies consist of a single, bulbous autozooid, with sometimes a few attached zooid buds and with kenozooidal basal processes. These muscular processes move constantly, and some have adhesive glands which allow temporary attachment to surrounding particles. Three species are known: *M. ambulans* (Figure 48.1b) from shallow *Amphioxus*-sands and shell gravels in the North Sea; *M. bulbosum* (Figure 48.1c) from shallow water off North Carolina; and *M. sandersi* from 805-811 m off the northeastern coast of the U.S.A. (d'Hondt, 1983; Hayward, 1985; d'Hondt and Hayward, 1981).

Family Nolellidae.--This family includes the genus *Nolella,* in which colonies are formed by nets or chains of discrete autozooids, connected by long, tubular extensions of body wall. Rhizoids are kenozooids which have the capacity to produce new zooid buds terminally. The subgenus *Franzenella* includes those species which have colonies with few zooids (4-25), which are not attached to a substratum, and which have elongated autozooids which may feed above the surface of the sediment. Only one of the three species, *N. (F.) limicola* (Figure 48.1a), which was originally attributed to *Monobryozoon,* is from shallow waters, where it is often found with *M. ambulans.* The other two species are *N. (F.) monniotae,* and *N. (F.) radicans,* both from between 4000-6000 m in the Atlantic (d'Hondt, 1983). *Nolella (F.) monniotae* is often associated with komokiacean foraminiferans, which form small cushion-like massses in the uppermost layers of abyssal oozes (Gooday and Cook, 1984).

Family Pachyzoontidae.--This family includes a single, monospecific genus, *Pachyzoon,* which has colonies forming somewhat similar cushions of contiguous autozooids. The inferred upper surfaces of *P. atlanticum* (Figure 48.1e) bear elongated autozooids which probably feed above the surface of the sediment. The opposing surfaces bear numerous rhizoids. The body walls of autozooids and kenozooids are adhesive, and colonies are covered with attached sediment particles. *Pachyzoon atlanticum* occurs from 800-1600 m in the Bay of Biscay (d'Hondt, 1983).

Family Beaniidae.--Among the Cheilostomata, the Suborder Anasca includes species in which the frontal body wall, which is continuous with the operculum, is uncalcified and flexible, although it may be overarched, and underlain by calcified structures. In the Family Beaniidae, most species have colonies which form loosely attached mats of discrete autozooids, anchored by cuticular basal rhizoids. The frontal body wall is often surrounded by long spines. *Beania proboscidea* (Figure 48.1f) has colonies comprising 1-3 autozooids with numerous rhizoids attached to particles of sediment. The elongated "proboscis"-like oral region suggests that autozooids feed above the surface of the sediment. *Beania proboscidea* occurs from 1457-1463 m off western New Zealand (Gordon, 1986).

Family Setosellinidae.--This family includes two genera, *Setosellina* and *Heliodoma,* both of which encrust minute substrata and have interzooidal avicularia with long, stabilizing, setiform mandibles. Autozooids and avicularia are interspersed in somewhat irregular spirals in *Setosellina,* which includes four species: *S. roulei* (Figure 48.1g), from 500-2330 m in the northeast Atlantic and the Indian Ocean; *S. capriensis,* from 50-250 m in the Mediterranean; *S. goesi,* from 20-480 m in the Gulf of Mexico and Caribbean; and *S. constricta,* from 18-60 m in the East Indies. In the monospecific genus *Heliodoma, H. implicata* has zooids budded laterally in paired, interlocking spirals. Colonies show some substratum preference and are usually found with *S. roulei* (Harmelin, 1977; Hayward and Cook, 1979).

Families Selenariidae (or Lunulitidae) and Cupuladriidae.--These are two unrelated, but highly convergent families having conical or discoid colonies which incorporate the original minute substratum particle within a mass of extrazooidal calcification. They both have regularly patterned avicularia with setiform supportive mandibles (Cook and Chimonides, 1983). All species are potentially meiofaunal at the earliest growth stages, but most of the 50 or more Recent forms exceed 2 mm in diameter at maturity. Only 4 species of the Family Selenariidae (or Lunulitidae, see Cook and Chimonides, 1986) are less than 1.5 mm when sexually mature. *Selenaria* includes species with radial budding and large, isolated avicularia. Distinctive male and female zooids form the outer zones of colony growth, and the avicularia sustain locomotion (Cook and Chimonides, 1978; Chimonides and Cook, 1981). *Selenaria initia* is found from 46-100 m in the Bass Strait, Australia (Cook and Chimonides, 1985a). The genus *Otionella* includes species with smaller avicularia which may occur in radial series. The mandibles are not as elongated as those of *Selenaria* and are not known to sustain colony locomotion; no distinctive male zooids occur. Both *Otionella auricula* and *O. minuta* (Figure 48.1i) are found with *S. initia* from the Bass Strait (Cook and Chimonides, 1985b). The genus *Helixotionella* has similar, small colonies, but the autozooids and avicularia are budded in paired spirals. Two of the three known species are meiofaunal when sexually mature; *H. spiralis* and *H. scutata* are found together from 80-140 m off western Australia (Cook and Chimonides, 1984b).

Division Cribrimorpha.—This group is included with either the Anasca or the Ascophora; it comprises forms in which the flexible frontal body wall is overarched by a shield of fused calcified spines. Only one monospecific genus, *Inversiscaphos*, is known to be meiofaunal. Colonies of *I. setifer* (Figure 48.1h) encrust minute substrata and have small avicularia with long, setiform stabilizing mandibles; the species occurs from 370-384 m off southeastern Africa (Hayward and Cook, 1979).

Suborder Ascophora.—In this group, the flexible part of the frontal body wall is concealed beneath a complete calcified shield, which usually has an overlying coelom beneath an investing cuticle. This coelom is in communication with the visceral coelom through pores, which allows the development of zooid buds frontally. All the purely interstitial ascophoran genera have colonies formed almost entirely of frontal buds, although members of some other genera, which pass only their earliest growth stages as meiofauna, develop in the more usual, distally budded manner (Cook and Lagaaij, 1976; Cook and Chimonides, 1985c).

Family Celleporariidae.—This family includes many species, some of which are massive, and nearly all of which are encrusting. One genus, *Sphaeropora*, has minute, globular colonies, which are frequently interstitial and are anchored among the sediment particles by a long, wide, turgid, extrazooidal rhizoid. The complex known as *"S. fossa"* (Figure 48.1d) includes several species, all from Australasia, some of which are meiofaunal at maturity (Cook and Chimonides, 1981a).

Family Orbituliporidae.—Consists of species which are all known, or inferred to be interstitial and anchored by rhizoids. The genus *Lacrimula* has pear-shaped colonies, anchored by a wide adapical rhizoid. *Lacrimula pyriformis* (Figure 48.1n) occurs from 100-700 m off southeast and east Africa (Hayward and Cook, 1979). The closely similar genus *Batopora* also has species with an adapical rhizoid. *Batopora nola* and *B. murrayi* (Figures 48.1l,k) occur from 376-805 m off southeastern Africa (Hayward and Cook, 1979).

Family Conescharellinidae.—This family includes several genera, all of which have colonies living anchored by rhizoids within sediments. Some species appear to spend their entire life history as interstitial forms, and some are meiofaunal in size range. Colonies of the genus *Conescharellina* are conical or globular. *Conescharellina africana* (Figure 48.1m) differs from others in its apical bulb, which is surrounded by small avicularia; it occurs from 100-800 m off southeastern Africa (Hayward and Cook, 1979).

Habitats and Ecological Notes

Meiofaunal bryozoans occur with other "sand fauna" species, which live on the surface, or within the upper 10 cm, of muddy sands, and medium to coarse shell sands and gravels, from only a few meters to abyssal depths. Bryozoans are now known from high energy environments as well as from deep water oozes, where they may be associated with thread-like masses of komokiacean foraminiferans. Many shallower water species are associated in a similar manner with the rhizoid masses and entangled sediments of larger "sand fauna" bryozoan colonies, such as *Parmularia* (Cook and Chimonides, 1981b; 1985c). To date, meiofaunal and other interstitial species have been reported from cold waters off Norway and Sweden as well as from warm waters in all oceans. It seems very probable that more will be reported as further sediments are examined. Nearly all the cheilostome genera mentioned here have fossil representatives in the Tertiaries of Europe, the Americas or Australasia. A few living species such as *Selenaria initia* and *Helixotionella spiralis* were first described as fossils and have apparently existed, virtually unchanged from Early Oligocene times to the present. Several purely fossil genera have colony forms analogous to those of Recent meiofaunal groups (Cook and Chimonides, 1981a; Cook and Voigt, 1986).

Methods of Collection and Extraction

The best samples are taken using an epibenthic sledge (Gray, 1971; Aldred et al., 1976). Ideally, living colonies may be separated by examination of sediments kept in shallow trays of aerated seawater.

Most "free living" colonies will be seen at or near the surface after a few hours, and some species of *Selenaria* will move towards light. The basal processes of *Monobryozoon* move continuously, but these colonies, together with the early growth stages of many other, like *Conescharellina,* and *Sphaeropora,* are all semi-transparent and difficult to see. Only detailed examination of every sand grain and very careful dissection of komokiaceans and rhizoid masses will reveal these colonies, together with those of the encrusting species.

Preparation for Taxonomic Study

All colonies may be relaxed in magnesium chloride solution and are best killed, after relaxation, by the very gradual addition of 5% formalin to the seawater (see also Gray, 1971). Preserved samples of sediment may be examined dry or in water. Dry sediments will retain only the calcified skeletons of cheilostomes and cyclostomes and will include an abundance of accumulated dead colonies. Preservation in alcohol tends to shrink and distort any ctenostomes present and the rhizoids of other species. They may be restored to shape by transferring washed specimens to a 10% solution of trisodium phosphate for 5-30 minutes. Alcohol with a little glycerin is a suitable preservative for all specimens; buffered formalin or propylene phenoxetol is suitable only for ctenostomes. For further notes on practical methods see Hayward and Ryland (1979).

Type Collections

Extensive collections of "sand fauna" species, including some meiofaunal types, are located at the British Museum (Natural History), London, UK; the Musée National d'Histoire Naturelle, Paris, France; the National Museum of Victoria, Melbourne, Australia; and the Western Australian Museum, Perth, Australia. Other type specimens have been reported to be housed in the American Museum of Natural History, New York, USA; the University of Uppsala, Uppsala, Sweden; and the Zoologisches Museum Berlin, Berlin, GDR.

References

Aldred, R.G., M.W. Thurston, A.L. Rice, and D.R. Morley
1976. An Acoustically Monitored Opening and Closing Epibenthic Sledge. *Deep-Sea Research*, 23:167-174.

Berge, J.A., H.P. Leinaas, and K. Sandoy
1985. The Solitary Bryozoan, *Monobryozoon limicola* Franzén (Ctenostomata), a Comparison of Mesocosm and Field Samples from Oslofjorden, Norway. *Sarsia*, 70:91-94.

Boardman, R.S., A.H. Cheetham, and P.L. Cook
1983. Introduction to the Bryozoa. Pages 3-48, 24 Figures in R.A. Robison, editor, *Treatise on Invertebrate Paleontology, Part G, Bryozoa Revised.* Boulder, Colorado and Lawrence, Kansas: Geological Society of America Inc. and Univeristy of Kansas.

Cheetham, A.H., and P.L. Cook
1983. General Features of the Class Gymnolaemata. Pages 138-195, 23 Figures in R.A. Robison, editor, *Treatise on Invertebrate Paleontology, Part G, Bryozoa Revised.* Boulder, Colorado and Lawrence, Kansas: Geological Society of America Inc. and University of Kansas.

Chimonides, P.J., and P.L. Cook
1981. Observations on Living Colonies of *Selenaria* (Bryozoa, Cheilostomata), II. *Cahiers de Biologie Marine*, 22:207-219.

Cook, P.L.
1965. Polyzoa from West Africa. The Cupuladriidae (Polyzoa, Anasca). *Bulletin of the British Museum (Natural History), Zoology*, 13(6):191-227.
1981. The Potential of Minute Bryozoan Colonies in the Analysis of Deep Sea Sediments. *Cahiers de Biologie Marine*, 22:89-106.

Cook, P.L., and P.J. Chimonides
1978. Observations on Living Colonies of *Selenaria* (Bryozoa, Cheilostomata), I. *Cahiers de Biologie Marine*, 19:147-158.
1981a. Morphology and Systematics of Some Rooted Cheilostome Bryozoa. *Journal of Natural History*, 15:97-104.
1981b. Early Astogeny of Some Rooted Cheilostome Bryozoa. Pages 59-64, 3 figures in G.P. Larwood and C. Nielsen, editors, *Recent and Fossil Bryozoa.* Fredensborg: Olsen and Olsen.
1983. A Short History of the Lunulite Bryozoa. *Bulletin of Marine Science*, 3(3):566-581.
1984a. Recent and Fossil Lunulitidae (Bryozoa, Cheilostomata) 1. The Genus *Otionella* from New Zealand. *Journal of Natural History*, 18:227-254.
1984b. Recent and Fossil Lunulitidae (Bryozoa, Cheilostomata) 2. Species of *Helixotionella* gen.nov. from Australia. *Journal of Natural History*, 18:255-270.
1985a. Recent and Fossil Lunulitidae (Bryozoa, Cheilostomata) 3. "Opesiulate" and Other Species of *Selenaria sensu lato.* *Journal of Natural History*, 19:285-322.
1985b. Recent and Fossil Lunulitidae (Bryozoa, Cheilostomata) 4. American and Australian Species of *Otionella*. *Journal of Natural History*, 19:575-603.
1985c. Larval Settlement and Early Astogeny of *Parmularia* (Cheilostomata). Pages 71-78, 4 figures in C.Nielsen and G.P. Larwood, editors, *Bryozoa: Ordovician to Recent.* Fredensborg: Olsen and Olsen.
1986. Recent and Fossil Lunulitidae (Bryozoa, Cheilostomata) 6. *Lunulites sensu lato* and the Genus *Lunularia* from Australasia. *Journal of Natural History*, 20:681-705.

Cook, P.L., and R. Lagaaij
1976. Some Tertiary and Recent Conescharelliniform Bryozoa. *Bulletin of the British Museum (Natural History), Zoology*, 29(6):317-376.

Cook, P.L., and E. Voigt
1986. *Pseudolunulites* gen. nov. (Bryozoa), a New Kind of Lunulitiform Cheilostome from the Upper Oligocene of Northern Germany. *Verhandlungen des Naturwissenschaftlichen Vereins in Hamburg* (NF), 28:107-127.

Gooday, A.J., and P.L. Cook
1984. An Association Between Komokiacean Foraminifers (Protozoa) and Paludicelline Ctenostomes (Bryozoa) from the Abyssal Northeast Atlantic. *Journal of Natural History*, 18:765-784.

Gordon, D.P.
1986. The Marine Fauna of New Zealand: Bryozoa: Gymnolaemata (Ctenostomata and Cheilostomata Anasca) from the Western South Island Continental Shelf and Slope. *Memoir of the New Zealand*

Oceanographic Institute, 95:1-121.

Gray, J.S.
1971. The Meiobenthic Bryozoa. Pages 37-39 in N.C. Hulings, editor, Proceedings of the First International Conference on Meiofauna. *Smithsonian Contributions to Zoology*, 76:1-205.

Hakansson, E., and J.E. Winston
1985. Interstitial Bryozoans: Unexpected Life from a High Energy Environment. Pages 125-134, 7 figures in C. Nielsen and G.P. Larwood, editors, *Bryozoa: Ordovician to Recent*. Fredensborg: Olsen and Olsen.

Harmelin, J.G.
1977. Bryozoaires du banc de la Conception (nord de Canaries). Campagnes Cinéca I du Jean-Charcot. *Bulletin du Muséum National d'Histoire Naturelle Paris, (3) Zoologie*, 341:1057-1074.

Hayward, P.J.
1985. Ctenostome Bryozoans. *Synopses of the British Fauna* (New Series) No. 33. 169pp. D.M. Kermack and R.S. Barnes, editors. Linnean Society, London and Estuarine and Brackish-water Sciences Association. Leiden, London, Cologne, and Copenhagen: Brill.

Hayward, P.J., and P.L. Cook
1979. The South African Museum's Meiring Naude Cruises, Pt. 9, Bryozoa. *Annals of the South African Museum*, 79(4):43-130.

Hayward, P.J., and J.S. Ryland
1979. British Ascophoran Bryozoans. *Synopses of the British Fauna* (New Series) No. 14. 312 pages. D.M. Kermack and R.S. Barnes, editors. Linnean Society, London and Estuarine and Brackish-water Sciences Association. London, New York, San Francisco: Academic Press.

d'Hondt, J.-L.
1975. Bryozoaires Cténostomes bathyaux et abyssaux de l'Atlantique Nord. *Documents des Laboratoires de Géologie de la Faculté des Sciences de Lyon*, H.S., 3(2): 311-333.
1983. Tabular Keys for Identification of the Recent Ctenostomatous Bryozoa. *Mémoires de le Institut Océanographique, Monaco*, 14: 1-134.

d'Hondt, J.-L., and P.J. Hayward
1981. Nouvelles récoltes des Bryozoaires Cténostomes bathyaux et abyssaux. *Cahiers de Biologie Marine*, 22: 267-283.

d'Hondt, J.-L., and T.J.M. Schopf
1984. Bryozoaires des grandes profondeurs recueillis lors des campagnes océanographiques de la Woods Hole Oceanographic Institution de 1961 a 1968. *Bulletin du Muséum National d'Histoire Naturelle, Paris*, (4) series 6A, number 4:907-973.

Monniot, F.
1971. Les Ascidies littorales et profondes des sédiments meubles. Pages 119-125, in N.C. Hulings, editor, Proceedings of the First International Conference on Meiofauna. *Smithsonian Contributions to Zoology*, 76:1-205.

Ryland, J.S.
1970. *Bryozoans*. London: Hutchinson University Library. 175 pages.

Winston, J.E., and E. Hakansson
1986. The Interstitial Bryozoan Fauna from Capron Shoal, Florida. *American Museum Novitates*, 2865:1-50.

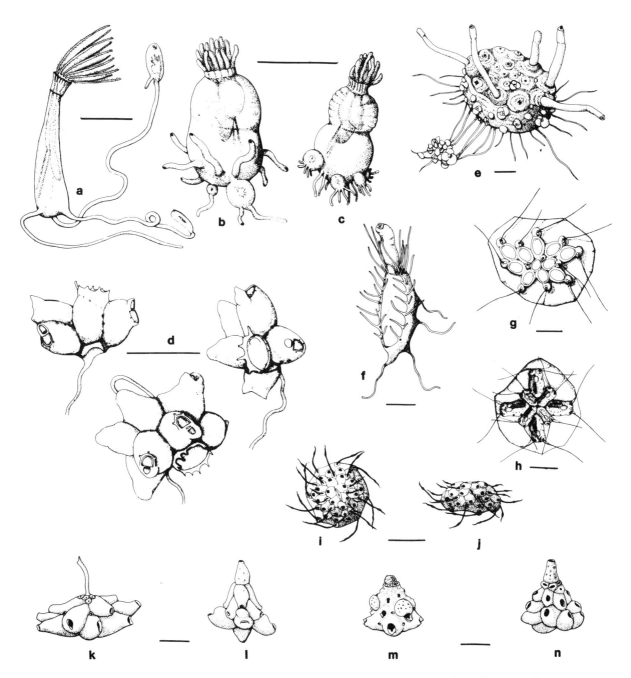

Figure 48.1.--Bryozoa: **a**, *Nolella (Franzenella) limicola* (Franzen), after Franzen; **b**, *Monobryozoon ambulans* Remane, after Remane; **c**, *Monobryozoon bulbosum* Ott, after Ott; **d**, *Sphaeropora "fossa"* Haswell, from Cook, reproduced by courtesy of the Systematics Association; **e**, *Pachyzoon atlanticum* d'Hondt, from specimen in British Museum (Natural History) (note adhering foraminiferans); **f**, *Beania proboscidea* Gordon, from specimen in British Museum (Natural History); **g**, *Setosellina roulei* Calvet, from Hayward and Cook, 1979; **h**, *Inversiscaphos setifer* Hayward and Cook, from Hayward and Cook, 1979; **i,j**, *Otionella minuta* Cook Chimonides, from specimen in the British Museum (Natural History); **k**, *Batopora murrayi* Cook, from Hayward and Cook, 1979; **l**, *Batopora nola* Hayward and Cook, from Hayward and Cook, 1979; **m**, *Conescharellina africana* Cook, from Hayward and Cook, 1979; **n**, *Lacrimula pyriformis* Cook, from Hayward and Cook, 1949; (**c,d,k-n** reproduced by courtesy of the South African Museum). (Scale = 0.5 mm.)

49. Entoprocta

Claus Nielsen

The Entoprocta is a small phylum of sessile or vagile, solitary or colonial metazoans. Each zooid has a globular to slightly flattened body with a U-shaped gut, a stalk (longer or shorter than the body) which rises from a branching stolon in most of the colonial species, and an almost closed "horseshoe-shaped" lophophore of ciliated tentacles. When slightly irritated, the zooids bend the stalk and perform the nodding movements which inspired the older name for the phylum (Kamptozoa). With stronger irritation, the zooids also curl the tentacles towards the middle of the horseshoe and a ring muscle at their bases contracts so that the tentacles become fully enclosed.

Entoprocts have brood protection and their eggs develop into feeding larvae while still attached to the parent zooid by a gelatinous stalk and surrounded by the tentacles. The larvae are trochophores and vary widely both in morphology and type of metamorphosis.

The solitary species constitute the family Loxosomatidae, and the only hitherto described interstitial-meiofaunal species belongs to the genus *Loxosoma*. In this genus, the stalk of the animal ends in a sucking disk. The species of this genus are generally associated with polychaetes and are found either attached to the worm itself or to the inner side of its tube.

Loxosoma isolata Salvini-Plawen (Figure 49.1) has been found in coarse sand (grain diameter about 0.5-1.0 mm) from 6-7 m depth in the northern Adriatic (Salvini-Plawen, 1968). An undescribed species has been reported by Thane-Fenchel (1970) from coarse sand slightly below the low water mark at the north coast of Key Biscayne, Florida, U.S.A.

Nearly any qualitative or quantitative collecting protocol conducted in a coarse sand habitat is adequate for collection of Entoprocta if they are present. Material can be extracted from the sediment best by initial relaxation or narcotization of the animals using a mixture of equal volumes of sea water and isotonic magnesium chloride in distilled water followed by the filtration of the supernatant through 42 μm mesh netting. The specimens will soon recover when returned to sea water.

It is essential to study entoprocts alive, or to first narcotize and fix them in an expanded condition; material fixed directly in formalin or alcohol is often very difficult to identify. Although the above-mentioned technique using magnesium chloride gives good results with many species, cocaine or stovain is generally easier to use and is more reliable. Fixation and storage for general purposes is best in 5% formalin in sea water.

References

Salvini-Plawen, L. v.
1968. Neue Formen im marinen Mesopsammon: Kamptozoa und Aculifera (nebst der für die Adria neuen Sandfauna). *Annalen des Naturhistorischen Museums in Wien*, 72:231-272.

Thane-Fenchel, A.
1970. Interstitial Gastrotrichs in Some South Florida Beaches. *Ophelia*, 7:113-138.

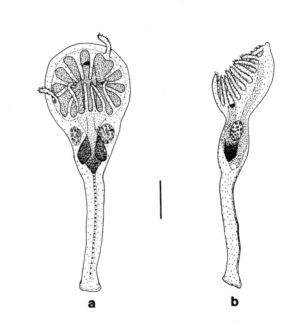

Figure 49.1.--Entoprocta: a, *Loxosoma isolata*, a half-way expanded specimen in frontal view; b, same, lateral view (after Salvini-Plawen, 1968). (Scale = 100 μm.)

50. Brachiopoda

Claus Nielsen

Brachiopods are benthic animals characterized by the presence of a dorsal and a ventral valve secreted by a pair of mantle folds whose edges are bordered with fine setae. The larger part of the mantle cavity is occupied by the lophopore, a more-or-less complicated arrangement of long, slender, ciliated tentacles. The articulate brachiopods are attached to a hard substratum by a short stalk which protrudes through a narrow hole in the ventral valve. The three other main groups of brachiopods, sometimes united by the name inarticulates, include: (1) the lingulids, with a long cylindrical stalk protruding between the valves and attached loosely in a soft substratum; (2) the discinids, which have a short stalk protruding through a notch in the ventral valve and attached to a hard substratum; and (3) the craniids, with the ventral valve cemented to the substratum.

Most brachiopods are free spawners and each of the four main groups has a characteristic larva. The minute species mentioned below have brood protection in the mantle cavity, where one or two larvae develop at a time.

The only brachiopod *described* as meiofaunal (Swedmark, 1967) is the articulate, *Gwynia capsula* (Jeffreys) (Figure 50.1). This species attains a size of 1.5 mm and has been found in the western part of the English Channel and along the western shores of the British Isles at depths of 15-45 m. *Gwynia capsula* usually attaches to stones or boulders, but some specimens have been found attached in the sheltered concavities of fragments of serpulid tubes in shell-gravel habitats.

A second species of articulate brachiopod, equally small, is *Argyrotheca bermudana* Dall. This species attaches to the undersides of rocks or corals in Bermuda. No information exists as to whether or not it has been collected during the normal course of dredging sediments, however. It is easily confused with the foraminiferan *Homotrona rubrum* which is of similar size and color.

Neither of the two brachiopod species mentioned is known to be less than 1 mm as a sexually mature adult, yet neither exceeds 2 mm. Whether or not these minute brachiopods should be considered meiofaunal or not is a matter of opinion, but they clearly represent the smallest members of this phylum.

A complicating factor is the fact that newly settled specimens of macrofaunal species are equally small, therefore, the sexual maturity of a specimen is important to determine before classifying it as a permanent member of the meiofauna.

Methods of Collection and Preparation of Specimens for Taxonomic Study

The sessile nature of the Brachiopoda creates collecting problems not normally found in most meiofaunal taxa. Mere relaxation of the specimen, although an advantage in the identification of the species, will not cause its release from its substratum. Where shell gravel has been a substratum under investigation for its meiofaunal components, only close inspection of the individual pieces of shell, gravel, or hard surfaces making up the habitat will reveal brachiopods should they be present. Since this procedure is rarely followed by most meiofaunal investigators, very little information has been published on this taxon.

Specimens should be relaxed with isotonic $MgCl_2$ or a suitable narcotic such as cocaine, if possible. Fixation in 10% formalin and subsequent storage in 70% alcohol is preferred. Identification is usually based on the characteristics of the valves as well as some internal soft parts.

References

Emig, C.C.
1986. Brachiopoda. Pages 518, 519 in Wolfgang Sterrer, editor, *Marine Flora and Fauna of Bermuda*, xxx + 742 pages. New York: John Wiley & Sons, Inc.

Swedmark, B.
1967. *Gwynia capsula* (Jeffreys), an Articulate Brachiopod with Brood Protection. *Nature*, 213:1151-1152.

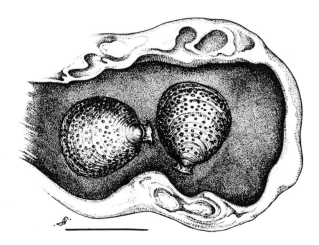

Figure 50.1.—**Brachiopoda**: *Gwynia capsula*, two adult specimens located in the concavity of a fragment of a serpulid worm (from Swedmark, 1967). (Scale = 1 mm.)

51. Aplacophora

M. Patricia Morse and Amelie H. Scheltema

Members of the class Aplacophora (solenogasters) are wormlike, bilaterally symmetrical, nontorted molluscs found in marine environments in all oceans. They range in length from less than 1 mm to 13 cm; most are elongate (length 2 to 30 times width), but a few are nearly round while others are extremely elongate and thread-like. Some burrow within soft mud bottoms or creep on the surface; others are found wrapped around hydroids and corals; still others are interstitial in habitat and found among coarse sand grains. The meiofaunal species described here are defined as 2 mm in length (as recorded in the literature) since the diameter of a specimen this long would easily allow its passage through a 1 mm sieve. Aplacophora exist most abundantly and in greatest diversity at depths of over 100 m. They are distinguished from most other worm-shaped invertebrates by their glistening coat of calcareous spicules (they share this with some marine gastrotrichs and a few marine turbellarians). The spicules are carried either erect or closely pressed to the body. The anterior end is rounded and lacks eyes and tentacles; posteriorly the body is rounded, pointed, or truncated, with a dorsal sensory organ shaped like a pit or groove. The spicules are embedded in a cuticle surrounding the body, are secreted by the underlying epidermis, and are diagnostic for species and often for higher taxa. A usually nonrasping radula with two teeth per row (distichous) is present and exhibits great morphological variability; it defines species and higher taxa. Each genus or family is marked by a particular body shape, which includes external details of the anterior oral area and posterior mantle cavity, or cloaca. Copulatory spicules, if present, define species. Knowledge of internal organization, particularly the salivary glands, reproductive organs, and epidermal cells, is usually necessary for classification of species but seldom for species identification.

Classification

Two subclasses, Chaetodermomorpha (=Caudofoveata) and Neomeniomorpha (=Solenogastres *sensu* Salvini-Plawen) are recognized (Scheltema, 1978).

They are distinguished from each other by the presence (Neomeniomorpha) or absence (Chaetodermomorpha) of a ciliated foot in a narrow ventral groove. Of the 262 named species, 17 species in 11 genera belong to the meiofauna. Species with special adaptations to live in the interstices of coarse sand are found only within the Neomeniomorpha and in relatively shallow water (intertidal to less than 60 m); all other meiofaunal aplacophorans have been taken in soft muds and silts and extend from shallow water to hadal depths of more than 7,000 m.

Neomeniomorpha creep on a mucous track formed by glands associated with the foot groove and usually, an anterior pedal pit. The spicules range from solid scales to elongate rods or needles that are often hollow. Anteriorly the head is slightly differentiated by an eversible vestibule above the mouth, within which are sensory protuberances; there may also be sensory stereocilia that project forward in crawling animals. Neomeniomorphs are obligate feeders on Cnidaria. They are hermaphroditic and many species have copulatory spicules.

A number of interstitial Neomeniomorpha has been mentioned in the literature, but only six species have been described. In 1886, Marion and Kowalevsky described *Lepidomenia hystrix* associated with a piece of coral in the Gulf of Marseille at 30 m. Swedmark (1956) collected additional specimens in coarse sand from the same locality and identified them as *L. hystrix*. The status of this species needs further study on organisms from new collections at the type-locality. Salvini-Plawen (1968) described a second species, *Biserramenia psammobionta*, from coarse sand at Terenez, near Roscoff. Morse (1979) described *Meiomenia swedmarki*, a species Swedmark had previously collected at Friday Harbor, Washington. In 1985, Salvini-Plawen described three new species and a new genus, *Meioherpia*, from Bermuda and North Carolina and designated family relationships of the interstitial species.

In shape, the interstitial neomeniomorphs are moderately elongate, with length 5 to 8 times width. The spicules of *Biserramenia* are hollow and needle-shaped; those of *Lepidomenia*, *Meiomenia*, and *Meioherpia* are solid. In the latter two genera they can be grouped according to location, with shape and type of bonding to the underlying cuticle considered as adaptations for an interstitial habitat. Spicules of the general body covering are flattened, oval to round scales loosely attached by a cuticular protuberance that fits into a depression in the scale. Elongate paddles and/or spines are scattered on the body surface, or are most numerous along the sides or around the posterior end; they work against the sand grains as an aid to locomotion by means of elaborate cuticular connections that allow movement in several directions. Small triangular spicules of the head also have elaborate connections; they meet forces from many directions when the organism everts the vestibule during crawling and when the mouth and vestibule are withdrawn on disturbance. Pedal spicules alongside the pedal groove show species differences. Triangular spicules surround the posterior dorsal sensory organ if it is present. The radula is distichous; the numbers of transverse rows range from 7 to 26, with 3 to 6 denticles on each tooth, often with one of the denticles larger than the others. *Biseramenia* differs in having 45 rows of rake-like teeth with 30 sharp denticles per tooth. Copulatory spicules occur in *Meiomenia* and *Meioherpia,* but not in *Biserramenia* or *Lepidomenia*. In *Meiomenia swedmarki* a large adhesive gland is found in the posterior end of the body.

Eight species of Neomeniomorpha in 6 genera are meiofaunal but are not known to be interstitial. Six of the species occur in the Subantarctic; three of them are poorly described from single specimens (*Nematomenia incirrata* Salvini-Plawen, 1978, *Lepidomenia harpagata* Salvini-Plawen, 1978 and *Pholidoherpia lepidota* Salvini-Plawen 1978); the other subantarctic species are *Nematomenia protecta* Thiele, 1913, *Harpagoherpia tenuisoleata* Salvini-Plawen, 1978 and *Pholidoherpia cataphracta* (Thiele, 1913). Two other species, *Rupertomenia fodiens* (Schwabl, 1955) and *Genitoconia atriolanga* Salvini-Plawen, 1967, are known from Scandinavian waters. Seven species have solid spicules in the form of scales and a thin cuticle, characters which place them in the order Pholidoskepia (Salvini-Plawen, 1978). The eighth, *Harpagoherpia tenuisoleata,* has solid spines. Further taxonomic relationships do not exist. However, these species fall in the realm of meiofaunal in the size definition used in this paper (2 mm in length).

The subclass Chaetodermomorpha is recognized by the lack of foot and foot groove; there is an anterior cuticular oral shield and a fringe of long spines around a posterior cloaca which contains a pair of ctenidia. Chaetoderms burrow slowly through silt and mud. The spicules are solid scales and are often ornamented. Shape of the body reflects the internal arrangement of gonad, digestive gland, and stomach. Chaetoderms are dioecious and lack copulatory structures. They are either carnivorous on foraminiferans or other small organisms or they feed on organic detritus.

Only one chaetoderm family, the Prochaetodermatidae, has representatives of meiofaunal size, with four of its eight described species averaging 2 mm or less in length. In this family a pair of large jaws can often be seen through the cuticle at the anterior end and the oral shield is divided. The body terminates in a slender, tail-like posterium that is 1/4 to 1/3 total body length; preserved specimens are often flexed dorsally into an arch. The spicules, which characterize species, are elongate, pointed scales with little or no ornamentation; they vary in size and morphology from anterior to posterior and from dorsal to ventral. The radula is small and distichous, with 8 to 12 rows of teeth which are similar among species; between each pair of teeth is a radular plate which is species specific in shape. The prochaetodermatid radula is unique among Aplacophora because it is capable of rasping.

Prochaetodermatidae are deep-sea organisms found at depths greater than 100 m in both the Atlantic (*Chevroderma scalpellum* Scheltema, 1985; *Prochaetoderma yongei* Scheltema, 1985, and Pacific Oceans (*Chevroderma whitlatchi* Scheltema, 1985) but not in Arctic or Antarctic Oceans. There are no particular morphological features other than size to differentiate meiofaunal from macrofaunal species.

Habitats and Ecological Notes

Interstitial neomeniomorphs are found in coarse substratum which may consist of mixed shell fragments and larger gravel, or it may be well-sorted coarse sand. Organic silt is often included in the substratum but sulfides are never found. Favorable collection sites are usually in areas of strong currents. Areas which have been described are generally offshore and organisms have been reported from depths to 59 m. Associated interstitial fauna reported are other molluscs, such as species of the opisthobranch orders Acochlidiacea and Nudibranchia, representatives of the annelid genus *Protodrilus,* and interstitial-sized sipunculans, holothurians, actinulids, and hydroids. Associated macrofaunal species reported are the bivalve *Glycymeris subobsoleta,* the polychaete *Polygordius,* and the sipunculan *Golfingia pugettensis.*

Meiofaunal, non-interstitial Neomeniomorpha and Prochaetodermatidae (Chaetodermomorpha) are all part of the deep-sea, level-bottom community, a large group of diverse organisms that belong to many phyla and live in cold-temperature muds under pressure (Hessler and Sanders, 1967). Among the Prochaetodermatidae, whose species, in some localities, are among the most numerous organisms, meiofaunal size may be correlated with a life-history pattern of rapid growth and maturity (Scheltema, 1987). No explanation has been offered for why a few species of non-interstitial Neomeniomorpha should be meiofaunal in size.

Methods of Collection and Extraction

Sites for collection of interstitial aplacophorans include subtidal and intertidal habitats where there is adequate pore space in the substratum. Additionally, there must be a continuous exchange of water either as a result of currents or of wave action. Favorable collection sites are those from which other interstitial mollusks have been described. The occurrence of sentinel macrofaunal taxa such as *Glycymeris* or *Amphioxus* might indicate a molluscan interstitial habitat. Sand can be collected from boats by dredging with an anchor dredge lined with a canvas bag which prevents water from passing through the sand as the dredge is pulled back into the boat and washing out the mollusks. In shallower habitats, buckets of sand can be collected by SCUBA divers; only the surface (5-15 cm) should be collected. This method is particularly effective near large rocks and in the spur and groove zones of coral reefs.

Interstitial solenogasters can be collected intertidally in special habitats. For example, at Crow Neck, Maine, sand samples can be taken from around boulders accessible only at extreme low tides. This is a region that has a tidal amplitude of 8-10 m and the low areas are not subjected to extreme changes in temperature for any critical length of time. On San Juan Island, Washington, an intertidal locality for *Meiomenia swedmarki* is Minnesota Reef. This area is exposed only at the lowest tides, and large kelps attached to boulders protect the underlying sand from extreme temperature changes. The solenogasters are found in the sand that collects around boulders when strong tidal currents run through the exposed area.

Moist sand can be left in tubs for several days before samples are taken from the upper layers, to which the solenogasters seem to migrate. Extracting the animals is done by elutriation. Sand is subsampled into a small bucket, seawater added and the mixture vigorously stirred. The seawater is rapidly decanted through a 42 µm nylon mesh screen from which the animals can be washed off with seawater into a Petri dish. Animals can be detected through a dissecting microscope and appear shiny opaque white in color.

The most satisfactory deep-sea collections are made with an epibenthic sled, which takes large samples over a wide, but undetermined, area (Hessler and Sanders, 1967); with a spade box corer, which takes an undisturbed quantitative sample (Hessler and Jumars, 1974); or from a deep submersible research vessel (i.e., submarine), which uses a variety of sampling methods under the collector's control at the site of collection (Grassle, 1980). Sieving at sea should be by elutriating, that is by running water through the sample which has been placed in a tall container and pouring the overflow gently through a screen (Sanders, Hessler, and Hampson, 1965).

Preparation of Specimens for Taxonomic Study

Organisms should first be studied *in vivo* to observe patterns of movement, distribution of spine-like scales, occurrence of anterior sensory bristles and length while crawling. One specimen should then be studied under a compound microscope slowly withdrawing the water so that camera lucida drawings of the various types of scales and their connection to the underlying cuticle can be made. Characteristics to be noted include presence or absence of a dorsoterminal sensory organ and copulatory spicules. Finally the buccal mass and the radula should be examined. The radula and copulatory spicules can be studied by first dissecting out the tissue in which they are embedded, placing the tissue on a depression slide, and dissolving the tissue in a solution of potassium hydroxide or common household bleach. The solvent is withdrawn by a pipette and the radula and spicules rinsed several times by adding and withdrawing distilled water with a pipette; a drop of glycerin is then added. Very small radulae or spicules can be transferred in a drop of glycerin to a slide for study with oil immersion microscopy.

It may not be possible to sort deep-sea samples alive; in that case the sample should be fixed in buffered 10% formalin and after 24 hours changed to 70 or 80% buffered alcohol. Specimens so fixed and preserved will be adequate for spicule and radula examination and for standard histologic preparations after they are decalcified with acetic acid.

Acknowledgments

Contribution Number 6697 from the Woods Hole Oceanographic Institution. Contribution Number 173, Marine Science Center, Northeastern University, Nahant (MPM).

References

Grassle, J.F.
1980. *In Situ* Studies of Deep-Sea Communities. In F.P. Diemer, F.J. Vernberg, and D.Z. Mirkes, editors, *Advanced Concepts in Ocean Measurements for Marine Biology*. Belle Baruch Library of Marine Science, No. 10. Columbia: University of South Carolina Press.

Hessler, R.R., and P.A. Jumars
1974. Abyssal Community Analysis from Replicate Box Cores in the Central North Pacific. *Deep-Sea Research*, 21:185-209

Hessler, R.R., and H.L. Sanders
1967. Faunal Diversity in the Deep-Sea. *Deep-Sea Research*, 14:65-78.

Marion, A.F., and A.O. Kowalevsky
1886. Organisation du *Lepidomenia hystrix*, Nouveau Type de Solenogastre. *Comptes Rendus Hébdomadaire des Séances de l'Académie des Sciences, Paris*, 103:757-759.

Morse, M.P.
1979. *Meiomenia swedmarki* gen. et sp. n., a New Interstitial Solenogaster from Washington, U.S.A. *Zoologica Scripta*, 8:249-253.

Salvini-Plawen, L.v.
1967. Neue Scandinavische Aplacophora (Mollusca, Aculifera). *Sarsia*, 27:1-63.
1968. Neue Formen im marinen Mesopsammon: Kamptozoa und Aculifera. *Annalen des naturhistorischen Museums in Wien*, 72:231-272.
1978. Antarktische und Subantarktische Solenogastres. *Zoologica* 128:1-315.
1985. New Interstitial Solenogastres (Mollusca). *Stygologia*, 1:101-108.

Sanders, H.L., R.R. Hessler, and G.R. Hampson
1965. An Introduction to the Study of Deep-sea Benthic Faunal Assemblages along the Gay Head-Bermuda Transect. *Deep-Sea Research*, 12:845-867.

Scheltema, A.H.
1978. Position of the Class Aplacophora in the Phylum Mollusca. *Malacologia*, 17:99-109.
1985. The Aplacophoran Family Prochaetodermatidae in the North American Basin, Including *Chevroderma* n.g. and *Spathoderma* n.g. (Mollusca: Chaetodermomorpha). *Biological Bulletin*, 169:484-529.
1987. Reproduction and Rapid Growth in a Deep-Sea Aplacophoran Mollusc, *Prochaetoderma yongei*. *Marine Ecology Progress Series*, 37:171-180.

Schwabl, M.
1955. *Rupertomenia fodiens* n.g. n.sp., Eine neue Lepidomeniide von der Südwestküste Schwedens. *Österreichische zoologische Zeitschrift*, 6:90-146.

Swedmark, B.
1956. Étude de la Microfaune des Sable Marin de la Région de Marseille. *Archives de Zoologie Expérimentale et Générale*, 93:70-95.
1968. The Biology of Interstitial Mollusca. *Symposium of the Zoological Society of London*, 22:135-149.

Thiele, J.
1913. Antarktische Solenogastren. *Deutsche Südpolar Expedition*. 14(2):35-65.

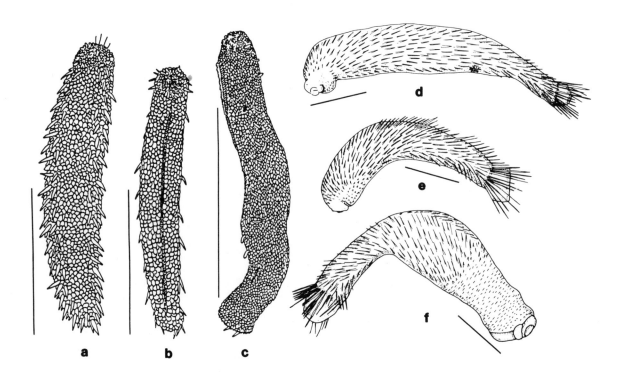

Figure 51.1.--Aplacophora. a-c, external appearance of three interstitial aplacophorans (subclass Neomeniomorpha): **a**, *Meioherpia stygalis* Salvini-Plawen, 1985; **b**, *Meiomenia swedmarki* Morse, 1979; **c**, *Meiomenia arenicola* Salvini-Plawen, 1985; **d-f**, external appearance of three meiofaunal aplacophorans (subclass Chaetodermomorpha): **d**, *Prochaetoderma yongei* Scheltema, 1985; **e**, *Spathoderma clenchi* Scheltema, 1985; **f**, *Chevroderma whitlatchi* Scheltema, 1985. Illustrations d-f from Scheltema, 1985 (courtesy *Biological Bulletin*). (Scale = 500 μm.)

52. Gastropoda and Bivalvia

Claude Poizat and Patrick M. Arnaud

Two major benthic molluscan classes, the Bivalvia and the Gastropoda, are commonly encountered in marine and freshwater sediments. Although nearly all of the benthic Bivalvia are temporary meiofauna in the early stages of their life history, and none fall within the size limits of the strict definition of meiofauna upon reaching sexual maturity, some commentary is appropriate because they may represent a significant food source for fishes and larger invertebrates in benthic habitats. Additionally, we should point out that several of the 85 families of Bivalvia such as the Lasaeidae, Nuculidae, Malletiidae, and Nuculanidae have representatives with maximum dimensions not much larger than 1-2 mm.

Juvenile bivalves are, nevertheless, a common occurrence in sediment samples and they can be confused with other bivalved benthic invertebrates such as the Ostracoda, and to a lesser extent, the smaller Brachiopoda. The typical shell of a bivalve mollusc consists of two symmetrical valves connected by an elastic ligament. In juveniles, the valves may be smooth or with concentric ridges or growth lines and a conspicuous umbo or beak. Often these valves are transparent or translucent so that some internal anatomical detail can be seen, thus adding characters (such as lack of jointed appendages as found in the Ostracoda or lophophore as found in the Brachiopoda, and often the appearance of distinct gills and the typical muscular foot) which separate this taxon from the other bivalved benthic invertebrates.

The identification of juvenile Bivalvia is so difficult that they are often counted with other "miscellaneous taxa." Nevertheless, a few species are listed from time to time in the literature, and in terms of the biomass they may be seasonally important. Spat of such species as *Macoma baltica* may be especially abundant as can be the case in spat of *Mytilus edulis*. The lack of meiobenthic representatives of adult Bivalvia may be related to their reproduction. As noted by Sellmer (1967) and others, a bivalve shell smaller than 3-4 mm cannot accommodate a gonad of sufficient size to produce enough eggs and still permit the organism to have a life history strategy which includes planktotrophy, the strategy of most Bivalvia. In all the smallest known species of Bivalvia, a small number of eggs are brooded, but the process requires a delicate balance in the allotment of space shared with the ever present gills.

The Gastropoda, on the other hand, are represented in the meiofauna as both temporary and permanent meiofauna. Most belong to various families of the Prosobranchia and Opisthobranchia. Relative to the total number of species in the Gastropoda, there are only a few meiofaunal representatives. There are about 57 species of Opisthobranchia and 44 species of Prosobranchia (Arnaud et al., 1986) listed as meiofauna. These species are, for the most part, poorly known and their descriptions are often inadequate. Because of their morphology, most meiofaunal gastropods occupy interstitial microhabitats. The following remarks concern only the Gastropoda.

Classification

The subclass Prosobranchia, order Mesogastropoda includes small organisms with either a very small operculated planorbid-like shell with a smooth surface (Omalogyridae, Figure 52.1a) or a curved tubular shell in the adult phase (Figure 52.1b), with or without a small coiled juvenile conch (Caecidae). Adult caecid prosobranchs may be confused with some temporary meiofauna such as the Scaphopoda (tusk shells) and tubicolous Polychaeta (*Ditrupa*). However, one must remember that both extremities of the caecid shell are closed (Figure 52.1b); the rear end of the adult shell is shut by a plug-septum and the mouth of the shell by an operculum. In contrast, the tubular shell of the scaphapod is more slender and open at both ends, and the increasing diameter of the *Ditrupa* shell abruptly narrows at the front opening. *Omalogyra* may also be confused with juvenile caecid prosobranchs (growth stage 1 of Folin), however, the protoconch (stage 1) of the caecid is smaller than the *Omalogyra* shell and subsequently acquires an uncoiled portion during growth.

The subclass Opisthobranchia correspond to an ill-defined group which includes the four following orders:

Order Cephalaspidea have a head shield, a reduced visceral hump with internal shell (Philinidae, Figure 52.1c) or generally without (Philinoglossidae, Figure 52.1d-g), a ciliated sole, lateral parapodia (Philinidae), with gill (Philinidae) or without (Philinoglossidae), with or without jaw and gizzard plates.

Order Ascoglossa have linear, sometimes flattened bodies (Platyhedylidae, Figure 52.1h), without palps (Platyhedylidae), with rhinophores (Stiligeridae, Figure 52.1i) or without (Platyhedylidae).

Order Acochlidioidea (Figure 52.1j-p) lack a shell, gill and cephalic disc, but are provided with one or two pairs of free head appendages (tentacles, rhinophores), axial hump, and subepidermal spicule formations. In this order sexes are separated (Microhedylidae) or not (Hedylopsidae); sometimes sperm transfer is by means of spermatophores.

Order Nudibranchia have cluster of dorsal ceratal processes with typical cnidosacs (Tergipedidae, Figure 52.1q) or without (Pseudovermidae (Figure 52.1r), acorn-shaped head devoid of appendages (Pseudovermidae) or with tentacles modified to a wide, bilobed velum and short cylindrical rhinophores (Embletoniidae, Figure 52.1q), ciliated pedal epidermis, no distinctly delimited foot (Pseudovermidaea), cerata in a single row on each side of the dorsum (*Embletonia*).

Two families of doubtful systematic position between the Opisthobranchia and the Pulmonata include hermaphroditic shell-less gastropods with pseudopulmonary and/or dermal respiration, with a well-delimited foot anteriorly differentiated into a kind of sucker (Smeagolidae, Figure 52.1t), or without a delimited foot (Rhodopidae, Figure 52.1s), without tentacles, without jaw, without eyes (Smeagolidae), or with one pair of subdermal eyes (Rhodopidae).

Meiobenthic opisthobranchs (Rhodopidae, Smeagolidae, Pseudovermidae) may be confused with Turbellaria, especially when in a preserved condition, and occasionally (Platyhedylidae and Philinoglossidae) can be confused with Nemertina which often fragment during fixation. However, the presence of a radula in the opisthobranchs, as well as the presence of spicules, rhinophores, and tentacles, does permit these molluscs to be separated from these non-molluscan taxa.

Many species have been described only on the basis of external morphology such as the cephalic shield, parapodia, gills, palps, oral veil, rhinophores, dorsal cerata, subdermal eyes, pigmentation of the tegument, visceral hump, foot sole, supplemented by a few details of the radula, jaws, gizzard plates and internal shell. Nowadays we are learning more about species through the use of transmission electron microscopy (TEM) and scanning electron microscopy (SEM). These techniques have helped achieve a more accurate description of taxonomic characters such as internal morphological structures which include alimentary canal, nervous system, sense organs, reproductive systems, excretory systems, heart, blood vessels, and hemocoel. The taxonomy of caecid gastropods is mainly based on the shell shape, diameter and sculpture (ribs and striae), the mouth of the shell (swollen or unswollen), operculum, plug-septum (angulate or mucronate), profile of the plug (flat, concave, triangular, etc.). The taxonomy of Omalogyridae is also mainly based on the shell (number and shape of whorls, depth of suture, and sculpture), and the operculum.

Habitats and Ecological Notes

Meiobenthic gastropods are presumed to be present in most sublittoral zones of the world's oceans provided the substrate is suitable. In western Europe, these gastropods live mainly in coarse sands (0.75-1.5 mm), very coarse sands (1.5-2 mm), and granules (2.0-4.0 mm) of high homogeneity (low to high sorting index) and frequently devoid or very poor in silt-clay particles and organic detritus (<5%, Poizat, 1981b, 1985b; Arnaud and Poizat, 1981). These kinds of sediments are the result of high energy to waves and currents. In such substrates, the pore space is extensive and ensures good drainage, resulting in well-oxygenated interstitial water. Generally, these interstitial gastropods can be considered as rheophilous species, and are useful indicators of clean, well-oxygenated, non-polluted habitats (Poizat, 1985a).

The interstitial gastropods exhibit a remarkable set of adaptations (Swedmark, 1964, 1968, 1971) to this environment. The Acochlidioidea have very well developed epidermal spicules and the caecids have a tiny curved shell, both ensuring mechanical protection. Others have adhesive glands and static sense organs which help maintain their retention in the habitat. These adaptations also include cutaneous fertilization and a lack of photic reaction in the larvae which thus remain in the interstitial water rather than being washed away by waves and currents.

In the Mediterranean Sea, where long term surveys have been conducted, meiobenthic gastropods exhibit various grain-size preferences (Poizat, 1981b; Arnaud and Poizat, 1981). Certain species such as *Hedylopsis spiculifera, Pontohedyle milaschewitschii, Philinoglossa helgolandica, Abavopsis latosoleata, Embletonia pulchra, Caecum auriculatum,* and *C. trachea* prefer gravel and very coarse sands. Other species such as *Unela glandulifera, Philine catena,* and *Caecum subannulatum* prefer a mixture of very coarse sands and coarse sands with a certain amount of clay particles and organic matter (<5%). A third category of very rare species such as *Pseudovermis papillifer* and *Platyhedyle denudata,* prefers a mixture of coarse and medium sands with clay particles and organic matter (up to 7-8%).

In tideless seas or in seas with very little tidal influence such as the Mediterranean and locally in the Skagerak, at the Bonden Island (outer Gullmarfjord (Poizat, 1980)), interstitial Gastropods are scanty in the intertidal zone; they become more abundant from 10 to 25 m; they progressively disappear from 25 m down to 50, sometimes 60 m.

In areas with a significant tidal range such as the English Channel, Irish Sea, and the North Sea, interstitial gastropods are found more frequently in the intertidal zone (low water mark) down to depths of 80-90 m, in sediment where the silt-clay and organic matter contents (up to 9-10%) are generally more important than in the Mediterranean substrates (Poizat, 1979; 1981a). As in other meiofaunal groups, the distribution of the gastropods is very patchy (Poizat, 1981a). The animals seem to be attracted by a common food or by some other mechanism such as mucous trails of the same species which, during reproductive periods, may have a significant value for the exchange of gametes.

Generally, horizontal and vertical distributions of these animals within a given sediment vary according to seasonal and nocturnal cycles, as well as to their abundance or density (Arnaud and Poizat, 1979; Poizat, 1980, 1983a,b, 1984). The mineralogical composition of the sediment is of major importance in the distribution of the species (Poizat, 1980). Interstitial gastropods are very rare or absent in detritic sediments along metamorphic or granitic shores (e.g., Skagerrak; Port-Cros Island, western Mediterranean) in spite of a convenient grain size, as a result of high density of the sand and gravel particles which are therefore hardly rearranged by waves or currents. The shape of the particles is also important in that certain species *(Embletonia pulchra)* prefer flat particles under which they hide for protection from the light (Poizat, 1980).

Methods of Collection

The very low density and patchiness of meiofaunal gastropods (about 1.5% of the total mesopsammon are opisthobranchs, and about 6% caecid prosobranchs; Poizat, 1981b; Arnaud and Poizat, 1981) make it impossible to accurately assess populations with small diameter corers. Sampling of large volumes of sand, up to 50 dm^3 per station, is compulsory. This is usually accomplished by SCUBA diving, using devices such as the "stainless-steel wedge" of Uhlig (1968) or by using dredges or special grabs. The dredge commonly used by the authors is the "Charcot" dredge (Poizat, 1978) but any dredge designed to collect the uppermost layers of sediment can be utilized in such collecting procedure. Grabs are notoriously difficult to use in coarse sediment where the majority of meiofaunal gastropods are found but the Smith-McIntyre grab (Smith and McIntyre, 1954), the Van Veen grab (Dybern et al., 1976) or the more sophisticated box corers are highly recommended. A significant factor in the sampling process is the minimization of disturbance, especially washing, of the sediment while in the collecting apparatus.

Methods of Extraction and Sorting

Sediment should be transferred to clean 10 dm^3 plastic buckets, preferably with a lid which avoids loss of material during transport, and evaporation during storage. Storage at low temperatures of 6° C and aeration of the seawater is recommended since gastropods quickly succumb to low oxygen tension. This is particularly important in warm or tropical regions. If only qualitative samples are sought, the "deterioration" or anaesthetization-decantation-sieving method is sufficient (see Chapter 9). For quantitative studies it is necessary to process the samples in a more precise manner. In general, we recommend a modification of the Uhlig seawater-ice technique (see Chapter 9). Live animals collected by this technique may be stained slightly in a solution of neutral red in sea water (1 g/l) for a few minutes to facilitate sorting under the stereomicroscope. Elutriation techniques should be avoided as a means of extracting the gastropods because it does not work well with caecid prosobranchs and most meiobenthic "hard" molluscs, and the interstitial opisthobranchs are damaged by excessive agitation of the sediment. The loss of important taxonomic characters such as cerata, rhinophores, or ruptured visceral humps make subsequent identification impossible. Similarly, fixation of otherwise unprocessed sediment will cause the loss of most of the opisthobranchs hidden in empty shells. Anaesthetizing prior to fixation is very difficult, particularly for samples exceeding 50 dm^3 because animals will not expand sufficiently and therefore identification is impossible. In general, meiobenthic gastropods must be extracted while alive.

Preparation of Specimens for Taxonomic Study

Living specimens should be isolated into small plastic boxes fitted with lids and filled with filtered seawater. These boxes should be of the dimensions and optical quality to allow proper observation under a microscope. Alternative methods include their transfer to the "rotative compression chamber" (see Chapter 9) described by Uhlig and Heimberg (1981) which has been developed for use with nearly all kinds of microscopes.

Anesthetization is necessary prior to the fixation of specimens for study. Ten percent magnesium chloride ($MgCl_2$) solution in seawater, nembutal (1 g/l seawater), or propylene phenoxetol (1% in seawater (Robillard, 1969) are all suitable for relaxing gastropods. Completely anaesthetized specimens should be fixed for histological study with Bouin's fixative, preferably at 60° C for 2-6 hours and if internal shells *(Philine)* or spicules (Acochlidioidea) are to be studied, fixation in 5% buffered formalin in seawater for 24 hours and transfer to 70% ethanol is recommended. Internal shells, spicules, jaws, gizzards, and radulae sometimes can be observed in living squeezed material, but is preferable to first dissolve the soft tissues in a solution of 0.5 to 1% sodium hydroxide (Fritchmann, 1960). The shell can be mounted in concavity slides according to Gisin's (1968) technique, and the radulae can be mounted in the manner described by Fritchmann (1960). Shell structure and radulae are best studied by SEM. The preparation of such material is described by Thompson and Hinton (1968) (see also Morse, 1976).

For a more precise study of internal morphology, TEM techniques should be employed. But any osmic fixation of interstitial opisthobranchs in relaxed conditions remains difficult, resulting most of the time in the contraction of the specimens in spite of their anaesthetization. Morse's technique (1976) which was used for the description of *Hedylopsis riseri* is proposed: the whole organisms are fixed in a 0.2 M phosphate buffered paraformaldehyde-gluteraldehyde fix followed by post-fixation in 2% phosphate buffered osmium tetraoxide.

Laboratory Culture

We feel that it is particularly important to study the life history, diet, and reproductive behavior of meiobenthic gastropods. Most species are easily maintained for several weeks and even months in Petri-dishes with glass beads or shell-sand provided, and with the seawater changed every two days. It is more satisfactory, however, to try to maintain the animals in a recirculated seawater aquarium of high capacity (up to 250-300 liters seawater) with rigidly controlled temperatures and under other conditions that closely approximate their natural habitat conditions (Poizat, 1978).

Such animals are kept in transparent plastic tubes (10 cm long, 1 cm diameter) with the ends fitted with 100 μm mesh nylon netting to retain the specimens. The tubes are laid on a shelly gravel layer at the bottom of completely dark vessels in running and intensively bubbled seawater. Under these conditions, it is easy to observe the animals, particularly if they have been slightly stained by neutral red, by placing the transparent tube "cages" under the microscope. Either in Petri dishes or in an aquarium, the interstitial opisthobranchs must not be kept together with starved interstitial copepods, a mistake which always results in the destruction by predation of these shell-less gastropods

Type Collections

The type material of various meiobenthic prosobranch and opisthobranch Gastropoda is found in the Muséum National d'Histoire Naturelle, Paris, France; National Museum of Natural History, Smithsonian Institution, Washington, D.C., USA; British Museum (Natural History), London, UK; Naturhistorisches Museum, Wien, Austria; Naturhistoriska Riksmuseum, Stockholm, Sweden; and the Dominion Museum, Wellington, New Zealand.

References

Arnaud, P.M., and C. Poizat
1979. Données écologiques sur des Caecidae (Gastéropodes Prosobranches) du golfe de Marseille. *Malacologia,* 18(1-2):319-326.
1981. Signification écologique de quelques gastéropodes Caecidae des cotes de Provence. *Haliotis,* 11:29-35.

Arnaud, P.M., C. Poizat, and L. v. Salvini-Plawen
1986. Interstitial Gastropods. Pages 153-176 in L. Botosaneanu, editor, *Stygofauna mundi.* Leiden: E.J. Brill.

Dybern, B.I., H. Ackefors, and R. Elmgren
1976. Recommendations on Methods for Marine Biological Studies in the Baltic Sea. *Baltic Marine Biologists,* 1:1-98.

Fritchmann, H.K.
1960. Preparation of Radulae. *Veliger,* 32:52-53.

Gisin, H.
1968. A Cavity Slide Technique for Preparing Permanent Fluid Preparations of Small Organisms. *Revue d'Ecologie et de Biologie du Sol,* 4:81-584.

Heunert, H.H., and G. Uhlig
1966. Erfahrungen mit einer neuen Kammer zu Lebendbeobachtung Beweglicher Mikroorganismen. *Research Film,* 5:642-649.

Morse, M.P.
1976. *Hedylopsis riseri* sp. n., a New Interstitial Mollusc From the New England Coast (Opisthobranchia, Acochlidiacea). *Zoologica Scripta* 5:221-229.

Luft, J.H.
1961. Improvements in Epoxy Resin Embedding Methods. *Journal of Biophysical and Biochemical Cytology,* 409-414.

Poizat, C.
1978. *Gastéropodes mésopsammiques de fonds sableux du golfe de Marseille: écologie et reproduction.* Université Aix-Marseille III, Thèse Doctorat Sciences, volume 1:1-301; Volume 2: figures 1-84.
1979. Gastéropodes mésopsammiques de la mer d'Irlande (Portaferry, Northern Ireland): écologie et distribution. *Haliotis,* 9(2):11-20.
1980. Gastéropodes opisthobranches mésopsammiques du Skagerrak (Suède occidentale): distribution et dynamique des populations. *Vie et Milieu,* 30(3-4):209-223.

1981a. Gastéropodes mésopsammiques de la mer du Nord (Robin Hood's Bay, U.K.): écologie et distribution. *Journal of Molluscan Studies*, 47:1-10.
1981b. Signification écologique de quelques gastéropodes opisthobranches mésopsammiques des côtes de Provence. *Haliotis*, 11:201-212.
1983a. Mesopsammic Opisthobranchs from the Provençal coast (Marseille Bouches-du-Rhône, France): Long term variations of the populations. *Journal of Molluscan Studies*, supplement, 12A:126-135.
1983b. Opisthobranches interstitiels: migrations nychthémérales, données préliminaires. *Haliotis*, 13:35-44.
1984. Seasonal Variations of Mediterranean Interstitial Opisthobranch Assemblages. *Hydrobiologia*, 118:83-94.
1985a. Interstitial Opisthobranch Gastropods as Indicator Organisms in Sublittoral Sandy Habitats. *Stygologia*, 1(1):26-42.
1985b. Inventaire préliminaire des gastéropodes opisthobranches mésopsammiques de l'archipel des Embiez (Var, France). *Vie Marine*, 6:1-5.

Robillard, G.A.
1969. A Method of Color Preservation in Opisthobranch Molluscs. *Veliger*, 11:289-291.

Sellmer, G.P.
1967. Functional Morphology and Ecological Life History of the Gem Clam, *Gemma gemma* (Eulamellibranchia: Veneridae). *Malacologia*, 5(2):137-223.

Smith, W., and A.D. McIntyre
1954. A Spring-loaded Bottom-sampler. *Journal of the Marine Biological Association of the United Kingdom*, 33:257-264.

Swedmark, B.
1964. The Interstitial Fauna of Marine Sand. *Biological Review*, 39:1-42.
1968. The Biology of Interstitial Molluscs. *Symposium of the Zoological Society of London*, 22:135-149.
1971. A Review of Gastropoda, Brachiopoda, and Echinodermata in Marine Meiobenthos. *Smithsonian Contributions to Zoology*, 76:41-45.

Thompson, T.E., and H.E. Hinton
1968. Stereoscan Electron Microscope Observations on Opisthobranch Radulae and Shell Sculpture. *Bijdragen tot de Dierkunde*, 38:91-92.

Uhlig, G.
1968. Quantitative Methods in the study of Interstitial Fauna. *Transactions of the American Microscopical Society* 87(2):226-232.

Uhlig, G., and S.H.H. Heimberg
1981. A New Versatile Compression Chamber for Examination of Living Microorganisms. *Helgoländer Meeresuntersuchungen* 34:251-256.

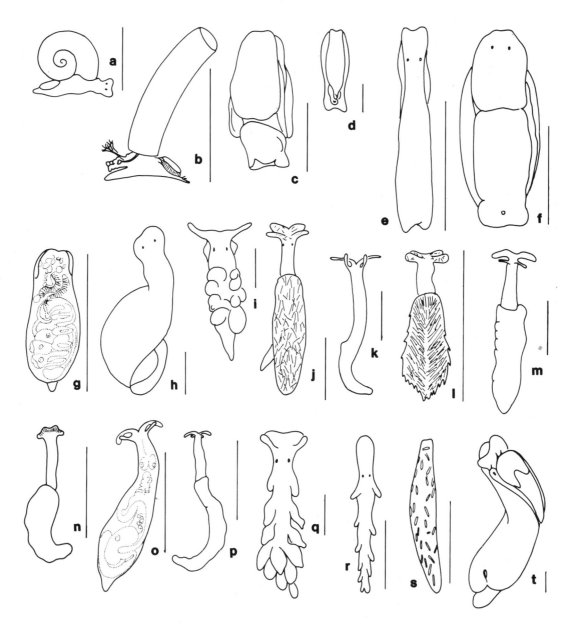

Figure 52.1.--Gastropoda: a, *Omalogyra atomus* (Philippi, 1841); **b,** *Caecum glabrum* (Montagu, 1803), (after Poizat); **c,** *Philine catena* (Montagu, 1803), (after Poizat); **d,** *Pluscula cuica* Marcus, 1953, ventral view, (from Marcus); **e,** *Philinoglossa helgolandica* Hertling, 1932, (after Cobo Gradin); **f,** *Abavopsis latosoleata* Salvini-Plawen, 1973, (after Poizat); **g,** *Sapha amicorum* Marcus, 1959, (after Marcus); **h,** *Platyhedyle denudata* Salvini-Plawen, 1973, (after Poizat); **i,** *Calliopaea bellula* d'Orbigny, 1837, (after Todd); **j,** *Hedylopsis spiculifera* (Kowalevsky, 1901), (after Gradin); **k,** *Pseudunela cornuta* (Challis, 1970), (after Challis); **l,** *Asperspina loricata* (Swedmark, 1968), (after Gradin); **m,** *Unela nahantensis* Doe, 1974, (after Doe); **n,** *Pontohedyle verrucosa* (Challis, 1970), (after Challis); **o,** *Microhedyle cryptophthalma* Westheide and Wawra, 1974, (after Westheide and Wawra); **p,** *Paraganitus ellynnae* Challis, 1968, (after Challis); **q,** *Embletonia pulchra* Alder and Hancock, 1844, (after Poizat); **r,** *Pseudovermis artabrensis* Urgorri, 1983, (after Cobo Gradin); **s,** *Rhodope veranyi* Kölliker, 1847, (after Odhner); **t,** *Smeagol manneringi* Climo, 1980, (after Climo). (Scale = 1 mm.)

53. Holothuroidea

David L. Pawson

Echinoderms are familiar and important members of magafaunas and macroinfaunas worldwide, from intertidal to abyssal depths. They are also represented in the meiofauna, mostly as temporary residents. Of the six extant classes (Asteroidea, Ophiuroidea, Echinoidea, Holothuroidea, Crinoidea, and Concentricycloidea (Baker et al., 1986)), only the Holothuroidea includes what appear to be permanent meiofaunal representatives. The list below undoubtedly includes some species that spend only a part of their life history in the meiofauna. Samples of meiofauna may occasionally contain numerous specimens of minute post-larval brittle-stars (Ophiuroidea) or starfish (Asteroidea), but these are only temporary residents of the meiofauna; they rapidly grow too large to qualify as meiofaunal components.

General summaries of the characteristics of the Holothuroidea may be found in Hyman (1955), Barnes (1980), and Pawson (1982). There are six orders, of which only one, the Apodida, includes meiofaunal taxa. Apodida comprises three families: Chiridotidae, Synaptidae, and Myriotrochidae, all of which have meiofaunal members. In general, the apodous holothurians are "derived" from the basic holothurian body plan. The simple body is worm-like and lacks any external manifestation of a water-vascular system (i.e., tube feet), apart from the circumoral tentacles. The tentacles are simple, digitate to pinnate. Respiratory trees are absent; apparently gaseous exchange takes place directly through the thin body wall. Dermal calcareous ossicles may be numerous, in the form of wheels or anchors and anchor plates; in a few species ossicles are absent. Early post-larval meiofaunal holothurians may retain their characteristic larval wheels for some time, even while adult-type ossicles, such as wheels or anchors and anchor plates, are developing.

Publications dealing with strictly meiofaunal holothurians are few indeed. Swedmark (1971), in an excellent review article, discussed the four species of meiofaunal holothurians known to him at that time. The list has not grown appreciably in the intervening years. In their seriously flawed review article, Salvini-Plawen and Rao (1986) mention five species. In this paper, seven species, all within the Order Apodida Brant, 1935, will be included.

Classification, Habitats and Ecological Notes

Family Synaptidae.--In synaptid holothurians the ossicles are chiefly anchors supported by anchor plates. The body walls are generally sticky to touch due to the presence of protruding anchor arms. There are approximately 12 genera and 130 species in this family; most occur subtidally, especially in the tropics.

Rhabdomolgus ruber Keferstein. This species, which lacks calcareous ossicles, is common in many parts of Europe in depths of 0-20 m (Menker, 1970; Schmidt, 1972). In an elegant paper, Menker (1970) discusses many aspects of the biology of *R. ruber*, and summarizes past work on this species, now the best-known interstitial holothurian. Adaptations to interstital life are conspicuous: *R. ruber* produces few eggs, lacks pelagic larval stages, is resistant to mechanical damage, and has mucus-laden, sticky tentacles which cling firmly to the substratum, making the animal very difficult to collect alive.

Leptosynapta minuta (Becher) has a distribution pattern similar to that of *R. ruber*. Unlike the latter species, *L. minuta* has well-developed ossicles in the form of anchors and oval anchor plates. Recent records of *L. minuta* include those of Salvini-Plawen (1972a,b) from the Mediterranean Sea off Livorno, Italy, at a depth of 3.5 m, and O'Connor (1981) from off the west coast of Ireland (he noted that all had a maximum of five young in the body cavity). Williams (1972) listed the species in a detailed study of interstitial fauna from Menai Bridge, Wales.

Labidoplax buskii M'Intosh is known from both sides of the Atlantic and also the Mediterranean in 18-540 m (Mortensen, 1927; Salvini-Plawen, 1972a,b, 1977; Pawson, 1967). This species has 11 tentacles, each with 3 distal digits; dermal ossicles are anchors and anchor plates, the latter equipped with distinctive narrow "handles." Little is known of its biology; Salvini-Plawen (1972b) included photographs of living specimens, and line drawings of ossicles.

Labidoplax media Oestergren occurs in the north Atlantic (Outer Hebrides, Norway) and was recently recorded by Salvini-Plawen (1977) from the Adriatic Sea and the Cote Vermeille of France at depths of 70-122 m. *L. media* differs from *L. buskii* in having 12 tentacles, each with 4 digits.

Family Chiridotidae.--In chiridotid holothurians the ossicles occur usually as six-spoked wheels, which may be scattered in the body wall or aggregated into small papillae. In addition, sigmoid or C-shaped rods may be present. A few species lack ossicles altogether. There are approximately 7 genera and 60 species; they occur from the intertidal zone to depths in excess or 3,000 m.

Trochodota furcipraedita Salvini-Plawen was described in 1972 from the mesopsammon off Livorno, Italy, in depths of 4-5 m. The body wall contains wheel ossicles and sigmoid hooks. There are ten tentacles with terminal forks. Live specimens may reach a length of 5 mm. The brief description of this species gives only a few details of its internal organization. *Trochodota havelockensis* Rao is a species inhabiting coarse sand mixed with fine shell-gravel (Rao, 1975) in the intertidal zone of the Andaman Islands. There are scattered wheels and sigmoid rods in the body wall. The ten tentacles have terminal forks. The body length is 2.5-3.0 mm. *Chiridota rotifera* (Pourtales). Engstrom (1980) described some specimens of *C. rotifera* found in the meiofauna at Key Biscayne, Florida, just below lowtide level. This species broods its young in its coelom (Clark, 1910); it is widely distributed throughout the Caribbean where it reaches lengths of as much as 40 mm or more, and typically lives on sand under rocks. The specimens studied by Engstrom were minute (1.5 mm long), and they apparently spent some considerable time as meiofaunal animals before reaching a larger size. Engstrom (1980) described the development and behavior of the species while it is a member of the meiofauna.

Family Myriotrochidae.--The myriotrochids have a unique wheel type, usually with numerous (more than 6) spokes. The wheels are never aggregated into papillae. Approximately 6 genera and 40 species of myriotrochids are known, and the family is more or less restricted to the deep-sea, at depths exceeding 500 m.

Several species of the typically deep-sea myriotrochids, by virtue of their small size, would seem to qualify as meiofaunal taxa. These holothurians have been studied in the Pacific and Indian Oceans (Belyaev and Mironov, 1982) and in the Atlantic (Salvini-Plawen, 1971; Gage and Billett, 1986; Pawson and Gust, in press); these papers should be consulted for further details on this group.

Meiofaunal Holothurians of Doubtful Status.-- *Psammothuria ganapatii* Rao (1968) was a new genus and new species described from Waltair, India. Engstrom (1980) and others have expressed doubt over the validity of this species (cited as *P. ganatii* by Engstrom) which has eight bifurcating tentacles and apparently no gonads, and which is obviously a juvenile, an "interstitial stage in another synaptid's life history" (Engstrom, 1980:93). A second species in this category includes *Myriotrochus geminiradiataus* Salvini-Plawen. The wheels of this form, as illustrated by the author (Salvini-Plawen, 1972), are typical "auricularia-wheels," common to numerous apodous holothurian larvae. Such wheels are illustrated and discussed by Pawson (1971). This "species" then is an early post-larva of some apodous holothurian, and has not yet developed its adult-type ossicles.

Meiofaunal holothurians can occur in a great variety of habitats. Some (with the exception of the deep-sea myriotrochids) have been recorded from "*Amphioxus*-sand" (see Menker, 1970), others from finer sediments. There seems to be no depth limitation, although most species have been recorded from less than 100 m.

Methods of Collection, Extraction, and Preservation

No special methods for collection or extraction of meiofaunal echinoderms have been developed. Like so many other meiofaunal organisms, the holothurians can cling tenaciously to a substratum by means of their sticky tentacles. A strong solution of magnesium sulfate will usually anesthetize holothurians and may cause them to release their hold on the substratum. Other anesthetizing agents such as propylene phenoxetol may also be helpful in relaxing specimens.

Preservation is best in 70% ethanol; buffered formalin is a useful fixative and short-term preservative.

Preparation of Specimens for Taxonomic Study

Taxonomic study of holothurians usually involves examination of the microscopic calcareous ossicles (when present). Small pieces of body wall tissue containing ossicles can be placed in a drop of liquid household bleach (sodium hypochlorite) on a glass microslide; the soft tissues dissolve leaving the ossicles. Permanent mounts can be made by drying the preparation, carefully irrigating it with distilled water, drying the slide, and adding a permanent mounting medium and a cover glass.

References

Baker, A.N., F.W.E. Rowe, and H.E.S. Clark
1986. A New Class of Echinodermata from New Zealand. *Nature*, 321:862-864.

Barnes, R.D.
1980. *Invertebrate Zoology*. Fourth edition, 1089 pages. Philadelphia: W.B. Saunders,

Belyaev, G.M., and A.N. Mironov
1982. The Holothurians of the Family Myriotrochidae (Apoda): Composition, Distribution and Origin. *Trudy Institute Okeanologii*, 117:81-120 [in Russian, with English summary].

Clark, H.L.
1910. Development of an Apodous Holothurian *(Chiridota rotifera). Journal of Experimental Zoology*, 9:497-516.

Engstrom, N.
1980. Development, Natural History, and Interstitial Habits of the Apodous Holothurian *Chiridota rotifera* (Pourtales, 1851) (Echinodermata: Holothuroidea) *Brenesia*, 17:85-96.

Gage, J.D., and D.S.M. Billet
1986. The Family Myriotrochidae Theel (Echinodermata: Holothuroidea) in the Deep Northeast Atlantic Ocean. *Zoological Journal of the Linnean Society*. 88(3):229-276.

Hyman, L.H.
1955. *The Invertebrates, Echinodermata*. vii + 763 pages. New York: McGraw-Hill Book Company.

Menker, D.
1970. Lebenszyklus, Jugendentwicklung und Geschlechtsorgane von *Rhabdomologus ruber* (Holothuroidea: Apoda). *Marine Biology, Berlin*, 6:167-186.

Mortensen, T.
1927. *Handbook of the Echinoderms of the British Isles*. 471 pages. London: Oxford University Press.

O'Connor, B.
1981. Some Echinoderms from the West Coast New to the Irish Fauna. *Irish Naturalist*, 20:247-249.

Pawson, D.L.
1967. *Protankyra grayi* New Species and *Labidoplax buskii* (McIntosh) from off North Carolina (Holothuroidea: Synaptidae). *Proceedings of the Biological Society of Washington*, 80:151-156.

1971. Second New Zealand Record of the Holothurian Giant Larva *Auricularia nudibranchiata* Chun. *New Zealand Journal of Marine and Freshwater Research*, 5:381-387.

1982. Holothuroidea. Pages 813-818 in S.P. Parker, editor, *Synopsis and Classification of Living Organisms*. New York: McGraw-Hill Book Company.

Pawson, D.L., and C. Gust
in press Myriotrochid Sea Cucumbers from the Atlantic Ocean (Echinodermata: Holothuroidea). *Smithsonian Contributions to the Marine Sciences*.

Rao, G.C.
1968. On *Psammothuria ganapatii* n.gen. n.sp., an Interstitial Holothurian from Beach Sands of Waltair Coast and Its Autecology. *Proceedings of the Indian Academy of Sciences*, 67:201-206.

1975. On a new Interstitial Species of *Trochodota* (Apodida, Holothuroidea) from Andamans, India. *Current Science*, 44:508-509.

Salvini-Plawen, L. v.
1972a. Zur Taxonomie und Ökologie mediterraner Holothuroidea-Apoda. *Helgoländer Wissenschaftliche Meeresuntersuchungen*, 23:459-466.

1972b. Die nordatlantische *Labidoplax buskii* (Holothuroidea- Synaptidae) in der Adria. *Zoologischer Anzeiger*, 188:301-304

1977. Caudofoveata (Mollusca), Priapulida und Apode Holothurien *(Labidoplax, Myriotrochus)* bei Banyuls und im Mittelmeer allgemein. *Vie et Milieu*, Series A, 27:55-81.

Salvini-Plawen, L. v., and G.C. Rao
1986. Echinoderma. Pages 701-704 in L. Botosaneanu, editor, *Stygofauna Mundi*. Leiden: E.J. Brill/Dr. W. Backkhuys.

Schmidt, P.
1972. Zonierung und jahreszeitliche Fluktuationen der interstitiellen Fauna in Sandstranden des Gebeits von Tromsö (Norwegen). *Mikrofauna des Meeresbodens*, 12:1-86.

Swedmark, B.
1971. A Review of Gastropoda, Brachiopoda and Echinodermata in Marine Meiobenthos. In N.C. Hulings, editor, Proceedings of the First International Conference on Meiofauna. *Smithsonian Contributions to Zoology*, 76:41-45.

Williams, R.
1972. The Abundance and Biomass of the Interstitial Fauna of a Graded Series of Shell-gravels in Relation to the Available Space. *Journal of Animal Ecology*, 41:623-646.

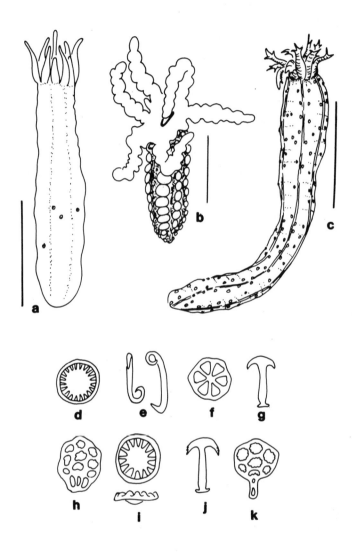

Figure 53.1.--Holothuroidea: **a**, *Myriotrochus geminiradiatus* Salvini-Plawen; this is a juvenile apodous holothurian, and may not belong to the Family Myriotrochidae (after Salvini-Plawen, 1972a; scale = 1 mm); **b**, *Chiridolta rotifera* (Portales) (after Engstrom, 1980; scale = 200 μm); **c**, *Labidoplax buskii* McIntosh (after Swedmark, 1971) (scale = 1 mm); **d-k**, dermal calcareous ossicles: **d**, wheel from body wall of *Myriotrochus geminiradiatus*, compare with Figure 53.1i; **e**, Sigmoid hooks from body wall of *Trochodota furcipraedita* (after Salvini-Plawen, 1972a); **f**, wheel from body wall of *T. furcipraedita* (after Salvini-Plawen, 1972a); **g**, anchor from body wall of *Leptosynapta minuta* (after Becher, 1906); **h**, anchor of *L. minuta* (after Becher, 1906); **i**, plan and profile views of wheel from body wall of holothurian larva *Auricularia nudibranchiata*, compare with Figure 53.1d, (after Pawson, 1971); **j**, anchor from body wall of *Labidoplax buskii* (after Salvini-Plawen, 1972b); **k**, anchor plate from body wall of *L. buskii* (after Salvini-Plawen, 1972b).

54. Tunicata

Françoise and Claude Monniot

Benthic tunicates or ascidians, commonly named "sea squirts," are solitary or colonial, living on all kinds of substrata at any latitude, but exclusively in marine water. The adult animals are sac-like, sometimes buried in sand or mud, but more commonly attached to rocks or any hard object. The body is covered with a cellulose tunic (the test) except for two openings, the inhalant buccal siphon and the exhalant atrial siphon. The water current enters the buccal siphon and passes through the branchial sac in which suspended food particles are trapped in mucus. Feces and gametes are discharged into the filtered water before it continues through the atrial siphon. The digestive tract and the hermaphroditic gonads are attached to the mantle. Sometimes the eggs are incubated inside the body. The larvae resemble small frog tadpoles and, like them, have a notochord, gill slits, and a dorsal hollow nerve cord. These organs are responsible for the inclusion of the Tunicata in the phylum Chordata. The notochord and the dorsal hollow nerve cord are lost during metamorphosis and only the gill slits remain. Adults have a unique nervous ganglion between the two siphons.

Colonial ascidians reproduce asexually by budding. The budding pattern varies according to the family. Solitary ascidians range in size from 400 μm to 40 centimeters, and colonies may reach several meters. The smallest forms, some of which belong to the meiofauna, are solitary. Larvae and young tadpoles of ascidians living on soft bottoms often exist in the littoral or abyssal meiobenthos.

Classification

Two classes of benthic tunicates are present in the meiobenthos. The Ascidiacea, as they are described above, and an exclusively deep-sea class, the Sorberacea.

Class Ascidiacea.--This class is divided into 3 orders: the Aplousobranchiata are colonial, not meiofaunal, with numerous small zooids in a common test, and larvae which are brooded internally; the Phlebobranchiata and Stolidobranchiata have simple and colonial forms, and are characterized by the presence of papillae or sinuses internal to the branchial sac. The major characters used to identify genera and species are the structure of the branchial sac and the digestive tract, the number and shape of the gonads, the presence of a kidney, endocarps, the mode of incubation, and the larval morphology.

Class Sorberacea.--These are solitary and have a hypertrophied oral aperture with prehensile lobes. They lack a branchial sac, and the esophagus opens into a large stomach which occupies most of the body. All have a large kidney. They are hermaphroditic and oviparous, but their development remains unknown. Their size varies between less than 1 mm to 15 mm according to the species. They are carnivores, feeding on many invertebrates including small crustaceans, nematodes, and polychaetes.

Characteristics of Meiobenthic Ascidiacea.

Very small ascidians may be found in two kinds of substrata: littoral coarse sand where they are interstitial, and deep ocean muds where they are epibenthic. The reduction of size of all meiobenthic ascidians is in keeping with the juvenile state of all organs except the gonads. Consequently, most meiobenthic ascidians probably evolved by neoteny, but some may have undergone regressive evolution.

Interstitial Ascidians.--These ascidians are generally tube-like with a strong test, or very soft and covered with sand particles (Figure 54.1j). Their size is less than the largest particles of the sediment in which they are found. They are not permanently attached, rather some are free and mobile, while others develop rhizoids at sexual maturity. Sand covered species can replace the sediment particles. Development is very slow; all the early developmental stages, which are typically condensed in macrobenthic ascidians of rocky substrata, are present and extended (Figure 54.1a-i) in interstitial forms. Branchial development progresses slowly until the gonads begin to appear in the warm season.

All meiobenthic Stolidobranchiata are incubatory and lack swimming tadpoles. The Phlebobranchiata have enlarged sperm ducts and oviducts and the release of gametes may be delayed. Interstitial ascidians are colorless unlike those living on hard substrates.

Abyssal Ascidians.—There are two principal body forms in abyssal ascidians. The first type has one or few strong ventral root-like rhizoids anchoring the individual into the mud; the second type has a rounded body with numerous soft, hairlike rhizoids covering most of the test, especially around the ventral surface. Foraminifera often make a dense crust on the test. The internal organs are simplified in some species, and reflect a true neoteny, whereas in others, simple and derived characters co-occur and suggest regressive evolution. Deep-water ascidians are always oviparous. The gonads appear at a very small size and the maturation of eggs and spermatozoa seems to be continuous. The larval stages are unknown.

All specimens are colorless. Both siphons are very short and do not protrude. They are never diametrically opposed, rather they subtend an angle of 120° or less. The body plan is so uniform that it is impossible to identify even the family without dissecting the specimens.

Habitats and Ecological Notes

Interstitial Ascidians.—These forms have never been collected in mud or sand with fine particles. Instead, they seem to require coarse, clean, well-oxygenated sand - in areas without decaying organic matter. They are eurythermous. Interstitial Ascidians are found from Norway to Senegal in the Atlantic, and from Spain to Greece in the Mediterranean Sea. One species has been collected in Patagonia.

Interstitial ascidians have never been found in coral sand, even when the sediment is appropriately coarse. The explanation for their absence seems to be that although the particles are large, they are flat and thin. As a result, they pack closely interstitial water flow is reduced, and ascidians are excluded. The best habitats to find interstitial ascidians are granitic or coarse shell sands with large ripple marks, under the breaking waves in channels with strong rip currents, and in giant sand waves such as occur on the west coast of France. Interstitial ascidians feed on unicellular organisms and detritus.

Deep-sea Ascidians.—These animals have been collected in all the oceans from the continental slopes to the deepest abyssal plains. They live patchily in the water-sediment interface. Except for very small forms, their size varies from 1 mm to 5 mm in diameter, but larger forms up to several cm can be found. All species have a wide distribution and many are cosmopolitan.

Like shallow-water ascidians, they shelter parasitic copepods in their branchial sac and digestive tract. They are microphagous.

Methods of Collection

Finding ascidians in sand requires sampling a large volume of the upper centimeter of sand collected over large areas. Any kind of dredge designed for such qualitative sampling is satisfactory. A toothed cutting edge on the dredge allows a better penetration into the sediment. A dredge mesh size of as much as 1 mm is effective. The collection of sand by SCUBA diving is necessary in areas between rocks or other restrictive barriers. Grabs are not generally useful since they do not take a large enough sample of just the upper horizon of sand.

In mud, two different devices may be used. Each tends to give different results even at the same station. The epibenthic dredge (Hessler and Sanders) gives good results on calcareous sediments, but it is less efficient on very soft radiolarian mud. The cod-end of the net within such a dredge should have a mesh size of no more than 0.5 mm even though the front part of the net may be constructed of 1.0 mm mesh net.

A beam trawl fitted with fine mesh (0.5 mm at the cod-end) also may be used. Anchor dredges or smaller dredges, such as the Higgins Meiobenthic Dredge, may also be very effective in stripping the uppermost centimeter of sediment from a given habitat.

Methods of Extraction

Samples of mud should be placed in large containers so that subsamples can be transferred to large plastic jars with a hole in the bottom and fitted with a tap to admit running sea-water and a spout at the top edge so that the water will overflow into a series of sieves. The material collected on each sieve should be fixed separately in 10% formalin. Animals are sorted using a dissecting microscope.

Samples of sand should be subsampled, transferred to containers of about 1 liter volume, vigorously stirred and decanted into a sieve of no larger than 250 μm. This procedure should be repeated several times for each subsample until all of the material has been processed. In the case of all materials, relaxed animals are much better for study. If living material is collected, crystals of menthol introduced into the seawater, before or after extraction, will enhance the taxonomist's ability to properly identify the specimens.

Preparation of Specimens for Taxonomic Study

The study of ascidians requires dissection and

the subsequent staining and mounting of various organs on microslides. Organs should be fixed for at least 48 hours in order to harden the tissues for adequate dissection. All of the following procedures are done using a dissecting microscope: (1) remove the tunic avoiding damage to the mantle (often in several layers). Do this using fine forceps, beginning around the siphons; (2) cut open the body wall along the ventral line, from one siphon to the other, following the endostyle; (3) using very small insect pins, pin the open body in a dissecting pan (with wax bottom); (4) stain the body with a solution of hemalum (0.2 g of hematein powder boiled 3 minutes in 100 ml of a saturated solution of alum); (5) rinse carefully with tap water; (6) remove the branchial sac by cutting all the dermatobranchial strands with iridectomy scissors; (7) separately pin the unfolded branchial sac; (8) stain once more if necessary; (9) dehydrate, replacing (twice) the water directly with 95% ethyl alcohol in the dissecting pan; (10) replace the ethyl alcohol with butyl alcohol; (11) mount on a slide in a resin (araldite).

Several specimens of the same species are necessary for a good identification; note that individual variations may possibly be important in the case of such soft tissues.

Type Collections

The collections of Millar are located in the British Museum (Natural History), London, UK. Our collections are deposited in the Museum National d'Histoire Naturelle, Paris, France.

References

Monniot C., and F. Monniot
1963. Présence à Bergen et à Roscoff d'Ascidies psammicoles du genre *Heterostigma*. *Sarsia*, 13:51-57.
1978. Recent Work on Deep-sea Tunicates. *Oceanography and Marine Biology Annual Reviews*, 16:181-223.

Monniot, F.
1961. Recherches sur les Ascidies interstitielles des gravelles à Amphioxus (2ème note). *Vie et Milieu*, 12:269-283.
1962a. Recherches sur les graviers à Amphioxus de la région de banyuls s/mer. *Vie et Milieu*, 13:232-322.
1962b. Présence à Roscoff d'une Ascidiidae interstitielle: *Psammascidia teissieri* n.g.n.sp. *Comptes Rendus de l'Academie des Sciences* (Paris), 255:2656-2658.
1962c. *Dextrogaster suecica*, n.g.n.sp., Ascidie interstitielle des graviers du Skagerrak. *Comptes Rendus de l'Academie des Sciences* (Paris), 255:2820-2822.
1964. *Polycarpa arnbackae* n.sp., Styelidae interstitielle des sables coquilliers de la cote Ouest de Suède. *Cahiers de Biologie Marine*, 5:27-31.
1965. Ascidies interstitielles des cotes d'Europe. *Mémoires du Museum National d'Histoire Naturelle* (Paris), séries A, 35:1-154, 11 plates.
1966. Les Ascidies interstitielles. 6ème congrès de biologie marine. *Veröffentlichungen des Instituts für Meeresforschung in Bremerhaven*, 2:161-164.

Weinstein, F.
1961. *Psammostyela delamarei* n.g.n.sp. Ascidie interstielle des sables à Amphioxus. *Comptes Rendus de l'Academie des Sciences* (Paris), 252:1843-1844.

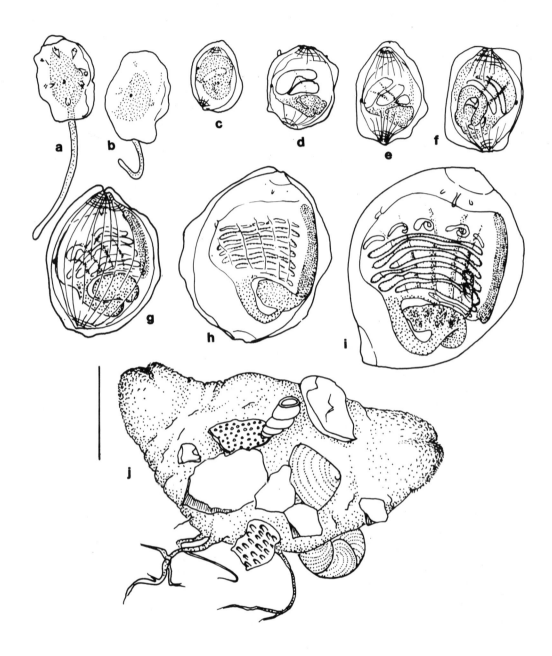

Figure 54.1.--**Tunicata:** a,b, *Heterostigma gonochorica*, tadpole stages; c-i, successive development stages seen through the tunic; j, *Heterostigma gonochorica*, adult. (Scale = 500 μm.)

Index

This index consists of two parts; a subject index and a taxonomic index. Boldface page numbers refer to illustrations.

Subject Index

Abiotic factors, 61-76
Abundance, marine fauna, 18-21, 24, 28, **27, 28**
Accretion dynamics, 68
Adenylates, total, 95, 108
Agglomerative classification strategy, 208
aggregated pattern, 200
Alpha-diversity, 222
Amino acids, 97
Amphioxus-sand, 68
Anaerobes, facultative, 191
    obligate, 191
Analysis, image, 149
    morphometric, 149
Analysis of variance, 129, 132
Ancestrula, 438
Anesthetic see Anesthetization,
Anesthetization, 115, 137, 149, 151, 154, 171
    ethanol, 137
    formalin, 137
    freshwater, 137
    magnesium chloride, 137, 138
Anoxic layers, 15
Antibiotics, 163
Antigen-antibody interactions, 186
ARIMA model, 226
ARMA model, 226, 227
Artificial sand system, 150, **151**
Ashfree dry weight, 79-81, 181, 182
Askö splitter, 141, **142**
Association matrices, 201
ATP, 79, 95, 96, 108
Aufwuchs, 47, 52, 122
Autocorrelation, 201, 225
Autocovariance, 201, 225
Autoradiography, 158
Autoregressive process, 226, 225
Bacteria, endosymbiotic, 184
Beach, slope, 24
    high energy, 24
Beta-diversity, 222
Binomial distribution, 200
Biomass, 18, 29, 30, 49, 53, 79, 95, 136
    determination, 181, 188, 189
        gravimetric, 181
        volumetric, 181
Biotic factors, 79-108
BMD, 197, 209
BMDP, 197
Bou-Rouch pump, 50-52, 115
Box corer, 117, 120, **119**

Bray and Curtis index, 209
Brillouin diversity index, 221
Broken stick model, 216
Broods, number of, 29
Bubbling technique, 137
Buffer, 135, 234
Calorimetry, 182, 192
Canberra metric, 206
Carbohydrates, anthrone method, 91
    MBTH method, 92, 93
    phenol sulfuric acid method, 92
Carbon, 182
Carbon dioxide, 42
Cartesian diver, 192
CEP (Cornell Ecology Programs), 197, 198
Chamber, agar, **147**
    compression, 147, 150, **148**
    continuous flow, **147**
    paraffin flow, 150, **147**
    sealed, **147**
Check list, 151
Chemicals, preparation of, 158, 159
Chitin, 102, 103
Chloroplastic pigment, fluorometric, 101
Chloroplastic pigments, spectrophotometric, 100
CHN analysis, 82, 83, 86, 182, 183
Chord distance, 202
Classification, 201, 202, 208
    agglomerative strategy, 208
    centroid, 206
    furthest neighbor, 228
    group average, 208
    hierarchical strategy, 208
    median, 208
    nearest neighbor, 208, 212
    weight average, 208, 209
Classification strategy, agglomerative-monothetic, 208
    agglomerative-polythetic, 208
    divisive, 208
CLUSTAN, 198, 209
Clustering, 208
Cobb slide frame, 152, **153**
Coefficient, similarity, 206
Collection, 238
    data, 239
    management, 238
Collector's curve, 212, 214
Colonization, 27, 32, 170
Communities, freshwater, 42-5
    gravel and rubble, lake shores, 46
    lake mud and sand, 47
    littoral sand, gravel and rubble, 46

littoral sand, mud and gravel, 46
mud margins, exposed, streams, 51
mud, rivers, 53
mud, shores, 46
psammolittoral (eu-), 41-46
sand, gravel and rubble, lotic, submerged, 52, 53
sand, gravel and rubble, streams, 50, 51
sand, streams, 51, 52
sandy margins, streams, 49
Community structure, 16
COMPCLUS, 198, 214
Complete-block sampling design, 130, **131**
Compression chamber, 147, 148, 150
Concentration, organisms, 134
CONDENSE, 198
Conditions, anaerobic, 41, 42
Conductivity, 72
Consumption rate, 187, 183
Contagious distribution, 200, 201
Container, 239, 240
Conversion factor volume/weight, 182, **183**
Core, 199
Core compaction, 63
Corer, box, 117, 120, **119**
    gravity, 63
    Hagge, **122**
    multiple, 63, **118**
    pole, 117, **118**
    single, 118
    SMBA, **118**
    syringe, **116**
Correlation, 207-209
Correlogram, 225
Counting, nanofauna, 235
Counting tray, organisms, 143
Covariance, 207, 208, 210
Cross-correlogram, 225
Cross-spectrum, 228
Crowding, 201
Culture, agnotobiotic, 161
    axenic, 161
    dishes, 162, **147**
    gnotobiotic, 161
    maintenance, 165
    monoxenic, 161
    polyxenic, 161
    techniques, 161-166
Culture dishes, 147
Curation, 238-242
Currents, 27
Cypris-larva, 370
Data, abundance, 202
    analysis, 197-231
    base management system (DBMS), 198
    Cycle Record, 198
    evaluation, 197-231
    matrix, 202, 203, 208
    processing, 197-231
    retrieval, 197
    storage, 197
Data base, 198
    field, 198
    file, 198
    matrix, 202
    record, 198
    standardization, 198
    taxonomic codes, 199
        NODC code, 199
        Rubin code, 199
Data standardization, 123
DATAEDIT, 198
dBaseIII+, 198
Decantation, 115, 137-140
DECORANA, 197, 211
Deep sea, 25
Definition, 11, 13-15
    absorption, 184, 188
    agnotobiotic, 161
    annual production, 189
    assimilation, 188
    assimilation efficiency, 192
    Aufwuchs, 47
    axenic, 161
    bacteriovore, 184
    carnivorous, 184
    cohort, 189
    cohort production, 189
    consumption, 188
    deposit feeding, 184
    detritivorous, 184
    documentation specimens, 239
    efficiency, ecological, 192
    eupsammolittoral, 40, 42
    eupsammon, 43
    euryphagous, 184
    excretion, 188
    feces, egested, 188
    food intake, 188
    formaldehyde, 134, 135
    formalin, 134, 135
    fungivore, 184
    gnotobiotic, 161
    herbivorous, 184
    holotypes, 239
    hydropsammon, 43
    hyperbenthos, 122
    hyporheic, 15, 39
    interstitial fauna, 12, 14
    lithophyton, 47
    macrobenthos, 11
    meiobenthon, 39
    meiobenthos, 11
    meiofauna, 11
    mesopsammon, 12
    microbenthos, 232
    monoxenic, 161
    mortality, 188
    nanobenthos, 12, 232

nanofauna, 13
omnivorous, 184
paratypes, 239
periphyton, 47
permanent, 11
phreatic, 39, 50
phytophagous, 184
polyxenic, 161
population, 126
predation, 184
production, 188
production efficiency, 192
respiration, 188
sample size, 123
scavenging, 184
support specimens, 239
survivorship, 188
suspension feeding, 184
temporary, 11
turnover rate, 30
zoophagous, 184
Demography, 176
Dendrogram, 202, 208
Density, 72
Density gradient centrifugation, 232, 236
Density gradient separation, 140
Descriptors, 202, 207-209
Desiccation, 23
Deterioration technique, 137
Detrended correspondance analysis, 211
Detritus, 29, 31, 79
Dispersal, 26, 27
Dispersion, 25
Dissection procedure, 463
Dissection technique, 372, 384, 399
Dissolved organic matter, 184, 187
Distance measures, 202, 202, 219
Distribution, differential, 173
    horizontal, 45, **26, 27**
    random, 200
Disturbance, 27, 171-175, **175**
    mimic, 172, 175
Disturber exclusion cages, 172
Diversity, 15, 16, 18, 22, 24, 25, 197
    information statistic, 219, 220
Diversity statistics, 217-242, 219
Divisive classification strategy, 208
DNA, 94, 95
Documentation, graphic, 147
    photographic, 147
DOM, 184, 187
Dominance index, 221, 223
Dredge, 115
Dry weight, 79, 80, 181, 182
Ecology, marine meiofauna, 18-38
Ecology of meiofauna, 16, 18-38, 39-60
Ecophysiology, 16
Efficiency, assimilation, 192
    ecological, 192

production, 192
Effluents, description, 61-76
    geography, 61
    geology, 61
    hydrography, 61, 63
    light, 61
    mineralogy, 63
    physiography, 61, 63
    solar illumination, 63
    terrestrial, 70
    turbidity, 63
Egg-ratio technique, 191
Eigenvalue, 211
Eigenvector, 211
Electrodes, 71-76, **74**
Electrometry, 71-76
Electrostatic charge, 372
Elutriation, 137-140, **139**
Emergence trap, 431, **122**
Energetic role, 16, 29, 30
Energetics, 166, 181-193
Energy, budget, 182, 188-192
    budget equation, 188
Energy charge, 79
Environment, 61
Enzyme, 186
Enzyme activity, 79
Equitability, 219, 223, 220
Erosion, 68
Ethanol, 240
Euclidean distance, 202, 223, 220
Eupsammolittoral, 40, 42, 49
Eupsammon, 43-45, 48, 49
Evenness, 22, 197, 219-224
Evolution, 16
Excretion, 192
Experiment, 16
    grazing, 187
Experimental, control, 169
    manipulation, 174
    technique, 169-178
Experiments, 32
    bacteria, 175
    feeding, 175, 176
    population growth, 175
    preference, 171
    tracer techniques, 175
Exposure, waves, 70
Extraction, life, 162
    specimens, life, 135, 137, 138
        phototactic, 137
        preserved, 140
        qualitative, 137
        quantitative, 137-141
Fast Fourier Transform (FFT), 228, 229
Faunal counts, 143
Feces production, 192
Fecundity, 29
Feeding, life, 162

mechanism, 184
Feeding preference, 26
Feeding techniques, tracer, 165
Field, in data bases, 198, 229
File, Header Record, 198
    in data bases, 198
Fixation, 134, 152-154, 233-235
    formalin, 134, 135
    glutaraldehyde, 135
    lactic acid, 378
    microwave, 135, 154
Flexible sorting strategy, 209
Flotation, 115, 137, 138, 140
Flow cytometry, 185
Food, 29, 70, 166
    intake, 184
    organisms, 164
    resources, **184**
Food web, **31**
Formaldehyde, 134, 135
Formalin, 134, 135, 234, 240
Fourier analysis, 227
Fractionation, size classes, 136
Freshwater shock see anesthetization, 151
Friedman rank-sums procedure, 131
Funnel trap, 366, 431
Furthest neighbor classification, 228
Fusing strategies, 208, 209
Gamma-diversity, 222, 209
Gas chromatography, 82
Geodesic metric, 206
Geometric series ranked abundance list, 216
Grab, Okean-50, 120
    Petersen, 120
    Van Veen, **121**
Gradient, 199, 211
Grain size, 15, 24, 61, 62, 64-66
    analysis, 64, 65
    distribution, 42, **66**
    median (Md), 65
    quartile (Q), 65
Granulometry, 63
Grazing experiment, 187
Groundwater, 62, 370
Groundwater see phreatic,
Group average classification, 208
Growth, 166, 175
    medium, 163
    rate, 188
Gut contents see stomach contents,
H-S slide frame, 153
Habitat, 18, 62
    classification, 39, 40
    lentic, 40
Habitats, lotic, 40
Harmonic terms, 208
Health safety, 135
Helcom system, 198
Hensen pipette, 141

Hierarchical classification strategy, 208
Higgins larva, 319, 320, **321**
Hill's diversity numbers, 220, 223
Histochemistry, 158
Histology, 153
History, 14-17, 40
    discovery, 14, 15, 40
    diversity, 15, 16
    ecology, descriptive, 15, 40
    ecology, experimental, 16
    laboratory studies, 15, 16
    phylogenetic consideration, 16
    systematics, 14, 15, 40
HPLC, 186, 187
Hydrogen sulfide, electrometry, 74
    titration, 74
Hydropsammon, 43, 45, 46, 48
Hygropsammon, 43, 45, 48
Hyporheal, 52
Hyporheic, 53
    communities, 40, 52
    water, 39, 52, 55
Identification, 240
Immunization, 186
Ingestion rate, 187
Interactions, biological, 171
International Association of Meiobenthologists, 14, 15
International Code of Zoological Nomenclature, 238
International Meiofauna Conference, 15
Interstitial fauna, 14, 15, 269
Irwin loop, 143, 149, **150**
Isotopes, radioactive, 165
Jaccard coefficient, 207, 208
Jensen splitter, 141, **142**
k-Dominance curve, 212, 214, 223
Kendall's rank correlation, 205, 207, 225
Kjeldahl method, 83-86
Kruskal-Wallis test, 129
Label, 240
Laboratory culture, 454
Lactic acid, 378
Lebensformtypus, 14
    meiobenthos, 14
    phreatic fauna, 15
    psammon, 15
    subterranean fauna, 15
Lefkovitch's partitioning method, 212
Length-weight regression, 190
Life history, 16, 171
Life history parameters, 29
    broods, number of, 29
    fecundity, 29
    life style, 26
    longevity, 29
    P/B ratio, 30
    rate of development, 29
    reproductive potential, 29
Light source, 146
Lipids, 79, 88

chloroform methanol method, 89
potassium hydroxide method, 89
Lithophyton, 47
Loadings, 211
Log series distribution, 217, 219
Log sheet, 61, **62**
Log-normal distribution, 217, 218, **218**
Lorenzen curve, 212, 214
Ludox, 140, 141, 162, 236
Macrobenthos, 11
Macrofauna, food web, 29, **31**
Magnesium chloride, 115, 120, 171, 236
Mandelbrot model, 218, **218**
Manhattan-metric, 206
Manipulation, experimental, 174
Mann-Whitney U test, 129
Maximum Entropy Spectral Analysis (MESA), 229
Meiobenthon, 39
Meiobenthos, 11, 270
    anomalies, 54-56
    densities, 39, 47, 48, 53, 54
    freshwater, 39-60
    rarities, 54-56
Meiofauna, 11
    as food, 30
    as predator, 30
    as prey, 30
    definition, 269, 270
    experimental, 169
    in food web, **31**
    marine, 18-38
    permanent, 11, 18
    size, 269
    temporary, 11, 18
Meiostecher, **121**
Mesh size, 12
Mesocosm, 174
Mesopsammon, 12, 269
Metabolism, 191
Microcosm, 171, 172
Microrespirometer, 192
    oxygen electrode, 192
Microscope, compound, 146
    inverted, 143, 147
    stereo-, 146
    video-, 285
Microscopy, scanning electron, 157
    transmission electron, 155
Microsplitter, 141
Microstrainer, 156, **157**
    syringe, 156
Microwave fixation, 135, 154
Microwave irradiation see fixation, microwave,
Migration, diurnal, 23
    horizontal, 23, **23**
    seasonal, 23
Mimic, disturbance, 172, 173
Minimum spanning tree, 212
Mortality, 188

Mounting, 235, 236
    living organisms, 149, 151, **150**
    permanent, 152
Mousetrap, 120, **121**
Mouth pipette, 143
Moving average, 226
Multivariate method, 209
Muramic acid, 79, 103, 104
Nanobenthos, 12, 232-237
Nanobiota, 245
Nanofauna, 13
Narcotization see anesthetization,
Negative exponential distribution, 200
Neutral models of species abundances, 217
Neyman distribution, 200
Niche preemption model, 216
Nomenclature, 238
Nonmetric multidimensional scaling, 211
NT-SYS, 198, 209
Nutrition, 29, 166
Objects, 206
Okean-50 grab, 120
ORDIFLEX, 198
Ordination, 197, 201, 203, 209, 211
    space partitioning, 212
Organic, carbon, 79, 81, 83
    matter, 32, 79
    nitrogen, 83
Organism, processing, 146-160
Oxygen, availability, 41, 63
    content, 62, 70
    diffusion rate, 62
    dissolved, 41, 42
    electrometry, 73
    titration (Winkler), 73
Oxygen consumption, 191
Oxygen depletion technique, 137, 138, 140
Oxygen tension, 24
P/B ratio, 30, 188-192, **189**
Paper-board slide, 372
Pasteur pipette, 149, 235
Patch, description, 201
    location, 201
    size, 201
Patchiness, 24, 39, 201
Pearson's correlation coefficient, 207
Percoll, 140, 162, 236
Periodogram, 227, 228
Periphyton, 47
Permanent mounts, CMC-10, 432
    Hyrax, 419
    Marc André, 426
Permeability, 62, 63, 67, 70
Permeameter, **67**
Pertubation, indicators, 32
Petersen grab, 120
pH, 41, 62, 70, 75, 76
    electrometry, 75, 76
    formalin, 135

indicator, 75
Phenoxetol, 135, 136, 235
Phototactic response, 115
Phreatic, communities, 40, 42
    fauna, 15, 53
    water, 39, 50
    zone, 55
Physiological efficiency, 175
Plain Language Record, in data bases, 199
Plastic slide, 372
Plate tectonics, 26
Poisson distribution, 199, 200
Pole corer, 117, **118**
Pollution, 16
    experiments, 173, 174
Population, statistical sense, 126
Population dynamics, 16
Pore water, in situ sampler, 68, 69, 71
    analysis, 71
    content, 23
    lance, 69
Porosity, 62, 63, 66, 67
Praniza stage, 398
Predation, 25, 30, 171, 173
Predator, 171, 172, 186
    exclusion, 172
    introduction, 172
Predator-prey interaction, 16
Preference, 16
    experiments, 171, 176
    feeding, 26
Preparation, cover glass, 235
Preparation of chemicals, 158, 159
Preservation, 134
    volume, 240
Prey, 186
Principal Component Analysis (PCA), 207, 209, 210, 211
Principal coordinate Analysis, 211
Processes, benthic, 29
Production, 29, 188-192
Production biomass ratio see P/B ratio, 188-192
Prospectus, 11-13
Protein, 79
    method of Bradford, 87, 88
    method of Lowry, 87
Protocol, 240
Psammolittoral, 40, 41, 44, 45, 49, 51, 55
Psammon, 15, 40, 42
Pseudoreplicated sampling design, 131, **131**
Q (mid-range diversity statistic), 219, 220
Q-mode, 202, 206
R-mode, 202, 207
Radioactive tracer, 187
Random pattern, 199, 200
Randomness, 200
Ranked species abundance curve, 212, 213
Rarefaction, 220
Reciprocal averaging, 211, 212
Recolonization, 15, 170, 173

chamber, 171
Record, in data bases, 198
Recruitment, 188
Redox, potential, 62, 63, 75
    potential discontinuity, 22, 23, 75
Reference diver, 192
Refractometry, 72
Remineralization, 29
Reproduction, 175
    rates, 166
Reproductive output, 191
Reproductive potential, 29
Resin embedding, 155, 156
Resin, epoxy, 249
Respiration, production, 189, 191
Respiration rates, 16
    experiments, 177
Rhodamine B, 143
RNA, 94, 95
Rose bengal, 143
Roundness, particles, 66
RPD, 22, 23, 75
Rubin codes, in data bases, 199
Salinity, 23, 24, 29, 62, 63, 72
    araeometry, 72
    conductivity, 72
    refractometry, 72
    titration, 73
Salt marsh, 25, **26**
Sample, life maintaining, 134
    processing, 134-143, **234**
    sieving, 136, 140
    size, **123**
    slicing, 134
    volume, 240
Sampling, aufwuchs, 122
    equipment, 115-123
    hyperbenthos, 122
    hypothesis testing, 128
    nanobenthos, 233
    plants, 121, 122
    pore water, 68, 69
    program design, 126-133
    qualitative, 115, 116
    quantitative, 116-123
    random, 127-129, 131, **127**, **131**
    sediment, 63, 115-121
    strategies, 126-133, 170
    stratified random, 127, **128**, **129**
    vertical sediment profile, 121, 122
Sand, baited, 171
SAS, 198
Scuba diving, 117
Seasonal differencing, 226
Seasonality, in time series, 225
Seawater ice (method) technique, 150, 162, 232, 453, **139**
Sediment, aerobic, 22
    anaerobic, 22
    defaunated, 171

types, 18
Series Header Record, in data bases, 198
Serology, **186**
Shannon-Wiener diversity index, 220, 221
Sieve, standard, 65
Sieving, 136
Similarity, 203, 206
Simpson's dominance index, 219-221
SIR, 198
Size, 269, 270
Size categories, 13
Size class, fractionation, 136
Size classes, 136
Sizing, specimens, 143
Skewness, 65
Sled, meiobenthic, **116**
Slide frame, Cobb, 152, **153**
    H-S, **153**
Slutzky-Yule effect, 226
Software package, 197, 209, 229
    BMD, 197
    BMDP, 197, 209
    CEP, 197, 198
    CLASP, 198
    CLUSTAN, 198, 209
    COMPCLUS, 198
    CONDENSE, 198
    DATAEDIT, 198
    DBaseIII+, 198
    DECORANA, 197, 211
    LOTUS 123, 198
    NT-SYS, 198, 209
    ORDIFLEX, 198
    SAS, 198
    SIR, 198
    SPSS, 197, 209
    TWINSPAN, 198, 212
Sorensen coefficient, 207-209
Space partitioning, 212
Spade corer see box corer,
Spatial pattern, 197, 199
    randomness, 200
Spearman's rank correlation, 205, 207
Species, anaerobic, 41
    facultative anaerobic, 41
Species abundance, analysis, 207, 212-218
    distribution, 197, 212, 215-223
    log-normal distribution, 217, 218, 218
    log-series distribution, 217
    negative exponential distribution, 216
    neutral models, 217
    Poisson log-normal distribution, 218
    ranked curve, 212, 213
    resource apportioning models, 216-218
    statistical models, 216-218
Species richness, 22, 219, 220
Specimens, documentation, 239
    holotypes, 239, 241
    paratypes, 239

    shipping of, 241
    support, 239
    voucher, 238, 241
Specimens per unit area, 143
Spectral analysis, 201, 224, 227, 228
Springs, 370
SPSS, 197, 209
Staining, 134, 143, 151-153
    chromosomes, 153
    Eosin B, 249
    nanofauna, 232, 235, 236
    Phloxine B, 432
    Sudan Black B, 249
    vital, 151
Standardization, in data bases, 198
Standing stock, 181, 183
Stationarity, 225
Statistical techniques, 197
Statistics, 197
Steedman's mixture, 135, 136
Stempel pipette, 141
Stomach contents, 184
    chlorophyll analysis, 185, 186
    serology, **186**
    tracer technology, 185
    visual examination, 185
Storage, 135
Student's T test, 129
Subdivision of diversity, hierarchical, 221-223
Submersible, 117
Subsampler, Askö splitter, 141, **142**
    Folsom splitter, 141
    Jensen splitter, 141, **142**
    microsplitter, 141
    Stempel pipette, 141
Subsampling, 123, 134, 141, 142, 236
Substrates, artificial, 171
    azoic, 171
Subterranean fauna, 15
Sucrose solution, 140
Sulfide system, 284
Surface tension, 240
Survivorship, 188
Suspension feeders, 32
Syringe, 68, 116
Taxonomic assistance, 240, 241
Taxonomic codes, in data bases
    NODC code, 199
    Rubin code, 199
Taxonomic identification, 240, 241
Taxonomy, 238-242
Temperature, 23, 24, 29, 40, 41, 62, 63, 70
Thermistor, 70
Thiobios, 284
Thomas distribution, 200
Tidal amplitude, 24
Tidal exposure, 15
Time series, 224-229
    differencing, 226

frequency-domain, 224
  seasonality, 225
  stationarity, 225
  time domain, 224, 226
Tolerance, 16
Tolerance experiments, 177
Tracer technique, 185, 187
  feeding, 165
Transformation, 210
Trap, **122**
Traps, emergence, 122, 177, 431
Trophic aspects, 184-187
Turnover rate, 30
TWINSPAN, 198, 212
Übersand technique, 276
Ultrasonic treatment, 136
Ultrastructure, 155
Van Veen grab, **121**
Variability, distribution, 23, **24**
  migration, 23
  spatial, 22
  temporal, 22, 28
Vials, 239, 240
Vital staining, 139, 151
Water, flow, 70
  saturation, 62, 63, 70
  table, 71
Water content, 15
Wave, exposure, 70
  recorder, 70
Wave action, 24
Wet weight, 79, 80, 182
Wilcoxon T test, 129
Zonation, horizontal, 23
  vertical, 22

Taxonomic Index

Abavopsis latosoleata, 452, **456**
Acantharea, 244
Acanthocyclops, 53
　　plattensis, 51
Acanthodasys, 304, **310**
Acantholeberis curvirostris, **369**
Acanthopriapulus horridus, 323
Acarina, 51-53, 428
Acarochelopodia, 418
　　cuneifera, **422**
Acaroidea, 418
Acaromantis, 418
　　subasper, **421**
Acarpomyxea, 243, 244
Acauloides ammisatum, 268, **272**
Achnanthes, 164
Acochlidiacea, 448
Acochlidioidea, 452, 454
Acoela, 273-275, 277, **280**
acoels see Acoela,
Acotylea, 274
Acrassea, 244
Acrobeles, 44
Acrocirridae, 334, **343**
Actinarctus doryphorus, **363**
Actinopoda, 244, 245
actinulid see Actinulida,
Actinulida, 268, 448
Adenophorea, 293
Adephaga, 430
Adinetidae, 312
Aeolosoma, 45, 334, 346
　　hemprichi, **348**
　　maritimum, 346
Aeolosomatidae, 345-348, **348**
Afronerilla, 334
Agathotanais hanseni, **407**
Agaue, 417
Agauopsis, 417
　　pteropes, **422**
Agnathiella beckeri, 283
Aktedrilus, 350-352
　　monospermathecus, **354**
Allogromiidae, 245, 251, **255**
allogromiids see Allogromiidae,
Allokoenenia, 425
Alona bessei, 55
　　phreatica, 55
Aloninae, 365, **369**
Amenophia peltata, **385**
Ammonia batavus, **257**
　　beccarii, 246
Ammotheidae, 423
Amoebae, 232, 243, 245, 247, **256**
Amphioxus, 449
Amphipoda, 39, 50, 51, 53, 55, 409-412, **43**, **412**

amphipods see Amphipoda,
Anaplostoma, **301**
Anarthrura simplex, **407**
Anarthruridae, 403
Anasca, 439-440
Anaspidacea, 389-390, **392**
anaspidacean see Anaspidacea,
Anaspididae, 389-390
Anchistropus minor, **369**
Ancorabolidae, 380, **387**
Ancorabolus mirabilis, **387**
Angeliera, 53
　　phreaticola, **401**
Anisonyches diakidius, **364**
Anisoptera, 429
Annelida, 39, 49, 52, 302, 332, 349, 428
annelids see Annelida,
Annulonemertes, 288, **291**
　　minusculus, 287
Anomalohalacarus, 418
　　minutus, **421**
Anopla, 287, 288, 289
anoplans see Anopla,
Anoplodactylus arescus, 423, **424**
　　evelinae, 423,
　　tarsalis, 424
Anoplopisione, 334
Anostraca, 365
Anthohydra, 268
Anthozoa, 266, 269
Anthuridea, 397, 398, **400**
Anticoma, **301**
Apharyngtus, 334
　　punicus, **342**
Aphropharynx, 335
Aplacophora, 18, 447-450, **450**
aplacophoran see Aplacophora,
Aplousobranchiata, 461
Apochela, 358
Apodida, 457
Apodopsyllus vermiculiformis, **387**
Apodotrocha, 334
　　progenerans, **342**
Apseudes setosus, **407**
　　spinosus, **407**
Apseudoidea, 403, **408**
apseudomorphan (Apseudomorpha), 402, 403
Arachinda, 423, 425
Archechiniscus marci, **364**
Archiannelida, 14, 25, 51, 54, 162, 170, 275, 302, 349, **54**
archiannelids (Archiannelida),
Archinemertina, 287, 288, **291**
archinemertines see Archinemertina,
Archotoplaninae, 274, **281**
Arcturidae, **400**
Arcturopsis rudis, **400**
Arenonemertes microps, 287
　　minutus, 287
Arenotrocha, 334

Argyrotheca bermudana, 445
Aricidea (Acesta) cerrutii, **344**
Armorhydra janowiczi, 269, **272**
Armorhydridae, 268
Artemia, 164
Arthrotardigrada, 358
Articulata, 332, 445
articulate see Articulata,
aschelminth see Aschelminthes,
Aschelminthes, 283, 328, 357
Ascidiacea see Tunicata,
ascidians see Tunicata,
Ascoglossa, 452
Ascophora, 440
Aselloidea, 397
Asellopsis, 192
   hispida, **386**
Asellota, 397, 398, **400, 401**
Aspelta, 43, 313
   egregia, **318**
Asperspina loricata, **456**
Aspidiophorus, 304, **311**
Aspidisca, 258
   costata, **265**
Aspidosiphon exiguus, 356, **356**
Asteroidea, 457
Asteropteron skogsbergi, 375
Astigmata, 418
Attheyella, 44
Auricularia nudibranchiata, 460
Aurila convexa, 376
Austrognathia microconulifera, 286
Axonolaimidae, 27
Axonolaimus, **301**
Bacteria, 29, 31, 46, 47, 163-166, 232
Baetidae, 429, **436**
Baetis, **436**
Bairdioidea, 371
Bairdioppilata, 375
Balcanella, 53
Basserolis kimblae, **400**
Bathychaetus, 334
Bathyconchoecia septemspinosa, 375
Bathydrilus, 351
   hadalis, 351
Bathylaimus, **301**
Bathynella, 53, **55**
   paranatans, **392**
   riparia, 51, 53
Bathynellacea, 51, 389, 390
bathynellaceans see Bathynellacea,
bathynellid see Bathynellidae,
Bathynellida, 53, 55
Bathynellidae, 389, 390, **392**
Bathynerilla, 334
Batillipes, 362
   bullacaudatus, 25
   dicrocercus, 25
   pennaki, 25

Batopora murrayi, 440, **443**
   nola, 440, **443**
Bdelloidea, 312, 315
bdelloids see Bdelloidea,
Beania proboscidea, 439, **443**
Beaniidae, 439
Beetles see Coleoptera,
Biserramenia, 448
   psammobionta, 447
Bivalvia, 451-456
Blepharisma, 258
   japonicum, **265**
Bodotriidae, 414
Bogidiella, 50, 53, 55
   brasiliensis, **55**
Bogidiellidae, 410
Boreohydra simplex, 267, **272**
Bothrioplanidae, 274, **281**
Brachionidae, 313
Brachiopoda, 11, 18, 445, 446, 451, **446**
brachiopods see Brachiopoda,
Brachycarida, 413
Brachycera, 430
Bradleya dyction, **376**
Bradyagaue, 418
   drygalski, **422**
Branchiopoda, 377, 445-446
Branchiura, 377
Brania, 333
   oculata, **341**
Brinkmaniellinae, 275, **282**
Bryceella, 313
   tenella, 314, **318**
Bryocamptus, 51, 53
Bryozoa, 438-443, **443**
Bulimina elongata, **257**
Bursovaginoidea, 283, **286**
Bythocythere turgida, **376**
caecid see Caecidae,
Caecidae, 451-453
Caecostenetroides ischitanum, **401**
Caecum auriculatum, 452
   glabrum, **456**
   subannulatum, 452
   trachea, 452
Caenorhabditis elegans, 158
Calabozoa pellucida, **401**
Calabozoidea, 397, 398, **401**
Calanoida, 49, 55-175, 186, 380
calanoids see Calanoida,
Calliopaea bellula, **456**
Campylaspis, 414
   minor, **416**
Campyloderes, 328
   macquariae, **331**
Candona candida, 375
Canthocamptus, 44
Canuella, 381
   perplexa, **386**

Canuellidae, 386
Capilloventer, 349
    atlanticus, 350, 351
Capilloventridae, 350, 351
Capitellidae, 349
capitellids see Capitellidae,
Caprellidea, 409
Carinina arenaria, 287, 288, **291**
    coei, 288
Carnacolaimus, **301**
Carphania fluviatilis, **364**
Carybdea alata, 267
    marsupialis, 267, **272**
Caryophyllidae, 269
Catenula, 51
Catenulida, 274, 283, **280**,
Catenulidae, 274
catenulids see Catenulida,
Cateria, 18, 328
    gerlachi, **331**
    submersa, **331**
Caudofoveata, 447
Celleporariidae, 440
Centroderes, 328
    spinosus, **331**
Centroderidae, 328
Cephalaspidea, 452
Cephalodasys, 303, **309**
Cephalodella, 43, 313
    compacta, 313, **318**
Cephalothrix, 287, 288, **291**
    germanica, 289
Ceramonematidae, 27
Ceratopogonidae, 46, 430, **437**
ceratopogonids see Ceratopoginidae,
Ceratopogoninae, 430
Cercopagidae, 366
Cervinia bradyi, **386**
Cerviniidae, **386**
Chaetodermomorpha, 447, 448, **450**
Chaetogaster, 45, 351
Chaetonotida, 302, 304
Chaetonotidae, 304, 305
chaetonotidans see Chaetonotida,
chaetonotids see Chaetonotidae,
Chaetonotus, 14, 46, 304, **311**
Chaetostephanidae, 322-324
Chaetostephanus praeposteriens, 323
Chaoborida, 48
Cheilostomata, 438, 439, 441
cheilostomes see Cheilostomata,
Chelon labrosus, 30
Chevroderma scalpellum, 448
    whitlatchi, 448, **450**
Chiridolta rotifera, 458, **460**
Chiridotidae, 457, 458
Chironex fleckeri, 267
Chironomida, 53
Chironomidae, 46, 48, 49, 428-432, **437**

chironomids see Chironomidae,
Chloroperlidae, 429, **436**
chlorophytes, 164
Chordariidae, 274
Chordodasys, 303
Chromadora lorenzeni, 294
Chromadoridae, 27, 293, 296
chromadorids see Chromadoridae,
Chromadorina germanica, 163, 165
Chrysomelidae, 430
Chydoridae, 366, **369**
chydorids see Chydoridae,
Chydorinae, 365, **369**
Ciliata see Ciliophora,
ciliates see Ciliophora,
Ciliopharyngiellidae, 275, **282**
Ciliophora, 39, 48, 139, 162, 166, 232, 233, 243, 258-265, 275, 304, 426, **233**, **265**,
Cirripedia, 370, 377
Cladocera, 39, 46, 48, 49, 52, 55, 365-369, **369**
Cladocopida, 371
Clamydotheca rudolphi, **375**
Clavarctus falculus, **363**
Clavodorum adriaticum, **341**
Cletodidae, **386**
Clitellata, 332, 345, 346, 349
Cnidaria, 266-271, 447, **272**
Coelogynoporidae, 274, **281**
Coleoptera, 56, 428-430, **437**
Collembola, 56, 426, 428, 429, **436**
Collotheca campanulata, 313
    ornata, 313
    voigti, 313
    wiszniewskii, 313, **318**
Collothecidae, 313
Colomastigidae, **412**
Colomastix, 409
    kapiolani, **412**
Colurella colurus, 313-315, **318**
    dicentra, 313
    grandiuscula, 313
    obtusa, 313
Colurellidae, 313
Combinata, 274
Comesomatidae, 27
Concentricycloidea, 457
Conchorhagae, 328
Conchostraca, 365
Condyloderes, 328
    multispinosus, **331**
Condylostoma, 258
    patulum, **265**
Conescharellina, 441
    africana, 440, **443**
Conescharellinidae, 440
Conophoralia, 283, **286**
Copepoda, 18, 19, 20, 21-23, 25, 26, 31, 39, 43, 46, 48, 50, 135, 140, 164, 170, 172, 174-177, 182, 377, 380-388, 426, 43, **385-388**
copepodid stages, **388**

copepods see Copepoda,
Copidoghnatinae, 418
Copidoghnatus, 418
Coralliotrocha, 334
Coronarctus tenellus, **364**
Corydalidae, 429
Cotylea, 274
Crasiella, 303, 309
Craspedacusta sowerbyi, 269, **272**
Cribrimorpha, 440
Crinoidea, 457
Crithionina sp., **256**
Crustacea, 162, 359, 399, 423
crustaceans see Crustacea,
Cryptorhagae, 328
Ctenocheilocaris, 377, 378
Ctenodrilidae, 334, **343**
Ctenodrilus, 334
    serratus, **343**
Ctenostomata, 438, 439, 441
ctenostomes see Ctenostomata,
Cubozoa, 266, 267
Cumacea, 413-416, **416**
cumaceans see Cumacea,
Cumella vulgaris, 414
Cumopsis, 414
Cupuladriidae, 439
Curculionidae, 430
Cushmanidea seminuda, **376**
Cyatholaimidae, 27
Cyclidium, 258
Cyclocypris globosa, **375**
Cycloleberis galatheae, **375**
Cyclopidae, 380
Cyclopinidae, 380
Cyclopoapseudidae, 403
cyclopoid see Cyclopoida,
Cyclopoida, 48-50, 52, 53, 55, 380
Cyclops agilis, 48
Cyclorhagae, 328
Cyclorhagida, 328
Cyclostomata, 439, 441
cyclostomes see Cyclostomata,
Cylindropsyllidae, **387**
Cylindropsyllus laevis, **387**
Cylindrotheca, 164
Cypretta foveata, **375**
Cypria, 48
    ophthalmica, **375**
Cyprideis lengae, **376**
    torosa, 30
Cypridina sinosa, **375**
Cypridoidea, 370, 371
Cypridopsis newtoni, **375**
Cypris maculosa, **375**
    marginata, **375**
Cytherelloidea bairdioppilata, **375**
    damericacensis, **375**
Cytheroidea, 371

Cytheromorpha fuscata, **376**
Cytheropteron abyssicolum, **376**
Dactylobiotus, 359
Dactylopodola, 302, 303, 305, **309**
Dactylopodolidae, 305
dactylopodolids see Dactylopodolidae,
Dalyelliidae, 275, **281**
Dalyellioida, 273, 275
Danielopolina orghidani, **375**
Daphniopsis, 366
Daptonema, **301**
Darcthompsonia fairliensis, **387**
Darcthompsoniidae, 380, **387**
Darwinula stevensoni, **375**
Darwinuloidea, 371
Dasydytes, 305, **311**
dasydytes see Dasydytidae,
Dasydytidae, 305
Dendrodasys, 303, **309**
Deontostoma timmerchioi, 293
Derocheilocaris, 377, 378
    typica, **379**
Desmodasys, 303, **310**
Desmodoridae, 27
Desmodorinae, 27
Desmoscolecidae, 27, 295
desmoscolecids see Desmoscolecida,
Desmoscolex, **300**
Diacyclops, 53
Diaphorosoma, 334
Diarthrodes aegideus, 26
    major, **385**
Diasterope schmitti, **375**
Diastylis bradyi, 414
    rathkei, 414
Diatoms, 29, 31, 164
Dichaetura, 305, **311**
Dichaeturidae, 304
Dicranophoridae, 313, 314
Dicranophorus, 43, 313
    hercules capucinoides, **318**
Digononta, 312
Dileptus, 258
    anser, **265**
    gigas, 48
Dinodasys, 303, **310**
dinoflagellates (Dinoflagellata), 258
Dinophilidae, 332, 334, **342**
Dinophilus, 334
    gyrociliatus, 158, 332, **342**
Diopatra, 173
Diosaccidae, **385**
Diplodasys, 304, **310**
Diplolaimella, 165
Diplolaimelloides, 192
Diplura, 426
Diptera, 39, 46, 50, 52, 428-432, 437
dipterans see Diptera,
discinids (Discinidae), 445

Discocephalus, 258
   ehrenbergi, **265**
Disparalona dadayi, **369**
Dissotrocha, 313
Ditrupa, 451
Diurella, 43
   pygocera, 314, 315, **318**
Diurodrilidae, 332, 334, **342**
Diurodrilus, 302, 332, 334
   westheidei, **342**
Dolerocypris fasciata, **375**
Dolichodasys, 303, 305, **310**
Dolichomacrostomidae, 274
Dolichomacrostominae, 274
Dolichopodidae, 430
Donacia, 430
Dorvillea, 334
   pacifica, **342**
Dorvilleidae, 334, **342**
Dorylaimida, 294, 298
Dorylaimus, 44
Dracograthis, 300
Draculiciteria, 304, **311**
Dunaliella, 164
   parva, 165
Duridrilus, 351
Dysteria, 258
   scutellum, **265**
Dytiscidae, 430
Echiniscidae, 357
Echiniscoidea, 358, 359
Echiniscoides heopneri, **364**
Echinocythereis dunelmensis, **376**
Echinoderes, 328
   coulli, **331**
   dujardinii, **331**
Echinoderida, 328
Echinoderidae, 328
Echinodermata, 359, 457
echinoderms see Echinodermata,
Echinoidea, 457
Echinolaophonte horrida, **387**
Echinotheristus, 300
Ectinosoma melaniceps, **386**
Ectinosomatidae, **386**
Ectopria, **437**
Eggerelloides scabrum, **256**
Eisothistos antarcticus, **400**
Elaphoidella, 44, 53
Eliberidens, 334
Elmidae, 430, **437**
Elosa, 43
   worallii, 314
Elphidium crispum, 245
   selseyensis, **257**
Embletonia, 452
   pulchra, 452, 453, **456**
Embletoniidae, 452
Empididae, 430, **437**

Encentrum, 43, 314
   axi, 314
   lineatum, 313, **318**
   marinum, 313, 315, **318**
   permutandum, 314
   villosum, 313
Enchytraeidae, 151, 349, 351
enchytraeids see Enchytraeidae,
Enhydrosoma buchholtzi, **386**
   propinquum, 26
Enopla, 287
Enoplidae, 27
Enoploides, **301**
Enoplolaimus, **301**
Enoplus, **301**
Enteromorpha, 163
Entocytheridae, 370
Entomobryidae, **436**
Entoprocta, 11, 18, **444**
entoprocts see Entoprocta,
Eophliantidae, **412**
Epacteriscidae, 380
Epactophanes, 44, 53
Ephelota gemmipara, **265**
Ephemerellidae, 429, **436**
Ephemeroptera, 46, 429, 431, **436**
ephemeropterans see Ephemeroptera,
Ephydridae, 430
Epicaridea, 397
epsilonematids (Epsilonematida), 295
Erignatha, 313
Erythraeidae, 418
Erythraeoidea, 418
Eubriidae, 430, **437**
Euchlanidae, 313
Euchlanis, 43
   arenosa, 313, **318**
Euchone, 336
Euchromadora, **301**
Eucladocera, 365, 366
eucladoceran see Eucladocera,
Eucypris afghanistanensis, **375**
Eucythere argus, **376**
Eudiplogaster paramatus, 165
   pararmatus, 293
Eugymnanthea psammobionta, 268, **272**
Eukalptorhynchia, 275
Eukoenenia, 425, 426
Eukoeneniidae, 425
Eumalacostraca, 409, 413
eumalacostran see Eumalacostra,
Euphysa aurata, 270
   ruthae, 270
Euplotes, 236, 258, **236**
Eurotatoria, 312, 313
Eurycercinae, 365
Eusarsiella cornuta, **375**
Eusyllis, 333
   homocirrata, **341**

Eutardigrada, 357-359
eutardigrade see Eutardigrada,
Exallopus, 334
Exogone, 333
    naidinoides, **341**
Fabricia, 334
Fabriciola, 334
Fauveliopsidae, 334, **343**
Fauveliopsis, 334
    brevis, **343**
Filitanais rebainsi, **407**
Filosea, 243, 244
Filospermoidea, 283, **286**
Flabellifera, 397, 398, **400**
flagellates (Flagellata), 46, 48, 139, 162, 166, 232, 243, 245, **233**
flatworms see Turbellaria,
Flexibacteria, 232, **233**
Florarctus salvati, **363**
Foraminifera, 25, 162, 164, 165, 243-247, 249, 250, 438, 439, **255, 257**
foraminiferans see Foraminifera,
Foraminiferida, 244, 245
Fragillaria, 164
Franzenella, 439, 443
Fungi, 232, **233**
Gammaridea, 409
Gastropoda, 18, 139, 451-456, **456**
gastropods see Gastropoda,
Gastrotricha, 11, 14, 16, 18, 22, 25, 39, 46-48, 50, 52, 135, 137, 139, 151, 161, 162, 164, 170, 177, 182, 240, 302-311, 334, 447, **43, 309-311**
gastrotrichs see Gastrotricha,
Geleia, 258
Genitoconia atriolanga, 448
Gephyrocuma, 414
Globigerina, **256**
Glycymeris, 449
    subobsoleta, 448
Gnathiidae, **400**
Gnathiidea, 397, 398, **400**
Gnathorhynchidae, 275
Gnathostenetroidae, **401**
Gnathostenetroidea, 398, **401**
Gnathostenetroides laodicense, **401**
Gnathostomaria, 284
Gnathostomula, 284
    tuckeri, 283, **286**
Gnathostomulida, 11, 14, 16, 18, 273, 275, 283-286, 302, **286**
gnathostomulids see Gnathostomulida,
Gnesiotrocha, 313
Gnosonesimidae, 274
Golfingia pugettensis, 448
Gomphidae, **436**
Goniadidae, 333, **343**
Goniadides, 333
    falcigera, **343**
Gonionemus vertens, 269, **272**
Gracilaria, 163
Graeteriella, 53

Graffillidae, 275, **281**
Grania, 349, 351
    pusilla, **354**
Granuloreticulosea, 244
Graptoleberus testudinaria, **369**
Greeffiella, 293, **300**
Gromia oviformis, 244
Gromiidae, 244
Gruberia, 258
Guernella, 365
Gwynia capsula, 445, **446**
Gymnamoebia, 243, 244
Gymnodorvillea, 334
Gymnolaemata, 438
Gynodiastylidae, 414, **416**
Gynodiastylis lata, **416**
Gyrinidae, 430
Habrotrichidae, 312
Halacarellus, 417
    subterraneus, **422**
Halacarida, 143
Halacaridae, 46
Halacaroidea, 18, 25, 48, 50, 51, 135, 177, 182, 417-422, 424-426, **421-422**
halacaroids see Halacaroidea,
Halacarus, 417
    rismondoi, **421**
Halalaimus, **300**
Halammohydra, 14
    coronata, 268
    octopodides, **272**
    schulzei, **272**
Halammohydrina, 266-269
Halechiniscus, 357
    flabellatus, **362**
Halectinosoma winonae, 26
Halichaetonotus, 304, **311**
Halicryptus, 324
Halicyclops, 26, 380
Halobiotus, 358
    arcturulius, **364**
halocyprid see Halocyprida,
Halocyprida, 371
Halolinda, 313
Halosbaena, 393
    acanthura, **396**
Haplognathia cf. rosacea, **286**
Haplopharyngida, 274, **280**
Haplophragmoides bradyi, **257**
Haplopoda, 365
Harpacticoid copepods see Harpacticoidae,
Harpacticoidae, 11, 22, 24, 25, 27, 29, 40, 44, 46, 48-53, 55, 161-163, 165, 170, 174, 176, 183, 184, 192, 377, 380-382, **43, 183, 385-388**
harpacticoids see Harpacticoidae,
Harpacticus, 381
    chelifer, **386**
Harpagoherpia tenuisoleata, 448
Hedylopsidae, 452
Hedylopsis spiculifera, 452, **456**

Hekodoma implicata, 439
Helicoprorodon, 258
Heliodoma implicata, 439
heliozoans see Heliozoea,
Heliozoea, 244, 245
Helixotionella scutata, 440
　　spiralis, 440
Helodidae, 430
Hemicytherura videns, 376
Hemidasys, 304
Hemiptera, 428
Heptageniidae, 429, **436**
Herpetocypris reptans, 375
Hesionidae, 333, **341**
Hesionides, 333
　　arenaria, 23
　　gohari, **341**
　　riegerorum, 55, 332
Hesionura, 333
　　laubieri, **343**
Heterocypris salina, 375
Heterodrilus, 350, 351
　　jamiesoni, **354**
Heterolaophonte, 380
Heterolepidoderma, 304, **311**
Heteronemertina, 287-289, **291**
heteronemertines see Heteronemertina,
Heteropodarke, 333
　　heteromorpha, **341**
Heterostigma fagei, 438
　　gonochorica, **464**
Heterotanais oerstedi, 407
Heterotardigrada, 357-359
heterotardigrades see Heterotardigrada,
Heteroxenotrichula, 304, **311**
Hexabathynella, 389, 390
　　halophila, **392**
Hirudinoidea, 349
holothurian see Holothuroidea,
Holothuroidea, 11, 18, 457-460, **460**
Homalorhagae, 328
Homalorhagida, 328
Homotrona rubrum, 445
Hoplonemertina, 287, 289, **291**
hoplonemertines see Hoplonemertina,
Hrabeiella periglandulata, 332
Hubrechtella dubia, 288
Huntemannia jadensis, 30
Hyadesiidae, 418
Hydra, 51
Hydracarina, 39
Hydracarinida, 48
Hydrachnellae, 46, 50, 418, **43**
Hydrachnoidea, 418
Hydrina, 267
hydroid see Hydrozoa,
Hydroptiidae, 430
Hydroptilla, **437**
Hydroptillidae, **437**

Hydrovolzioidea, 418
Hydrozoa, 18, 48, 182, 266, 267, 448
Hydryphantoidea, 418
Hygrobatoidea, 418
Hymenoptera, 428
Hyperammina, **256**
Hyperiidea, 409
hyphomicrobial bacteria, **233**
Hyphomicrobium, 232
Hypsibiidae, 359
Hypsibius, 45, 358, 359
　　convergens, **364**
　　itoi, 359
Hyssuridae, **400**
Hystricosoma, 346
　　chappuisi, 346, 347, **348**
　　insularum, 346
　　pictum, 346
Ichthydium, 46, 304, **311**
Ikosipodus, 334
Ilyocryptus spinifer, 369
Ilyocypris gibba, 375
Inanidrilus, 350, 351
inarticulates (Inarticulata), 445
Inflatana, 313
ingofiellids see Ingolfiellidea,
Ingolfiella, 53, 410, **43**
　　putealis, **412**
Ingolfiellidae, 55, 410
Ingolfiellidea, 409, **412**
Insecta, 51, 428-437, **436**, **437**
insects see Insecta,
Inversiscaphos, 440
　　setifer, 440, **443**
Ionosyllis, **341**
Ironidae, 293
Isobactrus, 418
　　hartmanni, **422**
Isochrysis galbana, 164
Isohypsibius, 45, 359
Isopoda, 22, 39, 50, 51, 53, 55, 182, 397-401, **43**, **400**, **401**
isopods see Isopoda,
Itaquascon, 359
Jaeropsidae, **401**
Jaeropsis beuroisi, **401**
Janaira gracilis, **401**
Janiridae, **401**
Janiroidea, 397, **400**, **401**
Kalliapseudidae, 403
kalliapseudids see Kalliapseudidae,
kalyptorhynch see Kalyptorhynchia,
Kalyptorhynchia, 273, 275, 289, **282**
Kamptozoa, 444
Kanaloa manoa, **412**
Karkinorhynchidae, 275
Karlingiinae, 274
Kentrophoros, 258
　　flavum, **265**
Kijanebalola, 305, *311*

Kinorhyncha, 11, 14, 18, 135, 137, 143, 182, 328-331, **331**
Kinorhynchus mainensis, **331**
Koeneniodes deharvengi, 426
Koinocystidae, 275
Komokiacea, 249, 251, 439, 440, **257**
komokiaceans see Komokiacea,
Koonungidae, 389
Krenosmittia, 56
Krithe bartonensis, **376**
Kytorhynchidae, 275, **282**
Labidoplax buskii, 457, **460**
    media, 457
Lacrimula pyriformis, 440, **443**
Lacrymaria, 258
    olor, **265**
Lagena, 244
    semistriata, **257**
Lagenidiopsis, 244
    elegans, **256**
Lagenina, **257**
Lanthus, **436**
Laophonte thoracica, **386**
Laophontidae, 386, **387**
Laophontodes bicornis, **387**
Lasaeidae, 451
Latona setifera, 369
Latonopsis occidentalis, 369
Lecane, 43, 46, 51, 313
    cornuta, 315
    mucronata, 313, **318**
    psammophila, 313, **318**
Lecanidae, 313
Lecithoepitheliata, 274, **280**
Leguminocythereis oertili, **376**
Lepadella psammophila, 313, 314, **318**
    triptera, 313
Lepidodasyidae, 303
Lepidodasys, 303, 304, **310**
Lepidodermella, 46, 304, **311**
Lepidomenia, 448
    harpagata, 448
    hystrix, 447
Lepidoptera, 428
Lepidostoma, **437**
Lepidostomatidae, **437**
Lepoarctus coniferus, **363**
Leptastacus macronys, **387**
Leptobathynella, **43**
Leptoceridae, 430
Leptocheliidae, 402, 403, **408**
Leptocythere levis, **376**
Leptognathiidae, 403, 404, **408**
Leptohalysis, **256**
Leptokoenenia, 425, 426
    gerlachi, 426
    scurra, 426, **427**
Leptolaimidae, 27, 293
Leptolaimus, **301**
Leptophlebiidae, **436**

Leptosomatidae, 27
Leptosynapta minuta, 457, **460**
Leuconidae, 414
Leuctra, **436**
Leuctridae, 429, **436**
Leydigia acanthocercoides, 369
Ligiarctus eastwardi, **362**
Limnocnida indica, 269
    tanganyicae, 269
Limnocythere sanctipatricii, **376**
Limnodriloides, 350, 351
    barnardi, **354**
Limnodriloidinae, 350
Limnohalacarinae, 418
Limnohydrina, 269
Limnoposthia polonica, 54, **54**
Limnosbaena, 393
    finki, 393
Lindia, 43
    janickii, 313
    tecusa, 313
Lindiidae, 313
lingulids (Lingulidae), 445
Linhomoeidae, 27
Litonotus, 258
    lamella, **265**
Lobatocerebridae, 334, **342**
Lobatocerebrum, 334
    psammicola, **342**
Lobosea, 243, 244
Lohmannella, 418
Lohmannellinae, 418
Loricifera, 11, 14, 18, 137, 143, 240, 314, 319-321, **321**
loriciferans see Loricifera,
Loxoconcha propunctata, **376**
Loxodes, 258, 259
    magnus, **265**
Loxophyllum helus, **265**
Loxosoma isolata, 444, **444**
Loxosomatidae, 444
Lunulitidae, 439
Maccabeus, 324
    cirratus, 323, 324
    tentaculatus, 323, 324, 327
Macoma balthica, 451
Macrobiotidae, 359
Macrobiotus, 45, 359
Macrochaeta, 334
    multipapillata, **343**
Macrocypris minna, 375
Macrodasyida, 302-309
Macrodasyidae, 303, 305
macrodasyidans see Macrodasyida,
macrodasyids see Macrodasyida,
Macrodasyoida, 16
Macrodasys, 303, **309**
Macrostomida, 274, 277, **280**
Macrostomidae, 274
macrostomids see Macrostomida,

Macrothricidae, 365, 366, **369**
macrothricidae see Macrothricidae,
Makrokylindrus, 414
Malacostraca, 377, 397, 402
malacostracans see Malacostraca,
Malletiidae, 451
Manayunkia, 334
    aestuarina, **343**
Maraenobiotus, 53
Marenda nematoides, 245
Maricola, 274
Marinellina, 305, **311**
    flagellata, 54, **54**
Marionina, 351, 352
Mastigophora, 243, 244, **233**
mastigophorans see Mastigophora,
Maxillopoda, 377
Megadasys, 303, 305, **309**
Megaloptera, 429, 430
megalopterans see Megaloptera,
Meganerilla, 334
    swedmarki, **344**
Meiodorvillea, 334
Meioherpia, 447, 448
    stygalis, **450**
Meiomenia, 448
    arenicola, **450**
    swedmarki, 447, 449, **450**
Meiopriapulus fijiensis, 322-324, **327**
Meiorhopalon arenicolum, 268, **272**
Mesodasys, 303, 304, **309**
Mesodinium, 258
Mesogastropoda, 451
Mesonerilla, 334
    ecuadoriensis, **344**
Mesostigmata, 418
Mesostominae, 275, **282**
Mesostygarctides orbiculatus, 362
Mesostygarctus intermedius, 362
Mesotardigrada, 358
Metachromadora vivipara, **24**
Metahadzia, 53
Metaingolfiella, 410
Metaingolfiellidae, 410
Metalinhomoeus, **301**
Metapseudidae, 403
Metaxypsamma, 333
Metidae, 380, **385**
Metis ignea, **385**
Microarthridion littorale, 24, 26, 30
Microcerberoidae, **401**
Microcerberoidea, 397, 398, **401**
Microcerberus, 55, **43**, **401**
Microcharon, 53, 55
Microcyclops pumilis, 51
Microcytherura nigrescens, 376
Microhedyle cryptophthalma, **456**
Microhedylidae, 14, 452
Microhydrula pontica, 269

Microlaiminae, 27
Microlaimus, **301**
Microparasellidae, 393, **401**
Microparasellus, 55
Microphthalmus, 333
    listensis, **341**
    sczelkowii, **341**
    similis, **341**
Microteliphonida, 425
Microthorax pusillus, **265**
Miliolina, **257**
Milnesium, 359
    tardigradum, **364**
Mirandotanais vorax, **407**
Mirocyclops, 53
mites see Halacaroidea,
Moina, 366
Monhystera, 44
Monhysteridae, 27, 293, 296
monhysterids see Monhysteridae,
Monimotrocha, 313
Monobryozoon, 14, 439, 441
    ambulans, 439, **443**
    bulbosum, 439, **443**
    sandersi, 439
Monobryozoontidae, 439
Monocelididae, 274, **281**
Monochrysis lutheri, 164
Monochus, 44
Monodella, 393, **55**
    atlantomaroccana, 393, **396**
    sanctaecrucis, 393, **396**
    texana, 393
Monodellidae, 393
Monogononta, 313
Monoposthia, **301**
Monostilifera, **291**
Monostyla, 43, 313, 315, **318**
Monotoplanidae, 274
Moraria, 53
Morimotoa, 56
multiflagellate, **233**
Multitubulatina, 304
Munna boecki, **400**
Munnidae, **400**
Musellifer, 305, **311**
Myersina, 43
Myersinella, 315
    tetraglena, 313
Myodocopa, 370, 371
Myodocopida, 370-372
myodocopids see Myodocopida,
Myriotrochidae, 457, 458, **460**
myriotrochids see Myriotrochidae,
Myriotrochus geminiradiatus, 458, **460**
Mystacocarida, 15, 18, 377-379, **379**
mystacocarids see Mystacocarida,
Mystides, 333
Mytilus edulis, 451

Naididae, 45, 48, 350, 351
naidids see Naididae,
Nais, 45, 350
Nannastacidae, 414, **416**
Nannastacus inflatus, **416**
Nannopus palustris, 26
Nanochloris, 164, 165
Nanoloricida, 319
Nanoloricidae, 319
Nanoloricus mysticus, 319, **321**
Narapa bonettoi, 351
Narapidae, 350, 351
narapids see Narapidae,
Narcomedusae, 268
nauplius stages, **388**
Nebalia, 312
Nemathelminthes, 328
Nematocera, 430
Nematoda, 11, 18-25, 27, 29, 39, 44, 46-53, 134, 135, 139, 161-166, 170, 172-177, 181-184, 192, 293-301, 349, 380, 426, 428, **43, 300, 301**
nematodes see Nematoda,
Nematomenia incirrata, 448
   protecta, 448
Nematoplanidae, 274, **281**
nemerteans see Nemertina,
Nemertina, 18, 55, 135, 275, 287-292, 452, **291, 292**
nemertines see Nemertina,
Nemertodermatida, 274, **280**
Neocentrophyes, 328
   satyai, **331**
Neocentrophyidae, 328
Neochromadora, **301**
Neocytherideis senescens, **376**
Neodasyidae, 304
Neodasys, 302, 304, 305, **311**
Neogossea, 305, **311**
Neomeniomorpha, 447-449, **450**
neomeniomorphs see Neomeniomorpha,
Neostygarctus acanthophorus, **362**
Neotanaidae, 403, **408**
neotanaids see Neotanaidae,
Nephtys, 29
Nerilla, 334
   parva, **344**
Nerillidae, 332, 334, **344**
Nerillidium, 334
   lothari, **344**
Nerillidopsis, 334
   hyalina, **344**
Neuroptera, 428
Nihotunga noa, **412**
Nihotungidae, 409, **412**
nihotungids see Nihotungidae,
Niphargopsis, 53
Niphargus, 53
Nitocra, 53, 174
   lacustris, 26
   spinipes, 174

Nitocrella, 44
Nitzschia, 164
Nodellum, **255**
   membranacea, **255**
Nolella, 439,
   limicola, 439, **443**
   monniotae, 439
   radicans, 439
Nolellidae, 439
Normanina, **257**
Notholca baikalensis, 313
   kozhovi, 313
   psammarina, 313
Notiphila, 430
Notodiaptomus caperatus, 55
Notodromas monacha, **375**
Notommata, 43
   bennetchi, 313
Notommatidae, 313
Notostraca, 365
Novaquesta, 334
   trifurcata, **343**
Nuculanidae, 451
Nuculidae, 451
Nudibranchia, 448, 452
Nymphonella lambertensis, 423
   lecalvezi, 423
Odonata, 429, **436**
Odontolaimus, 44
Oedicerotidae, 410, **412**
Ogdeniella maxima, **256**
Olavius, 350, 351
   geniculatus, **354**
Oligochaeta, 18, 45-48, 50-53, 140, 161, 177, 182, 275, 332, 345, 346, 349-354, 426, **354**
oligochaetes see Oligochaeta,
Oligochoerus limnophilus, 54
Olisthanellinae, 275, **282**
Omalogyra, 451
   atomus, **456**
Omalogyridae, 451, 452
Oncholaimidae, 27
Oniscoidea, 397
Opalinata, 244
Ophiuroidea, 457
Ophryotrocha, 192, 334
   gracilis, **342**
Ophryoxus gracilis, 48
Opisthobranchia, 14, 451, 452
opisthobranchs see Opisthobranchia,
Opisthoporata, 274
Optioservus, **437**
Orbiniidae, 334, **343**
Orbituliporidae, 440
Oribatei, 46, 418
Orthocladiinae, **437**
Orzeliscus belopus, 364
Ostracoda, 18, 22, 39, 46-53, 140, 161, 164, 182, 370-376, 451, **375, 376**

Ostracods see Ostracoda,
Otionella auricula, 440
  minuta, 440, **443**
Otohydra tremulans, 269
  vagans, 268, 269, **272**
Otomesostomidae, 274, **281**
Otonemertes marcusi, 287
otoplanid see Otoplanidae,
Otoplanidae, 274, 304, **281**
Ototyphlonemertes, 287-289, **292**
  americana, **292**
  antipai, **292**
  aurantiaca, **292**
  brevis, **292**
  brunnea, **292**
  cirrula, **292**
  erneba, **292**
  evelinae, **292**
  fila, 288, **291**
  macintoshi, **292**
  pallida, 288, **291**
Ougia, 334
Ovammina, **255**
Oxystomatidae, 27
oxystominids (Oxystominidae), 296
Oxytricha, 258
  elliptica, **265**
Pachyzoon, 438, 439
  atlanticum, 439, **443**
Pachyzoontidae, 439
Paedotrocha, 313
Pagurapseudes largoensis, **407**
Paijenborchella cymbula, **376**
Palaeocaridacea, 390
Palaeonemertina, 287, 288, **291**
palaeonemertines see Palaeonemertina,
Paleokoenenia, 425
  mordax, 425
palpigrades see Palpigradida,
Palpigradida, 425-427
Paludicola, 274
Pancarida, 393
Pantopoda, 423
Parabathynella motasi, **392**
Parabathynellidae, 390, **392**
Paracentrophyes, 328
  praedictus, **331**
Parachela, 358
Paracytheridea luandensis, **376**
Paracytheroma sudaustralis, **376**
Paradasys, 303, **310**
Paradicranophorus, 313
Paradicronophorus hudsoni, **318**
Paraganitus ellynnae, **456**
Paragnathia formica, **400**
Paraleptophlebia, 429, **436**
Paramesochra dubia, **385**
Paramesochridae, 385, **387**
Paramonohystera wieseri, 294

Paranais, 351
Paranerilla, 334
Paraonidae, 334, **344**
Parapodrilus, 334
  psammophilus, **342**
Pararotatoria, 312
Parastenocaris, 51, 53, **43**
  brevipes, 44
  starretti, 44
  texana, 52
Parastygarctus higginsi, **362**
Paratanaidae, 404
Paraturbanella, 303, 310
Parergodrilidae, 332-334, **343**
Parergodrilus heideri, 332, 334
Pargurapseudes largoensis, 402
Parmularia, 440
Parmursa fimbriata, **363**
Paronychocamptus wilsoni, 26
Parophryotrocha, 334
Parougia, 334
Patagonacythere tricostata, **376**
Paucitubulatina, 304
Pauropoda, 426
Pedipartia, 43, 315
  gracilis, 313
Peltidiidae, **385**
Peltidium purpureum, **385**
Peltoperla, **436**
Peltoperlidae, 429, **436**
Penaeoidea, 389
Penilia, 366
Peracarida, 393, 397, 402, 409, 413
peracaridans see Peracarida,
Perissocytheridea meyerabichi, **376**
Petitia, 333
  amphophthalma, **341**
Petrocha, 334
Pettiboneia, 334
  australiensis, **342**
Phaeodactylum, 164
Phaeodarea, 244
Phallodrilinae, 350
Phallodrilus, 350, 351
Phascolion, 355, 356, **356**
Philine, 454
  catena, 452, **456**
Philinidae, 452
Philinoglossa helgolandica, 452, **456**
Philinoglossidae, 452
Philodina, 313
Philodinavidae, 312
Philodinidae, 312
Philomedes globosus, **375**
Phlebobranchiata, 461
Phlyctenophora aff. *zealandica*, **375**
Pholidoherpia cataphracta, 448
  lepidota, 448
Pholidoskepia, 448

Pholoe, 333
   swedmarki, **343**
Pholoides, 333
Phoxichilidiidae, 423
Phreatocoidea, 397, 398
Phreatodytes, 56
   relictus, **437**
Phreatodytidae, 428, 430, **437**
phreatodytids see Phreatodytidae,
Phyllodocidae, 333
Phyllognathopus, 44, 53
   paludosus, 44
   viguieri, 51
Phyllopodopsyllus bradyi, **386**
Phylothalestris mysis, **386**
Phytomastigophorea, 244
Picrocuma poecilata, **416**
Pionosyllis, 333, **341**
Pisione, 334
   galapogoensis, **343**
Pisionella, 334
Pisionidae, 333, 334, **343**
Pisionidens, 334
Placus, 258
Plakosyllis, 333
   brevipes, **341**
Planodasyidae, 303
planodasyids see Planodasyidae,
Planodasys, 303, **309**
Plasmodiophorea, 244
Platycopida, 371
Platydasys, 304, 310
Platyhedyle denudata, 453, **456**
Platyhedylidae, 452
Platyhelminthes, 283
Platyhelminthomorpha, 283
Plecoptera, 429, 430, **436**
Plectus, 44
Pleocola limnoriae, **364**
Pleurocope dasyura, **400**
Pleurocopidae, **400**
Pleurodasys, 303, **309**
Pleuronema, 258
Pleuroxus denticulatus, **369**
Pliciloricus, 320
   grazilis, **321**
Pliciloridiidae, 320
Ploima, 313
Pluscula cuica, **456**
Podocopa, 370, 371
Podocopida, 371
Podonidae, 366
Polychaeta, 18, 25, 55, 140, 161, 164, 174, 182, 192, 275, 332-344, 349, 350, 426, 451, **341-344**
polychaetes see Polychaeta,
Polycladida, 273, 274
polyclads see Polydladida,
Polycopsis serrata, **375**
Polycystididae, 275

Polycystinea, 244
Polygordiidae, 333, 334, **344**
Polygordius, 334, 448, **344**
   madrasensis, **344**
Polymerurus, 304, **311**
   delamarei, **311**
Polyphaga, 430
Pomponema, **301**
Pontarachnidae, 418
Pontohedyle milaschewitschii, 452
   verrucosa, **456**
Porcellidiidae, **385**
Porcellidium viride, **385**
Porolohmannella, 418
   violacea, **421**
Potamodrilidae, 345-348, **348**
Potamodrilus, 346
   fluviatilis, 345, 347, **348**
Priapulida, 18, 322-327
priapulids see Priapulida,
Priapulopsis, 324, **327**
   cnidephorus, 322-324
Priapulus, 324
Pristina, 45
Proales, 313
   germanica, 314, **318**
   halophila, 314
   syltensis, 314
Proalidae, 313
Proasellus, 53, 55
   walteri, 55
Problognathia minima, **286**
Procephalothrix, 287, 288, **291**
Prochaetoderma yongei, 448, **450**
prochaetodermatid see Prochaetodermatidae, 448
Prochaetodermatidae, 448
Proichthydidae, 305
Proichthydioides, 305, **311**
Proichthydium, 305, **311**
Prokoenenia, 425
Prolecithoepitheliata, **281**
Prolecithophora, 273
Promesostomidae, 275, **282**
Promesostominae, 275, **282**
Propontocypris cedunaensis, **375**
Proporata, 274
Prorhynchidae, 274
Prorodon, 258
   teres, **265**
Proseriata, 274, 277, **281**
Prosobranchia, 451, 453
prosobranchs see Prosobranchia,
Prostigmata, 417, 418
Prostoma graecense, 55
Prostomatella, 288
   arenicola, 287, 288, **291**
Protista, 258
Protodorvillea, 334
Protodrilidae, 332, **344**

Protodriloidae, 332, 334
Protodriloides, 334
  chaetifer, 334, **344**
Protodrilus, 14, 334, 448
  adhaerens, **344**
  brevis, **344**
  helgolandicus, **344**
Protohydra caulleryi, 267
  leuckarti, 267
  psamathe, 267
Protozoa, 15, 31, 46-49, 161-164, 166, 232, 233
protozoans see Protozoa,
Protura, 426
Provorticidae, 275, **281**
Psammaspidae, 389, 390
Psammocephalus faurei, **265**
Psammodrilidae, 333, **343**
Psammodriloides, 334
  fauveli, **343**
Psammodrilus, 334
  balanoglossoides, **343**
Psammohydra nanna, 267, **272**
Psammonobiotidae, 243
Psammonobiotus communis, **256**
Psammorhynchidae, 275
Psammoriedlia, 334
Psammothuria ganapatii, 458
  ganatii, 458
Pselionemax, **300**
Pseudobradya pulchella, 26
Pseudocyclopidae, 380
Pseudostenhelia wellsi, 26
Pseudostomella, 304, **311**
Pseudostugarctus triungulatus, **362**
Pseudotanaidae, 403, **408**
Pseudotrocha, 313
Pseudoturbanella, 303, **310**
Pseudovermidae, 452
Pseudovermis artabrensis, **456**
  papillifer, 453
Pseudozeuxo belizensis, **407**
Pseudunela cornuta, **456**
Pterygocythereis jonesi, **376**
Ptycholaimellus ponticus, **24**
Ptychostomella, 304, **311**
Pulmonata, 452
Pusillotrocha, 334
  akessoni, **342**
Pycnogonida, 11, 423-424, **424**
pycnogonids see Pycnogonida, 423
Pycnophyes, 329
  greenlandicus, **331**
Pycnophyidae, 329
Questa, 334
  media, **343**
Questidae, 334, **343**
Quinquiloculina seminulum, **257**
radiolarians (Radiolaria), 245
Raiarctus colurus, **363**

Randiella, 351
  multitheca, **354**
Randiellidae, 350, 351
Raricirrus, 334
Raumahara dertoo, **412**
Remanella, 258
  margaritifera, **265**
Renaudarctus, 357, 358
  psammocryptus, **362**
Reophax, **256**
Retronectidae, 274
Rhabditis marina, 165, 293
Rhabdocoela, 273, 275, 277, **281**
Rhabdolaimidae, 293
Rhabdomolgus ruber, 457
Rhagidia, 425
Rhaphidrilus, 334
Rhaptapagis cantacuzeni, 269
Rheomorpha, 346
  neizvestnovae, 346, 347, **54**, **348**
Rhizopoda, 48, 243-245, 248
rhizopods see Rhizopoda,
Rhodacaridae, 418
Rhodatrachnidae, 418
Rhodomonas lens, 164
Rhodope veranyi, **456**
Rhodopidae, 452
Rhomboarctus thomassini, **363**
Rhombognathinae, 418
Rhyacodrilinae, 350
Rhynchotalona falcata, **369**
Rhynchothoracidae, 423
Rhynchothorax, 423
  philopsammum, 423, **424**
Richtersia, **300**
Ridgewayiidae, 380
Robertsonia propinqua, 26
Rotaliina, **257**
Rotaria citrina, 315
  rotatoria, 313, **318**
  tardigrada, 313
Rotifera, 14, 15, 18, 39, 40, 43, 46-48, 50-53, 302, 312-318, 357, **44**, **318**
rotifers see Rotifera,
Rugiloricus, 320
  carolinensis, **321**
  cauliculus, **321**
Rupertomenia fodiens, 448
Rutiderma rostrata, **375**
Sabatieria, **301**
  pulchra, **24**
Sabellidae, 334, **343**
Saccaminidae, **255**
Saccocirridae, 333, **344**
Saccocirrus krusadensis, **344**
  minor, **344**
Sacodiscus littoralis, **385**
Sapha amicorum, **456**
Sarcodina, 232, 233, 244, **233**

sarcodinans see Sarcodina,
sarcodines see Sarcodina,
Sarcomastigophora, 233, 243-257, **255**, **257**
sarcomastigophorans see Sarcomastigophora,
Sayciinae, 365
scaphapod see Scaphopoda,
Scaphopoda, 451
Scaptognathus, 418
    hallezi, **422**
Schizopera knabeni, 26
Schizorhynchidae, 275
Schroederella, 334
    pauliani, **343**
Scirtidae, 430
Sclerochilus semivitrens, **376**
Scleroperalia, 283, **286**
scorpions see Palpigradida,
Scottopsyllus pararobertsoni, **387**
Scyphozoa, 266
Seba, 409
    ekepuu, **412**
Sebidae, **412**
Secernentea, 293
Seison, 312
Selenaria, 439, 441
    initia, 440
Selenariidae, 439
Semnoderes, 328
    armiger, **331**
Semnoderidae, 328
Separata, 274
Seriata, 273
Serolidae, **400**
Setosellina, 439
    capriensis, 439
    constricta, 439
    goesi, 439
    roulei, 439, **443**
Setosellinidae, 439
Sialidae, 429
Sida crystallina, 365, **369**
Sididae, 365, 366, **369**
sidids see Sididae,
Sigalionidae, 333, **342**
Simognathinae, 418
Simognathus, 418
Siphonohydra, 268
    adriatica, 268
Siphonolaimidae, 27
Sipuncula, 11, 18, 355-356, **356**
sipunculan see Sipuncula,
Sipunculids see Sipuncula,
Smeagol manneringi, **456**
Smeagolidae, 452
Soldanellonys monardi, **421**
Soldanellonyx, 51, 418
solenogasters see Solenogastres,
Solenogastres, 447, 449
Solenopharyngidae, **282**

Sorberacea, 14, 461
Spaerosyllis centroamericana, **341**
Spathidium musicola, **265**
Spathoderma clenchi, **450**
Spelaeogriphacea, 413
Speocyclops, 53
Sphaerodoridae, 333, **341**
Sphaerolaimidae, 27
Sphaeropora, 440, 441
    fossa, 440, **443**
Sphaerosyllis, 333
Sphenoderes, 328
    indicus, **331**
Sphenotrochus, 266
    andrewianus, 269, 270, **272**
Spionidae, 334
Spirinia, **301**
Spiroplectammina biformis, **256**
Spirostomum, 258
springtails see Collembola,
Stenacron, **436**
Stenasellus, 53
Stenbelia bifida, 26
Stenetroidea, 397, 398
Stenhelia normani, **385**
Stenocypris major, **375**
Stenolaemata, 438
Stenothoidae, 409, **412**
stenothoids see Stenothoidae,
Stentor roeselli, **265**
Sternarthronidae, 425
Stiligeridae, 452
Stilocladius, **437**
Stolidobranchiata, 461
Strandesia bicornuta, **375**
Strationyidae, 430
Streblocerus serricaudatus, **369**
Streptosyllis, 333
Strombidium, 258
Stygarctus, 358
    granulatus, 25, **362**
Stygobromus, 53
    coloradensis, 51
    pennaki, 51
Stygocapitella subterranea, 334, **343**
Stygocarella pleotelson, **392**
Stygocaridacea, 55
Stygocarididae, 389, 390, **392**
Stylochaeta, 305, **311**
Stylocoronella riedli, 266, **272**
    variabilis, 267
Stylonychia putrina, **265**
Styraconyx, 358
    hallasi, 359
    kristenseni, **363**
Suctoria, **233**
Sweltsa, **436**
Syllidae, 333, **341**
Syllides, 333, **341**

Symphyla, 426
Synaptidae, 457
Syncarida, 55, 389-392, **43**, **392**
Synpseudes idios, **407**
Tabanidae, 430
Tachidiidae, **385**
Tachidius discipes, 30, **385**
Tanaidacea, 82, 402-408, 413, **407**, **408**
tanaidaceans see Tanaidacea,
Tanaidae, 402-404, **408**
Tanaidomorpha, 402-404, **408**
tanaidomorphan see Tanaidomorpha,
tanaids see Tanaidacea,
Tanarctus arborspinosus, 363
Tansanapseudes longiseta, **407**
Tantulocarida, 377
Tanypoinae, **437**
Tardigrada, 11, 18, 24, 25, 39, 45, 47, 48, 52, 53, 137, 143, 182, 240, 319, 357-364, **43**, **362-364**
tardigrades see Tardigrada,
Tegastes falcatus, **385**
Tegastidae, **385**
Telmatodrilinae, 350
Tergipedidae, 452
Terricola, 274
Terschellingia communis, **24**
  longicaudata, **24**
Testacea, 46
Testacealobosia, 243, 244
Testate, 232
Tetrakentron synaptae, **364**
Tetranchyroderma, 151, 304, **311**
Textulariina, **255**, **257**
Thalassarachna, 417
Thalassia, 163
Thalassoalaimus, **301**
Thalassochaetus, 55, 334
Thalestridae, **385**, **386**
Thaumastoderma, 304, **311**
Thaumastodermatidae, 303-305
thaumastodermatids see Thaumastodermatidae,
Thaumatocyprididae, 371
Theosbaena, 393
  cambodjiana, 393, **396**
Theristus blandicor, 294
Thermosbaena, 393
  mirabilis, 393, **396**
Thermosbaenacea, 55, 393-396, 413, **55**, **396**
thermosbaenaceans see Thermosbaenacea,
Thermosbaenidae, 393
Thiodasys, 303
Tholoarctus, 358
  natans, **363**
Thompsonula hyaenae, 165
Thoracostoma, **301**
thraustochytrid fungi, 232
Thulinia, 359
  ruffoi, **364**
Tigriopus, 164, 174

Tisbe, 174, 176
  furcata, **385**
  holothuriae, 163
Tisbidae, **385**
Tobrilidae, 293
Torrenticola, 51
Trachelocerca, 258
  tenuicolis, **265**
Trachelonema, 258
Tracheloraphis, 258
  phoenicopterus, **265**
Trichocera pygocera, 314, **318**
Trichocerca taurocephala, 315
Trichocercidae, 314
Trichoptera, 429-431, **437**
Trichosphaerium, 243
Trichotria, 43
Trigonostomidae, 275, **282**
Trilobodrilus, 334
  axi, **342**
Tripedalia cystophora, 267, **272**
Tripyla, 44
Tripylidae, 293
Trochodota furcipraedita, 458, **460**
  havelockensis, 458
Troglochaetus, 50, 334
  beranecki, 51, 53, 54, 332, **54**
Trogloleleupia, 410
Tubificidae, 349-351
tubificids see Tubificidae,
Tubificinae, 350
Tubificoides, 350, 351
Tubiluchidae, 322, 324
Tubiluchus, 324
  australensis, 323
  corallicola, 322, 323, **327**
  philippinensis, 323
  remanei, 323
Tubulanus pellucidus, 288
Tubulariidae, 268
Tubuliporata, 438
Tunicata, 14, 18, 438, 461-464, **464**
tunicates see Tunicata,
Turbanella, 303, **310**
Turbanellidae, 303
turbanellids see Turbanellidae,
Turbellaria, 16, 18, 29, 39, 46, 48-50, 52-54, 134-135, 137, 139, 161-162, 170, 172, 174, 182, 184, 246, 273-284, 289, 302, 447, 452, **280-282**
turbellarians see Turbellaria,
Turritellella laevigata, **257**
  shoneana, 245
Typhloplanida, 275, **282**
Typhloplanidae, 275, **282**
Typhloplaninae, 275, **282**
Typhloplanoida, 273, 275
Typosyllis, 333
  glarearia, **341**
Ulva, 163

Unela glandulifera, 452
    nahantensis, **456**
Urodasys, 303, **309**
Uronema, 259
Uronychia transfuga, **265**
Uropodidae, 418
Uropodinae, 418
Valvifera, 397, 398, **400**
Viscosia, **301**
Wandelia, 409
    wairarapa, **412**
Wandesia, **43**
Westheideia, 334
Wierzejskiella, 43, 51
    elongata, 313, **318**
Wigrella, 43, 313
    depressa, **318**
Wingstrandarctus corallinus, **362**
Xenodasys, 303, **309**
xenophyophores see Xenophyophoria,
Xenophyophoria, 243-245
Xenotrichula, 304, **311**
Xenotrichulidae, 304, 305, 307
xenotrichulids see Xenotrichulidae, 304
Xestoleberis baja, **376**
Yeast-like cells, **233**
Zaus spinatus, 177, **385**
Zoomastigophiorea, 244
Zostera, 163
Zygoptera, 429